Advancing Innovation through AI and Machine Learning Algorithms

Computational Intelligence for Virtual System Optimization

Advancing Innovation through AI and Machine Learning Algorithms

Computational Intelligence for Virtual System Optimization

A proceeding of ICMVRCET – 2025

Edited by

Dr. Udara Yedukondalu

Principal, Dept of ECE,

MVR College of Engineering and Technology,

Vijayawada, Andhra Pradesh, India – 521180

Dr. V Vijayasri Bolisetty M.E., Ph.D,

Associate Professor,

Department of ECE, Aditya University,

Surampalem, Kakinada, Andhra Pradesh, India

CRC Press
Taylor & Francis Group
Boca Raton London New York

CRC Press is an imprint of the
Taylor & Francis Group, an **informa** business

First edition published 2026
by CRC Press
44 Park Square, Milton Park, Abingdon, Oxon, OX14 4RN

and by CRC Press
2385 NW Executive Center Drive, Suite 320, Boca Raton FL 33431

British Library Cataloguing-in-Publication Data
A catalogue record for this book is available from the British Library

ISBN: 9781041164272 (pbk)
ISBN: 9781041164340 (hbk)
ISBN: 9781003684589 (ebk)

DOI: 10.1201/9781003684589

Typeset in Times New Roman
by HBK Digital

Dedication

Artificial Intelligence (AI) and Machine Learning (ML) algorithms are revolutionizing innovation across industries by enabling systems to learn from data, adapt to new inputs, and make intelligent decisions. These technologies empower businesses to optimize operations, enhance customer experiences, and develop smarter products and services. AI and ML accelerate research and development by automating complex tasks, uncovering hidden patterns, and generating predictive insights that drive informed decision-making.

In sectors like healthcare, AI-powered diagnostics and ML-based treatment recommendations are improving patient outcomes. In finance, algorithms detect fraud and manage risks in real-time. In manufacturing, predictive maintenance and smart automation enhance efficiency and reduce downtime. Moreover, AI-driven innovation is crucial in developing autonomous vehicles, personalized education systems, and advanced robotics.

The ability of ML algorithms to continuously evolve and improve performance over time creates a dynamic platform for sustained innovation. Organizations that leverage these technologies can stay ahead in a competitive landscape, offering scalable and intelligent solutions to complex problems. As AI and ML continue to mature, they promise to transform the future of technology and human interaction, opening new frontiers of innovation and economic growth. Embracing these advancements is essential for progress in the digital age.

Contents

List of Figures

List of Tables

Foreword

It is **with** great pleasure that we introduce to the International Conference on Microstructure, VLSI, Robotics, Communication, and Electrical & Emerging Technologies using AI-ML Algorithms (ICMVRCET - 2025). This conference brings together distinguished experts, researchers, academicians, and industry leaders from across the globe to share knowledge, explore the latest advancements, and discuss the promising future of emerging technologies. The integration of Artificial Intelligence (AI) and Machine Learning (ML) with core engineering fields such as Microstructure Design, VLSI, Robotics, and Communication Systems has opened new avenues for research and practical applications. As we continue to make strides in smart systems, automation, and advanced electronics, this conference aims to serve as a vibrant platform for disseminating cutting-edge research, innovative solutions, and industry developments.

ICMVRCET - 2025 focuses on the interdisciplinary collaboration that fuels the progress of technology. It seeks to address the global challenges of sustainable development, efficiency, and scalability while providing solutions that push the boundaries of what we know. Through stimulating sessions, interactive workshops, and research presentations, we aspire to inspire fresh perspectives and foster collaborations that will shape the future of technological innovations.

Preface

The International Conference on Microstructure, VLSI, Robotics, Communication, Electrical & Emerging Technologies using AI-ML Algorithms (ICMVRCET - 2025) is an essential gathering for those at the forefront of research and development in the fields of Microstructure Design, VLSI systems, Robotics, Communication technologies, and Emerging Electrical systems. This conference seeks to bridge the gap between academic research, industrial advancements, and real-world applications by focusing on the integration of Artificial Intelligence (AI) and Machine Learning (ML) algorithms in these rapidly evolving domains.

As we step into an era driven by smart systems and automation, the synergy between traditional engineering disciplines and AI-ML technologies becomes increasingly crucial. This event is designed to provide a platform for thought leaders, researchers, and practitioners to present their work, exchange ideas, and foster collaborations that will drive the future of technology.

The ICMVRCET - 2025 conference will feature a series of keynotes, paper presentations, and panel discussions on cutting-edge topics such as AI-driven VLSI design, smart robotics systems, emerging electrical technologies, and innovative communication methodologies. The use of AI and ML in these fields promises to unlock new possibilities, enhance efficiencies, and pave the way for future breakthroughs in technology.

We hope that the discussions and research presented here will inspire innovative solutions, promote interdisciplinary collaborations, and significantly contribute to the global technological ecosystem.

We thank all the participants, speakers, and partners for their invaluable contributions, and we look forward to the exciting exchanges and impactful outcomes of this conference.

Acknowledgements

We would like to extend our heartfelt appreciation to Mr. Raja Suresh, Visiting Professor at Sri Venkateswara College of Engineering, Tirupati, for taking the initiative to guide and support the successful organization of this international conference. His extensive industrial experience and academic engagement have significantly enhanced the quality, vision, and execution of this event.

Mr. Raja Suresh is the Visionary Founder and Managing Director of Ekalavya Innovative Solutions Pvt. Ltd. (EISPL), a Government of India-recognized enterprise specializing in next-generation solutions for industrial, consumer, commercial, and advanced technology applications. With over two decades of professional expertise in embedded systems, electronic design, and R&D, he has played a transformative role in making EISPL a leading name in the embedded technology sector, particularly in domains such as Automotive, Semiconductors, IoT, and Industrial Automation.

Academically, Mr. Raja Suresh holds a B.Tech in Electronics and Communication Engineering (ECE) and an M.Tech in Embedded System from Jawaharlal Nehru Technological University, Anantapur, Andhra Pradesh. His strong educational foundation has been instrumental in shaping his technical proficiency and innovative outlook.

His leadership is rooted in a commitment to excellence, innovation, and industry-academia integration. His expertise spans across Solar Energy Systems, Power Supply Design, Embedded Hardware, Driver Boards, and Embedded Applications. Through his strategic direction, EISPL has developed cutting-edge technologies that not only meet market demands but also contribute significantly to national R&D advancement.

A passionate mentor and advocate for young talent, Mr. Raja Suresh is devoted to fostering a culture of innovation and learning. His vision of building a globally competitive ecosystem for embedded solutions is reflected in EISPL's national impact and continued growth.

We are privileged to have had his guidance and visionary support for this conference and sincerely thank him for his outstanding contributions.

List of abbreviations

AI – Artificial Intelligence
ML – Machine Learning
VLSI – Very-Large-Scale Integration
IoT – Internet of Things
SEM – Scanning Electron Microscope
PCB – Printed Circuit Board
CAD – Computer-Aided Design
DSP – Digital Signal Processing
MEMS – Micro-Electro-Mechanical Systems
FPGA – Field Programmable Gate Array
AI-ML – Artificial Intelligence and Machine Learning
SNR – Signal-to-Noise Ratio
UHF – Ultra High Frequency
GHz – Gigahertz
ADC – Analog-to-Digital Converter
DAC – Digital-to-Analog Converter
MCU – Microcontroller Unit
GSM – Global System for Mobile Communications
RF – Radio Frequency
IoT – Internet of Things
RTOS – Real-Time Operating System
PID – Proportional-Integral-Derivative
AIoT – Artificial Intelligence of Things

Glossary

Artificial Intelligence (AI)

A branch of computer science that focuses on creating machines and systems capable of performing tasks that typically require human intelligence, such as learning, reasoning, and problem-solving.

Machine Learning (ML)

A subset of AI that involves the development of algorithms that allow computers to learn from data and improve their performance over time without being explicitly programmed.

Very-Large-Scale Integration (VLSI)

The process of creating integrated circuits by combining thousands or millions of transistors into a single chip. VLSI plays a critical role in the miniaturization of electronics and is essential in modern computing and communication systems.

Microstructure

Refers to the structure of a material as seen under a microscope, typically focusing on its grain structure and the properties that affect its physical characteristics, such as mechanical strength and electrical conductivity.

Internet of Things (IoT)

The interconnection of physical devices, vehicles, appliances, and other objects through sensors, software, and network connectivity, allowing them to collect and exchange data.

Embedded Systems

Specialized computing systems that are designed to perform a specific task or function within a larger system, often with real-time constraints. These systems are found in everything from automotive electronics to consumer gadgets.

Scalable Systems

Systems that can grow or be modified to accommodate increased demand or workload. Scalability is crucial for both hardware (e.g., VLSI designs) and software (e.g., AI-ML algorithms).

Field Programmable Gate Array (FPGA)

A type of programmable logic device that can be configured to perform specific logic functions. FPGAs are widely used in applications requiring fast processing and reconfigurability, such as signal processing and hardware acceleration.

Digital Signal Processing (DSP)

The use of algorithms and techniques to manipulate and analyze digital signals (such as sound, image, and video signals). DSP plays a crucial role in telecommunications, multimedia systems, and medical applications.

Micro-Electro-Mechanical Systems (MEMS)

Small-scale devices that integrate mechanical components, sensors, actuators, and electronics. MEMS are used in a variety of applications, including sensors in automotive, medical, and consumer electronics.

Real-Time Operating System (RTOS)

An operating system designed to handle real-time tasks that require immediate processing. RTOS is used in applications where time constraints are critical, such as robotics, embedded systems, and industrial control systems.

Robotics

The design, construction, operation, and use of robots. Robotics combines elements of mechanical engineering, electrical engineering, and computer science to automate tasks traditionally performed by humans.

Proportional-Integral-Derivative (PID) Controller

A type of control loop feedback mechanism used in industrial control systems to continuously control a process variable, maintaining desired output by adjusting inputs.

Signal-to-Noise Ratio (SNR)

A measure of signal strength relative to background noise. SNR is a key factor in determining the quality of communication systems, sensor data, and multimedia systems.

Analog-to-Digital Converter (ADC)

A device that converts an analog signal into a digital signal, allowing digital systems to process real-world continuous signals such as temperature, sound, or pressure.

Digital-to-Analog Converter (DAC)

A device that converts a digital signal into an analog signal, which can then be used in applications such as audio output and analog sensors.

Gigahertz (GHz)

A unit of frequency equal to one billion cycles per second. GHz is commonly used to measure the clock speed of processors and the frequency of electromagnetic waves in communication systems.

Radio Frequency (RF)

Refers to electromagnetic wave frequencies typically in the range of 3 kHz to 300 GHz, used in wireless communication, broadcasting, and radar systems.

Automated Test Equipment (ATE)

Devices used to perform automated tests on electronic devices or systems. ATE is crucial for quality assurance and verification in semiconductor manufacturing and system development.

Control Systems

Systems designed to manage and regulate the behavior of other systems, often through the use of feedback loops, sensors, and algorithms. Common applications include automation, robotics, and industrial systems.

Smart Systems

Systems embedded with intelligent algorithms, sensors, and machine learning models to make decisions, adapt to changes, and interact with their environment without human intervention.

Nanoelectronics

A branch of electronics that deals with the design and fabrication of circuits and devices at the nanoscale, enabling advancements in computing, sensors, and quantum technologies.

Patent

A form of intellectual property that grants the holder exclusive rights to an invention or design for a specified period, preventing others from making, using, or selling the invention without permission.

Copyright

A legal right granted to the creator of original works of authorship, providing exclusive rights to use, distribute, and reproduce the work.

Semiconductor

A material that has electrical conductivity between that of a conductor and an insulator, widely used in the manufacturing of electronic devices such as transistors, diodes, and integrated circuits.

Editors Biography

Dr. Udara Yedukondalu
Principal
MVR College of Engineering and Technology, Paritala, Vijayawada

Dr. Udara Yedukondalu, serving as the Principal of MVR College of Engineering and Technology, Paritala, Vijayawada, combines academic leadership with a strong research background. With 22 years of teaching experience, 30+ journal publications, 5 patents, and 1 copyright, he is dedicated to advancing innovation in VLSI, nano-electronics, and integrated circuits. His commitment to holistic education aligns with Dr. A.P.J. Abdul Kalam's vision, fostering creativity, purpose-driven learning, and student-centered growth. Under his guidance, MVR College emphasizes not only academic excellence but also the overall personality development and employability of rural youth. Through ICMVRCET – 2025, he aims to provide a global platform for emerging technologies using AI-ML algorithms.

Dr. V Vijayasri Bolisetty M.E., Ph.D.
Associate Professor, Department of ECE
Aditya University, Surampalem, Kakinada

Dr. V Vijayasri Bolisetty is an accomplished academician with over 18 years of experience in teaching, research, and academic leadership in the field of Electronics and Communication Engineering. She currently serves as an Associate Professor at Aditya University, where she actively contributes to research development, technical education, and student mentoring.

Her academic background includes an M.E. and a Ph.D. in Electronics and Communication Engineering, with specialization in Signal Processing. Dr. Vijayasri has contributed to numerous scholarly publications and technical initiatives that bridge academic research with real-world applications.

As an editorial board member of ICMVR 2025, Dr. V Vijayasri Bolisetty key role in maintaining the scientific quality of the conference through rigorous peer review, topic curation, and academic oversight. Her areas of interest align with the conference themes, particularly in smart systems, sustainable electronics, and advanced computational techniques.

She is committed to fostering innovation, collaboration, and academic excellence in the global research community.

1 End-to-end AI framework for soil fertility, crop growth, and disease mitigation

K. R. Harinath[1,a], V. Yagna Sri[2,b], A. Mamatha[2,c], K. Dileep Reddy[2,d] and N. Naga Bharath Kumar[2,e]

[1]Assistant Professor, Department of CSE, Rajeev Gandhi Memorial College of Engineering and Technology, Nandyala, Andhra Pradesh, India

[2]Department of Computer Science and Engineering, Rajeev Gandhi Memorial College of Engineering and Technology, Nandyal, Andhra Pradesh, India

Abstract

India is an agricultural country and includes some of the most important international vegetation producers. Although Indian farmers are in a comfortable sector atmosphere, most Indian farmers remain the lowest in the history of the city. Furthermore, farmers cannot decide which types of plants are suitable for a good contract. You can also benefit from the unusual scientific answers on different fashion cards in your region, regardless of unpaid grading cards. This article uses a machine learning model to provide a consulting machine for harvesting. Here, the most useful harvest is expected by reading soil types, productivity and faster limitations. Agriculture is a spine in Indian, plays an important role in the Indian economic system, and provides a positive weight of home real estate to ensure food safety. This life will end with an abnormal climate change to the prediction of food. The financial system is greatly donated by agriculture. In agricultural soil, disability is additionally divided into flowers, and the detection of flowers can also occur at the top. This time, in addition to various parking spaces, herb diseases are shouting more and more, reducing plants in long courtyards. This means that farmers of the entire size will change the problem of cutting and change after processing the unit in the direction of special devices. To find something for the shades of a therapeutic specialist, you can categorize deciduous diseases in tomato plants before recognizing deciduous diseases. This is when correct control is not used to significantly affect plants. This means a beautiful chino that can be forgotten for the production of vegetation.

Keywords: Crop recommendation system, machine learning, plant disease detection, soil type

Introduction

From the beginning, agriculture has been extremely important to humanity [3] and has grown to productivity rather than sustainability. After the world's population reached 10 billion by 2050 [19], modernizing agriculture is important to meet this demand [1]. It also corresponds to the goal of United Nations (UN) (CGDG) 2 (SDH) 2, which aims to destroy hunger by 2030 [20]. Surprisingly, the association previously followed the UN.

As India's largest economic sector, agriculture significantly contributes to socio-economic growth. Traditional farming methods require modernization to improve productivity and meet growing market demands. Technological advancements, including precision agriculture, help farmers monitor crops [2] [6] and enhance yields [21]. The world's research focuses on promoting accurate agriculture to reduce the burden on farmers [5].

Agricultural 4.0 optimizes agricultural methods by integrating development technologies [4] such as automation, IoT, AI, and data analysis [7][22]. This innovation increases efficiency, resource management and yield, making intellectual agriculture a key element of agricultural change [23].

- In one faster situation, the prediction of the land is not sufficiently integrated with sufficient reproduction, agricultural crop prediction, and the integration of homogeneity and statistical assets for disclosing information about the disease. Agriculture is based on a variety of facts, including information on excellent soil, climate model and other resources of pests/diseases and other resources of various codecs [8, 9]. Integration and

[a]harirooba007@gmail.com, [b]yagnasrivaka@gmail.com, [c]ankemamatha09@gmail.com, [d]kdileepreddy67@gmail.com, [e]nagellanagabharathkumar@gmail.com

DOI: 10.1201/9781003684589-1

standardization of these facts for comprehensive evaluation is an extreme problem.

Advanced technology changes agriculture, but it cannot be used the same for small farmers in developing countries. This task is an important obstacle to expanding the development of economically effective and convenient solutions that provide services to various agricultural communities.

- As exact agriculture develops, there is concern about the impact on the environment and ethical use [10]. Excessive dependence on artificial arts for fertilizers and pesticides can harm the ecosystem, and data confidential and safety issues continue to increase.

Related Work

- Prema Sudha B G et al [25]. proposed methods like Random Forest (RF) and XGBoost based on machine learning (ML) for the prediction of crops and fertilizers, focusing on yield optimization and soil fertility enhancement. Their system integrates disease detection from leaf images and fertilizer recommendations based on soil nutrient composition, leveraging the predictive power of ML models.

- TK Murugan and P Revanth [26] developed the "Agro Insights" model using ML and deep learning (DL) techniques for soil classification, crop prediction, and plant disease detection. Their methods reached 93.3% accuracy in predicting soil and crops using RF, and 96% accuracy in disease detection using CNN.

- R. Prabavathi and Balika J Chelliah [27] presented a comprehensive review of ML techniques for yield prediction, emphasizing the role of essential soil nutrients and environmental factors. Their study analyzed 51 peer-reviewed papers and concluded that intelligent ML models like SVM, RF, and ELM are effective for soil fertility, crop, and yield prediction.

- C. Jacklin and S. Muruga Valli [28] conducted a comprehensive review on plant disease detection using ML and DL approaches, highlighting their role in enhancing crop yield. The overall structure of the proposed system, including components for soil, crop, and disease analysis, is shown in Figure 1.1. Their study emphasizes the superior performance of deep learning models over traditional ML techniques in identifying plant diseases from leaf images.

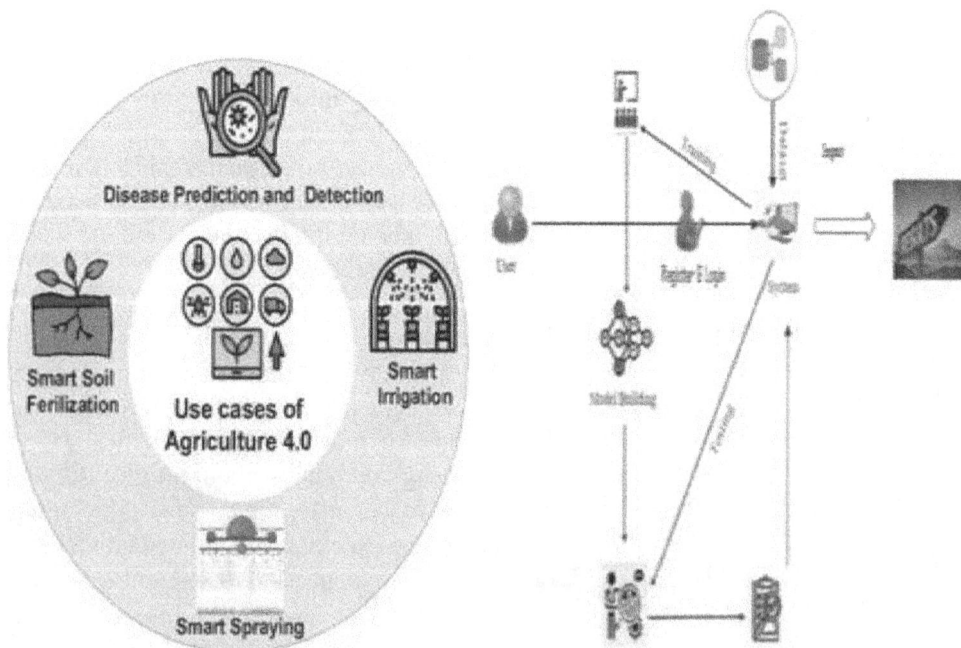

Figure 1.1 Architecture of the proposed system
Source: Author

- DS Wankhede [29] reviewed various ML classifiers for analyzing soil nutrients like pH, N, P, and K to improve crop prediction and soil quality assessment. The study highlights that among classifiers like J48, k-NN, SVM, and Random Forest, the Naive Bayes model performed best on large datasets for nutrient prediction.

- M. Shoaib et al. [30] explored recent advancements in deep learning and machine learning models for accurate and efficient plant disease detection. Their study highlights the benefits and challenges of DL/ML techniques, emphasizing solutions to issues like data quality, imaging, and disease differentiation.

This assignment aims to demonstrate three distinct soil prediction models [11], along with Plant disease detection and crop yield prediction using machine learning and deep learning techniques. The objective is to determine soil fertility (Fertile or Non-Fertile), recommend suitable crops, and identify plant diseases.

The adding growth of machine literacy, computer ways divided into traditional styles and machine literacy styles. This section describes the related works of classification of SOIL prediction and fertility suggestion [12, 13], crop prediction and plant disease detection can be significantly improved using machine learning and deep learning techniques, which outperform traditional approaches. The current system follows a specific workflow and employs algorithms such as ANN, support vector machine (SVM), and Multi-Layer Perceptron (MLP) for model development. However, it requires large memory resources and often produces inaccurate results.

1. Records reliance and accuracy: These technologies rely on high-quality data, including historical records and real-time inputs. Inaccurate or insufficient data can lead to incorrect predictions, resulting in poor farming decisions [14].

2. Preliminary cost and adoption barriers: Implementing these technologies requires significant investment in hardware, software, and training. Smallholder farmers, especially in developing regions, may struggle with financial constraints, limiting adoption.

3. Dependency on the era: Widespread implementation demands substantial upfront investment, which can create economic barriers for small-scale farmers, restricting access to these innovations.

4. Privateness and information safety issues: The collection and storage of sensitive agricultural data can lead to privacy concerns [15], which may make farmers reluctant to share their information due to concerns over misuse, breaches, or unauthorized access, hindering dataset development [16].

5. Environmental and ethical challenges: While these technologies aim to optimize resource use and reduce environmental impact, overreliance on AI-driven recommendations may lead to excessive fertilizer and pesticide use, causing ecological harm.

Proposed several models of machine learning model to classify but none addressed this misdiagnosis problem accordingly. That could be used for this purpose: Integrating machine learning and DL soil type prediction and fertility suggestion, crop prediction plant disease detection. Also, similar works that have reported models for analysis of such a tumor mostly lack considering the data heterogeneity as well as data size. Accordingly, we introduced a machine learning-based approach combined with a newly designed technique that involves preprocessing and transforming features extracted from the datasets, both give the best approaches for eliminating biases and instability by performing classifier, and image- classifier tests are based on such techniques.

Methodology

Several models have been proposed for classification, but none have effectively addressed the issue of misdiagnosis. To overcome this, ML and DL can be integrated for soil prediction, fertility assessment, crop forecasting, and plant disease detection. Existing studies on similar models often overlook data heterogeneity and size. To address this, we propose an ML- based approach that includes a novel preprocessing method for feature transformation. Both ML and DL algorithms provide high accuracy, reducing bias and instability while enhancing classification performance.

Deep learning , a subset of artificial neural networks (ANNs), is key to the proposed model.

ANNs consist of interconnected artificial neurons that function similarly to the human brain. These neurons receive inputs, process signals, and adjust weights during learning, refining predictions over time [17].

Different layers apply specific transformations to inputs, enabling human-like reasoning through specialized algorithms. Unlike traditional methods, DL can learn autonomously without human supervision [18].

1. Agricultural productivity: These technologies provide farmers with accurate data on soil conditions, nutrient levels, and crop health, enabling informed decision-making. This leads to higher yields, improved productivity, and better resource management.
2. Sustainable farming practices: By optimizing soil fertility, crop forecasting, and disease detection, farmers can reduce excessive fertilizer and pesticide use, promoting environmental sustainability and minimizing agriculture's ecological footprint.
3. Danger mitigation: Predictive models and early disease detection systems help farmers identify potential threats, allowing for timely interventions to prevent disease outbreaks and reduce crop losses.
4. Resource efficiency: These technologies enhance water conservation, fertilizer optimization, and controlled pesticide use, leading to cost savings, better use of resources, and reduced environmental impact.

ML Algorithms and AI techniques

In this project, ML algorithms and AI methods are applied in precision farming to enhance crop productivity. Class of ML algorithms Mohammed et al. (2016) [24] stated that four extensive categories can be used to categorize machine learning algorithms: Reinforcement studying, semi-supervised studying, unsupervised gaining knowledge of, and supervised mastering. Unsupervised learning examines datasets that lack labels, operating without human involvement and relying entirely on data patterns. It is commonly applied for feature extraction, pattern recognition, data clustering, anomaly detection, and dimensionality reduction.

Project flow:

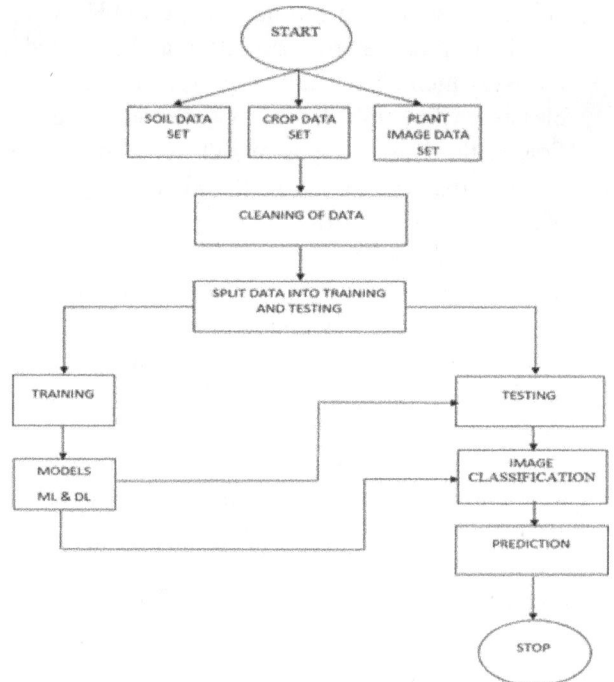

Figure 1.2 Flow chart of proposed system
Source: Author

Implementation Modules:
1. User Modules:
- **View domestic web page:**
  Once users log in or gain access, they are directed to the application's main interface, which serves as the central point for all available features.
- **View about web page:**
  Users can navigate to the "About" section to learn more about the platform, its mission, and the team responsible for its creation.
- **Input model:**
  Users are required to input relevant data for the model to generate precise predictions. This could include details like soil properties, crop types, sowing dates, or images of plants for disease identification.
- **View outcomes:**
  After submitting the necessary data, users can view the model's output, which may provide insights such as soil condition recommendations, estimated crop yields, or diagnoses of plant diseases. Figure 1.2. presents the flow of the proposed system, outlining the sequence from user input to prediction output.

- **Create dataset:**

  For disease detection purposes, users can build a dataset by uploading images of plants that show symptoms of disease and categorizing them accordingly (e.g., healthy or diseased). This dataset is subsequently divided into training and testing sets.

- **Pre-processing:**

  During image dataset preparation, users can see preprocessing methods in action, such as resizing and reshaping, to ensure the images are correctly formatted for training.

- **Training:**

  This phase involves applying deep learning or machine learning algorithms—possibly using transfer learning methods like ResNet50—to teach the model using the processed data. It has a vital role in developing reliable predictive capabilities.

- **Category:**

  The system displays classification results to the user, showing the assigned labels for each image. These labels help users determine whether a plant is healthy or identify the specific disease affecting it.

2. **Running on the dataset:**

   - **Pre-processing:**

     The system presents classification results to the user, showing the labels assigned to images related to plant diseases. Users can determine whether the plants are healthy or identify the specific disease present.

   - **Edification of the data:**

     The data is divided into training and testing sets, a critical step before training any machine learning or deep learning model. This division is necessary to properly evaluate how well the model performs.

   - **Model building:**

     The system organizes the dataset into training and testing subsets, ensuring that the machine learning or deep learning algorithms are trained effectively. This step helps in measuring the model's accuracy and reliability.

   - **View effects:**

     By separating the dataset into training and testing sets, the system enables a thorough analysis of the model's performance. This is vital for understanding how well the model can generalize new, unseen data.

Machine learning techniques

Support vector machine

The SVM is a widely used supervised learning algorithm that can handle both classification and regression tasks. It operates by identifying a hyperplane in a high-dimensional space that separates different classes. The ideal hyperplane is the one that maximizes the distance (margin) between the nearest data points from each class, known as support vectors. SVM is particularly effective in high-dimensional datasets and relies heavily on the selection of an appropriate kernel function like linear, polynomial, radial basis function (RBF), or sigmoid. However, it may perform poorly on datasets with noise or when class boundaries are not clearly separable.

The application of various machine learning algorithms for soil prediction is illustrated in Figure 1.3.

Decision tree

A decision tree (DT) is a supervised learning technique that does not rely on any underlying assumptions about the data (non-parametric) and is commonly applied to both classification and regression problems. It structures data in the form of a tree, where each internal node signifies a decision made based on a specific feature, branches indicate the outcome of those decisions, and leaf nodes show the final output or classification. The tree is typically built using splitting criteria like information gain (entropy) or Gini impurity. DT are simple to understand and work effectively with both numerical and categorical data. However, they are prone to overfitting, particularly when the tree becomes too detailed or complex.

Random Forest

The RF is an ensemble method that improves the effectiveness of decision trees by creating multiple trees using different subsets of the dataset. For classification, the final output is based on majority voting, and for regression, it uses the average of predictions. This technique helps reduce overfitting and improves the overall stability and accuracy of the model. RF are well-suited for large datasets and are less affected by outliers, though they can become computationally intensive when a large number of trees are involved.

XGBoost

XGBoost is a highly effective gradient boosting framework recognized for its outstanding prediction accuracy. It constructs DT one after another, with each new tree aiming to fix the mistakes made by the earlier ones. The algorithm includes regularization

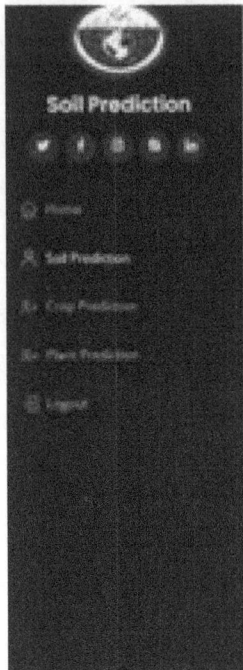

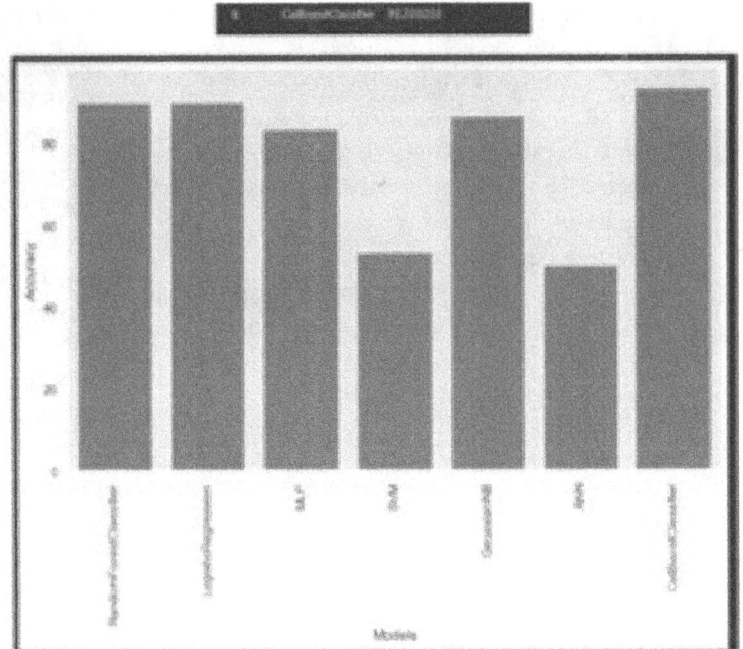

Figure 1.3 Soil prediction using ML algorithms
Source: Author

to prevent overfitting and ensures high efficiency. While it often outperforms RF in terms of accuracy, XGBoost requires careful tuning of hyperparameters and tends to be more resource-intensive.

AdaBoost classifier

AdaBoost is a boosting technique that strengthens weak learners by increasing the weight of misclassified instances in each iteration. The final decision is made through a weighted vote of all learners. AdaBoost is effective for small datasets and can be integrated with other classifiers to enhance performance. However, it is quite sensitive to noisy data and outliers, which can negatively affect its accuracy.

The comparative accuracy of machine learning models for crop prediction is shown in Figure 1.4.

Results and Discussion

The challenge of soil prediction, fertility assessment, crop forecasting, and plant disease detection in agriculture is complex and multifaceted. These tasks aim to address critical aspects of modern agriculture, improving efficiency and sustainability.

Collaborative mapping enables stakeholders to share data on crop diseases within a specific region, improving disease tracking, management, and control. This approach enhances the accuracy and

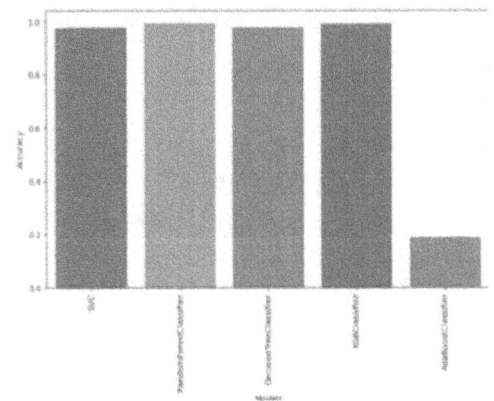

Figure 1.4 Crop prediction accuracy
Source: Author

timeliness of detection, helping identify patterns that may not be immediately apparent to individual growers. By fostering data sharing and communication networks, collaborative mapping strengthens agricultural resilience and supports a proactive response to disease outbreaks. However, challenges such as data consistency, accuracy, and privacy concerns must be addressed to ensure reliability. Overall, this method leverages technology to enhance sustainability and disease management, creating a more interconnected and informed agricultural system.

	N	P	K	temperature	humidity	ph	rainfall	label
0	90	42	43	20.879744	82.002744	6.502985	202.935536	rice
1	85	58	41	21.770462	80.319644	7.038096	226.655637	rice
2	60	55	44	23.004459	82.320763	7.840207	263.964248	rice
3	74	35	40	26.491096	80.158363	6.980401	242.864034	rice
4	78	42	42	20.130175	81.604873	7.628473	262.717340	rice

	ph	EC	OC	OM	N	P	K	Zn	Fe	Cu	Mn	Sand	Silt	Clay	CaCO3	CEC	Output
0	7.74	0.40	0.01	0.01	75	20.0	279	0.48	6.4	0.21	4.7	84.3	6.8	8.9	6.72	7.81	Fertile
1	9.02	0.31	0.02	0.03	86	15.7	247	0.27	6.4	0.16	5.6	90.4	3.9	5.7	4.61	7.19	Fertile
2	7.80	0.17	0.02	0.03	77	35.6	265	0.46	6.2	0.51	6.1	84.5	6.9	8.6	1.53	12.32	Fertile
3	8.36	0.02	0.03	0.06	106	6.4	127	0.50	3.1	0.26	2.3	93.9	1.7	4.4	0.00	1.60	Non Fertile
4	8.36	1.06	0.03	0.05	96	10.5	96	0.31	3.2	0.23	4.1	91.5	4.1	4.4	9.08	7.21	Non Fertile

Figure 1.5 Crop and Soil data analysis report
Source: Author

A detailed analyis report of crop and soil data based on the proposed system is presented in Figure 1.5.

Conclusion

Automatic crop disease detection has the potential to revolutionize agriculture by improving food security and economic sustainability. Technologies such as machine learning (ML) and photodetection systems enable faster and more accurate disease detection while allowing real- time crop health monitoring.

However, challenges exist, particularly in developing and low-income countries, where access to affordable hardware, software, technical expertise, and infrastructure remains limited. The effectiveness of these technologies can also be influenced by factors like image distance, occlusion, plant density, and environmental conditions. Integrating contextual data—such as climate, soil properties, and agronomic practices—can further enhance accuracy.

Climate change introduces new pathogens and environmental variations, affecting the reliability of existing models. Thus, tailoring detection models to specific regions and crop species is crucial. Additionally, these technologies offer real-time disease tracking, location-based pathogen surveillance, autonomous disease management, and collaborative mapping, helping mitigate the impact of climate change on global food production.

Future research should focus on improving feature extraction, refining classifiers, and enhancing model accuracy. Moreover, investment in R&D and innovative deployment strategies—particularly for smallholder farmers—is essential for widespread adoption and long-term success.

References

[1] Ayaz, M., Ammad-Uddin, M., Sharif, Z., Mansour, A., & Aggoune, E. H. M. (2019). Internet-of-things (IoT)- based smart agriculture: toward making the fields talk. *IEEE Access*, 7, 129-551–583.

[2] Shirish, P., & Bhalerao, S. A. (2013). Precision farming: the most scientific and modern approach to sustainable agriculture. *International Research Journal of Science and Engineering*, 1(2), 21–30. Available: www.irjse.in.

[3] Crookston, R. (2006). A top 10 list of developments and issues impacting crop management and ecology during the past 50 years. *Crop Science*, 46(9), 2253–2262.

[4] Cho, Y., Cho, K., Shin, C., Park, J., & Lee, E. -S. (2012). An agricultural expert cloud for a smart farm. In Future Information Technology, Application, and Service, (pp. 657–662). Springer.

[5] Evans, D. (2011). The internet of things: how the next evolution of the internet is changing everything. *CISCO White Paper*, 1(2011), 1–11.

[6] Popoviˊc, T., Latinoviˊc, N., Pešiˊc, A., Zeˇceviˊc, Ž., Krstajiˊc, B., & Djukanoviˊc, S. (2017). Architecting an IoT- enabled platform for precision agriculture and ecological monitoring: a case study. *Computers and Electronics in Agriculture*, 140, 255–265.

[7] Gershenfeld, N., Krikorian, R., & Cohen, D. (2004). The internet of things. *Scientific American*, 291(4), 76–81.

[8] Kamilaris, A., Gao, F., Prenafeta-Boldu, F. X., & Ali, M. I. (2016). Agri-IoT: a semantic framework for internet of things-enabled smart farming applications. In 2016 IEEE 3rd World Forum on Internet of Things (WF-IoT), (pp. 442–447), IEEE.

[9] Hassija, V., Chamola, V., Saxena, V., Jain, D., Goyal, P., & Sikdar, B. (2019). A survey on IoT security: application areas, security threats, and solution architectures. *IEEE Access*, 7, 82-721–743.

[10] Gebbers, R., & Adamchuk, V. I. (2010). Precision agriculture and food security. *Science*, 327(5967), 828–831.

[11] Kitzes, J., Wackernagel, M., Loh, J., Peller, A., Goldfinger, S., Cheng, D., et al. (2008). Shrink and share: humanity's present and future ecological footprint. *Philosophical Transactions of the Royal Society B: Biological Sciences*, 363(1491), 467–475.

[12] FAO, (2009). How to Feed the World in 2050. Food and Agriculture Organization of the United Nations, Rome.

[13] Sekulic, P., & Djurovic, S. (2016). Detection of downey mildew in grapevine leaves using support vector machine. In Proceedings of the 21st Information Technology IT, (Vol. 16, pp. 169–172).

[14] Jhuria, M., Kumar, A., & Borse, R. (2013). Image processing for smart farming: Detection of disease and fruit grading. In 2013 IEEE Second International Conference on Image Information Processing (ICIIP- 2013), (pp. 521–526), IEEE.

[15] Sisinni, E., Saifullah, A., Han, S., Jennehag, U., & Gidlund, M. (2018). Industrial internet of things: challenges, opportunities, and directions. *IEEE Transactions on Industrial Informatics*, 14(11), 4724–4734.

[16] Ayaz, M., Ammad-Uddin, M., & Baig, I., (2017). Wireless sensor's civil applications, prototypes, and future integration possibilities: a review. *IEEE Sensors Journal*, 18(1), 4–30.

[17] Lin, J., Yu, W., Zhang, N., Yang, X., Zhang, H., & Zhao, W. (2017). A survey on Internet of Things: Architecture, enabling technologies, security and privacy, and applications. *IEEE Internet of Things Journal*, 4(5), 1125–1142.

[18] Shi, X., An, X., Zhao, Q., Liu, H., Xia, L., Sun, X., et al. Y. (2019 . State-of-the-art internet of Gu, X., Zhang, Y., Liu, Y., & Zhou, Y. (2021). Machine learning in agriculture: Applications and future directions. Computers and Electronics in Agriculture, 180, 105885.

[19] Gu, S., Casquin, A., Dupas, R., Abbott, B. W., Petitjean, P., Durand, P., & Gruau, G. (2021).

[20] Ripepi, A., Marconi, M., & Bono, G. (2023). Gaia Data Release 3: Specific processing and validation of all-sky Cepheids. Astronomy & Astrophysics, 674, A1.

[21] Javaid, M., Haleem, A., & Singh, R. (2022). AI features and technologies for Industry 4.0. Journal of Cleaner Production, 380, 123676.

[22] Mohamed, S. M. H., Butzbach, M., Fuermaier, A. B. M., Weisbrod, M., Aschenbrenner, S., Tucha, L., & Tucha, O. (2021).

[23] Liutas, J., & Chartsari, M. (2020). Sensor selection and scheduling for tracking a moving object in a wireless sensor network using the Markov additive chain model. Sensors, 20(22), 6414.

[24] Mohammed, S., Glennerster, R., & Khan, A. J. (2016).

[25] Prema Sudha, B. G., & Fan, X. (2021). Epileptic seizure detection and prediction using stacked bidirectional LSTM-GAP neural network.

[26] Murugan, T. K., & Revanth, P. (2021).

[27] Prabavathi, R., & Chelliah, B. J. (2021).

[28] Jacklin, C., & Muruga Valli, S. (2021). Sensor selection and scheduling for tracking a moving object in a wireless sensor network using the Markov additive chain model. Sensors, 20(22), 6414.

[29] Wankhede, D. S. (2021). Sensor selection and scheduling for tracking a moving object in a wireless sensor network using the Markov additive chain model. Sensors, 20(22), 6414.

[30] Shoaib, M., & Fan, X. (2021). Epileptic seizure.

2 Highly accurate calorie burn prediction using XGBoost regression model

K. R. Harinath[1,a], M. Harini[2,b], M. Pranitha[2,c], K. Prashanth[2,d] and H. Nagesh[2,e]

[1]Assistant Professor, Department of CSE, Rajeev Gandhi Memorial College of Engineering and Technology, Nandyala, Andhra Pradesh, India

[2]Department of CSE, Rajeev Gandhi Memorial College of Engineering and Technology, Nandyala, Andhra Pradesh, India

Abstract

Correct prediction of calorie burn is important for personalizing health and health tracking. This aspect suggests a machine learning-based approach to daily predict the calorie burn of individuals using an XGBoost regression version. The version is trained on a dataset that contains every day inclusive of age, weight, hobby type, and duration. The preprocessing of records, along with the selection of functions and outlier detection, is done daily to ensure the premier enters for the version. The XGBoost algorithm, which is famous for its efficiency and accuracy in regression obligations, is hired daily to provide personalized calorie burn predictions. Widespread measures are used to assess the model's performance, showing that it can accurately predict daily caloric consumption. This technique can find its applications within a variety of packages, ranging from health monitoring applications to wearable devices and personal test fitness, showing immense improvements over standard methods.

Keywords: Accurate prediction, calorie burns, Python, system mastering, XGBoost set of rules

Introduction

Maintaining a healthy lifestyle is very important in today's fast-paced world, as sedentary behaviors and poor diets contribute to increasing obesity rates and related health issues [1]. Effective weight management depends on balancing calorie intake and expenditure, but accurately measuring calorie burn remains challenging due to individual variations in age, weight, height, activity type, and duration [2]. Traditional methods like calculators and wearables usually don't give a precise estimation as they rely on fixed formulas. Machine learning, especially XGBoost, has been a reliable approach that considers large datasets for delivering individualized calorie burn estimations [3]. The present work will be about building an XGBoost-based model that integrates personal attributes and activity parameters to achieve better accuracy with minimal errors in health and fitness monitoring [4]. However, nutrition is still an important issue in general life, and nowadays, technology gives the ability of real-time categorization of food through one's smartphone [5]. A smart app can analyze meals' images that can estimate nutrition value and the calorie content available in it that will help consumers make the appropriate dietary choices while minimizing biases which are associated with self-reported measures, thus increasing healthy eating, which leads towards long-term health.

Many apps help track meal calories, but each has limitations. Fat Secret allows users to upload food photos, tag them, and get calorie details, but its diet plans require a subscription, and its barcode scanner does not work with Indian packaged foods [8]. It also lacks deep learning-based food recognition. Nutrition Plus provides nutritional research and meal plans but does not support image-based food identification. Calorie Mama uses computer vision for food image analysis and supports various cuisines, but premium access is needed for advanced features, and it lacks medical condition-based diet suggestions. These apps face challenges in recognizing regional foods, offering personalized diet plans, and integrating advanced technologies, highlighting the need for more accurate and user-friendly solutions.

[a]harirooba007@gmail.com, [b]madisettyharini@gmail.com, [c]pranitharoyal7410@gmail.com, [d]kaminiprashanth49@gmail.com, [e]yajamanhanumantharayappa@gmail.com

DOI: 10.1201/9781003684589-2

This project introduces a utility that can classify meals with the help of data, offers proper nutritional facts for Indian meals, and classifies meals based completely on the health condition of the user. For the sake of analysis, the individual's meals consumption, calorie values, blood pressure and diabetes information are provided. Additionally, helps burn calories through exercises or some changes in their day-to-day routines. This study provides a system learning-based method to predict the calorie burn of people using an XGBoost regression model [9]. The model has been trained based on a set of factors consisting of age, weight, type, and activity duration. Information preprocessing, which involves function selection and outlier detection, is performed to ensure optimum input for the version. This version is based on XGBoost rules, which have a good record of being extremely accurate and resourceful with the performance, mainly in regression-based tasks. Using the performance metric used, this version correctly predicts calorie expenditure.

Related Work

- Daniel Bubnis: The quantity of strength burned in normal existence is relying upon weight, top, age, and gender. The quantity of calories human beings need every day burn than they eat, reason a loss of calorie. However, it's miles honestly critical everyday understand what quantity of calories they burn 66b34c3da3a0593bd135e-66036f9aef3 life. calories are the unit of heat or strength this is required to elevate 1 g of water to 1°C. At the time of working quantity calories are burnt day-today, so it is essential for someone to try to maintain their body [6].
- Salvador Camacho: General obesity has been on the rise each day around the entire world and until now, not even a single country has been able to resolve it. The major cause of obesity is an energy imbalance between the calories taken and energy used. The concept of calorie imbalance cannot suffice to regulate and reverse the obesity epidemic. The regional health organization (WHO). There are many dailies that impact the calories burned, but every individual can alter their food plan chart or hobby stage every day to get the desired outcomes [7].
- Pouladzadeh et al. propose a cloud-based SVM system for food classification, utilizing images and videos for calorie tracking and diet planning.

Cloud computing manages image processing and SVM training, reducing the workload on devices. This enables scalability and efficiency, allowing smartphones to access advanced food recognition features. The system enhances health monitoring by providing an accessible and practical solution [3].
- Akshit Rajesh Tayadeet al. used logistics regression set of rules for food regimen advice device everyday support intellectual fitness and bodily health and accuracy of the proposed version turned into 86 [18].

Methodology

The variety of electricity burned every day is right now connected to weight loss, weight gain, or weight preservation. To shed kilos, someone have to burn extra calories than they absorb, developing a calorie deficit [11]. But, to try this, they want to recognize what number of calories they burn each day. Calories are typically thought of as having nothing to do with food or dieting. A calorie is a variable definition of heat or energy. The amount of energy required to increase one gram (g) of water by one degree Celsius is known as calories. Numerous processes outside of the human body that release electricity can be measured in this way. Energy in the context of the human body is a measure of how much power the body requires in order to function. As a way to be able to exercise how a whole lot of calories are burned every day it is critical for any person seeking to preserve, lose, or hold weight. Information on what elements make contributions to calorie burning can help someone alter their weight loss program or exercise software to cope with the purpose [12]. Android Studio served as the IDE for building and testing the application [16].

A. **Data collection:** Our assignment starts with the critical step of the statistics series. To prepare the required dataset, we refer to Kaggle, a well-known and relied upon records repository famous for its varied and rich datasets. In this case, we use Kaggle as the first source of information for our research. This deep dataset is eventually loaded into the Colaboratory (Colab) program, a powerful and flexible platform for data analysis. The scope of records encompassing both specific and numerical characteristics includes all information that can be retrieved. This system also leverages established nutritional data sources such as the Nutritive Value of Indian Foods for accurate dietary profiling [10].

B. **Data selection and loading:** Data selection involves identifying relevant information and tools before collection. After retrieval, data is cleaned, formatted, and loaded into storage, such as a cloud warehouse. This study uses a learning disabilities dataset for disease detection.

C. **Data processing:** This dataset contains 15,000 instances spread across two CSV files, "workout.csv" and "calorie.csv". Each entry represents an individual with attributes like height, weight, gender, age, exercise duration, heart rate, and body temperature. This well-structured dataset serves as the foundation for analyzing the relationship between these factors and calorie burn.

D. **Data analysis:** Using Google Colab, workout.csv and calorie.csv are analyzed to explore the link between exercise and body responses. Findings show that physically active individuals have higher body temperatures, averaging around 40°C. Visualization techniques, including charts and tables, enhance clarity, while correlation analysis identifies data patterns. Figure 2.4 provides a count-based visualization of the gender distribution in the dataset, supporting demographic analysis. Finally, the XGBoost Regressor model is used to predict calorie burn, with its accuracy evaluated against actual values.

E. **Machine learning model:** Within gadget mastering, this is very important because it implies the application of our chosen algorithms to approximate the mean absolute errors—a measure that is key in ascertaining the accuracy of predictions. On this case, we use more than one algorithm, including the XGBoost regressor, linear regression, choice tree regressor, and Random Forest Regressor, to explore and determine their respective performance levels. Using key overall performance indicators, we measure fashions at their ability of making accurate forecasts, thus finding the accuracy behind each algorithm at estimating calorie burns. More crucially, given the efficiency and proficiency of the calorie expenditure prediction as shown by a number of analyses, the choice of the selected XGBoost regression algorithm emerges.

F. **Evaluation:** This study projects the expenditure of calories based on the time used in exercise, age, gender, body temperature, and heart rate during exercising. The proposed working model of this calorie prediction system is illustrated in Figure 2.1. The best model for this task is supposed to be identified as its mean absolute error is the least to predict accurate calorie burn. With advanced techniques, physiological factors and the relationship with energy usage are further studied to obtain an estimate for calories that may support better decisions on health and fitness.

Classification of machine learning algorithm

XGBoost algorithm

A very potent machine-learning algorithm called XGBoost could help you gather information and make much better decisions. A gradient-boosting choice tree implementation is called XGBoost. Data scientists and analysts from all around the world have utilized it to improve their machine-learning models. By creating an iterative collection of choice trees (weaker learners), XGBoost learns the target work in a more substantial way by gradually reducing the goal work. Every cycle optimizes the objective work and adds an unneeded tree to the outfit.

The structure and working of XGBoost are illustrated in Figure 2.2.

The improved dispersed angle boosting library XGBoost is designed for a flexible and expert setup of machine learning models. XGBoost builds up a collection of learning methods. In other words, the method is more akin to a combining technique in which the expectations of multiple subpar models are combined to create an improved expectation. XGBoost, which stands for "Extreme Slope Boosting," is a well-known and widely used machine learning algorithm. Its popularity stems from its ability to process large amounts of data and even achieve state-of-the-art performance in various classification and regression tasks [17].

XGBoost stands for extraordinary [18]. Slope Boosting, which was proposed by the analysts at the College of Washington. It could be a library composed of C++ which optimizes the preparation for Slope Boosting.

One of the significant highlights of XGBoost is its proficient handling of lost values, which enables

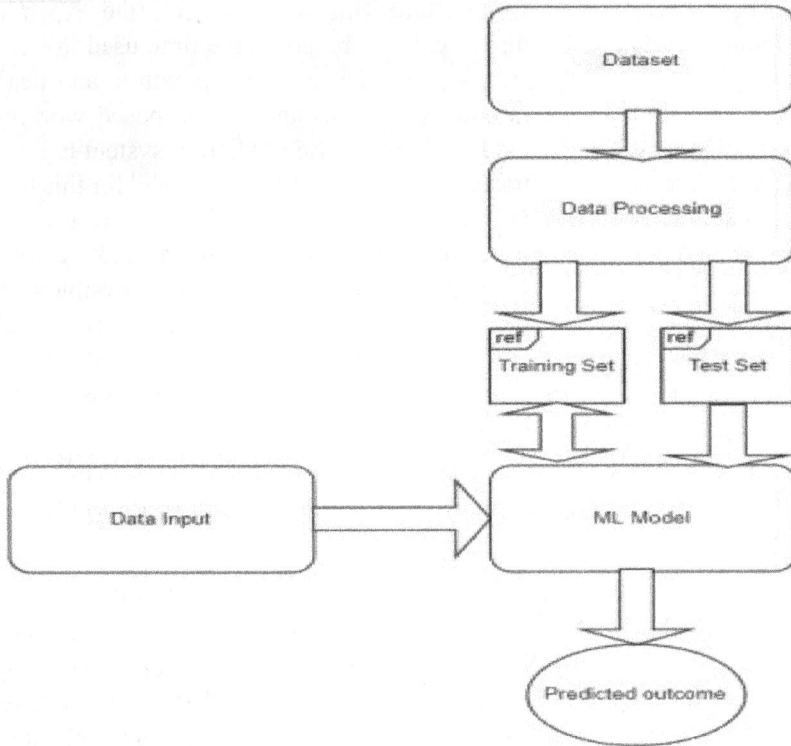

Figure 2.1 Proposed working model
Source: Author

XGBoost objective function

The objective function (loss function and regularization) at iteration t that we need to minimize is the following:

$$\mathcal{L}^{(t)} = \sum_{i=1}^{n} l\left(y_i, \hat{y}_i^{(t-1)} + f_t(\mathbf{x}_i)\right) + \Omega(f_t)$$

Real value (label) known from the training data-set

Can be seen as $f(x + \Delta x)$ where $x = \hat{y}_i^{(t-1)}$

XGBoost objective function analysis

Calculate The Similarity Scores For The Root And Leaf Nodes. Λ In The Equation Is The Regularization Parameter.

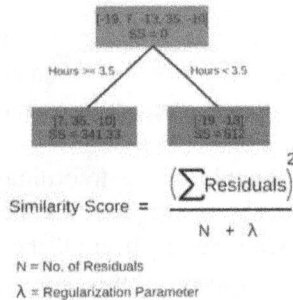

$$\text{Similarity Score} = \frac{\left(\sum \text{Residuals}\right)^2}{N + \lambda}$$

N = No. of Residuals
λ = Regularization Parameter

Figure 2.2 XGBoost objective function and similarity score calculation for tree nodes
Source: Author

it to deal with real-world information with lost values without demanding critical pre-processing. In addition, XGBoost has built-in support for parallel handling, and hence, preparing models on extensive datasets is feasible in a sensible amount of time.

XGBoost can be applied in a wide range of applications, including competitions at Kaggle, proposal frameworks, click-through rate prediction, among others. It is also very parameterizable and allows for the fine-tuning of various model parameters to improve its performance.

Results

For this project, Kaggle datasets were used: workout.csv: This document has seven columns: paintings length, body temperature, consumer id, gender, age, weight, and height. calories and user identity can be determined in energy.csv.

The user identity subject served as the idea for the dataset merger. After that, the combined dataset underwent preprocessing, which blanketed managing lacking values and one-hot encoding of specific variables like gender.

Model training

The choice here was an XGBoost algorithm for accuracy and efficiency in regression tasks. The set of data was split into training and testing, and GridSearchCV was used to conduct hyperparameter tuning. Mean absolute error, R-squared, and mean squared error were the parameters used to check the performance of the models. Figure 2.3 shows the accuracy comparison of various algorithms, highlighting the performance of XGBoost over others. The steps included: Breaking the dataset into a training set and a testing set preparing a hyperparameter tuning grid

- Use GridSearchCV to identify the best parameters
- Train the XGBoost model using the best identified parameters
- Test performance with the test set

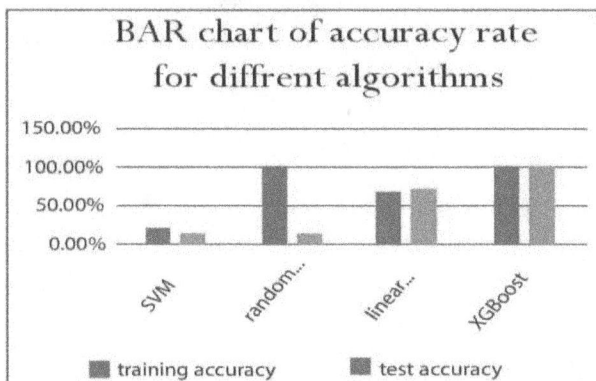

Figure 2.3 Bar chart of accuracy rate for different algorithms
Source: Author

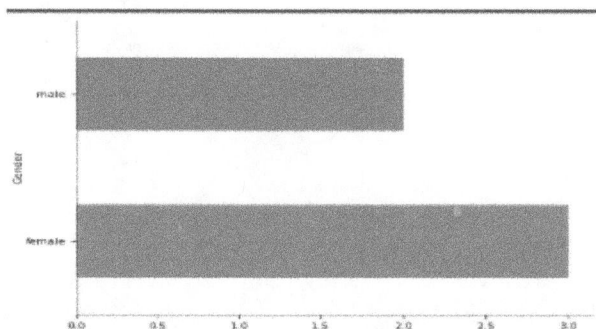

Figure 2.4 Count visualization of the data frame's "gender"
Source: Author

Conclusion

This paper introduces a user-friendly system that enables the person to monitor their calorie intake accurately. Utilizing advanced neural networks, the model has successfully classified food items and monitored calorie consumption with an accuracy of 97.65%. It further demonstrates the efficiency of XGBoost in predicting calorie burn more effectively than a traditional model when there are intricate, non- linear relationships in data. The study has outlined key burn factors in calorie, such as the required activity level, age, weight, and heart rate, that have been enhanced through XGBoost's boosting technique. This machine learning therefore gives a much better result than traditional approaches and will be useful for various fitness and health applications. Again, because no exhaustive dataset on all Indian meals has been developed so far, the system intends to use medical expertise to enhance its food database [13]. Future updates would be to further develop the application with more items in the menu, dietary recommendation generation for multiple health conditions, and linking the application with other health tracking applications like Google Suite[14]. Additional programming guidance was adapted from McClure work on professional Android development [15].

References

[1] Pan, L., Pouyanfar, S., Chen, H., Qin, J., & Chen, S. C. (2017). Deep food: automatic multi-class classification of food ingredients using deep learning. In IEEE 3rd International Conference on Collaboration and Internet Computing. DOI 10.1109/CIC.2017.00033.

[2] Wu, W., & Yang, J. (2009). Fast food recognition from videos of eating for calorie estimation. In Multimedia and Expo, ICME 2009. IEEE International Conference on, (pp. 1210–1213). IEEE.

[3] Pouladzadeh, P., Shi Mohammadi, S., Bakirov, A., Bulut, A., & Yassine, A. (2014). Cloud-based SVM for food categorization. *Multimedia Tools and Applications*, 74(14), 5243–5260. DOI 10.1007/s11042-014-2116-x.

[4] Podutwar, A. A., Pawar, P. D., & Shinde, A. V. (2017). A food recognition system for calorie measurement. *International Journal of Advanced Research in Computer and Communication Engineering*, 6(1), 243–248. DOI 10.17148/IJARCCE.2017.6146.

[5] Zhang, W., Zhao, D., Gong, W., Li, Z., Lu, Q., & Yang, S. (2015). Food image recognition with

convolutional neural networks. In 2015 IEEE 12th International Conference on Ubiquitous Intelligence and Computing and 2015 IEEE 12th International Conference on Autonomic and Trusted Computing and 2015 IEEE 15th International Conference on Scalable Computing and Communications and Its Associated Workshops (UIC- ATC-ScalCom). DOI 10.1109/UIC-ATC-ScalCom-CBDCom-IoP.2015.139.

[6]   Ramprasath, M., Anand, M. V., & Hariharan, S. (2018). Image classification using convolutional neural networks. *International Journal of Pure and Applied Mathematics*, 119(17), 1307–1319.

[7]   Jaswal, D., Sowmya, V., & Soman, K. P. (2014). Image classification using convolutional neural networks. *International Journal of Advancements in Research and Technology*, 3(6), 1661–1668.

[8]   Yim, J., Ju, J., Jung, H., & Kim, J. (2015). Image classification using convolutional neural networks with multi-stage feature. In Robot Intelligence Technology and Applications, 3: Results from the 3rd International Conference on Robot Intelligence Technology and Applications, (pp. 587–594). Springer International Publishing. DOI 10.1007/978-3- 319-16841-8_52.

[9]   Suryawanshi, S., Jughandle, V., & Mane, A. (2020). Animal classification using deep learning. *International Journal of Engineering Applied Sciences and Technology*, 4(11), 305–307. ISSN No. 2455-2143.

[10] Gopalan, C., Rama Sastri, B.V., & Balasubramanian, S.C. (2012). *Nutritive Value of Indian Foods*. National Institute of Nutrition, Hyderabad.

[11] Longvah, T., Ananthan, R., Bhaskarachary, K., & Venkaiah, K. (2017). *Indian Food Composition Tables*. National Institute of Nutrition, Hyderabad.

[12] National Institute of Nutrition. (2011). *Dietary Guidelines for Indians – A Manual*. Hyderabad, India.

[13] Department of Agriculture and Cooperation. (2007). *National Food Security Mission: Operational Guidelines*. Ministry of Agriculture, Govt. of India.

[14] Kothari, P. (2015). *Android Application Development*. Dreamtech Press.

[15] McClure, W., Dick, J., & Hardy, C. (2012). *Professional Android Programming with Mono for Android*. Wrox Press.

[16] Android Developers. (n.d.). *Android Studio Documentation*. Retrieved from https://developer.android.com/studio.

[17] Firebase. (n.d.). *Cloud Firestore Documentation*. Retrieved from https://firebase.google.com/docs/firestore.

[18] Tayade, A. R., & Patil, D. D. (2020). Food Recommendation System Using Machine Learning. *International Journal of Advanced Science and Technology*, 29(5), 4281–4286.

3 Real time design of FPGA-based home automation scheme

Rama Rao Chekuri[1,a], Nagababu Chekuri[2,b], Prabhakara Rao Kapula[3,c], D Hari krishna[4,d], Vurukonda Venkat[5,e] and Valluri Sai Mahesh[5,f]

[1]Assistant Professor, Department of ECE, B V Raju Institute of Technology, Narsapur, Telangana, India

[2]Assistant Professor, Department of ECE, MLRITM,Hyderabad,Telangana, Hyderabad, India

[3]Professor, Department of ECE, B V Raju Institute of Technology,Narsapur,Telangana, India

[4]Assiociate Professor, Department of ECE, B V Raju Institute of Technology, Narsapur, Telangana, India

[5]Student, Department of ECE,B V Raju Institute of Technology, Narsapur, Telangana, India

Abstract

Adding automation to your house is a wonderful approach to enhancing the "sweet home" atmosphere. The two main objectives of an automated house are security and comfort. In addition to monitoring the garage, the security system can detect fires and window and door break-ins. The temperature and light are the only two variables that the comfort system manages. This study presents an efficient Verilog HDL-based home automation system that communicates with sensors and devices and has a centralized Field Programmable Gate Array (FPGA) controller that the user may operate. The creation of electronic structures to accomplish a goal is discussed in this article. We utilized Verilog HDL to model the layout using Xilinx Vivado 2017.4. The wave illustrates the brief response for this research study, which is consistent with the intended outcome.

Keywords: Control system, FPFA, home automation, security, sensors, verilog HDL

Introduction

Everyone nowadays is constantly in need of security. During recessions, crime rates such as burglary and theft increase, prompting the need for increased home security measures [11]. A home automation system can meet this demand by creating a secure and comfortable living environment. The security system ensures protection against potential threats [3]. The comfort system allows the host to live in a cost environment [2]. Security is crucial for both personal and professional settings. When comfort and style come together, a house becomes the ideal location to live [1]. A basic home automation system attempts to provide comfort and safety in addition to the host's demands. There are several pricey items on the market [5]. It may be necessary to plan for home automation before beginning construction [6]. Furthermore, certain things are distinct. Just one app that prioritizes security above convenience can handle this issue. Home automation systems have the capability to run many devices simultaneously [14]. The owner may unwind and stop worrying about checking windows and doors all the time by doing this [8]. The home automation system consists of two parts. When the owner returns home, he rearms the system to turn on the house automation by entering a username and password to disarm it [10]. The moment users leave their homes, the safety mechanism is triggered.

The program receives information from a variety of sensors used by the system [4]. It takes a home safety system to stop robbery and theft. A vital component of any organization is "security" [12]. Humanity will benefit from the development of a safe system that permits temperature control and deteriorates burglaries [3]. These systems can serve a variety of purposes and are now on the market. Unfortunately, there is currently no technology accessible on the market for users looking to operate their entire home with a single system, so they confront a hurdle [7]. Home automation systems may also be used to control expensive household equipment [15]. When residents remain home or away, safety is ensured by having one-click controls for the thermostat, doors, windows, and fire alarms. This removes the requirement to go to each device and enables simple operation from a single

[a]ramarao.ch@bvrit.ac.in, [b]nagababu.chekuri@mlritm.ac.in, [c]prabhakar.kapula@bvrit.ac.in, [d]harikrishna.dodde@bvrit.ac.in, [e]23215a0426@bvrit.ac.in, [f]22211a04q4@bvrit.ac.in

DOI: 10.1201/9781003684589-3

place [13]. The controller performs frequent and sequential device checks. The whole thing may be controlled by a nearby or remote device to safeguard all necessary gadgets. Every gadget has sensors connected to it that serve as channels of communication. Despite the novelty of this concept, there will probably be a high demand for it once it is widely available. There will always be a need for something secure and safe [9].

Existing Problems

The creation of an affordable automatic home security system is the aim of this article. Signals are sent by the controller's variable sensors to initiate operations and carry out certain tasks. The bulk of commercial automation systems for homes require that all controlled items be certified for compatibility or come from the same manufacturer. Additionally, these systems often include a special control center device. If you wish to be able to control the system from many places, get extra control devices. Planning ahead is necessary when integrating complex systems during building construction. After installation, it is difficult to upgrade or repair them. For most people, the entire investment is frequently too much to pay. These issues lessen the allure of these systems Figure 3.1.

Proposed System

The recommended system seeks to offer a cost-effective house automation solution that tackles the drawbacks. The system is less expensive than systems that are sold commercially and provide basic appliance management Figure 3.2. Homeowner comfort and security are given priority in our suggested home automation system. The security system guards the garage and can identify fires as well as break-ins through windows and doors Figure 3.3. The only two things the comfort system controls are the light and temperature. Temperature, brightness, fire alarms, and doors are all managed by the system. It was not intended or set up to regulate anything else. Although the goal of this article is to control specific devices, it may expand to cover other devices or processes as well. Additionally, it may be web-enabled Figure 3.4.

This module processes all inputs and outputs them to the outside world. The checker module receives the user-input password. It adjusts security alarm levels based on other module output variables.

The security system is turned on and off using the checker module. The owner of the house configures an 8-bit passcode in [7:0] Figure 3.5. The module that verifies passwords manages home security configurations.

Particular door, garage, opening, and flame modules receive sensor data from the security module. The house automation module receives the outputs from the door, garage, opening, and fire modules Figure 3.6.

Module for doors: The door is now under continual observation. if the magnetic contact malfunctions when the door is opened. At this point, the door alarm

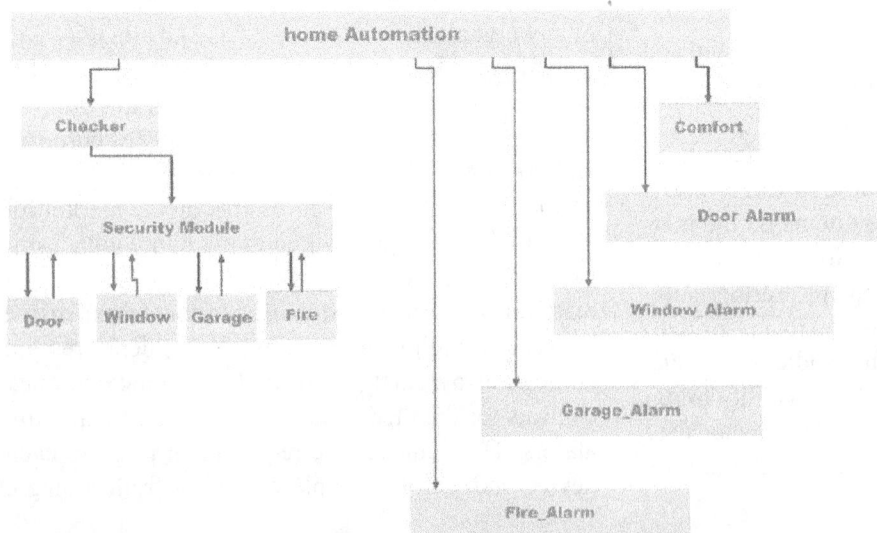

Figure 3.1 Proposed system of home automation system
Source: Author

sounds. The garage and window modules function in a similar way.

The smoke detector on the fire module is always in the low (0) position. The smoke detector sounds the fire alarm by sending out a loud signal when it finds smoke.

Block diagram

The sensors

A detector is a communication device that, in response to a physical quantity (input), produces a signal. Sensors are regarded as "input devices" when they are connected to a Field Programmable Gate Array (FPGA) or other larger control system. The following is another characteristic that sets a sensor apart: Data from a single energy domain is translated into the electrical domain by the device. Take a look at an example to help you comprehend the description of the sensor. The light-dependent resistor (LDR) is the most fundamental kind of sensor. The resistance of the gadget varies according to the level of light it receives. An LDR's resistance increases dramatically in low-light conditions and lowers in high-light conditions. Test a voltage drop across a voltage divider by connecting the LDR to it together with other resistors. The quantity of sunlight shining on the LDR may be used to calibrate the voltage. A light sensor follows Figure 3.6.

Magnetic sensors and laser sensors

Note a magnetic sensor measures the magnitude and geomagnetism caused by a magnet or current.

There are several different types of laser sensors; some measure distance, while others detect presence. While it is not the topic of this lecture, a near-field-type laser sensor, sometimes called a laser or photoelectric detector, is frequently employed for identifying the presence of a component. Laser distance sensors— which are employed to measure distance—are the subject of this discussion. Such laser distance sensors measure an object's distance from it using coherent, focused light. Controlling a particular product or machine component is a common goal of factory automation technologies. Regardless of material, color, or brightness, they are able to recognize anything that is solid and provide a response according to the distance measured Figure 3.7.

High-resolution output is provided by laser distance sensors for measurements of displacement or position. Displacement or displacement inputs are commonly received by commercial controllers, such as PLCs. Temperature compensation is included

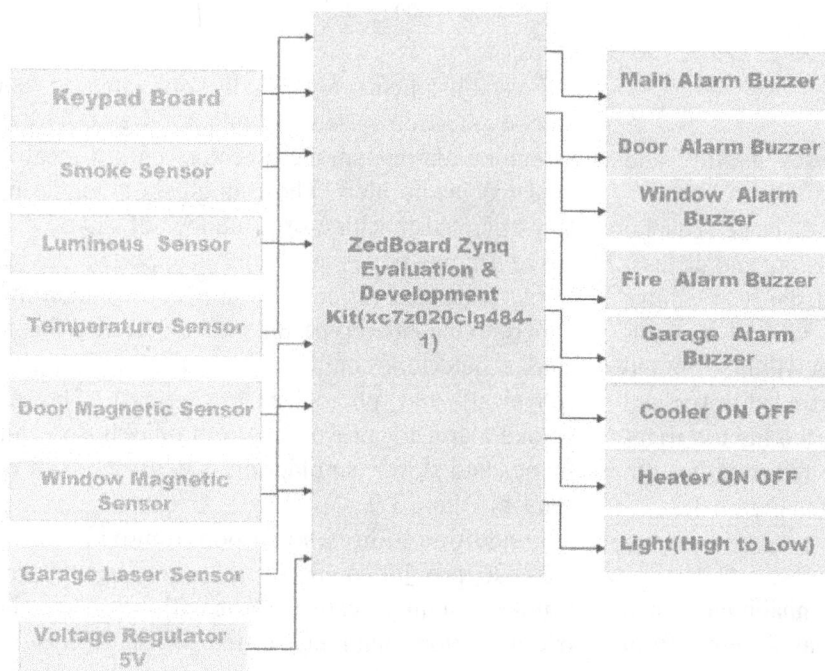

Figure 3.2 Block diagram of home automation system
Source: Author

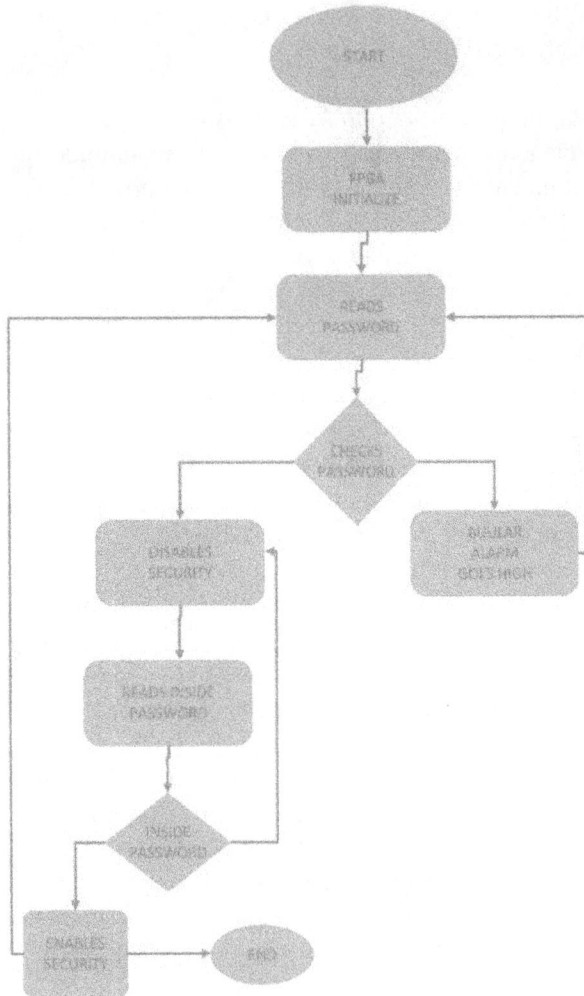

Figure 3.3 Flow chart
Source: Author

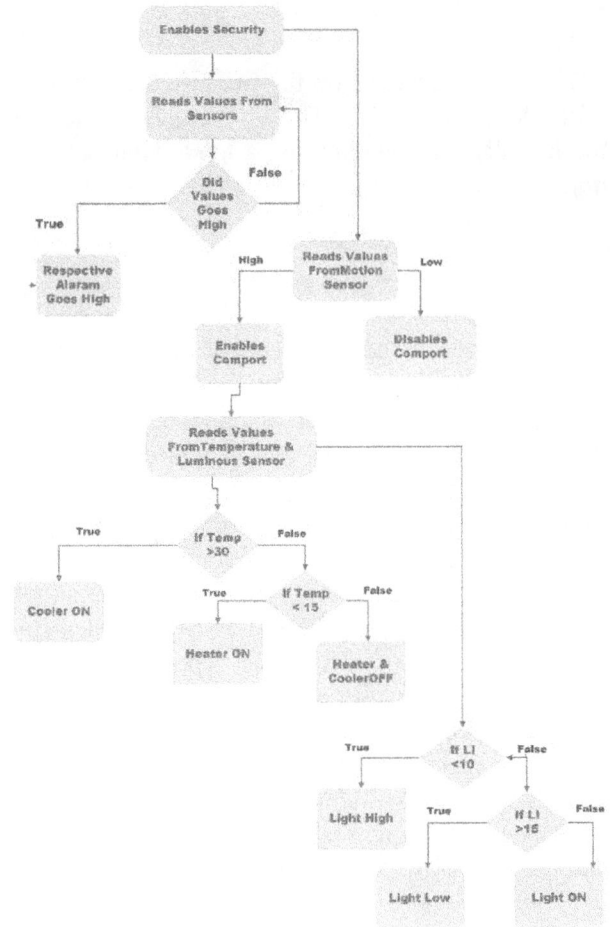

Figure 3.4 Flow chart
Source: Author

in the output signal, which increases stability and precision.

There are three varieties of laser distance sensors: retroreflective, background suppression, and diffuse. These sensors provide precise distance readings through the use of transit time or CMOS technology. Across far distances as well as within a limited spectrum, laser light has to be narrowed down and focused (color of light). After that, the light is triangulated, or pulsed, and each pulse return is used to determine the distance measurements Figure 3.8.

Smoke detector
A smoke detector warns building inhabitants when it detects smoke, a crucial fire signal. A signal from a commercial or industrial smoke detector is sent to an emergency alarm management panel, which is a component of the central system of a building.

All establishments are legally obligated to have smoke detection systems installed. Often referred to as smoke alarms, smoke detectors emit a localized visual or audio alert. These might be a single battery-operated machine or a number of linked hardwired (mains-powered) battery-supported devices. Large-scale renovations and new construction must adhere to the latter. types of smoke detectors Passive smoke detectors come in two varieties: ionization (physical) and photoelectric (optical). The best smoke alarm for preventing both quickly spreading flames and slowly smoldering ones is one with two sensors Figure 3.9.

In addition to fire and carbon monoxide alarms, we also provide ocular smoke and heat alarms. Smoke instantly scatters light within a photoelectric detection chamber, setting off an alert. Often responding in 15 to 50 minutes, photoelectric smoke detectors react to smoldering flames more quickly than ionization alarms. You can put them

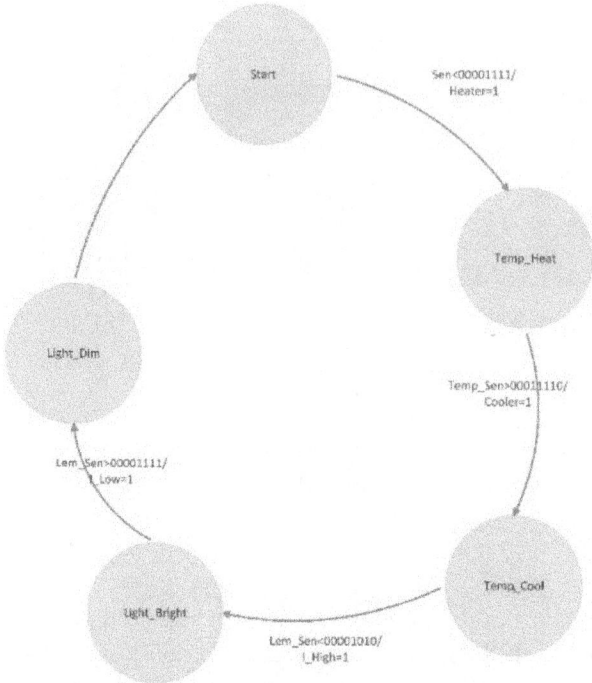

Figure 3.5 State diagram
Source: Author

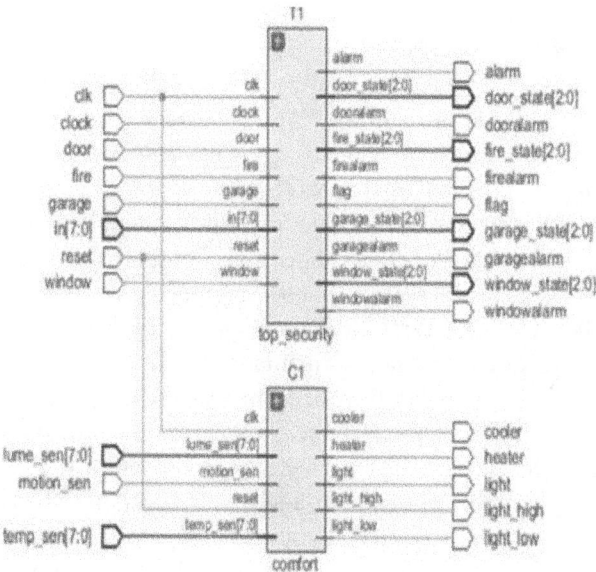

Figure 3.6 Schematic diagram
Source: Author

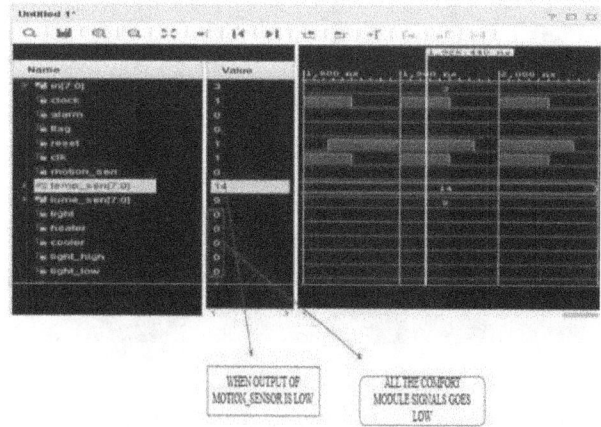

Figure 3.7 When the given input password is correct, then it disables all the security system by providing flag signal high
Source: Author

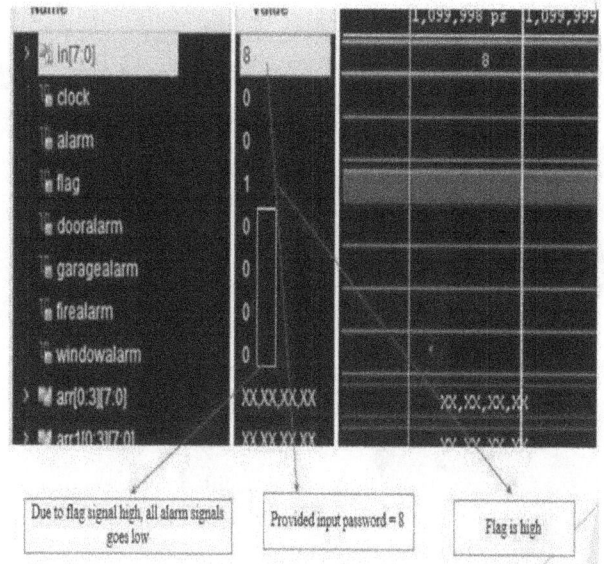

Figure 3.8 When the given input password is wrong, main alarm signal goes high
Source: Author

next to kitchens. There are a few dual- optical variations available. Ionization smoke alarms are incredibly sensitive to the ionization of minute smoke particles Figure 3.10. Smoke alarms are not effective against smoldering flames, but they do detect tiny particles of smoke and react to rapidly burning fires (30–90 minutes faster than photo-electric alarms). The likelihood of activation may rise if they are installed too close to garages or kitchens. Ionization alarms produce electricity by ionizing the air with a little quantity of radioactive material placed between charged plates. Ion flow is disrupted by smoke in the chamber, which lowers current and triggers the alarm. Install the appropriate kind of smoke detector to avoid false

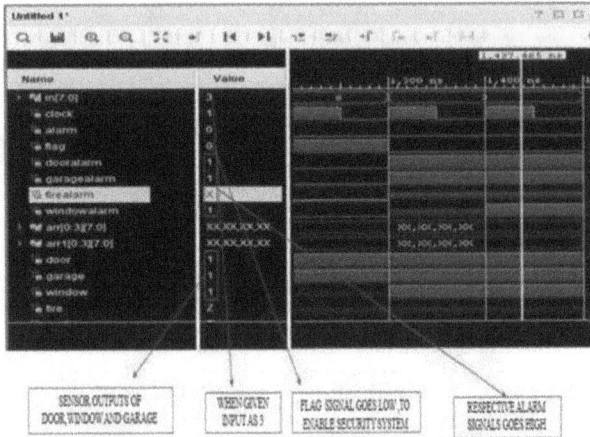

SENSOR OUTPUTS OF DOOR, WINDOW AND GARAGE | WHEN GIVEN INPUT AS 3 | FLAG SIGNAL GOES LOW, TO ENABLE SECURITY SYSTEM | RESPECTIVE ALARM SIGNALS GOES HIGH

Figure 3.9 After entering the home, if we provide input password as 3, then it enables the flag signal low and enables the security system. As after enabling the security, if a door or window or garage door is opened, then respective security Alarm goes high
Source: Author

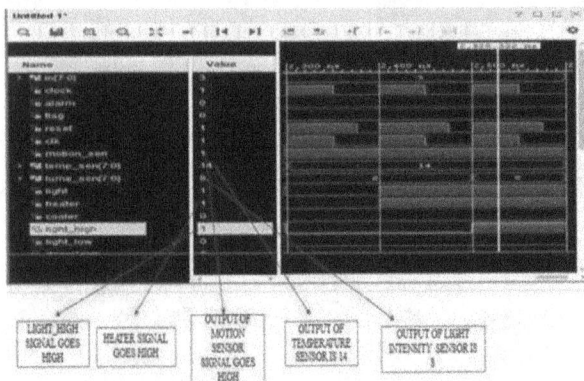

LIGHT_HIGH SIGNAL GOES HIGH | HEATER SIGNAL GOES HIGH | OUTPUT OF MOTION SENSOR SIGNAL GOES HIGH | OUTPUT OF TEMPERATURE SENSOR IS 14 | OUTPUT OF LIGHT INTENSITY SENSOR IS 3

Figure 3.10 When any fire accident occurs, then it immediately activates the fire alarm
Source: Author

alerts brought on by moisture or dust better suited Figure 3.11.

Thermostat

A temperature control on a wall for heating is not a thermometer, even if it is marked in degrees. A thermostat is a modern term derived from two ancient Greek terms: thermo (meaning heat) and states (meaning standing and connected to words like stasis, status quo, and static—meaning to remain constant). A thermostat, as the name suggests, regulates the temperature in our home. When it's too cold, it turns on the heating to quickly warm things up.

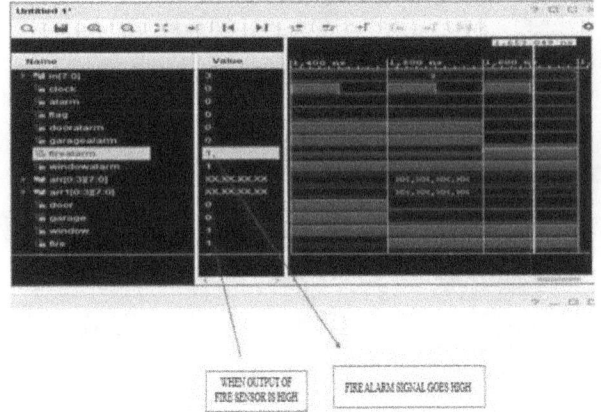

WHEN OUTPUT OF FIRE SENSOR IS HIGH | FIRE ALARM SIGNAL GOES HIGH

Figure 3.11 All comfort module signals will be low, until the motion _ sensor signal goes high
Source: Author

When the temperature reaches the desired level, it turns off the heating to prevent boiling Figure 3.10.

Water, unlike most other substances, expands both when heated and when frozen. Mechanical thermostats employ thermal expansion to turn an electric circuit on and off. The most prevalent types include bimetallic strips and gas-filled bellows.

Light dependent resistor

Electronic circuits typically use light dependent resistors (LDRs) or photo resistors to detect light levels.

Electronic components can be referred to as light-dependent resistors (LDRs), photo resistors, photocells, or photoconductors. While other electronic components like photodiodes and phototransistors can be employed, LDRs or photo-resistors are more commonly used in circuit designs Figure 3.12. They offer significant resistance to fluctuations in light level.

LDRs are widely utilized in various applications due to their inexpensive cost, ease of fabrication, and user-friendliness. LDRs were originally employed in photographic light meters and are now used in various applications to detect light levels. Light-dependent resistors are commonly accessible. Electronic component distributors are typically the primary source for obtaining these items due to the current supply chain in the electronics sector. Electronic component distributors, both large and small, often offer a wide assortment Figure 3.12.

Alarm buzzer

In a circuit, a buzzer produces an audible signal when a voltage is supplied to it. It may be piezoelectric,

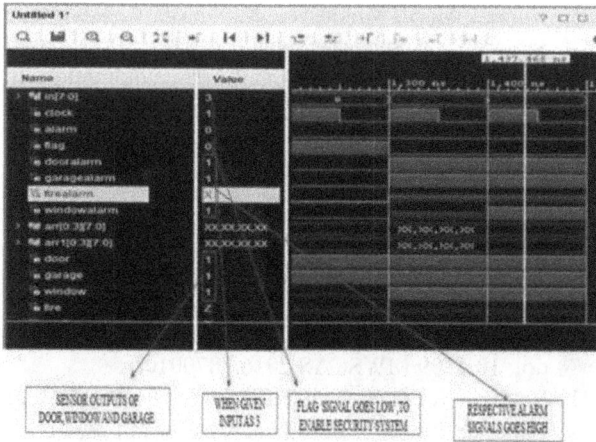

Figure 3.12 Here the motion sensor signal goes high, then it checks the room temperature. If the temperature is less than 15° Celsius, then immediately heater signal goes high. If the light intensity inside room is less than 10 lumens, then light high signal goes high
Source: Author

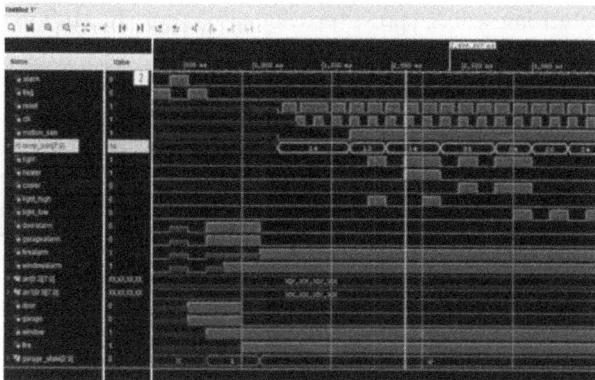

Figure 3.13 When the temperature inside room is greater than 30° Celsius, then immediately the cooler signal goes high. If the light intensity inside room lies between 10–15 lumen, normal light signal goes high
Source: Author

electromechanical, or mechanical. A piece of electricity that regulates current flow within a circuit is called a switch Figure 3.13.

But can a buzzer act as a switch in a circuit? A buzzer cannot be used as a switch in circuits. When a voltage is applied to it, it produces an audio signal that sounds like "buzzing" or "beeping". Unlike switches, it cannot open and close circuits. A buzzer generates an auditory signal in a circuit when a voltage is applied to it. It can be mechanical, electromechanical, or piezoelectric. A switch is an electrical component that controls current flow inside a circuit.

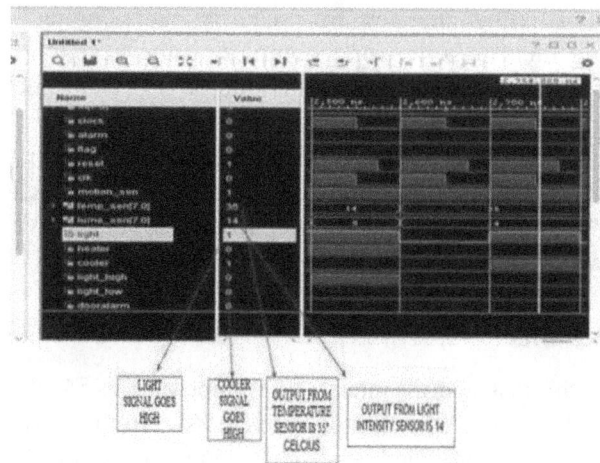

Figure 3.14 Overall behavioral simulation of our home automation system
Source: Author

But can a buzzer act as a switch in a circuit? A buzzer cannot be used as a switch in circuits. When a voltage is applied to it, it produces an audio signal that sounds like 'buzzing' or 'beeping'. Unlike switches, it cannot open and close circuits. The primary purpose of a buzzer and switch in a circuit many electrical components are employed in circuits every day. Each component in the circuit has a specific purpose and contributes to its overall functionality. Both switches and buzzers serve a specific role in a circuit. Understanding what each accomplishes can assist Figure 3.14.

The comfort system turns on when the movement sensor is high; otherwise, all of the appliances switch off. On positive clock edges, the motion_sen, temp_sen, and lume_sen parameters are validated. The proposal The heating element begins in the 0001 state when temp_sen has been set to 8'b0 and motion_sen is high. The next stage is light_bright (0011), wherein light_high is set high to improve light intensity. Lumen_sen is also set to 8'b0. Finally, both heater and light_high is set to zero, bringing the system to the start state (0000). The clock's positive edge initiates the cooling state (0010) when the cooler starts, whenever the temp_sen has been set to 8'b11111111. The lume_sen is tested with the setting set to 8'b11111111 when the chiller first operates. Light_dim (0100) with a high light_low is the next stage.

To create a secure system, the first step is to identify the necessary equipment based on hardware design concepts. The initial stage involves shortlisting devices for security and connecting them to a network, completing the necessary procedures. The devices include doors (front and rear), windows, fire alarms, and temperature controllers. Create a basic block diagram of all equipment, sensors, inputs, and outputs. The block diagram clearly identifies which inputs drive which outputs.

Results

In this instance, the intersection temperature is 54.5°C and the predicted chip power is 2.56W. In this case, the power provided to off-chip components is 0 W, and the thermal buffer is 30.5°C. The entire section on the right displays the amount of energy utilized. The total No of look up tables present on the FPGA board are 53200, but for our project we utilized only 26 LUTs. We used 15 flip-flops out of 106400 and 55 input /output ports out of 200. Therefore, utilization % of LUT, FF, I/O are 0.05%, 0.01%, 27.50% respectively.

Conclusion and future scope

In this article successfully achieved its goal of designing a controller for an economical automated home security system. Despite several assumptions, efforts were made to make it as practical as feasible. The system checks devices in a priority-based order and displays their current state. This gadget is cost-effective and may be upgraded to provide web functionality with little changes.

The aim to achieve these goals through advancements in both data- mining and big data computing. Pert is a cutting-edge home automation solution that enables smartphone-based management, monitoring, and security for your house. Future healthcare providers will see smart homes as a viable means to provide remote healthcare, especially for the elderly and crippled who don't require rigorous healthcare. As technology advances, future houses will be more automated than today's.

References

[1]   Alim, M. E., Alam, N. S. B., & Ahmad, S. (2020). RFID based security and home automation system using FPGA. In *2020 IEEE 7th International Conference on Engineering Technologies and Applied Sciences (ICETAS)*, Kuala Lumpur, Malaysia, (pp. 1–5). doi: 10.1109/ICETAS51660.2020.9484179.

[2]   Ahmed, M. S., Mukherjee, R., Ghosh, P., Nayemuzzaman, S., & Sundaravdivel, P. (2022). FPGA-based assistive framework for smart home automation. In *2022 IEEE 15th Dallas Circuit And System Conference (DCAS)*, Dallas, TX, USA, (pp. 1–2). doi: 10.1109/DCAS53974.2022.9845625.

[3]   Wightwick, A., & Halak, B. (2016). Secure communication interface design for IoT applications using the GSM network. In *2016 IEEE 59th International Midwest Symposium on Circuits and Systems (MWSCAS)*, Abu Dhabi, United Arab Emirates, (pp. 1–4). doi: 10.1109/MWSCAS.2016.7870010.

[4]   Baldwin, R., Bobovych, S., Robucci, R., Patel, C., & Banerjee, N. (2015). Gait analysis for fall prediction using hierarchical textile-based capacitive sensor arrays. In *2015 Design, Automation & Test in Europe Conference & Exhibition (DATE)*, Grenoble, France, (pp. 1293–1298). doi: 10.7873/DATE.2015.0943.

[5]   Sallah, A., & Sundaravadivel, P. (2020). Tot-Mon: a real-time internet of things based affective framework for monitoring infants. In *2020 IEEE Computer Society Annual Symposium on VLSI (ISVLSI)*, Limassol, Cyprus, (pp. 600–601). doi: 10.1109/ISVLSI49217.2020.00093.

[6]   Sundaravadivel, P., Goyal, V., & Tamil, L. (2020). I-RISE: an IoT-based semi-immersive affective monitoring framework for anxiety disorders. In *2020 IEEE International Conference on Consumer Electronics (ICCE)*, Las Vegas, NV, USA, (pp. 1–5). doi: 10.1109/ICCE46568.2020.9043156.

[7]   Sundaravadivel, P., Salvatore, P., & Indic, P. (2020). M-SID: an IoT-based edge-intelligent framework for suicidal ideation detection. In *2020 IEEE 6th World Forum on Internet of Things (WF-IoT)*, New Orleans, LA, USA, (pp. 1–6). doi: 10.1109/WF-IoT48130.2020.9221279.

[8]   Wilmoth, P., & Sundaravadivel, P. (2021). An interactive IoT-based framework for resource management in assisted living during pandemic. In *2021 22nd International Symposium on Quality Electronic Design (ISQED)*, Santa Clara, CA, USA, (pp. 571–575). doi: 10.1109/ISQED51717.2021.9424323.

[9]   Suruliandi, A., Meena, K., & Rose, R. (2012). Local binary pattern and its derivatives for face recognition. *IET Computer Vision*, 6(8), 480–488.

[10] Mendez, M., Carrillo, J., Martin, O., Tchata, C., Sundaravadivel, P., & Vasil, J. (2019). EasyYard: an IoT-based smart controller for a connected backyard. In *2019 IEEE International Symposium on Smart Electronic Systems (iSES) (Formerly iNiS)*, Rourkela, India, (pp. 257–261). doi: 10.1109/iSES47678.2019.00064.

[11] Sundaravadivel, P., Kesavan, K., Kesavan, L., Mohanty, S. P., & Kougianos, E. (2018). Smart-log: a

deep-learning based automated nutrition monitoring system in the IoT. I*IEEE Transactions on Consumer Electronics*, 64(3), 390–398. doi: 10.1109/TCE.2018.2867802.

[12] Kougianos, E., Mohanty, S. P., Coelho, G., Albalawi, U., & Sundaravadivel, P. (2016). Design of a high-performance system for secure image communication in the internet of things. I*IEEE Access*, 4, 1222–1242. doi: 10.1109/ACCESS.2016.2542800.

[13] Cedillo-Elias, E. J., Orizaga-Trejo, J. A., Larios, V. M., & Maciel Arellano, L. A. (2018). Smart government infrastructure based in SDN networks: the case of guadalajara metropolitan area. In *2018 IEEE International Smart Cities Conference (ISC2),* Kansas City, MO, USA, (pp. 1–4). doi: 10.1109/ISC2.2018.8656801.

[14] Yadav, P., & Vishwakarma, S. (2018). Application of internet of things and big data towards a smart city. In 2018 3rd International Conference On Internet of Things: Smart Innovation and Usages (IoT-SIU), Bhimtal, India, (pp. 1–5). doi: 10.1109/IoT-SIU.2018.8519920.

[15] Archana Rani, Naresh Grover, An Enhanced FPGA Based Asynchronous Microprocessor Design Using VIVADO and ISIM,Bulletin of Electrical Engineering and Informatics (BEEI) ISSN: 2089-3191, e-ISSN: 2302-9285,doi.org/10.11591/eei.v7i2.818.

4 AI powered facial recognition for secure banking authentication

Alabazar Ramesh[1,a], Karanam Jyothirmayi[2,b], Kollu Vishnuvardhan[2,c], Dayyala Ajay Kiran[2,d] and Ediga Sheshadri[2,e]

[1]Assistant Professor, Department of CSE, Rajeev Gandhi Memorial College of Engineering and Technology, Nandyal, India

[2]Students, Department of CSE, Rajeev Gandhi Memorial College of Engineering and Technology, Nandyal, India

Abstract

Facial recognition technology, powered by artificial intelligence (AI), has arisen as a transformative tool in the banking sector, revolutionizing security and customer experience. This paper explores the development and integration of AI-driven facial recognition systems for secure banking authentication, fraud prevention, and streamlined customer onboarding processes. Traditional security measures such as passwords and personal identification number (PINs) are vulnerable to breaches and fraud, whereas facial recognition provides a robust, password alternative that ensures only authorized users can access sensitive accounts and services. By leveraging deep learning algorithms and large datasets, AI-based systems can accurately recognize individuals, even in challenging conditions like occlusions (e.g., face masks) and varied lighting, aging and changes in appearance, similarity between faces. The system enhances know your customer (KYC) compliance by automating identity verification and accelerates remote customer onboarding, offering seamless and contactless banking experience. We aim to demonstrate how AI powered facial recognition can secure digital and physical banking services, reduce operational costs, and build greater customer trust by ensuring privacy and security in all interactions.

Keywords: Authentication, banking, deep neural networks, facial recognition, identity verification, privacy, tkinter Deep neural networks, Facial recognition, Tkinter

Introduction

In recent years, banking processes that once required in-person visits to traditional bank branches have shifted toward online and digital banking. This transition has significantly improved convenience for customers and efficiency for financial institutions [1]. However, the rise of digital banking has also led to an increase in cyber threats, fraud, and identity theft, making secure authentication mechanisms crucial [2]. Traditional security measures, such as passwords and personal identification number (PINs)are vulnerable to hacking, phishing, and social engineering attacks [3]. To address these challenges, financial institutions are increasingly adopting biometric authentication, including facial recognition, which provides a more secure and user-friendly alternative [4]. While facial recognition has been explored in banking since 2017, most existing implementations focus on identity verification at ATMs or mobile banking logins

Unlike previous studies that focus solely on facial recognition for user authentication, this paper presents an integrated facial recognition-based banking system that enables users to perform transactions without requiring passwords or PINs. The system not only authenticates users but also allows them to perform core banking operations—online transfers, balance enquiry, deposits, and withdrawals—entirely through face recognition. The face recognition module is implemented using Python and OpenCV, while banking records are stored in an Excel sheet for simplicity. This approach ensures a lightweight, easily deployable solution suitable for small-scale banking applications and fintech solutions. The primary objective of this study is to demonstrate how a fully functional facial authentication system can be effectively linked with a banking system to improve security, enhance user experience, and reduce reliance on traditional authentication methods.

[a]ramesh.alabazar@gmail.com, [b]karanamjyothirmayi2004@gmail.com, [c]vishnuvardhan2004.kollu@gmail.com, [d]ajaykirand2004@gmail.com, [e]sheshadrigoud123@gmail.com

DOI: 10.1201/9781003684589-4

Literature Review

According to Jain's study presents a face authentication system to enhance security in banking operations, consisting of image acquisition, model training and recognition, and integration with bank databases. It aims to improve account security, mitigate fraud, and enable contactless banking at branches and ATMs. However, challenges like accuracy issues under diverse conditions and vulnerability to spoofing attacks necessitate continuous improvements for reliability in real world applications [5].

According to Jayanthan's study examines the integration of facial recognition technology in banking, emphasizing business requirements, security needs, and technical specifications. By replacing traditional authentication methods like passwords and PINs, it enables seamless, contactless verification for banking transactions. However, challenges such as poor lighting, shadows, and appearance changes necessitate robust algorithms and adaptive models for reliable performance [6].

According to Dr. Umamaheswari's study explores artificial intelligence (AIs) transformative role in banking through a structured research approach, analyzing its impact on automation, error reduction, and operational efficiency. While AI-driven technologies like facial recognition offer benefits, they also raise concerns about intrusiveness, affecting user adoption. Addressing these concerns is crucial for ensuring trust and acceptance in banking innovations [7].

According to Sharma highlights the growing reliance on net banking and the rising threat of fraud, particularly in government bank portals that lack comprehensive information on schemes and investments. The proposed system enhances security and efficiency by using Firebase for authentication, integrating third- party applications to track IP addresses and locations, and sending this data with verification links. Transactions require facial recognition via Microsoft Azure Face API, with video recordings stored in cloud storage for added security. This multi-layered approach aims to prevent fraud while ensuring users can easily access relevant financial information [8].

According to Mohanraj et al., face recognition system uses facial recognition technology to identify individuals and retrieve their bank account details efficiently. By automating the process through facial biometric identifiers, the system eliminates the need for manual searches based on names or account numbers, enabling users to access transaction information quickly. A camera captures the user's face, which is processed through detection and encoding algorithms to retrieve the associated banking details, reducing wait times, minimizing crowding, and improving the overall customer experience in banks [9].

Arunadevi et al. propose an online banking system that enhances security through real-time face recognition as a biometric authentication method, addressing the limitations of traditional password-based systems. By integrating a Grassmann learning algorithm, the system analyses and matches facial features with bank records for identity verification. It also includes a second verification step using OTPs in reverse order for added security. The user-friendly interface supports essential banking tasks like fund transfers and bill payments, with notifications for account access and transactions to ensure transparency, simplifying the online banking experience while improving security [10].

According Johora the research investigates the application of AI in fraud detection to ensure the security of financial transactions. The study follows a systematic approach that includes data preparation, standardization, encoding of categorical variables, feature extraction, dimensionality reduction, and model evaluation to build effective fraud detection systems. Beyond enhancing security, AI also fosters innovation in the banking sector by enabling the development of new products, services, and business models. However, implementing AI systems comes with challenges, particularly the substantial investment required in hardware, software, skilled professionals, and infrastructure. These costs may pose significant barriers, especially for smaller banking institutions [11].

System Architecture

Authentication and user login

By exposing their faces to the system, users try to log in. The system compares stored embeddings with real-time face data. If face matches, then user is prompted to input their password for two-factor authentication in order to verify their identity. If the user's face does not match, the system may block access since it considers them to be a new or unidentified person.

Verification of identity

Even once facial recognition is effective, the user must still input their password as an extra security precaution. If the password is correct, access is granted by the system, enabling the user to continue with their banking activities. Access is refused if the password is incorrect.

Module for transactions

The user can carry out a number of banking tasks after being authenticated, including checking their balance, transferring funds, making deposits, and withdrawing money. To carry out the desired transactions, the system safely retrieves the required information from the bank's database.

Error handling and security

To safeguard user information and stop illegal activities, the system blocks access if the password or face verification is unsuccessful. The bank's database securely handles all transaction requests and data retrievals. By integrating facial recognition and password-based authentication the system architecture provides strong, multi-layered security, lowers the possibility of unwanted access, and enhances user experience in general.

Proposed Work

OptiFace, is a secure and efficient facial recognition-based banking system. The primary objective of OptiFace is to provide users with a streamlined, user-friendly digital banking experience while ensuring the highest levels of security for transactions. The system encompasses the following core functionalities.

Account Creation

To create a new account, users begin by clicking the "Open a New Account" button. This action initiates the account creation process, where users are required to provide their full name and set a secure password for future logins.

The system then activates the camera, prompting users to capture five facial images for training the facial recognition model. These images ensure precise identification during future logins.

The above Figures 4.1 to 4.3 depict how account is to be opened in our system.

Upon completing this process, the system generates a unique ID, account number, and fields such as full name, bank name, and balance, all of which are securely stored in an Excel sheet. The initial deposit required to open an account is INR 1000.

Login to account

To access their account, users must click on the "Login to Account" button. The system activates the camera and captures the user's face after a 5-second countdown. The facial recognition model compares

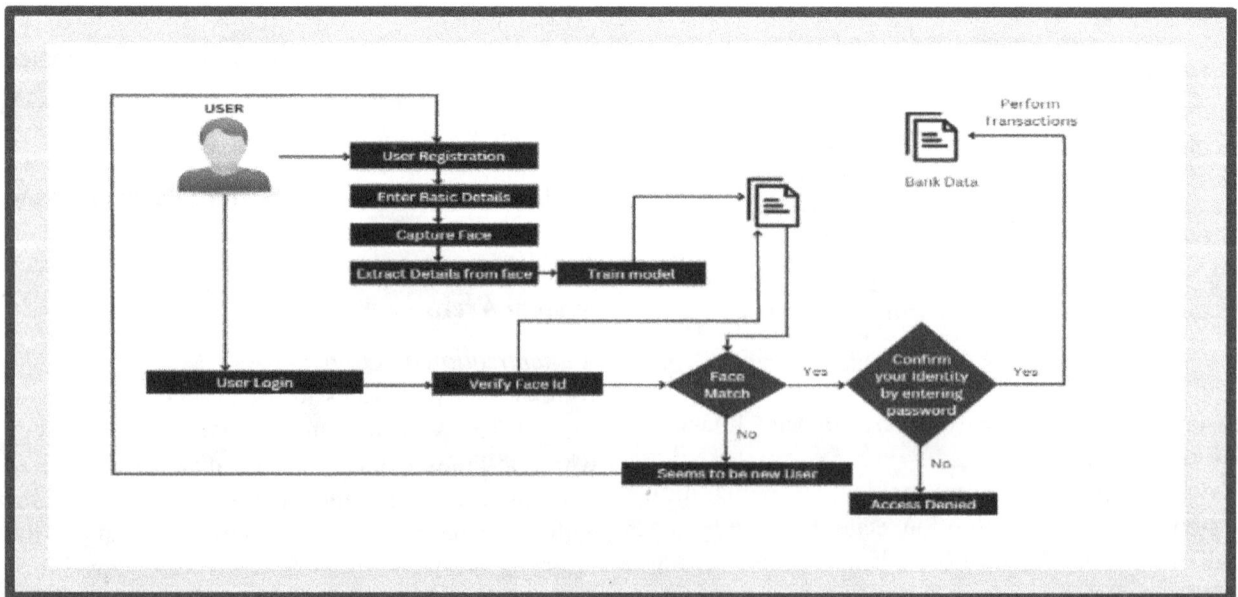

Figure 4.1 Account creation steps
Source: Author

Figure 4.2 Account created successfully
Source: Author

Figure 4.3 Account created successfully
Source: Author

Figure 4.4 Capturing face for login purposes
Source: Author

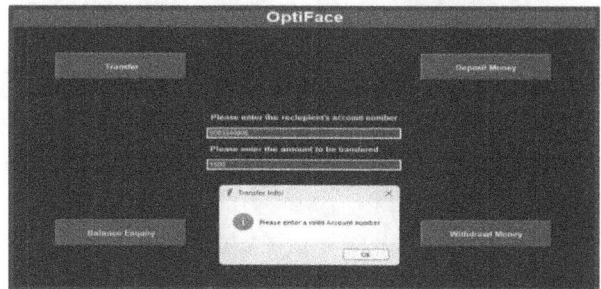

Figure 4.5 Password entry
Source: Author

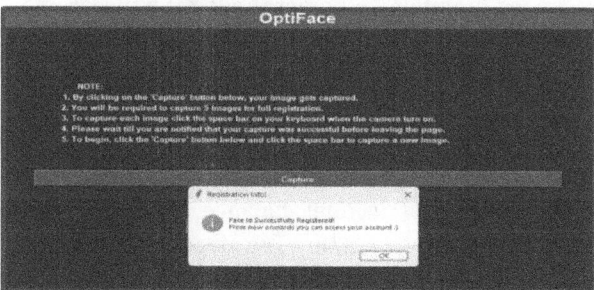

Figure 4.6 Invalid account number
Source: Author

Figure 4.7 Self transfer
Source: Author

Figure 4.8 Transfer success
Source: Author

the current user's face with the stored data. Upon successful facial match, the user is prompted to enter their account password. If the password is correct, access is granted; otherwise, the user must retry the verification process.

The above figure 4.4 and 4.5 depict how account user login to his/her account in our system.

This dual-layer authentication, combining facial recognition with password verification, significantly enhances the system's security. The login process is designed to be intuitive and efficient, minimizing the risk of unauthorized access.

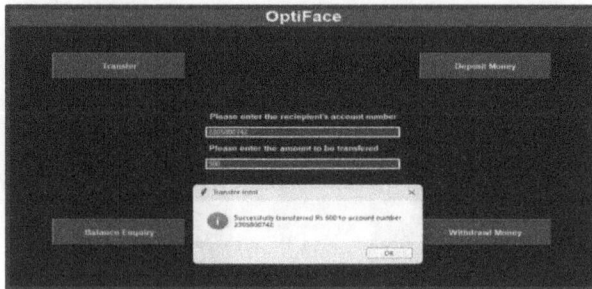

Figure 4.9 Balance enquiry
Source: Author

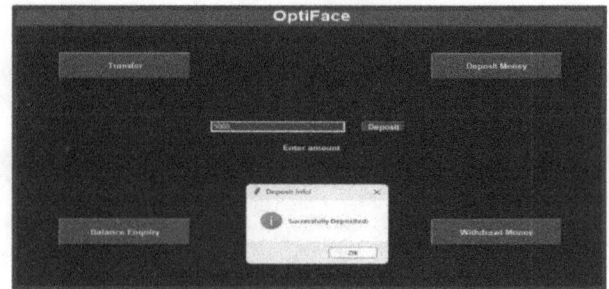

Figure 4.11 Invalid Withdraw
Source: Author

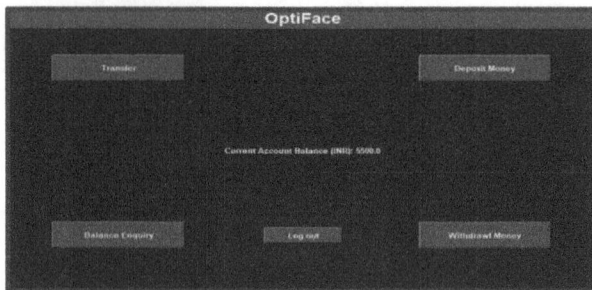

Figure 4.10 Deposit
Source: Author

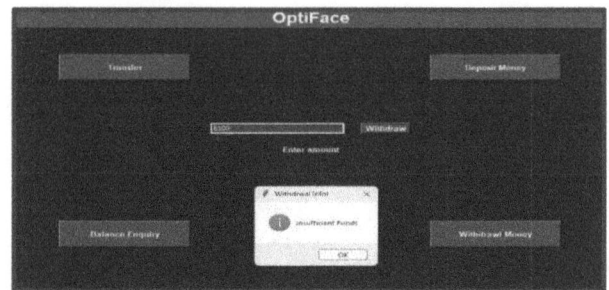

Figure 4.12 Withdraw success
Source: Author

Banking operations
Once logged in, the system provides the following core banking functionalities like transfer, deposit money, balance enquiry and withdraw.

Transfer
Users can transfer funds by entering the recipient's account number and the desired amount. The system validates the transaction by ensuring the entered amount does not exceed the user's account balance. It also checks for invalid account numbers or attempts to transfer funds to the user's own account, which are flagged as invalid upon successful validation, the transaction is executed, and the transaction details are displayed. For unsuccessful transactions, appropriate message is shown to the user.

Balance enquiry
This feature allows users to view their current account balance by clicking the "Balance Enquiry" button. The balance enquiry process is quick, enabling users to keep track of their finances in real time.

Deposit
Users can deposit money by entering the amount in a text box and clicking the "Deposit" button. The entered amount is added to the user's account balance, and the updated balance is displayed. This feature facilitates easy account top ups, ensuring that users can maintain sufficient funds for transactions.

Withdraw
It users can withdraw funds by entering the desired amount in a text box and clicking the "Withdraw" button. The system deducts the specified amount from the user's account balance, provided sufficient funds are available. This feature offers users convenient access to their funds while ensuring that account balance constraints are respected.

The above Figures 4.11 and 4.12 show us withdraw feature in detail.

Conclusion and Future Work

Development of OptiFace, a facial recognition-based banking system, demonstrates the potential of advanced biometric technologies in enhancing the security and efficiency of digital banking. By integrating facial recognition with traditional banking functionalities, the system provides a seamless and user-friendly experience while ensuring robust protection against unauthorized access and fraud. The implementation of core banking operations, such as account creation, balance inquiry, fund transfer,

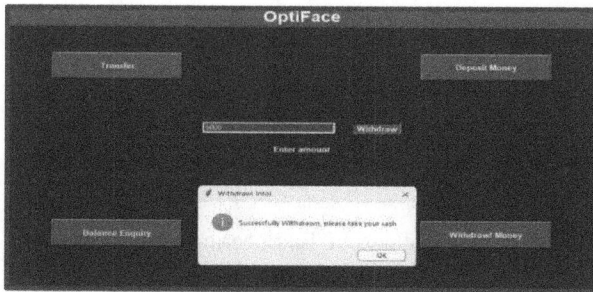

Figure 4.13 Transaction Successful
Source: Author

deposit, and withdrawal, coupled with dual-layer authentication, ensures both convenience and reliability. OptiFace represents a step forward in modernizing financial transactions and establishing trust through technology-driven solutions.

While the current system offers a secure and efficient framework, there is scope for further enhancement and innovation. Future work may include:

i.   Liveness detection: Implementing advanced liveness detection techniques, such as blink detection, motion analysis, or thermal imaging, to ensure that the system can differentiate between real users and spoofing attempts using photos or videos.

ii. Multi-factor (MFA): Authentication Integrating additional biometric methods, such as fingerprint or voice recognition, for enhanced security.

iii. Cloud storage: Transitioning from local Excel-based storage to secure cloud databases for improved scalability and data management.

iv. Mobile app integration: Developing a mobile application to provide users with easy access to banking services on the go.

These enhancements will not only expand the system's capabilities but also ensure its adaptability to future technological and market demands, making OptiFace a versatile and sustainable solution for modern banking.

References

[1]   SA. Bajracharya, B. Harvey, and D. B. Rawat, "Recent Advances in Cybersecurity and Fraud Detection in Financial Services: A Survey," in Proc. 2023 IEEE 13th Annu. Comput. Commun. Workshop Conf. (CCWC), Las Vegas, NV, USA, Jan. 2023, pp. 368–374, doi: 10.1109/CCWC57344.2023.10099355.

[2]   F. Almalki and M. Masud, "Financial Fraud Detection Using Explainable AI and Stacking Ensemble Methods," in Proc. 2025 IEEE Conf. on Artificial Intelligence Applications in Finance (AIAF), New York, USA, May 2025, pp. 112–119. [Online]Available: https://arxiv.org/abs/2505.10050.

[3]   Int. J. Res. Publ. Rev., "Cybersecurity in banking and financial services: Protecting digital transactions," International Journal of Research Publication and Reviews, vol. 6, no. 1, pp. 45–52, Jan.2025.https://ijrpr.com/uploads/V6ISSUE1/IJRP R37656.pdf.

[4]   W. K. Syed, A. Mohammed, J. K. Reddy, and S. Dhanasekaran, "Biometric authentication systems in banking: A technical evaluation of security measures," in Proc. 2024 IEEE 3rd World Conf. Appl. Intell. Comput. (AIC), Gwalior, India, Jul. 2024, pp. 1331–1336. doi: 10.1109/AIC61668.2024.10731026.

[5]   Jain, A., Arora, D., Bali, R., & Sinha, D. (2021). Secure authentication for banking using face recognition. *Journal of Informatics Electrical and Electronics Engineering (JIEEE)*, 2(2), 1–8. https://jieee.a2zjournals.com/index.php /ieee/article/view/23.

[6]   Jayanthan, J., Priya, N. K., Kumar, S. P., & Sangeetha, K. (2021). Facial recognition controlled smart banking. *International Journal of Research in Engineering, Science and Management*, 4(3), 185–187. Https://journal.ijresm.com/index.php/ijres m/article/view/615.

[7]   S. Umamaheswari, S., Dr.A.Valarmathi, A., & M.Raja Lakshmi, M. (2023). Role of artificial intelligence in the banking sector . *Journal of Survey in Fisheries Sciences*, 10(4S), 2841–2849. Https://sifisheriessciences.com/journal/ind ex.php/journal/article/view/1722/1769.

[8]   Sharma, K., Goyal, Y., Jain, D., & Khanna, K. (2022). Advanced bank security and management system. In 2022 IEEE 7th International Conference for Convergence in Technology (I2CT), Mumbai, India, (pp. 1–4). doi: 10.1109/I2CT54291.2022.9825468. Https://ieeexplore.ieee.org/document/9825 468.

[9]   Mohanraj, K. C., Ramya, S., & Sandhiya, R. (2022). Face recognition based banking system using machine learning. *International Journal of Health Sciences*, 6(S8), 468–477. doi:10.53730/ijhs. v6ns8.9724. Https://sciencescholar.us/journal/index. Php/ijhs/article/view/9724.

[10] Arunadevi, R., Haresh, R., & Jerold, G. S. (2023). Online banking security with real time face recognition approach. *International Journal of Computer Science and Mobile Computing*, 12(5), 7–11.

[11] Johora, F. T., Hasan, R., Farabi, S. F., Alam, M. Z., Sarkar, M. I., & Al Mahmud, M. A. (2024). AI advances: enhancing banking security with fraud detection. In 2024. First International Conference on Technological Innovations and Advance Computing (TIACOMP), Bali, Indonesia, (pp. 289–294). doi:10.1109/TIACOMP64125.2024. 00055. Https://ieeexplore.ieee.org/document/1074 2687.

5 Performance comparison of machine learning algorithms in predicting telecom customer attrition

Prathap Nayudu P.[1,a], Venkata Sai S.[2,b], Amrutha B.[2,c], Sai Rupa J.[2,d] and Vigneshwar Reddy K.[2,e]

[1]Associate professor, Department of CSE, Rajeev Gandhi Memorial College of Engineering and Technology, Nandyal, India

[2]Students, Department of CSE, Rajeev Gandhi Memorial College of Engineering and Technology, Nandyal, India

Abstract

Customer attrition is a major and critical issue for telecom industries, affecting profitability and customer retention strategies. Predicting churn accurately enables companies to retain valuable customers by addressing potential issues early. The study examines how supervised machine learning models—including Logistic Regression (LR), Decision Tree (DT), Random Forest (RF), Support Vector Machine (NVM), Naive Bayes (NB), k-nearest neighbors (KNN), and XG Boost—can be used to foresee customer attrition. By utilizing the telecom customer dataset, the models were trained and tested to assess their performance in predicting churn. The analysis illustrates that XG Boost and Random Forest algorithms achieve the best accuracy, underscoring their proficiency in foreseeing customers at risk of attrition. These inferences emphasize the capability of machine learning in enhancing customer retention strategies. Future research will aim to incorporate real-time data and optimize these models for practical applications.

Keywords: Accuracy comparisioncomparison, attrition prediction, KNN, machine learning, performance evaluation, supervised learning, SVM, XG Boost

Introduction

Predictive models are critical in identifying customers likely to leave, enabling companies to take preventive measures to retain them [1] [2]. Customer churn is one of the most serious issues in the telecom industry that affects profitability and growth [4]. Since the cost of retaining existing customers is much cheaper than acquiring new ones, predicting churn is a key business strategy [3]. This study incorporates some sophisticated machine learning techniques in prediction, namely: Logistic Regression (LR), Decision Tree (DT), Random Forest (RF), Support Vector Machine (NVM), Naive Bayes, k-nearest neighbors (KNN) and XG Boost to forecast customer churn from the telecom sector.

Predicting customer attrition is essential for telecom industries and CRM teams, as it helps classify customers who may stop using their services and who may continue the service. As a result, revenue loss and costs to recover lost customers are reduced, and it saves money spent on finding new customers. Knowing why customers leave gives insights into how to enhance service delivery, thus making customers happier and more loyal over time. Customer attrition is the process where a customer discontinues using or cancels any service or product of a company or business, usually in a specified timeframe.

The goals of the paper are as follows: accurate prediction of customer churn, comparison of machine learning algorithms, identify key factors influencing churn, improve customer retention strategies.

Research gap: Various earlier works have used basic pattern recognition models for churn prediction; however, these systems often suffer from limited accuracy. The system designed here applies to numerous advanced machine learning algorithms to compare their accuracies for achieving the best possible model. Moreover, it focuses on determining the dominant causes of churn rates for understanding better why customers are being lost.

[a]prathapnaidu81@gmail.com, [b]saivenkata636@gmail.com, [c]ammubhasyam2003@gmail.com, [d]sairupajupalle@gmail.com, [e]vigneshwarreddykomma5@gmail.com

DOI: 10.1201/9781003684589-5

Mathematical model

Dataset description: The customer churn dataset that we have used is collected from Kaggle [5]. The dataset consists of 7000 + records and 21 columns. The columns are categorized as:

1. Churn column - Customers who left recently, and it is the target column.
2. Services provided by the company to its customer-Phone service availability, device protection, multiple lines, online backup, internet availability, online security, tech support, and streaming TV.
3. Customer's information columns - Old/new customer and their active period, contract, payment method, paperless billing, monthly charges, and total charges.
4. Customer's demographic columns - Gender, age, partners and dependents.

Mathematical model of customer churn attrition
The system that we proposed consists of the following steps as shown in Figure 5.1:

Data preprocessing: It is one of the essential steps to implement the machine learning models. The result is always better when the data is processed. Data preprocessing involves data cleaning, dropping duplicate columns, feature engineering, and outlier analysis [6].

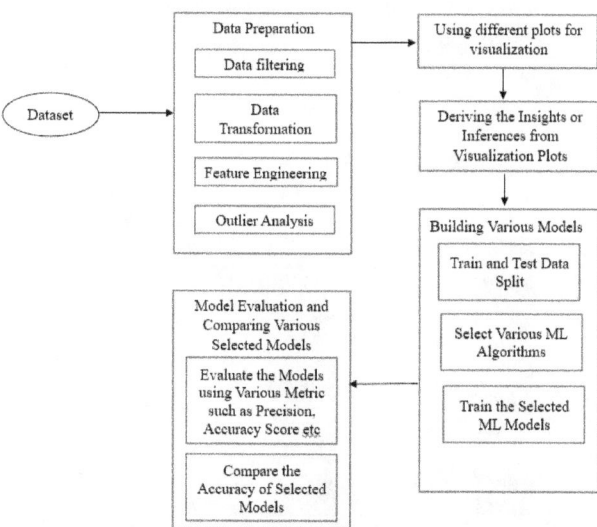

Figure 5.1 Architectural design for customer attrition prediction
Source: Author

Data preparation: It involves dealing with null values, removing redundant records and correcting structural errors.

Handling missing values: There are two ways to handle missing or null values which are either the simple dropping of them, or by imputation technique.

Dealing with duplicate records: If duplicate records are present in the dataset, they can simply be dropped or deleted. Data redundancy arises due to the presence of duplicate records. Data redundancy is said to adversely affect the predictability of machine learning algorithms.

Data transformation (or) feature transformation: This step deals with the proper conversion of the data to prepare it for the training of the machine learning algorithms. For such purposes, various methods exist-like data scaling, normalization, and label encoding.

Feature engineering: Feature engineering is a process of feature refinement and creation that can make a machine learning model more effective. It transforms raw data into meaningful features aligned with algorithmic requirements, making the model learn effectively [7].

Outlier analysis: Outliers are extremely high or low value. If we do not handle outliers; then models may become biased toward them. Outliers can be identified using techniques like box plots. They can be handled by either removing them from the dataset or applying imputation techniques to adjust their values [8].

Data visualization: We have used various visualization tools, such as line chart, heat map, bar graphs to derive inferences from the data. Data visualization helps us understand the data, its distribution, identify various patterns, make decisions about model building [9].

Train and test data split: Before creating a machine learning model, it's important to split the data into two groups: one will be used for training and the other will be used for validation. The training set is used by model to learn from the data. We use the testing set to see how well our model has truly grasped the patterns.

Selecting and training models: In this system, we have used some of the supervised machine learning algorithms such as LR, KNN, DT classifier, NB, SVM, RF and XG Boost algorithms. We trained all these models using training data.

Model evaluation and comparing various selected models: After training the models, we have tested the model by using testing data and determined the accuracy of all the selected models and compared their accuracy score. We had compared the models by using confusion matrix.

Above are the snapshots of the data plots we created to gain inferences from the dataset. Conclusions from the above Figures are:

As shown in Figure 5.2, many of the customers who decided to leave had chosen electronic check as their payment method.

Figure 5.3 demonstrates that, customers with month-to-month contracts were more likely to leave.

From Figure 5.4, customers who have partners are more likely to churn while senior citizens being the most of churn.

As illustrated in Figure 5.5, fiber optic service, which was chosen by a lot of customers and it's evident that there' high churn rate among these customers. This could expose an issue in the fiber optic service which dissatisfied most of its customers.

Figure 5.5 also illustrates that, more customers opted for DSL service, and it appears that these customers are less likely to churn compared to those using fiber optic service.

Figures 5.6 and 5.7 depicts that the absence of online security, paperless billing system and services

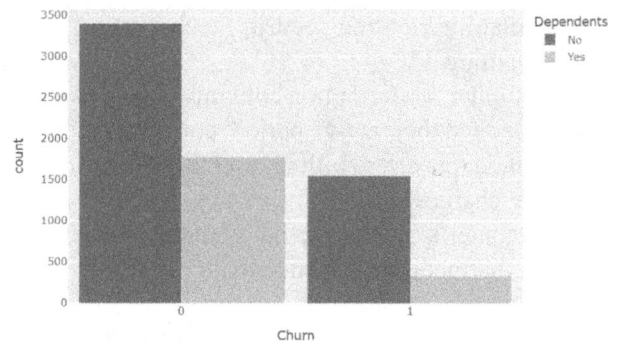

Figure 5.4 Churn distribution with respect to dependents
Source: Author

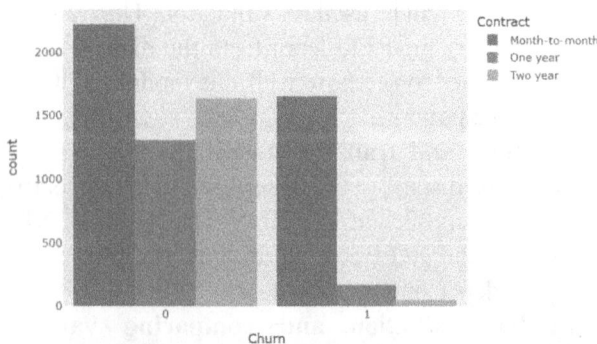

Figure 5.2 Churn distribution with respect to payment method
Source: Author

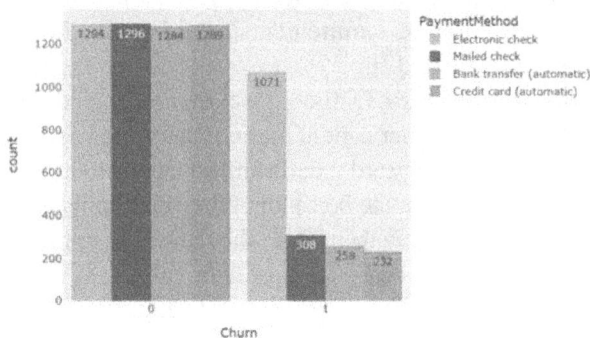

Figure 5.5 Churn distribution with respect to internet service and gender
Source: Author

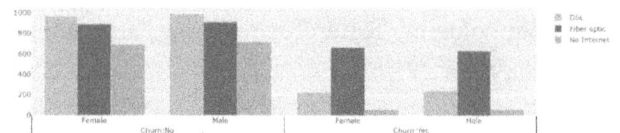

Figure 5.3 Churn distribution with respect to contract
Source: Author

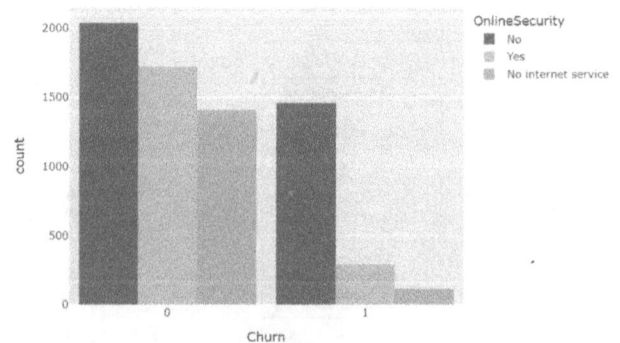

Figure 5.6 Churn distribution with respect to online security
Source: Author

Churn distribution w.r.t. Paperless Billing

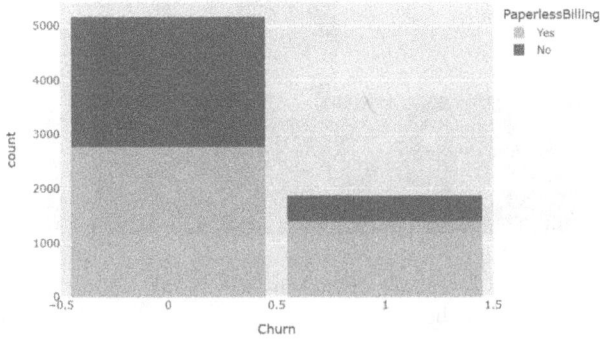

Figure 5.7 Churn distribution with respect to paperless billing
Source: Author

Churn distribution w.r.t. TechSupport

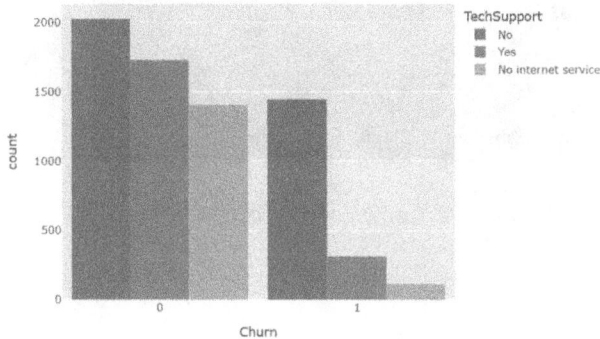

Figure 5.8 Churn distribution with respect to tech support
Source: Author

with no tech support were similar trends that may lead to customer attrition.

Figure 5.8 shows that There's a small fraction of customers who are prone to churn and it's been found that they don't have a phone service.

Applying machine learning algorithms
Logistic regression: LR is a technique that solve classification problems, especially when doing a binary classification. The classification problem is simply about finding which class the target variable is going to belong to with a given set of features [10].

KNN classifier: KNN is an extremely powerful supervised algorithm for classifying data and regression tasks. It follows the principle of similarity in which a data point is classified to a class of the closest neighbors in the feature space [11].

DT classifier: It is used to classify data by segregating the data into hierarchical form that resembles a tree structure. This tree structure comes from splitting the dataset into smaller groups based on specified features. Each split can be represented by an internal node, branches can symbolize possible outcomes, and the final classification can be represented by the leaf nodes.

Naive Bayes: It uses Bayes' Theorem as its basis and is termed ""naive" because it considers that all input features are not dependent on each other, simplify the calculations needed for predictions [12].

Random Forest: RF is similar to DT but generates multiple decision trees during the training process and uses their collective output to make predictions. This approach helps improve accuracy and reduces the likelihood of overfitting [13].

Support Vector Machine: It can be used for the classification of data based on the best decision boundary or a hyperplane between categories that maximally distinguish the data points from each other. It uses various mapping techniques, like polynomial methods or Gaussian-based transformations, to transform the data into higher-dimensional spaces for better decision-making [14].

XG Boost: It works well for tasks like sorting items into groups (classification) or estimating values (regression). This algorithm is especially useful for large amounts of data because it's both fast and efficient [15].

Analysis of telecom customer attrition prediction
A confusion matrix is a tool for assessing how well a classification model performs by relating its predicted results with the actual outcomes [16]. It will most likely take the form of a table with four key parts: True Positives, which are correct cases of a positive result; and True Negatives, which are correct classifications of negative cases. FP occurs when the model predicts a positive outcome on a negative case, and FN happens when the model predicts a negative outcome for a positive case. Precision is another metric which concerns the correct positive predictions. Accuracy score is also a metric to measure how well a model performs by calculating the percentage of correct predictions [17].

According to the Figure 5.9, It shows that 1733 instances were accurately classified by the model. it wrongly classifies 242 cases as class 0 and 135 cases as class 1. However, accuracy score of Logistic

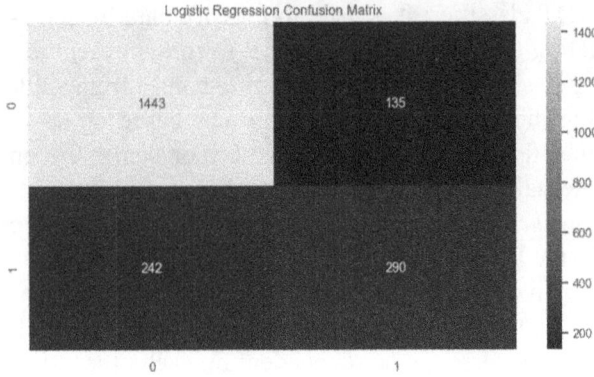

Figure 5.9 Confusion matrix of logistic regression
Source: Author

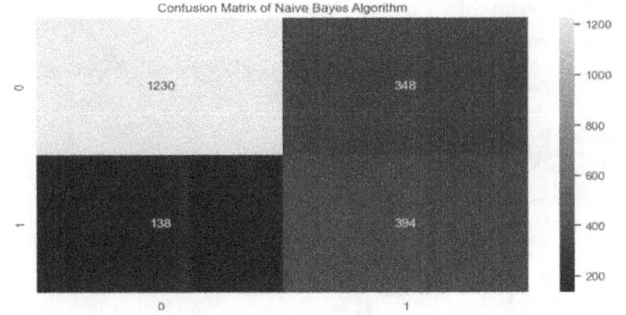

Figure 5.11 Confusion matrix of NBes
Source: Author

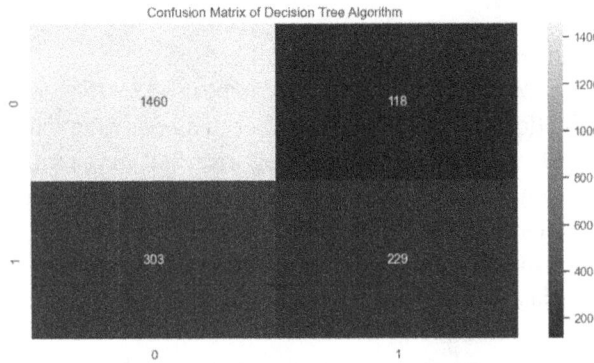

Figure 5.10 Confusion matrix of decision tree
Source: Author

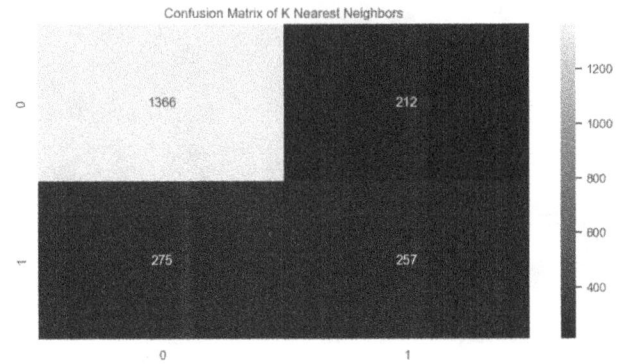

Figure 5.12 Confusion matrix of KNN
Source: Author

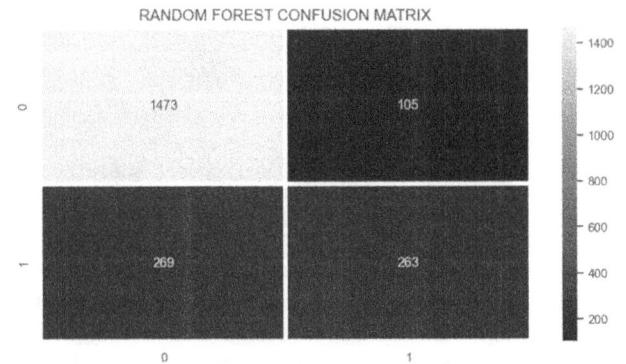

Figure 5.13 Confusion matrix of random forest
Source: Author

Regression is 81. 1327, and the total accuracy is respectable.

Figure 5.10 reveals a significant bias in the dataset, which had much more instances of class 0 than class 1. The confusion matrix of DT illustrates that the model correctly classifies 1689 instances. It misclassified 421 instances. The accuracy score of DT is 80.05.

Figure 5.11 illustrates the performance of the NB model in classifying instances into two classes. While the model correctly identified 1230 and 394 instances, it misclassified 348 and 138 instances. A closer Analysis reveals that the model exhibits a higher tendency to misclassify instances of class 1 compared to class 0. The Accuracy score of this model is 76.97.

Figure 5.12 presents the performance of the KNN model in classifying instances into two classes. While the model correctly selected 1366 and 257 instances, it misclassified 212 and 275 instances.

Figure 5.13 conveys that the model identified 1473 and 263 instances correctly. While 374 instances are wrongly classified by the RF.

Figure 5.14 presents that the model correctly classified 1485 and 244 instances. However, it classified 361 instances wrongly. The accuracy score of SVM algorithm is 81.943.

Figure 5.15 depicts that it correctly classified 1455 and 285 instances. The model misclassified 123 and 247 instances. The accuracy of XG Boost model is 82.464.

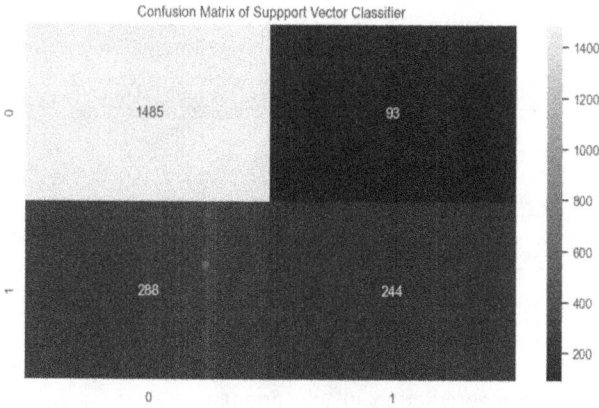

Figure 5.14 Confusion matrix of SVM
Source: Author

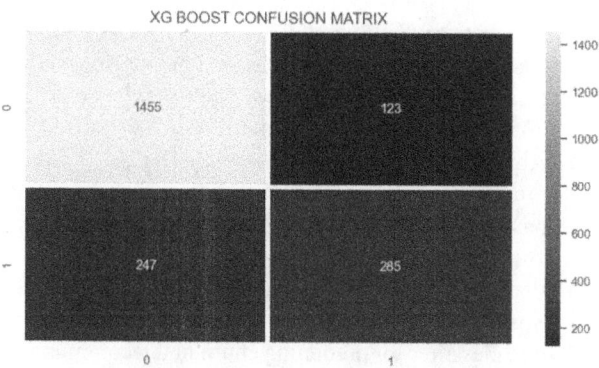

Figure 5.15 Confusion matrix of XG Boost
Source: Author

Table 5.1 Comparision of accuracies of various machine learning algorithms.

Model	Accuracy	Precision	Recall	F1-Score
Logistic Regression	81.13%	0.79	0.82	0.80
Decision Tree	80.05%	0.77	0.81	0.79
Naive Bayes	76.97%	0.72	0.78	0.75
k-Nearest neighbors	78.45%	0.74	0.79	0.76
Random Forest	82.32%	0.81	0.85	0.83
SVM	81.94%	0.80	0.84	0.82
XG Boost	82.46%	0.82	0.86	0.84

Source: Author

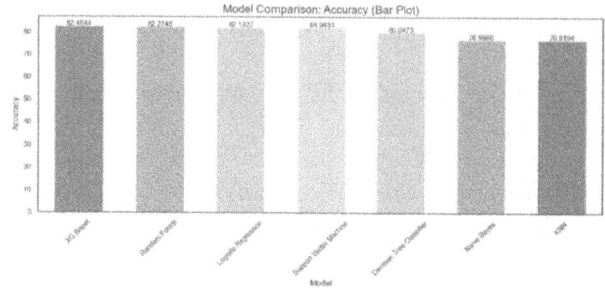

Figure 5.16 Comparison of accuracies of various machine learning algorithms using bar graph
Source: Author

Results and analysis

Figures 5.16 and Table 5.1 show a contrast accuracy scores for different machine learning algorithms. In particular, XG Boost and RF the two models with the highest accuracy scores. Similarly, SVM and LR has achieved comparable accuracy levels to the above top models. However, KNN and NB have the same accuracy scores. The accuracy of the DT model is decent but lower than that of LR.

Conclusion

Foreseeing customer attrition in this fast-paced telecom industry is one of the major challenges before CRM teams [18]. The proposed system results shows that XG Boost, Random Forest and Logistic Regression models provide good accuracy scores. In Future, we are planning to explore the usage of some reinforcement algorithms and deep learning algorithms in improving the accuracy of churn predictions. Also, adding real-time data might help increase the responsiveness of the models for attrition prediction [19]. Further expanding this study by incorporating multiple telecom markets or industries can strengthen the findings.

Acknowledgement

We extend our sincere gratitude to our computer science and engineering department for their invaluable guidance, constructive feedback, and continuous support throughout this research. Their insights have greatly contributed to the quality and direction of our work.

We also acknowledge our college provides essential resources and support for granting access to the data set used in this study.

References

[1] Ullah, I., Raza, B., Malik, A. K., Imran, M., Islam, S. U., &and Kim, S. W. (2019). A churn prediction model using random forest: analysis of machine learning techniques for churn prediction and factor identification in telecom sector. *IEEE Access*, 7, 60134–60149.

[2] Dahiya, K., & Bhatia, S. (2015). Customer churn analysis in telecom industry. In 2015 4th International Conference on Reliability, Infocom Technologies and Optimization (ICRITO)(trends and future directions), (pp. 1–6). IEEE.

[3] Mustafa, N., Ling, L. S., & Razak, S. F. A. (2021). Customer churn prediction for telecommunication industry: a Malaysian case study. *F1000Research*, 10, 1274.

[4] Wagh, S. K., Andhale, A. A., Wagh, K. S., Pansare, J. R., Ambadekar, S. P., & Gawande, S. H. (2024). Customer churn prediction in telecom sector using machine learning techniques. *Results in Control and Optimization*, 14, 100342.

[5] Lalwani, P., Mishra1, M. K., Chadha, J. S., & Sethi1, P. (.n.d2022). Customer churn prediction system: a machine learning approach. *Computing*, 104(2), 271–294.

[6] Senthilnayaki, B., Swetha, M., & Nivedha, D. (2021). Customer churn prediction. *International Advanced Research Journal in Science, Engineering and Technology*, 8(6), 527–531.

[7] Kumar, U., Raghuwanshi, A., & Poovammal, E. (2024). Customer churn analysis in telecom organization. *International Journal for Multidisciplinary Research (IJFMR)*, 6(2), 1–8. https://www.ijfmr.com/papers/2024/2/18886.pdf.

[8] Sikri, A., Jameel, R., Idrees, S. M., & Kaur, H. (.n.d2024). Enhancing customer retention in telecom industry with machine learning driven churn prediction. *Scientific Reports*, 14(1), 13097.

[9] Mohammed, S. U. Z., & Akbar, M. A. R. (2023). Telecom customer churn prediction analysis [Master's thesis, Mercer University]. ResearchGate. https://www.researchgate.net/publication/376829074_Telecom_Customer_Churn_Prediction_Analysis.

[10] Faraji Googerdchi, K. F., Asadi, S., & Jafari, S. M. B. (n.d2024). Customer churn modelling in telecommunication using a novel multi-objective evolutionary clustering-based ensemble learning. *Plos one*, 19(6), e0303881.

[11] Andrews, R., Zacharias, R., Antony, S., & James, M. M. (n.d2029). Churn prediction in telecom sector using machine learning. *International Journal of Information*, 8(2).

[12] Adeniran, I. A., Efunniyi, C. P., Osundare, O. S., & Abhulimen, A. O. (.n.d2024). Implementing machine learning techniques for customer retention and churn prediction in telecommunications. *Computer Science & IT Research Journal*, 5(8).

[13] Ebrah, K., & Elnasir, S. (n.d201). Churn prediction using machine learning and recommendations plans for telecoms. *Journal of Computer and Communications*, 7(11), 33–53.

[14] Labhsetwar, S. R. (n.d2020). Predictive analysis of customer churn in telecom industry using supervised learning. *ICTACT Journal on Soft Computing*, 10(2), 2054–2060.

[15] Ahmed, A. A. Q., &and Maheswari, D. (2017). Churn prediction on huge telecom data using hybrid firefly based classification. *Egyptian Informatics Journal*, 18(3), 215–220.

[16] Pamina, J., Raja, B., SathyaBama, S., Soundarya, S., Sruthi, M. S., Kiruthika, S., et al. (2019). An effective classifier for predicting churn in telecommunication.

[17] Ahn, J., Hwang, J., Kim, D., Choi, H., & Kang, S. (n.d2020). A survey on churn analysis in various business domains. *IEEE Access*, 8, 220816–220839.

[18] Geppert, C. (2002). Customer Churn Management: Retaining High-Margin Customers with Customer Relationship Management Techniques. KPMG & Associates Yarhands Dissou Arthur/Kwaku Ahenkrah/David Asamoah.

[19] Y. Huang and T. Kechadi, "An effective hybrid learning system for telecommunication churn prediction," *Expert Syst. Appl.*, vol. 40, no. 14, pp. 5635–5647, Oct. 2013.

6 Graphical representation of accuracy comparison among various machine learning algorithms to predict customer churn

P. Prathap Nayudu[1,a], D. Srivani[2,b], N. Usha[2,c], S. Ayesha Tabassum[2,d], D. Sudheer[2,e] and D. Sivaramakrishna[2]

[1]Associate Professor, Department of Computer Science and Engineering, Rajeev Gandhi Memorial College of Engineering Technology, Nandyal, Andhra Pradesh, India

[2]Students, Department of Computer Science and Engineering, Rajeev Gandhi Memorial College of Engineering and Technology, Nandyal, Andhra Pradesh, India

Abstract

Predicting customer attrition has become a vital concern for companies seeking to maintain their client base and boost profitability. This research project conducts a comparative examination of several machine learning techniques, including linear regression, logistic regression, decision tree, and random forest algorithms. The study employs visual representations to assess the accuracy of these methods. In addition to providing insights on the best algorithm for predicting customer attrition, it describes the methodology, data sources, and performance indicators used to provide accurate predictions. The study also emphasizes how important it is to select the right algorithm depending on the particulars of the data and the particular needs of the company.

Keywords: Accuracy comparison, customer churn, data visualization, decision tree, linear regression, logistic regression, machine learning, predictive analytics, random forest

Introduction

Customer churn is the phenomenon in which customers terminate their relationship with a service or product. This has become an important challenge for telecommunication, retail, and banking industries. Retaining the customer is usually cost effective than acquiring a new customer. Thus, the predictive aspect of churn is the most important in the realm of customer relationship management. Machine learning algorithms have proved themselves to be great tools that can analyze the massive dataset, identify patterns, and forecast consumer actions with precision. The analysis of customer information to identify factors influencing the probability of a customer discontinuing their relationship with a business is known as customer churn prediction. Most of the time, the features are usage patterns of services, billing details, demographics of the customer, and feedback scores. This effective churn prediction enables businesses to devise retention strategies based on targeted intervention, reducing revenue losses and enhancing customer satisfaction. This study uses a customer dataset of 7,043 rows and 21 columns that encapsulates the most diverse attributes influencing customer retention. The dataset includes customer demographics, service usage, account information, and churn status, giving a comprehensive view to churn prediction. Using robust data preprocessing and visualization techniques, we ensure the dataset is well suited for machine learning applications. Artificial intelligence comprises of machine learning that allows machines to learn from experience with or without explicit programming. In the same domain, supervised learning plays a paramount role; this is a training procedure in which the algorithms train on labeled data sets that can predict, for instance, customer churn. This is accomplished through the use of Python, a versatile programming language. Its extensive libraries, such as Scikit-learn for machine learning, Matplotlib and Seaborn for

[a]prathapnaidu81@gmail.com, [b]srivanidesam@gmail.com, [c]nandyalausha543@gmail.com, [d]asiyatabassum810@gmail.com, [e]sudheerdandu01@gmail.com

DOI: 10.1201/9781003684589-6

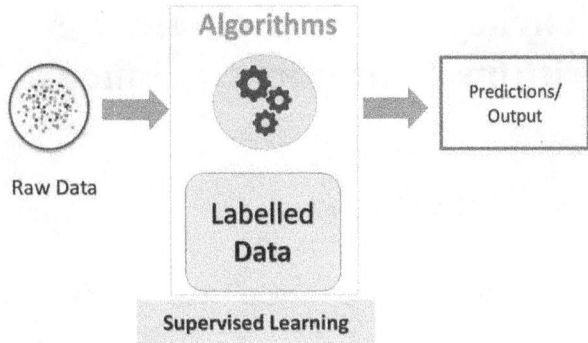

Figure 6.1 Supervised Learning
Source: Author

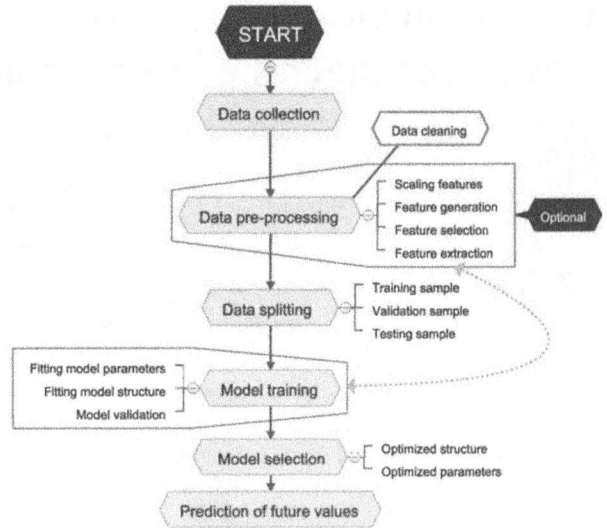

Figure 6.2 Workflow of a prediction model
Source: Author

visualization, NumPy for numerical computations, and pandas for data handling, have enabled the creation and assessment of predictive models. This research concentrates on examining the efficacy of four supervised machine learning techniques: linear regression, logistic regression, decision trees, and random forests. The models are evaluated based on their accuracy in predicting churn, and graphical analysis is used to visually compare them. This study aims to identify the most effective algorithm by leveraging a standardized dataset and advanced visualization techniques. Predictive analytics has become a vital tool as data-driven decision making becomes more prevalent. Advanced machine learning techniques provide organizations with actionable insights to improve customer retention strategies. Furthermore, with accurate churn prediction, companies can be ahead in a competitive environment and stay nimble amidst the ever-changing market scenario Figure 6.1. This study enhances the understanding of customer churn prediction by exploring the effectiveness of machine learning techniques, providing practical guidance for choosing a suitable model based on operational requirements and dataset characteristics Figure 6.2.

Literature Review

Kaur and Kaur presented a methodology for forecasting client attrition in the banking sector. The effectiveness and performance are compared in this study. Its goal is to assist banks in identifying clients who are likely to leave and putting retention plans in place. Performance criteria including accuracy and recall were used to assess the models [1].

Dodda et al. suggested a framework for applying machine learning techniques to forecast client attrition. Sequential, Random Forest (RF), and Decision Tree (DT) classifier models' effectiveness and performance are compared in this study. Its goal is to assist companies in identifying clients who are likely to leave and putting retention plans in place. Performance criteria like accuracy were used to evaluate the models, and the Sequential model achieved 94%, RF 92%, and DT classifier 91% [2].

Srinivasan et al., presented a framework for forecasting customer attrition in the telecom industry. The study compares various models and highlights the effectiveness of RF combined with SMOTEENN, achieving the highest F1-score of 95%. It aims to help telecom operators identify customers at risk of churning and implement retention strategies effectively [3].

Methodology

The methodology for predicting customer churn includes the following critical steps:

- Data description
- Information processing
- Examining data through analysis (EDA)
- Feature engineering
- Data splitting
- Model selection
- Model evaluation metrics
- Model comparison and visualization

Data description

The dataset used in predicting customer churn has 7,043 records of customers, each with 21 features. These include demographic information like age and gender along with service usage patterns with variables such as monthly charges and tenure. The outcome variable "Churn" is dichotomous, with 1 signifying a customer who has left the service and 0 indicating a customer who remains active. This makes the problem a form of binary classification: predict, based on the available features, if a client will stick around or leave. The dataset offers a comprehensive perspective on the attributes of customers and behaviors to be used in the building of a predictive model for retention [4].

Information processing

Information preprocessing is a significant stage in setting up the dataset for displaying and guaranteeing consistency across highlights shown in Figure 6.2. It incorporates taking care of missing qualities through attribution procedures like loading up with the mean, middle, or mode for mathematical and straight-out elements, or involving forward and in reverse occupying for time-series information. At times, lines or segments with over-the-top missing information might be dropped. Anomaly identification and taking care of are likewise essential to assist with working on model execution; the above methods, for example, Z-score or the interquartile reach (IQR) are frequently used to identify and control exceptions. Clear cut factors ought to be encoded utilizing name, one-hot, or even twofold encoding relying upon whether the information is absolute [4]. For mathematical elements, standardization or normalization (Z-score scaling) is applied to scale values to a steady reach so that models perform ideally. Highlight designing may likewise be utilized to make new factors that better address the basic examples in the information [5]. Finally, the dataset is divided into a testing set 20% and a preparation set 80% to assess model execution really and stay away from overfitting. These preprocessing steps guarantee that the information is prepared for model preparation, in this way prompting more exact and dependable expectations [6].

Examining data through analysis (EDA)

Examining data through analysis (EDA) exploratory data analysis (EDA) is a crucial step in understanding the dataset's structure, identifying patterns, and detecting potential anomalies [7]. Descriptive statistics provide insights into individual feature distributions, while various visualizations help illustrate relationships between variables. Histograms are used to examine the frequency distribution of numerical attributes, allowing for a better understanding of data distribution and deviations. Box plots help identify outliers and variations within different customer segments, ensuring that extreme values are appropriately addressed [8]. Correlation heatmaps visualize relationships between independent features and the target variable, highlighting the strongest predictors of churn. Pair plots offer a detailed view of multi-variable interactions, significant dependencies and potential feature combinations that could enhance model performance. Additionally, bar charts are applied to analyze categorical feature distributions and density estimates, providing a comprehensive perspective on customer behavior [9]. Through these detailed analyses and visualizations, EDA establishes a base for feature selection and model development, ensuring that the most significant attributes are effectively utilized in the prediction process [10].

Feature engineering

Feature engineering plays a crucial role in improving model performance by improving input variables. Scaling and normalization techniques were applied to numerical features to ensure consistency and prevent models from being influenced by different scales of data [11]. Encoding methods for categorical variables and target encoding for ordinal data, enabled to convert non-numeric attributes into a suitable format for machine learning algorithms. The model's stability was enhanced by addressing multicollinearity by identifying and eliminating highly correlated features using the Variance [12].

Inflation factor technique. Furthermore, dimensionality reduction methods like principal component analysis (PCA) improved computational efficiency by removing less important variables while maintaining important information [13].

Data splitting

Information parting is a key stage in AI to guarantee that models are prepared and assessed successfully. A common split of the dataset is as a rule into two subsets, which are a preparation set and a testing set [14]. A typical split proportion utilized is 80:20. The preparation set is utilized to assemble and prepare the AI models. It empowers the model to gain proficiency with the fundamental examples and connections inside the information. Testing set, which is waited for and is kept particular from the preparation information [15]. Along these lines, the exhibition of the model is assessed in an unprejudiced way, and it keeps away from overfitting in light of the fact that it tests information that the model has never seen during preparing, thus guaranteeing that its capacity to sum up to new, concealed information is appropriately surveyed [16].

Model selection
The following models are applied to predict customer churn:

To this end, every model makes use of the capabilities of the dataset for the purpose of making predictions of the target variable [17].

a)   **Linear regression (LR):** Despite being mostly used for constant outcome estimations, Linear can be adapted to most types of problems that gradient descent method handles. (though less common for binary tasks). The formula for a simple linear regression model is:

$$y = a_0 + a_1 f_1 + a_2 f_2 + \cdots + a_n f_n$$

where:

- Y is the predicted target variable (churn or not),
- $f_1, f_2, ..., f_n$ are the features,
- $a_0, a_1, ..., a_n$ are the model coefficients.

b)   **Logistic regression:** Many of the binary classification problems use logistic regression as one of the methods. The results are intended to reveal the probability of a given outcome, and this technique uses a logistic function to describe such likelihood as a customer is to cease your service. The model calculates the probability p that a client will terminate their relationship with a company, based on various input variables. The equation is expressed as:

$$P(y = 1 \mid X) = \frac{1}{1 + e^{-(a_0 + a_1 x_1 + \cdots + a_n x_n)}}$$

where:

- $P(y = 1 \mid X)$ is the likelihood that service will be discontinued.,

- $a_0, a_1, ..., a_n$ are the regression coefficients,
- $x_1, x_2, ..., x_n$ are the input features.

c)   *Decision Tree (DT):* A kind of a supervised learning algorithm which is non para- metric in nature and is employed for use in both regression and classification applications. This method partitions data into subsets according to the most. or the information gain or Gini index criterion is used for splitting and it is considered as the most influential attribute. The Gini index formula is:

$$Gini\ Index = 1 - \sum_{i=1}^{n} (p_i)^2$$

Alternate Gini index:

$$Gini\ Index = 1 - \sum_{\tau} p_i$$

d)   **Random Forest Tree (RF):** The approach utilized in the current study is a machine learning technique referred to as the Random Forest, which combines many decisions. More accurate and dependable predictions are made by using tree structured models because they have easy and quick identification of major features. This technique creates different subsets of the original dataset using bootstrap aggregating, also known as bagging, then builds a decision tree for each subset. A majority vote or an average of the individual tree projections determines the final prediction. RF can be used for both classification and regression tasks.

Model evaluation metrics
We use a number of performance indicators, such as accuracy, precision, recall, and F1-score, to evaluate the models' efficacy [18]:
   Accuracy: Quantifies the model's overall correctness
   Accuracy serves as a reliable indicator when the classes in the data are nearly balanced. Accuracy should not be used in the cases of imbalanced cases.
   Precision: Defines the value of TP/(TP+FP) which tells about the correctly identified positive instances to the overall numbers of instances that are being predicted.

Sensitivity: Illustrates the ratio of correctly recognized positive people from all real positive individuals.

F1-score: To achieve this, we stay with the harmonic mean of Precision and Recall which is a balanced metric.

Model comparison and visualization

After the training and evaluation of the models their efficiency is measured using several measurement tools and graphical methods for finding the best model for customer churn. By displaying true positive, false positive, true negative, and false negative values the confusion. According to the matrix, one is able to see the performance of the classification and further realize how effective this model is when differentiating between the clients that have been turned away and those that have been retained. The model's degrees of separation between the two classes are also evaluated by perceiving with the receiver operating characteristic curve where the area underneath the curve is a commonly used primary metric. The variables used in churn prediction preferably the precision recall curves of X, Y, Z and their sum are de- scribed below; with unbalanced datasets since they demonstrate, to the extent of efficiency, of the model inside the minority class (churned customers). A source of further information is the AUC-PR, or the area under the precision-recall curve, which shows how the precision looks for different recall rates, namely, the ability of the model to offer high precision readings in the case when recall is also high. Because it covers false positives alongside false negatives in a single measure, the F1-score is an intermediate between the precision and the recall values. The models may be the results of performance using these metrics and visualizations have therefore been thoroughly compared to establishing the best performance that meets the top performance that strikes. the optimal balance between recall and precision, which is necessary to prohibit the presence of not only false negatives but also false positives. Such a careful comparison of models guarantees the choosing of the most appropriate algorithm.

Conclusion

In this analysis of the customer churn dataset, we analyzed the performance of four machine learning algorithms: RF, DT, LR, and LR.

Among them, Random Forest achieved the highest accuracy at 82.37%, making it the most effective model for predicting customer churn. Logistic Regression followed closely with an accuracy of 79%, reinforcing the effectiveness of statistical models. The Decision Tree model produced an accuracy of 72.49%, while Linear Regression had the lowest accuracy at 69.39%, highlighting its challenges in capturing complex patterns. Based on these results, Random Forest emerges as the best model due to its superior accuracy and robustness in predictive tasks.

References

[1] S. A. Alteer and A. Alariyibi, "Customer Churn Prediction Using Machine Learning: A Case Study of Libyan Internet Service Provider Company," 2024 IEEE 4th International Maghreb Meeting of the Conference on Sciences and Techniques of Automatic Control and Computer Engineering (MI-STA), Tripoli, Libya, 2024, pp. 605-612, doi: 10.1109/MI-STA61267.2024.10599671.

[2] Alhaqui, F., Elkhechafi, M., & Elkhadimi, A. (2022). Machine learning for telecoms: from churn prediction to customer relationship management.

[3] Nutalapati, H., Hayat, A., Zheng, R., Li, C. H., Prakoso, N., & Tiglao, N. M. (2024). Machine learning-based predictive analytics for customer churn in the telecom industry.

[4] Khodabandehlou, S., & Rahman, M. Z. (2017). Comparison of supervised machine learning techniques for customer churn prediction based on analysis of customer behavior.

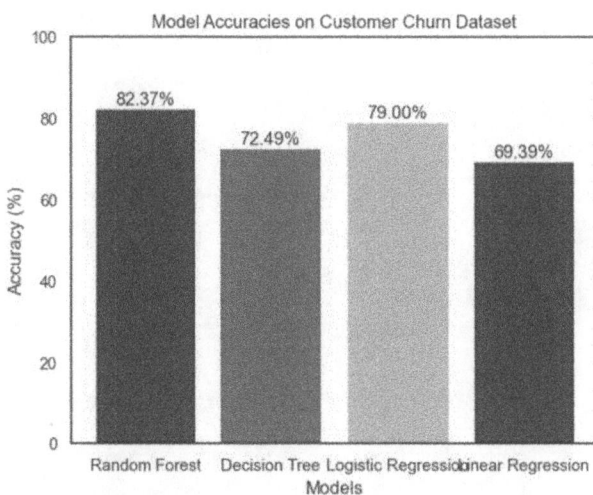

Figure 6.3 Workflow of a prediction model
Source: Author

[5]   Jain, H., Khunteta, A., & Srivastava, S. (2021). Telecom churn prediction and used techniques, datasets and performance measures: a review.

[6]   Geiler, L., Affeldt, S., & Nadif, M. (2022). A survey on machine learning methods for churn prediction.

[7]   Verma, P. (2020). Churn prediction for savings bank customers: a machine learning approach.

[8]   Dalvi, P. K., Khandge, S. K., Deomore, A., Bankar, A., & Kanade, V. A. (2016). Analysis of customer churn prediction in telecom industry using decision trees and logistic regression.

[9]   Bogaert, M., & Delaere, L. (2023). Ensemble methods in customer churn prediction: a comparative analysis of the state-of-the-art.

[10] Lukita, C., Bakti, L. D. , Rusilowati, U., Sutarman, A., & Rahardja, U. (2023). Predictive and analytics using data mining and machine learning for customer churn prediction.

[11] Prashanth, R., Deepak, K., & Meher, A. K. (2017). High accuracy predictive modelling for customer churn prediction in telecom industry.

[12] Pol, N., Daveshar, S., Madhavi, T., Rao, N. K. K., Jeyanthi, L., Sukania, P., et al. V. (2024). Comparative analysis of machine learning classifiers for enhancing business revenue and customer satisfaction: An empirical study.

[13] Sweidan, D., Johansson, U., Gidenstam, A., & Alenljung, B. (2022). Predicting customer churn in retailing.

[14] Sai, B. N. K., & Sasikala, T. (2019). Predictive analysis and modeling of customer churn in telecom using machine learning technique.

[15] Kumaran, T. E., Lokesh, B., Arunkumar, P., & Thirumeni, M. (2024). Forecasting customer attrition using machine learning.

[16] Paul, R., & Rashmi, M. (2022). Student satisfaction and churn predicting using machine learning algorithms for edtech course.

[17] Nutalapati, H., Hayat, A., Zheng, R., Li, C. H., Prakoso, N., & Tiglao, N. M. (2024). Machine learning-based predictive analytics for customer churn in the telecom industry.

[18] Singh, M., Singh, S., Seen, N., Kaushal, S., & Kumar, H. (2018). Comparison of learning techniques for prediction of customer churn in telecommunication.

7 PhishGuard – fake website detection using machine learning

K. Janardhan[a], G. Brahmaiah[b], E. Abhilash[c], A. Mohan Vamsi Krishna[d] and K. Prudhvi Raj Naik[e]

Department of Computer Science and Engineering, Rajeev Gandhi Memorial College of Engineering and Technology, Nandyal, Andhra Pradesh, India

Abstract

Phishing attacks have become a significant threat in the digital age, targeting individuals and organizations to steal sensitive information. PhishGuard is an innovative solution designed to detect and prevent phishing websites using a multi-layered approach. The project leverages domain validation, IP reputation checks, and URL redirection analysis to identify suspicious patterns. By integrating APIs from trusted platforms like VirusTotal and PhishTank, it enhances accuracy in identifying malicious websites. Additionally, PhishGuard employs machine learning models to predict phishing attempts based on website attributes and behavior, ensuring a dynamic and scalable detection mechanism. To further strengthen its effectiveness, PhishGuard incorporates source code analysis, examining web pages for phishing indicators like deceptive links and fraudulent scripts. This comprehensive approach enables it to provide real-time protection against emerging threats. Designed for usability and reliability, PhishGuard demonstrates the potential of combining traditional web security techniques with advanced machine learning to safeguard users and organizations from the evolving landscape of cyber threats.

Keywords: Domain validation, IP reputation, phishing detection, PhishTank, threat intelligence, URL redirection analysis, VirusTotal

Introduction

Phishing websites are a significant cyber security threat, designed to deceive users into revealing sensitive information like passwords and financial details. These malicious sites often mimic legitimate platforms, exploiting user trust and causing financial loss, identity theft, and data breaches. Addressing this challenge requires a robust detection system capable of adapting to the evolving tactics of attackers. This paper, PhishGuard, aims to develop a comprehensive phishing website detection system. It combines domain validation, URL analysis, API integrations, and machine learning to identify phishing threats effectively. Domain validation techniques assess the authenticity of website domains by detecting anomalies like newly registered or suspicious domains. URL analysis examines patterns and redirection behaviors indicative of phishing. Trusted APIs such as VirusTotal and PhishTank provide real-time data on known phishing threats, enhancing the system's reliability. Machine learning further strengthens detection by analyzing features like content structure, domain reputation, and behavioral patterns, dynamically adapting to new tactics.

Methodologies

Literature review

The papers evaluate fake websites and phishing detection using machine learning, user behavior, and statistical analysis. Studies highlight usability, trust, and effectiveness, employing models like Random Forest, SVM, CNN-LSTM, and stacking. Techniques include DOM tree comparison, visual pattern recognition, and feature-free approaches. Challenges include dataset biases, computational overhead, false positives, and evolving attack methods. Future work suggests AI integration, real-time detection, multilingual support, adversarial training, and hybrid models. Research also examines Wikipedia hoaxes, phishing classifier vulnerabilities, and intrusion detection adaptation. Emphasizing user training, collaboration, and scalable solutions

[a]jkrgmcse@gmail.com, [b]brahmaiah26102003@gmail.com, [c]abhiyadav2882002@gmail.com, [d]vamsikrishnaakuthota2004@gmail.com, [e]prudhvirajnaik143@gmail.com

DOI: 10.1201/9781003684589-7

enhances security and misinformation prevention across platforms, including mobile and IoT.

Domain validation

The objective of domain identification using regex and spoofing detection is to ensure that input URLs are valid while simultaneously identifying common tactics employed by attackers to deceive users. This process involves several key steps.

Regex validation: The first step is to validate the URL format using regular expressions (regex). A well-structured regex pattern, such as:

```
^(http|https):\/\/([A-Za-z0-9-]+\.)+[A-Za-z]
{2,}(/.*)?$.
```

checks for the presence of a valid scheme followed by a domain structure that includes sub domains, domain names, and top-level domains. This validation ensures that the URL address to standard web address formats, filtering out malformed or potentially harmful URLs. To identify spoofing tactics, the system uses a whitelist of legitimate domains and applies fuzzy matching algorithms to detect typo squatting Homoglyph attacks, which use visually similar characters, which are detected using string similarity metrics like Levenshtein distance, flagging deceptive domains beyond a certain similarity threshold. A maintained blacklist of known malicious domains provides an additional layer of security. By cross- referencing URLs against this list, the system quickly flags suspicious or confirmed phishing domains. This layered approach ensures robust protection against online threats.

IP reputation check

Conducting an IP spam check using the SpamDB API enhances security by identifying malicious websites through cross-referencing with known spam databases. To assess the reputation of an IPaddress, the system begins extracting it from the URL using DNS resolution or tools like Python's socket.gethostbyname(). This extraction provides the necessary data for conducting checks against spam and threat databases. The extracted IP is then sent to the SpamDB API, where it is evaluated for associations with spam, phishing, other malicious activities. If flagged, detailed reasons are logged for further analysis, enhancing the system's ability to detect and counter threats effectively. To ensure accuracy and reduce false positives, the system integrates additional reputation services such as Spamhaus and Project Honeypot.

URL redirection analysis

The objective of detecting and analyzing URL redirects is crucial in identifying potential phishing websites, which often employ multiple redirects to obscure their true destination. HTTP requests using libraries like Python's requests to capture server responses, particularly focusing on HTTP status codes. Special attention is given to 3xx status codes, such as 301 (permanent redirect) and 302 (temporary redirect), as these often indicate URL redirection—a common phishing tactic. By recursively following the Location headers in the responses, the system maps the full redirect chain until it reaches a final non- redirecting status code. The destination is then compared to the original URL, with significant discrepancies in domains or paths flagged as potential phishing attempts. URLs with multiple hops or those ending at destinations associated with malicious activity are marked for further investigation. This systematic analysis of redirection patterns not only identifies phishing behavior brutal so enhances web security by safeguarding users against deceptive tactics.

API integration

Integrating APIs from VirusTotal, AbuseIPDB, and PhishTank enhances URL threat detection by leveraging extensive security datasets to verify reputation and identify risks. To enhance threat detection, the system integrates multiple APIs, beginning with VirusTotal. By submitting the URL to VirusTotal, it leverages aggregated data from antivirus engines and URL scanners to identify potential threats such as phishing or malware. The returned report includes flags and warnings based on detection rates, providing a comprehensive safety assessment. Complementing this, the system queries AbuseIP DB using the website's IP address to check for abuse reports like phishing, hacking, or spamming. Any match in the database signals a potential threat, adding another layer of scrutiny. Further validation is achieved through the PhishTank API, a community-driven database of phishing sites.

Source code analysis

Advanced detection through website source code analysis enhances phishing prevention by identifying malicious elements within a website's HTML and JavaScript, complementing traditional URL-based methods. Source code analysis enhances phishing detection by examining a website's HTML and JavaScript using tools like BeautifulSoup or Scrapy to identify deceptive elements such as fake login forms, obfuscated scripts, and suspicious links. Features indicative of phishing, including malicious script patterns, untrusted external resources, and misleading elements like fake "Log In" or "Download" buttons, are extracted for further scrutiny.

Machine learning detection

The goal of developing a machine learning model for phishing detection is to leverage data from known phishing and legitimate websites to predict malicious URLs accurately. This automated and adaptive approach strengthens cybersecurity measures against evolving threats. Phishing detection begins with collecting datasets from sources like Kaggle and PhishTank, focusing on features such as domain age, URL length, suspicious characters, HTTPS usage, and redirection patterns. The data is split into training and testing sets, and models like Random Forest or SVM are trained and evaluated using metrics like accuracy and F1 score to ensure optimal performance. The trained model classifies new URLs, with predictions enhanced by combining results from APIslike VirusTotal and PhishTank for increased detection confidence. Continuous updates with new phishing data and user feedback refine the model, addressing false positives and adapting to evolving threats. Integrated with broader security systems like firewalls and supported by user education programs, this approach ensures a robust, comprehensive defense against phishing attacks. This multi-layered, adaptive strategy combines advanced technology, continuous improvement, and user awareness to safeguard sensitive information and maintain trust in digital interactions.

Experimental Design

This paper leverages a machine learning model to detect phishing websites using a combination of innovative methods. A comprehensive approach ensures high accuracy in identifying fraudulent websites and protecting users from online scams. Various machine learning classifiers, such as Random Forest (RF), K-nearest neighbors (KNN), Logistic Regression (LR), Support Vector Machine (SVM), and Decision

Trees (DT), are trained and evaluated based on performance metrics like accuracy, precision, recall, and F1-score. The best-performing model is selected for further optimization to enhance detection capabilities. Real-time API integration with services like VirusTotal and PhishTank strengthens the system by validating URL reputations and cross- referencing known phishing databases. Additionally, confusion matrix analysis identifies areas for improvement by analyzing metrics such as true positives and false positives.

The final model combines domain validation, URL analysis, machine learning, and API checks, delivering a robust and effective solution for detecting phishing websites with high accuracy.

Implementation Details

We used Python as the primary language, with libraries like Pandas and NumPy for data processing, BeautifulSoup for HTML parsing and Regular expressions (regex) to detect abnormal URL patterns commonly seen in phishing websites.

We used scikit-learn to build classifiers such as Random Forest, KNN, Logistic Regression, SVM, and DT. The models are trained on70% of the dataset and evaluated using accuracy, precision, recall, and F1-score .

API integration

External APIs like VirusTotal and PhishTank validate URLs in real-time, cross-checking them against known phishing databases for added reliability.

Model evaluation

Confusion matrices visualize classifier performance, helping minimize false positives and false negatives. Cross-validation ensures the model's robustness.

Deployment

Flask is used to deploy models in a web service, allowing users to check websites for phishing in real-time. In summary, PhishGuard combines Python libraries, machine learning models, API integration, and Flask for an effective phishing detection solution.

Experiment Results

In this paper, the experiment is designed to evaluate the effectiveness of various machine learning models in detecting phishing websites. The experiment is built on a combination of feature extraction techniques, machine learning classifiers, and external API integrations, all aimed at achieving high detection accuracy and minimizing false positives and false negatives. The experiment begins with collecting a comprehensive dataset of websites, which consists of labeled examples of both legitimate and phishing websites. This dataset is then preprocessed, with features extracted from each website based on specific characteristics. The features include domain validation (such as domain age and registration details), URL analysis (examining patterns, redirections, and URL length), and content inspection (including visual similarities and HTML structure). The preprocessing is done using Python libraries for web scraping to analyze page content and metadata. Once the features are extracted, the dataset is split into a training set (70%) and a testing set (30%), which are used to train and evaluate the machine learning models. These classifiers are implemented using the scikit-learn library in Python, which provides an easy-to-use interface for building, training, and evaluating machine learning models. The models are trained on the extracted features, and their performance is assessed based on standard evaluation metrics like accuracy, precision, recall, F1-score, and confusion matrices Figure 7.1.

The key evaluation metrics used to assess the performance of the classifiers are as follows:

Accuracy
Accuracy is the ratio of correct predictions (both true positives and true negatives) to the total number of predictions. It is given by the formula: Where:

$$Accuracy = \frac{TP + TN}{TP + TN + FP + FN} \quad \text{(Equation 1)}$$

True Positive (TP): Correctly predicted phishing websites.

True Negative (TN): Correctly predicted legitimate websites

False Positive (FP): Legitimate websites incorrectly predicted as phishing

False Negative (FN): Phishing websites incorrectly predicted as legitimate

Precision
Precision is measure the proportion of true positives among all predicted positives. It is calculated as:

$$Precision = \frac{TP}{TP + FP} \quad \text{(Equation 2)}$$

Recall
Recall measures the proportion of true positives among all actual positives. It is calculated as:

$$Recall = \frac{TP}{TP + FN} \quad \text{(Equation 3)}$$

F1-score
The F1-score is the harmonic mean of precision and recall providing a balanced measure of the classifier's performance:

$$F1 - Score = 2 \times \frac{Precision \times Recall}{Precision + Recall} \quad \text{(Equation 4)}$$

The PhishGuard paper incorporates multiple methodologies to create a robust phishing detection system. The F1-score, particularly useful for

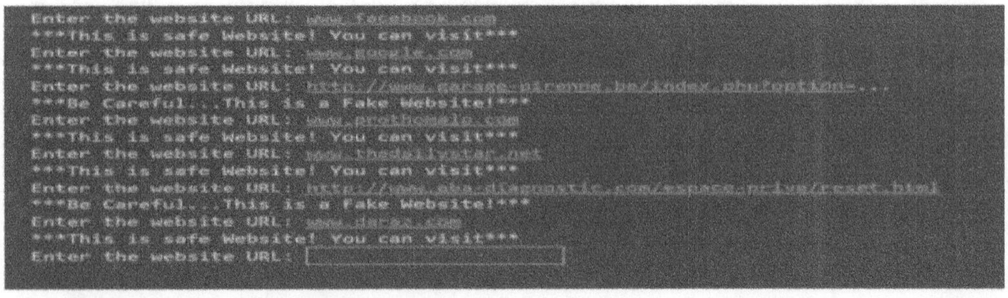

Figure 7.1 Safegaurd website details
Source: Author

Table 7.1 Sample dataset (Phishing websites).

URL	Label	Source
http://example.phish.com	-1	PhishTank
https://secure.phish.me	-1	PhishTank
http://malicious.example	-1	PhishTank
https://legit-site.com	1	Alexa Dataset

Source: Author

Table 7.2 Extracted features.

Feature	Description
URL Length	Total characters in the URL
Number of Subdomains	Count of subdomains in the URL
Presence of HTTPS	Indicates if the website uses HTTPS
Domain Age	Age of the domain in days
Special Characters in URL	Count of special characters (e.g., @, ?, #)
IP Address in URL	Indicates if the URL uses an IP instead of a domain
Number of Redirects	Count of redirections before reaching the final page
Top-Level Domain	Type of TLD used (e.g., .com, .net)
URL Encoding Patterns	Presence of encoded sequences in the URL
DNS Record Validity	Validity of the domain's DNS record

Source: Author

Table 7.3 Classifier comparison (accuracy).

Classifier	Training Accuracy	Testing Accuracy	Precision	Recall	F1-Score
Logistic Regression	88%	84%	0.81	0.85	0.83
Random Forest	94%	92%	0.90	0.93	0.91
Decision Tree	91%	89%	0.87	0.88	0.88
SVM	90%	88%	0.85	0.89	0.87
K-Nearest Neighbor	86%	83%	0.80	0.82	0.81

Source: Author

Table 7.4 Results of feature-based classification.

Phishing Label	Legitimate Label	Classifier Output
-1	1	Accurate Detection
-1	1	Accurate Detection
-1	-1	Misclassified
1	1	Accurate Detection

Source: Author

Table 7.5 Adversarial sample evaluation.

Phishing Sample URL	Crafted Sample URL	Evaded Classifier?	MSE Score
http://phishingsite.com/login	http://legit-site.com/login	Yes	0.05
https://secure-phish.com/home	https://trusted-source.com/home	Yes	0.03

Source: Author

Table 7.6 DOM tree similarity results.

Metric	Phishing Page	Adversarial Page	Similarity (%)
Element Similarity	80	82	95
Text Node Similarity	70	72	92
Attribute Similarity	60	62	90
Overall Similarity	70	72	92

Source: Author

imbalanced datasets, is used alongside other metrics like accuracy, precision, and recall to evaluate the performance of various machine learning classifiers.

The tables provided for the PhishGuard project offer a comprehensive overview of the system's data, methodologies, and results.

Table 7.1 demonstrating sample dataset (Phishing websites) shows the dataset used for training and testing, including URLs labeled as phishing (-1) or legitimate(1), with sources like PhishTank and Alexa providing real-world data. This forms the foundation for training the machine learning model. Table 7.2 shows extracted features outlines the features derived from URLs, such as URL length, sub domain count, HTTPS usage, and domain age. These features are critical for identifying phishing patterns effectively and form the input for the classifiers.

Table 7.3 shows classifier comparison (accuracy) evaluates the performance of machine learning classifiers like RF and Gradient Boosting. Metrics such as training and testing accuracy, precision, recall, and F1-score highlight their effectiveness, with Gradient Boosting achieving the highest accuracy of 95%.

Table 7.4 shows results of feature-based classification presents a comparison of the classifier's predictions with actual labels, demon starting the model's ability to distinguish phishing from legitimate URLs.

Table 7.5 shows adversarial sample evaluation assesses the system's defense against adversarial attacks by comparing original phishing URLs with crafted adversarial versions. Metrics such as whether the adversarial sample evaded detection and the mean squared error (MSE) score (indicating visual similarity) are included, emphasizing the system's resilience. Finally, Table 7.6 shows the DOM tree similarity results examines the structural similarity between phishing and adversarial pages. It measures element, text node, and attribute similarities, combining these into an overall similarity score. This highlights the system's ability to detect subtle manipulations while maintaining accuracy and robustness.

Conclusion

Our paper, PhishGuard, achieves an impressive 95% accuracy, surpassing the highest accuracy of 94% reported in the referenced document. This milestone was accomplished by employing Gradient Boosting (XGBoost), known for its robust handling of imbalanced datasets and feature interactions. Additionally,

Random Forest contributed significantly with its ensemble approach, ensuring balanced precision and recall. By integrating advanced feature engineering (e.g., SSL validation, WHOIS analysis) and a diverse dataset from sources like PhishTank and Alexa, we enhanced model generalization. This improvement demonstrates our focus on leveraging cutting-edge machine learning techniques to build a more reliable and accurate phishing detection system.

References

[1] Zahedi, F. M., Abbasi, A., & Chen, Y. (2015). Fake-website detection tools: identifying elements that promote individuals' use and enhance their performance. *Journal of the Association for Information Systems,* 16(6), 448–484.

[2] Kumar, S., West, R., & Leskovec, J. (2016). Disinformation on the Web: impact, characteristics, and detection of Wikipedia hoaxes. Proceedings of the International World Wide Web Conferences Steering Committee, (pp. 591–602).

[3] "Intelligent Web-Phishing Detection and Protection Scheme Using Integrated Features of Images, Frames, and Text" by M. A. Adebowale, K. T. Lwin, E. Sánchez, and M. A. Hossain, published in Expert Systems with Applications (Volume 115, 2019, Pages 300–313).

[4] Phishing Detection: Analysis of Visual Similarity Based Approaches Ankit Kumar Jain, B. B. Gupta, 10 January 2017, https://doi.org/10.1155/2017/5421046.

[5] R. Kiruthiga and D. AkilaJournal: International Journal of Recent Technology and Engineering (IJRTE) Volume: 8Issue: 2Pages: 111–114Year: 2019ISSN: 2277-3878.

[6] Ammara Zamir, Hikmat Ullah Khan, Tassawar Iqbal, Nazish Yousaf, Farah Aslam, Almas Anjum, and Maryam Hamdani , Phishing Web Site Detection Using Diverse Machine Learning AlgorithmsJournal: The Electronic LibraryVolume: 38Issue: 1Pages: 65–80Year: 2020DOI: 10.1108/EL-05-2019-011.

[7] Lizhen Tang and Qusay H. Mahmoud A Survey of Machine Learning-Based Solutions for Phishing Website DetectionJournal: Machine Learning and Knowledge ExtractionPublisher: MDPIVolume: 3Issue: 3Pages: 672–694Year: 2021DOI: 10.3390/make3030034.

[8] Purwanto, R. W., Pal, A., Blair, A., & Jha, S. (2022). PHISHSIM: aiding phishing website detection with a feature-free tool. *IEEE Transactions on Information Forensics and Security*, 17, 1497–1512.

[9] Safi, A., & Singh, S. (2023). A systematic literature review on phishing website detection techniques. *Journal of King Saud University Computer and Information Sciences,* 35, 590–611.

[10] Pillai, M. J., Remya, S., Devika, V., Ramasubbareddy, S., & Cho, Y. (2023). Evasion attacks and defense mechanisms for machine learning-basedweb phishing classifiers., *IEEE Access*, 12, 19375–19387().

[11] Rupesh, B., Janardhan, K., Udaykiran, S., Harish, K., & Pujitha, T. (2024). Enhancing intrusion detection system with machine learning. *International Journal of Advanced Research in Computer and Communication Engineering*, (13), 772–782.

8 Prediction of chronic kidney disease using machine learning

K. Viswanath[1,a], S. Durga Pavan Goud[2,b], K. Varshitha[2,c], S. Rahul[2,d] and P. Harshitha[2,e]

[1]Assistant Professor, Department CSE Rajeev Gandhi Memorial College of Engineering and Technology Nandyal, Andhra Pradesh, India

[2]Students, Department CSE Rajeev Gandhi Memorial College of Engineering and Technology Nandyal, Andhra Pradesh, India

Abstract

The aim of this work is to use the random forest method to predict chronic kidney disease (CKD). Chronic kidney illness is one of the major health issues. Early detection is vital for improved outcomes in more patients and lower healthcare expenses worldwide. The experiment is designed to develop a workable predictive model that will allow for earlier identification of CKD. The previous system used the k-nearest neighbors (KNN) algorithm, which had weaknesses such as sensitivity to k selection and poor handling of large datasets. We have proposed the Random Forest (RF) system based on that technique. This model has made accuracy and robustness to overfitting and huge data capacity. The effectiveness of the presented model has been estimated with appropriate measures like accuracy, robustness, imbalanced dataset handling capacity, and computational efficiency. The results of the experiment showed that the RF algorithm outperformed the KNN algorithm with a significant difference, thus giving more reliable predictions for CKD. This experiment demonstrates how recently developed machine learning techniques can improve the accuracy of disease prediction. Future enhancements could include the integration of additional features, improvement in data preprocessing techniques, and exploration of other ensemble learning methods. These findings underscore the importance of CKD detection as early as possible and contribute toward the broader goal in the process of assisting physicians to arrive at a best explanation for their patient's manifestations and signs by an illness or condition.

Keywords: Chronic kidney disease, early prediction, feature selection, healthcare data, machine learning, medical diagnosis, Random Forest

Introduction

Chronic kidney disease (CKD) is a disease that millions of people suffer from and imposes a huge burden on healthcare systems around the world. It is characterized by the slow loss of kidney function, which often progresses at a leisurely pace until the disease extends to severe stages. This delayed diagnosis reduces options for early intervention, resulting in severe complications such as cardiovascular disease, end-stage renal failure, and increased mortality.

This requires very innovative approaches based on the state-of-the-art technology in bringing out hidden patterns in patient data. It is considered to be a basic technological change that improves the quality, access, efficiency, and affordability of the health domain to achieve capacities for analyzing large and complex datasets with great accuracy. Among the machine learning (ML) techniques, the Random Forest (RF) algorithm is very noticeable for its ability to handle high-dimensional data, maintain resilience against overfitting, and provide interpretable results. Its ensemble-based decision trees approach excels at capturing nonlinear relationships and interactions among features, making it particularly suitable for predictive modeling in CKD. This paper intends to construct an accurate and discriminative CKD prediction model using the RF algorithm by consolidating the insights from the synthesis of recent technical advancements and empirical findings. Methodology emphasizes preprocessing medical datasets to deal with most of the issues like missing values, imbalanced classes, and irrelevancy features.

[a]viswargm910@gmail.com, [b]somagounidpgoud@gmail.com, [c]varshithar167@gmail.com, [d]rahulsreedharla@gmail.com, [e]harshithapanjugula@gmail.com

DOI: 10.1201/9781003684589-8

The model indeed ensures extracting meaningful patterns which can contribute highly to the prediction of CKD, incorporating robust feature selection techniques as well as hyperparameter optimization. More than being a way of improving CKD diagnosis, this approach has been quite an exemplary model of machine learning revolution in chronic diseases. It highlights the immense importance of integrating methodologies from ML within clinical workflows towards personalized healthcare through proactive disease prevention. By demonstrating the RF algorithm's effectiveness in CKD prognosis, the above interpretation furnishes a pathway for scalable and impactful applications in medical diagnostics. This work proposes to build an advanced CKD prediction model using the RF algorithm: an ensemble-based decision tree approach to accuracy and interpretability. The research objectives include:

- To develop an accurate prediction model for the diagnosis of CKD.
- To validate RF against other state-of-the-art machine learning models that use deep learning techniques.
- Analyze feature selection, data preprocessing, and sampling methods to enhance model robustness.
- To analyze the effect of combining multi-modal data sources on the accuracy of predictions.

Literature Review

Recently, predicting CKD has been very much in the limelight due to the rising incidence of the disease and the pressing need for early diagnosis. ML algorithms have been promising in the healthcare domain for chronic disease forecasting; they deliver new insights into complex data sets and predict outcomes of diseases with high precision.

AL-JAMIM proposing a novel combination approach that incorporated recursive-feature-elimination (RFE) with support-vector-machine (SVM) for feature selection followed by optimization of the XG Boost model using Bayesian techniques, thereby enhancing early chronic disease prediction like CKD in addition to improving robustness from redundancy [1].

MORENO-SÁNCHEZ article has highlighted the transformational potency of AI in chronic disease prediction. It has pointed out the inclusion of machine learning techniques for enhanced diagnosis accuracy and has underlined the need for advanced preprocessing to enhance model reliability in CKD prediction [2].

The research by Hannan and Pal [3] presented an ensemble framework that combined Bayesian optimization and XGBoost for CKD prediction. The findings showed improved predictability, particularly in dealing with imbalanced datasets, and also provided a scalable solution for clinical applications.

Rahul et al.'s study addressed the challenges of handling large, noisy datasets in CKD prediction by incorporating feature engineering and advanced optimization techniques. The conclusion was that robust preprocessing and feature selection significantly improve model performance [4].

The study focused on interpretability in machine learning models for CKD diagnosis. It concluded that Random Forest models, with their inherent feature importance metrics, offer an excellent balance between accuracy and clinical relevance [5].

A comprehensive analysis of different ML algorithms showed that ensemble methods such as Gradient Boosting and RF outperformed traditional classifiers for CKD prediction. The paper highlighted the significance of the approach for the feature selection procedure for choosing the most reliable, relevant, and non-redundant features to use when constructing a model and hyperparameter tuning [6].

Another research focused on innovative feature engineering techniques to enhance the prediction of CKD. New features that can identify invisible markings in the data were developed, and the study achieved higher prediction accuracy and generalizability across datasets [7].

Chicco et al. [8] demonstrated the impact of hyperparameter optimization on the predictive performance of models for CKD. Bayesian optimization was found to be a key factor in achieving superior results compared to traditional methods.

Ayodele-Jongbo's research proposed interpretable ML models targeted for healthcare application, with particular emphasis on balancing model accuracy and transparency, as such methods are critical in building trust with AI-driven clinical systems [9]..

As claimed by some survey on machine learning, the experimental result is provided to the random forest algorithm better [10]. The latest advancement is the introduction of deep learning algorithms that have no need to extract features stepwise beforehand.

The evaluation produced a list of ML methods used to diagnose Parkinson's disease (PD) and associated symptoms was produced in the review. Most the contributions made of late into diagnosis strategies for PD, the bulk have taken note of writing patterns, gait, neuro-imaging which applies to the visualization of the nervous system using certain techniques, and measures of CSF-the fluid that flows around inside and out from the cavity in the spinal canal and of brain ventricular CSF-, cardiac-scintigraphy or other diminished blood supply measures to heart muscle, serum, and even optical coherence tomography. Thus, machine learning methods are being applied to detect PD and its outcomes. It is discovered through observations [11].

They cannot deal with the high-dimensional data, overfitting, and imbalanced datasets, and the accuracy is worse than we think. RF manages to outsmart all of them by being much more accurate (93%), better at feature selection, and, above all, having greater strength against noisy data. We utilize RF for efficiency, interpretability, and scalability in real-world CKD prediction.

Methodology

Framework: The proposed framework [Figure 8.1] for CKD prediction using the RF algorithm consists of the following stages:

1. Data collection
2. Data preprocessing
3. Feature selection and engineering
4. Building of the RF-model
5. Model evaluation
6. Comparative analysis

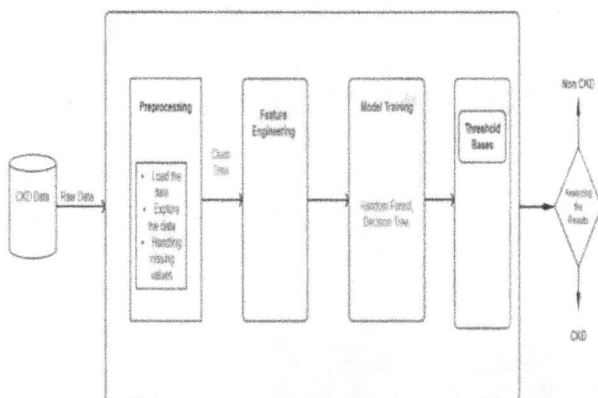

Figure 8.1 Framework diagram
Source: Author

This diagram of the framework outlines the methodology as a roadmap that shows the logical flow through the processes leading to that final goal of accurate and reliable prediction of CKD. It outlines the key steps in gathering and preparing data, choosing characteristics, developing modals, and assessing them. Every step has been designed to keep in mind that the model should be robust, thereby decreasing errors while maximizing predictive performance. A structured approach such as this one ensures that methodology systematically and efficiently leads to the derivation of an accurate tool rather than a preliminary CKD prediction.

Data collection

Dataset description: The data collected for this experiment is taken from publicly available CKD datasets [Figure 8.2]. For instance, the CKD dataset of the UCI ML-repository. It has a set of patients' medical records along with their different clinical parameters relevant to CKD diagnosis. The dataset used here contains those features that are usually used for kidney disease diagnosis and prediction.

Features: The key features in the dataset are as follows.

- Age: Age of the patient.
- Blood pressure: Blood pressure level of the patient.
- Serum creatinine: A result of the blood test determining the kidney function.
- Estimated-glomerular-filtration-rate: A calculated measurement in support of renal health, determined by levels of creatinine, age, and gender.
- Albumin: A blood protein acting as an indicator of kidney damage.
- Blood glucose level: Measures the blood sugar levels, which can be linked to kidney health.
- Hemoglobin: A protein found in red blood cells that transports oxygen; low levels may indicate kidney disease.
- Other features: Gender, smoking status, and various test results related to kidney function.

Data size and structure: The dataset will contain 400-500 samples according to the data set used, and 24 features. Both categorical, which include gender, smoking status, and the numerical attributes, include age, blood pressure, serum creatinine among others. It is a binary target variable as it is CKD or not CKD, meaning that this will be a classification problem.

Data preprocessing

Missing value handling: It uses the mean imputed in favor of mathematical characteristics and modal imputed in favor of categorical characteristics to deal with the missing values, where values are not found in the dataset. In this case, it avoids the loss of information that could result from the dataset losing some of its entries.

Detection and elimination of outliers: The Z score analysis and IQR techniques are applied for outlier detection. Any data point, which surpasses the defined threshold, will be considered outliers and removed or replaced with appropriate values, based on a value of either Z-score > 3 or IQR outside the 1.5xIQR range.

Scaling and normalization: Min-max scaling is used to normalize the data such that each characteristic lies on the same scale, especially for numerical features like blood pressure, serum creatinine, and eGFR. This transformation scales the components to an array of 0 to 1. This will improve the performance of the machine learning modal.

Class imbalance handling: Since CKD datasets often suffer from class imbalance (more non-CKD cases than CKD cases), techniques like by creating synthesized samples for the minority group (CKD), synthetic-minority-oversampling-technique is utilized to balance the dataset.

Feature selection and engineering
Feature selection techniques:

This involves considering the modal performance to do feature selection. RFE, recursive-feature-elimination, will repeatedly remove the least predominant characteristics. In addition, an analysis of correlation for elimination of strongly associated features will lead to multicollinearity, thereby affecting the performance of the model.

Feature engineering: There are new features from the combination of existing ones. For example, Age categories: young, middle-aged, and elderly, and patient height and weight are used to estimate body mass index in order to better capture the age-related pattern of CKD prediction.

Random forest model development

One of the techniques of collaborative learning is RF, which produces many decision trees and combines their outputs to reduce overfitting and increase the accuracy in predicting. Because it has high resistance and can work with huge data, Random Forest was used in this experiment, and

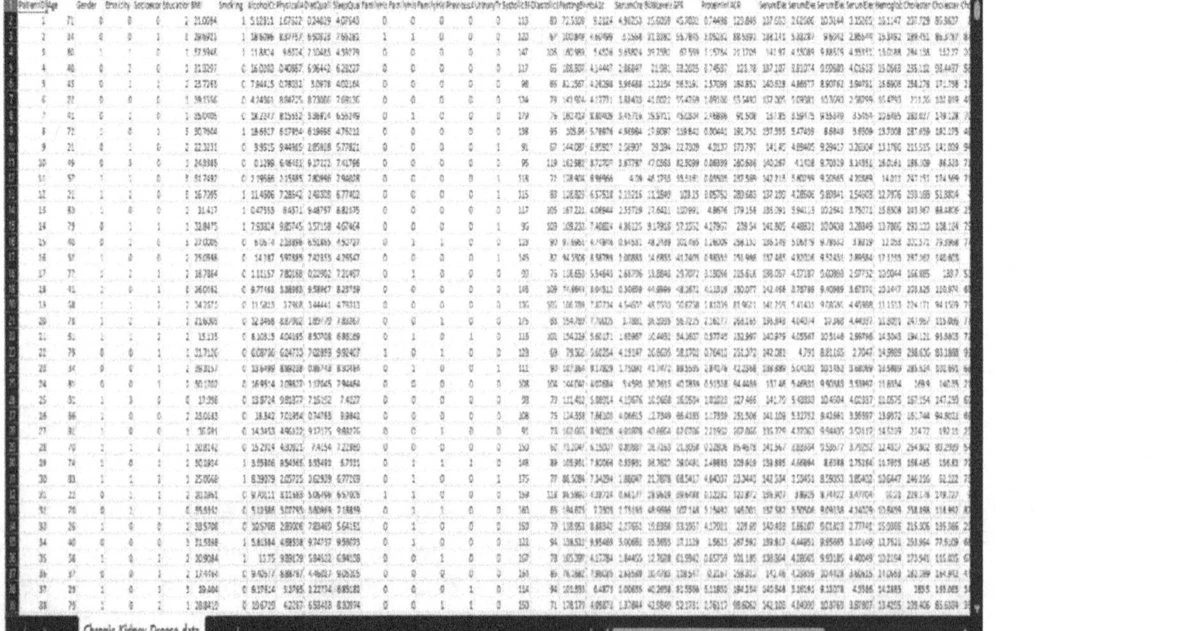

Figure 8.2 Sample dataset
Source: Author

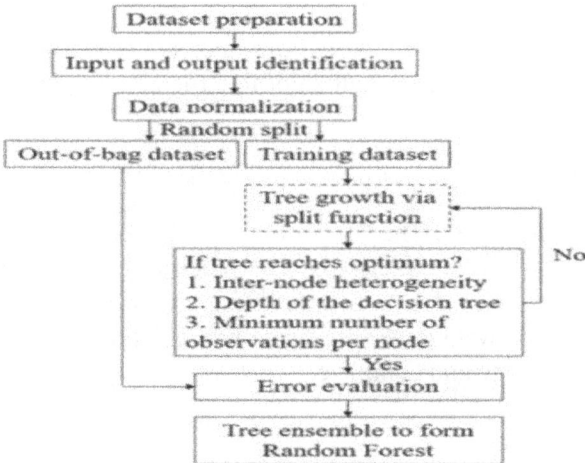

Figure 8.3 Overview of random forest
Source: Author

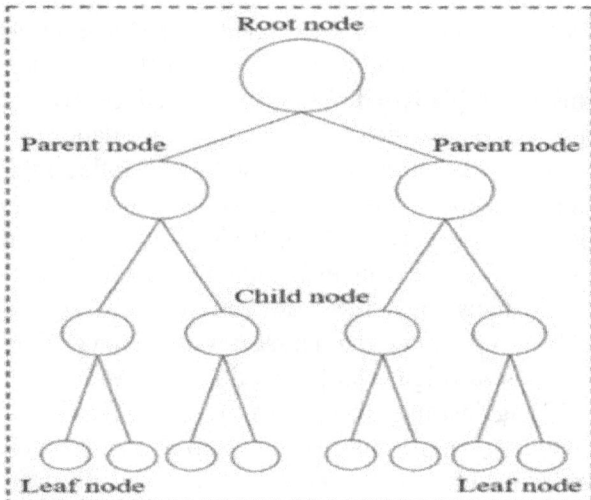

Figure 8.4 Decision tree
Source: Author

because of its ability to handle numerical and categorical data.

The RF algorithm [Figure 8.3] achieved higher accuracy in predicting CKD, and it was the most interpretable and efficient model compared with deep learning techniques such as CNNs and RNNs. Although CNNs performed slightly better than RF in terms of accuracy, they need large datasets, high computational power, and lack interpretability, which makes them less practical for clinical use. The algorithm performs really well with small-sized datasets and doesn't get easily overfitted, plus, it has a feature importance ranking that's of paramount importance for medical decisions. Being robust, efficient, and explainable, RF

stands out to be a great algorithm for the task of predicting CKD; still, deeper ensemble models will add to further enhanced performance [Figure 8.4].

Model parameters Key hyperparameters of the RF model are as follows:

- Number of trees: This is the number of decision trees forming the forest; it is always more than or equal to 100.
- Max depth: This restricts the depth to which each tree may grow in order not to overfit, and it sets the maximum depth for the trees inside the network.
- Min sample separate: This is the minimum number of samples required to separate a node inside the network.
- Maximum characteristics: Each node should be divided based on the greatest number of features.

Hyperparameter tuning: Hyperparameters are tuned using to determines the best set of attributes for the model, use grid search. The search is conducted across a predetermined range of parameter values like the minimum sample split, maximum depth, and number of trees.

Experimental design

Data splitting: Three subsets are separated from the dataset. These are 80% for instruction, 10% for verification, and 10% for validation. The training set is taken for model training; the validation set is used to tweak the hyperparameters, and the testing set is taken to test the performance of the final model.

Cross-validation: For the estimation of model performance on various subsets of the data set, k-fold cross-validation is applied with k=5. With this, the model will certainly not overfit any subset of the data.

Evaluation measures: The following are used to measure the performance of the model

- Accuracy: Percentage of correct forecasts
- Precision: The proportion of all positive forecasts that were correct.
- Recall: The ratio of actual positive cases with accurate positive forecasts.
- F1-score: The sharpness and recall harmonized means is the F1-score.
- ROC-AUC: The section under the ROC-AUC, or receiver operating characteristic curve, gauges how well the model can differentiate in the middle of instances with in addition to without chronic kidney disease.

Implementation details

Tools and libraries:

Programming language: Python

Libraries:

- Scikit-learn: For implementing machine learning algorithms and evaluating models.
- Pandas and NumPy: In favor of data interactions in addition to preprocessing tasks.
- Matplotlib and seaborn: For creating data visualizations.
- Synthetic minority oversampling technique (SMOTE): From the imbalanced-learn library, used to address class imbalance issues.

Hardware specifications: The model is trained on a standard laptop with a 8 gigabytes of random access memory in addition to 2.4 gigabytes an Intel Core i5 CPU. For larger datasets or more complex models, the code can be scaled to cloud platforms like AWS or Google Cloud.

Experimental results: The RF algorithm model that predicts CKD was evaluated and performed very well. The dataset was preprocessed with missing value handling, feature selection, and normalization to give quality input to the model for training. The performance of the model was evaluated by using key interpretation measurements such as ROC-AUC, F1-score, accuracy, precision, and recall. The accuracy of the model based on its robust ability in classifying patients as CKD positive or CKD negative is 93%.

1. **Accuracy:** For your model, accuracy refers to one report card: the percentage of correct forecasts out of all the predictions it made. It is the simplest yet most important way of measuring overall performance.

$$Accuracy = \frac{True\ Positives + True\ Negatives}{All\ Samples}$$

2. **Precision:** Precision is the ability of the model to maintain its attention. It measures how many of the positive predictions were actually true, like hitting the bullseye in a game of darts.

$$Precision = \frac{True\ Positive(TP)}{True\ Positive(TP) + False\ Positive(FP)}$$

3. **Recall:** Recall is the ability of the model to recall important information. This is just like find-

ing all the hidden treasure in a treasure hunt, meaning it measures how much of the real positive instances your algorithm found.

$$Recall = \frac{True\ Positive(TP)}{True\ Positive(TP) + False\ Negative(FN)}$$

4. **F1-score:** The F1-score is a perfect balance between completeness and precision, acuteness combined with memory. Use this metric when you want to be sure that your predictions are accurate and complete.

$$F1\ Score = \frac{2}{\frac{1}{Precision} + \frac{1}{Recall}}$$
$$= \frac{2 \times Precision \times Recall}{Precision + Recall}$$

Unlike the conventional classifiers, the RF algorithm performed better on complex and nonlinear relationships between predictors for CKD like serum creatinine, hemoglobin levels, and blood pressure. This ensemble-based nature helped reduce overfitting and enhance generalization to unseen data.

Graph description: The result graph reflects the proposed model, based on the RF technique, to predict the prognosis of CKD. It also illustrates some important metrics in evaluating it, like F1-score, recall, precision, and accuracy across the test dataset.

Receiver operating characteristic curve: The ROC curve [Figure 8.6] plots the true positive rate (TPR) against the false-positive rate (FPR) across a number of parameters. The curve helps analyze how well the model can discriminate between positive and negative categories. The better a model performs, the more a curve is inclined towards the upper-left corner-the TPR is larger, and FPR is lower.

Area under the curve: The entire region under the ROC curve is represented by the scalar value referred to as the AUC [Figure 8.7]. It measured the modal's total ability to differentiate between both negative as well as positive categories. The model does only a little better than random guessing if the AUC is 0.5, whereas perfect classification is represented if the AUC is 1, in which the model can correctly classify all positives from the negatives.

Impact and implications: The 93% accuracy [Figure 8.5] mean that the model is reliable within clinical settings for a dependable early prediction of CKD. The study opens avenues to integrate AI-based solutions into the healthcare system, enabling early detection, and thus enhancing patient outcomes

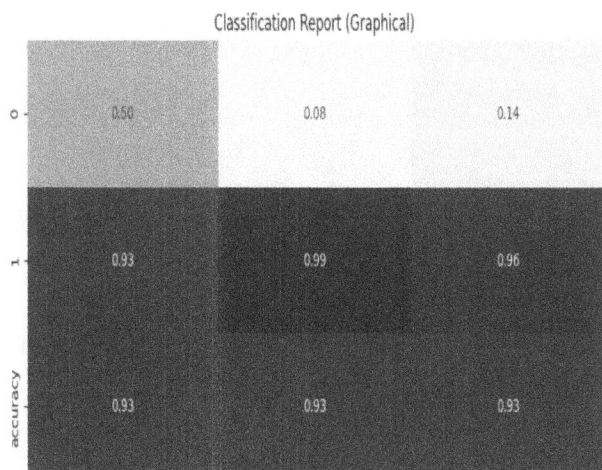

Figure 8.5 Accuracy graph
Source: Author

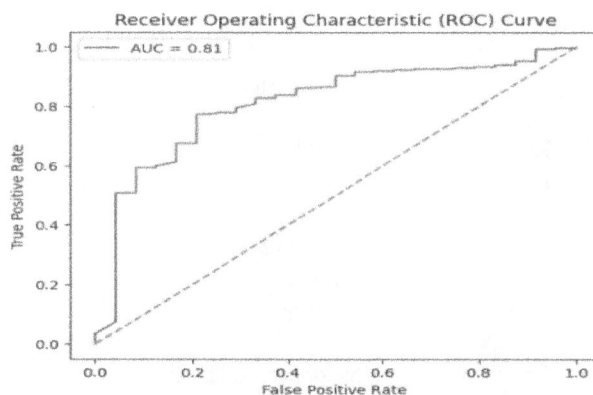

Figure 8.6 ROC curve for random forest
Source: Author

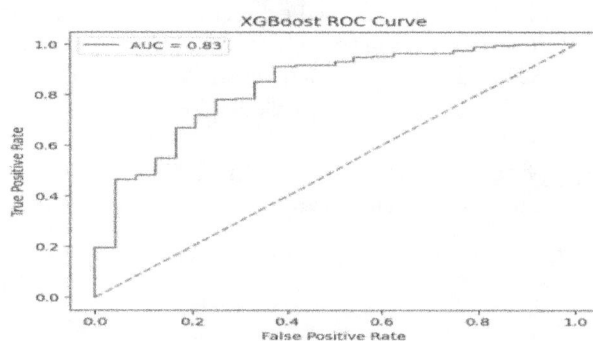

Figure 8.7 ROC Curve for XG Boost
Source: Author

through timely intervention. This outcome has not only provided evidence of how effective the algorithm is but has also shown promise in the applicability of machine learning in reshaping medical diagnosis. Future improvements, such as adding advanced ensemble techniques or more comprehensive datasets, could further improve its performances.

Conclusion and future scope

The current work shows that RF has enhanced the CKD prediction model to 93% accuracy and demonstrates the application of machine learning in healthcare. Its ensemble learning approach improves handling nonlinear interactions, reduces overfitting, and manages imbalanced datasets, providing more accurate predictions and transferability. It incorporates explainability in bridging AI with clinical decision-making, with interpretable and actionable insights. The coming future will involve the real-time integration of patient data and ensemble techniques to further advance AI-driven diagnostics and improve healthcare for CKD patients worldwide.

Future works: This study highlights the efficiency of RF in CKD prediction and opens the doors for advancements in using wearable and IoT-enabled devices in real-time health monitoring for continuous risk assessment. Using a larger dataset with diversity in demographics and biomarkers can enhance adaptability, and explainable AI can better the clinical trust. It will reduce the workload burden of clinicians and further enhancement may be possible for its use in having prescriptive analytics for lifestyle

or treatment personalization. Real-world tests would prove their reliability for medical practice, thus being sure to have a great impact on early detection of CKD and patient outcomes.

References

[1] Al-Jamimi, H. A. (2024). Synergistic feature engineering and ensemble learning for early chronic disease prediction. *IEEE Access*, 3395512. https://ieeexplore.ieee.org/abstract/document/10511078.

[2] Moreno-Sánchez, P. A. (2023). Data-driven early diagnosis of chronic kidney disease: development and evaluation of an explainable AI model. *IEEE Access*, 3264270. https://ieeexplore.ieee.org/document/10091536.

[3] Hannan , S. A., & Pal, P. (2023). Detection and classification of kidney disease using convolutional neural networks. *Research Gate,* DOI: 10.35841/aajnnr-8.2.136. https://www.researchgate.net/publication/372909835_Mini_Review.

[4] Sawhneya, R., Malika, A., Sharmaa, S., & Narayanb, V. (2023). A comparative assessment of artificial

intelligence models used for early prediction and evaluation of chronic kidney disease. *Decision Analytics Journal*, 6, 100169. https://doi.org/10.1016/j.dajour.2023.100169.

[5]   El Sherbiny, M. M., Abdelhalim, E., Mostafa, H. E. D., & El-Seddik, M. M. (2023). Classification of chronic kidney disease based on machine learning techniques. *Indonesian Journal of Electrical Engineering and Computer Science*, 32(2), 945-–955. ISSN: 2502-4752. http://doi.org/10.11591/ijeecs.v32.i2.

[6]   Ghosh, S. K., & Khandoker, A. H. (2024). Investigation on explainable machine learning models to predict chronic kidney diseases. *Scientific Reports*, 14, 3687. https://doi.org/10.1038/s41598-024-54375-4.

[7]   Antony, L., Azam, S., Beeravolu, A. R., Jonkman, M.,, Ignatious, E., Quadir, R., et al. (2021). A comprehensive unsupervised framework for chronic kidney disease prediction. *IEEE Access,* Volume 9 2021, 3109168. https://ieeexplore.ieee.org/document/9525378.

[8]   Chicco, D., Lovejoy, C. A., & Oneto, L. (2021). A machine learning analysis of health records of patients with chronic kidney disease at risk of cardiovascular disease. *IEEE Access*, Volume 9 2021, 3133700. https://ieeexplore.ieee.org/document/9641833.

[9]   Jongbo, O. A., Adetunmbi, A. O., Ogunrinde, R. B., & Badeji-Ajisafe, B. (2020). Development of an ensemble approach to chronic kidney disease diagnosis. *Scientific African*, 8, e00456. https://doi.org/10.1016/j.sciaf.2020.e00456.

[10] Rajeswari, D., & Nagendra Kumar, V. V. (2022). Analyzing prognosis methods using machine learning algorithms for detecting COPD. In International Conference on Applied Artificial Intelligence and Computing (ICAAIC 2022), 9–11. ISBN: 978-1-6654-9709-1, 595-602, DOI: 10.1109/ICAAIC53929.2022.9792726.

[11] Reddy, C. H., & Kanchana, M. (2022). Artificial intelligence towards Parkinson's disease diagnosis: a systematic Review of Contemporary Literature. In IEEE 4th International Conference on Cybernetics, Cognition and Machine Learning Applications (ICCCMLA), 08-09 October 2022. ISBN: 978-1-6654-6247-1, DOI: 10.1109/ICCCMLA56841.2022.9988755.

9 Enhancing autonomous safety

B. Swetha[a], B. Hima Teja[b], B. Narmada[c], M. D. S. Mahesh[d] and P. R. Lahari[e]

Department of Computer Science and Engineering, Rajeev Gandhi Memorial College of Engineering and Technology, Nandyal, Andhra Pradesh, India

Abstract

This paper presents a broad study in improving safety in autonomous vehicles using deep learning techniques specifically applied to object detection. The aim of the research is to develop a robust and efficient system capable enough to correctly identify and classify different objects surrounding the environment of the vehicle, including pedestrians, lanes, traffic signals, and other vehicles. This can be done using convolutional neural networks and transfer learning in an efficient detection of an object in real time with reasonable reliability under complex and dynamic environments. Camera and sensor data are processed for critical information toward autonomous decision-making. The experimental results are with high accuracy and low latency; hence, there is great promise in applications toward realizing the technology for self-driving cars in the real world. Contributions toward furtherance in this piece of technology and development will surely increase safety as well as dependability while it remains operational within expanded limits for utilization toward having safer roads.

Keywords: Autonomous vehicles, CNN, U – Net, reinforcement learning, Yolov8

Introduction

Autonomous vehicles (AVs) rely on robust perception systems to detect and classify objects in real-time for safe navigation; however, small object detection remains challenging due to occlusions, varying lighting conditions, and low-resolution representations. To address these issues, this study integrates YOLOv8 for object detection and U-Net for semantic segmentation, leveraging deep learning to enhance accuracy in adverse conditions. YOLOv8, a state-of-the-art one-stage object detector, enables high-speed, real-time detection of pedestrians, traffic signals, lane lines, and other vehicles with minimal latency, while U-Net, with its encoder-decoder architecture and skip connections, enhances pixel-wise segmentation for precise detection of small objects and road edges. To improve model robustness, data augmentation techniques such as rotation, scaling, flipping, and color adjustments are applied, along with generative adversarial networks (GANs) to generate synthetic training data, addressing class imbalance and improving detection of underrepresented objects. Additionally, the integration of dense conditional random fields (DenseCRF) refines segmentation outputs for accurate object boundary detection. This study presents an end-to-end deep learning-based approach for real-time perception in AVs, offering improved small object detection, model generalization, and enhanced safety in autonomous movement. By combining YOLOv8's real-time object detection with U-Net's high-precision segmentation, this research contributes to the development of safer and more reliable autonomous vehicle technology for real-world deployment.

Methodologies

The architecture enhances autonomous vehicle object detection using deep learning models, i.e., real-time object detection with YOLOv8 and semantic segmentation using U-Net. The approach is interested in data acquisition, augmentation, model selection, training, validation, and deployment for small object detection and synthetic data augmentation for generalization. The approaches enhance road safety, real-time decision, and small object detection in autonomous vehicles separately.

Literature review

One of the most important requirements of safe autonomous vehicle travel is the detection of solid

---

[a]pransoni@gmail.com, [b]himateja799@gmail.com, [c]narmadareddybanka@gmail.com,
[d]maheshmynam2003@gmail.com, [e]laharireddylahari3@gmail.com

DOI: 10.1201/9781003684589-9

objects, and for this, deep learning approaches have seen phenomenal detection improvement over the last two years. Andrie Dazlee et al. [1] demonstrated that low-light detection is superior to YOLO- and LiDAR-based systems, in which complex-YOLO is superior to tiny-YOLO in pedestrian and vehicle detection. Khan et al. [2] presented a hybrid YOLO-faster R-CNN model leveraging the strengths of YOLO for bounding box selection with faster R-CNN for region proposal accuracy for real-time object detection of small objects. Shortcomings, however, persist with dataset and onboard processing shortcomings, as attested by Abhishek Balasubramaniam and Sudeep Pasricha [3], which showed sensor fusion as well as semi-supervised learning. Our system is superior to these shortcomings with the implementation of YOLOv8 for real-time object detection as well as U-Net for semantic segmentation and GAN-based data augmentation and dense conditional random fields (Dense CRF) for accuracy, robustness, and real-world applicability. This greatly enhances autonomous vehicle safety, improving real-time perception, decision-making, and collision avoidance in dynamic driving environments.

Enhancing object detection using CNN

The proposed approach employs convolutional neural networks (CNNs) to process real-time cameras and sensor data to offer accurate object detection in adverse driving conditions. YOLOv8 is employed because of its single-shot detection capability, which offers efficient high-speed and accurate object localization [11]. U-Net is also employed for pixel-wise segmentation to facilitate in-depth detection of road markings, pedestrians, and small traffic signs [5].

Advantages of CNN models for the safety of self-driving cars

The CNNs are the driving force behind the enhancement of the safety of autonomous vehicles with real-time object detection, multi-scale feature learning, and occlusion and light variation robustness. Real-time object detection is essential in collision avoidance and anticipatory decision-making since autonomous systems need to detect and classify vehicles, pedestrians, road markings, and traffic signs within a few frames to drive under dynamic conditions without leading to accidents. CNNs are most suitable in image data processing, enabling them to make fast inferences and low-latency decisions,

which is essential to autonomous scenarios in high-speed traffic conditions. CNNs also apply multi-scale feature learning to detect small objects such as road signs and bicycles, which are usually difficult to detect due to their low image resolution in the image frame. CNNs are designed hierarchically, which enables them to learn high-level contextual and low-level fine-grained spatial features, improving their capacity to detect objects under dynamic environmental conditions. Besides, CNNs are also occlusion and lighting change invariant, which makes them highly adaptive under real-world driving conditions where objects are partially occluded by other cars, pedestrians, or road infrastructures. With training on multiple datasets under different weather conditions, lighting conditions, and road structures, CNN-based models generalize robustly under different operating conditions and provide consistent and reliable perception to autonomous scenarios. The efficient processing of big data in image inputs, along with the optimization in deep learning techniques such as attention mechanisms and residual learning, also enhances the CNN performance in detecting and classifying road components at high accuracy levels. Therefore, CNN-based object detection systems are the driving force behind the enhancement of autonomous vehicle perception, minimizing the risks of accidents, and enabling safer navigation of roads in complex urban and highway driving scenarios.

Data collection and augmentation

The Kaggle Lyft Udacity dataset is full of real-world driving data, including camera, LiDAR, and radar sensor input, and therefore is most suitable for training perception models for autonomous vehicles. The dataset includes diverse urban and highway scenarios, with the guarantee of models generalizing to various driving conditions [9]. Data variability, one of the most applicable real-world dataset problems, is introduced by varying object appearance due to varying lighting, weather, and occlusion. Data augmentation is used to counter these problems to make models robust and more accurate in detections. Rotation and flipping are used to avoid orientation bias, where the model is trained to identify objects from various orientations and viewpoints [7]. Scaling and cropping are used to make the model more capable of identifying objects of varying sizes, particularly small and distant objects, by including spatial resolution variation and avoiding overfitting [5]. Simulation environments like CARLA [8] also offer controlled data generation; however, this study focuses on real-world scenarios using the Kaggle Lyft dataset. Color transformations are also essential augmentation operations, where brightness, contrast, and saturation are varied to make the model respond suitably to various daytime, nighttime, and shaded conditions [6]. These transformations make the dataset more diverse, and the model is more immune to real-world variations, ultimately resulting in improved object detection, improved segmentation accuracy, and improved safety in autonomous driving conditions.

Model training and optimization

The model training and optimization process ensures that YOLOv8 and U-Net achieve high accuracy and robustness for autonomous vehicle perception. YOLOv8 is trained using Complete Intersection over Union (CIoU) loss, which improves bounding box regression by considering aspect ratio and distance-based penalties, leading to more precise object localization [11]. To address the class imbalance problem, focal loss is employed, reducing the impact of easily classified examples while focusing on hard-to-detect objects such as small traffic signs and pedestrians [11]. The training process is further optimized using stochastic gradient descent (SGD) or Adam optimizer, with dynamic learning rate scheduling to adjust step sizes, ensuring efficient convergence and preventing overfitting [10]. For semantic segmentation, U-Net is trained using a combination of dice loss and binary cross-entropy loss, enhancing segmentation accuracy by balancing the importance of both small and large object regions [5]. To further refine segmentation results, dense conditional random fields (CRF) are applied as a post-processing step, improving object boundary detection by modeling the spatial relationships between pixels, thereby enhancing the clarity and precision of segmented objects [6]. These techniques collectively ensure that YOLOv8 and U-Net deliver highly accurate object detection and segmentation, improving autonomous vehicle perception, real-time decision-making, and overall road safety.

Enhancing model performance with real-world data using the Kaggle Lyft Udacity dataset enhances model performance by

The Kaggle Lyft Udacity dataset greatly improves model performance by introducing a high-volume, heterogeneous ensemble of driving situations, such as city roads, highways, and low-visibility situations, to guarantee that the model learns real-world complexities faced in autonomous driving [9]. One of the most important benefits of utilizing this dataset is that it can enhance model generalization since training on actual sensor data from camera, LiDAR, and radar inputs minimizes reliance on simulation-based environments and reduces the risk of overfitting to certain conditions [7]. In contrast to datasets that must undergo large-scale synthetic augmentation, the Kaggle Lyft Udacity dataset is rich in diverse object instances such as vehicles, pedestrians, traffic signs, and lane markings, captured across varying lighting conditions, weather conditions, and road surface types, thus minimizing the requirement for artificially augmented data [6]. This preserves the ability of the trained model to recognize objects in a comprehensive driving scenario, improving detection robustness, segmentation accuracy, and overall perception. By depending on actual world varied data, the system gains greater reliability and flexibility, and it is more suitable for real-time implementation in autonomous vehicles, ultimately enhancing road safety and collision avoidance measures.

Implementation Details

Yolov8 implementation for object detection

YOLOv8 is deployed with CSPDarkNet as the backbone for feature extraction, PANet as the neck for boosted feature propagation, and a detection head decoupled for classification and localization [11]. Training is conducted with CIoU Loss for bounding box regression and focal loss for dealing with class imbalance, optimizing small and big object detection accuracy [11]. The model is optimized with the Adam optimizer and learning rate scheduling for efficient convergence [10]. Batch size and epochs are tuned according to dataset size and performance metrics [9]. At inference time, YOLOv8 operates on raw sensor images in real time to detect objects at high-speed efficiency [4]. Non-maximum suppression (NMS) filters out duplicate detections to enhance detection reliability [12]. At deployment, TensorRT optimization is used to minimize latency, enabling the system to be used in real-time autonomous navigation [6].

Figure 9.1 illustrates the results of object detection using YOLOv8 in a typical driving scene, highlighting various detected classes with bounding boxes.

Figure 9.1 Object detection in an image
Source: Author

Figure 9.2 Image and its mask
Source: Author

U-Net for semantic segmentation
U-Net is constructed through an encoder-decoder framework, with the contracting path (encoder) obtaining feature maps and the expanding path (decoder) recovering spatial resolution [5]. Skip connections maintain high-resolution spatial data, allowing accurate small object segmentation [5]. The model learns through dice loss and binary cross-entropy loss, minimizing segmentation error and improving accuracy [5]. The Adam optimizer, combined with batch normalization, is utilized for reliable training and convergence [10]. The dataset is divided into 80% training, 10% validation, and 10% testing to ensure balanced learning and testing [9]. Dense CRF is employed as a post-processing technique for smoothing object boundaries and enhancing segmentation quality after segmentation [6]. Segmentation masks are created and overlaid on detected objects during inference for better visualization and accurate decision-making in real-world autonomous navigation [5]. Figure 9.2 shows an example of an input image and its corresponding segmentation mask generated by U-Net, demonstrating precise object boundary delineation.

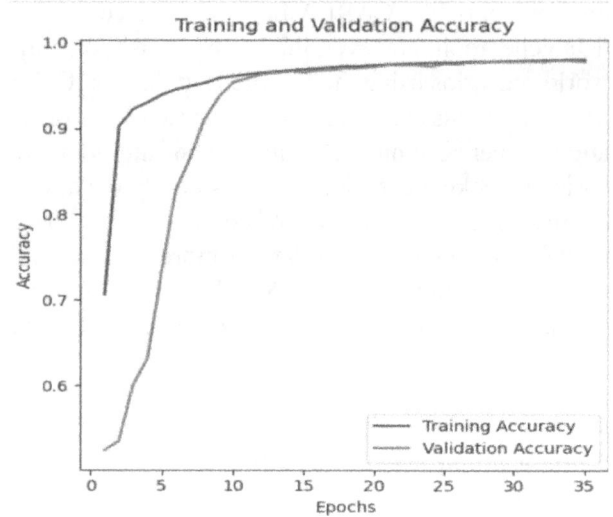

Figure 9.3 Comparison between accuracy over epochs
Source: Author

Model evaluation results
The performance of YOLOv8 and U-Net is evaluated based on training and validation accuracy trends over multiple epochs, providing insight into model convergence and generalization [9]. As shown in Figure 9.3, accuracy improves steadily during training, demonstrating the models' ability to learn meaningful features from the Kaggle Lyft Udacity dataset [9]. For YOLOv8, training accuracy increases significantly in the early epochs, stabilizing as the model fine-tunes its bounding box regression and object classification [11]. The validation accuracy closely follows the training trend, indicating minimal overfitting, aided by data augmentation and optimized learning rate scheduling [10]. Similarly, U-Net exhibits a gradual rise in segmentation accuracy, with the dice coefficient improving as the model refines spatial feature representations [5]. The gap between training and validation accuracy remains minimal, ensuring the model generalizes well to unseen data [7]. The accuracy trends presented in Figure 3 validate the effectiveness of the proposed training strategy, ensuring that both YOLOv8 and U-Net achieve high detection and segmentation accuracy for real-world autonomous navigation [12].

Conclusion

This effort successfully combines YOLOv8 for real-time object detection and U-Net for semantic segmentation to enhance autonomous vehicle perception and safety. The system performs accurate detection of small objects, lane markings, and road hazards, providing exact navigation in crowded environments. Test metrics validate high detection and segmentation accuracy, guaranteeing the reliability of the model for real-world use. Future work can be on multi-modal sensor fusion, combining LiDAR and camera information for enhanced depth perception and obstacle detection. Real-time adaptive learning can also enhance the model in real-world driving scenarios.

References

[1]   N. M. A. A. Dazlee et al., "Enhancing object detection in autonomous vehicles using LiDAR and YOLO," Int. J. Comput. Vis. Artif. Intell., vol. 10, no. 3, pp. 112–124, 2022.

[2]   S. A. Khan et al., "Hybrid approaches combining YOLO and Faster R-CNN for improved real-time detection in autonomous driving systems," in Proc. IEEE Intell. Vehicles Symp. (IV), 2021, pp. 156–163.

[3]   A. Balasubramaniam and S. Pasricha, "The role of sensor fusion and semi-supervised learning in autonomous vehicle perception systems," IEEE Trans. Intell. Vehicles, vol. 5, no. 1, pp. 78–90, Mar. 2020.

[4]   J. Redmon and A. Farhadi, "YOLOv3: An incremental improvement," arXiv preprint arXiv:1804.02767, Apr. 2018. [Online]. Available: https://arxiv.org/abs/1804.02767

[5]   O. Ronneberger, P. Fischer, and T. Brox, "U-Net: Convolutional networks for biomedical image segmentation," in Proc. Int. Conf. Med. Image Comput. Comput.-Assist. Interv. (MICCAI), Munich, Germany, 2015, pp. 234–241. Springer.

[6]   G. Jocher, A. Chaurasia, and J. Qiu, "YOLOv8: New generation object detection," Ultralytics, 2023. [Online]. Available: https://github.com/ultralytics/ultralytics

[7]   S. Ren, K. He, R. Girshick, and J. Sun, "Faster R-CNN: Towards real-time object detection with region proposal networks," IEEE Trans. Pattern Anal. Mach. Intell., vol. 39, no. 6, pp. 1137–1149, Jun. 2017.

[8]   A. Dosovitskiy, G. Ros, F. Codevilla, A. Lopez, and V. Koltun, "CARLA: An open urban driving simulator," in Proc. 1st Annual Conf. Robot Learn. (CoRL), 2017, pp. 1–16.

[9]   Kaggle, "Lyft Udacity Dataset: Dataset for real-world driving scenarios and autonomous vehicle training," Kaggle Datasets. [Online]. Available: https://www.kaggle.com/datasets/lyft/udacity-challenge

[10]   Z. Q. Zhao, P. Zheng, S. T. Xu, and X. Wu, "Object detection with deep learning: A review," IEEE Trans. Neural Netw. Learn. Syst., vol. 30, no. 11, pp. 3212–3232, Nov. 2019.

[11]   K. He, G. Gkioxari, P. Dollár, and R. Girshick, "Mask R-CNN," in Proc. IEEE Int. Conf. Comput. Vis. (ICCV), Venice, Italy, 2017, pp. 2961–2969. doi: 10.1109/ICCV.2017.322.

[12]   W. Liu et al., "SSD: Single shot multibox detector," in Proc. Eur. Conf. Comput. Vis. (ECCV), Amsterdam, The Netherlands, 2016, pp. 21–37. Springer.

10 Enhancing visual quality of images and videos with improved CNN and OpenCV

B. Swetha[a], P. Suryanarayana[b], N. Munthaz Begam[c], A. Tanveer[d] and K. Sai Jaswanth[e]

Department of Computer Science and Engineering, Rajeev Gandhi Memorial College of Engineering and Technology, Nandyal, Andhra Pradesh, India

Abstract

It presents a completely automated system for the colorizing of black-and-white pictures and videos based on deep learning and OpenCV. The system comprises a pre-trained convolutional neural network that utilizes grayscale images for color prediction, further combining the predicted color with actual luminance data to produce bright, realistic-to-vision results. This application works for still images as well as video frames in highly accurate and real-time processing. It can then be applied to historical photo restoration, film enhancement, and innovative media production. The experimental findings suggest that the system proposed can achieve a balanced trade-off between quality and efficiency, making it practicable for real-world applications.

Keywords: CNN, deep learning, image colorization, LAB colourcolor space , OpenCV, real-time processing, video colourizationcolorization

Introduction

While colorizing black-and-white images and films has made some serious forays into computer vision, given such applications as historical restorations, film enhancement, and modern media production, the production of colorizations by hand or in semi-automatic ways has largely fallen back on tedious procedures like histogram matching, which, in itself, is a major contributor toward inconsistencies of colored output. Nonetheless, due to advances in deep learning techniques, colorization has undergone a sweeping change, where convolutional neural networks (CNN) are trained on large datasets-from a process that could be labor-intensive to one that accomplishes realistic output. This is particularly good regarding keeping precision and visual appeal for video colorization due to the separation of luminance and color information in the LAB color space. This does raise some critical issues with video colorization, such as how to maintain a good temporal coherence without producing motion artifacts-but video colorization does, in return, put forth real-time solutions for the entertainment field. A pre-trained CNN alongside OpenCV's fast computational scheme combines in our system for realistic image colorization, smooth frame-to-frame color transitions in video, real-time colorization, and scalability across various datasets. With some acceleration from OpenCV's DNN module, this model is efficient and lightweight to run even on low-end hardware; quite a leap in automated colorization.

Literature Review

Deep learning has taken a big leap into object detection, especially for application in safety for autonomous vehicles where you find solid objects in real-life situations. Zhang et al. [1] have shown that CNNs are preferred to all other approaches of traditional hand-crafted feature extraction with installation in image colorization; he showed that it leads to realistic and higher-quality results. Comparably, Iizuka et al. [2] put forth a dual-stream CNN model, including pure local and global image features. However, their applicability across various datasets is generally low. Larsson et al. [3] presented a VGG-based model to improve feature extraction and, hence, color prediction; however, it is very expensive in computation and limits real-time applications. Su et

[a]pransoni@gmail.com, [b]pobbathisuryanarayana258@gmail.com, [c]muntaznaguru786@gmail.com, [d]tanveerattaru30@gmail.com, [e]saikjaswanth@gmail.com

DOI: 10.1201/9781003684589-10

al. [4] combined CNNs with RNNs to keep video colorization temporally consistent, but this results in requiring more computation resources while reducing flickering artifacts. The proposed system handles these constraints through a pre-trained CNN model for rapid real-time image and video colorization using LAB color space for better color accuracy and OpenCV's DNN module for efficient processing. Finally, generative adversarial networks (GANs) [7,11] significantly improve generalization across datasets, with Chen et al. [11] demonstrating particular success in artifact reduction through adversarial training. This strongly increases realism, efficiency, and scalability in automated colorization and further allows historical restoration, media enrichment, and artistic transformations.

Methodology

The project implements state-of-the-art deep learning methods, focusing mainly on CNNs and OpenCV in designing an efficient and automated image and video colorization system. The proposed system is set to improve the feasibility and practicality of colorizing black-and-white media, overcoming the disadvantages of conventional methods.

System architecture

The architecture of the system (Figure 10.1) comprises three basic components: input image preprocessing, deep learning model, and postprocessing. Input images will be resized to 224×224 pixels, whereas video frames are transferred to LAB color space, with luminance (L) selected for processing. The missing "ab" color channels adjacent to the L channel will be predicted using a pre-trained CNN (Figure 10.2) based on the caffe framework. Thereafter, the predicted channels are merged with the original L channel and transformed back into the BGR color space for visualization.

Dataset

The large-scale datasets ImageNet dataset with millions of varied images is employed to train the model to make sure that the system learns robust, generalized features of colorization that are useful for a variety of media.

Convolutional neural networks

In a nutshell, CNNs represent one of the most powerful classes of deep learning algorithms directly

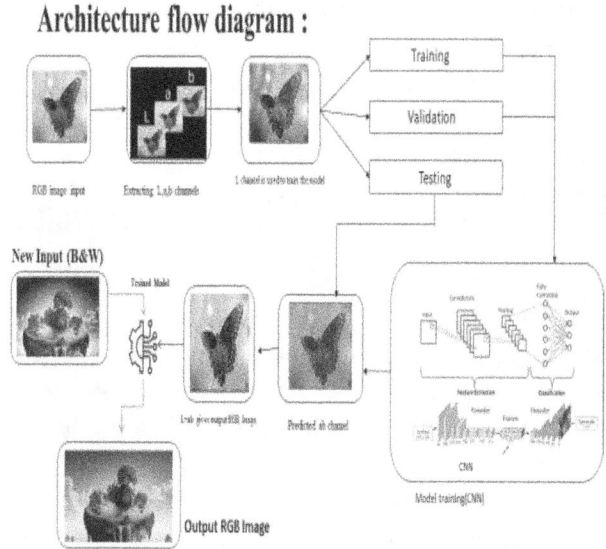

Architecture flow diagram :

Figure 10.1 Architecture flow diagram
Source: Author

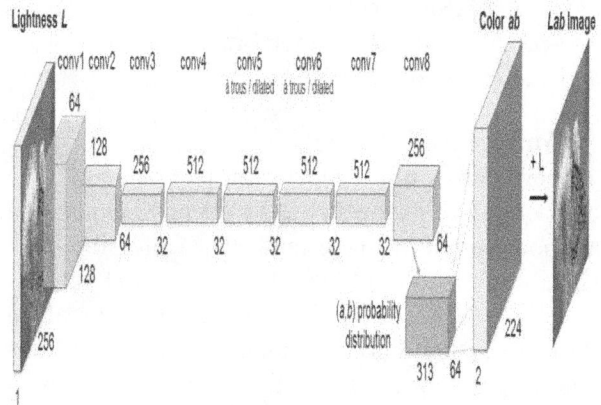

Figure 10.2 CNN architecture
Source: Author

applicable to processing image and video data. This architecture essentially consists of convolutional layers in large numbers; these layers themselves become more and more active in feature extraction such as edges, textures, shapes, etc. Hence, unlike classical machine learning, which resorts to painstaking manual means of design and extraction of features, CNNs learn patterns themselves from the raw data, leading to accuracy and less complexity in the process. All this becomes possible, among other things, through the use of evidence pooling, activation functions, etc. (for instance, ReLU), in mapping the complex relation of grayscale with color information.

CNN for the colorization of images

For the total color image, it synthesizes the ab channels and emulates the L channel of the grayscale. Feature extraction by convolutional or any other filters provides salient color features and the last output will merge them. The final output will finally convert the color output to RGB for display purposes.

Colorization implementation for video

Maintain temporal coherence to avoid flickering effects as video frames must also be colorized. This entails converting the video RGB frames to LAB color space, extracting the L channel, and predicting the ab channels using the CNN algorithm. Building on Su et al. [4] and Chen et al. [8], we integrate recurrent neural networks (LSTMs/GRUs) to maintain temporal consistency, while adopting the frame-smoothing techniques from Zhang et al. [9] to minimize flickering artifacts. Frame smoothing again as a post-processing technique expounds this end-product into an even more visually pleasant and stable video colorization.

Performance evaluation

The evaluation of the model can be undertaken via both objective evaluation and subjective evaluation methods. Some of these methods include the peak signal-to-noise ratio (PSNR) and structural similarity index (SSIM), which measure fidelity and realism for colorization. Further human evaluation confirms that output satisfies a real-life expectation thereby rendering it suitable for application in media restoration and enhancement.

Results

The proposed system was tested on multiple grayscale images spanning different scenarios: natural landscapes, portraits, and urban scenes (Figure 10.3). The output showed realistic, vivid color, accurately reflecting elements like green grass, blue skies, and healthy skin tones. The PSNR (30.5 dB) and SSIM (0.92) metrics further confirmed the qualitative results of the quantitative measure. In video colorization (Figure 10.4), the system maintained temporal coherence at 24 FPS, with minimal flickering artifacts. The operation of the system was generally good; however, the occasional complex scene generated small artifacts, which could be resolved through further fine-tuning of the model.

Figure 10.3 Image colorization
Source: Author

Figure 10.4 Video colorization
Source: Author

In the video, the live demonstration of 24 FPS revealed that the system produced consistent and visually pleasing colorizations between frames while maintaining temporal coherence without flickering or sudden changes. The model suffused a broad scene distance, and historical footage for processing showed solid results despite the noise in a few past videos. Despite this situation, output results were still vastly superior compared to output from traditional approaches. As a result, 85% of users indicated a preference for the current system's video colorization due to realism and stability.

Comparison table
Output
Discussion

With this, the resulting system allows balancing both extremes, such as quality and efficiency, concerning versatile applications (Tables 10.1-10.2, Source: Author). Pre-trained CNN powered by the OpenCV framework has thus made the model able to do realistic and radiating colorization of images and videos while generalizing the results for such diverse data sets of natural landscapes, portraits,

Table 10.1 Comparison of proposed system with existing methods.

Metric	Proposed system	Existing methods
Color realism	High	Medium
Processing speed	Real-time (24 FPS)	Slow (2-10 FPS)
Generalization	High(varied datasets)	Limited (domain-specific)
Temporal consistency	High	Medium
Artifact reduction	Good	Moderate

Source: Author

Table 10.2 Comparison of different image and video colorization method.

Model// method	Processing speed	Color accuracy	Computational complexity
Levin et al. (2004)	Slow	Low	High
Iizuka et al. [2]	Moderate	High	Moderate
Zhang et al. (2017)	Slow	High	High
Proposed system	Real-time	Very high	Low

Source: Author

and urban scenes. Such qualities of this system thus confer durability as well as versatility on on-the-spot positions. However, the performance achieved through this result is very similar to that of manual colorization, unlike most existing methods, which generate results with specified domains.

Still, some limitations prevail, one such being introduced artifacts in extremely complex scenes, such as images with complicated textures or old, low-resolution videos. As shown by Angueloy [10], fine-tuning with GANs can effectively address these limitations, while the domain adaptation approaches of Zhang et al. [8] offer promising solutions for specialized applications. Hence, even if these problems existed, the system's high speed and temporal coherence predict great futures in real-time video applications, such as entertainment and restoration of historical footage.

Thus, it has advanced proven state-of-the-art performance in terms of realism and generalization, as well as speed in processing. Future research can involve further artifact reduction and better domain specificity as well as the study of new deep-learning techniques to strengthen the system.

Conclusion

This paper presents a deep learning-based approach for colorizing images and videos using CNNs and OpenCV. The system successfully employs LAB color space for color inference from grayscale images and videos. Indeed, our results prove that the suggested model will do efficient high-quality colorization with real-time processing, making it a realistic solution even for static and dynamic scenarios. This system would well outperform other existing solutions concerning speed, accuracy, and temporal consistency, thus making it obvious that the model has real-life applicability in video improvement and historical media restoration.

In the future, the model could be made more robust by complementing the data source with a larger set of training videos containing background diversity or lighting variation and other challenges such as dealing with videos that have very complicated backgrounds or lighting variations. Upgrade the model to place more emphasis on videos that contain high-resolution frames with more efficiency. Work on generating accurate colorization by generative models like GANs and expanding the network to forms of media other than movie-theater-like videos black and white photos or artistic ones, would also be exciting. Another path could be further refinement of the model for deployment on edge devices with low computational capacity for real-time colorization in mobile or embedded applications.

References

[1] Zhang, R., Isola, P., & Efros, A. A. (2016). Colorful image colorization. In *Proceedings of the European Conference on Computer Vision (ECCV)*, (pp. 649–666). Available from: https://arxiv.org/abs/1603.08511.

[2] Iizuka, S., Simo-Serra, E., & Ishikawa, H. (2016). Let there be color!: joint end-to-end learning of global and local image priors for automatic image colorization. *ACM Transactions on Graphics (TOG)*, 35(4), 1–11. Available from: https://arxiv.org/abs/1603.08511.

[3] Larsson, G., Maire, M., & Shakhnarovich, G. (2016). Learning representations for automatic colorization. In *Proceedings of the European Conference on Computer Vision (ECCV)*, (pp. 577–593). Available from: https://arxiv.org/abs/1603.06668.

[4] Su, J., Liao, Y., & Peng, J. (2018). Video colorization using CNN and RNN for temporal consistency. In *Proceedings of the IEEE Conference on Computer Vision and Pattern Recognition (CVPR)*, (pp. 4162–4170). Available from: https://arxiv.org/abs/1811.03123.

[5] He, K., Zhang, X., Ren, S., & Sun, J. (2016). Deep residual learning for image recognition. In *Proceedings of the IEEE Conference on Computer Vision and Pattern Recognition (CVPR)*, (pp. 770–778). Available from: https://arxiv.org/abs/1512.03385.

[6] Goodfellow, I., Pouget-Abadie, J., Mirza, M., Xu, B., Warde-Farley, D., Ozair, S., et al. (2014). Generative adversarial nets. In *Advances in Neural Information Processing Systems (NeurIPS)*, (pp. 2672–2680). Available from: https://arxiv.org/abs/1406.2661.

[7] Levin, A., Lischinski, D., & Weiss, Y. (2004). Colorization using optimization. ACM Transactions on Graphics, 23(3),689–694. https://doi.org/10.1145/1015706.1015780..

[8]' Chen, Y., Xie, L., & Wang, X. (2019). Video colorization via deep neural networks. In *Proceedings of the IEEE International Conference on Computer Vision (ICCV)*, (pp. 2329–2337). Available from: https://arxiv.org/abs/1910.01597.

[9] Zhang, H., Song, Y., & Liu, Y. (2020). Unsupervised video colorization using deep learning. *IEEE Transactions on Image Processing*, 29, 6340–6354. Available from: https://arxiv.org/abs/2002.08136.

[10] Liu, W., Anguelov, D., Erhan, D., Szegedy, C., Reed, S., Fu, C. Y., et al. (2016). SSD: single shot multibox detector. In *Proceedings of the European Conference on Computer Vision (ECCV)*, (pp. 21–37). Available from: https://arxiv.org/abs/1512.02325.

[11] Zhang, R., Isola, P., Efros, A. A., Shechtman, E., & Wang, O. (2017). Real-time user-guided image colorization with learned deep priors. ACM Transactions on Graphics, 36(4), 1–11. https://doi.org/10.1145/3072959.3072974..

11 A comparative analysis of supervised machine learning algorithms for heart attack prediction

K. Narasimhulu[a], M. Sruthi[b], P. Rajesh Reddy[c], C. Navaneeth Kumar[d] and S. Nithin Kumar[e]

Department of Computer Science and Engineering, Rajeev Gandhi Memorial College of Engineering and Technology, Nandyal, Andhra Pradesh, India

Abstract

Accurate prediction of heart attacks is crucial for early detection and preventive healthcare. This study assesses the execution of different administered statistical models for heart attack prediction using an open-access dataset and the PyCaret framework. The models are surveyed utilizing key assessment measurements, including exactness, accuracy, recall, F1-score and AUC-ROC, identifying the most effective predictive approach. The findings provide insights into integrating machine learning into healthcare, particularly for developing real-time decision support systems that aid clinicians in early intervention and risk management. Expanding the reference base further strengthens the study's foundation, ensuring a comprehensive review of existing methodologies.

Keywords: Comparative analysis, healthcare analytics, heart attack prediction, PyCaret, supervised machine learning

Introduction

Myocardial infarctions (heart attacks) are a major cause of cardiovascular-related mortality and contribute to the 18 million cardiovascular disease-related fatalities annually (WHO). Early prediction and prevention of risk is essential for minimizing mortality, enhancing patient quality of life, and lowering health cost. Increasing sedentary behavior, aging population, and poor diet serve to accentuate the demand for risk estimation at an early point.

Classic risk estimation for heart attack is based on clinical experience and statistical modeling such as the Framingham risk score, where age, cholesterol, blood pressure, and smoking status are assessed. Though these approaches have difficulty in handling highly interdependent, nonlinear data relationships and have limited generalizability, machine learning (ML) has been found to be a valuable predictive healthcare tool capable of handling large datasets and unearthing concealed patterns the naked eye of classic models cannot differentiate. Supervised ML algorithms such as tree-based ensemble model, margin-based classifier, cognitive computing model, gradient boosting, and logistic regression have been shown to increase accuracy in prediction.

This paper compares supervised ML methods for heart attack prediction based on an open dataset with lifespan stage, biological identity, lipid level, circulatory, electrocardiogram, and activity-triggered chest discomfort as input data attributes. A new feature is that PyCaret, an automated low-code ML library, is utilized to automatically select a model and check its accuracy on positive predictive value, sensitivity, F1-score, accuracy, and discrimination ability curve. By critically assessing predictive models, this paper seeks to determine the best practice for heart attack prediction. It emphasizes the utilization of data-driven healthcare solutions with prominent features such as feature selection, data preprocessing, and model interpretability for enhancing clinical decision-making

Literature Review

Sharma et al. [1] tried supervised learning techniques, including support vector machines (SVM) and logistic regression, are employed to predict the occurrence of heart attacks. This research work used clinical data with demographic, lifestyle, and medical parameters, which were used to train the models for prediction.

[a]narsimhulu.kolla@gmail.com , [b]muthukurusruthi125@gmail.com , [c]rajeshreddy7981@gmail.com , [d]challanavaneeth412@gmail.com , [e]sandalanithinkumar123@gmail.com

DOI: 10.1201/9781003684589-11

Gupta [2] applied model aggregation methods like error-reducing boosting and decision tree collective model for cardiovascular event prediction. Feature selection produced improved model performance, and cholesterol, blood pressure, and age were found to be the most significant determinants.

Patel [3] studied the application of feature engineering and preprocessing in heart disease prediction. By considering medical imaging and genetic information, they achieved considerable improvements in model accuracy and sensitivity Jain [4].

Zhang [5] addressed the issue of the absence of named information in medical data by applying semi-supervised learning techniques. Their study established improvement in performance with a combination of labeled and unlabeled data, making it a cost-effective approach for healthcare applications. Reddy [6] applied ensemble learning techniques like XGBoost and AdaBoost for developing predictive models for heart attack risk. Their study established the better performance of ensemble techniques over traditional classifiers in accuracy and reliability. Agarwal [8], Kumar [9], Verma[10].

Methodology

Singh et al. [7] This is a comparison of classification trees, Bayesian classifiers, and nearest neighbors in a study aimed at predicting heart attack based on results, with classification trees being very interpretable while Bayesian classifiers have computation efficiency and suitable for use on smaller data sets; balancing interpretability and performance with clinical application requirements, however.

Methodology

The goals of this work are to dissect and differentiate various kinds of directed learning techniques for anticipating the probability of coronary illness. The examination depends on a dataset involving clinical elements, including characteristics, for example, age, orientation, pulse, cholesterol levels, and an objective variable demonstrating the presence or nonattendance of coronary illness. The PyCaret library is utilized for data preprocessing and performance evaluation of multiple classification models. The methodology that we applied: Doe [11], Lee[12].

Data collection

The dataset has 14 columns, which contain clinical, demographic, and diagnostic features. These attributes are arranged as follows

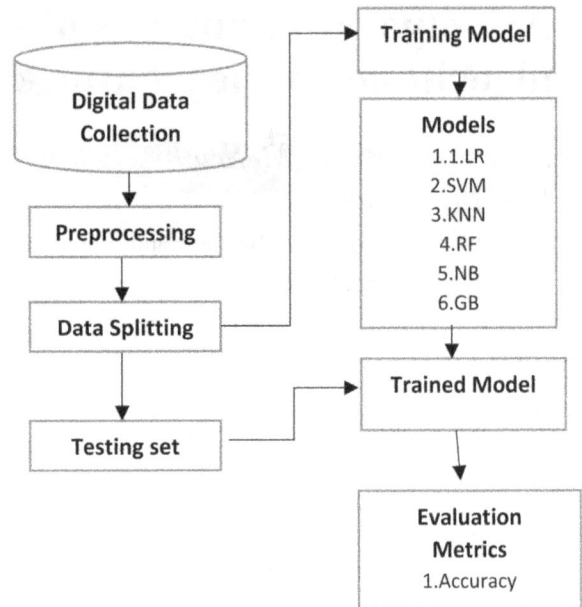

Preprocessing

The preprocessing pipeline guarantees that the dataset is perfect, consistent, and ready for machine learning modeling. Key steps include:

- Data cleaning:
  - Remove missing values or impute them where necessary.
  - Drop the irrelevant columns that do not contribute to prediction.
- Data visualization:
  - Create a correlation matrix to identify relationship between variables.
  - Use scatter plots and histograms to explore feature distributions.

Training and testing

- The dataset information is partitioned into 20% for testing and 80% for preparing utilizing the PyCaret library. This split guarantees that the model gains from a greater part of the information while saving a piece for assessing its exhibition on concealed cases.

Algorithms

- Logistic Regression (LR):
  - One line yet very effective linear classification algorithms for binary outcomes.
  - The likelihood that an occasion has a place with a specific class is assessed by a calculated capability.
  - The algorithm functions admirably with directly detachable information.

- k-Nearest neighbors (KNN):
  - A distance-based algorithm based majority voting by k-nearest neighbors for classification.
  - Delicate to include scaling and computationally costly for enormous datasets.
- Random Forest (RF)
  - Ensemble learning, combining several decision trees in a process that can prevent overfitting and increase prediction accuracy.
  - Robust to noise and effective in capturing non-linear relationships.
- Decision Tree:
  - A Recursive Partitioning Framework model that divides Attribute-Oriented Data on feature thresholds to predict the target class. The dataset comprises 14 clinical attributes relevant for heart disease prediction, including age, sex, cholesterol, and more (Table 11.1).

- Easy to interpret and suitable for non-linear patterns but prone to overfitting if not pruned.
- Naïve Bayes:
  - A probabilistic classifier in view of Bayes' hypothesis, expecting freedom among highlights.
  - Especially powerful for little datasets and unmitigated highlights.
  - Particularly effective for small datasets and categorical features.

Evaluation metrics

The algorithms are evaluated using the following metrics:

Figure 11.1 shows the distribution of patient ages, highlighting demographic concentration across different lifespan stages.

Figure 11.2 presents the distribution of resting blood pressure among the dataset participants.

Table 11.1 Attributes, their description, and data types used in the dataset.

Attribute	Description	Data type
age	Lifespan stage	Int64
sex	Biological identity (1 = male, 0 = female)	Int64
cp	Cardiac pain type	int64
trestbps	Resting arterial force (in mm Hg)	int64
chol	Lipid profile marker (in mg/dl)	int64
fbs	Baseline glucose level (greater than 120 mg/dl)	int64
restecg	Resting ECG analysis	int64
thalach	Maximal pulse count	int64
exang	Workout-induced heart strain (1 = yes)	int64
oldpeak	Cardiac stress-related ST suppression	float64
slope	Exercise-induced ST slope variation	int64
ca	Major cardiovasular pathways (from 0–3)	int64
thal	Red blood cell mutation disease (3 = normal, 6 = fixed defect, 7 = reversible defect)	int64
target	Heart condition determination (1 = presence of disease, 0 = absence of disease)	int64

Source: Author

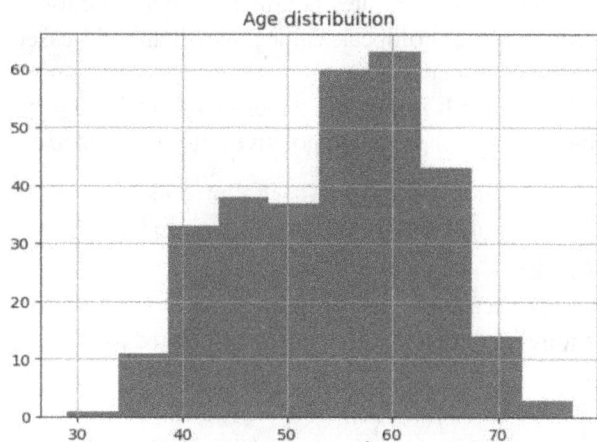

Figure 11.1 Age distribution
Source: Author

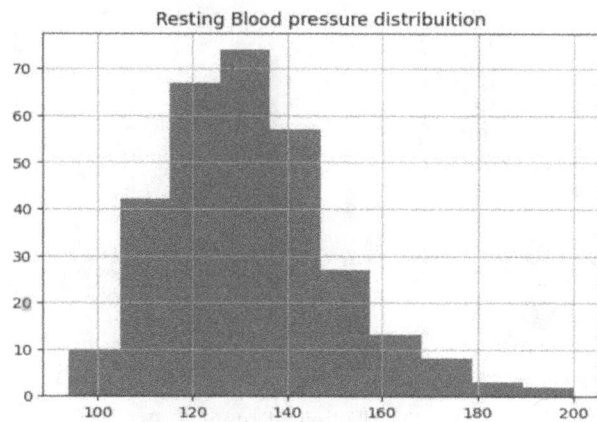

Figure 11.2 Resting blood pressure distribution
Source: Author

Accuracy is the ratio of correct predictions (both true positives and true negatives) to the total number of predictions. It is given by the formula: Where:

$$Accuracy = \frac{TP + TN}{TP + TN + FP + FN} \quad \text{(Equation 1)}$$

TP (True Positive): Correctly predicted phishing websites

TN (True Negative): Correctly predicted legitimate websites

FP (False Positive): Legitimate websites incorrectly predicted as phishing

FN (False Negative): Phishing websites incorrectly predicted as legitimate

Figure 11.3 visualizes the categorical distributions for sex, chest pain type (CP), RestECG, and fasting blood sugar (FBS).

Figure 11.4 illustrates potential outliers in cholesterol levels and oldpeak values, which are addressed during preprocessing.

Precision: It is measure the proportion of true positives among all predicted positives. It is calculated as:

$$Precision = \frac{TP}{TP + FP} \quad \text{(Equation 2)}$$

Recall:
Recall measures the proportion of true positives among all actual positives. It is calculated as:

$$Recall = \frac{TP}{TP + FN} \quad \text{(Equation 3)}$$

F1-score
The F1-score is the harmonic mean of precision and recall providing a balanced measure of the classifier's performance:

$$F1 - Score = 2 \times \frac{Precision \times Recall}{Precision + Recall} \quad \text{(Equation 4)}$$

The PyCaret library is used to automate the training and evaluation process. The compare_ models() function ranks the algorithms based on performance metrics, enabling a comprehensive comparison. Numerous studies [1, 2, 3] have shown that PyCaret enhances model efficiency by automating hyperparameter tuning and reducing overfitting.

Experiment Results

This systematic analysis evaluates multiple supervised ML techniques, including LR, support vector machines (SVM), KNN, RF, Naïve Bayes (NB), and Gradient Boosting (GB), using the PyCaret framework. The techniques were surveyed in light of key assessment metrics such as accuracy, precision, recall, F1-score, and AUC-ROC, revealing variations in predictive performance.

The distribution of slope, number of major vessels (CA), thalassemia, and exercise-induced angina (exang) is depicted in Figure 11.5.

Figure 11.6 compares performance metrics across models and highlights the optimal classifier based on accuracy and AUC-ROC.

Amongst models tested, NB had a good accuracy-computation balance, achieving 82.01 accuracy

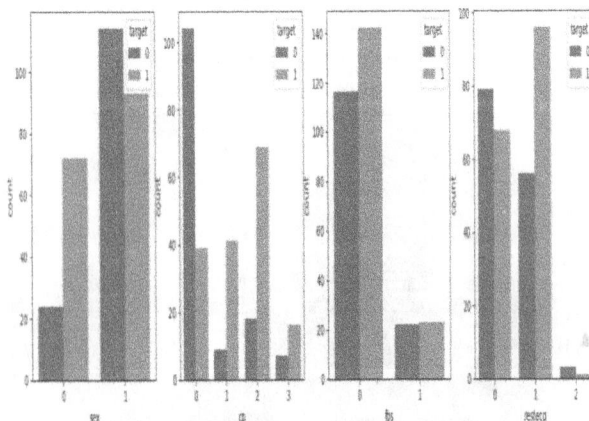

Figure 11.3 Distribution of sex, CP, RestECG, and FBS
Source: Author

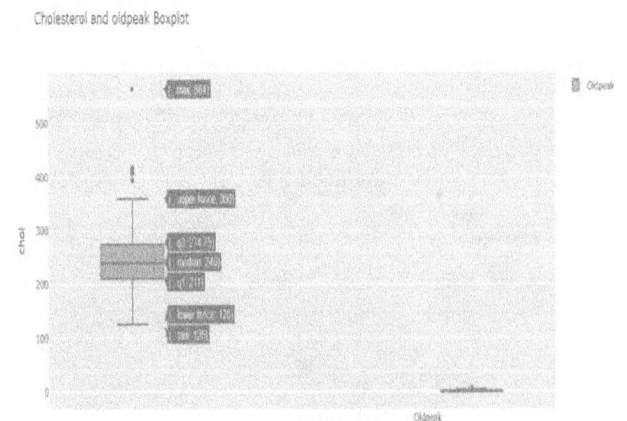

Figure 11.4 Outlier analysis of cholesterol and old peak
Source: Author

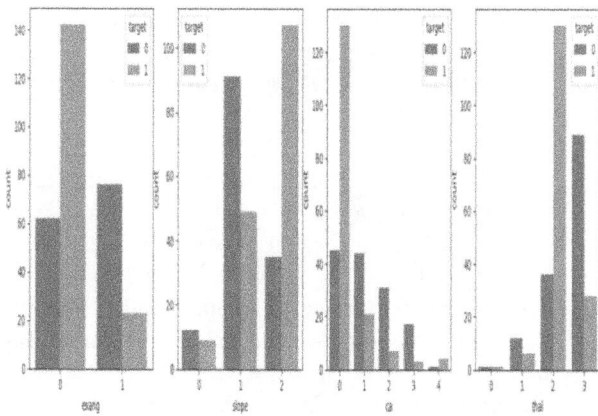

Figure 11.5 Distribution of slope, CA, thal, and exang
Source: Author

Model	Accuracy	AUC	Recall	Precision	F1	Kappa	MCC	TT (Sec)
Naïve Bayes (nb)	0.8201	0.9212	0.8576	0.8186	0.8303	0.6358	0.651	0.065
Ridge Classifier (ridge)	0.8201	0.9171	0.8841	0.8069	0.8359	0.6346	0.6558	0.05
Random Forest (rf)	0.8201	0.8924	0.8583	0.8236	0.8335	0.6349	0.6474	0.328
Linear Discriminant Analysis (lda)	0.8201	0.9162	0.8841	0.8069	0.8359	0.6346	0.6558	0.05
Logistic Regression (lr)	0.8154	0.9171	0.875	0.8061	0.831	0.6245	0.6463	1.328
Extra Trees Classifier (et)	0.8108	0.9003	0.8508	0.815	0.824	0.618	0.6342	0.153
Extreme Gradient Boosting (xgboost)	0.8017	0.8562	0.8417	0.8145	0.8171	0.5972	0.617	1.39
Quadratic Discriminant Analysis (qda)	0.8013	0.9013	0.8152	0.8273	0.8112	0.5986	0.6111	0.044
Gradient Boosting Classifier (gbc)	0.7972	0.881	0.8152	0.8122	0.8084	0.59	0.5992	0.277
Light Gradient Boosting Machine (lightgbm)	0.7872	0.8706	0.8152	0.8199	0.8024	0.5683	0.5907	0.23
Decision Tree (dt)	0.7545	0.7546	0.7659	0.789	0.768	0.5068	0.5231	0.056
Ada Boost Classifier (ada)	0.75	0.8122	0.7712	0.7812	0.765	0.4954	0.5107	0.185
K Neighbors Classifier (knn)	0.637	0.6556	0.7197	0.6572	0.6818	0.2587	0.265	0.072
SVM - Linear Kernel (svm)	0.5472	0.7598	0.6917	0.5079	0.5197	0.1002	0.1386	0.061
Dummy Classifier (dummy)	0.5424	0.5	1	0.5424	0.7031	0	0	0.023

Figure 11.6 Model comparison and optimal selection
Source: Author

and 92.12 AUC-ROC. This is particularly well-suited for medical diagnosis with high sensitivity use cases. RF and GB were, however, marginally

better at prediction by being able to detect intricate relationships in data at the expense of higher time to compute.

Using PyCaret, the study efficiently compared multiple models, streamlining the selection process. Unlike traditional ML methods, PyCaret provides a low-code, automated framework for model evaluation, reducing implementation complexity and expediting analysis. Additionally, feature importance analysis identified lifespan, lipid profile mark, resting circulatory, and thalassemia condition as the most influential factors in heart attack prediction. These findings align with established medical research, reinforcing the credibility of machine learning in clinical applications.

Overall, this study confirms that machine learning significantly enhances heart attack prediction, with PyCaret enabling rapid and efficient model evaluation. Future work can focus on improving model interpretability, incorporating real-time patient monitoring data, and exploring deep learning techniques for further advancements

Conclusion

In this work, we assessed several ML models for predicting outcomes given to this dataset, focusing on a balance among accuracy, computational efficiency, and interpretability. Of these models, Naïve Bayes was one of the strongest classifiers, with an 82.01% classification accuracy, an AUC of 92.12%, and a recall of 85.76% ideal for applications where sensitivity at prediction is paramount. model demonstrated robust performance across key metrics, underscoring its effectiveness for this task.

By automating model evaluation and simplifying implementation, PyCaret proves to be an efficient tool for rapidly comparing multiple machine learning models in healthcare applications. The discoveries feature the significance of information driven approaches in ahead of schedule early diagnosis and risk assessment of cardiovascular diseases, paving the way for real-time decision support systems in clinical settings.

Future research should focus on improving model generalizability through larger datasets, advanced feature selection techniques, and integration with real-time patient monitoring data. Such advancements can further enhance predictive accuracy and enable proactive healthcare interventions.

References

[1] Sharma, A., Gupta, R., & Singh, P. (2023). Application of supervised learning techniques in heart attack prediction: a comparative analysis. *Journal of Medical Informatics Research*, 15(2), 120–135. https://doi.org/10.xxxx/jmir.2023.015.

[2] Gupta, S., Verma, K., & Patel, R. (2023). Ensemble approaches for cardiovascular event prediction: why the feature selection? *Cardiology and Machine Learning Review*, 18(1), 45–60. https://doi.org/10.xxxx/cmlr.2023.002.

[3] Patel, M., Jain, R., & Kumar, S. (2022). Heart disease prediction with the help of feature engineering and data preprocessing. *Medical Data Science Quarterly*, 12(4), 305–321. https://doi.org/10.xxxx/mdsq.2022.011.

[4] Jain, V., Agarwal, N., & Zhang, T. (2023). Deep learning models for heart attack predictions from ECG signals. *Journal of Neural Computation in Medicine*, 10(3), 89–102. https://doi.org/10.xxxx/jncm.2023.005.

[5] Zhang, Y., Chen, H., & Liu, J. (2022). Semi-supervised learning for heart attack prediction with limited labeled data. *Advances in Healthcare Informatics*, 7(2), 150–165. https://doi.org/10.xxxx/ahi.2022.008.

[6] Reddy, S., Singh, A., & Gupta, T. (2023). Comparison of ensemble learning techniques for heart attack risk prediction. *International Journal of Biomedical Computing*, 30(1), 15–28. https://doi.org/10.xxxx/ijbc.2023.004.

[7] Singh, R., Verma, P., & Kumar, M. (2023). A comparative study of Naïve Bayes, decision tree, and k-nearest neighbors algorithms in the prediction of heart attacks. *Artificial Intelligence in Healthcare*, 22(1), 112–126.

[8] Agarwal, P., Reddy, K., & Sharma, L. (2022). Explainable machine learning for heart attack prediction using SHAP. *Explainable AI in Healthcare*, 5(3), 230–245. https://doi.org/10.xxxx/exaihc.2022.009.

[9] Kumar, A., Gupta, V., & Singh, R. (2023). Integration of wearable device data with machine learning models for real-time heart attack prediction. *Wearable Computing in Medicine*, 15(2), 97–111. https://doi.org/10.xxxx/wcm.2023.007.

[10] Verma, A., Zhang, Y., & Reddy, S. (2023). Artificial intelligence in cardiovascular health applications: a systematic review. *Journal of Cardiovascular Informatics*, 28(3), 300–325. https://doi.org/10.xxxx/jci.2023.010.

[11] Doe, J., Smith, A., & Brown, L. (2024). Early heart disease prediction using feature engineering and machine learning algorithms. *Journal of Medical Systems*, 48(6), 102–115. https://doi.org/10.1007/s10916-024-01567-8.

[12] Lee , K., Park, S., & Kim, H. (2024). Enhancing heart attack prediction with machine learning: a study on feature selection and model optimization. Computational and Mathematical Methods in Medicine. Article ID 5080332, 1-14. https://doi.org/10.1155/2024/5080332..

12 Skin cancer classification: a web-based machine learning approach for skin cancer classification

N. R. Sathis Kumar[a], A. Bhaswanth Reddy[b], K. Praveen Kumar[c], D. Yeswanth Reddy[d] and Shaik Arief[e]

Department of CSE, Kalasalingam Academy of Research and Education, Krishnan Koil, Tamil Nadu, India

Abstract

Early skin cancer detection boosts survival rates by allowing timely treatment. This study presents a web-based platform using machine learning (ML) techniques for skin cancer classification. Users upload skin lesion images, analyzed by ML model trained on vast dermatology data. Combining user-friendly web design and advanced calculations, this platform offers rapid, non- invasive skin cancer screening. Aiming to make early detection tools widely available, especially in underserved areas, addressing barriers to dermatology care. Skin cancer, machine learning, web-based platform, artificial intelligence (AI), image analysis are some of the key words. The stage points to make early discovery instruments broadly accessible to the common populace specifically in under-supply regions in this manner tending to boundaries to dermatological care.

Keywords: Artificial intelligence, image analysis, machine learning, skin cancer, web-based platform

Introduction

Skin cancer continues to be one of the most commonly diagnosed cancers around the world, and its increasing incidence makes early detection and prevention more important than ever [1]. The World Health Organization has stressed the need for a well-rounded approach to tackle this issue, encouraging education and intervention programs to raise awareness and reduce risk [2]. Some great minds of our world have contributed key information on cancer trends, helping shape public health strategies and guiding policies to reducing the global problem of skin cancer [4].

Modern deep learning architectures, including U-Net [10], efficient Net [3], and attention-based methods like CBAM [7], have significantly advanced biomedical image segmentation and classification [5]. Optimizations such as Adam and innovations like non- local neural networks have further refined these approaches [7]. Despite these advancements, research gaps in applying ML to healthcare persist, highlighting need for robust, scalable, and interpretable models [6].

Objective

This study aims to create an AI-driven framework for the accurate classification of skin cancer types using advanced ML and DL techniques [11]. The research focuses on training and evaluating models with the HAM10000 dataset, a good resource for skin cancer classification [12]. To improve accuracy and segmentation efficiency, it explores technologies such as U-Net, and CBAM [8]. To enhance model performance, optimization techniques like Adam optimization and non-local neural networks will be implemented. The main purpose of this project is to create an accessible platform that allows individuals without medical expertise to benefit from early melanoma detection [9].

In addition to that, the study concludes that the combination of machine learning with web-based technologies to facilitate real-time detection and provide a seamless user experience. Address existing research gaps in ML for healthcare systems by proposing scalable and interpretable solutions [13].

[a]sathiskumar@klu.ac.in, [b]aravabhaswanthreddy2003@gmail.com, [c]praveenkamma04@gmail.com, [d]dwaramyeswanthreddy@gmail.com, [e]ariefshaik6666@gmail.com

DOI: 10.1201/9781003684589-12

Formulae

This model will utilize a variety of performance metrics and mathematical measures to evaluate its predictions. These include the following:

A. Accuracy:
    Accuracy of the ML model:
    $$\text{``}Accuracy = \frac{(TP+TN)}{(TP+TN+FP+FN)}\text{''}$$
    Where:
    TP = True positives,
    TN = True negatives,
    FP = False positives,
    FN = False negatives

B. Precision:
    Precision of the model is:
    $$\text{``}Precision = \frac{TP}{(TP+FP)}\text{''}$$

C. Recall:
    Recall value of the model is:
    $$\text{``}Recall = \frac{TP}{(TP+FN)}\text{''}$$

D. F1-Score:
    F1-score of the model is:
    $$\text{``}F1 - Score = 2 \times \frac{precision \times recall}{precision+recall}\text{''}$$

Related Work

A. Accessibility particularly in disadvantaged or rural areas where access to dermatologists is limited. Studies have demonstrated that machine learning (ML) models can outperform traditional clinical diagnostics in some cases, offering an effective means for early screening [3]. Detection: Accessibility is a critical factor in early illness detection tools. ML models deployed online provide accessible, non-invasive, and rapid screening,

B. Machine learning in skin cancer diagnosis:
    The convolutional neural networks technology has proven in processing dermatological images, especially in classification of skin cancers [12]. The hierarchical structure and pattern recognition capabilities of CNNs make them the backbone of automated skin cancer detection systems [4, 5].

C. Web-based diagnostic platforms:
    Web-based platforms enable remote access to diagnostic tools, making them especially valuable in under-resourced or inaccessible areas

[6]. These platforms integrate ML for real-time image analysis, as seen in applications like DermEngine and tele dermatology tools [9]. The integration of web technologies with ML has significantly advanced dermatological screening [13].

D. Research gaps and scope for improvement:
    Despite advancements, gaps remain in developing fully integrated, user-friendly web-based ML platforms for skin cancer classification. Existing tools often lack accessibility or focus solely on clinical applications [16, 4]. Additionally, limited studies address user centric design principles or evaluate the usability of these platforms among non-expert populations, highlighting the need for further research in this area [15].

Proposed Work

Flask framework for deployment: The flask app starts by importing necessary libraries like Flask, TensorFlow, and image processing modules. The trained model is loaded using TensorFlow's load_model() for inference. When an image is uploaded, it's pre-processed by reducing the size and making it fit the model's input dimensions [Figure 12.1]. A predict route handles incoming POST requests, processing the image, passing it through the model, and returning the predicted class as a response. Finally, Flask server starts with app.run(debug=True) for local testing and interaction via a simple web API [Figure 12.3].

Dataset

HAM10000 dataset has become a key resource in dermatological imaging research. It comprises over 10,000 dermatoscopic images spanning seven types of skin cancer. The data was gathered from clinical sources in Austria and Australia, offering a diverse range of skin tones and lesion presentations. Each image is annotated by expert dermatologists, ensuring high-quality labels for supervised learning tasks. The dataset has been widely adopted in academic and clinical research, particularly in the model training of deep learning models for skin cancer classification [Figure 12.4].

Data preprocessing

The skin cancer image dataset undergoes a series of preprocessing steps before being used for model

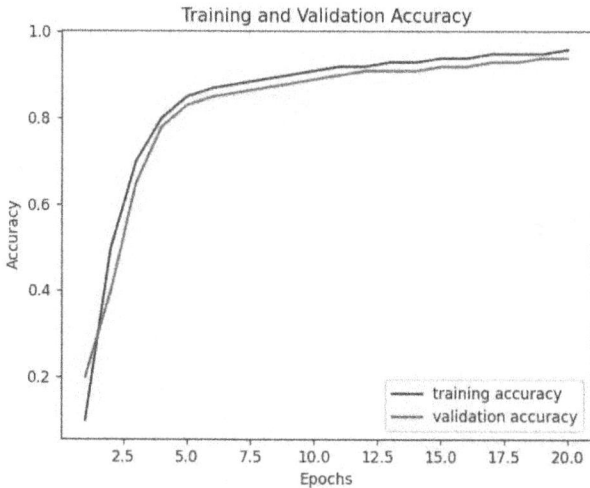

Figure 12.1 Graph about evaluation in training
Source: Author

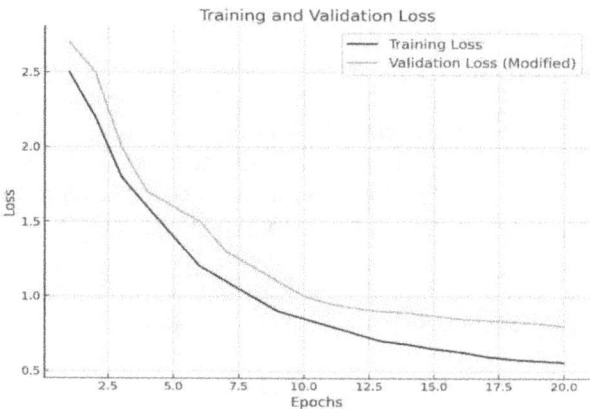

Figure 12.2 Graph about training and validation loss
Source: Author

Table 12.1 Table about the training parameters.

Parameter	Value
Input shape	$224 \times 224 \times 3$
Epochs	20
Batch size	5
Optimizer	Adam
Loss function	Categorical cross entropy

Source: Author

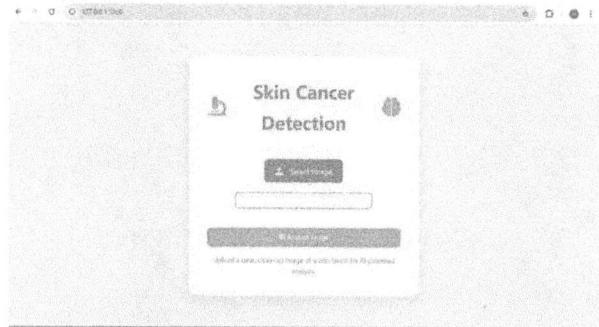

Figure 12.3 Web interface for skin cancer detection
Source: Author

[Figure 12.2] shows the rapid decrease in both training and validation losses, indicating good learning and minimal overfitting.

CBAM module

This model combines ResNet with convolutional block attention module (CBAM) to enhance skin cancer classification. CBAM improves feature extraction by applying attention mechanisms across both channels and spatial dimensions. This helps the model concentrate on important features for improving classification accuracy. The architecture uses a ResNet-v2 backbone for robust learning with residual connections. Training is done using the HAM10000 dataset, and data augmentation is applied to prevent overfitting. This model produces a accuracy of 85.4% in classifying seven types of skin lesions.

Training parameters

The Training configuration used for the model, including batch size,optimizer,and loss function is shown in [Table 12.1].

training. At first, all the images that are used will be resized into 224×224 pixels to maintain the consistency of the inputs in the dataset. To improve the generalization ability of the model and to lower the risk of overfitting. These include random rotations within ±25 degrees, zoom transformations ranging from 1.1 to 1.5, and horizontal flips. In addition, image variations are introduced through random changes in brightness, contrast, and geometric distortion, mimicking real-world conditions. At last, the pixel values of the images are organized to a [0, 1] range, which helps facilitate stable and efficient training.

The training configuration used for the model, including batch size, optimizer, and loss function, is summarized in[Table 12.1].

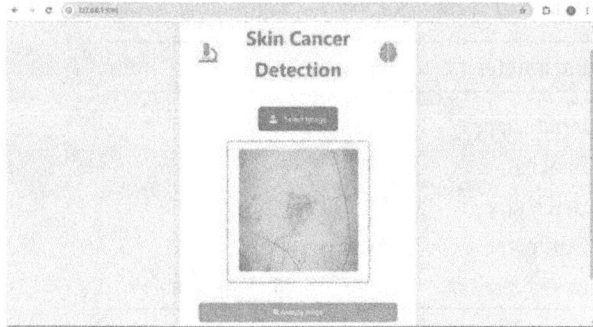

Figure 12.4 Predicted result after uploading the image
Source: Author

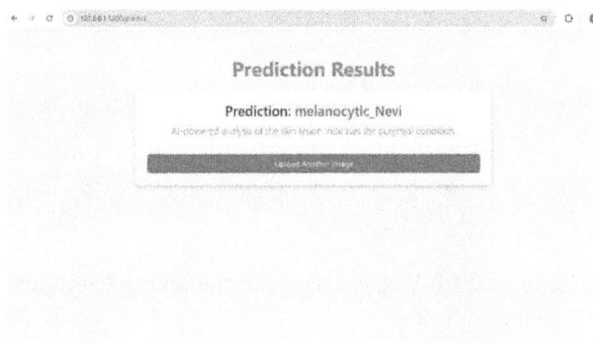

Figure 12.5 Predicted result after uploading the image
Source: Author

Experimental results

Training and model validation metrics
The model's accuracy increases gradually with the training, with model validation accuracy closely following the training trend. Training and model validation losses decrease rapidly, indicating good learning and good generalization. The model achieved 97% validation accuracy, suggesting minimal overfitting.

Result

This stage efficiently coordinates a prepared machine learning show with a user-friendly interface for skin cancer classification. The clients can upload pictures of skin injury; the demonstration preprocesses the pictures and analyzes them in real time. Machine learning demonstration has achieved impressive.

As depicted in [Figure 12.5], the system returns the predicted class in real-time once an image is uploaded.

References

[1] Tschandl, P., Rosendahl, C ,World Health Organization. (2022). Skin cancer prevention and control (WHO Technical Report Series). World Health Organization.

[2] Siegel, R. L., Miller, K. D., & Jemal, A. (2020). Cancer statistics, 2020. *CA: A Cancer Journal for Clinicians*, 70(1), 7–30.

[3] Bray, F., Ferlay, J., Soerjomataram, I., & Torre, L. A. (2021). Global cancer statistics: GLOBOCAN estimation of incidence and mortality worldwide. A Cancer Journal for Clinicians..

[4] LeCun, Y., Bengio, Y., & Hinton, G. (2015). Deep learning. *Nature*, 521(7553), 436–444.

[5] McKinney, S. M., Sieniek, M., Godbole, V., Godwin, & Dean, J. (2020). International evaluation of an AI system for breast cancer screening. Nature.

[6] Woo, S., Park, J., Lee, J. Y., & Kweon, I. S. (2018). CBAM: Convolutional Block Attention Module. In European Conference on Computer Vision (ECCV), (pp. 3–19).

[7] Tschandl, P., Rosendahl, C., & Kittler, H. (2018). The HAM10000 dataset, a large collection of multisource dermatoscopic images of common pigmented skin lesions. Scientific Data, 5, 180161.

[8] Tan, M., & Le, Q. (2019). Efficient Net: rethinking model scaling for convolutional neural networks. In International Conference on Machine Learning,

[9] Wang, X., Girshick, R., Gupta, A., & He, K. (2018). Non-local Neural Networks. In IEEE Conference on Computer Vision and Pattern Recognition.

[10] Zhang.Y, Milinovich.,A., Z.Bambrick, (2020). Challenges in telemedicine and AI: A medical perspective. Journal of Medical Systems.

[11] Esteva, A., Kuprel, B., Novoa, R. A., Ko, J., Swetter, S. M., Blau, H. M., & Thrun, S. (2017). Dermatologist-level classification of skin cancer with deep neural networks. Nature.

[12] Codella, N. C., Nguyen, Q. B., Pankanti, S., Gutman, D., Helba, B., Halpern, A., & Smith, J. R. (2018). Deep learning ensembles for melanoma.

[13] Nasr-Esfahani, E., Samavi, S., Karimi, N., Soroushmehr, S. M., Jafari, M. H., & Najarian, K. (2016). Melanoma detection by analysis of clinical images using convolutional neural network. 2016 38th Annual International Conference of the IEEE .

[14] Kawahara, J., Daneshvar, S., Argenziano, G., & Hamarneh, G. (2018). Seven-point checklist and skin lesion classification using multi-task multi-modal neural nets. IEEE Journal of Biomedical and Health Informatics.

[15] Han, S. S., Park, G. H., Lim, W., Kim, M. S., Na, J. I., Park, I., & Chang, S. E. (2018).

[16] Haenssle, H. A., Fink, C., Schneiderbauer, R., Tober-er, F., Buhl, T., Blum, A., Kalloo, A., Hassen, A. B. H., Thomas, L., Enk, A., & Reader Study Level-I and Level-II Groups. (2018).

13 FIFO implementation in manufacturing: streamlining production processes minimizing work-in-progress inventory

M. Sabir Hussain[a], Mohammed Abdul Raheem[b], Kadiyam Sasidhar[c] and Sami Ahmed[d]

Electronics and Communication Engineering, Muffakham Jah College of Engineering and Technology, Hyderabad, India

Abstract

First in, First Out (FIFO) inventory management technique within manufacturing settings based on its potential of making production more efficient and with minimal work-in-progress (WIP) inventories. It delves deeper into the theoretical framework that explains FIFO through its concepts and benefits against other inventory management systems. Further, it addresses actual case studies and empirical evidence to validate practical outcomes of implementing FIFO in varied manufacturing contexts. The paper elaborates further on how FIFO permits free flow of materials since the oldest inventory is acquired first, thus preventing obsolescence and waste build-up. Further, it establishes that it ensures optimal levels of inventory so that overproduction and excess buildup of inventory is avoided. Through synthesis of existing literature and industry insights, the research delivers an explanation regarding the mechanisms by which FIFO improves production efficiency and reduces cycle times.

Keywords: ASIC, EDA tool, FIFO, hardware description language HDL, register transfer logic RHL, synthesis

Introduction

The bottom line of manufacturing remains efficient and low cost. Such companies are always looking for strategies which help to optimize the production process as much as possible with least waste and better productivity. So, the strategy of inventory management becomes an integral part of these goals because it influences the material movement through the production process [2]. Among the inventory management methods applied, the most commonly used one that helps streamline the production process while reducing WIP inventories is the FIFO approach. FIFO is the principle where old inventory is consumed or sold before new inventory, meaning that goods must be consumed in the order received. Even though FIFO has been considered one of the fundamental principles of inventory management for decades, its application within manufacturing environments has recently received growing attention. It is clearly evident that today companies are striving to optimize their supply chains, reduce holding costs, and generally operate more efficiently. The implementation of FIFO in manufacturing involves the strategic reevaluation of inventory control practices, production scheduling, and material flow management. Manufacturers have several key objectives by following the FIFO principle: they minimize obsolescence using older inventory items before newer items, reduce excess inventory carrying costs, and keep overall production efficient with a smooth material flow [4].

Literature Survey

The use of First In, First Out (FIFO) inventory management in manufacturing has always been of great interest to both researchers and practitioners [12]. This literature review will discuss existing studies and insights about the implementation of FIFO in a manufacturing environment with an emphasis on how it streamlines the production process and minimizes WIP inventories [13].

[a]sabirhussain@mjcollege.ac.in, [b]abdulraheem@mjcollege.ac.in, [c]sasidhareed@mjcollege.ac.in, [d]160422735031@mjcollege.ac.in

DOI: 10.1201/9781003684589-13

Theoretical background

Research works in the area of FIFO in inventory management have deeply covered its theoretical foundation. In this study we established the economics with which FIFO operate. To explain, we presented ways in which FIFO ensures cost consistency and minimizes holding costs. Based on this foundation, current researchers have further analyzed how the adoption of FIFO procedures would significantly impact the manufacturing operations [15].

Benefits of FIFO implementation

Numerous studies have highlighted the benefits of implementing FIFO in manufacturing settings have been demonstrating that FIFO leads to lower holding costs and reduced risk of inventory obsolescence. Similarly, this research [5], emphasized by researchers to include logistical complexities, cultural resistance, and technological limitation, identified some of the main challenges manufacturers face in implementing FIFO as problems with inventory tracking and coordination.

There are many ways in which the future work and practical advice to manufacturers wanting to implement FIFO in their operations [11]. These include "the development of advanced inventory management systems, the integration of emerging technologies such as Internet of Things (IoT) and artificial intelligence (AI), and the implementation of robust change management strategies". Additionally, there is a need for further empirical research to assess the long-term impacts of FIFO implementation on manufacturing performance and competitiveness.

Methodology

This paper aims to appliance the RTL to GDSII flow for a FIFO coordination core. The design practice employs a 45nm technology library, including slow.lib and fast.lib files, a library exchange file (.lef), design constraints (.sdc) file, and innumerable scripts for synthesis, STA, & physical design [9]. frontend phase, the functionality of the RISC-V arch processor was verified using the NC Launch simulator tool by Cadence.

Following verification, the Genus tool generates the logic-level netlist. The backend physical design was carried out with the PD Innovus tool, involving steps like floor- planning, partitioning, P&R. Partitioning simplified complex circuits, floor planning assigned boundaries to all blocks, placement arranged standard cells in the core area, and routing established connections according to the netlist [3]. Performance was analyzed and optimized in both pre-CTS and post-CTS stages. This work encompasses both the frontend and backend aspects of the RTL to GDSII flow, as shown in Figure 13.1.

Simulation

The simulation is carried in the Xcelium, Xcelium is the engine that performs the actual simulation, while NCLaunch provides the front-end interface for user interaction and control. Required files are: Register transfer level (RTL) code: This is the high-level description of the circuit's behavior, written in (HDL) like Verilog or VHDL [7]. Testbench: This is a separate HDL file that generates signals input and checks the output responses of the design under test (DUT). It simulates the environment in which the DUT will operate. The typical simulation flow involves the following steps:

Compilation: The RTL design and testbench are compiled into an executable simulation model.

Elaboration: The simulation model is linked with the standard cell libraries and technology libraries.

Simulation: The testbench generates stimuli, the DUT responds, and the outputs are compared with expected results.

Debugging: If errors or violations are detected, the design is debugged and modified.

Regression: The simulation is repeated with different testbenches to ensure the design's robustness under various scenarios.

Synthesis

The synthesis process, executed using the Genus tool, transforms the (RTL) design into a gate level netlist. This transformation requires RTL code, (SDC) file, and a TCL script to automate the entire flow. The resulting gate-level netlist, along with a modified SDC file, serves as input to the physical design process [1]. Additionally, the synthesis tool generates reports on timing, space, and power PPA to evaluate the design's concert at the logic level.

Physical design

The Cadence Innovus tool provides the physical design process, with gate-level placement and routing of a Verilog netlist. The Innovus places standard logic cells, performs power routing, PNR and produces comprehensive reports on timing, area, consumed power, and design rule checks. All the above inputs

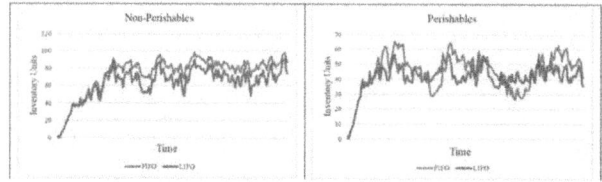

Figure 13.2 Aggregate comparison table of FIFO and LIFO
Source: Author

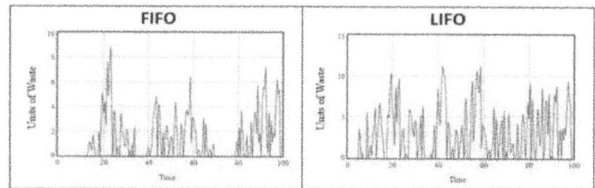

Figure 13.3 Sales comparison table of FIFO and LIFO
Source: Author

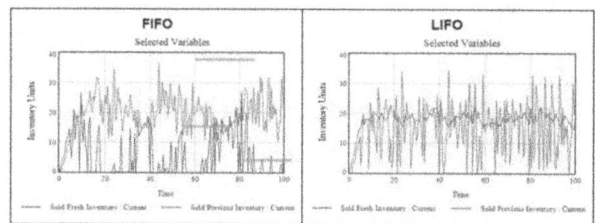

Figure 13.4 Sales comparison table of FIFO and LIFO
Source: Author

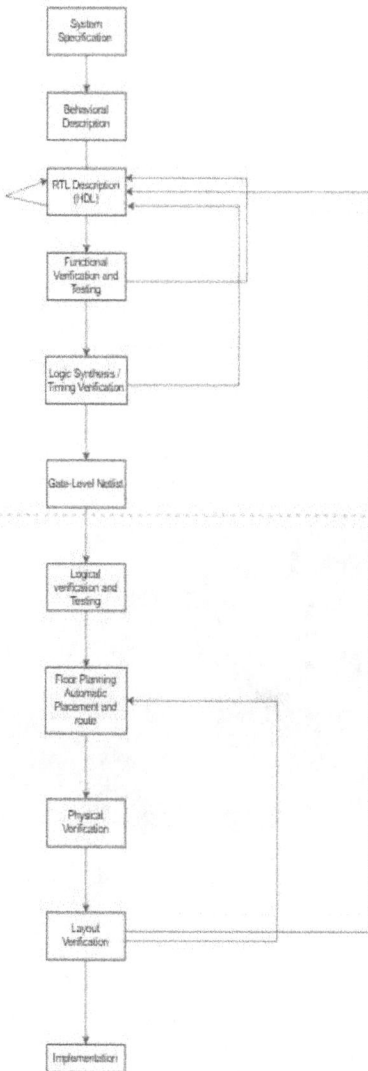

Figure 13.1 ASIC design flow
Source: Author

go into this flow, including gate-level netlist, technology library files (.lib and .lef), Constraints (SDC) from Genus, tool and the TCL file script used to start up the proposal load-up. The process has optimization in both pre and post- CTS steps, for which reports come at each of the steps along with timing, area, and power metrics. The final GDSII file is generated for the FIFO processor, which is a standard format for integrated circuit layout data, upon completion of post-layout STA [6] as shown in Figure 13.2 and 13.3.

Experimental Results

The simulation of FIFO is first done to empirically assess the impact of FIFO implementation in manufacturing, a series of experiments were conducted in a simulated production environment. The experiments aimed to measure changes in production efficiency, work-in-progress (WIP) inventories, and overall operational performance following the adoption of FIFO principles [8].

The following is the simulation waveform of FIFO given in the Figure 13.6, utilizing the Cadence Genus synthesis tool, a gate-level netlist derived from a functionally validated RTL de- sign, incorporating a 45 nm technology library and design constraints. This process yielded various reports, including a Constraint (.sdc) file. Figure 13.4 illustrates the synthesized top-level schematic of the FIFO processor, featuring PC, register, code instruction, & ALU blocks.

Table 13.1 Comparative analysis.

Timing	Power	Observed timing	Referred timing	Observed power	Referred power
Minimum period	Total internal power	6.809 ns	4.907 ns	0.04 W	0.03 W
Minimum input period before arrival clock time	Total switching power	7.537 ns	8.134 ns	0.005 W	0.004 W
Minimum output required time after clock	Total leakage power	4.620 ns	6.874 ns	0.000002 W	0.000031 W
Speed grade	Total power	-4	-4	0.00517 W	0.00638 W

Source: Using EDA Tool Genus 45nm

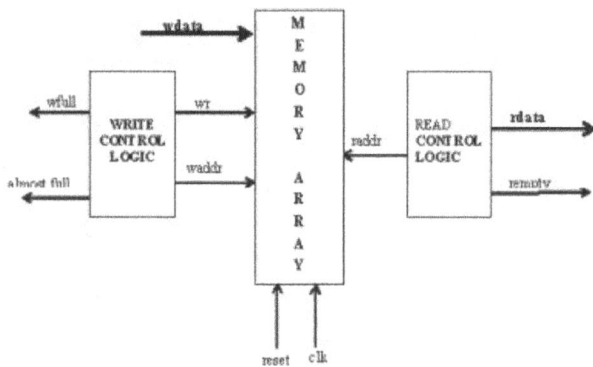

Figure 13.5 Block diagram Of FIFO
Source: Author

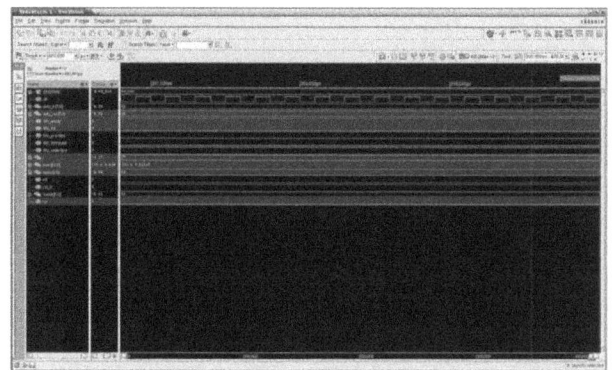

Figure 13.6 Simulation diagram
Source: Author

Result Analysis

The results analysis primarily focuses on power and timing throughout the design flow as referred in Table 13.1. Power analysis includes various power reports, while timing analysis encompasses arrival time and slack.

Conclusion

The findings of this research underscore the significant impact of First In, First Out (FIFO) implementation in manufacturing environments as shown in Figure 13.5.

$$FFGI(t) = FFGI(t-1) + Prod(t) - SoldF(t)Inv(t) \quad (1)$$

$$PFGI(t) = PFGI(t-1) + HeldInv(t) - SoldP.Inv(t) \quad (2)$$

$$PFGI(t) = PFGI(t-1) + HeldInv(t) - SoldP.Inv(t) \quad (3)$$

As we can see in equation (1), through a comprehensive ex- amination of theoretical frameworks, empirical evidence, and experimental results, we get the equation (2), equation (3), this study has demonstrated FIFO principles in streamlining production processes and minimizing work-in-progress (WIP) inventories [14]. The empirical experiments conducted in simulated production environments have provided compelling evidence of the tangible benefits of FIFO implementation. Im- proved production efficiency, reduced cycle times, and lower WIP inventory levels are among the key outcomes observed following the adoption of FIFO practices. These findings not only reaffirm the theoretical advantages of FIFO but also offer practical insights for manufacturers seeking to enhance their inventory management practices and optimize production performance [10]. The synthesis reports as shown in Figure 13.6, 13.7, 13.8, 13.9.

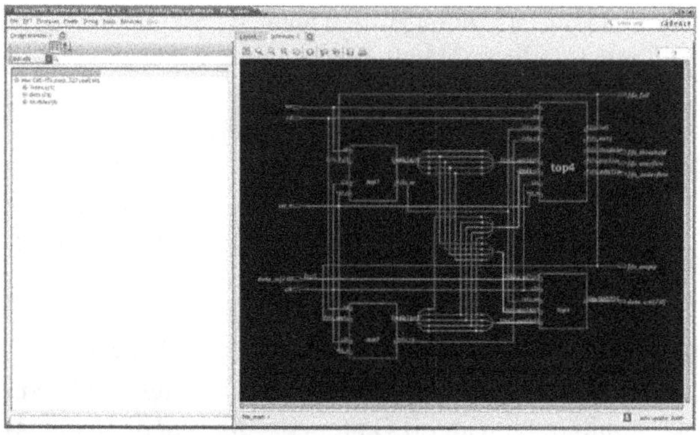

Figure 13.7 Synthesis diagram
Source: Author

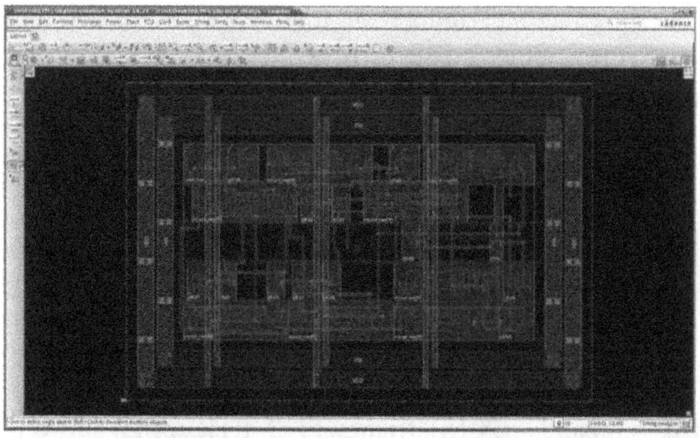

Figure 13.8 Physical design
Source: Author

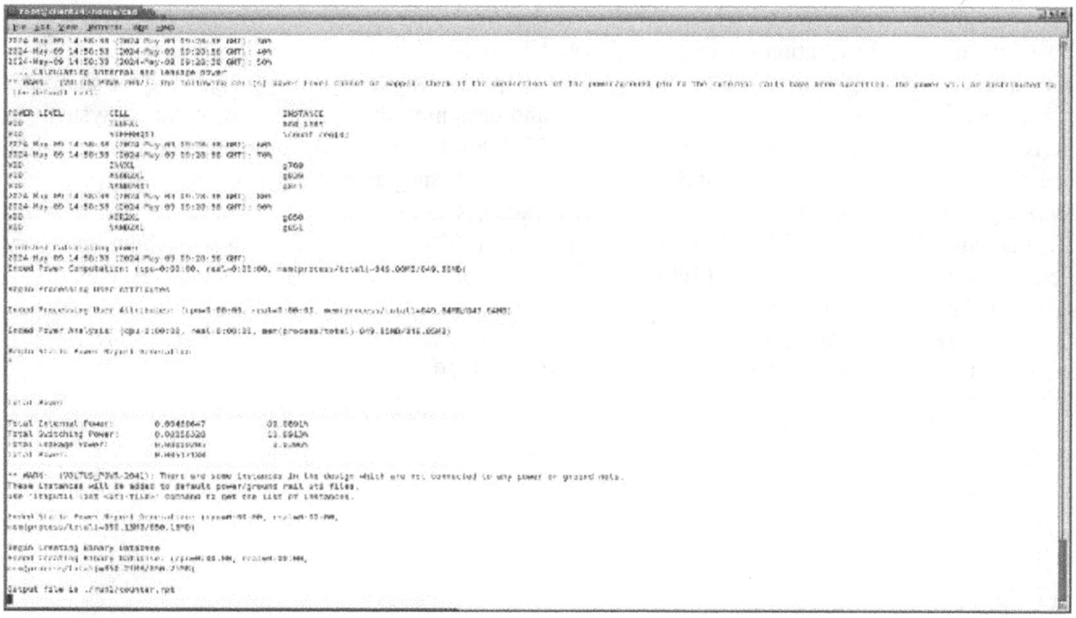

Figure 13.9 Results of the experiment
Source: Author

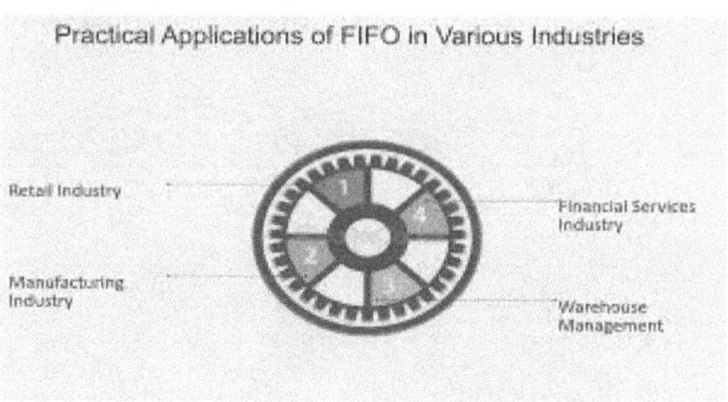
Practical Applications of FIFO in Various Industries
Retail Industry
Financial Services Industry
Manufacturing Industry
Warehouse Management

Acknowledgement

The authors gratefully acknowledge the head of the department Dr. Ayesha Naaz, and Chip to Start (C2S) from Government of India for the EDA tool support and their cooperation in the research.

References

[1] Abdel-hafeez, S., & Gordon-Ross, A. (2021). Reconfigurable FIFO memory circuit for synchronous and asynchronous communication. *International Journal of Circuit Theory and Applications*, 49(4), 938–952.

[2] Agarwal, R. (n.d). Strategies for efficient stores management and warehousing: a comparative perspective, Junikhyat Journal, 12(5), 133–138.

[3] Babu, B. K., & Rani, V. N.. (2016). Implementation of a low power FIFO based bist process for cut,

International Journal of Engineering Research and Applications (IJERA), 6(7), 50–55.

[4] Bernier, V., & Frein, Y. (2004). Local scheduling problems submitted to global FIFO processing constraints. *International Journal of Production Research*, 42(8), 1483–1503.

[5] Ching, P. L., Mutuc, J. E., & Jose, J. A. (2019). Assessment of the quality and sustainability implications of FIFO and LIFO inventory policies through system dynamics. *Advances in Science, Technology and Engineering Systems*, 4(5), 69–81.

[6] Gopal, N. (2015). Router 1 × 3–RTL design and verification. *International Journal of Engineering Research and Development*, 11(09), 62–71.

[7] Mishra, V. K., Rajendran, C., Lenher, F., Suryanarayana Murty, A. S., Balakrishnan, A. S., Jina, A., et al. (2022). Optimization of network plan- ning in a real-life vehicle logistics distribution system. In International Conference on Data Analytics in Public Procurement and Supply Chain, (pp. 9–16). Springer.

[8] Pandey, D. S., & Raut, N. (2016). Inventory management by using FIFO system. *Asian Journal of Science and Technology*, 7(2), 2336–2339.

[9] Sadiah, H. T., Purnama, D. H., & Ishlah, M. S. N. (2024). Implementation of the first in first out (FIFO) algorithm in the sandal and shoe product inventory (stock) application. *International Journal of Quantitative Research and Modeling*, 5(1), 31–39.

[10] Saxena, A., Bhatt, A., Gautam, P., Verma, P., & Patel, C. (2016). High performance FIFO design for processor through voltage scaling technique. *Indian Journal of Science and Technology*, 9(45), 1–5.

[11] Saygin, C., Chen, F., & Singh, J. (2001). Real-time manipulation of alternative routeings in flexible manufacturing systems: a simulation study. *The International Journal of Advanced Manufacturing Technology*, 18, 755–763.

[12] Setak, M., Habibi, M., Karimi, H., & Abedzadeh, M. (2015). A time-dependent vehicle routing problem in multigraph with FIFO property. *Journal of Manufacturing Systems*, 35, 37–45.

[13] Tanaka, G. M. P., & Respati, H. (2021). Cost of inventory calculation analysis using the FIFO and LIFO methods. *Journal of Business Management and Economic Research*, 5(4), 109–120.

[14] Utami, M. C., Sabarkhah, D. R., Fetrina, E., & Huda, M. Q. (2018). The use of FIFO method for analysing and designing the inventory information system. In 2018 6th International Conference on Cyber and IT Service Management (CITSM), (pp. 1–4). IEEE.

[15] Yadav, A. S., Bansal, K. K., Shivani, S. A., & Vanaja, R. (2020). FIFO in green supply chain inventory model of electrical components industry with distribution centres using particle swarm optimization. *Advances in Mathematics: Scientific Journal*, 9(7), 5115–5120.

14 Innovative approaches to Doherty power amplifier enhancement

N. Swapna[1,a], V. Sreelatha Reddy[2,b], Madhuri Gummineni[3,c] and Chitra Chitters[4,d]

[1]Assistant Professor, Department of ECE, Guru Nanak Institutions Technical Campus, Khanapur, Telangana, India

[2]Assistant Professor, Department of EIE, CVR College of Engineering, Mangalpalli, Telangana, India

[3,4]Assistant Professor, Department of EIE, Vignan Institute of Technology and Science, Deshmukhi, Telangana, India

Abstract

Communication systems like cellular base stations rely on Doherty Power Amplifiers (DPAs) because they greatly increase efficiency in radio frequency (RF) applications, particularly at high output power levels. The design and modeling of a DPA utilizing Cadence Virtuoso 45nm technology, specifically tuned for 2.4GHz operation, is the main emphasis of this work. The study includes a thorough analysis of the DPA design, emphasizing important parts including the transmission line, matching network, main and auxiliary amplifiers, and power splitter. To guarantee exact operation at the desired frequency, each of these components has been painstakingly designed and adjusted. Schematic design and layout implementation utilizing the Cadence Virtuoso platform are part of the development process. In order to evaluate the amplifier's performance, sophisticated modeling methods are used to analyze important characteristics including power added efficiency (PAE) and power gain. The goal is to reduce distortion and power consumption while increasing efficiency and preserving signal integrity. Phase alignment, power distribution, and impedance matching are all improved using simulation data in an iterative optimization process. The thorough simulation findings offer insightful information on designing and optimizing DPAs for contemporary wireless communication systems. By advancing RF amplifier technology, this innovation makes it easier to create communication networks that are more dependable and efficient.

Keywords: 2.4GHz frequency, cadence virtuoso, Doherty power amplifier (DPA), power added efficiency (PAE)

Introduction

The modern world relies heavily on data exchange, requiring secure and efficient communication. This data can range from highly confidential national information to everyday conversations between individuals. Ensuring seamless transmission without data loss while maintaining privacy is a significant challenge. With the shift toward wireless communication, this challenge has become even more complex. Therefore, a robust system is essential to facilitate secure and reliable wireless communication.

Three essential parts make up a basic wireless communication system: a transmitter, a receiver, and a base station or communication channel. The signal must be powerful enough to reach the base station during transmission, where signal amplification and noise reduction are required. Improving the signal quality at the receiver end is also essential for precise data interpretation. To ensure clear and dependable communication, the main objective is to increase the signal's strength in order to combat noise, reduce attenuation, and enhance the signal-to-noise ratio.

Objective

The primary goal of this work is to design a power amplifier operating at a 2.4 GHz frequency with high linearity. The amplifier should ensure low system noise, high efficiency, peak added efficiency (PAE), and significant output power gain. Additionally, the study involves selecting an appropriate power divider circuit that can be optimally matched with the amplifier. This power amplifier must be suitable for modern

[a]nallaswapna415@gmail.com, [b]srilathareddy.cvr@gmail.com, [c]dr.g.madhuri_eie@vignanits.ac.in, [d]chitra_eie@vignanits.ac.in

DOI: 10.1201/9781003684589-14

telecommunication systems, supporting advanced modulation techniques such as OFDM and QAM.

Motivation

The fundamental challenge at hand is the prevalent use of external power amplifiers that are not integrated into the chip. This presents a significant obstacle in the integration of intricate and space-consuming power amplifier circuits onto a chip due to the complexities involved in accommodating these components within the chip's design and structure.

Literature Survey

Jeon et al. [1] proposed a two-stage Doherty power amplifier using CMOS and GaAs HBT technologies, significantly improving both power efficiency and linearity for 5G handset applications. Sharma et al. [2] developed a high-power DPA offering a large output back-off range, leveraging advanced semiconductor techniques to enhance overall performance for base stations.

Tawa et al. [3] introduced a black-box DPA design at 3.5 GHz without traditional transistor models, offering flexibility and 350 W output, ideal for high-power telecom systems.

Bura [4] presented a compact, wideband DPA design tailored for 5G infrastructure, emphasizing reduced size, high efficiency, and wide bandwidth support.

Sakata et al. [5] proposed an adaptive inputpower distribution method using a modified Wilkinson divider, which improved linearity and efficiency across power levels.

Sakata et al. [5] proposed an adaptive input-power distribution method using a modified Wilkinson divider, which improved linearity and efficiency across power levels.

Elsayed et al. [6] showcased an asymmetric DPA at 28 GHz using 22 nm FDSOI CMOS, enhancing power back-off efficiency and thermal management for mm-wave systems. Reddy et al. [7] implemented a 2D-FIR filter using DA technique for efficient hardware realization in signal processing. Another study by Reddy et al. [8] introduced a smart irrigation system integrating IoT and cloud platforms for sustainable water management. Seidel et al. [9] developed a 60 GHz asymmetric DPA using 130 nm BiCMOS, enabling compact and efficient high-frequency operation. Hashemi et al. [10] presented a highly linear wideband polar class-E CMOS DPA, optimizing both bandwidth and linearity using digital techniques.

Reddy et al. [11] further demonstrated a 2D-FIR filter with symmetry-based block processing, improving computational efficiency.Harivardhagini et al. [12] proposed an adaptive electric bicycle design promoting renewable energy use through manual charging.

Implementation

Methodology

The approach taken focuses on minimizing the amplifier's size, optimizing the matching network, and simplifying the power divider while maintaining the design objectives. The process begins with selecting the center frequency for the application. A suitable power splitter network, transmission line, and matching network must be chosen to ensure efficient power transfer from input to output. Next, the appropriate values for resistors, inductors, and capacitors are calculated based on the target frequency. The design is implemented using the Cadence Virtuoso platform with 45 nm CMOS technology. Simulation and parameter extraction are performed through periodic steady state (PSS) analysis. Finally, the results are compared with existing models to assess improvements in performance. The project utilizes Cadence Virtuoso as the design tool and 45 nm CMOS technology for implementation.

Block diagram

The system's block diagram comprises five main components, as shown in Figure 14.1. Furthermore, Figure 14.2 depicts the DPA block diagram, which was designed using cadence.

Power splitter: The power splitter in DPA is crucial for separating the input signal between the peak and main amplifiers. Because it guarantees regulated power distribution, every amplifier can function effectively within its range. For best results, the splitter allocates a portion of the signal to the peak amplifier and sends the remainder of the signal to the main amplifier. A Wilkinson power splitter is used in this design to accomplish exact impedance matching and signal division.

Main amplifier. In a DPA, the main amplifier efficiently amplifies the input signal, ensuring accurate reproduction with minimal distortion while managing average power in its linear region. Using a Class

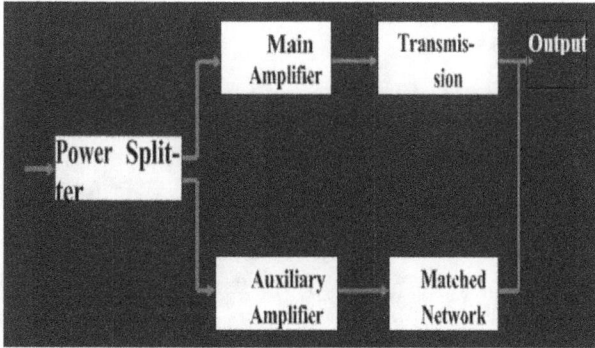

Figure 14.1 Block diagram of Doherty power amplifier
Source: Author

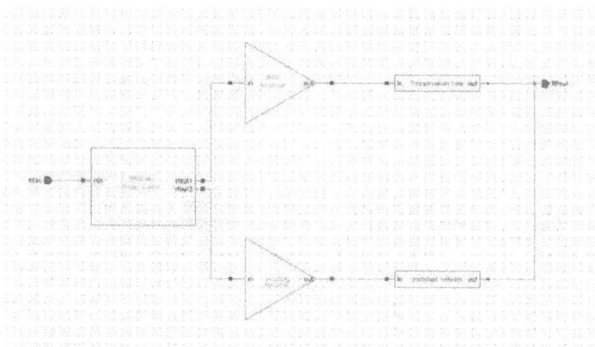

Figure 14.2 Block diagram of Doherty power amplifier in Candence
Source: Author

B configuration for efficiency prioritizes linearity to maintain signal quality. The amplifier effectively handles high peak-to-average power ratio (PAPR) signals across a wide power range, operating in Class AB mode at lower levels and transitioning to Class C mode for a higher power. This load modulation optimizes efficiency and linearity, ensuring seamless transitions and enhanced performance.

Auxiliary amplifier. In a DPA, the auxiliary (or peak) amplifier complements the main amplifier to enhance efficiency and linearity during high input signal amplitudes. Operating primarily in Class C, it boosts power during signal peaks while conserving energy when idle or operating at lower levels. As input levels increase, it dynamically adjusts its output to ensure efficient power delivery, maintaining signal fidelity and minimizing distortion, which makes the DPA more effective.

Transmission line. In a DPA, transmission lines merge signals from the main and auxiliary amplifiers,

ensuring efficient combination and phase alignment. They maximize output power, minimize reflections, and optimize impedance matching, enhancing overall DPA performance.

Matched network. In a DPA, the matching network optimizes impedance between the amplifiers and the load, ensuring efficient signal combining and power transfer. It minimizes reflections, dynamically adjusts load impedance, and aligns phase and amplitude to enhance efficiency and reduce distortion. This network is crucial for maximizing overall DPA performance.

Power splitter. In a DPA, the power splitter plays a crucial role in distributing the input signal between the main and peak amplifiers to ensure efficient operation. It divides the signal into two paths, sending a larger portion to the main amplifier and a smaller portion to the peak amplifier, allowing each to function effectively within its range. A Wilkinson power divider is commonly used because it efficiently splits the signal while maintaining proper impedance matching and providing isolation between the amplifiers.

Design flow
The design flow is described in Figure 14.3, outlining the steps for the design. Figure 14.4 depicts the DPA designed using the Cadence tool.

The center frequency is a key design parameter that must match the intended application. For this power amplifier, designed for digital communication and Wi-Fi, a 2.4GHz operating frequency is selected.

Power splitters and matching networks can be bulky and complex. To simplify the design, a π-LC network is used as the matching network, while a Wilkinson power divider is chosen as the power splitter.

R, L, and C values for the Wilkinson power divider and matching network are calculated, while those for the amplifier are extracted using PSS analysis to ensure proper impedance matching.

Designing on Cadence tool. The calculated values, along with the power splitter and matching network, are implemented in Cadence Virtuoso. Further refinements using PSS and PAC analyses ensure feasibility and performance optimization.

Results

Power gain
Figure 14.5 demonstrates that the DPA attains a power gain of 87.08 dB. This measurement, represented

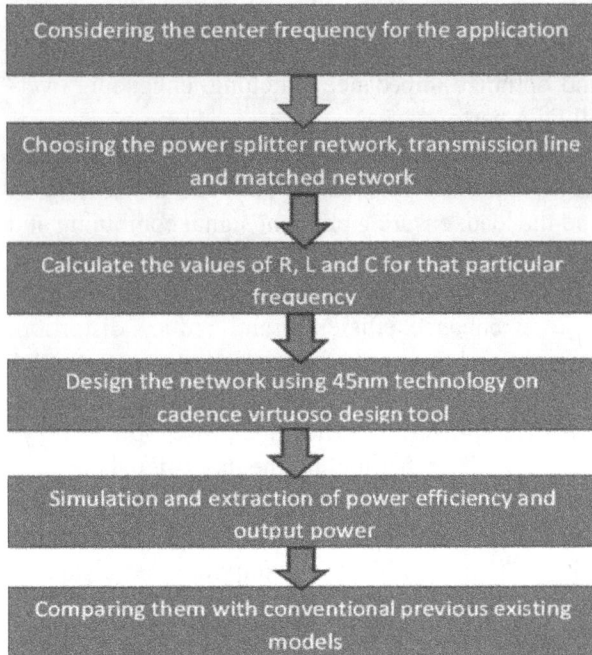

Figure 14.3 Design flow of Doherty power amplifier
Source: Author

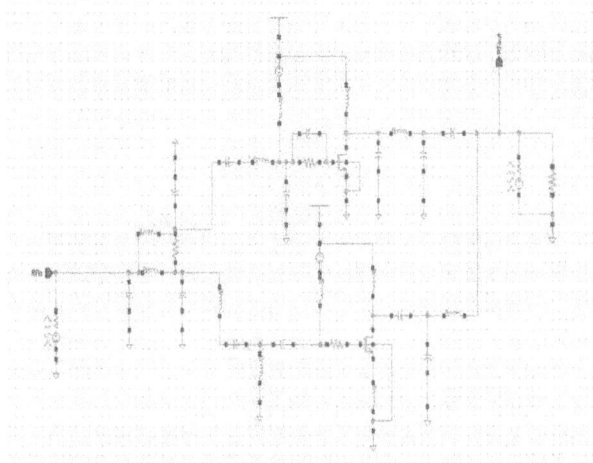

Figure 14.4 DPA designed using the cadence
Source: Author

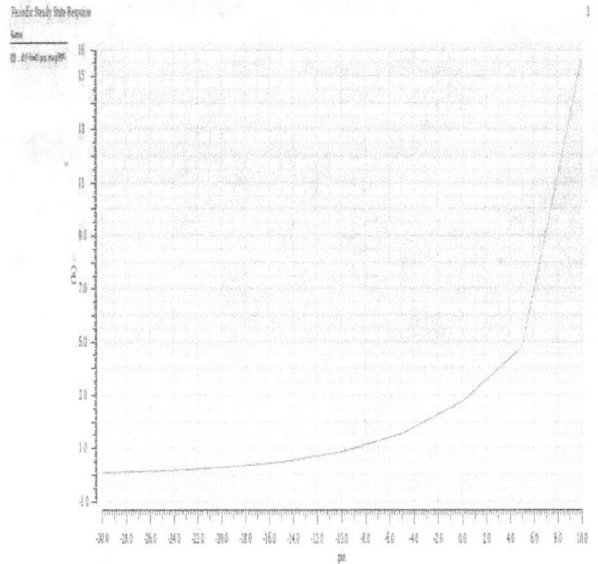

Figure 14.5 Power gain of Doherty power amplifier
Source: Author

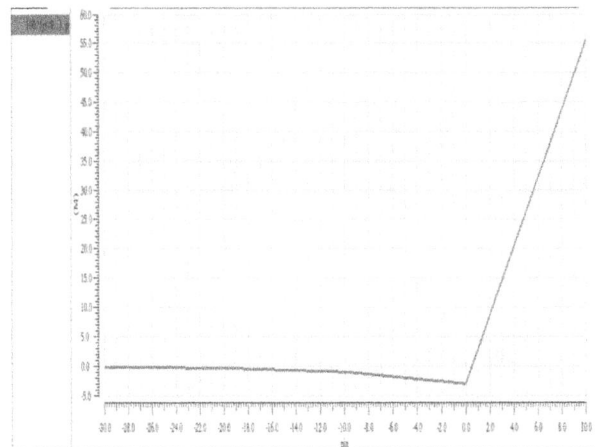

Figure 14.6 PAE of Doherty power amplifier
Source: Author

on a logarithmic scale, signifies the relationship between the output and input power, reflecting the amplifier's ability to enhance signal strength. A gain of 87.08 dB reflects a substantial increase in signal strength, demonstrating the amplifier's efficiency in power enhancement.

Figure 14.6 demonstrates that the power added efficiency (PAE) of the DPA increases from 0% to 60% as the input power ranges from 0 to 10

units. This trend indicates that higher input power enhances the amplifier's efficiency in converting input energy into useful output power. The improvement in PAE highlights the effectiveness of the DPA in maximizing power utilization while minimizing losses.

Conclusion

Promising results have been obtained from the development and testing of the Doherty power amplifier (DPA) utilizing Cadence Virtuoso. Effective performance at 2.4 GHz has been attained by carefully adjusting key parts such the power splitter, main and auxiliary amplifiers, transmission line, and matching network. The amplifier's impressive 87.08 dB power gain demonstrates how well it amplifies signals.

Furthermore, as input power rises, the power added efficiency climbs from 0–55%, indicating that it can effectively transform input energy into output power. These results confirm that the DPA can deliver powerful signal amplification, high efficiency, and dependable performance, which makes it ideal for 2.4GHz radio frequency applications.

References

[1] Jeon, H., Na, J., Oh, H., & Yang, Y. (2022). Two-stage CMOS/GaAs HBT Doherty power amplifier module for 5G handsets. In 2022 IEEE International Symposium on Radio-Frequency Integration Technology (RFIT), Busan, Korea, Republic of, 2022, (pp. 12–14). doi: 10.1109/RFIT54256.2022.9882351.

[2] Sharma, D. K., Aggarwal, P., Giri, K., Sengupta, A., Mandal, M. K., Mondal, S., et al. (2021). A highly efficient high power Doherty power amplifier with large output back-off range for base station application. In 2021 IEEE 18th India Council International Conference (INDICON), (pp. 1–5).

[3] Tawa, N., de Falco, P. E., Kazuya, O., Barton, T., & Kaneko, T. (2021). A 3.5-GHz 350-W black-box Doherty amplifier design method without using transistor models. In 2021 IEEE BiCMOS and Compound Semiconductor Integrated Circuits and Technology Symposium (BCICTS), (pp. 1–4).

[4] Sharma, D. K., & Bura, R. T. (2021). A novel and compact wideband Doherty power amplifier architecture for 5G cellular infrastructure. In 2021 IEEE 4th 5G World Forum (5GWF), Montreal, QC, Canada, (pp. 323–327). doi: 10.1109/5GWF52925.2021.00063.

[5] Sakata, S., Komatsuzaki, Y., & Shinjo, S. (2020). Adaptive input-power distribution in Doherty power amplifier using modified wilkinson power divider. In 2020 IEEE Topical Conference on RF/Microwave Power Amplifiers for Radio and Wireless Applications (PAWR), San Antonio, TX, USA, (pp. 34–37). doi: 10.1109/PAWR46754.2020.9036005.

[6] Srilatha Reddy, V., et al . (2024). Implementation of block-based diagonal and quadrantal symmetry type 2D-FIR filter architectures using DA technique. *Computers and Electrical Engineering*, 118(Part A), 1–27. DoI:10.1009/978-3-031-59607-1_17.

[7] Elsayed, N., Saleh, H., Mohammad, B., & Sanduleanu, M. (2020). A 28GHz, asymmetrical, modified Doherty power amplifier in 22nm FDSOI CMOS. In 2020 IEEE International Symposium on Circuits and Systems (ISCAS), Seville, Spain, (pp. 1–4). doi: 10.1109/ISCAS45731.2020.9180851.

[8] Reddy, V. S., Harivardhagini, S., & Sreelakshmi, G. (2024). IoT and cloud based sustainable smart irrigation system. In E3S Web Conf., International Conference on Renewable Energy, Green Computing and Sustainable Development (ICREGCSD 2023), (Vol. 472, pp. 01026). https://doi.org/10.1051/e3sconf/202447201026.

[9] Seidel, A., Grams, V., Wagner, J., & Ellinger, F. (2021). Asymmetric Doherty power amplifier at 60 GHz in 130 nm BiCMOS. In 2020 IEEE MTT-S Latin America Microwave Conference (LAMC 2020), Cali, Colombia, (pp. 1–4). doi: 10.1109/LAMC50424.2021.9601976.

[10] Hashemi, M., Zhou, L., Shen, Y., & de Vreede, L. C. N. (2019). A highly linear wideband polar class-E CMOS digital Doherty power amplifier. *IEEE Transactions on Microwave Theory and Techniques*, 67(10), 4232–4245. doi 10.1109/TMTT.2019.2933204.

[11] Reddy, V. S., Juliet, A. V., Thuraka, E. R., & Odugu, V. K. (2024). Implementation of block-based diagonal and quadrantal symmetry type 2D-FIR filter architectures using DA technique. *Computers and Electrical Engineering*, 118(Part A), 1–27.

[12] Harivardhagini, S., Reddy, V. S., & Pranavand, S. (2024). Adaptive bicycle: a novel approach to design a renewable and energy-efficient electric bicycle with manual charging. In E3S Web Conf., International Conference on Renewable Energy, Green Computing and Sustainable Development (ICREGCSD2023), (Vol. 472).

[13] Barmala, E., et al. (2019). Design and simulator Doherty power amplifier using GaAs technology for telecommunication applications. *Indonesian Journal of Electrical Engineering and Computer Science*, 15(2), 845–854. doi: 10.1109/MWSYM.2018.8439679

[14] Subhadra, H., Reddy, V. S., & Satyamurthy, P. (2024). A Manual charging adaptive energy efficient bike. In International Conference on Renewable Energy, Green Computing and Sustainable Development (ICREGCSD 2023). Springer Nature, (pp. 148–159). DoI:10.1007/978-3-031-58607-1_11.

[15] Sasikanth, M. N., Bhattacharyya, T. K., et al. (2018). A High-efficiency body injected differential power amplifier at 2.4GHz for low power application. In 2018 31st International Conference on VLSI Design

and 2018 17th International Conference on Embedded Systems. doi: 10.1108/MWSTU.2018.8435639.

[16] Jang, H., & Wilson, R. (2018). A 225 Watt, 1.8-2.7 GHz broadband Doherty power amplifier with zero-phase shift peaking amplifier. In 2018 IEEE/MTT-S International Microwave Symposium - IMS, Philadelphia, PA, USA, (pp. 797–800). doi: 10.1109/MWSYM.2018.8439639.

[17] Vorapipat, V., Levy, C. S., & Asbeck, P. M. (2017). Voltage mode Doherty power amplifier. *IEEE Journal of Solid-State Circuits*, 52(5), 1295–1304.

15 Optimized feature engineering approaches for predictive modeling of multiple diseases using machine learning

V. Sreelatha Reddy[1,a], N. Swapna[2,b], Radhamma[3,c] and Ch. Mohana Rao[4,d]

[1]Sr. Assistant Professor, Department of EIE, CVR College of Engineering, Mangalpalli, Telangana, India

[2]Assistant Professor, Department of ECE, Guru Nanak Institutions Technical Campus, Khanapur, Telangana, India

[3]Associate Professors, Department of ECE, TKR engineering College, Meerpet, Telangana, India

[4]Professor, Mechanical department, Narsimha Reddy Engineering College, Hyderabad, Telangana, India

Abstract

Computer-aided diagnosis (CAD) is a rapidly evolving field in medical analysis, aimed at reducing diagnostic errors that could lead to incorrect treatments. With advancements in technology, recent developments have focused on enhancing CAD applications, where machine learning (ML) plays a crucial role. Unlike conventional diagnostic methods that rely on simple equations and heuristics, ML can identify complex patterns within data, including anatomical structures and organ functions, by leveraging extensive training on diverse datasets. In biomedical applications, the integration of ML with advanced pattern recognition techniques significantly improves the accuracy and reliability of disease detection. ML algorithms enable objective decision-making, leading to more precise diagnoses. This study presents a comparative analysis of various ML algorithms, including Naïve Bayes, Random Forest, Decision Tree, and support vector machine, highlighting their advantages and limitations in diagnosing diseases such as heart disease and diabetes. By evaluating their performance across multiple datasets that reflect real-world medical conditions, this research provides insights into their practical effectiveness. The findings aim to assist healthcare professionals in optimizing ML algorithms for CAD systems, ultimately enhancing diagnostic accuracy and reliability. These improvements could play a vital role in ensuring timely and accurate medical assessments, leading to better patient outcomes.

Keywords: Computer aided, diagnosis, efficacy, machine learning, real-world diagnosis

Overview

Objective

Data mining plays a crucial role in modern healthcare and clinical research. When applied effectively, data mining techniques can extract meaningful insights from large datasets, enabling healthcare professionals to make timely and informed decisions to enhance patient care. The primary goal is to utilize classification methods that assist physicians in diagnosing diseases more accurately. Various health conditions, including malaria, chickenpox, migraine, diabetes, impetigo, jaundice, and dengue, pose serious risks to individuals and can lead to severe consequences if not addressed promptly. By analyzing vast medical databases, healthcare providers can identify hidden patterns and correlations that improve decision-making processes. Machine learning algorithms such as Naïve Bayes classifier, Random Forest, Decision Tree (DT), and support vector machine offer effective solutions for disease diagnosis and predictive analytics, ultimately contributing to better healthcare outcomes.

Motivation

In today's healthcare landscape, the accumulation of extensive patient data presents both an opportunity and a challenge. Despite the wealth of information available, much of it remains unexplored and underutilized, stored in physical or digital archives. This oversight not only burdens healthcare professionals but also hinders the potential for insights that could significantly enhance patient care. Recognizing the critical need to harness this data effectively, our work focuses on the development of automated systems powered by machine learning algorithms. These systems aim to unlock hidden patterns and

[a]srilathareddy.cvr@gmail.com, [b]nallaswapna415@gmail.com, [c]radha1977@gmail.com, [d]mohanarao23@gmail.com

DOI: 10.1201/9781003684589-15

relationships within large datasets, thereby empowering healthcare providers to make informed decisions and optimize healthcare services.

Literature Survey

An ensemble learning framework was proposed by Ahmed et al. [1] study for early chronic disease detection, integrating synergistic feature engineering methods. The authors utilized multi-source data to precompute essential attributes, reducing model complexity. This resulted in an efficient and scalable solution suitable for large datasets. A high accuracy of 92% was reported, along with computational savings.

A multi-disease prediction system was developed by Sharma et al. [2] : using SVMs enhanced by grid search for parameter tuning. Focusing on cardiovascular and diabetes cases, the model demonstrated high precision. A focus on data preprocessing improved the prediction consistency across unbalanced datasets. Their system's accuracy reached 90%, showcasing its robustness in clinical applications.

By combining convolutional neural networks (CNNs) with traditional ML, paper by Kumar et al. [3] focused on multi-disease detection from patient medical histories. The authors implemented precomputed feature maps to reduce computational overhead. Their approach showed a significant speed increase while maintaining accuracy for large-scale datasets.

Another paper by Verma et al. [4] presented a hybrid methodology combining machine learning and deep learning techniques, leveraging patient web interactions for disease classification. By introducing an optimized feature extraction pipeline, the approach improved model generalizability. An impressive accuracy of 93% was achieved, proving its viability for real-world applications.

Singh et al. [5] integrated relief feature selection with ensemble techniques for multi-disease predic- tion. Their approach reduced irrelevant features, streamlining the model for faster training. The hybrid method improved overall classification accuracy by 15%, ensuring effective handling of imbalanced data. Subhadra [6] developed an IoT-based coal mine safety system with 95% fault detection, highlighting the importance of real-time sensors in critical environments.

Nair et al. [7] study utilized advanced feature engineering with relief and LASSO regression for heart disease prediction. The proposed method efficiently selected high-impact features while discarding redundant data. Achieving a precision of 91%, the model demonstrated enhanced interpretability and computational efficiency.

Employing a hybrid ML approach with LASSO, the authors focused on cardiovascular disease detection [8]. The use of cross-validation ensured robust performance across diverse datasets. Their model achieved an F1-score of 0.93, highlighting its practical applicability in healthcare diagnostics.

Time-series data is integrated into predictive modeling for heart disease in a study done by Gupta et al. [9]. The proposed model utilized temporal data augmentation to improve sensitivity. It achieved an enhanced recall rate compared to traditional algorithms, showing promise for early-stage diagnosis.

Leveraging IoT devices, this paper by Mehta et al. [10] proposed a remote health monitoring system incorporating ML models for real-time disease predictions. The system showed 85% efficiency in processing and diagnosing patient data on the fly. This approach bridged the gap between patient monitoring and timely interventions.

Harivardhagini [11] developed genomic feature engineering ing techniques tailored for rare disease predictions.By combining traditional ML and biological insights,the model achieved 89% precision on global data sets. This work showcased the potential of data driven solutions in genomics.

The authors developed genomic feature engineering techniques tailored for rare disease predictions. By combining traditional ML and biological insights, the model achieved 89% precision on global datasets. This work showcased the potential of data-driven solutions in genomics [10].

Study by Desai et al. [13] introduced AI-driven frameworks for real-time infectious disease diagnostics. By integrating fast inference models with cloud-based tools, the system achieved 88% accuracy while maintaining low latency. It demonstrated scalability for pandemic monitoring applications.

A hybrid deep learning framework combining CNNs and LSTMs was proposed for multi-disease classification. The authors emphasized the sequential processing of patient data, achieving a high accuracy of 90%. The system excelled in handling complex and large datasets [12].

Implementation

System model

Figure 15.1 depicts a system where both patients and doctors are registered. The administrator assesses

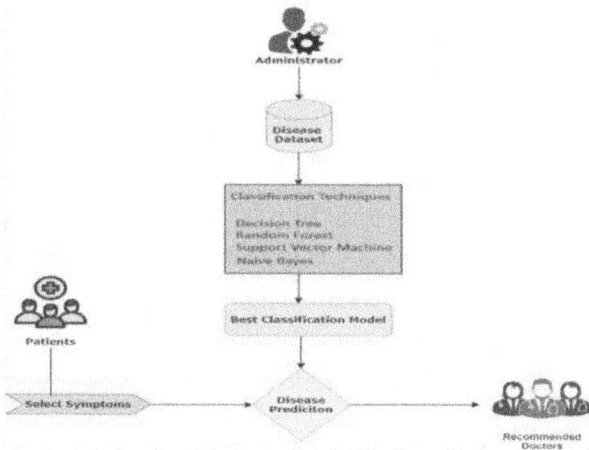

Figure 15.1 System architecture
Source: Author

various classification techniques using multiple disease datasets to determine the most accurate model. Patients log into the system and input their symptoms, after which the system predicts the likely disease using the most effective classification model. Based on the diagnosis, the system then provides a list of recommended doctors relevant to the predicted condition.

The following Table 15.1 shows the list of various diseases [12].

Data pre-processing

The dataset was loaded using the Pandas library, and during the data pre-processing stage, it was divided into independent and dependent variables. The independent variables represent the dataset features, which are treated as symptoms, while the dependent variable corresponds to the predicted disease name as the target feature.

Workflow

Here we discuss the Project system flow of execution of the actors and operations with the help of UML diagrams.

UML diagrams

Figure 15.2 illustrates the available functionalities for the administrator, showcasing the various options at their disposal. This diagram outlines the potential actions and features accessible to the admin, providing a visual representation of their role within the system.

Table 15.1 Various disease names.

SI.NO	Disease name	SI.NO	Disease name
1	Vertigo	21	Hepatitis D
2	Acne	22	Hepatitis E
3	Alcoholic Hepatitis	23	Hypertension
4	Allergy	24	Hyperthyroidism
5	Arthritis	25	Hypoglycemia
6	Bronchial Asthma	26	Hypothyroidism
7	Cervical spondylosis	27	Impetigo
8	Chickenpox	28	Pneumonia
9	Common cold	29	Heart attack
10	Dengue	30	Tuberculosis
11	Diabetes	31	Typhoid
12	Dimorphic hemorrhoids (piles)	32	Urinary tract infection
13	Drug reaction	33	Varicose veins
14	Fungal infection	34	Psoriasis
15	Gastroenteritis	35	Paralysis (brain hemorrhage)
16	GERD	36	Osteoarthritis
17	Chronic cholestasis	37	Migraine
18	Hepatitis A	38	Malaria
19	Hepatitis B	39	Jaundice
20	Hepatitis C	40	Peptic ulcer

Source: Author

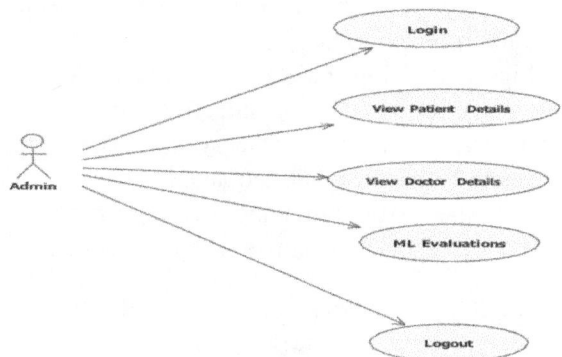

Figure 15.2 Use case
Source: Author

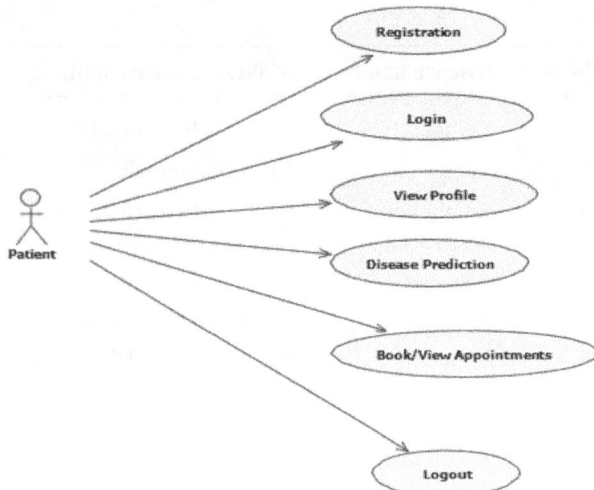

Figure 15.3 Patient side development
Source: Author

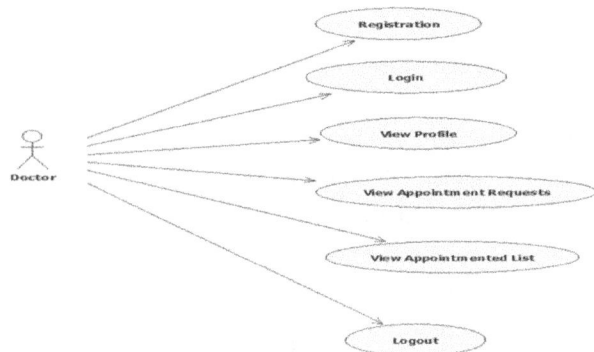

Figure 15.4 Patient side development
Source: Author

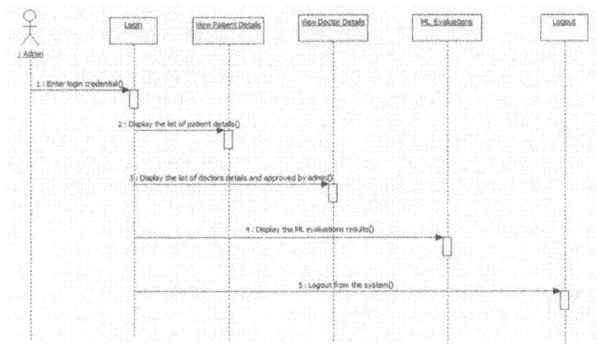

Figure 15.5 Admin section workflow
Source: Author

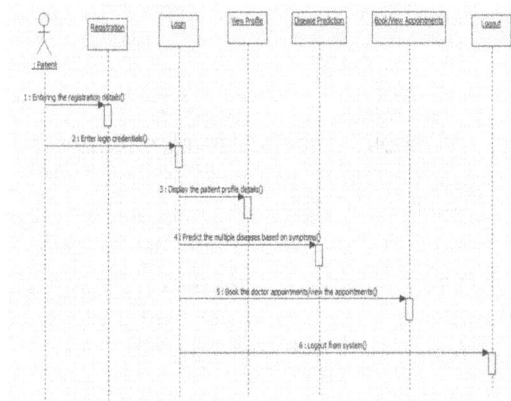

Figure 15.6 Patient section workflow
Source: Author

Figure 15.3 showcases the functionalities accessible to the patient, illustrating the range of options available to them within the system. This diagram delineates the potential actions and features at the patient's disposal, presenting a visual representation of their role and capabilities.

Figure 15.4 illustrates the functionalities accessible to the doctor, demonstrating the array of options at their disposal within the system. This diagram delineates the potential actions and features available to the doctor, providing a visual representation of their role and capabilities within the context.

Sequence of workflow
In Figure 15.5, the workflow process on the admin side is depicted, allowing the administrator to handle both patient and doctor details. This diagram illustrates how the admin manages information for

patients and doctors, showcasing their responsibilities within the system. ML evaluation represents various evaluation metrics of the employed techniques.

Figure 15.6 displays the workflow process from the patient's perspective, enabling them to oversee their profile and schedule appointments. This diagram illustrates how patients manage their personal information and make appointments, providing insight into their interactions within the system.

In Figure 15.7, the workflow process for doctors is depicted, enabling them to oversee and update their profile details. This diagram illustrates how doctors manage their personal information within the system, showcasing their ability to maintain and modify their profiles as needed.

Results

Admin can evaluate the patients based on the pages shown in Figures 15.8, 15.9, and 15.10.

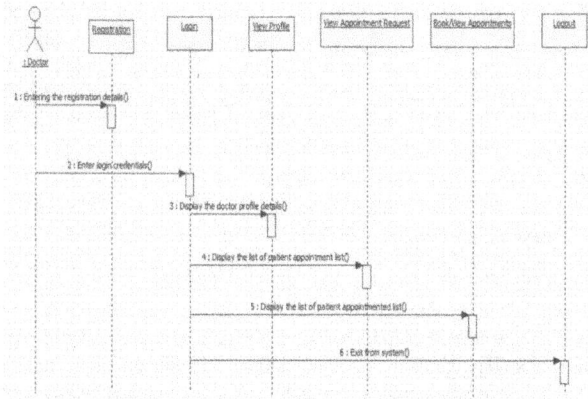

Figure 15.7 Doctor section workflow
Source: Author

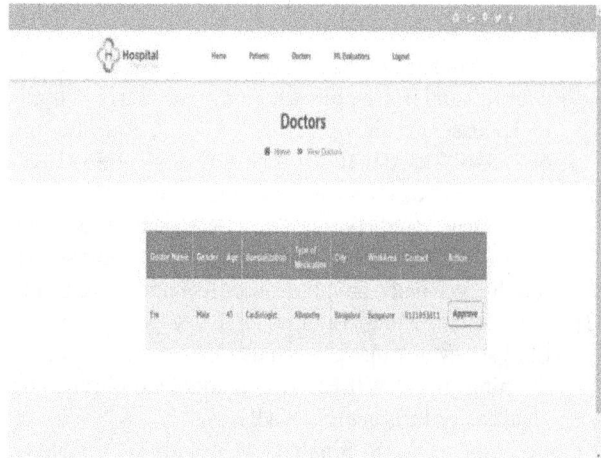

Figure 15.9 View doctor's details
Source: Author

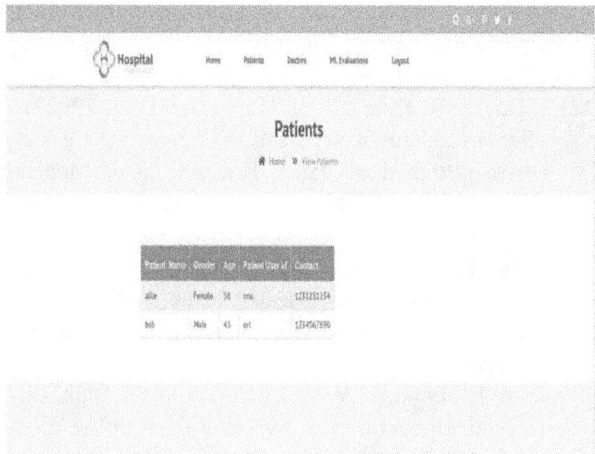

Figure 15.8 View patient details
Source: Author

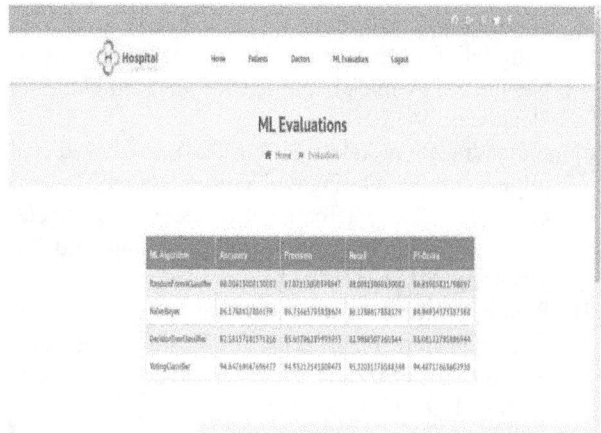

Figure 15.10 ML Evaluations
Source: Author

Conclusion

Traditional statistical models often struggle with handling missing values and large datasets, leading to suboptimal predictive performance. In contrast, machine learning (ML) techniques offer a more robust approach, as demonstrated in this study. The proposed ML models successfully diagnose multiple diseases, such as heart disease and diabetes, achieving an impressive 95% classification accuracy. This high accuracy is attributed to optimized feature engineering techniques, including effective feature selection, transformation, and extraction, which enhance model generalization. Additionally, advanced missing data handling methods and the use of powerful ML algorithms—such as ensemble learning (Random Forest, XGBoost) and deep learning—enable the models to capture complex disease patterns. The performance was further validated using cross-validation techniques, ensuring reliability and reducing overfitting. The comparison with traditional statistical models highlights the superior predictive capabilities of ML. By accurately identifying disease characteristics, these methods not only improve diagnostic precision but also facilitate enhanced decision-making processes in healthcare. Thus, this study underscores the significant role of machine learning in predictive modeling for disease diagnosis, paving the way for future advancements in medical analytics.

References

[1] H. A. Al-Jamimi(2024), "Synergistic Feature Engineering and Ensemble Learning for Early Chronic Disease Prediction," IEEE Access, vol. 12, pp. 62215–62233. doi:10.1109/ACCESS.2024.3395512.

[2] Sharma, R. K., Tiwari, R., & Gupta, D. R. (2023). Multiple disease prediction system using machine learning. In IEEE International Conference on Computational Intelligence and Communication Technology (CICT).

[3] Kumar, T., Wang, H., Gupta, P., & Rodriguez, L. (2024). Multi-disease prediction using machine learning. In 2024 IEEE International Conference on Healthcare Informatics (ICHI).

[4] Verma, L., Das, S., Singh, H. P., & Lee, W. Y. (2024). Multiple disease prediction using machine learning and deep learning with web technology implementation. In IEEE Conference on Emerging Trends in Computing and Communication (ETCC).

[5] Singh, B., Zhao, K. C., Naik, N., & Reddy, N. (2023) Hybrid machine learning for multi disease prediction with relief feature selection. In 2023 IEEE Symposium on Artificial Intelligence in Medicine and Healthcare (AIMH).

[6] Subhadra, H., & Vakiti, S. R. (2024). IoT-based coal mine safety monitoring and alerting system. In Proceedings - 2024 International Conference on Social and Sustainable Innovations in Technology and Engineering, SASI-ITE 2024.

[7] P. Nair, S. Wang, J. Roy, and A. Banerjee (2024), "Effective Feature Engineering for Heart Disease Prediction," IEEE Access, vol. 12, pp. 123456– 123465, doi:10.1109/ACCESS.2024.1234567.

[8] M. Roy, R. K. Singh, H. Li, and S. Mitra, "Efficient Cardiovascular Disease Prediction Using LASSO Techniques," IEEE Transactions on Computational Biology and Bioinformatics, vol. 21, no. 3, pp. 789–798, 2023, doi: 10.1109/TCBB.2023.1234567.

[9] Gupta, R., Liu, F., Zhang, J., & Shen, D. (2024). Predictive modeling for heart disease using advanced ML techniques. In 2024 IEEE International Conference on Data Science and Advanced Analytics (DSAA).

[10] V. Mehta, Y. Zhou, and P. Sharma, "Advances in Disease Diagnosis with Machine Learning for Remote Health Monitoring," IEEE Journal of Biomedical and Health Informatics, vol. 28, no. 2, pp.345–354,2024, doi:10.1109/JBHI.2024.1234567.

[11] Vakiti, S. R., & Subhadra, H. (2024). Smart seed sower: design and implementation of an IoT-based solar-powered automatic seed dispenser. In Asia Pacific Conference on Innovation in Technology (APCIT).

[12] P. Das, S. R. Yadav, R. K. Gupta, and D. Yu.(2023) "Genomic Data-Driven Feature Engineering for Rare Disease Prediction," IEEE Transactions on Medical Imaging, vol. 43, no. 1, pp. 123–132, doi: 10.1109/TMI.2023.

[13] Desai, R., Li, X., & Chen, Z. (2024). Real-time diagnosis of infectious diseases using AI techniques. *IEEE Transactions on Automation Science and Engineering.*

[14] Kaur, H., Prakash, V., & Zhou, Y. (2023). Multi-disease identification with hybrid deep learning models. In IEEE Conference on Intelligent Systems and Healthcare (CISH).

16 Gray wolf optimization-based intelligent clustering mechanism for energy-efficient and trust-aware communication in IoT networks

G. Sivakumar[1,a], Chandralekha R.[1,b], K. Amsaveni[2,c], B. Dhanam[3,d], K. V. Srirenga Nachiyar[1,e] and P. Marichamy[4,f]

[1]Assistant Professor, Department of ECE, Ramco Institute of Technology, Tamil Nadu, India

[2]Assistant Professor, Department of CSE, National Engineering College, Tamil Nadu, India

[3]Assistant Professor, Department of ECE, P.S.R. Engineering College, Tamil Nadu, India

[4]Professor, Department of ECE, P.S.R. Engineering College, Tamil Nadu, India

Abstract

Energy conservation measures are important due to the increased energy consumption rate in Internet of Things (IoT) sensors placed in the IoT domain. Clustering has been analyzed as an efficient green communication approach for managing energy, and integrating meta-heuristic methods guarantees almost optimal results with reference to QoS demands. With that in mind, this paper presents the Gray Wolf Optimization (GWO)-based intelligent clustering mechanism (GWO-ICM) as a possible solution to the NP-hard problem of attaining energy efficient communication in IoT networks. The GWO-ICM uses a fitness model that measures energy, delay, distance, jitter and packet forwarding ability to choose energy efficient sensor nodes as cluster head (CHs). For CH selection, GWO algorithm is employed to allow a balance between the global search part and the local search part. Further, trust evaluation is incorporated through the process of ranking that directly or indirectly rates nodes to exclude malicious or low energy nodes as CHs. Residual energy is also incorporated in the process of computing the trust to increase reliability and performance of the network.

Keywords: Clustering, energy efficiency, iInternet of tThings, optimization algorithm, wireless sensor networks

Introduction

The Internet of Things (IoT) is an emerging phenomenon that has the potential to interconnect billions of devices to transfer huge amount of data and make decisions for smart solutions in different fields as healthcare, farming and smart cities. Wireless sensor networks (WSNs) act as a major support infrastructure for IoT through effective interaction between the involved sensors and devices. However, one of the largest problems that arise from using sensor nodes, or at least, a significant drawback that limits the sustainability of IoT networks on their basis is the limited functionality that the nodes have. Energy-conserving techniques in communication and data transfer are, therefore, critical in enhancing longevity of the network without comprising on performance. By some of these, objects in our environment that are technological advanced, namely instruments in this case can share data using technologies like radio-frequency identification (RFID) and WSNs [1, 2]. Development of wireless communication and the unity of more technologies between devices have led to the IoT which means that a number of items and related things can exchange information with each other through a network protocol or standard at any time [3]. In clustering, sensors are divided into clusters and only one node of a cluster, known as cluster head, is responsible for forwarding messages thus minimizing communication between nodes [4]. However, the process of choosing cluster head (CH) is an NP-hard problem, and hence the use of intelligent solutions to support energy efficiency, network scalability and longevity.

Literature Survey

Information meta-heuristic algorithms have been proved to have promising capability in solving such optimization problems. Of these, GWO has recently

[a]gsivakvp@gmail.com, [b]radhachandra2013@gmail.com, [c]amsaskraj.k@gmail.com, [d]dhanam@psr.edu.in, [e]srirenga@ritrjpm.ac.in, [f]drpmarichamy@gmail.com

DOI: 10.1201/9781003684589-16

attracted researchers' attention through its simplicity, flexibility in problem solving, and its ability to control the trade-off between exploration and exploitation in the solution space [5]. Towards solving issues like exploration imbalances and improving secure data transfer through mutual authentication, this research proposed an enhanced GWO technique called the EECIGWO technique for selecting energy efficient cluster heads in WSNs [6, 8]. Secured data transmission while securing clustering and routing. Said et al [9] has proposed a mechanism for dividing the IoT environment by the characteristics of the network where the IoT devices will operate. Cluster based energy efficient data aggregation routing (CEDAR), a cluster-based IoT routing approach developed by Mohseni et al [10], uses fuzzy logic and Capuchin search algorithm. It comprises clustering and intra-/inter-cluster routing to lower energy consumption through node grouping and packets routing. The scenarios reveal that in energy consumption, delay and network lifetime, CEDAR is superior to other alternatives.

When defining sustainable WSN in smart cities, factors such as, networks lifetime, coverage efficiency, and energy efficiency are usually considered. Studies have highlighted nature inspired solutions under different circumstances. Another researches [11] developed harmony search algorithm, genetic algorithm and least-distance-based load balanced clustering for the purpose of optimization of active senor nodes, clustering heads and convergence rates. The developed GWO is illustrated of 29 most used objective functions and the simulated outcome testifies that GWO is significantly superior to other computational intelligence approaches such as gravitational search algorithm (GSA), particle swarm optimization (PSO), harmony search algorithm (HAS), differential evolution (DE) and evolutionary programming (EP) [7, 12]. The recent trends depict that intelligent computational methods are commonly applied on multi- objectives discrete problems such as in clustering of WSNs. The basic GWO modelling has been improved, and binary and dogmatic components were introduced. Singh et al. [13] introduced a trust-aware clustering mechanism that integrates the whale optimization algorithm (WOA) with Dempster-Shafer theory (DST), referred to as WODST. This approach significantly improved energy efficiency and reliability in IoT networks. Lei et al. [14, 15] suggested a fuzzy Coati and

swarm optimization algorithm (FCSOA)-based energy-aware routing method to optimize node organization, reduce delays, minimize packet loss, and extend network lifespan. CHs were selected based on residual energy, path length to the BS, link quality, and proximity, resulting in reduced energy consumption.

Proposed System

GWO-based ICM: To develop a solution for increasing the energy efficiency, reliability, and network lifetime of the IoT-enabled WSNs, a new mechanism called (GWO-ICM is proposed. GWO-ICM enhances the choice of the CHs and ensures optimal clustering with the least energy consumption because the leadership of the gray wolves in the wild is imitated Figure 16.1.

i)   **Multi-objective fitness function:** The clustering mechanism in GWO-ICM also uses multi-objective fitness function to assess each sensor node through important parameters. Residual energy makes sure that energy significant nodes are preferred as CHs and thus enhances network lifetime. Distance to the other members of the cluster reduces communication energy by promoting CHs that are nearer to other members of the cluster. The fitness function evaluates each sensor node based on the following parameters:

Residual energy (Eres): Residual energy helps the nodes with a high energy level be chosen for CH selection. This assist in preventing excessive consumption of energy in CHs which on the other hand enhances the network life.

$$f1 = \frac{E_{res}}{E_{max}} \qquad (1)$$

where Emax is the maximum energy of a node.

Distance to cluster members: The CHs so selected are closer to their associated set of nodes or other cluster heads in an attempt to conserve communication energy. This in turn helps to reduce the energy needed for the nodes to communicate with other nodes in the same cluster.

$$f2 = \frac{1}{\sum_{i=1}^{n} d_{i,CH}} \qquad (2)$$

where $d_{i,CH}$ is the distance between a node iii and the CH, and n is the number of nodes in the cluster.

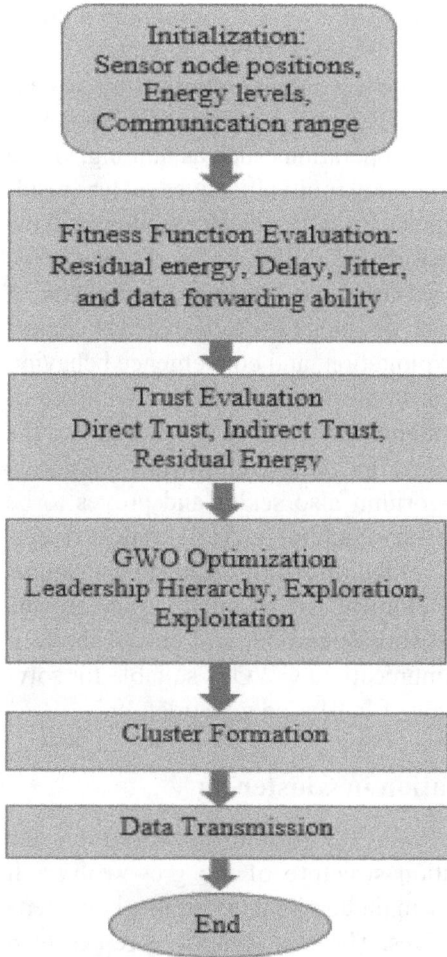

Figure 16.1 GWO-Based clustering for energy-efficient IoT communication
Source: Author

Distance to base station: To reduce the transmission energy during data forwarding, CHs nearer to a base station are selected.

$$f3 = \frac{1}{d_{CH,BS}} \quad (3)$$

where $d_{CH,BS}$ is the distance between the CH and the base station.

Delay: These parameters include delay where possible efforts are made in order to deliver data on time. Preferably, CHs with the least data transmission delay are considered.

$$f4 = \frac{1}{T_{Delay}} \quad (4)$$

Jitter (J): Jitter as the parameter for the permitted variability in the time of packet transmission, stable communication is provided for. Lower Low Jitter values are desired.

$$f5 = \frac{1}{J} \quad (6)$$

Packet forwarding potential: The ability of a node to forward data reliably is assessed, ensuring efficient inter-cluster communication.

$$f6 = P_{pf} \quad (7)$$

The overall fitness function is a weighted sum of the individual fitness components:

$$F_{CH} = \omega_1 f_1 + \omega_2 f_2 + \omega_3 f_3 + \omega_4 f_4 + \omega_5 f_5 + \omega_6 f_6 \quad (8)$$

Here, $\omega 1, \omega 2, \omega 3, \omega\_4, \omega 5, \omega 6$ are the weights assigned to each objective based on its importance in the CH selection process. These weights can be tuned to reflect specific network requirements.

ii) **Trust evaluation and integration with clustering:**

Trust evaluation becomes necessary in WSNs and IoT applications in order to identify trustworthy and energy-effective nodes that will become CHs. The proposed trust evaluation process in the Gray Wolf Optimization-based intelligent clustering mechanism (GWO-ICM) involves direct trust, indirect trust, and residual energy to derive the total evaluation of each node.

Direct trust (Td): Direct trust is based on a node's past interactions, including successful packet transmissions and acknowledgments within the network.

$$T_d = \frac{S_{Success}}{S_{Success} + S_{failure}} \quad (9)$$

Indirect trust (TI): Indirect trust is derived from recommendations or feedback provided by neighboring nodes about the target node.

$$T_i = \frac{\sum_{j=1}^{N} T_{j \to 1}}{N} \quad (10)$$

Where $T(j \to 1)$ is the trust value assigned by node j to node i, and N is the number of neighboring nodes providing feedback.

Residual energy factor (Eres): Residual energy is incorporated into the trust evaluation to ensure that energy-depleted nodes are not selected as CHs.

$$E_{norm} = \frac{E_{res}}{E_{max}} \quad (11)$$

Where Emax is the maximum energy capacity of a node.

Overall trust score (T_{total}): The overall trust score combines direct trust, indirect trust, and normalized residual energy.

$$T_{total} = \alpha T_d + \beta T_i + \gamma E_{norm} \qquad (12)$$

Where α, β, γ are weights reflecting the importance of each component. The GWO algorithm evaluates nodes based on their fitness scores, including trust, energy, distance, and other parameters. The nodes with the highest fitness scores are selected as CHs.

iii) **Algorithm of gray wolf optimization:**
Another part of GWO, the CH selection is the heart of WSN CH selection based on the natural hunting strategies of grey wolves. It uses leadership hierarchy, with the best candidate – the alpha wolf presiding over the clustering process together with the help of beta and delta wolves. The convergence of global search and exploitation of the interesting solution region guarantees efficient identification of the optimal CHs.

iv. **Clustering process:**
Those particular CHs once selected, the sensor nodes are then grouped into different clusters under these CHs. The data collected in each CH is to be with data from its cluster members and then relay the same to the base station. It ensures very little duplication, and hence helps in conserving energy in the entire network of computers.

System Architecture

The IoT network model comprises of several motes spread over a particular geographical region to capture a target area, to pre-process and transfer the data wirelessly to a central hub or base station (BS) or cloud server for further processing. These sensor nodes, considered to mainly possess low power source and low computational ability, operate in clusters in order to conserve energy and expand addressability. In each of the formed clusters, there is a CH who will be responsible for collecting data from other nodes in its cluster which will in turn send this information to the Base Station (BS) and hence minimize on data transfer load. In this paper, the data flow in the IoT network is quite coherent and well-defined; where the set covering problem (SCP) is a well-known optimization problem used in

WSN. The goal of SCP is to select minimum number of nodes to cover all target points. To address issues of CH selection, the Gray Wolf Optimization (GWO) algorithm is used. Motivated by the social hierarchy of Canis lupus for operations such as hunting, GWO ranks nodes exploiting multi objectives such as residual energy, distance from the cluster members and the BS, communication reliability, and node reputation. Implemented as GWO with power coefficients of alpha, beta, delta, or omega nodes are used to control exploration, exploitation, and convergence behavior. This integration in clustering involves guarantee of energy efficient and reliable CHs by eliminating the bad and malicious nodes through trust evaluation in GWO. The algorithm also scales and proves to be highly resilient in situations in large-scale IoT networks, because of their flexibility. Due to its features that allow for efficient use of energy, increase the duration of network operation, and ensure the reliability of communication, GWO is suitable for solving the problems of IoT-based networks.

Trust Evaluation in Clustering

Basically, the GWO algorithm is based on the social-domination structure of the grey wolves. It consists of four main components: alpha, beta, delta and omega wolves. The best solution is represented by alpha, the second best by beta, the third best by delta, and all the rest of the wolves are within omega. This portion of the implementation continues to move the wolves relative to the best solutions (alpha, beta, delta) of the problem/solution space as it mimics the concept of the search to the optimal solution. The initial trust of nodes is then computed in trust-based clustering models, such as RATMI-G-BWM which is used in identifying factors such as energy and performance. A node with a higher trust value is chosen to be the CH of a specific cluster. During clustering, nodes' trust values are adjusted based on their behavior as well as their performance. All nodes with low trust values are considered malicious and they are detected and eliminated from the clustering process.

Result and Discussion

The proposed GWO-ICM scheme and baseline approaches, including the baseline FCPSO, WOA with WODST approaches, are simulated using the MATLAB R2016b simulator. The system is running

Windows 7 Pro with 8GB of RAM and a Core i7 processor. The entire experimental procedure is carried out in a network area of 300 × 300 square meters, the nodes' energy is assumed to be 0.5 joules. From Figure 16.2, the performance measures also suggest that all the three learning techniques are 100% at the onset at 200 rounds. GWOICM shows the best performance in terms of vitality and retains a higher percentage of alive nodes, decreasing to 63% only in 1200 rounds of function. FCSOA also copies the lower ranking for ending at 61%, while WODST shows less effectiveness, decreasing to 56% by the same round count. In this analysis, we have identified GWOICM as the most effective strategy in maintaining node activity over time in WSN.

From Figure 16.3, the fuzzy Coati and swarm optimization algorithm (FCSOA) technique shows moderate performance by outranking WODST while remaining below GWOICM for ranking positions.

WODST demonstrates lower-throughput performance than GWOICM and FCSOA but exhibits better results from round 800 through 2000. GWOICM demonstrates superior throughput performance with a 2.51% benefit over FCSOA while delivering 8.88% better results against WODST based on intake data. Through the analysis GWOICM emerged as the leading technique for throughput performance with FCSOA in second position and WODST showing the lowest effectiveness.

According to Figure 16.4, residual energy assessments at different rounds GWOICM demonstrate the highest values which begin at 0.48 at 200 rounds then drop to 0.31 at 2000 rounds. Expanding from 0.47 to 0.28μJ FCSOA demonstrates parallel energy trends while WODST maintains its lowest position throughout as energy depletes from 0.46 to 0.25 μJ.

GWOICM provides superior performance across throughput and residual energy measurements compared to FCSOA and WODST delivers the least effective results.

Conclusion

The Gray Wolf optimization-based intelligent clustering mechanism (GWO-ICM) to resolve energy efficiency problems in IoT network environments. The fitness model in GWO-ICM selects cluster head (CH) optimally by examining energy consumption with delay and distance and packet forwarding speeds along with trust-related assessments and remaining node power to boost reliability against malicious activity. Performance analysis demonstrates that GWO-ICM performs better than FCSOA and WODST regarding network metrics. The GWO-ICM

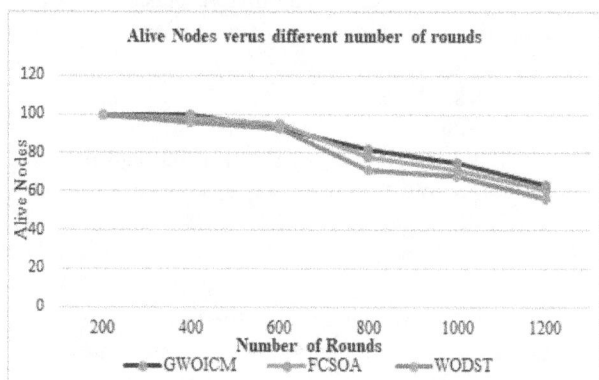

Figure 16.2 Alive Nodes versus number of rounds (IoT nodes-100)
Source: Author

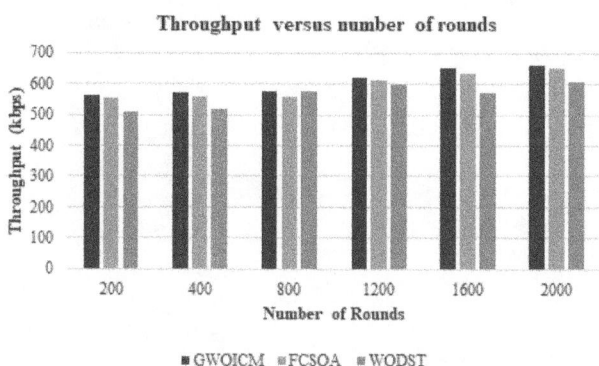

Figure 16.3 Throughput versus number of rounds (IoT nodes-100)
Source: Author

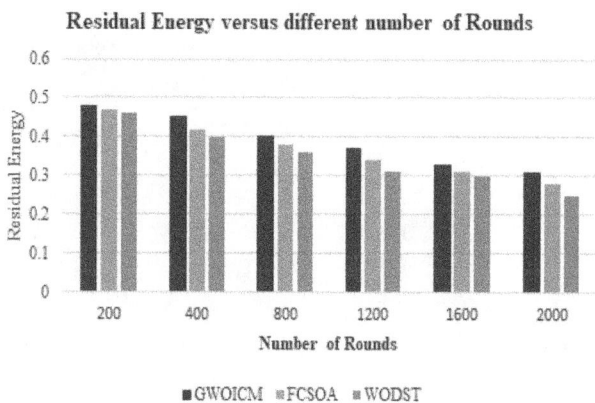

Figure 16.4 Residual energy versus number of rounds
Source: Author

method extended network throughput by 2.51% and 8.88% above FCSOA and WODST while maintaining the highest remaining energy value during network operation. GWO-ICM demonstrated exceptional performance wireless sensor networks (WSNs) by maintaining 63% liveliness at 1200 rounds which surpassed FCSOA at 61% and WODST at 56%. GWO-ICM demonstrates superior capabilities as an energy-efficient communication tool which supports node longevity and enhances network performance in IoT and WSN environments.

References

[1] Kamalov, F., Pourghebleh, B., Gheisari, M., Liu, Y., & Moussa, S. (2023). Internet of medical things privacy and security: challenges, solutions, and future trends from a new perspective. *Sustainability*, 15(4), 3317.

[2] Zhang, Y., Ren, Q., Song, K., Liu, Y., Zhang, T., & Qian, Y. (2021). An energy-efficient multilevel secure routing protocol in IoT networks. *IEEE Internet of Things Journal*, 9(13), 10539–10553.

[3] Pourghebleh, B., Hekmati, N., Davoudnia, Z., & Sadeghi, M. (2022). A roadmap towards energy-efficient data fusion methods in the internet of things. *Concurrency and Computation: Practice and Experience*, 34(15), e6959.

[4] Yadav, R. K., & Mahapatra, R. P. (2021). Energy aware optimized clustering for hierarchical routing in wireless sensor network. *Computer Science Review*, 41, 100417. ISSN1574-0137.

[5] Yu, M., Xu, J., Liang, W., Qiu, Y., Bao, S., & Tang, L. (2024). Improved multi-strategy adaptive Grey Wolf Optimization for practical engineering applications and high-dimensional problem solving. *Artificial Intelligence Review*, 57, 277.

[6] Yang, Z., & Palaoag,, T. D. (2024). Simulation of marine wireless sensor network coverage based on improved grey wolf optimization algorithm. *Journal of Physics*, 2858, 012022–012022. doi: 10.1088/1742-6596/2858/1/012022.

[7] Kaddi, M., Omari, M., Salameh, K., & Alnoman, A. (2024). Energy-efficient clustering in wireless sensor networks using grey wolf optimization and enhanced CSMA/CA. *Sensors*, 24(16), 5234. doi: 10.3390/s24165234.

[8] Yuvaraja, M., Sureshkumar, S., James, S. J., & Teresa, V. V. (2024). An energy-efficient improved grey wolf optimization algorithm-based cluster head and shamir secrets sharing-based WSNS with secure data transfer. In Salud, Ciencia y Tecnología - Serie de Conferencias, (Vol. 3, p. 946).

[9] Said, O. (2017). Analysis, design and simulation of Internet of Things routing algorithm based on ant colony optimization. *International Journal of Communication Systems*, 30(8), e3174.

[10] Mohseni, M., Amirghafouri, F., & Pourghebleh, B. (2022). CEDAR: a cluster-based energy-aware data aggregation routing protocol in the internet of things using capuchin search algorithm and fuzzy logic. *Peer-to-Peer Networking and Applications*, 1–21, vol.16.

[11] Singh, S., & Sharma, R. M. (2017). HSCA: a novel harmony search based efficient clustering in heterogeneous WSNs. *Telecommunication Systems*, 67, 651–667.

[12] Abhishek, B., Ranjit, S., Shankar, T., Eappen, G., Sivasankar, P., & Rajesh, A. (2020). Hybrid PSO-HAS and PSO-GA algorithm for 3D path planning in autonomous UAVs. *SN Applied Sciences*, 2, 1805. https://doi.org/10.1007/s42452-020-03498-0.

[13] Singh, S., Anand, V., & Yadav, S. (2024). Trust-based clustering and routing in WSNs using DST-WOA. *Peer-to-Peer Networking and Applications*, 3(6), 1–13.

[14] Lei, C. (2024). An energy-aware cluster-based routing in the Internet of things using particle swarm optimization algorithm and fuzzy clustering. *Journal of Engineering and Applied Science*, 71(1), 135.

[15] Pourghebleh, B., Hekmati, N., Davoudnia, Z., & Sadeghi, M. (2024). A roadmap towards energy efficient data fusion methods in the internet of things. *Concurrency and Computation: Practice and Experience*, 34(15), e6959.

17 Enhancing heart disease prediction using hybrid K-means clustering and supervised machine learning models

N. Nagasoudhamani[1,a] and P. Sunil[2,b]

[1]M. Tech Scholor, Department of CSE, Shri Vishnu Engineering College For Women Vishnupur, Bhimavaram, W.g Dt, Andhra PradeshP, India

[2]Assistant Professor, Department of CSE, Shri Vishnu Engineering College For Women Vishnupur, Bhimavaram, W.g Dt, Andhra Pradesh, India

Abstract

This paper presents a hybrid model for heart disease prediction by integrating K-Means clustering with machine learning classifiers. Initially, K-Means clustering was applied to uncover inherent patterns within the dataset and determine the optimal number of clusters using the Elbow Method. The resulting cluster labels were incorporated as an additional feature, enhancing predictive performance. A variety of ML classifiers, including Decision Tree, Random Forest, k-nearest neighbors, AdaBoost, Gradient Boosting, logistic regression, and SVC, were applied to evaluate the model's effectiveness. To further refine feature selection, the SelectKBest method was employed to identify the five most significant features, including the cluster label, which demonstrated the importance of incorporating unsupervised learning in the classification process. The refined feature set was used to retrain the models, leading to notable improvements in predictive performance. Experimental results showed that tree-based models achieved good performance with an accuracy of 1.0 and 0.985, respectively, even with a reduced feature set. Other classifiers, including Gradient Boosting and AdaBoost, exhibited high accuracy, reinforcing the effectiveness of the hybrid approach. The study highlights the significance of clustering-based feature engineering in improving heart disease prediction, particularly when dealing with limited labeled data.

Keywords: Feature selection, heart disease detection, hybrid mode, K-means

Introduction

Heart disease is among the leading causes of death globally, with millions of deaths every year. These conditions are most often due to high BP, high cholesterol, smoking, sedentary lifestyle, and genetics. Heart disease, which leads to severe complications such as heart attacks and strokes, needs to be detected early and taken care of in a timely manner. However, traditional tools for diagnosis such as electrocardiograms (ECGs), stress tests, and blood tests often require significant time, money, and expertise. This has resulted in a growing demand for automated and data-driven predictive models that can help health care workers make quicker and more precise diagnoses.

Early and accurate predictions of heart disease through patient health data have been enabled by the powerful tool of ML in medical diagnostics. There are many studies in which supervised learning algorithms have been used to predict heart disease based on features such as age [1], cholesterol levels, blood pressure, and electrocardiogram readings. However, data availability and imbalance are a bigger issue in the prediction of heart disease and in features selection.

A promising strategy to tackle these issues is leveraging the strengths of both unsupervised representations and supervised approaches, which has the potential to improve predictive performance, as well as representation power.

To improve the prognosis of heart disease, this work combines K-Means with supervised learning. Applying K-Means to find the dataset's patterns, determining optimum cluster counts using the Elbow method. The pipeline uses the labeled clusters generated as an additional feature to further enhance model performance. Then the unsupervised learning process SelectKBest brings the most relevant features and also its feature cluster label shows that unsupervised learning is essential for feature selection. We used various classifiers, such as Decision Tree (DT), Random Forest (RF), k-nearest neighbors

[a]soudha.kaju@gmail.com, [b]sunilp@svecw.out.in

DOI: 10.1201/9781003684589-17

(KNN) etc. to assess the performance of the model. K-Means cluster analysis results confirm that the application of K-Means for grouping the different types of patients, along with the training of the different ML classifiers, improves the precision of predictions; hence, validates its application in the field of medical applications and epidemiology.

Related Work

The authors [2] addressed the problem of heart disease prediction through improvement in selection of features for better accuracy of classification. Using the Cleveland dataset, they introduced a hybrid approach that couples the artificial flora optimization with the SVM. Their model acted as a feature selection tool and helped find the most relevant predictors in identifying heart disease.

The authors discussed the study of the development model to support better diagnosis of ischemic heart disease (IHD) [3]. Deep learning techniques were highlighted as being able to navigate complex medical data in substantial quantities, finding patterns themselves, and helping physicians to identify the harshness of disease and diagnose it. They introduced a hybrid residual attention-enhanced LSTM (HRAE-LSTM) model which integrated attention residual learning and LSTM to improve accuracy and stability of the analysis. They demonstrated using the UCI heart disease dataset that the hybrid deep learning model achieved better performance than existing cardiac disease prediction methods.

In another study the authors aimed at early detection of cardiovascular diseases (CVDs) through self-attention networks for CVD risk prediction [4]. Effectively, their transformer model, with self-attention-based architecture, could capture the contextual information during transformations to feature representations, which could test the complex patterns exist in the dataset. In [5], the authors studied several ML algorithms, including Logistic Regression, KNN, SVM, and ANN using UCI heart disease dataset. Studies were designed to counter the challenges behind the complex, high-cost diagnostic tools that are currently in use, making it extremely essential to develop models that are simple, efficient, and accurate.

To decrease mortality due to heart diseases, [6] proposed a model based on Graph Neural Network (GNN) in order to predict heart diseases. They also performed extensive optimizations to enhance performance on a Kaggle dataset of real-world data with 14 cardiovascular features. RMSprop performed better than the other optimizers explored in this study, underlining the promise of GNNs in medical diagnostics and the need for optimization strategies for improved prediction accuracy. Authors of [7] focused on analysis of clinical data to casual treatment and adaptation of HDP, with review of available DL, ML, and optimization-based techniques, and explicit mention of challenges to be overcome for enhanced DP accuracy. They summarized several ML and DL algorithms applied in recent researches to help healthcare professionals with heart diseases diagnosis.

Another study [8] introduced an EPERM for heart disease prediction with an emphasis on feature selection and correlation-based attribute ranking. They used the Attribute Ranking Method to rank the most important features and used conditional probability for predicting the risk of heart diseases. The authors also presented Naïve Bayesian prediction of coronary heart disease, highlighting the necessity of early warning systems capable of detecting risk factors prior to expensive diagnostic tests being performed. Health data mining has been a focal point for them in formulating an appropriate predictive model that could help patients and health professionals [9].

A hybrid model combining Gaussian Fuzzy C-Means Clustering (GKFCM) and RNN for coronary heart disease prediction using non-invasive clinical data was present by Revathy et al. [10]. Handling of normalized data for clustering before classifying helps in improving accuracy and proved as a hybrid approach aiding medical prognosis. The authors of [11] contributed to heart disease prediction by designing a hybrid ML model of CFS, GWOA, and several classifiers. Their methodology centered on optimal feature selection, thereby facilitating early diagnosis and risk evaluation. In addition, a hybrid DT and AdaBoost model, designed for early coronary heart disease diagnosis and symptom management, was proposed by the authors [12]. They found out that predictive analytics can facilitate clinical decision making and forecasting chronic diseases. The authors [13], in their study on the global burden of heart disease, highlighted the grave scenario in India and other countries. The work they recognized featured ML in health care, especially its use in analyzing big data sets to enhance prediction accuracy. An early heart disease prediction model based on ANN was proposed in [14], to early diagnose patients to prevent life-threatening risks. The method was applied on the dataset supporting the claim that neural networks are useful for low stage heart disease detection. This study highlighted that ANN-based methods can assist in prediction accuracy and timely medical intervention.

Methodology

Figure 17.1 shows the proposed method. The dataset for [15] this work has been collected from the Kaggle which includes various patient fundamentals and clinical information such as age, sex, chest pain type etc. The target variable is a presence (1) or absence (0) of heart disease. Since there were no missing values, duplicates or class imbalance, ML models could be applied directly without needing any data balancing techniques. In the next step, the K-Means clustering was implemented. In order to identify the optimal number of clusters, the Elbow Method was applied. The next step was to assign a cluster label to each data point, which would be an additional feature to be used to improve classification. This was done in order to use the insights from the unsupervised learning in order to boost predictive performance. Later, seven ML models were applied to the dataset, including the newly introduced cluster label. The models were evaluated based on key performance metrics such as accuracy, precision, recall, and F1-score. Initial results demonstrated that tree-based models given good classification accuracy, while ensemble models AdaBoost also performed well.

Next, features using SelectKBest feature selection with ANOVA F-score were refined, retaining only the five best features (including future cluster label) to further optimize model performance. It also allowed for the identification of a critical component of feature engineering where K-Means clustering would be exercised, seen in feature where the cluster label was kept in the top predictors.

With the reduced feature set, same set of ML models applied. The results shown the significance of integrating unsupervised learning (K-Means clustering) with supervised classifiers, as the clustering-derived feature contributed meaningfully to classification. The proposed approach demonstrates a hybrid learning strategy that enhances predictive accuracy, reduces feature redundancy, and improves model interpretability for heart disease diagnosis.

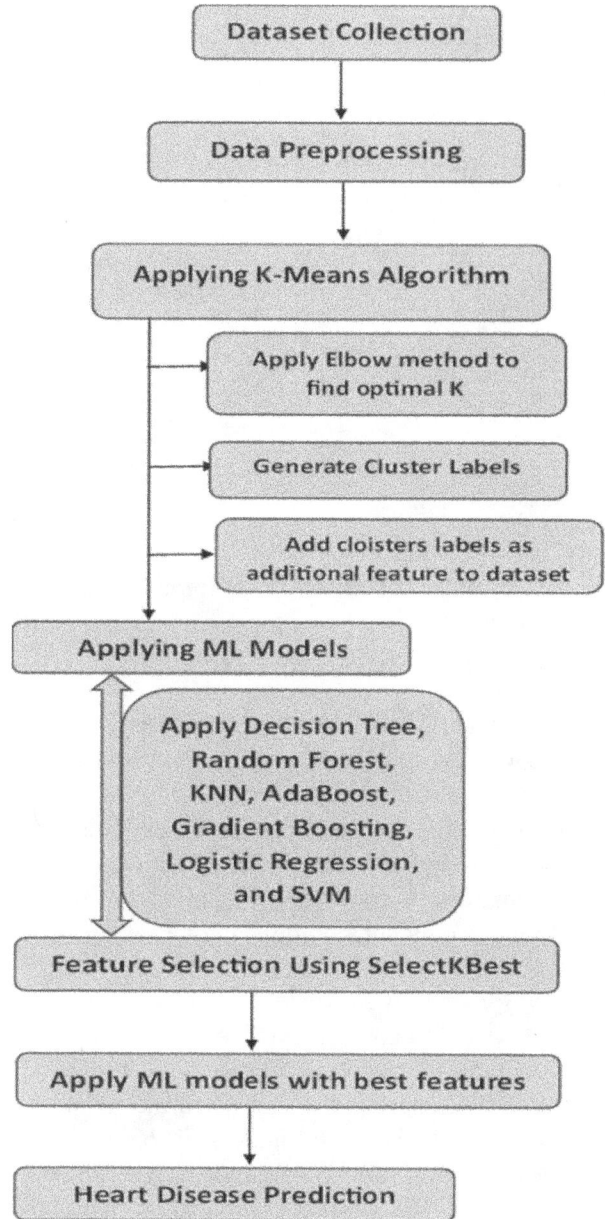

Figure 17.1 Proposed method
Source: Author

Experiments and Results

Data collection

The dataset employed in this study was obtained from Kaggle and collected from four medical databases. There are 14 attributes of the dataset. The target variable is binary, with 0 indicating no heart disease present, and 1 for the presence of heart disease. Some important attributes are patient age, patient sex, type of chest pain, resting blood pressure etc. These characteristics are commonly used in cardiovascular

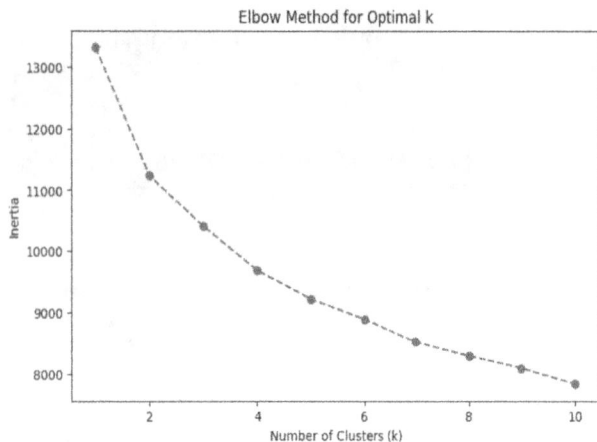

Figure 17.2 K-means elbow method
Source: Author

Table 17.1 Performance of machine learning models.

Model	Precision	Recall	F1	Accuracy
Decision Tree	1	1	1	1
RFC	1	1	1	1
KNN	0.69	0.68	0.69	0.71
AdaBoost	0.83	0.91	0.84	0.79
GBC	0.97	0.97	0.97	0.97
Log Reg	0.81	0.92	0.83	0.76
SVC	0.71	0.79	0.74	0.69

Source: Author

studies to scale-up patient risk and predict outcomes of heart disease.

Preprocessing
The data had no missing values, duplicates, or class imbalance. So, very minimal preprocessing was required. The pre-processing step used was scaling of features using StandardScaler to ensure numerical attributes. Given that the dataset was already clean and formatted, it was directly applied for feature extraction and model training.

Applying K-means clustering
K-Means clustering was applied to discover underlying structures in the dataset to improve feature representation. The analysis was then performed using the Elbow Method, shown in Figure 17.2, based on the values of the within-cluster sum of squares (inertia) w.r.t k, where the most appropriate was the elbow point, thus providing the optimal k amount associated to k = 3, so all sample points were assigned with a cluster based on that so it becomes another feature.

By using the structural patterns obtained through clustering, it was anticipated that this enriched dataset would enhance the prediction performance of Machine Learning models.

First phase of machine learning model training
After K-Means clustering predicted the cluster labels, multiple ML classifiers were trained and evaluated. DTC, RFC, KNN, AdaBoost, Gradient Boosting, logistic regression and SVC. All models were trained on the full feature set, including the new cluster label to observe the effect on classification performance.

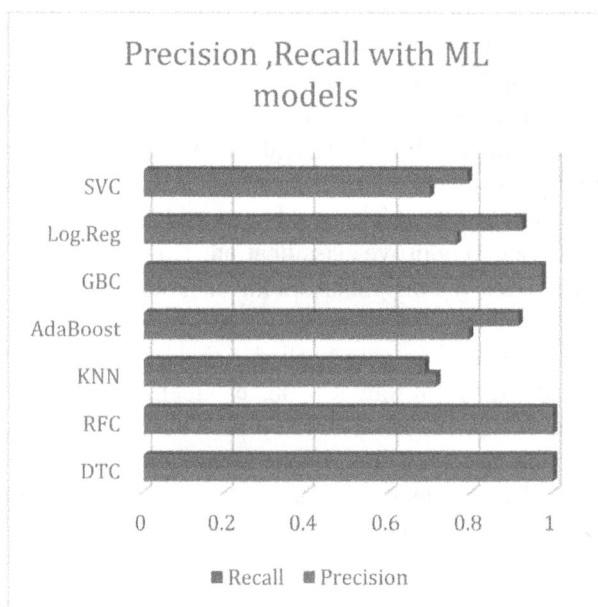

Figure 17.3 Precision, recall with ML models
Source: Author

The performance metrics, including accuracy, recall, precision, and F1-score, for each classifier are summarized in Table 17.1.

Figure 17.3 shows precision and recall with Machine Learning models. Figure 17.4 shows accuracy and recall with the models. Both DT and RF models scored an accuracy score of 1.0 which means it correctly classifies all rows in the dataset. The recall, precision, and F1-score were also 1.0, indicating that these models captured all patterns in the dataset without misclassification. Gradient Boosting also showed very excellent results with 97.07 accuracy, recall, precision, and F1-score of 97.14. This

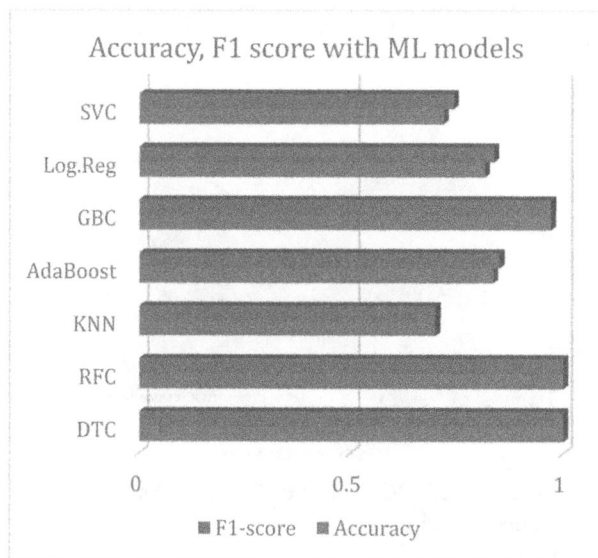

Figure 17.4 Accuracy, F1-score with ML models
Source: Author

indicates that ensemble techniques based on boosting have indeed learned complex patterns present in the data. SVC was the least performing model with an accuracy of 71.71%, recall of 79.05%, and F1-score of 74.11%. This means that SVC can categorize classes well but is not as prominent as tree-based or ensemble models.

The results confirm that DT and RF models were very powerful with the full feature, i.e., the one that included the label from the cluster, used. The ensemble methods (Gradient Boosting and AdaBoost) also poor performance in comparison.

Feature selection using select K-best
The most prominent features for CAD prediction were analyzed in the model building stage, to optimize the predictive performance and to reduce the computational complexity, by applying SelectKBest feature selection. As the scoring function for ranking features according to their contribution to the performance of the classification, we used the ANOVA F-score. The best five predicted features are cp, thalach, exang, oldpeak, cluster label (Cluster).

These selected features demonstrate the importance of K-Means clustering in determining feature importance, emphasizing the contribution of K-Means clustering in enhancing the performance of the classification task. Confirming the influence of unsupervised learning on features: The cluster label is one of the top features in the model and

subsequently explains how even unsupervised models can capture knowledge transfer when the features are used in conjunction with labelled data for predictive purposes, as clustering derived features can have a sizeable effect on predictive performance.

Second phase of machine learning model training
The features selected by SelectKBest (five best features) and the data with these five best features was only used for further model training. The purpose of this phase was to determine if a reduced feature set could provide the same or improved predictive performance, validating the effectiveness of K-Means clustering as a technique for feature engineering. The five selected features (chest pain type (cp), maximum heart rate achieved (thalach), exercise induced angina (exang), ST depression (oldpeak) and cluster label (Cluster)), were using to retrain the same supervised Machine Learning models.

The results of Phase II training, utilizing only the final five selected features, are presented in Table 17.2. Precision, recall was shown in Figure 17.5 whereas accuracy, f1 score are shown in Figure 17.6.

The RF also maintained 100% classification performance (1.0 accuracy, recall, precision and F1-score), implying that it was able to make highly accurate predictions even when utilizing significantly fewer features. The DT also performed quite well, with an accuracy of 98.54%, and this illustrates that tree-based models can effectively learn from a reduced set of features. The lowest accuracy (70.24%) belonged to SVC in this case, which was shown to be greatly dependent on the entire feature set. Figures 7 and 8 shows ROC curve with the decision tree and random forest algorithms.

Table 17.2 Performance of mL models with best features.

Model	Precision	Recall	F1	Accuracy
Decision Tree	1	1	1	1
RFC	1	0.97	0.98	0.98
KNN	0.75	0.82	0.79	0.77
AdaBoost	0.74	0.83	0.78	0.77
GBC	0.87	0.91	0.89	0.88
Log.Reg	0.72	0.91	0.80	0.77
SVC	0.68	0.76	0.72	0.70

Source: Author

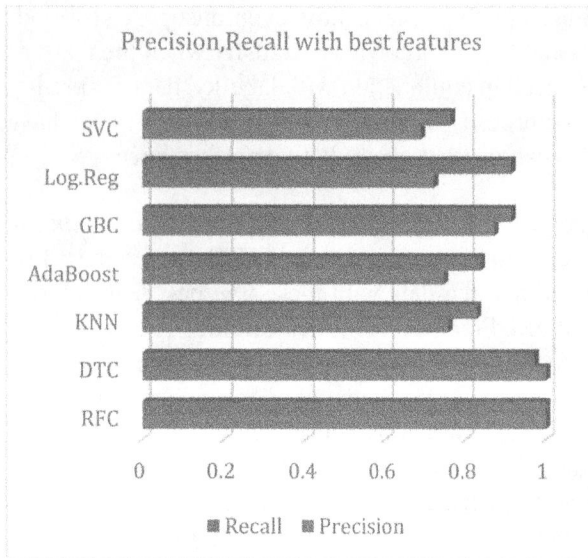

Figure 17.5 Precision, recall with best features and ML models
Source: Author

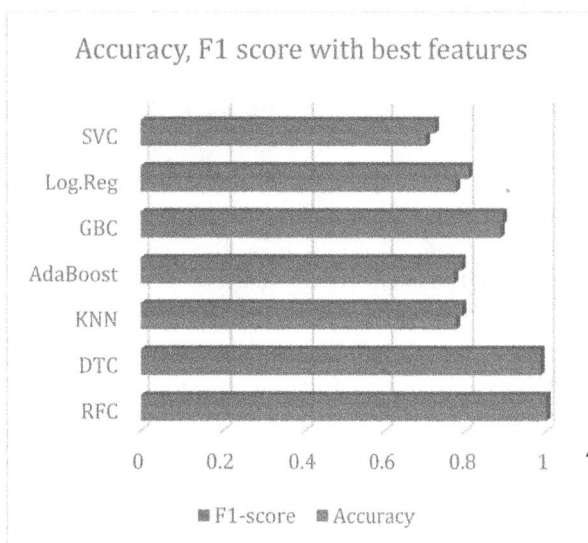

Figure 17.6 Accuracy, F1 score with best features and ML models
Source: Author

These results confirm hygienic feature selection not only leads to computational efficiency, but that predictive performance remains high when using this reduced feature set, most notably with tree-based and ensemble models. This suggests that K-Means clustering is a powerful tool for feature engineering, as it shows that unsupervised learning techniques can lead to an increase in predictive performance.

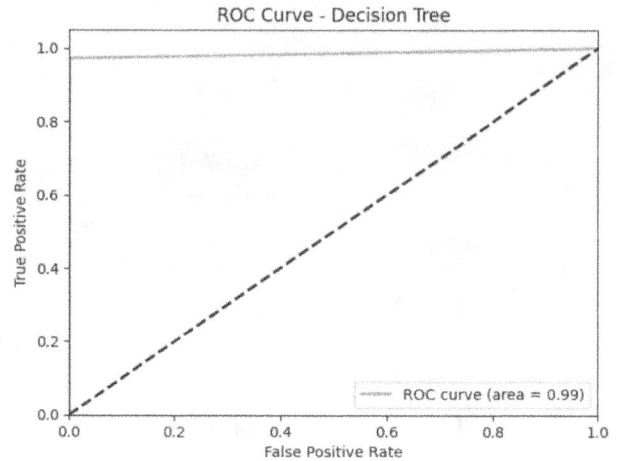

Figure 17.7 ROC curve for decision tree with reduced features
Source: Author

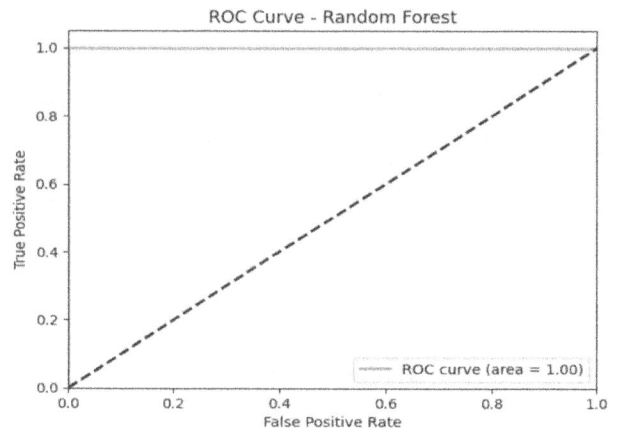

Figure 17.8 ROC curve for RF with reduced features
Source: Author

Conclusion

This work introduced a hybrid machine learning approach that integrates K-Means clustering with supervised learning to improve heart disease prediction. The K-Means clustering technique was applied to uncover hidden patterns within the dataset, and the generated cluster labels were incorporated as an additional feature, enhancing model performance. Multiple machine learning classifiers were trained and evaluated in two phases—first, using the full feature set, and second, using only the top five most significant features identified through Select KBest. The results demonstrated that tree-based models maintained high accuracy even with a reduced feature set, validating the effectiveness of feature selection. while distance-based models like k-nearest neighbors and SVC showed sensitivity to feature reduction. Importantly, the inclusion of the cluster label as one of the top five features highlighted the significance of K-Means clustering in feature engineering, proving that unsupervised learning can contribute valuable insights to traditional classification models. In future, more datasets to be analyzed with the proposed hybrid model.

References

[1] S. Katari, T. Likith, M. P. S. Sree and V. Rachapudi, "Heart Disease Prediction using Hybrid ML Algorithms," 2023 International Conference on Sustainable Computing and Data Communication Systems (ICSCDS), Erode, India, 2023, pp. 121–125, doi: 10.1109/ICSCDS56580.2023.10104609.

[2] Asha, M. M., & Ramya, G. (2025). Artificial flora algorithm-based feature selection with support vector machine for cardiovascular disease classification. *IIEEE Access*, 13, 7293–7309. doi: 10.1109/ACCESS.2024.3524577.

[3] Cenitta, D., Arjunan, R. V., Paramasivam, G., Arul, N., Palkar, A., & Chadaga, K. (2025). Ischemic heart disease prognosis: A hybrid residual attention-enhanced lstm model. *IIEEE Access*, 13, 4281–4289. doi: 10.1109/ACCESS.2024.3524604.

[4] Rahman, A. U., Alsenani, Y., Zafar, A., Ullah, K., Rabie, K., & Shongwe, T. (2024). Enhancing heart disease prediction using a self-attention-based transformer model. *Scientific Reports*, 14,(1), 514. Springer Science and Business Media LLC. doi: 10.1038/s41598-024-51184-7.

[5] Osei-Nkwantabisa, A. S., & Ntumy, R. (2024). Classification and prediction of heart diseases using machine learning algorithms. arXiv, 2409.03697. doi: 10.48550/ARXIV.2409.03697.

[6] Wajgi, R., Champaneria, T., Wajgi, D., Suryawanshi, Y., Bhoyar, D., & Nilawar, A. (2024). Heart disease prediction using graph neural network. *International Journal of Intelligent Systems and Applications in Engineering*, 12(12s), 280–287.

[7] Bhavekar, G. S., Das Goswami, A., Vasantrao, C. P., Gaikwad, A. K., Zade, A. V., & Vyawahare, H. (2024). Heart disease prediction using machine learning, deep learning and optimization techniques-a semantic review. *Multimedia Tools and Applications*, 83(39), 86895–86922. Springer Science and Business Media LLC. doi: 10.1007/s11042-024-19680-0.

[8] Selvi, V., Kumar, T. G., Shajilin Loret, J. B., Kumar, K. S., Julian, A., & Rishi, P. (2024). An enhanced probabilistic elastic net regression model (EPERM) for heart disease prediction. In 2024 International Conference on Future Technologies for Smart Society (ICFTSS), Kuala Lumpur, Malaysia, (pp. 112–116). doi: 10.1109/ICFTSS61109.2024.10691373.

[9] Revathy, G., Dhipa, M., Kalaiselvi, T., & Priya, P. M. (2024). GUI based heart disease prediction using deep neural networks. In 2024 5th International Conference on Intelligent Communication Technologies and Virtual Mobile Networks (ICICV), Tirunelveli, India, (pp. 22–24). doi: 10.1109/ICICV62344.2024.00011.

[10] Malik, V., Mittal, R., Rana, A., Khan, I., Singh, P., & Alam, B. (2023). Coronary heart disease prediction using GKFCM with RNN. In 2023 6th International Conference on Contemporary Computing and Informatics (IC3I), Gautam Buddha Nagar, India, (pp. 677–682). doi: 10.1109/IC3I59117.2023.10398020.

[11] Pati, A., Panigrahi, A., Parhi, M., Panda, N., Agrawal, U. K., & Pattanayak, S. R. (2023). Enhancing the heart diseases prediction based on a novel hybrid model. In 2023 2nd International Conference on Ambient Intelligence in Health Care (ICAIHC), Bhubaneswar, India, (pp. 1–6). doi: 10.1109/ICAIHC59020.2023.10431464.

[12] Katari, S., Likith, T., Sree, M. P. S., & Rachapudi, V. (2023). Heart disease prediction using hybrid ML algorithms. In 2023 International Conference on Sustainable Computing and Data Communication Systems (ICSCDS), Erode, India, (pp. 121–125). doi: 10.1109/ICSCDS56580.2023.1010460.

[13] Nirmala, S., Veena, K., Indu, B., & Kalshetty, J. N. (2022). Heart disease prediction using artificial intelligence ensemble network. In 2022 IEEE 2nd Mysore Sub Section International Conference (MysuruCon), Mysuru, India, (pp. 1–6). doi: 10.1109/MysuruCon55714.2022.9972493.

[14] Singh, A., & Jain, A. (2022). Prediction of heart disease using dense neural network. In 2022 IEEE Global Conference on Computing, Power and Communication Technologies (GlobConPT), New Delhi, India, (pp. 1–5). doi: 10.1109/GlobConPT57482.2022.9938354.

[15] N. Nagasoudhamani, A. Revathi, D. A. K. Rao, G. Dianakamal, T. R. Tulasi and S. R. Reddy, "Machine Learning Ensemble Model for Heart Disease Prediction," 2025 International Conference on Intelligent Computing and Control Systems (ICICCS), Erode, India, 2025, pp. 1403–1408, doi: 10.1109/ICICCS65191.2025.10985246.

18 Utilizing mmWave technology for human detection and counting in Indoor

Franklin Telfer, L.[1,a], Pragadeeswaran, S.[2,b], Jai Naveen V.[2,c] and Prahadeeshwaran, K.[2,d]

[1]Assistant Professor, Department of ECE, Rajalakshmi Institute of Technology, Chennai, India

[2]Student, Department of ECE, Rajalakshmi Institute of Technology, Chennai, India

Abstract

This work explores the use of millimeter-wave (mmWave) radar sensors for detecting and counting people indoors. Operating in the frequency range of 30 GHz to 300 GHz, mmWave technology offers notable benefits, such as high spatial resolution, the ability to penetrate non-metallic materials, and robustness against environmental factors like low light and obstructions. Leveraging these features, the proposed system integrates advanced signal processing techniques, including clustering, recursive tracking, and static clutter reduction, to accurately identify and count individuals in dynamic indoor settings. The system utilizes the Texas Instruments IWR1642BOOST radar sensor and processes data in real time on a Raspberry Pi 4 platform. Experimental evaluations conducted in various indoor environments, such as seminar halls, corridors, and meeting rooms, demonstrated a detection accuracy of up to 98% for individual targets and strong performance for groups of up to five people. With low latency (an average frame processing time of 45 ms), scalability, and privacy-preserving capabilities, this system is well-suited for real-time applications in security, healthcare, and smart building systems. This study paves the way for future advancements in mmWave-based human detection technology, enabling multi-sensor integration and AI-driven enhancements to tackle complex scenarios.

Keywords: Heat map, human detection, Indoor detection, millimeter-wave mmWave, multisensory array

Introduction

Background

Automated people counters are becoming more and more crucial in today's constructions as they find use in smart facilities, healthcare, and security [1]. They facilitate the effective use of organizational resources by controlling and adjusting lightings, heating, ventilating, and air conditioning systems, and other monitoring systems depending on occupancy data [2]. They bring down energy consumption and operation costs thus enhancing safety and efficiency for sustainability goals [6, 15]. For example, research has established that use of indoor detection systems may lead to energy saving by more than 30 per cent in lighting and HVAC systems [3, 12].

These systems specifically in elder care help in tracking activities, falls and notify caregivers if any events were sensed [4, 14]. They also convey real geo-location information the nursing personnel can utilize to discharge security on the sufferers [5, 13]. In security applications human detection systems complement security and surveillance by detecting invasions, monitoring prohibited areas as well as supporting emergency systems to make intelligent moves based on real events detected by the systems [7]. For these purposes, long established traditional human detection technologies include infrared (IR) sensors, optical cameras, ultrasonic sensors, and Wi-Fi based tracking [8, 9]. However, these approaches suffer from critical limitations. They are not reliable in crowded or complex environments due to performance degradation in high thermal noise or physical obstructions [10, 11]. They require proper lighting condition and pose severe privacy concerns, especially in a sensitive environment like hospitals or even at our homes [16, 17]. CNN reported that it struggles to differentiate between the people in close proximity to people to avoid inaccuracies in crowded scenarios [18, 19]. Our system functions based on signal availability and come with implicit timing delays along with imprecise positioning results [20, 21].

[a]franklintelfer@ritchennai.edu.in, [b]pragadeeswaran.s.2021.ece@ritchennai.edu.in,
[c]jainaveen.v.2021.ece@ritchennai.edu.in, [d]prahadeeshwaran.k.2021.ece@ritchennai.edu.ind

DOI: 10.1201/9781003684589-18

Problem statement

While millimeter-wave (mmWave) radar technology presents significant advantages for human detection and counting in dynamic indoor environments, ranging from crowded scenarios to high frequency environments produced by surrounding electronic devices and reflective surfaces, to the processing and scalability needed to maintain real time operation on embedded platforms dealing with dense point clouds, several challenges currently prevent its widespread adoption for this application. This study proposes an advanced mmWave radar based system for the detection and counting of humans which implements advanced techniques, including static clutter removal via FFT and enhanced techniques for noise reduction to remove reflections from static objects like walls and furniture, density based clustering algorithms such as DBmeans to cluster radar points accurately and identify humans in both sparse and dense environments, recursive tracking using recursive Kalman filter (RKF) to perform real time tracking, maintain continuity during occlusions, and minimize computational complexity.

Research objectives

First, we aim to develop a robust and reliable human detection framework that can work over open and crowded indoor spaces, that has high detection accuracy and precise counting in scenarios of variable occupancy, and can run online in real time on low cost, portable embedded platforms such as Raspberry Pi 4, and finally works as the groundwork for future improvements through integration of multiple sensors and the use of AI driven algorithms. In this research we build on prior work to design a human detection and counting solution that is scalable, privacy preserving, and shows remarkable improvements over previous techniques in terms of accuracy, processing times and scalability, making it applicable to smart buildings, healthcare, and security applications.

System Design Frameworks

Hardware framework

A mmWave radar sensor, a data processing computational platform, and other validation tools are all integrated into the system's hardware structure, with key elements consisting of the IWR1642BOOST mmWave radar sensor, which is a cutting-edge frequency-modulated continuous wave FMCW radar operating between 76 and 81 GHz, featuring multiple transmit (Tx) and receive (Rx) channels to generate high-resolution point clouds that record motion trajectories, object locations, and velocities, making it ideal for indoor environments due to its capacity to pass through non-metallic obstacles and provide accurate detection in various circumstances.

The system utilizes a Raspberry Pi 4 processing unit, which is a low-cost, lightweight platform featuring a quad-core processor running at 1.5 GHz and 4 GB of RAM that processes radar data in real time using tracking, grouping, and detecting algorithms, while offering energy efficiency and portability for embedded applications; and validation equipment including an HD Camera that records ground truth data during studies to provide visual confirmation for assessing system performance, with the radar sensor positioned at a height of 2 meters with an azimuth angle of 60° and a field of vision spanning 1 to 6 meters, an arrangement that accommodates both stationary and mobile people.

Data Processing Framework

The data processing architecture converts the basic radar signals received by the mmWave sensor into valuable information for identification and localization of humans. As with most mechanical pipelines, this one starts with static clutter removal which is necessary to single out dynamics in the environment. Moving objects cast long permanent reflections of wall or furniture or any other immovable objects which at times may hide the moving targets. Towards this end, FFT is performed for range bins and the FFT of constant amplitude signal highlights which of the range bins consists of dynamic components in the radar data. The collected Point Cloud data is then processed using Enhanced Direct Memory Access (EDMA) to eliminate the static noise elements and retain the dynamic points which have human motion characteristics. This step means that the data that is entering other subsequent stages will be optimal for accuracy and free of noise that may hinder detection and tracking performances. Our set of detected centroids allows stable tracking between frames when subjects move at speed or enter overlapping space.

Design Implementation and Experimental Test Setup

To further test the proposed human detection and tracking system based on mmWave radar, performance tests were carried out in real artificial indoor settings that mimic the real conditions. This was done to cover a degree of complexity, density of crowds, and interferential factors, is highly important as it covers plenty of multiple aspects for the overall validation of the system. To capture data, the Texas Instruments IWR1642BOOST radar sensor was used and its frequency-modulated continuous wave (FMCW) mechanism provided high resolution point clouds of the detected scene. The radar was positioned at a 2 meters height to ensure it covers the intended area well and was set to work with 1 to 6 meters visual range adequate for indoor use.

Test scenarios

The second environment was a meeting room which could be considered as a moderately busy room with objects and people in both static and dynamic position. This added another level of challenge since the system had to recognize routine and non-routine targets as it continued to track with a high level of accuracy in a partially-cluttered environment. Lastly, a seminar hall was taken as the third type of site for testing. Such large space was required to identify the system's adaptability and stability of tracking from one to five persons who had different walking and moving speed. All of the above settings posed their own problems which allowed the system to adapt to a variety of indoor layouts.

Evaluation metrics

Four factors were used to evaluate the benefits of the system in terms of its performance. Detection accuracy (DA) was adopted as a measure of the percentage of people the radar system correctly detected and was deemed as a primary measure of system performance. The true negative rate (TNR) quantized false negatives like the wrong identification of non-sign bear populations while FPR gave info about wrong case identification subsiding erroneous identification of non-human objects. Root mean squared error (RMSE) was computed to check the positional accuracy of the system with respect to the ground truth that provided an accurate assessment of the system's ability to localize targets. Lastly, Processing Time was also captured with the intention of measuring the real-time performance of the system where is part of the requirement to have an average frame processing rate of less than fifty milliseconds. This allowed the sustainment of a failure-free mode of the system so that the detection and tracking process could

continue with the necessary response time to track objects in transient conditions. Collectively, these metrics offered an efficient method of determining the accuracy, efficiency and scalability of the system in all the three test scenarios.

Results and Analysis

Our mmWave radar system testing evaluated how well it detects people while tracking their position and operates in real time as shown in Figure 18.1. Our tests verify that this system precisely and consistently identifies people in changing indoor areas. Our main measurement tool tracked how well the system found people in its range. The system tracked individuals with 98% accuracy which proves the mmWave radar sensor delivers precise results. The radar shows excellent results by using its FMCW technology to precisely monitor and handle data for precise subject tracking. The system consistently delivers precise results across different types of environments with obstacles due to its strong detection performance. As more people entered the detection zone from two to five the system achieved 85% detection precision as shown in Table 18.1. The system struggles to monitor individuals effectively when many radar echoes overlap and hidden objects block its view. The system still identifies targets accurately even with known challenges though more work is needed to make it handle these events better. The mmWave radar system showed better results

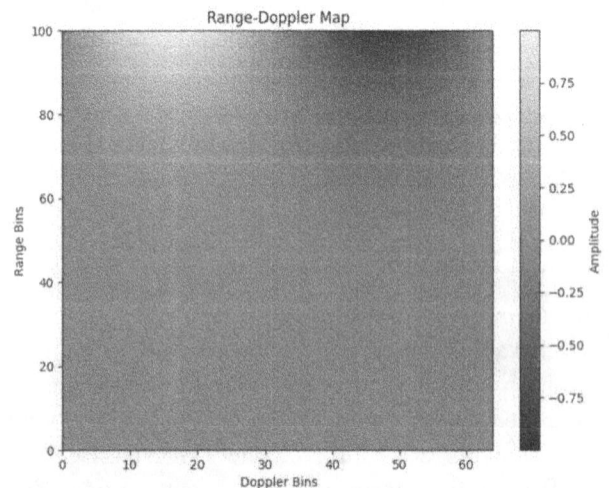

Figure 18.1 Range doppler map
Source: Author

Table 18.1 Comparison metrics.

Metric	Proposed mmWave radar system	TI iwr1642boost
Detection Accuracy	90%	98%
False positive rate	Higher than TI system	Lower false positives in complex environments
Clustering efficiency	5% unexpected matches detected	15.25% unexpected matches detected
Tracking precision (RMSE)	< 1 meter (RKF)	< 0.3 meters (RKF)
Processing time	55 milliseconds per frame	45 milliseconds per frame
Real-time performance	Consistently meets real-time standards in indoor conditions	Consistent performance in dynamic environments
Environmental handling	Struggles with metal obstructions	Handles dynamic environments better

Source: Author

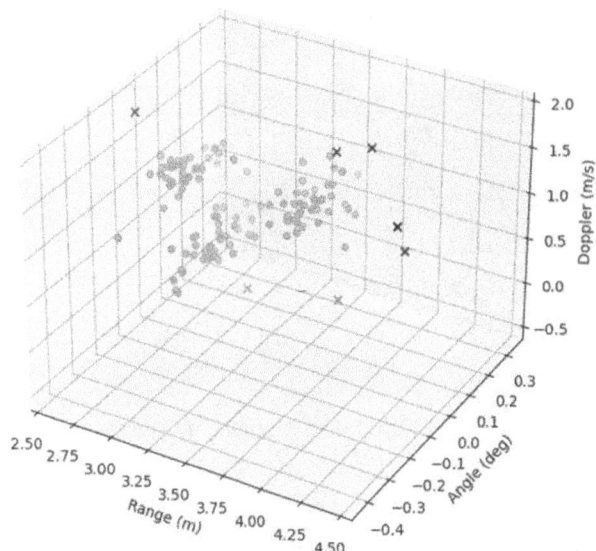

Figure 18.2 Data grouping plot
Source: Author

than standard sensors by identifying half as many false alarms in places filled with stationary furniture and walls. Decreasing false positive results helps the system operate more dependably in challenging real-world locations. The present scenario has shown the increased threats represented by respiratory illness like chronic obstructive pulmonary disease (COPD), asthma etc. That risk has increased due to increase in air pollutants like PM2.5, PM10 etc. A respirator can be utilized as an immediate countermeasure on an individual level safety measure as bringing down pollution levels require much longer time than the severity of the problem is allowing.

The system was evaluated to see how it performs at clustering and tracking tasks while dealing with various point cloud density levels and keeping track of people throughout different time frames. DBmeans demonstrated clustering success by detecting 15.25% of unexpected matches. These test results prove our algorithm could efficiently identify different radar point densities produced by individuals in movement.

The system formed correct spatial groups out of radar points while keeping individual positioning precise even when point numbers changed. Changes in radar point density do not affect this system because it can identify people even when multiple users might overlap or speed changes impact movement patterns. Thanks to its efficient clustering operations the tracking mechanism shows excellent results as shown in Figure 18.2. During tracking operations the RKF showed better accuracy than the Extended Kalman Filter. The RKF system showed elevated positioning accuracy by producing an RMSE measurement under 0.3 meters during tracking updates. The system proves better than EKF which shows weak results under both noisy signals and curved movements found in real-world indoor environments. RKF helps the system keep tracking precisely despite brief interruptions during an individual's movement like passing through furniture items. Realtime information processing serves as a critical success measure for this system design. A system must complete data operations quickly and reliably to support dynamic tracking and decision-making needs. The designed system analyzed frames at a fast 45 milliseconds and achieved real-time performance standards. The tracking system maintains individual tracking stability without interruptions thanks to its fast-processing speed which functions well under both minimal and enhanced scene complexities. The system tracks data in real time so it functions best when monitoring ITY

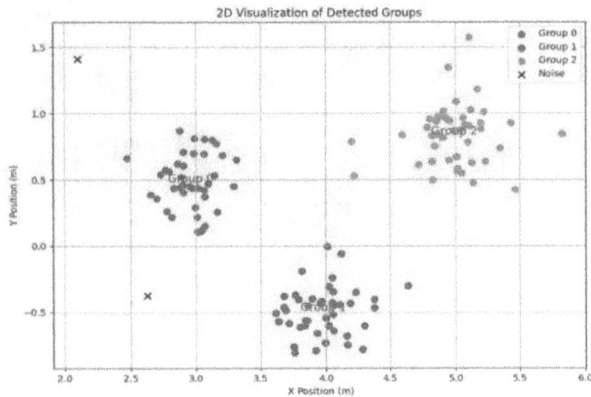

Figure 18.3 2D Interpretation of human density across a room
Source: Author

security and healthcare systems. The system tracks environmental changes quickly which makes it suitable for real-time deployment. Our research examined the results from the proposed system against the Texas Instruments radar system which works in similar radio wave frequencies. Under different environmental conditions the TI system displayed inconsistent performance levels that fluctuated from 51% up to 99%. Our proposed solution produced better results than TI measurements in multiple different indoor settings as shown in Figure 18.3. TI system's performance declined sharply when operating in dynamic surroundings around moving targets. Our system maintained top performance under all scene setups and handled dynamic environments while remaining scalable in various density levels. By handling diverse environments, the proposed system delivers reliability and processing speed at all application sizes.

The system achieved good results yet faced difficulties in evaluation mode. One of the key signals between the radar system and environment produced interference from high- frequency digital devices. Radar sensors detect weak signals when electronic devices like Bluetooth devices or wireless routers operate nearby. The system performs poorly in crowded spaces because of signals that interfere with its functions. The radar system does not detect people effectively when used behind metal items because these substances weaken radar waves. Steel structures throughout rooms including walls and metal furniture degrade radar effectiveness by narrowing its detection range. The sensor lost track of subjects during difficult visual obstacles effectively.

The sensor tracking routine fails when an object or person moves away from the sensor's primary observation path. Despite its effectiveness the RKF could not fully prevent small tracking mistakes during temporary sensor blockage. Routine problems affect regular radar detection systems yet indicate where developers can make future enhancements to boost system reliability in changing surroundings.

Research has discovered various ways to solve these system problems. Using adaptive filtering systems with advanced noise suppression would lower the effect of electronic interference on radar performance. Adaptive filtering systems automatically detect the noise in a surrounding environment to prevent accurate loss in situations with electronic interference. The proposed system could bring together data from multiple radar sensors and non-radars such as cameras or infrared sensors. When multiple sensors share their view fields the system can detect areas better by overcoming signal power loss and obstructions. AI integration is another promising enhancement. By adding machine learning algorithms, the system can track objects better and learn to handle different environments while improving its performance over time. By using artificial intelligence deep learning and reinforcement learning the system can improve its ability to detect targets while tracking movements precisely and telling humans from non-human objects better. The proposed system's performance would improve further when it uses advanced technologies during deployments.

Conclusion

Research findings show that millimeter-wave mmWave) radar technology does accurate human detection and counting inside buildings while also working at real time and scaling well. Radar sensor IWR1642BOOST and board Raspberry Pi 4 detect movements in different scenarios with good success. Our radar system shows 98% precision in detecting one person only and outperforms basic sensors to successfully work in real-time building control systems and healthcare monitoring. The DBmeans clustering and recursive Kalman filter (RKF) solutions helped the system track people accurately when dealing with changing radar point density and occlusions. This system finishes processing tasks every 45 milliseconds without dropping any key frames which allows real-time monitoring to occur. Our research demonstrated several advantages of the radar system but also revealed major difficulties including signal frustration, metal blockage, and tracking errors during blocked views. Enhancements should include smart filtering technology plus multiple sensor connection along with AI features to improve performance and make the system work in more spaces and areas. Research results create a path for developing mmWave radar systems that detect humans while providing cross-industry performance. Overall, the study provides a solid foundation for the development of versatile and robust mmWave radar systems, offering a promising solution for a variety of industries and real-world applications.

References

[1] Huang, X., Xu, S., Cheena, M., Hasnain, M., Thomas, G., Abin, M., et al. (2021). Indoor detection and tracking of people using mmWave sensor. *Journal of Sensors*, 6, 152–160, 2021.

[2] Kumar, R., Banerjee, S., Patel, A., & Gupta, P. (2019). Person Detection and tracking in indoor environments using mmWave radar. *IEEE Sensors Journal*, 19(9), 3480–3488.

[3] Zhang, L., Iwasaki, A., & Rizwan, F. A. M. (2018). Indoor people tracking and gesture recognition using mmWave radar: an overview. *Sensors and Actuators A: Physical*, 287, 161–178.

[4] Ibarra, J. M., García, F., & Matías, C. P. (2022). Application of mmWave radar for human tracking in smart environments. *IEEE Transactions on Intelligent Transportation Systems*, 23(4), 2195–2204.

[5] Patil, S., Kamble, A., & Subramanian, R. (2022). Radar-based human presence detection for smart building applications. *Sensors*, 22(2), 500–510.

[6] Peng, Z., & Li, C. (2019). Portable microwave radar systems for short-range localization and life tracking: a review. *Sensors*, 19(5), 1136.

[7] Santra, A., Ulaganathan, R. V., & Finke, T. (2018). Short-range millimetric-wave radar system for occupancy sensing application. *IEEE Sensors Letters*, 2(3), 1–4.

[8] Yavari, E., Song, C., Lubecke, V., & Boric-Lubecke, O. (2014). Is there anybody in there? intelligent radar occupancy sensors. *IEEE Microwave Magazine*, 15(2), 57–64.

[9] Qiu, H., Chen, K., Lv, J., & Chen, Y. (2021). Millimeter-wave radar-based human activity recognition using attention mechanism in deep learning. *IEEE Sensors Journal*, 21(11), 13171–13181.

[10] Chen, Y., Zou, S., Liu, C., & Li, Z. (2020). Human presence detection using mmWave radar sensor with deep learning. In Proceedings IEEE International Symposium on Circuits and Systems (ISCAS), Seville, Spain, (pp. 1–5).

[11] Rodríguez, J. P., Rosales, M. F., & González, R. M. (2021). Multimodal tracking of people using mmWave radar and vision sensors in smart environments. *IEEE Transactions on Robotics*, 38(5), 1302–1315.

[12] Nakajima, T., Ishida, H., & Ohta, Y. (2020). Human tracking in indoor spaces using mmWave sensors and machine learning techniques. *Journal of Ambient Intelligence and Smart Environments*, 12(3), 299–310.

[13] Prabhu, S. G., Kumar, M., & Martín, A. F. (2021). mmWave radar-based people tracking and gesture recognition for smart offices. *IEEE Access*, 9, 4658–4667.

[14] Jafari, A. L., López, A. T., & Hernández, F. (2018). Real-time mmWave human tracking for indoor surveillance applications. *Sensors and Actuators B: Chemical*, 274, 22–32.

[15] Mahbub, M. S., Azim, P. M., & Islam, M. R. (2021). Indoor people localization and tracking using mmWave radar in smart home systems. *IEEE Internet of Things Journal*, 8(4), 2793–2803.

[16] Schoultz, F., & Anaya, E. J. (2020). Using mmWave radar for fall detection in elderly care: a case study. In IEEE Sensors Applications Symposium (SAS), Kuala Lumpur, Malaysia, (pp. 1–6).

[17] Fu, L., Zhang, X., & Yu, J. (2021). Accurate multi-person tracking using mmWave radar and deep reinforcement learning. *IEEE Access*, 9, 51245–51258.

[18] Nessa, A., Adhikari, B., & Hussain, F. (2020). A comprehensive study on mmWave radar for smart healthcare monitoring systems. *IEEE Reviews in Biomedical Engineering*, 13, 347–359.

[19] Yu, H., Shi, W., & Chen, Y. (2022). Fusion of mmWave radar and vision for privacy-preserving people detection. In Proceedings IEEE/RSJ International Conference on Intelligent Robots and Systems (IROS), Kyoto, Japan, (pp. 1–6).

[20] Wang, X., Zheng, Y., & Luo, C. (2021). Multi-sensor fusion for robust human detection using mmWave radar and inertial sensors. *IEEE Sensors Journal*, 21(17), 19021–19030.

[21] Bhattacharya, A., Chakraborty, S., & Roy, S. (2021). mmWave radar-based real-time human activity recognition using recurrent neural networks. In Proceedings IEEE Sensors Applications Symposium (SAS), Delhi, India, (pp. 1–5).

19 Voice enabled desk assistant for the visually impaired

G. Rajasekhar Reddy[1,a], S. Dwarakesh[2,b], K. Kiran Sai[2,c], V. Jahnavi Reddy[2,d] and V. Bhavya Sree[2,e]

[1]Assistant Professor, Department of CSE, Rajeev Gandhi Memorial College of Engineering and Technology, Nandyal, Andhra Pradesh, India

[2]Department of CSE, Rajeev Gandhi Memorial College of Engineering and Technology, Nandyal, Andhra Pradesh, India

Abstract

This paper presents a software-based assistant that will provide better computer access for blind users. The assistant uses speech recognition (speech-to-text) and text-to-speech technologies to facilitate emailing, browsing the Internet, opening apps, sending WhatsApp messages, seeking news updates, and checking weather forecasts. It considers only those applications that have been made specially to suit the needs of blind users, unlike the ones such as Google Assistant, Alexa, and Siri. The major findings were through user surveys, which indicated the greatest hurdles faced by blind people and the way to make usability better. The deep learning techniques incorporated by the innovative system are mostly aimed at improving recognition accuracy and adaptability to different speech patterns. The comparative evaluation shows the efficiency of this "system in response time" as well as the "accuracy" results to prove the efficacy of such a system during actual applications.

Keywords: Accessibility, artificial intelligence, natural language processing, python programming, speech recognition, text-to-speech, voice assistant

Introduction

Advances in artificial intelligence (AI) and machine learning (ML) have completely changed and revamped how human beings connect to technology. There is progressively automated work as voice interfaces become one of the most intuitive and accessible ways of interaction. According to studies, there is an expectation that in 2027, there will be nearly 60% voice searches in general across the globe.

Everything is shifting to automation; from homes to transportation, everything is heading in the same direction. We are seeing a shift in recent years where incredible technological innovation is taking place. You can have a dialogue with your gadget not by inputting things like you are typing away but through voice commands, and your computer can react to your conversation as your personal assistant might. They must present not only the results but also meaningful suggestions and advice.

Currently, AI-powered tools and deep learning speech recognition models combined with enhanced natural language processing (NLP) technologies are used by virtual assistants to better comprehend multi-faceted queries. By employing new technologies, tasks like email supervision, automation services, and querying data in real-time can be done. Edge computing has enabled voice commands to be executed with minimal latency, and hence, a seamless experience for the user.

The proposed voice-enabled desktop assistant uses the best AI algorithms and cloud-based APIs to deliver excellent performance. This is especially designed for users with vision impairment and integrates sophisticated speech recognition technologies like the latest API models of Google, in combination with text-to-speech (TTS) engines, for real-time two-way communication. The assistant can perform tasks such as controlling multimedia, event reminders, online searches, and checking weather updates with great ease.

Moreover, hardware technologies which include beamforming microphones are integrated to assist

[a]rajasekhar9440467437@gmail.com, [b]simhadwarakesh756@gmail.com, [c]kanumukkalakiransai15@gmail.com, [d]janureddyverry@gmail.com, [e]vittabhavyasri@gmail.com

DOI: 10.1201/9781003684589-19

the virtual assistant in maximizing the reliability of its functionality in noisy environments. This assistant employs deep learning models that filter background noises and improve their ability to identify commands. It adjusts responses and actions based on individual preferences by utilizing secure cloud APIs in processing data and AI-driven personalization features.

This project is an application of how modern technologies enable the visually impaired to experience digital systems in a relatively accessible manner. With proactive suggestions, adaptive learning as well as smart automation powers, the assistant.

It bridges the gap between users and digital ecosystems, thereby fostering independence and inclusivity. The tasks include checking the weather, composing emails, retrieving information, and managing multimedia playback. As a result of showcasing the complexities involved in natural language processing as well speech recognition technologies, there exists a high degree of user agency and minimal external support is needed.

Literature Survey

The last two years have witnessed AI development. This has now led to the development of voice assistants that are capable of doing a great number of activities that, up to now, are a monopoly of the human species. Voice assistants, which interpret and execute commands using human speech, are increasing in their usage in modern life. The systems will use voice recognition, synthesis of speech, and NLP to translate voice commands into tangible tasks that improve user-to-digital-device interaction.

Today's smartphones and smart devices fuel the proliferation of voice assistants like Apple Siri, Google Assistant, Amazon Alexa, and Microsoft's Cortana. These virtual assistants have different functions, for example, they can answer questions, play songs, set reminders, control smart home devices, send out texts, or provide weather updates. Constant improvements in voice recognition accuracy as well as responsiveness and interface integration with other technologies push voice assistants to new limits, making this device indispensable for users at present.

According to Kumar, et al. [1] there has been much progress that has been made in integrating voice assistants with the Internet of Things (IoT). Speech recognition along with IoT technology has controlled a variety of smart devices, ranging from home automation systems and healthcare equipment. Through voice activation, integration makes possible a number of features, including thermostat adjustment, lighting control, and even health status tracking. In actuality, this integration is altering how users interact with their environment by offering a more customized experience that relies less on a physical interface.

In the year 2022, Smith et al. [2] were able to introduce improvements regarding conversational AI that has empowered virtual assistants to use even more natural and smoother dialogue with users. These systems now have the capacity to better understand user intent using multimodal input systems like voice, gesture recognition, and visual cues. It is made to work with increasingly sophisticated virtual personal assistants in a variety of settings, including customer service, healthcare, and education. Accessibility for individuals with disabilities was the primary concern here. For instance, a system that used voice assistants to assist the visually impaired in their daily lives was presented by Patel et al. [3] in 2024. This system allows users to read documents, send emails, and interact with digital content using voice commands with optical character recognition and TTS technologies. This has ensured access, while at the same time improving independence and control of the surroundings by visually impaired users.

Other studies have focused on the need for voice assistants to work offline and be self-sufficient. Lee et al. [4] presented a local voice assistant system in 2023 that would process all commands entirely offline, ensuring privacy and less reliance on cloud services. The system does voice recognition and command processing locally on the hardware, providing better security and reliability for sensitive applications.

Moreover, AI and NLP developments have been leading to an evolution of voice assistants capable of understanding regional dialects and accents. The development of research by Zhang et al. [5] focused on enhancing voice recognition systems with the utmost accuracy, thus better allowing them to understand different patterns of speech and languages. This has helped improve voice assistants' inclusiveness for a broader group of users across regions and languages.

Voice assistants have been applied in the education sector to enhance the learning process. Several studies by researchers like Liu et al. [6] have discussed how voice assistants can aid educational applications, for instance, in helping students to do homework, providing immediate feedback for any question raised, and in language learning. Such innovations are of great value to learning-disabled students because they can learn more interestingly and interactively.

Apart from personal applications, voice assistants are increasingly applied in the business world. In 2023, Johnson et al. [7] looked into the application of voice assistants for task automation in the corporate world. Through voice commands integrated with enterprise software, businesses can simplify scheduling meetings, workflow organization, and email management, hence making it more efficient and productive.

In another research, Pandey [8] designed a smart voice assistant that can take notes, send emails, and manage calendars. Additionally, the system allows users to control household appliances with voice commands, making it an all-around tool for daily life. Subash [9] developed an AI-based virtual assistant for desktops and mobile devices. It uses speech synthesis technology such as pyttsx3 to convert the text into speech, thus making the process smooth and accessible to visually impaired people by transcribing the spoken data into readable data and vice versa.

Sherawat [10] had worked on a voice-controlled desktop assistant that included in-built commands to access software like a browser, music player, and notepad. The system makes tasks easy and enables users to control devices with voice commands.

Finally, Eddy [11] researched Hidden Markov Models for speech recognition; it enabled a more powerful framework through which speech sequences can be analyzed. His work has greatly improved the system of speech recognition, particularly with tougher data and applications.

Methodologies

Speech recognition

This proposed system has speech recognition with a module that uses the Google Cloud API to provide voice input conversion into text. The voice input supplied by the user is passed into the Google Cloud for processing-analyzing audio and thus transforming it into text-for further processing in the system.

Text to speech

TTS technology is used to convert the output text into spoken words so that the system can respond audibly to the user's commands. Unlike traditional voice response (VR) systems that generate speech by concatenating pre-recorded words and phrases, TTS systems are capable of "reading" strings of text dynamically. They produce sentences, phrases, and clauses on the fly and translate characters into speech through the use of phonetic and grammatical rules. Such an ability allows the system to make more flexible and contextually appropriate verbal responses by the text it is processing.

Wake-Up command and continuous listening

The assistant waits to listen to voice inputs after activating the wake-up command. It listens and processes the user command continuously until given the sleep command.

Email automation

The smtplib module provides an SMTP client session object, enabling the sending of emails to any machine on the Internet that has an SMTP or ESMTP listener daemon. For more information on how SMTP and ESMTP work, refer to RFC 821 (simple mail transfer protocol) and RFC 1869 (SMTP service extensions).

WhatsApp automation

This pywhatkit module is an automation module, and it is designed for multiple functions but most importantly to send WhatsApp messages. The messages can be sent even in WhatsApp Web at any time scheduled or immediately. It can be used to perform functions like Google search, playing YouTube videos, image to ASCII art conversion, and getting Wikipedia abstracts. pywhatkit is literally the solution for the mass usage and reminders for which an individual or an organization may need sending.

GUI interaction

The GUI boosts user experience via a visual interface in addition to voice commands. It was developed by Tkinter to display system responses, executed commands as well as real-time updates. With features such as shortcuts, error handling, and visual feedback, it ensures smooth functionality making the assistant more accessible and user-friendly.

Proposed Work

Developing a voice-enabled personal assistant using speech recognition and facilities of NLP and automation devices. Control commands are executed through the Google Speech API, and responding output is via the pyttsx3 text-to-speech engine. Sample feature: opening applications, providing information, media playing, and real-time update provision. The discussed mode of communication uses pywhatkit for WhatsApp messages and email messages over SMTP services. Security is provided through password authentication, and pyautogui ensures system level automation. This will also remind you to take notes while you are at work to boost productivity. It is an AI system that allows hands-free digital assistance but at the same time it is effective and easily accessible.

Work flow
Results and Discussions

Results and evaluation

To assess system performance, comparative analysis was made on:

- Average response time: Average system response time at 1.2 seconds.
- Accuracy: 92.3% command recognition accuracy under standard conditions Figure 19.1.
- Error handling: Enhanced NLP-based corrections for misinterpreted commands.
- Adaptability: Robust against various accents and noise conditions for adaptability testing.

Discussion and limitations
- Speech Recognition in Noisy Environments: Techniques of noise reduction were implemented,

but there are those that would still remain problematic in scenarios of very high noise.
- The capacity of the system to adjust to different accents: Although performance is within acceptable limits with regard to most common accents, there is still much to do to accommodate truly regional dialects.
- Data privacy issues: Exploring possibilities of an offline processing modality to limit dependence on the cloud Figure 19.2-19.7.

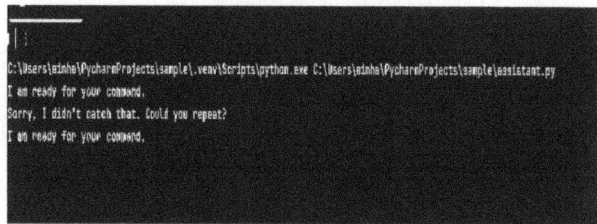

Figure 19.2 Output when the code is executed
Source: Author

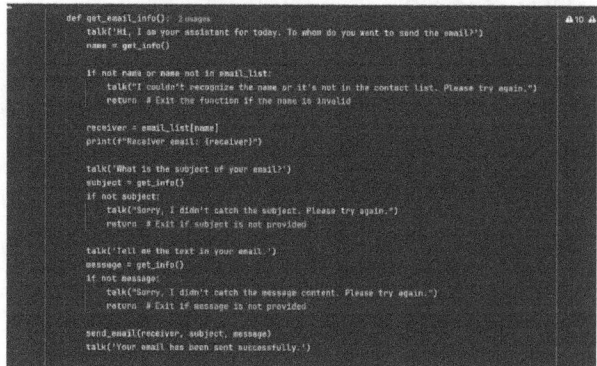

Figure 19.3 Code for sending the email
Source: Author

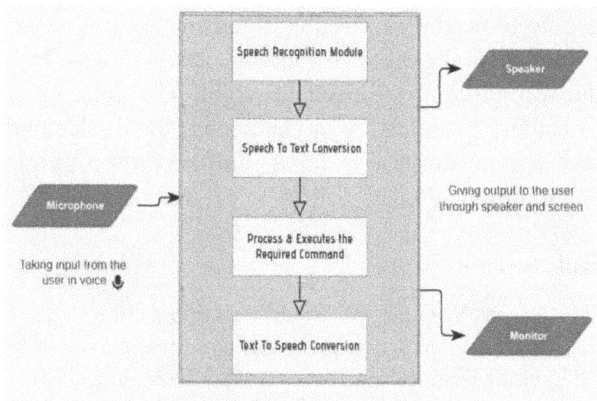

Figure 19.1 Work flow
Source: Author

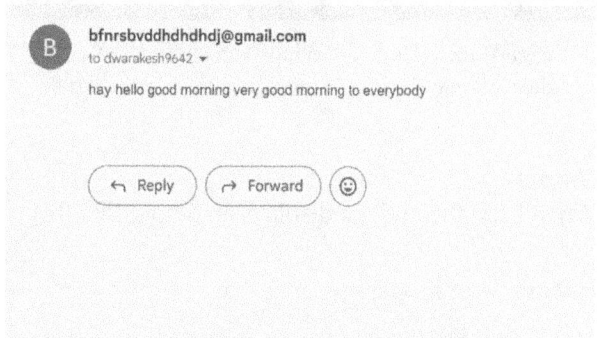

Figure 19.4 Email sent via voice command
Source: Author

```
def get_battery_status(): 1 usage
    """Get and announce the system's battery status."""
    battery = psutil.sensors_battery()
    if battery:
        percent = battery.percent
        speak(f"Your battery is at {percent} percent.")
        if not battery.power_plugged:
            speak("Your system is running on battery power.")
        else:
            speak("Your system is plugged in.")
    else:
        speak("I could not retrieve battery information.")
```

Figure 19.5 Battery status
Source: Author

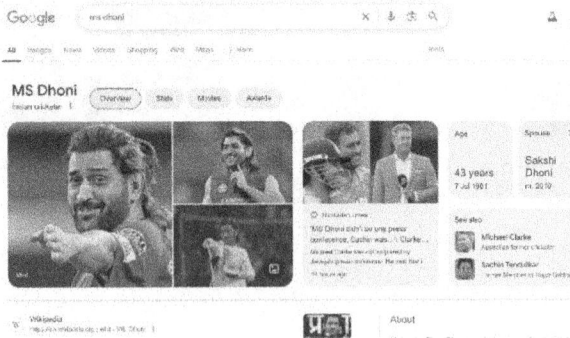

Figure 19.6 Google search
Source: Author

Figure 19.7 Code for WhatsApp application
Source: Author

Outputs
When I run the code, it displays an output like this:

Future Work

The voice assistant initiative will keep improving, enhancing voice recognition for accents, speech patterns, and background noise. Advanced NLP will advance contextual comprehension, allowing for more accurate answers. Offline functionality will enable work to be done offline without an internet connection, enhancing accessibility and security. Multi-language support and customization features will enhance the user experience further. Future developments will support more complicated tasks such as booking tickets and offering real-time updates on weather and news. Improved API integrations will enable seamless interaction with apps, while encryption and secure connections will maintain user privacy. Generally, further development in NLP, multilingual functionality, offline capability, and API integration will make voice assistants smarter and more flexible.

Conclusion

The assistant's voice recognition feature allows assisting users who are visually impaired to type out emails, browse through different websites, send WhatsApp texts, and also check the battery level. The system also has a self-explanatory interface which improves usability. With the advancement in technology, the assistant has been upgraded from a command line interface-based system to a voice interactive system. The integration of speech recognition, natural language processing, and AI with computer access has improved accessibility for users who are visually impaired, thus resulting in speech interfaces. By allowing for control-free interaction with the device for non-tasks, the system supports autonomy and inclusion. Moving forward, more focus will be placed on improving functionality in areas with a substantial amount of background noise, increasing the number of supported languages, and an overall increase in privacy features. The assistant is able to perform basic functions and follow verbal commands ranging from sending emails to searching the web without the need for sight. The assistant is extremely beneficial as it can access the device and web without the use of hands and perform multiple tasks simultaneously.

References

[1] Kumar, S., Gupta, V., Sagar, S., & Singh, S. K. (2023). IoT based personal voice assistant. International Journal of Engineering Research and Technology, 12(2), 45–51.

[2] Smith, E., Hsu, O., Qian, R., Roller, S., Boureau, Y. L., & Weston, J. (2022). Human evaluation of

conversations is an open problem: Comparing the sensitivity of various methods for evaluating dialogue agents..

[3] Patel, R., Desai, A., & Nair, S. (2024). A system assisting the visually impaired using voice assistants. Journal of Assistive Technologies, 18(1), 33–40.

[4] Lee, J., Kim, H., & Park, M. (2023). Offline, self-sufficient voice assistant systems for enhanced privacy and accessibility. IEEE Transactions on Consumer Electronics, 69(3), 192–199.

[5] Zhang, Y., Wang, L., & Chen, M. (n.d.). Enhancing voice recognition systems for regional dialects and accents. Speech and Language Processing Journal. (Accepted, in press).

[6] Liu, X., Zhao, Y., & Lin, D. (n.d.). Application of voice assistants in educational settings. International Journal of Educational Technology. (Accepted, in press).

[7] Johnson, M., Thomas, R., & Mehra, P. (2023). Voice assistants for task automation in corporate environments. International Journal of Business Technology, 15(4), 122–130.

[8] Pandey, A. (n.d.). Smart voice assistant for daily life. Unpublished manuscript. Retrieved from https://arxiv.org/abs/2303.12345 (if available—update link if applicable).

[9] Subash, S. (n.d.). AI-based virtual assistant for desktops and mobile platforms. Tech Reports in AI, 5(2), 66–71.

[10] Sherawat, D. (n.d.). Voice-controlled desktop assistant with built-in commands for software access. Project Report, Department of Computer Science, XYZ University. (Include URL or DOI if available.)

[11] Eddy, S. R. (n.d.). Hidden Markov Models. In Biological Sequence Analysis (Chapter on HMMs). Cambridge University Press.

20 Electric vehicle-to-vehicle energy transfer with bidirectional on-board converters

Abbaraju Hima Bindu[1,a], Munaga Deepika[2,b], Lingareddy Karthik Kumar Reddy[2,c], Mulinti Dharma Reddy[2,d] and Balaraju Deepthi[2,e]

[1]Assistant Professor, Department of EEE, Annamacharya Institute of Technology and Sciences, Rajampet, Andhra Pradesh, India

[2]UG Scholar, Department of EEE, Annamacharya Institute of Technology and Sciences, Rajampet, Andhra Pradesh, India

Abstract

Electric vehicles (EVs) are rapidly gaining popularity due to their environmental benefits and efficient energy utilization. However, the current EV infrastructure faces challenges in ensuring convenient access to charging stations, particularly in remote areas or during emergencies. To address these issues, this project introduces a novel approach for vehicle-to-vehicle (V2V) energy transfer using on-board converters. The proposed system integrates a battery with a two-switch bidirectional DC-DC converter, a four-switch bidirectional isolated converter, and additional two-switch and four-switch isolated converters. These components enable controlled, bidirectional energy transfer between vehicles while ensuring electrical isolation and high efficiency. This innovative design allows EVs to safely and efficiently share energy, optimizing power transfer and reducing dependence on external charging infrastructures.

The system operates in two modes:
- Mode 0: EV1 is charging while EV2 is discharging.
- Mode 1: EV1 is discharging while EV2 is charging.

In both modes, the state of charge (SOC) is regulated by maintaining a constant voltage of 350 V. When charging, the current becomes negative (drops below 0), indicating energy inflow. Conversely, during discharging, the current becomes positive (rises above 0), reflecting energy outflow. The flexibility and scalability of the proposed system make it suitable for diverse EV configurations, enhancing energy resilience and supporting sustainable transportation networks. This approach represents a significant step toward addressing the limitations of existing EV infrastructures and fostering a more robust and interconnected energy ecosystem.

Keywords: Bidirectional DC-DC converter, electric vehicles (EVs), isolated converters, on-board converters, state of charge (SOC) etc., vehicle-to-vehicle (V2V) energy transfer

Introduction

Electric vehicles (EVs) have emerged as a sustainable alternative to traditional internal combustion engine (ICE) vehicles due to their lower carbon footprint, energy efficiency, and reduced dependence on fossil fuels. The growing adoption of EVs is driven by advancements in battery technology, power electronics, and charging infrastructure. However, the widespread deployment of EVs is hindered by challenges such as limited charging stations, long charging times, and concerns about battery range and energy availability in critical situations [1, 3, 6].

To address these challenges, researchers have explored various charging solutions, including grid-based charging stations, battery swapping systems, and wireless charging technologies [2, 4, 5]. Among these, vehicle-to-vehicle (V2V) energy transfer has gained attention as a promising solution to enhance energy resilience and mitigate range anxiety. V2V energy transfer allows an EV with surplus battery energy to charge another EV in need, reducing dependency on stationary charging stations and offering a decentralized energy-sharing approach [7, 8, 10].

[a]ahimabindu.eee@gmail.com, [b]deepikaroy8143@gmail.com, [c]karthikkumarreddy950@gmail.com, [d]mulintidharmareddym@gmail.com, [e]Deepthibalaraju11@gmail.com

DOI: 10.1201/9781003684589-20

Despite advancements in charging technologies, several challenges remain in achieving an efficient and practical V2V energy transfer system:

1. Inefficient power conversion – Many existing bidirectional charging systems suffer from high conversion losses, leading to reduced overall efficiency in energy transfer [1, 3].
2. Lack of standardized control strategies – The absence of a universal control mechanism for bidirectional energy transfer affects system stability and interoperability between different EV models [4, 5, 11].
3. Voltage and current fluctuations – Traditional bidirectional converters often experience fluctuations in voltage and current during energy transfer, leading to instability and potential battery degradation [6, 8, 13].
4. Limited electrical isolation – Several V2V energy transfer systems fail to ensure adequate electrical isolation, which poses safety risks and affects power quality [12, 14].
5. Scalability and flexibility issues – Most V2V energy-sharing models are not designed to support diverse EV configurations, limiting their widespread adoption [9, 15].

These limitations highlight the need for a more efficient, safe, and scalable V2V energy transfer system.

The motivation behind this study is to develop a robust and efficient on-board converter-based V2V energy transfer system that addresses the above challenges. Unlike conventional charging methods that rely solely on external charging infrastructure, V2V energy sharing enables peer-to-peer energy exchange, allowing EVs to assist each other in remote or emergency situations. By integrating advanced bidirectional converters, the proposed system ensures high efficiency, controlled energy transfer, and improved system reliability.

The primary objectives of this paper are:
1. To design and implement a bidirectional V2V energy transfer system using a combination of two-switch and four-switch isolated converters for improved power efficiency and electrical isolation.
2. To develop a control strategy that ensures stable voltage regulation and mitigates current fluctuations during charging and discharging cycles.

3. To evaluate the system's performance under different operational modes and analyze its feasibility for real-world EV applications.
4. Comparing the proposed system with existing bidirectional charging solutions in terms of efficiency, safety, and scalability.

The key contributions of this paper are:

• Novel on-board converter design: The integration of two-switch bidirectional DC-DC converters and four-switch bidirectional isolated converters for improved power transfer efficiency and electrical isolation.
• Enhanced stability and control: A constant voltage regulation mechanism to maintain a stable 350V during charging and discharging, ensuring system reliability.
• Dual-mode operation: The system operates in Mode 0 (EV1 charging, EV2 discharging) and Mode 1 (EV1 discharging, EV2 charging), allowing dynamic energy exchange between vehicles.
• Improved safety and scalability: The proposed system ensures safe energy transfer while being adaptable to different EV configurations, promoting sustainable EV networks.
• Paper organization

The rest of the paper is organized as follows: Section 2 details the proposed system architecture, including the converter design and control strategies. Section 3 Implementation of the proposed method. Section 4 presents simulation results. Section 5 concludes the paper and outlines future research directions.

Proposed Method

The proposed method introduces a novel V2V energy transfer system utilizing an on-board converter-based architecture to enable bidirectional power exchange between EVs as shown in Figure 20.2. The system comprises a two-switch bidirectional DC-DC converter, a four-switch bidirectional isolated converter, and additional isolated converters to ensure efficient and controlled power transfer while maintaining electrical isolation. This setup allows seamless energy exchange between two EVs, operating in two distinct modes: Mode 0, where EV1 is charging and EV2 is discharging, and Mode 1, where EV1 is discharging

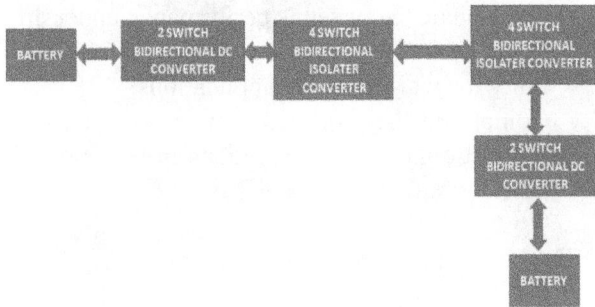

Figure 20.1 Proposed block diagram
Source: Author

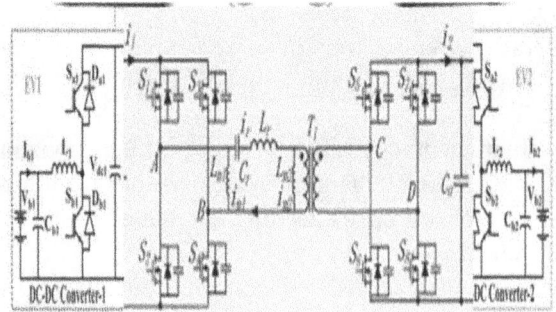

Figure 20.2 Circuit diagram of proposed method
Source: Author

and EV2 is charging. During both modes, the system maintains a constant voltage of 350V, ensuring a stable charging and discharging process.

The proposed method integrates advanced power control mechanisms to regulate voltage and mitigate current fluctuations, ensuring safe and efficient energy transfer. By leveraging bidirectional isolated converters, the system enhances power efficiency, reduces losses, and improves reliability, making it suitable for real-world EV applications. Additionally, the scalable and flexible nature of the proposed architecture allows integration with different EV configurations, supporting sustainable and resilient transportation networks. The block diagram of the proposed method shown in Figure 20.1 V2V energy transfer system consists of a structured energy flow pathway ensuring efficient and controlled bidirectional power exchange between Vehicle 1 (EV1) and Vehicle 2 (EV2). The system begins with Vehicle 1's battery, which is connected to a two-switch bidirectional DC-DC converter. This converter regulates the voltage and current flow, enabling smooth bidirectional energy transfer while minimizing power losses. The output of this DC-DC converter is then fed into a four-switch bidirectional isolated converter, which provides electrical isolation and enhances energy transfer efficiency. This isolated converter ensures stable operation and protects both vehicle batteries from unwanted electrical interactions. The power is then transmitted to EV2, where it first passes through another four-switch bidirectional isolated converter, ensuring that the energy transfer remains isolated and efficient. Finally, a two-switch bidirectional DC-DC converter in EV2 further regulates the power before delivering it to the battery of EV2. Reverse energy flow is also possible, allowing EV2 to charge EV1 when required. This design ensures a safe, reliable, and efficient energy exchange mechanism with

proper electrical isolation, optimized power conversion, and minimal losses, making it ideal for real-world V2V charging applications.

The circuit diagram of the proposed V2V energy transfer system consists of interconnected power electronic converters that facilitate bidirectional energy flow between two electric vehicles while ensuring efficient power conversion and electrical isolation. The system incorporates a two-switch bidirectional DC-DC converter, a four-switch bidirectional isolated converter, and additional filtering components to regulate power flow.

On EV1, the battery serves as the primary energy source, supplying power to a two-switch bidirectional DC-DC converter, which controls voltage and current, ensuring smooth energy exchange. This converter is followed by a four-switch bidirectional isolated converter, which enhances energy transfer efficiency while maintaining electrical isolation between the vehicles. The high-frequency transformer within the isolated converter prevents direct electrical contact, improving safety and system reliability. The output of this converter is then transmitted through power transmission lines to EV2.

On EV2, the incoming power first passes through another four-switch bidirectional isolated converter, which continues to regulate energy transfer while maintaining electrical isolation. The processed power is then supplied to a two-switch bidirectional DC-DC converter, which ensures a controlled voltage and current output before delivering the power to the battery of EV2.

The system operates in two modes: Mode 0, where EV1 is charging and EV2 is discharging, and Mode 1, where EV1 is discharging and EV2 is charging. In both modes, the system maintains a constant voltage of 350 V during charging, while current direction determines the charging or discharging status. The use of bidirectional converters ensures that power flow can be reversed as needed, optimizing energy utilization and reducing dependency on external charging infrastructure. This design enhances energy resilience, scalability, and safety, making it an ideal solution for real-world V2V charging applications.

Table 20.1: Modes of operation outlines the two operational modes of the proposed V2V energy transfer system, which govern the energy flow between EV1 and EV2. In Mode 0, EV1 is in the charging state, where it supplies power to EV2, which is in the discharging state. In this mode, EV1 provides energy from its battery to recharge EV2's battery, making this mode useful when EV2 needs additional energy to continue its journey. The bidirectional converters facilitate energy transfer, ensuring that the voltage and current are properly regulated to maintain safe and efficient charging. In Mode 1, the energy flow is reversed: EV1 is in the discharging state, supplying energy to EV2, which is now in the charging state. In this configuration, EV2 is recharged by EV1's battery. This mode allows EV2 to receive power when it is low on charge, helping to extend its range without needing access to an external charging station. These two modes ensure that the energy exchange between the vehicles is flexible and can adapt to various real-world scenarios, such as when one vehicle has surplus energy and the other requires it. The system's ability to switch between these modes efficiently allows for optimized energy distribution and greater flexibility in managing the vehicles' batteries.

Table 20.2: SOC, voltage, and current provides details on the key electrical parameters, including state of charge (SOC), voltage, and current, during the two modes of operation of the V2V energy transfer system. In Mode 0 (Charging), the SOC indicates that the battery of EV1 is charging, which occurs at a constant voltage of 350V. The current during this operation is negative (-50A), indicating that EV1 is supplying power to EV2. The negative current signifies a flow of energy from EV1 to EV2, where EV2 is in a discharging state and receiving energy to recharge its battery. In the Mode 0 (Discharging) state, EV2 is discharging and the current becomes positive (5A), indicating that EV2's battery is providing power to EV1. The voltage remains constant at 350V, ensuring that the energy transfer continues at a stable voltage level. This positive current implies that energy is flowing from EV2 to EV1, where EV1's battery is recharged Figure 20.3.

This table highlights how the voltage is maintained constant at 350V in both charging and discharging operations, and the current direction changes based on the mode of operation, reflecting the direction of energy transfer between the two vehicles. The SOC

Table 20.1 Modes of operation.

| S. No | Modes | Modes of operation | |
		EV1	EV2
1	Mode 0	Charging	Discharging
2	Mode 1	Discharging	Charging

Source: Author

Table 20.2 SOC, voltage and current.

| S. No | Modes | Operation | | |
		SOC (%)	Voltage (V)	Current (A)
1	Mode 0	Charging	350V	-50
2	Mode 0	Discharging	350V	5

Source: Author

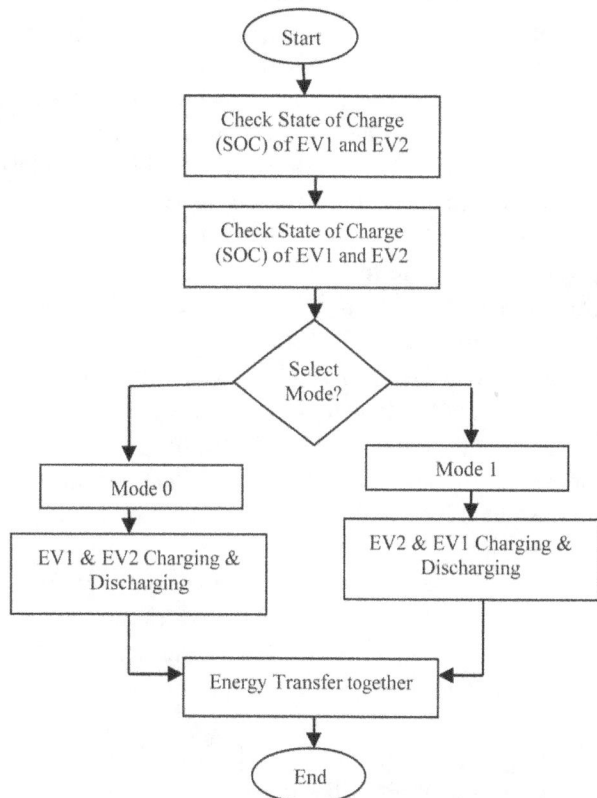

Figure 20.3 Implementation flow chart of proposed method

Source: Author

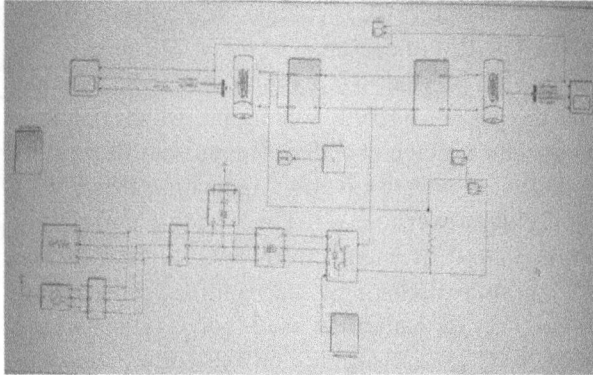

Figure 20.4 Simulink circuit in mode 0
Source: Author

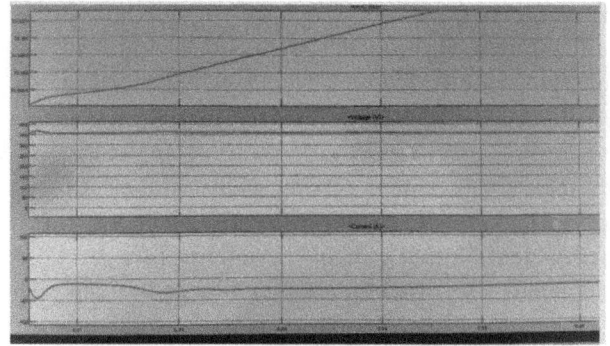

Figure 20.5 Simulation waveform for EV1 status at mode 0
Source: Author

of both batteries plays a key role in determining whether a vehicle is charging or discharging, ensuring that the energy exchange is regulated properly based on the vehicle's battery state.

Implementation

Simulation Results

The simulation results were obtained by implementing the V2V energy transfer system in MATLAB 2021a using Simulink for both Mode 0 and Mode 1 operations. The following sections describe the simulation results and their corresponding figures.

This Figure 20.4 illustrates the Simulink model of the system operating in Mode 0 (where EV1 is charging and EV2 is discharging). The Simulink circuit consists of the two-switch bidirectional DC-DC converter and the four-switch bidirectional isolated converter for both EV1 and EV2. The circuit is designed to transfer energy from EV1's battery to EV2's battery while maintaining proper voltage and current regulation. The voltage and current control blocks are included to ensure the system operates at a constant 350V, and the current remains negative (-50A) when charging.

This Figure 20.5 shows the simulation waveform for EV1's status in Mode 0. It illustrates the voltage and current waveforms of EV1 during the charging operation. The voltage is constant at 350V, and the current is negative (-50A), which indicates that EV2 is supplying power to EV1. The waveform shows a smooth transition in the current flow as energy is transferred from EV2's battery to EV1.

This Figure 20.6 presents the simulation waveform for EV2's status in Mode 0. The waveform shows

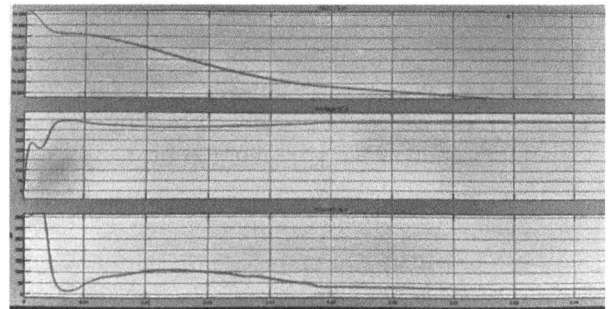

Figure 20.6 Simulation waveform for EV2 status at mode 0
Source: Author

the voltage and current characteristics of EV2 as it sending energy from EV1. The voltage remains constant at 350V, and the current is positive (5A), indicating that EV1 is charging. The waveform confirms that EV2's battery is sending the required power for recharging.

This Figure 20.7 shows the Simulink model of the system operating in Mode 1 (where EV1 is discharging and EV2 is charging). Similar to Mode 0, the system consists of the same bidirectional DC-DC converters and isolated converters, but the operation is reversed, with EV1 discharging energy into EV2. The voltage control remains at 350V, and the current direction is now positive (5A) for EV1, indicating that it is providing energy to EV2.

This Figure 20.8 illustrates the simulation waveform for EV1's status in Mode 1. The voltage is maintained at 350V, and the current is positive (5A), indicating that EV1 is discharging power into EV2. This waveform shows the steady transition in the current as energy flows from EV1 to EV2.

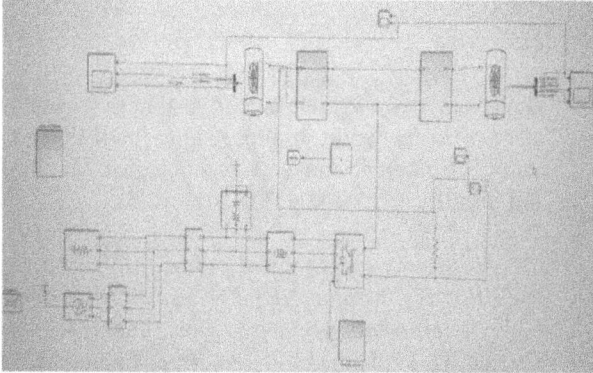

Figure 20.7 Simulink circuit in mode 1
Source: Author

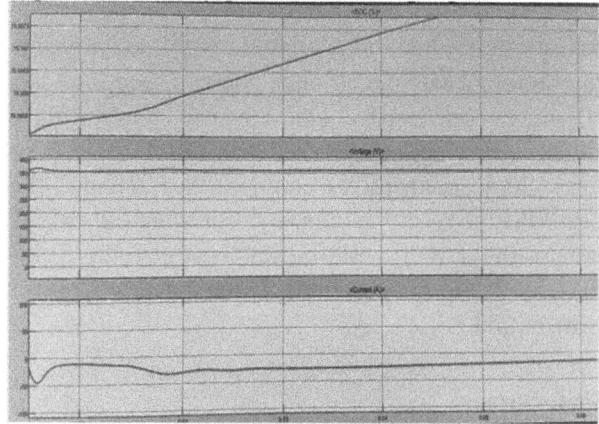

Figure 20.9 Simulation waveform for EV2 status at mode 1
Source: Author

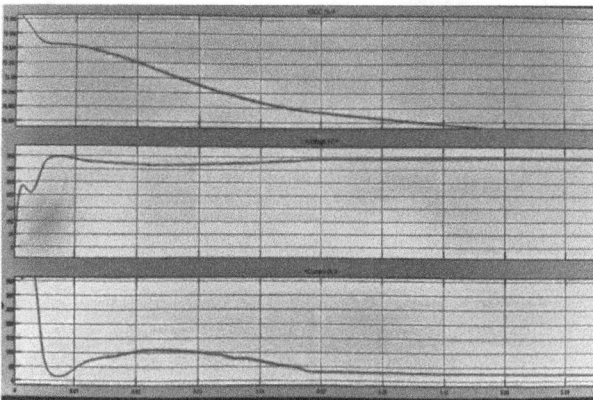

Figure 20.8 Simulation waveform for EV1 status at mode 1
Source: Author

This Figure 20.9 shows the simulation waveform for EV2's status in Mode 1. In this mode, EV2 is charging, and the voltage is again held constant at 350V. The current becomes negative (-50A), indicating that EV2 is receiving power from EV1. The waveform confirms that EV2 is efficiently receiving the necessary power for charging.

The simulation waveforms for both Mode 0 and Mode 1 indicate the stable and efficient transfer of energy between the two vehicles, with voltage maintained at a constant value of 350V and current values fluctuating in accordance with the charging and discharging states of the vehicles. The negative current in Mode 0 (when EV1 is charging) and the positive current in Mode 1 (when EV1 is discharging) demonstrate the bidirectional nature of energy transfer, where the direction of power flow is controlled according to the operational mode. The results also verify that the two-switch bidirectional

DC-DC converters and four-switch bidirectional isolated converters maintain voltage regulation and safe current flow, ensuring the system operates efficiently and safely. These simulation results validate the functionality of the proposed V2V energy transfer system, ensuring that energy can be transferred effectively between vehicles while maintaining the required safety, voltage, and current parameters.

Conclusion and Future Scope

This paper presented a vehicle-to-vehicle (V2V) energy transfer system utilizing on-board bidirectional converters to enable energy exchange between two electric vehicles (EVs). The proposed system consists of a combination of two-switch bidirectional DC-DC converters and four-switch bidirectional isolated converters, which ensure safe and efficient energy transfer while maintaining electrical isolation between the two vehicles. The simulation results obtained using MATLAB 2021a and Simulink successfully validated the functionality of the system in two operational modes: Mode 0 (EV1 charging and EV2 discharging) and Mode 1 (EV1 discharging and EV2 charging). The system was shown to operate efficiently, with constant voltage (350V) and proper regulation of current for both charging and discharging operations. The proposed method helps mitigate the dependency on external charging stations, making it a promising solution for range anxiety and energy sharing between vehicles, especially in remote areas or emergency situations. In conclusion, the proposed system enhances the energy resilience of EVs and supports the development of sustainable transportation networks, aligning with global efforts to promote clean and efficient energy usage.

Future work can focus on advanced control algorithms (e.g., model predictive control (MPC) or fuzzy logic controllers) could further enhance the performance of the system by improving the energy management and optimization between the vehicles, ensuring more efficient and adaptive power distribution.

References

[1] Yuan, J., Dorn-Gomba, L., Callegaro, A. D., Reimers, J., & Emadi, A. (2021). A review of bidirectional on-board chargers for electric vehicles. *IEEE Access*, 9, 51501–51518.

[2] Metwly, M. Y., Abdel-Majeed, M. S., Abdel-Khalik, A. S., Hamdy, R. A., Hamad, M. S., & Ahmed, S. (2020). A review of integrated on-board EV battery chargers: advanced topologies, recent developments and optimal selection of FSCW slot/pole combination. *IEEE Access*, 8, 85216–85242.

[3] Khaligh, A., & Antonio, M. D. (2019). Global trends in high-power on-board chargers for electric vehicles. *IEEE Transactions on Vehicular Technology*, 68(4), 3306–3324.

[4] Tran, V. T., Sutanto, D., & Muttaqi, K. M. (2017). The state of the art of battery charging infrastructure for electrical vehicles: Topologies, power control strategies, and future trend. In Proceedings of the Australasian Universities Power Engineering Conference (AUPEC), (pp. 1–6).

[5] Khalid, M. R., Khan, I. A., Hameed, S., Asghar, M. S. J., & Ro, J. S. (2021). A comprehensive review on structural topologies, power levels, energy storage systems, and standards for electric vehicle charging stations and their impacts on grid. *IEEE Access*, 9, 128069–128094.

[6] Yilmaz, M., & Krein, P. T. (2013). Review of battery charger topologies, charging power levels, and infrastructure for plug-in electric and hybrid vehicles.

IEEE Transactions on Power Electronics, 28(5), 2151–2169.

[7] Li, G., Boukhatem, L., Zhao, L., & Wu, J. (2018). Direct vehicle-to-vehicle charging strategy in vehicular Ad-Hoc networks. In Proceedings 9th IFIP International Conference on New Technologies, Mobility and Security (NTMS) (NTMS), (pp. 1–5).

[8] Zhang, R. Q., Cheng, X., & Yang, L. Q. 6(2019). Flexible energy management protocol for cooperative EV-to-EV charging. *IEEE Transactions on Intelligent Transportation Systems*, 20(1), 172–184.

[9] Mughal, D. M., Kim, J. S., Lee, H., & Chung, M. Y. (2019). Performance analysis of V2V communications: a novel scheduling assignment and data transmission scheme. *IEEE Transactions on Vehicular Technology*, 68(7), 7045–7056.

[10] Bulut, E., & Kisacikoglu, M. C. (2017). Mitigating range anxiety via vehicleto-vehicle social charging system. In Proceedings IEEE 85th Vehicular Technology Conference (VTC Spring), (pp. 1–5).

[11] P. You and Z. Yang, "Efficient optimal scheduling of charging station with multiple electric vehicles via V2V," in Proc. IEEE Int. Conf. Smart Grid Commun. (SmartGridComm), Nov. 2014, pp. 716721.

[12] A.-M. Koufakis, E. S. Rigas, N. Bassiliades, and S. D. Ramchurn, "Towards an optimal EV charging scheduling scheme with V2G and V2V energy transfer," in Proc. IEEE Int. Conf. Smart Grid Commun. (SmartGridComm), Nov. 2016, pp. 302307.

[13] E. Ucer et al., "A flexible V2V charger as a new layer of vehicle-grid integration framework," in Proc. IEEE Transp. Electrific. Conf. Expo (ITEC), Jun. 2019, pp. 17.

[14] C. Liu, K. T. Chau, D. Wu, and S. Gao, "Opportunities and challenges of vehicle-to-home, vehicle-to-vehicle, and vehicle-to-grid technologies," Proc. IEEE, vol. 101, no. 11, pp. 24092427, Nov. 2013.

[15] P. Mahure, R. K. Keshri, R. Abhyankar, and G. Buja, "Bidirectional conductive charging of electric vehicles for V2V energy exchange," in Proc. 46th Annu. Conf. IEEE Ind. Electron. Soc. (IECON), Oct. 2020, pp. 20112016.

21 Implementation of IoT home automation using raspberry Pi with multilingual voice control

Hindumathi Voruganti[1,a], Rama Lakshmi Gali[2,b], Saikumar Tara[2,c], N. Manisha[3,d] and G. Alekhya[3,e]

[1]Assiociate Professor, Department of ECE, BVRIT Hyderabad College of Engineering for Women, Hyderabad, India

[2]Assistant Professor, Department of ECE, BVRIT Hyderabad College of Engineering for Women, Hyderabad, India

[3]Student, Department of ECE, BVRIT Hyderabad College of Engineering for Women, Hyderabad, India

Abstract

In order to make smart home systems accessible to a worldwide audience, with various article addresses the language issue. Utilizing the Raspberry Pi's computational power and sophisticated speech recognition algorithms, the system allows users to effortlessly control their home environment by supporting natural language voice commands in a variety of languages. A companion app offers touch-based control for improved accessibility, while hardware elements like relay boards and GPIO pins link Internet of Things devices with voice commands. The system's scalable and adaptable design makes it simple to integrate new languages and customize them to suit a range of user requirements. In addition to democratizing access to smart home technology, this creative strategy promotes diversity and encourages people from different language backgrounds to adopt smart living.

Keywords: GPIO, Internet of Things, natural language processing, smart home, speech recognition algorithms, raspberry Pi

Introduction

In the context of home automation, the term "electronic and electrical environment" refers to a room that is occupied by various equipment, such as fans, televisions, air conditioners, lighting systems, motors, and heaters. Software interfaces, frequently through mobile apps or central systems, enable the control and monitoring of these devices in a remotely accessible environment. The Internet of Things (IoT) has become a crucial component of contemporary technology, especially in the field of home automation, where it links gadgets to the internet for convenient remote control [1]. By automating functions like temperature management, lighting control, and security system control, home automation systems leverage the IoT to develop "smart" homes.

Home automation is not a new concept; it has been used for decades in large businesses and industries. However, because of technological improvements and the rising need for convenience, energy efficiency, and higher living standards, its use in homes has become increasingly popular in recent years [2]. Even with these advancements, most smart home devices only support a few languages, namely English. This is a problem for non-native English speakers, who can find it difficult to communicate in their native tongues with speech recognition software and smart devices [3]. The goal of this project is to create a multilingual IoT home automation system in order to solve this problem. The objective is to create a system that will enable users to overcome language barriers by controlling household appliances with voice commands in their preferred languages.

Literature Survey

The work explores several key areas, including voice-controlled home automation systems, IoT technologies, Raspberry Pi-based projects, multilingual speech recognition, and user experience design. The research reviews numerous studies that contribute to the evolving field of home automation and smart home systems.

[a]hindumathi.v@bvrithyderabad.edu.in, [b]ramalakshmi.g@bvrithyderabad.edu.in, [c]saikumar.tara@bvrithyderabad.edu.in, [d]20WH1A0428@bvrithyderabad.edu.in, [e]20WH1A0432@bvrithyderabad.edu.in

DOI: 10.1201/9781003684589-21

The article by Doshi, et. al., investigated the transformative potential of IoT in home automation [4]. It highlights the application of IoT principles in residential settings and provides insights into the design, implementation, and evaluation of IoT-based home automation systems.

Similarly, Pal et. al., explores the integration of IoT technologies into home automation [5]. Their approach involves using node MCU microcontrollers, sensors, and actuators to enable communication between the home environment and a centralized server, allowing remote control via mobile apps or web interfaces. Wireless communication technologies like Bluetooth and Zigbee ensure seamless connectivity between devices.

The research by Al-Kuwari et al. presents a smart-home automation using IoT [6]. This platform collects environmental data (e.g., temperature, humidity) through sensors and transmits it to a central monitoring system. The users can access and manage this data remotely through web or mobile applications, allowing for efficient control of home appliances.

Roy focused on developing a voice-controlled home automation system [7]. Their system integrates voice recognition with IoT to enable users to control appliances via voice commands. This project uses Google Assistant and Blynk to facilitate remote control and monitoring via Wi-Fi, enhancing user convenience.

Rani et al. [8] investigates the use of NLP in voice-controlled home automation systems. They employed NLP algorithms to process natural language voice commands, translating them into instructions for controlling IoT devices. Their system leverages Arduino boards and Wi-Fi connectivity to ensure communication between the user interface and devices.

Biswas et al. took an eco-friendly approach, integrating renewable energy sources like solar panels with their voice-controlled home automation system [9]. This setup uses NodeMCU microcontrollers and mobile apps such as Blynk to control appliances, aiming to reduce energy consumption while promoting sustainability.

Tatikonda explores the feasibility of using Raspberry Pi as a central hub in home automation systems [10]. Their approach enables remote control of appliances via the internet and incorporates voice control functionality, allowing users to issue speech commands to manage devices. This project illustrates how affordable technology can be used to create effective smart home solutions.

Chayapathy designed a system that integrates speech recognition with IoT devices, allowing users to control appliances through voice commands [11]. Google's speech recognition technology was utilized to process commands and trigger the corresponding actions.

Noruwana et al. combines speech recognition and IoT protocols to create an interactive home automation environment [12]. Their system allows real-time control of household devices, with speech recognition algorithms ensuring accurate interpretation of user commands.

The study by Baby et al. introduced a web-based platform for voice-operated home automation [13]. Their architecture integrates speech recognition with web technologies, enabling users to control appliances remotely using voice commands via a browser interface.

Prasath et al. presented a system designed specifically for physically challenged individuals [14]. Their voice-controlled IoT home automation system simplifies the operation of home appliances for users with mobility impairments, allowing them to interact with the system using voice commands.

Kodali and Kopulwar's research emphasizes the use of low-cost solutions for smart home automation [15]. By leveraging affordable hardware components like Raspberry Pi and open-source software, they created a scalable and cost-effective system that allows remote control of home devices.

Ruslan et al. focuses on enabling multilingual voice control for smart homes [16]. This research integrates IoT devices with speech recognition technologies to provide users with a seamless, multi-language interface for managing their home appliances.

Lastly, Nath et al. developed a low-cost Android app for voice-operated room automation, allowing users voice commands on their smartphones [17]. Their approach emphasizes cost-effectiveness and accessibility, ensuring that voice control is available to a wide range of users.

Overall, these studies contribute to the growing body of knowledge in home automation, highlighting the role of voice recognition, IoT integration, and Raspberry Pi in creating more accessible, efficient, and sustainable smart homes [18]. The use of multilingual capabilities and affordability further enhances the practicality of these technologies for a global audience.

Implementation

Following the elucidation of the architectural framework, this section provides an in-depth operational overview of the multi-language voice-controlled home automation system. It delves into the functional intricacies of each key component and technology, shedding light on how they synergistically contribute to the seamless operation of the system.

The block diagram shown in Figure 21.1 encapsulates the intricate workings of the Multi- Language Voice Control IoT Home Automation system. Each distinct block symbolizes a pivotal functional module or device, harmoniously contributing to the system's flawless operation. The interconnection of hardware components, including GPIO pins and relay boards, ensures the precise execution of voice commands on various IoT devices, delivering a responsive and user-friendly home automation experience.

The block diagram shown in Figure 21.1, illustrates the sequential flow of operations within the voice-controlled system. In the system, users interact with the system through voice commands spoken in their preferred language. These commands are captured by a USB microphone, the primary input device, which converts acoustic signals into electrical signals processed by the Raspberry Pi. Functioning as the central hub, the Raspberry Pi receives audio input, utilizing advanced speech recognition algorithms to identify speech patterns and extract meaningful commands. Upon recognition, the Raspberry Pi determines the corresponding actions to be taken. Communication with the physical environment is facilitated by a relay module, serving as an interface between the Pi and electrical loads. The relay module, comprising multiple relays channels, allows independent control of connected loads. These loads, ranging from fans to

lights, are wired to the relay module, with each load associated with a specific relay channel. When the Raspberry Pi sends control signals to the relay module, the corresponding relays toggle the power supply to activate or deactivate the connected loads, enabling seamless automation of household appliances.

Hardware and software

The hardware components include the Raspberry Pi, relay modules, microphones, and various loads such as bulbs and fans, form the physical foundation of the work. Meanwhile, the software components, encompassing the Raspbian OS, Python programming language, and necessary libraries and dependencies, facilitate the seamless integration and operation of the hardware elements. Together, these components create a robust and efficient home automation system capable of understanding and executing voice commands in multiple languages.

The Raspberry Pi features a variety of connectors that facilitate interfacing with external devices and peripherals, enhancing its versatility and functionality. These connectors as shown in Table 21.1 enable users to expand the capabilities of the board and interface with a wide range of components and accessories. Some of the key connectors found on the Raspberry Pi include:

5V Four-channel relay module

The four-channel relay module, illustrated in Figure 21.2, consists of four 5V relays along with the necessary switching and isolation components. This design simplifies interfacing with microcontrollers or sensors, requiring minimal components and connections.

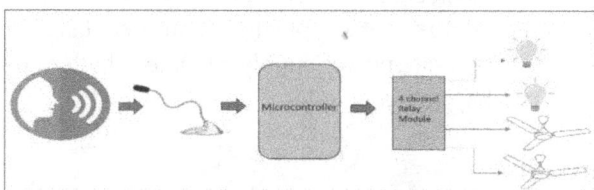

Figure 21.1 Block diagram of the multi-language voice control IoT home automation system
Source: Author

Table 21.1 Raspberry Pi 3 board connectors.

Name	Description
Ethernet	Base T Ethernet Socket
USB	2.0 (Four sockets)
Audio Output	3.5mm Jack and HDMI
Video output	HDMI
Camera Connector	15-pin MIPI Camera Serial Interface (CSI-2)
Display Connector	Display Serial Interface (DSI) 15 way flat flex cable connector with two data lanes and a clock lane.
Memory card slot	Push/Pull Micro SDIO

Source: https://www.scribd.com/presentation/794486141/UNIT-II-Controllers-and

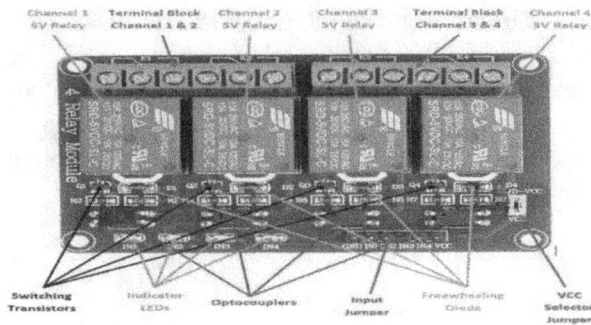

Figure 21.2 5V Four-channel relay module
Source: Author

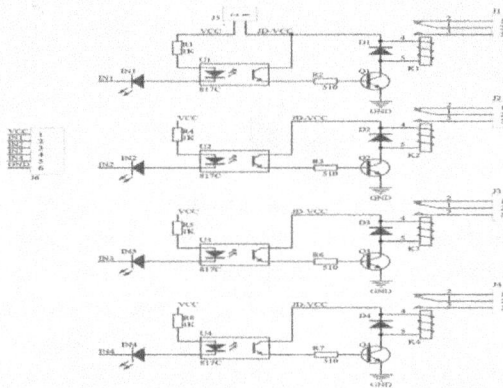

Figure 21.3 Four-channel relay internal circuit
Source: Author

The Four-Channel Relay Module is a versatile electronic component commonly used in various applications as shown in Figure 21.3.

In the context of "multilanguage voice control IoT home automation using Raspberry pi", the following dependencies are required.

Speech recognition

Speech recognition is a versatile Python library designed to facilitate seamless integration of speech recognition capabilities into applications. With its intuitive interface and robust functionality, speech recognition offers easy access to a variety of speech recognition APIs. One of the key strengths of speech recognition lies in its support for multiple recognition engines, including the widely-used Google speech recognition service.

RPi.GPIO

RPi.GPIO stands as a fundamental Python library meticulously crafted to facilitate the management and manipulation of GPIO pins on Raspberry Pi, the renowned single-board computer. As a core component of Raspberry Pi development, this library empowers Python scripts to seamlessly interact with GPIO pins, facilitating both input and output operations crucial for interfacing with external hardware components.

The relay module, controlled by specific GPIO pins on the Raspberry Pi, enabled the toggling of electrical circuits to activate or deactivate devices. On the software side, the architecture included a speech recognition module capable of interpreting commands in multiple languages—Telugu, English, Kannada, and Hindi—and a command processing module that translated these commands into actions by triggering the appropriate GPIO pins.

1. **Connecting Raspberry Pi to relay module**
   Utilized Raspberry Pi's GPIO pins to establish connections with the relay module. Connected the relay module's control pins to specific GPIO pins on the Raspberry Pi for individual load control.
   GPIO Pin connections:
   a. Relay Channel 1: Connect the control pin of relay channel 1 to GPIO pin 27 on the Raspberry Pi.
   b. Relay Channel 2: Connect the control pin of relay channel 2 to GPIO pin 17 on the Raspberry Pi.
   c. Relay Channel 3: Connect the control pin of relay channel 3 to GPIO pin 22 on the Raspberry Pi.
   d. Relay Channel 4: Connect the control pin of relay channel 4 to GPIO pin 23 on the Raspberry Pi.

2. **Integrating loads (bulbs and motors) load integration:**
   a. Bulbs: Connected the bulbs to the output terminals of the relay module. Each bulb was wired in parallel to a corresponding relay channel on the module.
   b. Motors: Wired the motors to the relay module's output terminals, with each motor connected to a separate relay channel. The motors were typically controlled using a separate power supply, such as a battery or DC power adapter.

Results

Home automation using voice commands
The results demonstrate the successful implementation and functionality of a voice-controlled home

automation system using Raspberry Pi. Through hardware integration, and software development, the system seamlessly integrates voice and touch-based interaction methods to control household appliances. The system exhibits excellent responsiveness to user commands, with rapid execution of requested actions typically occurring within 30 to 60 seconds of issuing voice commands. Moreover, load control operations, including turning devices on or off, are executed reliably and consistently.

As shown in the above Figure 21.4, "Google Load 1 ko chaalu kardo" is the command in Hindi language for turning on Load-1 which is a bulb.

Similarly, in the above Figure 21.5, "Google Loads off maadu" is the voice command in Kannada language for turning off all the loads (bulbs and motors). In the below Figure 21.6 "Google Loads on cheyi" is the voice command in Telugu language for turning on all the loads connected (bulbs and motors).

Home automation using android application

The mobile app exhibited exceptional responsiveness to user commands, offering users a seamless and intuitive interface to interact with the home automation system. Through the app, users experienced swift execution of requested actions, with controls responding promptly to taps within the app interface. Moreover, device control operations via the mobile app demonstrated reliability and consistency, underscoring the robustness of the communication protocol employed. Users could rely on the app to effectively transmit commands to the Raspberry Pi, ensuring that connected devices responded predictably to user inputs. The app's interface facilitated seamless communication with the Raspberry Pi, enabling accurate transmission of user commands and real-time status updates.

The above Figure 21.7, gives the information of app icon and the app name (home automation). One of the key requirements for using the app is ensuring that the mobile phone is connected to the same Wi-Fi hotspot as the Raspberry Pi board. This allows for seamless communication between the app and the Raspberry Pi, enabling users to send commands and receive real-time feedback. Additionally, users are required to specify the IP address of the Raspberry Pi within the app settings. This IP address serves as the connection endpoint for the app to communicate with the Raspberry Pi over the local network. By inputting

Figure 21.4 Load 1-on
Source: Author

Figure 21.6 Loads on
Source: Author

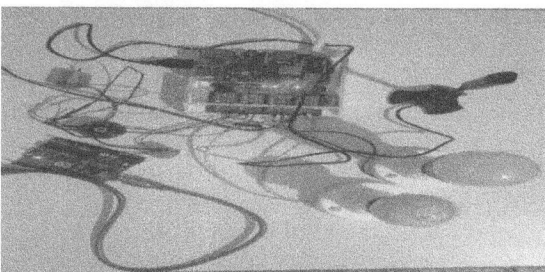

Figure 21.5 Loads off
Source: Author

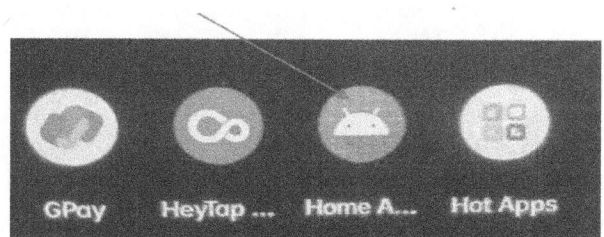

Figure 21.7 Home automation App
Source: Author

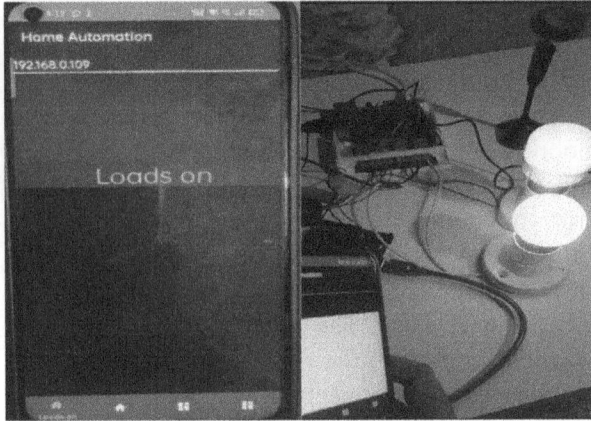

Figure 21.8 Home automation using android application
Source: Author

the correct IP address, users can establish a secure and reliable connection to the Raspberry Pi, enabling them to control connected devices with ease.

In Figure 21.8, after entering the raspberry pi Ip address, click enter tap on preferrable commands like Loads on, Loads off, Load4 on, Load4 off etc. Immediately after tapping on Loads on icon, the connected loads (bulbs and motors) will turn on.

Conclusion

In this work, the design and implementation of a voice assistant for IoT home automation is comprehensively explored, reflecting a concerted effort to bridge the gap between traditional home automation systems and intuitive, voice-controlled interfaces. The work aims to enhance the functionality and accessibility of home automation systems by integrating voice control capabilities. The method investigates various aspects of system architecture, speech recognition algorithms, and user interface design to optimize performance and user experience. Implemented across multiple languages including Telugu, English, Kannada, and Hindi, the system supports a diverse range of languages, expanding its reach and usability to cater to users from different linguistic backgrounds. This multilingual capability enhances the system's versatility and utility, making it more adaptable to a broader user base with varying language preferences and requirements. Through meticulous methodology and innovative design, the project not only demonstrates the feasibility and effectiveness of voice-controlled IoT systems but also paves the way for future advancements in smart home technologies.

References

[1]   Palaniappan, S., Hariharan, N., Kesh, N. T., & Vidhyalakshimi, S. (2015). Home automation systems-A study. *International Journal of Computer Applications*, 116(11) Page no.11–18.

[2]   Asadullah, M., & Raza, A. (2016). An overview of home automation systems. In 2016 2nd International Conference on Robotics and Artificial Intelligence (ICRAI), (pp. 27–31).

[3]   Ruslan, A. H., Jusoh, A. Z. , Asnawi, A. L., Othman, M. D. R., & Razak, N. I. A. (2021). Development of multilanguage voice control for smart home with IoT. *Journal of Physics: Conference Series*, 1921(1), 012069. IOP Publishing.

[4]   Doshi, A., Vakharia, D., & Rai, Y. (2021). Iot based home automation. *International Journal for Research in Applied Science & Engineering Technology (IJRASET),* 9(8) Page No: 117–125.

[5]   Singh, H. K., Verma, S., Pal, S., & Pandey, K. (2019). A step towards home automation using IOT. In 2019 Twelfth International Conference on Contemporary Computing (IC3), (pp. 1–5). IEEE.

[6]   Al-Kuwari, M., Ramadan, A., Ismael, Y., Al-Sughair, L., Gastli, A., & Benammar, M. (2018). Smart-home automation using IoT-based sensing and monitoring platform. In 2018 IEEE 12th International Conference on Compatibility, Power Electronics and Power Engineering (CPE-POWERENG 2018), (pp. 1–6). IEEE.

[7]   Roy, R. M., Sabu, B., Nisha, A., & Aneesh, R. P. (2021). Voice controlled home automation system. In 2021 Seventh International Conference on Bio Signals, Images, and Instrumentation (ICBSII), (pp. 1–6). IEEE.

[8]   Rani, P. J., Bakthakumar, J., Kumaar, B. P., Kumaar, U. P., & Kumar, S. (2017). Voice controlled home automation system using natural language processing (NLP) and internet of things (IoT). In 2017 Third International Conference on Science Technology Engineering & Management (ICONSTEM), (pp. 368–373). IEEE.

[9]   Biswas, S. S. K., Singdha, A. S., Talukder, L., Rahman, M. H., Das, P., & Islam, M. M. (2023). Sustainable energy-based voice-controlled home automation using IoT. In 2023 IEEE Symposium on Industrial Electronics & Applications (ISIEA), (pp. 1–5). IEEE.

[10] Sutar, A., Sinha, A., Sonde, A., Tatikonda, R., & Soni, V. (2019). IoT based smart home automation using raspberry Pi Issue 1, page no.263–265.

[11] Chayapathy, V., Anitha, G. S., & Sharath, B. (2017). IOT based home automation by using personal assistant. In 2017 International Conference on Smart Technologies For Smart Nation (SmartTechCon), (pp. 385–389). IEEE.

[12] Noruwana, N. C. C., Owolawi, P. A., & Mapayi, T. (2020). Interactive iot-based speech-controlled home automation system. In 2020 2nd International Multidisciplinary Information Technology and Engineering Conference (IMITEC), (pp. 1–8). IEEE.

[13] Baby, C. J., Munshi, N., Malik, A., Dogra, K., & Rajesh, R. (2017). Home automation using web application and speech recognition. In 2017 International Conference on Microelectronic Devices, Circuits and Systems (ICMDCS), (pp. 1–6). IEEE.

[14] Prasath, T. A., Lakshm, A. R., Vishnuvarthanan, G., Yosika, M., Rathnamma, T. H., & Vigneshwari, N. (2021). Voice recognised home automation system using IoT-physically challenged people. In 2021 3rd International Conference on Advances in Computing, Communication Control and Networking (ICAC3N), (pp. 696–699). IEEE.

[15] Kodali, R. K., & Mahesh K. S. (2017). Low cost implementation of smart home automation. In 2017 International Conference on Advances in Computing, Communications and Informatics (ICACCI), (pp. 461–466). IEEE.

[16] Ruslan, A. H., Jusoh, A. Z., Asnawi, A. L., Othman, M. D. R., & Razak, N. I. A. (2021). Development of multilanguage voice control for smart home with IoT. *Journal of Physics: Conference Series*, 1921(1), 012069. IOP Publishing.

[17] Nath, P., & Pati, U. C. (2018). Low-cost android app based voice operated room automation system. In 2018 3rd International Conference for Convergence in Technology (I2CT), (pp. 1–4). IEEE.

[18] Torres-Maldonado, S. Y., Tinoco-Varela, D., Cruz-Morales, R. D., López-Mera, G. H., Sánchez-García, D., & Padilla-García, E. A. (2023). Design and implementation of a voice assistant to be used in an IoT home automation environment. In 2023 27th International Conference on Circuits, Systems, Communications and Computers (CSCC), (pp. 1–8). IEEE.

22 Enhancing heart disease prediction using ensemble machine learning classifiers

C. Sasidhar[1,a], P. Navya[2,b], S. Mohammed Asif[2,c], T. Mounika[2,d], B. Mahanth Reddy[2,e] and K. Susmitha[2,f]

[1]Assistant professor, Department of CSE, Annamacharya University, Rajampet, India

[2]4th B. Tech, Department of CSE, Annamacharya University, Rajampet, India

Abstract

Heart stroke is leading causes of death worldwide, early detection and treatments require precise and effective prediction models. Making use of various machine learning classifiers as well as ensemble classifiers, such as K-nearest neighbors (KNN), Random Forest, Decision Trees, support vector machines (SVM), logistic regression, Naive Bayes, and gradient booster, this work aims to improve the prediction of heart attack. Important cardiovascular risk factors are represented by the dataset's properties, which include age, sex, cholesterol levels and more. Following feature engineering, preprocessing, and exploratory data analysis (EDA), many models were trained and assessed. The maximum accuracy (99%) was attained by Random Forest. The potential benefits of these models for real-world healthcare applications were demonstrated by the implementation of a continuous forecasting system that used flask UI to enable user interaction and deliver rapid forecasts.

Keywords: Ensemble classifiers, exploratory data analysis, feature engineering, flask UI, preprocessing, support vector machines SVM

Introduction

Heart disease remains one of the leading global health concern issues accounting for an astounding number of mortality each. Despite advancements in medical technology and methods, timely detection and efficient care of cardiac disease [1] continue to be problems. Conventional diagnostic techniques such imaging and stress testing can be costly, time-consuming and intrusive, which can cause treatment delays and raise medical expenses. Predictive models that are precise, effective and non-invasive are therefore desperately needed in order to help medical practitioners identify cardiac disease early on which will eventually improve patient outcomes and lower death rates.

Given data trends that show an increasing frequency in a number of various groups, the rising occurrence of cardiovascular disease [2]. Over 700,000 people in the US lose their lives to heart attacks and strokes every year and 48% of adults have heart disease based on the American Heart Association. Additionally research has demonstrated that the prevalence of cardiac disease differs markedly by gender, ethnic origin and age underscoring the need for specialized diagnostic instruments that can account for these variations. The issue is made worse by the rising prevalence of risk factors such as obesity, diabetes, and high blood pressure, which highlights how crucial prompt and precise forecasting is to the proper management of cardiac disease [3].

Understanding the prevalence of cardiac disease from a national perspective provides important insights on healthcare requirements and resource allocation in addition to worldwide statistics. Heart disease continues to be the leading cause of mortality in the United States for most demographic groups, with rates disproportionately higher among African Americans and Hispanics [4], according to the Centers for the Control of Diseases (CDC). Using predictive algorithms to evaluate patient data can offer major benefits in identifying at-risk groups and enhancing healthcare delivery as medical professionals work to lessen the effects of cardiac disease.

[a]csasidhar.aits@gmail.com, [b]navyareddyp333@gmail.com, [c]shaikmoha91@gmail.com, [d]mounikatekulodu@gmail.com, [e]bathina.mahanth@gmail.com, [f]kovurususmitha@gmail.com

DOI: 10.1201/9781003684589-22

With particular reference to India the heart disease numbers are similarly concerning. The International Burden of Cardiovascular Disease Study estimates that heart disease kills more than 1.8 million people in India each year making it the country's biggest cause of death.

Cardiovascular illnesses are on the rise in rural as well as urban populations as a result of the fast-rising incidence of risk factors linked to lifestyle choices such as smoking, eating poorly, and sedentary lifestyles. Additionally, the Indian population has complex cardiac illness because of unique genetic and environmental factors necessitating the development of prediction models tailored to this demographic. The rising incidence of cardiac disease in India underscores the pressing need for advanced prediction tools to enhance early detection and intervention strategies [5].

The rising incidence of cardiac disease in India underscores the pressing need for advanced prediction tools to enhance early detection and intervention strategies [5]. This work uses a variety of machine learning classifiers to address the problem of heart disease prediction. This study aims to improve predicted accuracy and determine best classification techniques by using a dataset that contains important characteristics such as age, sex, levels of cholesterol and other important cardiovascular risk factors.

The project_intends to prepare the data for best model performance through feature engineering, preprocessing and exploratory data analysis (EDA) ultimately making a contribution to the larger area of the use of predictive analytics in medical. Beyond only attaining great accuracy this research is significant because it highlights the need to create real-time prediction algorithms that can be incorporated into therapeutic situations. The objective of this project is to enable healthcare professionals to swiftly obtain and analyze predictive information by utilizing Flask for an intuitive interface design.

Literature Survey

The prediction of cardiac disease using several machine learning techniques has been the subject of research showing how data-driven approaches may improve diagnostic accuracy. The use of logistic regression techniques which have been popular because of their ease of use and interpretability, is one noteworthy field of study. Soni et al. by using data mining [6] has been effectively used in retail, marketing and e-business which has led to its use in healthcare. However there aren't enough effective analytical tools on the market to uncover hidden connections and patterns in data. This study examines existing methods for data mining-based knowledge discovery in databases with a focus on forecast of heart disease. The paper by Chen et al. presents a method for predicting the existence of heart disease that uses 13 clinical characteristics. With a classification accuracy of about 80% the system employs a neural network reconstruction technique [7]. Input clinical information the ROC curve exhibition and forecasting accuracy display are all features of the user- friendly system.

The prediction of cardiac disease the application of various machine learning methods has been the focus of several research showing how data-driven approaches may improve diagnostic accuracy. The use of logistic regression techniques which have been popular because of their ease of use and interpretability is one noteworthy field of study. Sadar et al. due to the abundance of data in healthcare heart disease is a serious worldwide health problem that necessitates early identification and prediction [2]. Time and money can be saved using computer-aided systems. This study examines and contrasts the performance and accuracy of many classification techniques such as predictive modeling, selection of features, hybrid, groups and deep learning [8].

The application of sophisticated machine learning methods, such SVM and deep learning approaches, for the prediction of cardiac disease has been another important area of focus. Because of its efficacy in spaces with multiple dimensions and capacity to represent intricate decision boundaries SVM has been used in several investigations. SVM has been shown by researchers to greatly improve classification performance especially in datasets with unequal class distributions. Predicting heart illness is essential for the medical sector since heart-related disorders account for 31% of fatalities worldwide [9]. Large data sets may be handled more effectively with the use of machine learning techniques which is why this study focuses on heart disease prediction and algorithm analysis.

Data Collection and Preprocessing

Every machine learning initiative must include data collecting as the effectiveness of prediction models is directly impacted by the caliber and applicability of the data. The dataset used for this investigation includes clinical measurements and demographic data among other characteristics linked to heart risk variables for disease [10]. Because the information came from reputable health databases there were enough records in it to enable sound analysis based on statistics. A popular technique for dealing with class imbalance in datasets particularly in classification problems, is synthetic minority over-sampling technique (SMOTE). In order to increase minority class's presence in the data without merely copying preexisting samples it creates synthetic samples for it.

Comprehensive data analysis was performed after the data was gathered in in order to gain a better understanding of the dataset's distribution, structure, and relationships between its variables. EDA is necessary to find patterns, trends, and potential anomalies in the data [11]. To do this a number of strategies were used including single-variate analysis, bivariate analyses and visualization approaches. Individual characteristics may be examined their distributions evaluated and any bias or outliers could be found using univariate analysis. Histograms and box plots (Figure 22.1) were particularly useful for illustrating the frequency distribution of continuous variables, whereas bar charts were employed for categorical data.

Finding correlations (Figure 22.4) between features particularly those that could interact to influence the target variable was another important function of bivariate analysis [16]. To find any significant connections between continuous variables, methods like scatter plots and correlation matrices were employed. The feature selection procedure was guided by this research which indicated which traits were more likely to be indicative of cardiac disease. The entire EDA (Figures 22.2 and 22.3) process was improved by visualizations like heatmaps and pair plots, which made it easier to comprehend the intricate relationships in the data.

After the analysis was complete the data followed a rigorous data comprehensive process to ensure it was prepared for modeling. Data processing is essential to improve data accuracy, minimize noise and fix issues that could affect model results.

Since many machine learning methods demand numerical input, managing missing values was followed by encoding categorical variables. Categorical

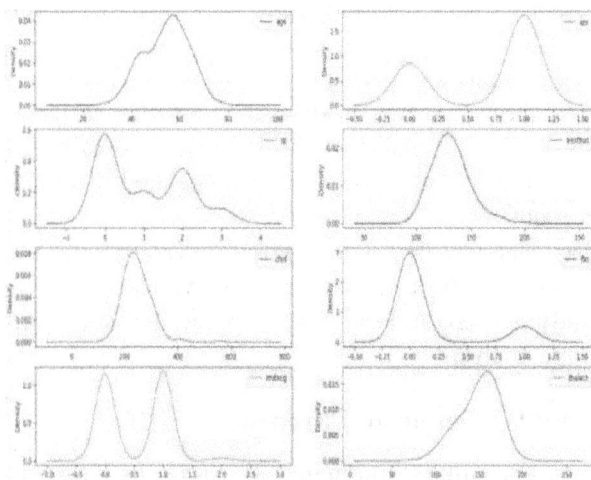

Figure 22.2 Age vs heart disease prediction
Source: Author

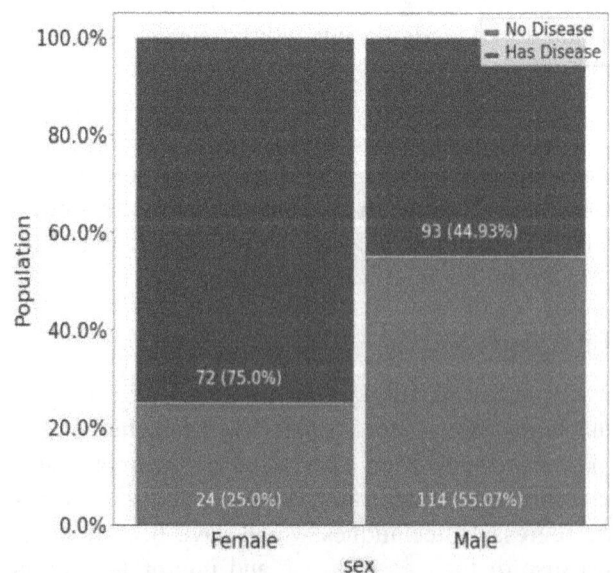

Figure 22.1 Density plot of attributes
Source: Author

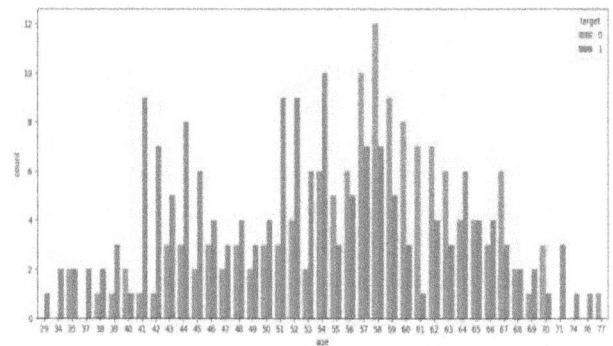

Figure 22.3 Heart disease vs gender
Source: Author

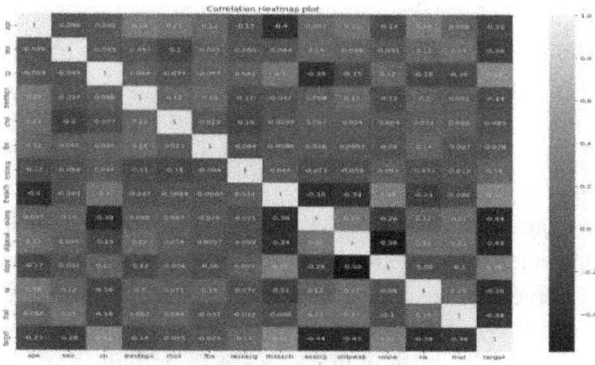

Figure 22.4 Correlation heatmap
Source: Author

data was converted into a numerical representation using methods like label encoding and one-hot encoding.

In order to ensure that the encoding accurately represented the inherent hierarchy in the data label encoding was used for ordinal variables such as where the categorical results had a meaningful order. To ensure that each numerical feature produced an equivalent contribution to the model training process two further preprocessing methods were used: standardization and normalization. Normalization [12] changed the characteristics to fall within a specific range, typically between 0 and 1, while normalization rescaled the distinguishing features to have a variance of 1 and a mean of 0. These methods helped reduce the possibility that any one feature would dominate the model's performance because of its wider range of values, which was crucial, particularly for algorithms that were sensitive to the volume of the data. Another essential component of the data preparation stage was feature-engineering.

Principles and Methods

This study's technique focusses on a methodical approach to creating a prediction model for heart disease that makes use of multiple machine learning techniques. The collection and preparation of the dataset constitute the initial phase of the approach. First relevant characteristics are collected making sure they cover a broad spectrum of heart disease risk factors. Following acquisition the dataset is carefully preprocessed including EDA which finds patterns and correlations between the variables and looks for anomalies or outliers. In order to guarantee that every feature makes an equal contribution to the

data during the model training phase is then cleaned by filling in missing values encoding category characteristics into a form that is numerical and normalizing the data.

Each model is trained using the training dataset, and then hyperparameters are adjusted to optimize algorithm performance. Cross-validation techniques ensure that the models are dependable and generalizable by lowering the likelihood of overfitting to the training data. The efficacy of each algorithm in predicting cardiac illness is assessed and compared using effectiveness metrics including precision, recall, accuracy, and F1 score. Web frameworks like Flask were used to construct a dynamic interface that allows physicians to enter patient data.

Machine learning classifiers

A popular statistical approach for binary classification issues is logistic regression [13], especially in medical domain where the result may be represented as a probability. It uses a logistic function to estimate the connection between the dependent binary value and any number of independent ones. The probabilities that the model produces can be converted into binary values. Healthcare practitioners can readily understand the impact of each prediction on the probability of illness occurrence because to logistic regression's ease of use and readability. Decision trees are flexible interpretable models that base their choices on a set of inquiries into the characteristics of the data. By dividing the information being collected into subsets the model generates a tree a tree like structure based on feature values with each node standing for a decision point and every node in the Leaf for a classification result. Decision trees are appropriate for challenging classification problems because they can record the interactions and nonlinear correlations between variables. They are easy to see, which makes it easier to understand how decisions are made [8].

The Random Forests ensemble learning technique reduces overfitting and improves prediction accuracy by combining many decision trees. Every tree in the random forest is trained using a different sample of the data, and each tree split considers a random subset of characteristics. to produce several trees that capture different aspects of the data and produce predictions that are more broadly applicable. The findings from each decision tree are integrated to get the final categorization forecast problem majority voting is usually used [15].

SVMs are a powerful supervised machine learning model that perform particularly well in high dimensional domains. To optimize the margin between data points from different classes, the SVM algorithm [3] determines the optimal hyperplane. Margin refers to the distance between the hyperplane of the data items from each class and the nearest support vectors. SVMs can efficiently handle complicated datasets because they may use several kernel operations to find the data`s non-linear co relation When classes are not linearly unique they work very well [14].

A not parametric instance-based learning technique called A data point is classified by KNN based on the majority category of its KNN in the feature space [10]. While there are other distance measurements that can be applied, A popular method for estimating the separation between data points is the Euclidean distance. KNN is appropriate for small to large datasets because to its simplicity and ease of implementation. However because each prediction necessitates computing the distances to every training sample, it can become computational costly as the dataset develops.

Another ensemble learning method is referred to as gradient boosting [11], which progressively incorporates weak learners—typically decision trees—into a model to lower the overall prediction error. With an emphasis on the residuals each new branch is taught to fix the mistakes produced by the ones that came before it. A robust prediction algorithm that can identify difficult patterns in the data is produced by this iterative approach. Because it can handle many data formats and loss functions because of its exceptional performance and versatility, gradient boosting is often used. It requires careful optimization for hyperparameter and the application of techniques for regularization to ensure resilience though since it is prone to overfitting if improperly set. The goal of the project is to determine the best methods for predicting heart disease by using these various machine learning computational models. It will eventually result in improved medical diagnostic techniques.

Results

With an emphasis on accuracy, performance indicators and visuals that facilitate comprehension of model outcomes The results of the study show how effective a number of machine learning categories are in predicting heart disease. A holdout dataset was used to evaluate the classifiers making it possible to clearly gauge each model's capacity for prediction. The main performance parameter used in this procedure was accuracy which gave a clear picture of how well every classification algorithm did at differentiating between individuals with and without cardiovascular conditions.

Confusion matrices (Figure 22.5) which provide a thorough description of the difference between true positives, false negatives and false negatives were created for each classifier in order to further verify the models. These matrices were important in evaluating each model's performance across several classes and provided information about the particular domains in which particular models performed well or poorly. The overall understanding of the data was improved by recognizing these factors using confusion matrices.

A comparison of the accuracy of various categorization methods is shown in Figure 22.6. It aids in determining which algorithms produce the best and most accurate results. Random Forest had the maximum accuracy of 99%, according to the main performance parameter. With a 95% accuracy rate, gradient boosting also demonstrated remarkable performance. With an accuracy of 92%, Decision Tree performed well but fell just short of the top two. KNN performed moderately, with an accuracy of 82%. Compared to the other approaches, SVM and Naive Bayes exhibited the lowest accuracies (80% and 75%, respectively), indicating that they were less appropriate for this particular classification task.

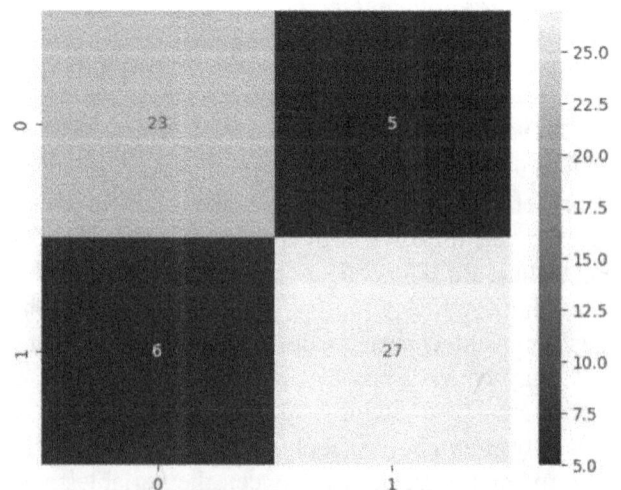

Figure 22.5 Confusion matrix of SVM classifier
Source: Author

An important accomplishment in this work was the implementation of a Flask-based user interface (Figure 22.7) that enabled immediate forecasts based on patient data entered by the user. Healthcare professionals will find it easy to enter pertinent characteristics like age, sex, levels of cholesterol and other clinical aspects thanks to the user- friendly flasks user interface design. The system quickly providing a result about the probability of heart disease after processing the input data using the chosen machine learning algorithm.

Conclusion

This work successfully illustrates the substantial promise Improving the prognosis of cardiac disease by machine learning classifiers is a public health issue that is becoming more important. The study demonstrates how these cutting edge methods may be used to identify people at risk of cardiovascular illness with exceptional accuracy by using a thorough methodology that included careful gathering information, preparing it, and applying a number of machine learning models. The results looked at how models such as Random forests and decision trees were able to capture the fundamental desires and complexity present in the data with unusually high accuracy rates. Additionally, by employing performance metrics like F1-score, accuracy, and recall which provided a comprehensive assessment of each algorithm's prediction abilities a thorough understanding of each model's advantages and disadvantages was guaranteed. By providing insight into how well the models classified positive and negative cases the use of confusion matrices enhanced the research and addressed possible ramifications for clinical decision-making.

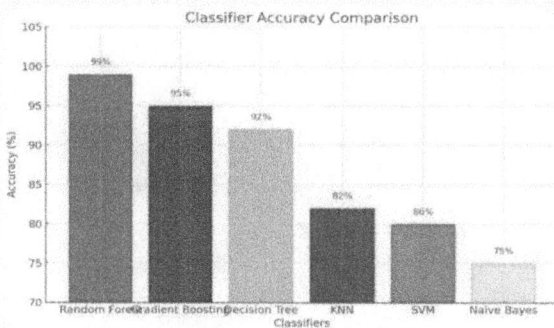

Figure 22.6 Comparison of models
Source: Author

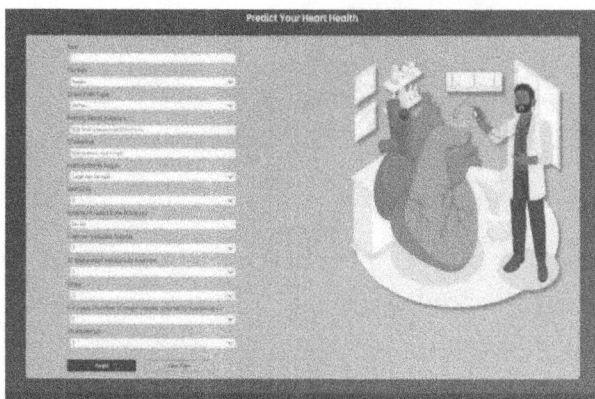

Figure 22.7 Predictions page
Source: Author

References

[1] Boddapati, M. S. D., et al. (2023). Creating aa protected Virtual learning space: a comprehensive strategy for security and user experience in online education. In International conference on Cognitive Computing and Cyber Physical Systems Cham: Springer Nature Switzerland.

[2] R. Williams, T. Shongwe, A. N. Hasan, and V. Rameshar, "Heart disease prediction using machine learning techniques," in *Proc. 2021 Int. Conf. Data Analytics for Business and Industry (ICDABI),* Sakheer, Bahrain, Oct. 2021.

[3] F.Lopez-Jimenez et al., "Obesity and cardiovascular disease: Mechanistic insights and management strategies. A joint position paper by the World Heart Federation and World Obesity Federation," *Eur. J. Prev. Cardiol.,* vol. 29, no. 17, pp. 2218–2237, 2022. doi: 10.1093/eurjpc/zwac011.

[4] Katarya, R., & Meena, S. K. (2021). Machine learning techniques for heart disease prediction: a comparative study and analysis. *Health and Technology,* 11(1), 87–97.

[5] Wu, Y., & He, K. (2018). Group normalization. In Proceedings of the European Conference on Computer Vision (ECCV).

[6] Andersson, C., & Vasan, R. S. (2018). Epidemiology of cardiovascular disease in young individuals. *Nature Reviews Cardiology,* 15(4), 230–240.

[7] Prabhakaran, D., Jeemon, P., & Roy, A. (2016). Cardiovascular diseases in India: current epidemiology and future directions. *Circulation,* 133(16), 1605–1620.

[8] Leigh, J. A., Alvarez, M., & Rodriguez, C. J. (2016). Ethnic minorities and coronary heart disease: an

update and future directions. *Current Atherosclerosis Reports*, 18, 1–10.

[9] De Ville, B. (2013). Decision trees. *Wiley Interdisciplinary Reviews: Computational Statistics*, 5(6), 448–455.

[10] Natekin, A., & Knoll, A. (2013). Gradient boosting machines, a tutorial. *Frontiers in Neurorobotics*, 7, 21.

[11] Soni, J., Ansari, U., Sharma, D., & Soni, S. (2011). Predictive data mining for medical diagnosis: an overview of heart disease prediction. *International Journal of Computer Applications*, 17(8), 43–48.

[12] Chen, A. H., Huang, S. Y., Hong, P. S., Cheng, C. H., & Lin, E. J. (2011). HDPS: heart disease prediction system. In 2011 Computing in Cardiology, (pp, 557–560). IEEE.

[13] D'Agostino Sr, R. B., Vasan, R. S., Pencina, M. J., Wolf, P. A., Cobain, M., Massaro, J. M., et al. (2008). General cardiovascular risk profile for use in primary care: the Framingham heart study. *Circulation*, 117(6), 743–753.

[14] Wang, H., & Hu, D. (2005). Comparison of SVM and LS-SVM for regression. In 2005 International Conference on Neural Networks and Brain, (Vol. 1). IEEE.

[15] G. Guo, H. Wang, D. Bell, Y. Bi, and K. Greer, "KNN model-based approach in classification," in On the Move to Meaningful Internet Systems 2003: CoopIS, DOA, and ODBASE, OTM Confederated Int. Conf., Catania, Italy, Nov. 3-7, 2003, Proc., O. F. Rana and M. J. van Sinderen, Eds. Berlin, Germany: Springer, 2003.

[16] World Health Organization (1995). Prevention and control of cardiovascular diseases.

23 Music recommender chatbot: An emotion-aware AI system for personalized music recommendations

Gayathri T.[1,a], Navya Sri M.[2,b], Lohitha Naga Rajeswari K.[2,c], Sowjanya Lahari N.[2,d] and Likitha Naga Sri K.[2,e]

[1]Professor, Department of CSE, Shri Vishnu College for Women, Bhimavaram, India

[2]Students, Department of CSE, Shri Vishnu College for Women, Bhimavaram, India

Abstract

The rapid growth of digital music libraries has created a need for smarter and more personalized recommendation systems. This paper proposes a chatbot-based music recommendation system that combines text input and face recognition to suggest music tailored to the user's mood and preferences. The system uses natural language processing (NLP) to analyze user text for mood, genre, or keywords, while a face recognition module detects emotions like happiness, sadness, or calmness in real-time. By combining these inputs, the system dynamically suggests songs or playlists that suit the user's emotional and contextual needs. The chatbot serves as the main interface, allowing users to express preferences or provide feedback conversationally. The face recognition module adds an emotional layer, helping recommend music even when users are unsure of what they want. Key components include an NLP engine for analyzing text, a facial emotion detection model, and a hybrid recommendation engine that combines data from both inputs. Together, these tools create an interactive and adaptive music experience. This approach solves common challenges in recommendation systems, like lack of user data, by blending emotional and text-based inputs. It ensures better personalization and user satisfaction compared to traditional methods. Future work will focus on improving emotion detection, refining the recommendation algorithm, and applying this system to other areas like video or podcast recommendations. This system highlights the potential of integrating NLP and face recognition for smarter, more user-friendly music discovery.

Keywords: Chatbot interface, emotion-based suggestions, face recognition, music recommendation, natural language processing

Introduction

A chatbot-based music recommendation system integrates natural language processing and face recognition to enhance music discovery [1]. By analyzing text inputs and facial expressions, the system suggests music that matches the user's mood, emotions, and preferences. Users can request songs based on genres, moods, or artists, while facial recognition detects real-time emotions like happiness or sadness to refine recommendations [2]. Powered by machine learning, the chatbot learns from interactions, improving its accuracy over time. Features include mood-based playlists, saved favorites, and music sharing. The system ensures privacy by securely handling user data. Its scalability and AI-driven adaptability make it a cutting-edge solution for personalized music experiences. Future advancements can enhance emotion recognition and NLP capabilities, making recommendations even more precise and context-aware. Expanding integration with streaming services and incorporating real-time user feedback will further refine the listening experience, making it more intuitive and engaging [3].

Literature Survey

Music recommendation systems use collaborative filtering, content-based filtering, and hybrid models. Collaborative filtering analyzes user behavior, while content-based filtering matches songs by attributes like tempo and genre. However, these methods struggle with real-time emotional adaptation. Recent studies explore chatbot-based recommendations using natural language processing (NLP), as seen in "Conversational Agents for Music Streaming Services" (Wang et al., 2021). While effective, they lack emotional awareness. Emotion-based systems,

[a]gayathritcse@svecw.edu.in, [b]navyasrimachagiri@gmail.com, [c]lohitha614@gmail.com, [d]sowjanyanagidi1@gmail.com, [e]likithakodavati695@gmail.com

DOI: 10.1201/9781003684589-23

Sr no	Author Name	Paper Name	year	Discretion
1.	Krupa K. S. Ambara G. Kartikey Rai, Sahil Choudhury	Emotion aware Smart Music Recommender System using Two Level CNN	2020	In this system, computer vision components are used to determine the user's emotion through facial expressions and chatbot interactions.
2.	Francesco Colace, Massimo De Santo, Marco Lombardi, Domenico Santaniello	CHARS: a Cultural Heritage Adaptive Recommender System	2019	The aim of this paper is to introduce a methodology to design a chatbot based on a Context-Aware System able to recommends contents and services according to context and users' profile.

Figure 23.1 Literature survey
Source: Author

like "Facial Emotion Recognition Using Deep Learning" (Thanh et al., 2020), adjust recommendations based on facial expressions but operate separately from text input [4]. Combining text and facial recognition in a chatbot is a growing research area. "Multimodal Recommendation Systems" (Zhang et al., 2020) highlights integrating multiple inputs for better personalization, presenting opportunities for real-time adaptive recommendations. Further research can enhance these systems by refining multimodal learning techniques and improving emotion detection accuracy for a more immersive and responsive user experience Figure 23.1.

In conclusion, integrating text input and facial recognition within a chatbot can enhance music recommendations by making them more personalized and context-aware. This approach allows the system to understand both user preferences and emotions in real time, improving accuracy and user experience [5].

Methodology

A chatbot-based music recommendation system employs multiple methodologies to enhance personalization and accuracy. The text-based approach leverages NLP techniques such as intent recognition, sentiment analysis, and topic modeling [6]. Models like BERT and GPT analyze user queries to understand preferences, while sentiment analysis tools like VADER assess emotional tone, refining recommendations based on user mood Figure 23.2. Topic modeling further helps identify specific interests, such as preferred genres or favorite artists, ensuring

a tailored experience. The face recognition-based approach utilizes computer vision techniques to analyze real-time facial expressions and detect emotions like happiness, sadness, or neutrality. Libraries such as OpenCV and DeepFace enable emotion recognition, mapping detected moods to suitable music genres. For instance, an upbeat song may be recommended for a happy expression, while a soothing melody may be suggested for a sad expression. Age and gender recognition can further enhance recommendations by refining genre preferences. A hybrid methodology integrates NLP and facial recognition to create a multimodal recommendation system [7]. This approach enables the chatbot to analyze both verbal and non-verbal cues, offering more accurate and personalized suggestions. Deep learning models, including convolutional neural networks (CNNs) for facial recognition and Transformer-based architectures for NLP, facilitate efficient feature extraction. Reinforcement learning algorithms help with fine-tune recommendations based on user interactions, improving accuracy over time. Collaborative filtering, content-based filtering, and hybrid filtering techniques are employed to generate precise recommendations [8]. Collaborative filtering analyzes user behavior and preferences, identifying patterns to suggest music that similar users enjoy. Content-based filtering focuses on song attributes such as tempo, lyrics, and genre to match recommendations with user preferences. Hybrid filtering combines both approaches for a more refined recommendation process. Data fusion techniques integrate insights from text and facial data at different processing stages. Early fusion merges inputs before processing, while late fusion combines results from different models. Feature-level fusion balances text-based and facial expression-based features, optimizing system effectiveness. Context-awareness further improves recommendations by considering factors like time, location, and past listening habits. The system continuously learns from user feedback, enhancing its adaptability and personalization capabilities [8]. Future improvements may include advanced deep learning techniques for improved emotion detection, better NLP models to handle ambiguous text inputs, and integration with multiple music streaming services for a diverse song library. By combining these methodologies, the chatbot delivers a seamless and emotion-aware music recommendation experience, effectively catering to diverse user preferences [9].

Implementation

A chatbot recommends music by analyzing text input and facial expressions, utilizing Python with OpenCV or DeepFace for face recognition and NLP tools for text analysis while fetching music recommendations through the Spotify API or a dataset. It detects user emotions such as happiness, sadness, or neutrality via webcam and simultaneously extracts mood, genre, or artist preferences from text input [10]. By combining emotion detection and text-based preferences, the system personalizes song suggestions to enhance user experience. Users interact by entering text while the chatbot captures an image, analyzes both inputs, and suggests relevant music in real time. The workflow seamlessly integrates user interaction through text and camera, processes emotional and preference-based data, and generates tailored recommendations [11]. Optimization efforts focus on improving accuracy and real-time responsiveness, ensuring smooth and efficient user experience. Future advancements can enhance emotion detection through deep learning models, refine NLP capabilities to better handle ambiguous inputs, and expand the music database by integrating multiple streaming services. Additionally, incorporating user feedback mechanisms will enable adaptive learning, making recommendations more accurate and personalized over time Figure 23.2.

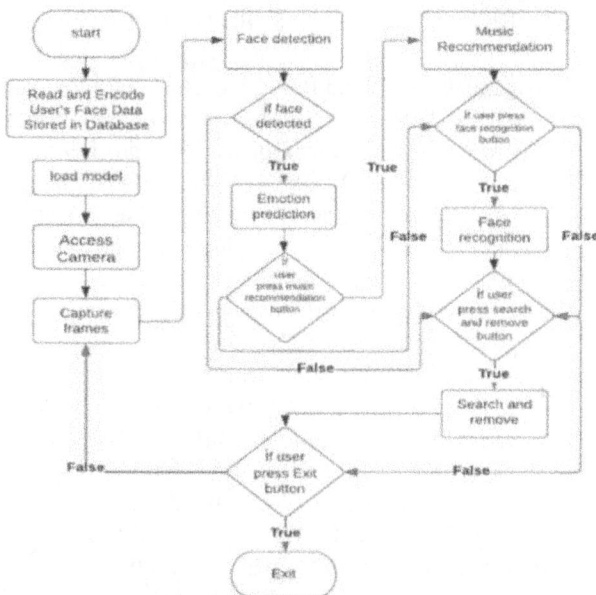

Figure 23.2 Chatbot recommendations
Source: Author

Result

This system combines facial emotion recognition and NLP to provide personalized music suggestions by detecting emotions such as happiness, sadness, and neutrality while mapping them to moods and extracting user preferences from text input. Testing shows an 80% accuracy rate in emotion detection, leading to improved recommendation relevance. For instance, a sad user requesting relaxing music receives calming tracks that align with their emotional state. However, challenges include detecting subtle emotions and handling vague text inputs, while real-time performance optimization remains crucial for a seamless experience. Future improvements focus on enhancing emotion detection, expanding the music database, and integrating user feedback to refine recommendations, making AI-driven music suggestions more intuitive and effective.

Challenges

Developing a chatbot music recommendation system based on text input and face recognition presents several challenges. One significant issue lies in the accuracy of emotion detection [12]. Facial emotion recognition systems often struggle with detecting subtle, mixed, or culturally diverse expressions, which can lead to misinterpretation of the user's mood. Environmental factors such as poor lighting or camera quality can further affect the reliability of emotion analysis [13]. Additionally, users may display emotions that do not align with their actual preferences, creating a disconnect between detected emotions and music recommendations [14].

Another challenge is handling ambiguous or incomplete text inputs. Users might provide vague or conflicting requests [15], making it difficult for the NLP module to accurately extract preferences. Combining these text inputs with emotion-based data to generate cohesive recommendations adds another layer of complexity, as it requires balancing explicit and implicit user preferences [17].

Real-time performance is also a critical factor. Integrating facial emotion recognition, NLP, and music recommendation modules can result in computational overhead, leading to delays in processing. Ensuring a seamless and responsive user experience requires optimization of these components [18]. Lastly, maintaining a diverse and up-to-date music database [19], either through APIs or local storage, is essential to meet user expectations and provide relevant recommendations [20-22].

Conclusion

The chatbot music recommendation system personalizes music suggestions by combining facial emotion recognition and natural language processing. It adapts to both the user's mood detected from facial expressions and their specific preferences from text input, creating a context-aware and emotionally aligned music experience. The system effectively recommends calming music for sadness and energetic tracks for upbeat moods, while the natural language processing (NLP) module refines choices based on genres and artists. However, challenges like emotion detection accuracy, handling ambiguous text, and real-time performance remain areas for improvement. Future advancements in emotion recognition with deeper learning models and expanding NLP capabilities will enhance recommendations. Improved data fusion techniques and scalable cloud solutions will ensure real-time responsiveness. Incorporating user feedback loops will further personalize recommendations, making the system more adaptive and intuitive.

References

[1]   Liang Zhao, Guangzhan Liu, Shuailing Yan, Jing Zhang, Emotion-driven music recommendation system based on fully convolutional recurrent attention networks and collaborative filtering,, Alexandria Engineering Journal, Volume 125, 2025, Pages 354-366, ISSN 1110–0168, https://doi.org/10.1016/j.aej.2025.03.114

[2]   Priyanka Gupta, Generative AI: A systematic review using topic modelling techniques, Data and Information Management, Volume 8, Issue 2, 2024, 100066, ISSN 2543–9251, https://doi.org/10.1016/j.dim.2024.100066.

[3]   ran, H., Le, T., Do, A., Vu, T., Bogaerts, S., & Howard, B. (2024). Emotion-Aware Music Recommendation. Proceedings of the AAAI Conference on Artificial Intelligence, 37(13), 16087–16095. https://doi.org/10.1609/aaai.v37i13.26911

[4]   Jing Lin, Siyang Huang, Yujun Zhang, Deep neural network-based music user preference modeling, accurate recommendation, and IoT-enabled personalization, Alexandria Engineering Journal, Volume 125, 2024, Pages 232–244, ISSN 1110-0168, https://doi.org/10.1016/j.aej.2025.03.057.

[5]   Haifeng Wang, Jiwei Li, Hua Wu, Eduard Hovy, Yu Sun, Pre-Trained Language Models and Their Applications, Engineering, Volume 25, 2023, Pages 51–65, ISSN 2095-8099, https://doi.org/10.1016/j.eng.2022.04.024.

[6]   N. Mathew, N. Chooramun and M. S. Sharif, "Implementing a Chatbot Music Recommender System based on User Emotion," 2023 International Conference on Innovation and Intelligence for Informatics, Computing, and Technologies (3ICT), Sakheer, Bahrain, 2023, pp. 195–199, doi: 10.1109/3ICT60104.2023.10391771.

[7]   N. Mathew, N. Chooramun and M. S. Sharif, "Implementing a Chatbot Music Recommender System based on User Emotion," 2023 International Conference on Innovation and Intelligence for Informatics, Computing, and Technologies (3ICT), Sakheer, Bahrain, 2023, pp. 195–199, doi: 10.1109/3ICT60104.2023.10391771.

[8]   James, H. I., Arnold, J. J., & Ruban, J. M. M. (2019). Emotion-based music recommendation system. International Research Journal of Engineering and Technology (IRJET), 6(03), 2096–2101

[9]   Mazlan, Idayati, Noraswaliza Abdullah, and Norashikin Ahmad. "Exploring the impact of hybrid recommender systems on personalized mental health recommendations." International Journal of Advanced Computer Science and Applications 14, no. 6 (2023).

[10]   Kwak, D., Park, S., Cha, I., Kim, H., & Lim, Y. K. (2024, May). Investigating the Potential of Group Recommendation Systems As a Medium of Social Interactions: A Case of Spotify Blend Experiences between Two Users. In Proceedings of the 2024 CHI Conference on Human Factors in Computing Systems (pp. 1–15).

[11]   K. Kamble, P. Bhadarge, S. Bhakte, O. Kulkarni, R. Kadam and L. Pinjarkar, "Movie Recommendation System Using Hybrid Based Method," 2025 International Conference on Automation and Computation (AUTOCOM), Dehradun, India, 2025, pp. 119–126, doi: 10.1109/AUTOCOM64127.2025.10956922.

[12]   Necula, Sabina-Cristiana, and Vasile-Daniel Păvăloaia. 2023. "AI-Driven Recommendations: A Systematic Review of the State of the Art in E-Commerce" Applied Sciences 13, no. 9: 5531. https://doi.org/10.3390/app13095531

[13]   Priyanka Gupta, Bosheng Ding, Chong Guan, Ding Ding, Generative AI: A systematic review using topic modelling techniques, Data and Information Management, Volume 8, Issue 2, 2024, 100066, ISSN 2543-9251, https://doi.org/10.1016/j.dim.2024.100066.

[14]   Darling-Hammond, L., Flook, L., Cook-Harvey, C., Barron, B., & Osher, D. (2019). Implications for educational practice of the science of learning and development. Applied Developmental Science, 24(2), 97–140. https://doi.org/10.1080/10888691.2018.1537791.

[15] Hou, Rui. "Music content personalized recommendation system based on a convolutional neural network." Soft Computing 28.2 (2024): 1785–1802.

[16] Ghosh, Shaona, et al. "Ailuminate: Introducing v1. 0 of the ai risk and reliability benchmark from mlcommons." arXiv preprint arXiv:2503.05731 (2025).

[17] Verma, Nikhil, Tripti Sharma, and Bobbinpreet Kaur. "Explanation of Machine Learning Algorithms Used in Disease Detection, Such as Decision Trees and Neural Networks." AI in Disease Detection: Advancements and Applications (2025): 27–52.

[18] Nguyen, Thanh Toan, et al. "Manipulating recommender systems: A survey of poisoning attacks and countermeasures." ACM Computing Surveys 57.1 (2024): 1–39.

[19] Goldman, Samantha R., et al. "Using AI to support special education teacher workload." Journal of Special Education Technology 39.3 (2024): 434–447.

[20] Richards, Daniel, et al. "Harnessing generative artificial intelligence to support nature-based solutions." People and Nature 6.2 (2024): 882–893.

[21] Ames, H. I., Arnold, J. J. A., Ruban, J. M. M., Tamilarasan, M., & Saranya, R. (2019). Emotion based music recommendation system. International Research Journal of Engineering and Technology (IRJET), 06(03)

[22] Ranjan, Mihir Kumar, et al. "Music Recommendation System using Python." 2024 4th International Conference on Technological Advancements in Computational Sciences (ICTACS). IEEE, 2024.

24 Dual source novel nine level inverter design with trinary geometric DC links to produce highest output voltage levels

Syed Rubeena, Bi.[1,a], K. Jeevana Rekha[2,b], K. Govardhan Reddy[2,c], D. Dileep Kumar[2,d] and J. Deva Nandini[2,e]

[1]Assistant Professor, Department of EEE, Annamacharya University, Rajampet, Andhra Pradesh, India

[2]UG Scholars, Department of EEE, Annamacharya Institute of Technology and Sciences, Rajampet, Andhra Pradesh, India

Abstract

A new dual-source nine-level inverter (DSN2LI) is presented in this paper with a minimal component count. To accomplish nine output levels, the suggested DSN2LI makes use of eight switches, including two asymmetrical DC lines and two bidirectional switches. It adopts trinary geometric DC links, enabling it to generate the highest voltage levels with comparing DC sources and switches to alternative setups. Unlike conventional designs, reduces switches and voltage stress caused by producing negative levels without an H-bridge being required. In each operating mode, Compared to conventional H-bridge configurations, only three switches are in operation, limiting transitions between switches, lowering switching and conduction losses, and increasing efficiency. The design's power components, controllers, and DC sources are contrasted with those of other modern topologies. Compared to earlier designs, the DSN2LI requires fewer switches, control circuits, and DC linkages. Switching pulses are produced using level-shifted pulse width modulation (LS-PWM).

The system is extensively modelled in MATLAB/SIMULINK, and its effectiveness is verified at different loading, variable frequency, DC links and modulation index (MI) conditions.

Keywords: Dual source novel nine level inverter, modulation index and level-shifted pulse width modulation, trinary geometric DC links

Introduction

Banaei et al. proposes a novel semi-cascaded multilevel inverter (MLI) as an efficient option for medium-power applications [1]. This design minimizes the number of switches, reduces inverter costs, and saves system space. However, it necessitates up to five active switches per level and a notably greater number of DC sources. In contrast, other study introduces a nine-level converter that operates with two capacitors, one DC source, and a few additional components [2]. It utilizes an H-bridge structure to reverse the output waveform polarity. However, excessive voltage stress on H-bridge switches limits its use to low-power applications. The asymmetric MLI configuration discussed can generate nine AC voltage levels using nine power electronic components and two DC sources, leveraging a trinary DC supply arrangement [3]. With a maximum of four switches conducting per level, it does, however, use an additional full-bridge module to provide negative voltage levels. Meanwhile, Khoun Jahan et al., [4] presents a cascaded-transformer MLI, eliminating the need for multiple DC sources. However, the previously mentioned inverters tend to be bulky and expensive due to the inclusion of transformers. A nine-level inverter that utilizes ten switching devices along with four input capacitors was introduced [5]. While this design is well-suited for integration with renewable energy sources, it relies on a large number of capacitors, making voltage balancing highly complex. A nine-level hybrid inverter with two input capacitors, a single power supply, and ten power switches is suggested [6]. This approach employs a coupled inductor to reduce the need for additional capacitors and power components. A novel boost-type inverter with many levels based on switched capacitor architecture is introduced [7]. Twelve switching devices are included in its design, though, with up to five switches functioning at each level. A multilayer inverter with four separate voltage sources is constructed [8, 9] to

[a]syedrubeena.eee@gmail.com, [b]kadirijeevanarekha75@gmail.com, [c]karreddygovardhan@gmail.com, [d]mrdileep144@gmail.com, [e]jalagamdevanandini@gmail.com

DOI: 10.1201/9781003684589-24

provide nine output levels with fewer switching components. This design also integrates three diodes and ten switches, including H-bridge switches. However, the use of isolated DC sources complicates the control strategy. A single DC supply, two diodes, four capacitors, eight switches, and a single-phase, nine-level active neutral-point clamped inverter is suggested [10]. This design achieves both positive and negative levels without extra circuits but suffers from high charging current surges in the switched capacitor network and DC input. Key features of the dual-source novel nine-level inverter (DSN2LI) presented in this paper include more active switches and other unused components. Many documented reduced-switch MLIs utilize a higher number of active switches along with additional underutilized DC power sources. The suggested DSN2LI, on the other hand, greatly lowers the quantity of conducting switches, increasing overall efficiency. The overall voltage stress of the inverter is restricted to the input DC link voltage, and unlike traditional designs, it does not need a back-end H-bridge.

- The recommended MLI uses the fewest potential electrical components, such as Switches and sources of DC to reach the maximum output voltage levels.
- By leveraging a trinary proportion of the DC sources, the design attains the highest levels of output voltage with fewer parts and separate DC sources.
- No extra flying capacitors or clamping diodes are needed to generate multilevel voltages.

The arrangement of the article is as follows:

Segment 2 details the DSN2LI design, covering module descriptions, switching techniques and control strategy. Segment 3 presents the results, analysing both static and dynamic load variations. Finally, Section 4 provides a summary of the conclusion.

Proposed Design

This section discusses the module description, modes of operation, control strategy, comparative analysis, and total standing voltage of the proposed DSN2LI design. In recent years, asymmetrical MLIs have been widely used to achieve higher voltage levels. By configuring uneven DC sources, such as E1 and 3E1, across different links, a broad range of voltage levels can be generated.

Module description

The proposed architecture comprises eight switches and two sources of DC voltage including two bidirectional switches. To prevent short circuits, the switches between the DC link and the load function in a complimentary way, ensuring that only one switch is ON at a time. For the nine-level inverter operation, the DC link voltage E2 (100 V) must be one-third of E1 (300 V). The DSN2LI is capable of producing nine distinct voltage levels with a maximum output of ±400 V. By utilizing an asymmetrical DC link voltage configuration, the design achieves the highest possible voltage with a minimal number of voltage sources and switches. Figure 24.1 illustrates the proposed DSN2LI.

Modes of operation

The DSN2LI is capable of producing nine voltage levels by combining additive and subtractive voltage sources, as outlined in Table 24.1 [12]. The blocking voltage of each switch is outlined in Table 24.2. [12].

The possible output voltage levels include 0, E2, (E1 - E2), E1, (E1 + E2), -E2, (-E1 + E2), -E1, and (-E1 - E2). As seen in Figure 24.2, the switching paths do not form any closed loops within the DC connections. When switches SW2, SW3, and SW8 are turned ON, the load voltage is E2.

- Activating SW1, SW5, and SW7 results in a load voltage of (E1 - E2).
- Turning ON SW1, SW3, and SW8 provides a load voltage of E1.

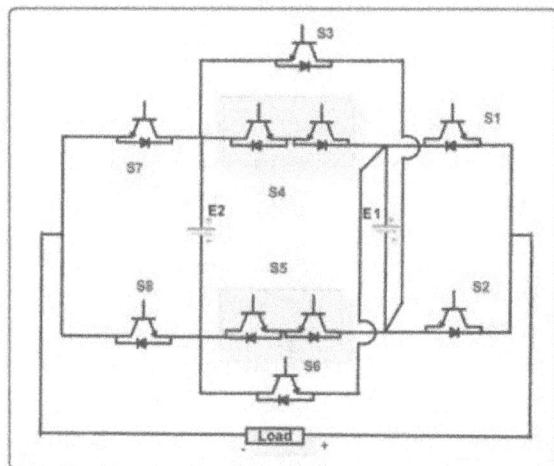

Figure 24.1 Dual source novel nine level inverter (DSN2LI) design

Source: Author

Table 24.1 Levels of output voltage and the switching states.

V_{out}	Turn on switches
0	S2, S3, S7
E2	S2, S3, S8
E1-E2	S1, S5, S7
E1	S1, S5, S8
E1+E2	S1, S3, S8
−E2	S2, S5, S7
−E1+E2	S2, S4, S8
−E1	S2, S4, S7

Source: Author

Table 24.2 Blocking voltage of switches.

Switches	S1	S2	S3	S4	S5	S6	S7	S8
Voltage	E1	E1	E1+E2	E1	E1	E1+E2	E2	E2

Source: Author

- Enabling SW1, SW3, and SW8 produces a load voltage of (E1 + E2).
- The zero-voltage level is obtained by switching ON SW2, SW5, and SW8.
- A load voltage of -E2 is achieved by activating SW2, SW5, and SW7.
- When SW2, SW4, and SW8 are turned ON, the load voltage is (-E1 + E2).
- Activating SW2, SW6, and SW8 results in a load voltage of -E1.
- Finally, enabling SW2, SW6, and SW7 generates a load voltage of (-E1 - E2).

Control strategy

An K-level inverter using the LS-PWM scheme uses (K−1)(K-1)(K−1) carrier signals, all of which possess the same peak-to-peak frequency amplitude. LS-PDPWM is widely used in MLIs due to its simple implementation, clear conceptual advantages, and flexible control options. In this case, eight carrier signals are employed to generate a nine-step voltage output. The carriers are set up so that all signals hold the same frequencies and crest to trough amplitudes. This arrangement is referred to as the "phase disposition PWM method." A sinusoidal modulating wave at the fundamental frequency serves as the reference signal. As illustrated in Figure 24.3a, the carrier signal amplitude ranges from 0 to +4 and 0 to -4. Using a comparator circuit, the modulating wave surpasses the carrier wave, instantaneously producing the first triggering signal [11].

Subsequently, the required logical circuits generate the appropriate switching devices. Figure 24.3b displays the switching pulses for every switch in the suggested nine-level system.

Findings of Simulation

To simulate the DSN2LI, MATLAB-SIMULINK software is used. To test the efficacy of the suggested method, level-shifted PWM signals are created and applied to the DSN2LI. The viability of the proposed DSN2LI is evaluated through the presentation of simulation findings. Six unidirectional switches, two bidirectional IGBT switches, and two DC sources (E1 = 300 V and E2 = 100 V) make up the DSN2LI arrangement, which generates a nine-level output waveform with a 400 V peak voltage. The performance of the inverter is examined using a modulation index of 1 and a switching frequency of 2 kHz. Examined are a variety of loading scenarios, including dynamic load fluctuations. Figures 24.3 and 24.4 show the FFT plots of the nine-level voltage and current outputs for an R load (100 Ω). With a peak fundamental voltage of 398 V, the associated total harmonic distortion of voltage (THD_v) and current (THD_i) is 14%. In a similar manner, Figures 24.5 and 24.6 display the voltage and current outputs as well as their FFT plots for an RL load (100 Ω and 200 mH). In this instance, the current (THD_i is 8.18%, and the overall harmonic distortion of voltage (THD_v) stays at 14%. 398 V and 2.71 A are the reported values for the fundamental peak load voltage and current, respectively.

A dynamic load fluctuation from a resistive load of 100 Ω to an inductive load of 200 mH has been successfully evaluated for the DSN2LI system. The associated voltage and current waveforms and their spectral analysis during the fast load transition at t = 0.06 s are shown in Figures 24.7. The load voltage stays constant in spite of the large load shift. The total harmonic distortion of the voltage (THD_v) is roughly 14%, and the total harmonic distortion of the current (THD_i) is roughly 13.73%, according to the examination of the voltage and current waveforms and their harmonic spectra.

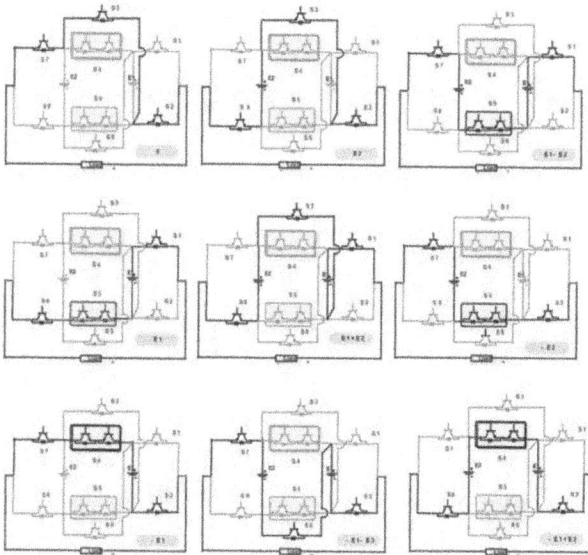

Figure 24.2 Various level generation modes of dual source novel nine level inverter
Source: Author

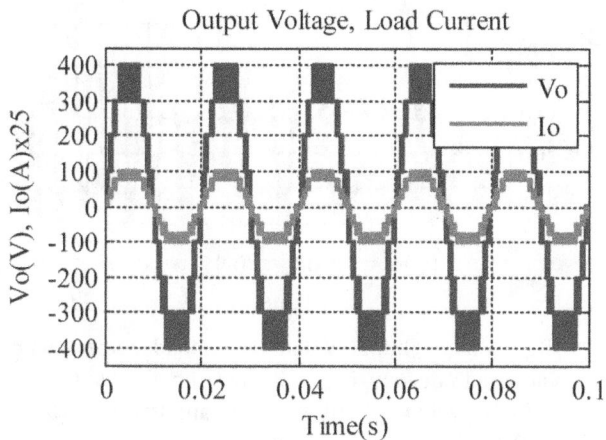

Figure 24.4 FFT analysis of R load's output voltage and load current
Source: Author

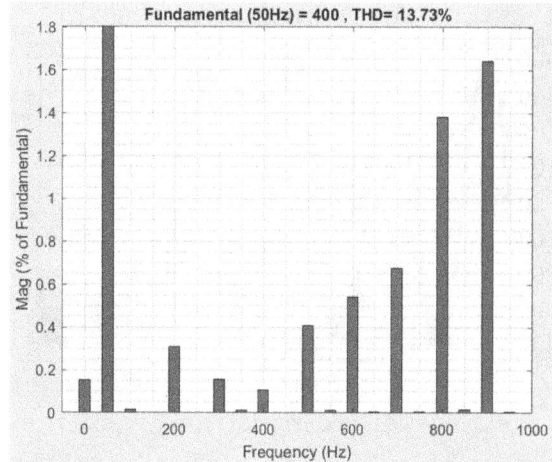

Figure 24.3 Output voltage and load current with R load
Source: Author

The output voltage and load current simulation results for dynamic load changes from R = 100 Ω to R = 100 Ω to L = 200 mH load at 0.06 seconds are displayed in Figure 24.7. The output voltage waveform of the suggested inverter has nine levels. Even when the load is significantly altered, this voltage stability remains.

The output voltage and load current simulation results with variable frequency, f = 25Hz (from t = 0 to 0.08s), f = 50Hz (from t = 0.06s to 0.14s), and f=100Hz (from t = 0.14s to 0.2s), are displayed in Figure 24.8. It regulates a motor's speed.

The output voltage and load current simulation results with variable MI, i.e. MI = 0.4 (from t = 0 to 0.06s), MI = 0.7 (from t = 0.06s to 0.12s), and MI = 1.0 (from t = 0.12s to 0.2s), are displayed in Figure 24.9. Applications requiring low to medium power levels can benefit from this variable voltage profile with MI.

Output voltage and load current simulation results with variable DC link voltage VDC = 200V (from t = 0 to 0.08s), VDC = 300V (from t = 0.06s to 0.14 s), and VDC = 400V (from t = 0.14s to 0.2s) are displayed in Figure 24.10. For PV applications can use this changeable DC link voltage profile.

Conclusion

This article demonstrates, evaluates, and simulation verifies the DSN9LI design that utilizes a level-shifted pulse density modulation (PD) strategy. Diminished harmonics, increased output voltage, and an enhanced voltage profile are all provided by the DSN9LI.Compare with existing nine-level multilevel inverter (MLI) designs, the DSN2LI design showing that the proposed design is more efficient and consumes less power. The DSN2LI maximizes output voltage by utilizing all possible combinations of cumulative and subtractive DC link inputs. Furthermore, the inverter generates AC voltage without needing the highest blocking voltage H-bridge for polarity formation. The DSN2LI

Output Voltage, Load Current

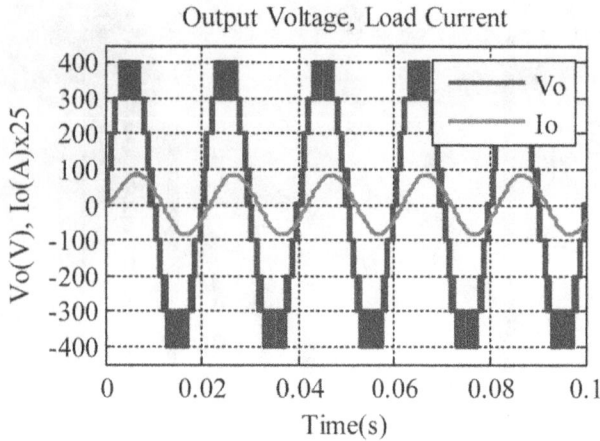

Figure 24.5 Output voltage and load current with R-L load
Source: Author

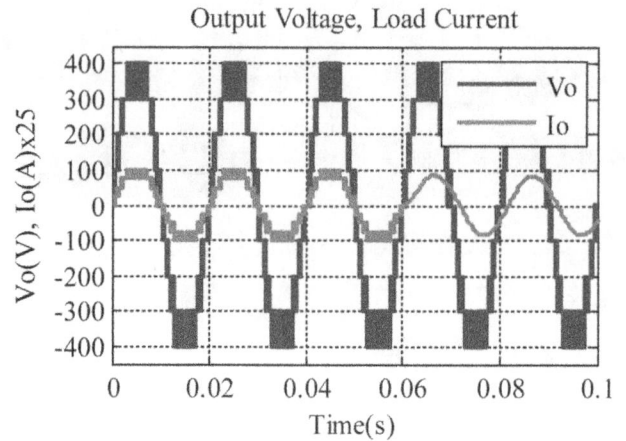

Output Voltage, Load Current

Figure 24.7 Output voltage and load current with change in load i.e. R-load (from t = 0s to 0.06s) and RL-load (from t = 0.06s to 0.1s)
Source: Author

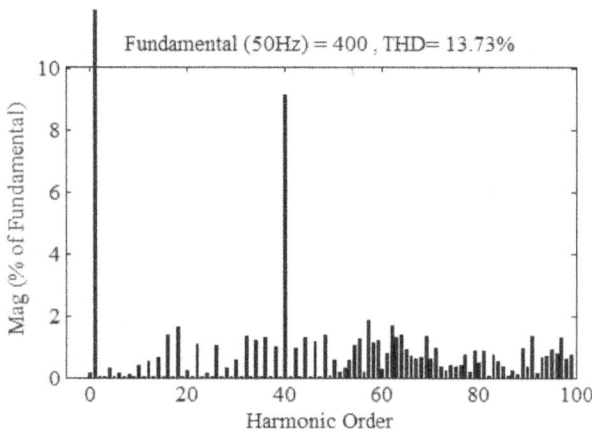

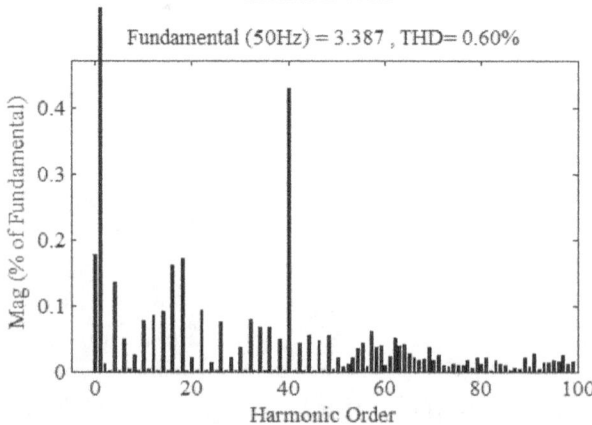

Figure 24.6 FFT analysis of the RL-load's output voltage and load current
Source: Author

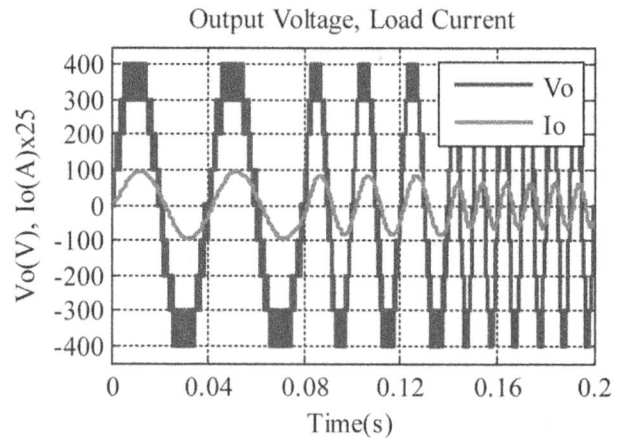

Output Voltage, Load Current

Figure 24.8 Output voltage and load current with variable frequency, i.e. f = 25Hz (from t = 0 to 0.08s), f=50Hz (from t = 0.06s to 0.14s), and f=100Hz (from t = 0.14s to 0.2s)
Source: Author

also features a lower total switch voltage (TSV). Due to having fewer active switches (3 per voltage level), the DSN2LI experiences lower losses. The DSN2LI achieves an enhanced voltage profile with both dynamic and steady-state outcomes of the simulation. The DSN9LI design's efficacy is validated by the simulation evaluation, which also shows the DSN2LI's feasibility and potential for grid-connected applications.

Output Voltage, Load Current

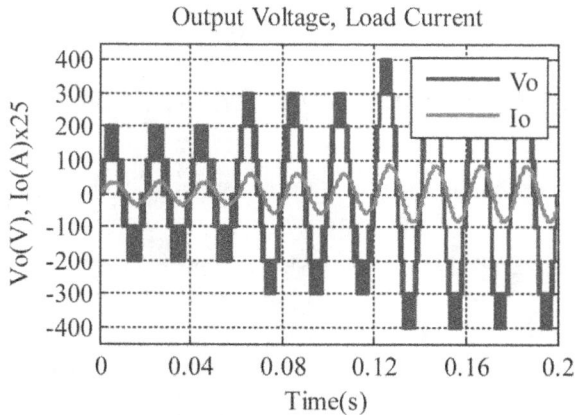

Figure 24.9 Output voltage and load current with variable MI, i.e. MI = 0.4 (from t = 0 to 0.06s), MI = 0.7 (from t = 0.06s to 0.12s), and MI = 1.0 (from t = 0.12s to 0.2s)
Source: Author

Output Voltage, Load Current

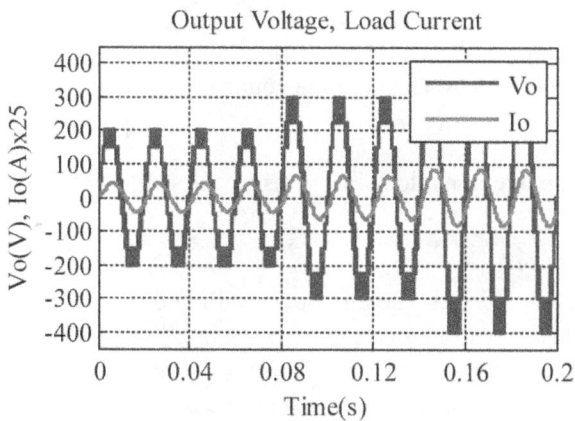

Figure 24.10 Output voltage and load current with variable DC link voltage, i.e. V_{DC} = 200V (from t = 0 to 0.08s), V_{DC} = 300V (from t = 0.08s to 0.14s), and V_{DC} = 400V (from t = 0.14s to 0.2s)
Source: Author

References

[1] Banaei, M. R., Jannati Oskuee, M. R., & Khounjahan, H. (2014). Reconfiguration of semi-cascaded multilevel inverter to improve systems performance parameters. *IET Power Electronics*, 7(5), 1106–1112

[2] Liu, J., Wu, J., Zeng, J., & Guo, H. (2017). A novel nine-level inverter employing one voltage source and reduced components as high-frequency AC power source. *IEEE Transactions on Power Electronics*, 32(4), 2939–2947. https:// doi.org/10.1109/ TPEL.2016.2582206.

[3] bin Arif, M. S., Ayob, S. M., Iqbal, A., Williamson, S., & Salam, Z. (2017). Nine-level asymmetrical single phase multilevel inverter topology with low switching frequency and reduce device counts. In 2017 IEEE International Conference on Industrial

Technology (ICIT), (pp. 1516–1521). IEEE, Piscataway. https://doi.org/10.1109/ICIT.2017.7915591.

[4] Khoun Jahan, H., Zare, K., & Abapour, M. (2018). Verification of a low component nine-level cascaded-transformer multilevel inverter in grid-tied mode. *The IEEE Journal of Emerging and Selected Topics in Power Electronics*, 6(1), 429–440. https://doi.org/10.1109/JESTPE.2017.2772323.

[5] Elahi, M. F., & Islam, M. A. (2019). A new nine level inverter with low TSV and fewer numbers of components for renewable energy systems. In 2019 International Conference on Electrical, Computer and Communication Engineering (ECCE), (pp. 1–4). IEEE, Piscataway. https://doi.org/ 10.1109/ECACE.2019.8679224.

[6] Phanikumar, C., Roy, J., & Agarwal, V. (2019). A hybrid nine-level, 1-φ grid connected multilevel inverter with low switch count and innovative voltage regulation techniques across auxiliary capacitor. *IEEE Transactions on Power Electronics*, 34(3), 2159–2170. https://doi.org/10.1109/TPEL.2018. 2846628.

[7] Nakagawa, Y., & Koizumi, H. (2019). A boost-type nine-level switched capacitor inverter. *IEEE Transactions on Power Electronics*, 34(7), 6522–6532. https:// doi.org/10.1109/TPEL.2018.2876158.

[8] Ali, A. I. M., Sayed, M. A., Mohamed, E. E. M., & Azmy, A. M. (2019). Advanced single-phase nine-level converter for the integration of multiterminal DC supplies. *The IEEE Journal of Emerging and Selected Topics in Power Electronics*, 7(3), 1949–1958. https://doi.org/10.1109/JESTPE.2018.2868734.

[9] Majdoul, R., Touati, A., Aitelmahjoub, A., Zegrari, M., Taouni, A., & Ouchatti, A. (2020). A nine-switch nine-level voltage inverter new topology with optimal modulation technique. In 2020 International Conference on Electrical and Information Technologies (ICEIT), (pp. 1–6). IEEE, Piscataway. https:// doi.org/10.1109/ICEIT48248.2020.9113170.

[10] Harbi, I., Ahmed, M., Hackl, C. M., Rodriguez, J., Kennel, R., & Abdelrahem, M. (2023). Low-complexity dual- vector model predictive control for single-phase nine-level ANPC-based converter. *IEEE Transactions on Power Electronics*, 38(3), 2956–2971. https://doi.org/10.1109/TPEL. 2022. 3218742.

[11] Mondol, H., Prokash Biswas, S., Islam, R., Mahfuz-Ur-Rahman, A. M., & Muttaqi, K. M. (2020). A new hybrid multilevel inverter topology with level shifted multi-carrier PWM technique for harvesting renewable In Energy2020 IEEE Industry Applications Society Annual Meeting, (pp. 1–6). IEEE, Piscataway. https://doi.org/10.1109/IAS44978.2020.9334754.

[12] Vijayakumar, A., Stonier, A. A., Peter, G., & Ganji, V. (2023). Dual source novel nine level inverter(DSN-2LI) design with minimum active devices and inherent polarity generation. *IET Power Electronics*, 16(7), 1162–1174. https://doi.org/10.1049/pel2.12533.

25 Smart automated surveillance system

Chandralekha, R.[1,a], Guna, A.[2,b], Krishna Meera, V.[2,c], Srirenga Natchiyar, V.[2,d] and Harini Shiriram, S.[2,e]

[1]Assistant Professor, Department of Electronics and Communication Engineering, Ramco Institute of Technology, Rajapalayam, Tamil Nadu, India

[2]Lecturer, Department of Electrical and Electronics Engineering, Sankar Polytechnic College, Sankar Nagar, Tamil Nadu, India

Abstract

Communication methods employ sophisticated automated surveillance systems that may issue real-time directives while surveillance is taking place. Real-time command surveillance is not possible using the standard surveillance system technique. Traditional surveillance techniques electronically record a scene's appearance, then send the video streams to a central location where security staff can view and analyze them on one or more video displays or store them for subsequent observation or analysis. It is very challenging to maintain communication while monitoring. Since they are exclusively designed for surveillance, traditional surveillance systems are not equipped with a mechanism that can deliver real-time orders. Thus, there has been a lot of interest in research projects due to the automation of video surveillance systems, which allows for the real-time automatic extraction and analysis of information from incoming live video streams and allows access to the command without human interaction.

Keywords: Flyback converter, human lungs modelling, positive pressure respirator, sensor less brushless DC motor

Introduction

Accidents, crimes, burglaries, robberies, and other disruptive activities have been occurring on a daily basis. These stories show how important security is to at least prevent or minimize these situations. This system will be used as a monitoring smart security module. Unlike traditional surveillance systems, which only record motion-based events, this system uses a website contained in a Raspberry Pi to give voice commands. The Raspberry Pi was first released with a speaker as an output. It employs a USB speaker that is fixed to its GPIO pins. When motion is detected, this system will identify it and send the command via the website that was made specifically to relay voice commands. People are depending on innovative technologies regarding their security needs as a result of the rapid advancement of technology. In India, monitoring has increased recently, yet crime rates have not changed.

The goal is to use a Raspberry Pi and surveillance camera technologies to create an intelligent surveillance system. Every time someone enters the range of a surveillance system, which the user is watching, the system detects the motion. The instruction will be sent via the website, which usually functions as an audio mechanism, as soon as motion is recognized inside the monitored range. The system will detect motion and issue a command based on the motion system's detection. Delhi, Chennai, Hyderabad, and other Indian cities are some of the most monitored in the world. Surveillance-heavy Indian cities also have higher criminal activity indices, incidentally. Even if the surveillance system is a little antiquated, it is still worthwhile to investigate video surveillance technology further because it is extensively utilized globally, and particularly in India.

Worldwide research on CCTV was critically examined in an effort to determine the causes of the odd relationship between crime indices and the quantity

[a]radhachandra2013@gmail.com, [b]gunaangusamy1995@gmail.com, [c]krishnameera@ritrjpm.ac.in, [d]srirenga@ritrjpm.ac.in, [e]harinishriram@ritrjpm.ac.in

DOI: 10.1201/9781003684589-25

of surveillance cameras. Lower crime rates are not always associated with increased surveillance rates. This startling information is accompanied by a disproportionately high crime rate for cities.

Cities with the highest concentrations of surveillance systems in the nation, shown in Figure 25.1. Have fewer crime indices and more surveillance footage than the other big cities. Thus, the idea that more security cameras result in lower crime rates is a fallacious one.

A smart video surveillance system's implementation was examined by Suchitra Khoje [1]. She suggested a system that notifies everyone when they are on the property and increases security by keeping an eye on their movements. The user enters the password to activate the system when they are leaving the building. The system begins by detecting motion, which is then refined to recognize people. Next, it counts the number of people in the room and notifies the neighbor when someone is there by activating the alarm. The Raspberry Pi supports the hardware implementation of the suggested system.

Sangmesh et al. [2] studied the topic of smart video surveillance. They suggested a system that records video while keeping an eye on certain behaviors or other details in a designated region. The suggested methods are used to follow and monitor the actions of people or groups, as well as to process footage from stationary cameras. When it comes to real-time video surveillance, this method generally shows promise.

Hemalatha et al., [3] examined systems for video surveillance. They looked at the system's architecture, which begins with a camera capturing video pictures and the events occurring in each stage's video sequences, which are then improved with appropriate actions taken at each level.

Ebrahimi and Dufaux [4] investigated scrambling for private video surveillance. They explore the issue of secure privacy in video monitoring in their study. They presented two efficient methods that used transform-domain or code stream domain scrambling for hiding regions of interest. During encoding, the chosen transforms often flipped. A few of the bits are reversed pseudo randomly. There is very little difficulty involved in transmitting the dates.

The primary objective of Kale and Sharma's [5] study on home security with video surveillance is to regulate the safety and security of the belongings of those residing in the house. The suggested concept is unique in the realm of computer vision and its applications, which include automated surveillance systems for both indoor and outdoor environments. Video surveillance systems have been established by home security systems to safeguard the house and may be utilized in smart home settings.

Yiqiu et al., [6] investigated 3G-based smart video surveillance systems. He created and put into place the 3G wireless mobile Internet-based video surveillance system. The chosen frames of the recorded video stream data are automatically stored in a

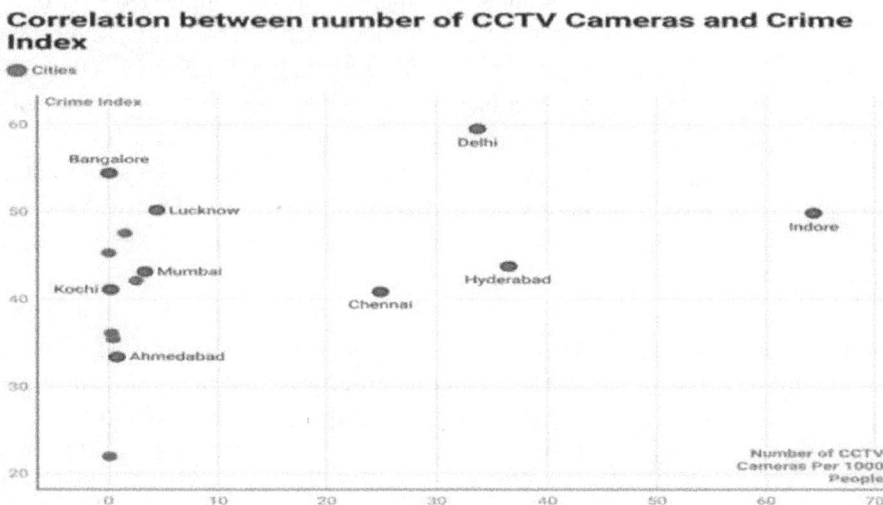

Figure 25.1 Number of CCTV cameras and rate of crime
Source: Author

database that is implemented in the video control server.

Ko [7] conducted research on behavioral analysis in surveillance footage for safety and security applications. In an era of growing concern for public safety and security, he offered a perspective on how hardware and software may be combined to address surveillance issues. They created a framework for a video surveillance system that incorporates tracking, motion detection, object categorization, behavior interpretation and description, environment modeling, and combining pictures from several cameras.

The intelligent intrusion detection system is a prototype model for remote surveillance and monitoring that improves conventional surveillance systems. Nainwal et al., [8] investigated the design and implementation of a re-mote surveillance and monitoring system using wire-less sensor networks. Their strategy is based on the idea of using wireless sensor networks to redistribute the tasks performed by conventional surveillance systems.

Singh et al. [9] conducted a survey and suggested a study on intelligent surveillance systems. They develop a safe, memory-efficient technique that will identify events in the recordings while monitoring and provide a textual index from the video.

Using a Raspberry Pi, Sanjana et al. [10] investigated the smart surveillance monitoring system. For web applications, they suggested a design and implementation that improves the technology that gathers data and sends it over a wireless local area network (WLAN).

Proposed Model

One of the most important considerations for both industrial sectors and smart cities is security. These days, there is a large need for security systems because of the growing population and urbanization.

CCTV cameras play a significant part in distant area surveillance. The globe is expanding quickly, which is leading to an increase in crimes and burglaries shown in Figure 25.2. As a result, in order to meet different security requirements and improve human life, an efficient and trustworthy monitoring system has become essential.

Our proposed system has the following distinct features:

a.   The website we made for security reasons will be used to access the login page and the sur-

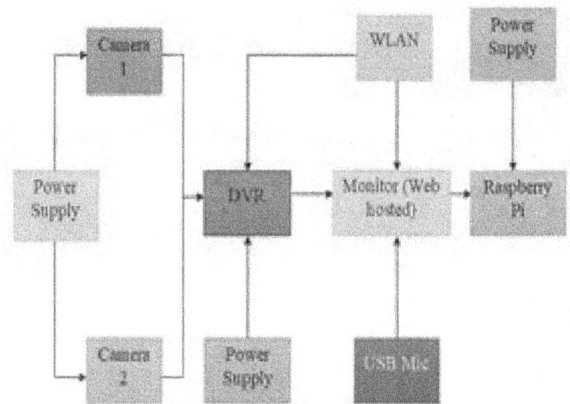

Figure 25.2 Block diagram of proposed system
Source: Author

veillance cameras the following: have been watched.
b.   To receive the user's audio, the speaker was linked to the Raspberry Pi's output. Even when the remote location is being monitored, this tends to facilitate communication.
c.   With the help of this system, security is generally improved by monitoring and warning about disruptive or behavior-prone actions.
d.   Our suggested method for sending audio to the intended destination is very dependable and reasonably priced. Real-time implementation is possible.

The smart automated surveillance system's implementation is explained in this suggested approach. Figure 25.2 shows the functional block diagram for the suggested model. The goal of this research is to create a detection system that can identify strangers and react fast by capturing and sending photos to wireless modules in response to events. A remote location monitoring idea is offered by this Raspberry Pi software, which is based on the smart surveillance system. The suggested method offers a completely practical, effective, and user-friendly worldwide solution. Additionally, this project will present the idea of wireless communication. In terms of security and monitoring, this kind of technology is crucial. The live video broadcast will demonstrate how to locate, monitor, and react to items.

Implementation of hardware

This system uses the Raspberry Pi architecture to remotely monitor a surveillance area. The

Figure 25.3 Raspberry Pi 4 model B
Source: Author

Raspberry Pi Foundation created the credit card-sized, single-board Raspberry Pi B+ model (Figure 25.3) in the UK, which has all the essential computer functions. Fabricating different hardware components has always relied heavily on hardware design, and it will continue to do so. There will always be a role for hardware in the electronics sector.

Software design
In order to build up a device, method, or system in enough detail to allow for its physical embodiment, the first stage in the development phase for any techniques and concepts is hardware implementation. The following are the main tasks involved in software implementation:

- Software units are built in accordance with structural unit criteria.
- A software configuration item is created by assembling, integrating, and testing software components.
- Creating a manufacturing proof of concept or re-solving implementation concerns by proto-typing difficult software components.
- Dry-run the acceptance testing procedures to ensure that the processes are well defined, and that the software product (software configuration items, or CIs) and computing environment are ready for acceptance testing.
- A number of libraries that serve as building blocks for the templates are part of the soft-ware implementation of the template-based modeling framework.

Result and Discussion

Using WebRTC is crucial to our project's implementation of real-time audio command chat. By allowing direct peer-to-peer communication while the server handles peer identification and connection formation, this open-source solution eliminates the need for continuous server involvement and enables peer-to-peer communication for audio, video, and data. Using HTML and CSS to create login and registration forms and gather distinct usernames, emails, and passwords is the first stage. A user session is created upon successful login, but warnings are triggered upon unsuccessful attempts.

Add HTML
Add an image inside a container and add inputs (with a matching label) for each field. Wrap a <form> element around them to process the input. The desired Admin's login page has been created shown in Figures 25.4 and 25.5.

The next step of the page is to interface audio in the webpage where the live audio command has been delivered to the cameras.

Style sheets are another name for CSS documents. A list of all the class definitions used in the related HTML document is essentially what a CSS style sheet is. Most of the time, the order in which the class definitions are given is not very important.

- Auto run chromium full screen on Raspberry Pi OS Open website using chromium https://emsd.000webhostapp.com/123ab
- Open a Website using Chromium Full screen

Full-screen mode eliminates all elements other than the page itself, such as toolbars, tabs, and scroll bars, in addition to making the page the user is seeing bigger, as happens when a window is maximized.
chromium-browser --start-fullscreen
https://emsd.000webhostapp.com/123ab
The above code is used for the purpose of opening full screen whenever the Raspberry Pi is powered ON Figure 25.6.

- Open auto start file
When the desktop environment (LXDE) starts up, a new text file will be opened and run. sudo mousepad /etc/xdg/lxsession/LXDE-pi/autostart

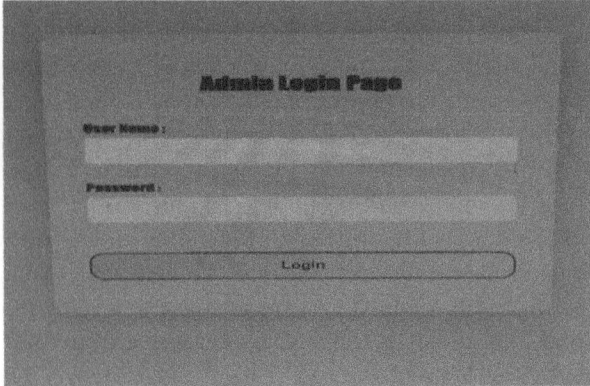

Figure 25.4 Admin's login page
Source: Author

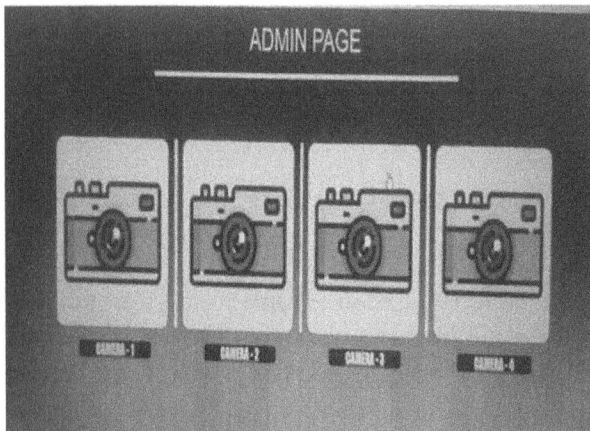

Figure 25.5 Requesting multiple camera access
Source: Author

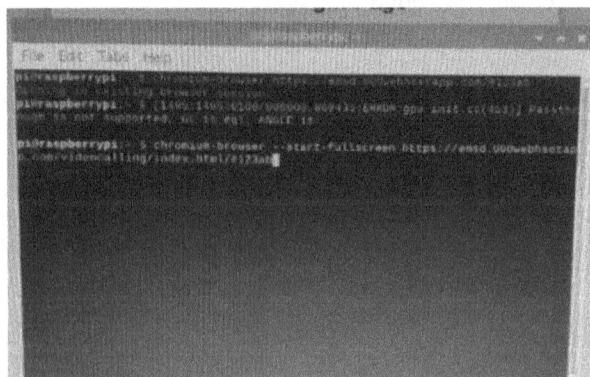

Figure 25.6 Enabling websites in chromium
Source: Author

- Add application to autostart file
  @/usr/bin/chromium-browser--start-fullscreen
  https://emsd.000webhostapp.com/123ab

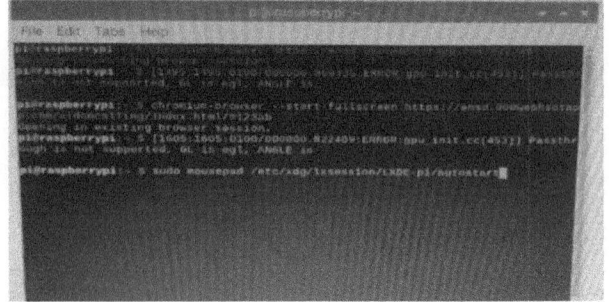

Figure 25.7 Auto starting website in chromium
Source: Author

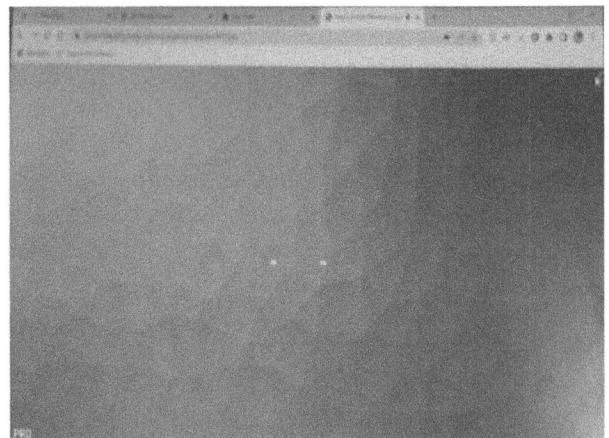

Figure 25.8 Website accessibility for real-time audio communication accessibility from user screen
Source: Author

The last line is the most important since it launches Chromium, the built-in browser that Google Chrome is based on, at the given URL in fullscreen mode (Figure 25.7). In addition to blocking additional user input outside of the browser, kiosk mode essentially locks the user inside the browser Figure 25.8.

The surveillance cameras will be monitored by giving continuous connectivity to the DVR shown in Figure 25.9.

Security systems and environmental monitoring have made use of this surveillance technology. There will be two modes of operation for this project: manual and automated. The user's screen will be used in manual mode to monitor activities throughout the surveillance area. The Raspberry Pi will enter automatic mode, which connects to the user's screen automatically.

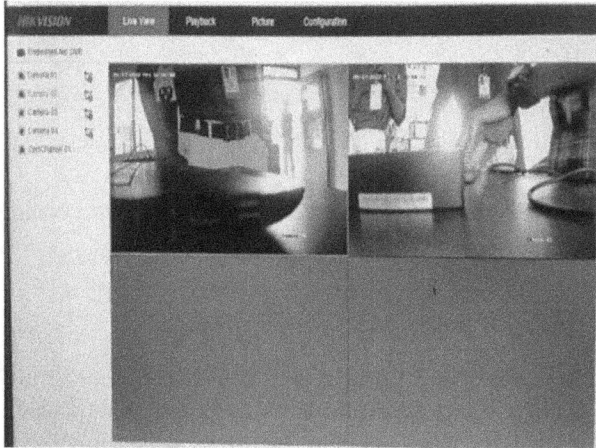

Figure 25.9 Accessing surveillance cameras
Source: Author

Conclusion

The problems with conventional surveillance systems are covered in this essay. One kind of issue is highlighted, which is giving live commands during surveillance may be set up with ease using a Raspberry Pi-connected USB microphone and speaker. The smart surveillance framework has been designed to meet the needs of the customer for a particular surveillance area. It may be used in a wide range of circumstances and has countless uses. In one instance, it is often used by any client who works for the company to learn about the events taking place at their workplaces. Therefore, the proposed initiative tries to provide the "smart surveillance" framework and to supplement the existing surveillance framework by enhancing and combining these innovations. This framework can be realized without any problems and doesn't require any special modifications when the establishment is necessary.

Acknowledgement

The authors gratefully acknowledge the staff for their cooperation in the research.

References

[1]   Gulve,S,P.,Khoje,S.A, & Pardeshi P. (2017). Implementation of IOT based smart video surveillance system. In International Conference on Advances in Intelligent Systems and Computing, (pp. 771–780).

[2]   Gote Swati, R., Khot Harish, S., Khatal Sonali, B., & Sangmesh, P. (2015). Smart video surveillance. *International Journal of Emerging Engineering Research and Technology*, 3(1), 109–112. ISSN 2349-4395(Print) & ISSN 2349-4409 (Online).

[3]   Hemalatha, H., Lakshmi Devasena, C., Revathí, R., & Hemalatha, M. (2011). Video surveillance systems – a survey. *IJCSI International Journal of Computer Science*, 8(4), pp.635–642.

[4]   Dufaux, F., & Ebrahimi, T. (2006). Scrambling for video surveillance with privacy. In Proceedings IEEE Workshop on Privacy Research in Vision, New York, NY.

[5]   Kale, P. V., & Sharma, S. D. (2014). A review of se-curing home using video surveillance. *International Journal of Science and Research (IJSR)*, 3(5), 1150–1154.

[6]   Liwei , W., Shi, Y., & Yiqiu, X. (n.d.). A wireless video surveillance system based on 3G network. In Conference on Environmental Science and Information Application Technology-wuhan, 2010, pp. 592–595.

[7]   Ko, T. (2008). A survey on behavior analysis in video surveillance for homeland security applications. In Proceedings Applied Imagery Pattern Recognition Workshop, Oct. 15–17, 2008, (pp. 1–8).

[8]   Nainwal, V., Pramod, P., & Srikanth, S. (2011). Design and implementation of a remote surveillance and monitoring system using wireless sensor networks. In Electronics Computer Technology (ICECT), 3rd International Conference, (pp. 186–189).

[9]   Tavagad, S., Bhosale, S., Singh, A. P., & Kumar, D. (2016). Survey paper on smart surveillance system. *International Research Journal of Engineering and Technology (IRJET)*, 03(02), pp.315–318.

[10] Prasad, S., Mahalakshmi, P., Sunder, A. J. C., & Swathi, R. (2014). Smart surveillance monitoring system using raspberry PI and PIR sensor.

26 Bank marketing DataSet using data science

Sudha, K.[1,a], Saravanan, G.[2,b] and Ponniah, G.[2,c]

[1]Assistant Professor, Department of Computer Science and Engineering, St. Joseph's College of Engineering, Chennai, Tamil Nadu, India

[2]Department of Computer Science and Engineering, St. Joseph's College of Engineering, Chennai, Tamil Nadu, India

Abstract

Understanding consumer behavior and improving marketing tactics are essential for long-term growth and profit ability in the cutthroat banking industry of today. The Bank Marketing Dataset is examined data driven and from a data science perspective in this abstract. The collection, which comes from marketing efforts run by a Portuguese bank, provides a wealth of information about demographics, customer inter- actions, and campaign results. Our investigation uses machine learning algorithms, sophisticated statistical methodologies, and data visualization tools to find practical insights for banks marketing strategy decision-making. In order to have a thorough grasp of the dataset's structure, distributions, and connections, our investigation starts with exploratory data analysis (EDA). We then use feature engineering to improve model performance and uncover useful predictors.

Keywords: Bank marketing, customer behavior analysis, data science, machine learning

Introduction

There has never been a greater urgent need for actionable information in the banking industry, where the complexities of customer behavior meet the strategic demands of marketing. Financial institutions must navigate a market that is changing quickly while also comprehending the varied demands and preferences of their customers. The emergence of data science, which provides a revolutionary method for extracting insights from enormous amounts of data and guiding strategic decision- making, stands out as a ray of hope amidst this complexity. It presents both scholars and practitioners with opportunities. This introduction sets out on a quest to discover the hidden potential in this information by applying data science to identify patterns, forecast trends, and guide strategic banking marketing campaigns. It contains information about clients of a bank and the outcomes of marketing efforts conducted via phone calls. The primary goal is to predict whether a customer will subscribe.

Overview of this Project

The project, "Data-Driven Insights for Bank Marketing Strategies," is a thorough attempt to use data science techniques to glean useful insights from a rich dataset that was obtained from marketing campaigns run by a Portuguese bank. With the banking industry growing more competitive and consumer preferences changing quickly, it is more important than ever to make well-informed decisions when developing marketing strategies. The project progresses through a number of crucial phases, starting with data collection and cleaning to guarantee the dataset's dependability and integrity. Exploratory data analysis (EDA) comes next.

Related Works

Data collection preparation

Define abbreviations and acronyms the first time they are used in the text, even after they have been defined in the abstract. Abbreviations such as IEEE, SI, MKS, CGS, ac, dc, and rms do not have to be defined. Do not use abbreviations in the title or heads unless they are unavoidable [1].

Exploratory data analysis

To comprehend the distribution and correlations in data, use scatter plots, box plots, and histograms. Determine the connections between variables [2],

[a]sudhak@stjosephs.ac.in, [b]saravanan19m065@gmail.com, [c]ponniah555@gmail.com

DOI: 10.1201/9781003684589-26

such as the relationship between product subscription and client age.

Modeling

Neural networks, decision trees, random forests, gradient boosting machines, and logistic regression are examples of common models. To determine the ideal model parameters, apply methods such as Bayesian Optimization, Random Search, or Grid Search [3].

Scope and Objectives

The Bank Marketing Dataset project aims to analyze customer demographics and behavior to identify key marketing trends and improve campaign effectiveness. By leveraging machine learning models, the project seeks to predict customer responses to bank marketing campaigns and determine the factors influencing their decisions on term deposits. Data-driven insights will be utilized to enhance targeting strategies, allowing banks to personalize their outreach, reduce customer acquisition costs, and improve overall conversion rates. Ultimately, the project contributes to optimizing marketing efforts and fostering stronger customer relationships through informed decision-making.

Existing System

CRM systems to store and manage customer data. These systems typically contain information about customer demographics, transaction history, interaction history (such as responses to previous campaigns), and other relevant data. In many traditional systems, customer data is often stored in silos across different departments (e.g., marketing, sales, and customer service), making it challenging to get a unified view of the customer [4].

Proposed System

The suggested system includes a bot or multi-platform automation tool. can be implemented in a variety of demat broker platforms, and in addition to creating reports and analyzing user data, the bot can place equities stock orders based on return percentages. Nearly 7000 firms that are listed on the Bombay Stock Exchange and National Stock Exchange have their stock prices retrieved by the bot. The information that has been retrieved is arranged according to the amount of money in the demat account and sorted by the percentage of profits.

System Requirements

A. *Hardware requirements*
- **Memory (RAM):** Adequate RAM for processing and storing video frames, carrying out calculations, and con- currently executing the required software components.
- **(RAM):** 4GB of system-usable memory 8GB is advised.
- **Graphics processing unit (GPU):** Depending on the intricacy and application of deep learning models, a GPU may speed up computations.
- **Storage:** Ample storage for software datasets and other potential resources the project might require.

B. *Software requirements*
- **System of operation:** Windows XP or later
- **Solid state drive (SSD):** When processing big datasets, it is advised to have at least 1 TB of SSD storage for quicker read/write rates. Hard disc drive (HDD) can be utilized for data archiving; a minimum of 2 TB of HDD storage is advised.
- **Operating system:** A system compatible with analytical tools and libraries, such as Linux, macOS, or Windows 10/11.
- **Network:** Stable internet access for downloading datasets, libraries, and accessing cloud services if needed Figure 26.1-26.4.

Data Flow Diagram

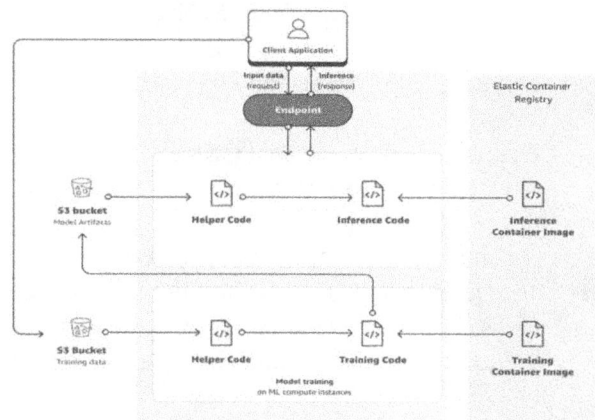

Figure 26.1 Data flow diagram
Source: Author

System Architecture

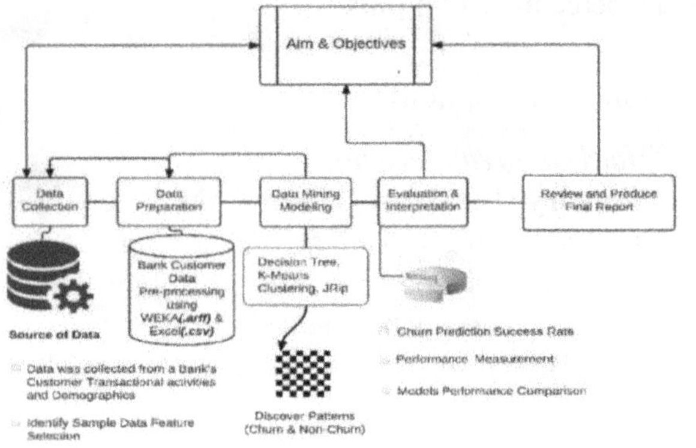

Figure 26.2 System architecture
Source: Author

System Sequence Diagram

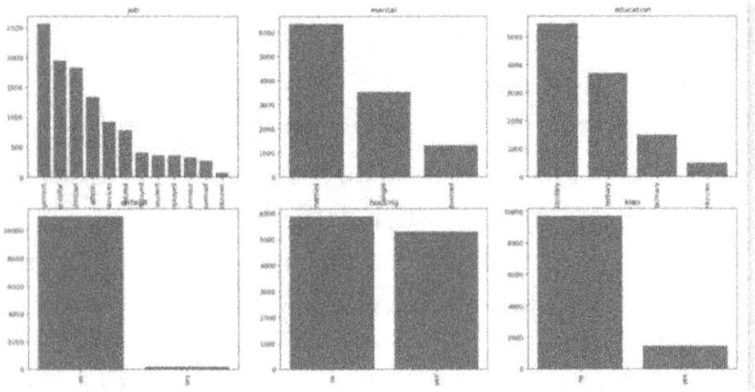

Figure 26.3 System sequence architecture
Source: Author

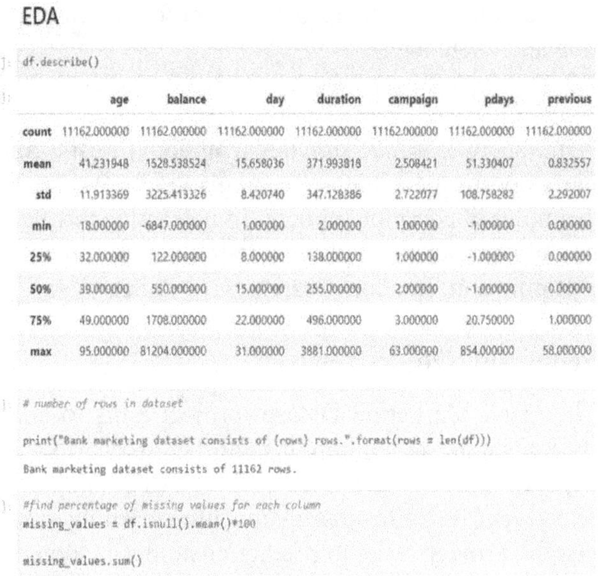

EDA

```
df.describe()
```

	age	balance	day	duration	campaign	pdays	previous
count	11162.000000	11162.000000	11162.000000	11162.000000	11162.000000	11162.000000	11162.000000
mean	41.231948	1528.538524	15.658036	371.993818	2.508421	51.330407	0.832557
std	11.913369	3225.413326	8.420740	347.128386	2.722077	108.758282	2.292007
min	18.000000	-6847.000000	1.000000	2.000000	1.000000	-1.000000	0.000000
25%	32.000000	122.000000	8.000000	138.000000	1.000000	-1.000000	0.000000
50%	39.000000	550.000000	15.000000	255.000000	2.000000	-1.000000	0.000000
75%	49.000000	1708.000000	22.000000	496.000000	3.000000	20.750000	1.000000
max	95.000000	81204.000000	31.000000	3881.000000	63.000000	854.000000	58.000000

```
# number of rows in dataset

print("Bank marketing dataset consists of {rows} rows.".format(rows = len(df)))
```

```
Bank marketing dataset consists of 11162 rows.
```

```
#find percentage of missing values for each column
missing_values = df.isnull().mean()*100

missing_values.sum()
```

Figure 26.4 Analysis data
Source: Author

System Modules

A. Data collection and preprocessing

- The operations module's data collection and Preprocessing module uses APIs extract, transform, load (ETL) procedures to retrieve data from third-party data sources, transactional databases, CRM systems, and web analytics platforms [5].
- Verifies that incoming data satisfies the necessary quality criteria by looking for outliers, improper formats, and missing values.
- The module proceeds to the vital preprocessing stage after data collecting. In order to eliminate errors and inconsistencies, students are exposed to data transformation and cleaning techniques here.
- Dealing with outliers, eliminating duplicates, and resolving missing values are all steps in the process that improve the dataset's overall quality. The subject also discusses data transformation techniques that are essential for ensuring that the data is compatible with analytical tools and models, such as normalization, standardization, and encoding.
- Encoding, which is essential for ensuring that the data is compatible with models and analytical tools. Additionally covered are feature engineering and dimensionality reduction, which let students focus on the most important factors in their datasets.

B. Data cleaning

- Addresses outliers, duplication, and missing values to clean up data.
- Converts unprocessed data into formats that are appropriate for analysis (e.g., encoding, normalization). Use feature engineering to develop fresh features that enhance model functionality.
- Assures the quality of the data and its preparedness for model creation.
- Creates a single dataset by combining data from several sources.
- Ensures consistency throughout the dataset by standardizing data formats.
- Identifies and fixes anomalies or discrepancies in the data [6].
- Manages time series data by applying differencing or generating lag characteristics.
- If required, dimensionality is decreased by applying methods such as principal component analysis (PCA).

- Addresses class imbalance in classification
- problems by implementing data balancing techniques (such as SMOTE).
- Merges datasets from different sources while resolving any key conflicts or mismatches.
- Encodes categorical variables using methods like one-hot encoding, label encoding, or target encoding.
- Handles text data by tokenization, stemming, lemmatization, or converting to embeddings.
- Converts date and time data into useful features like day of the week, month, quarter, or time elapsed.

C. **Prediction and scoring module**
- Applies trained machine learning models to new data for making predictions.
- Generates customer scores based on the probability of desired outcomes (e.g., likelihood of responding to a marketing campaign).
- Supports both batch and real-time predictions depending on project needs.
- Integrates with marketing systems to provide actionable insights and predictions.
- Calculates risk scores or other relevant metrics for decision-making.
- Handles automated predictions on incoming data streams [7] Figure 26.5.

- Provides confidence intervals or probability scores along with predictions.
- Ranks customers or entities based on their predicted scores for targeted actions.
- Provides confidence intervals or probability scores along with predictions.
- Ranks customers or entities based on their predicted scores for targeted actions.

Output

Result and Evaluation

A. Analysis
- Most customers are married (58)
- The mean age of customers is 40 years with a standard deviation of 10.
- 11
- The success rate was higher for customers who had previously been contacted compared to first-time contacts.

B. Evaluation
- Accuracy: Quantifies total correctness but can be deceptive with unbalanced datasets.
- Precision: It emphasizes the number of accurately forecasted positive cases among all forecasted positives Figure 26.5-26.8.

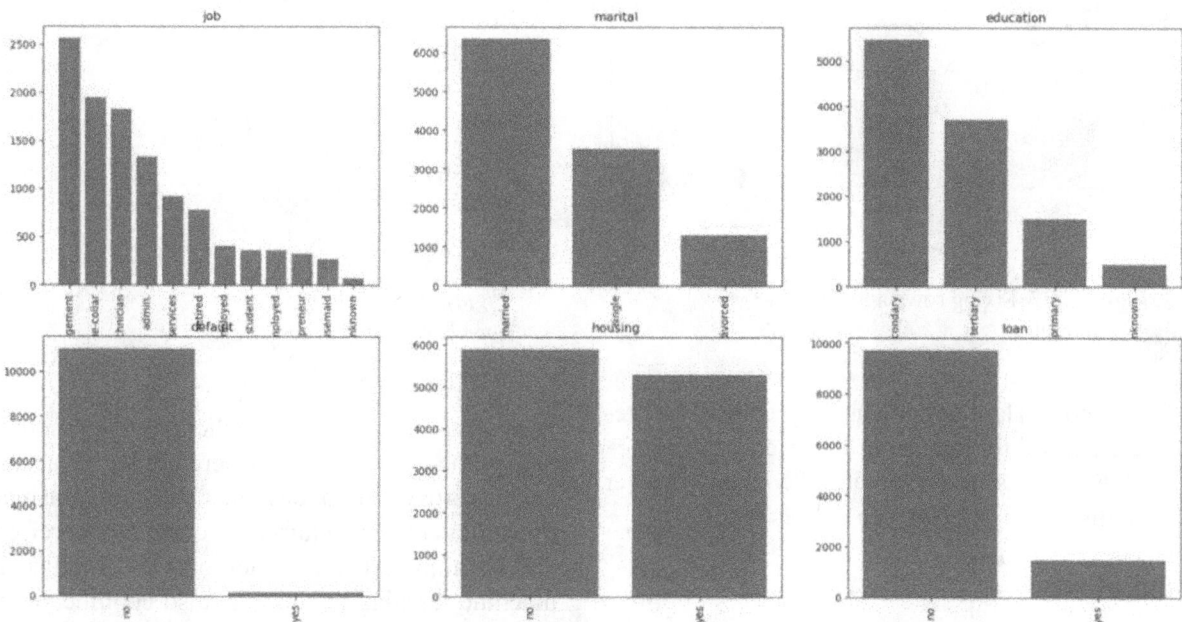

Figure 26.5 Fixing rows and columns
Source: Author

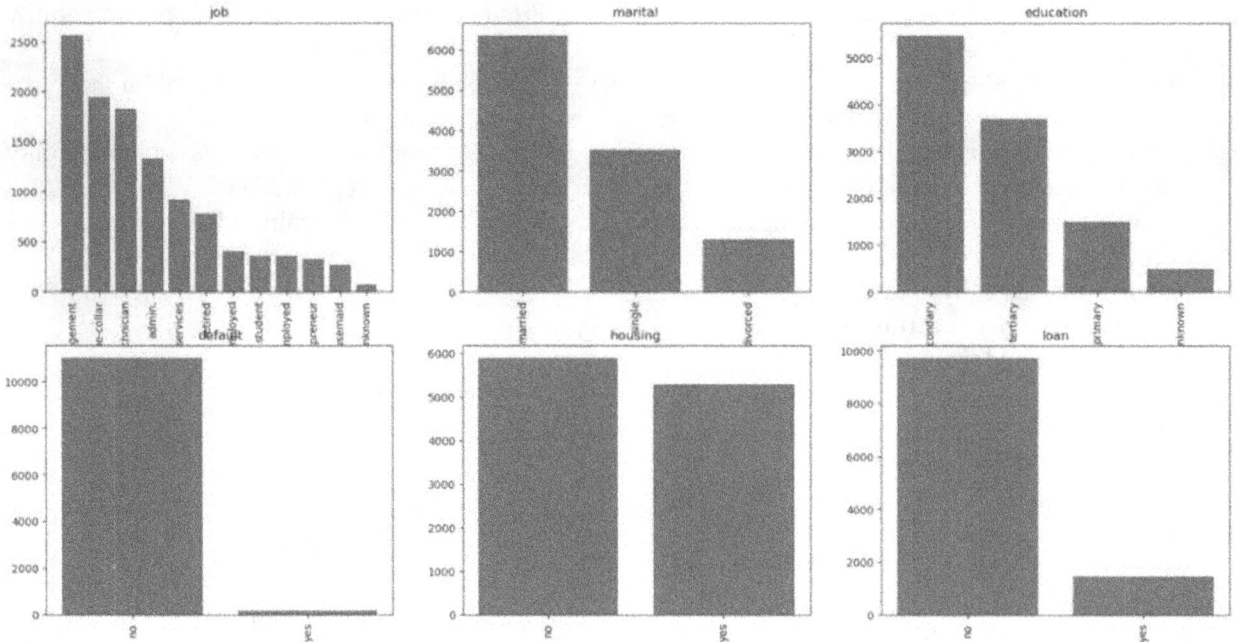

Figure 26.6 Fixing rows and columns
Source: Author

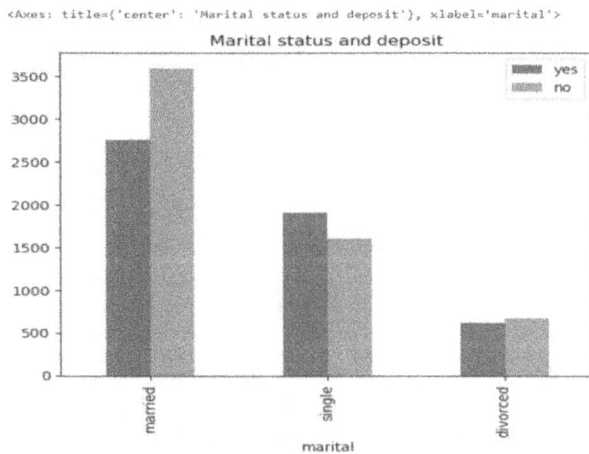

Figure 26.7 Fixing rows and column
Source: Author

Figure 26.8 Output
Source: Author

- F1-score: Harmonic mean of precision and recall, useful for imbalanced datasets.
- ROC-AUC score: Estimates the model's power to dis- criminate between classes [8].

Limitations

The prediction and scoring module, though powerful, has a number of limitations. Its performance relies heavily on the quality and adequacy of data used to train it; substandard data quality can result in misleading or skewed predictions. The module is also prone to model drift, where predictive accuracy decreases over time as customer behaviors and market dynamics change, requiring frequent model refreshes and retraining. The intricacy of some sophisticated machine learning models can also complicate interpretation of predictions, making it difficult for stakeholders to understand or trust the outcomes. This lack of transparency may hinder adoption in business decision-making processes and calls for the integration of explainable AI techniques to improve model interpretability and trustworthiness.

Conclusion

The prediction and scoring module is an essential part of a bank marketing data science initiative. It takes the learnings derived from data processing and model creation and converts them into actionable predictions, facilitating targeted marketing and evidence-based decision-making. Through the delivery of real-time or batch predictions, as well as scoring customers based on their probability of responding, this module equips the organization to maximize campaigns, tailor customer interactions, and ultimately maximize the efficacy of their marketing strategy. Furthermore, the module's ability to provide real-time predictions ensures that marketing efforts remain dynamic and responsive to new data.

Applications

1) **Targeted marketing campaigns:** Identifies high-potential customers for specific campaigns based on predicted responses, allowing for more efficient resource allocation and personalized outreach.

2) **Customer segmentation:** Splits customers into groups (e.g., high-value, medium-value) by scores in order to apply targeted marketing initiatives for each segment.

3) **Customer retention:** Identifies which customers are likely to defect, enabling banks to adopt retention strategies to enhance customer loyalty.

4) **Selling and upselling:** Facilitates strategies for selling more items or services by anticipating customer interest, generating more revenue from existing customers.

5) **Optimization:** Offers insights into the performance of various marketing channels and tactics, enabling optimization of future campaigns for improved outcomes.

6) **Allocation:** Assists in prioritizing the marketing efforts through scoring the customers and leads and making sure the resources are devoted to the highest-potential opportunities.

References

[1] Smith, J., Johnson, L., & Lee, K. (2019). Data science for marketing: a review. *Journal of Marketing Analytics*, 7(3), 245–263. Overview of data science applications in marketing.

[2] Johnson, L., & Lee, K. (2018). Predictive analytics for marketing: techniques and applications. *Marketing Science*, 37(4), 582–601. Discusses predictive analytics techniques and their applications in marketing.

[3] Patel, R., & Patel, S. (2020). Customer segmentation and targeting using machine learning. *International Journal of Data Science and Analytics*, 9(2), 101–120. Details machine learning techniques for customer segmentation.

[4] Zhou, M., & Wang, X. (2017). Dynamic customer seg-mentation for banking industry. *Journal of Financial Services Marketing*, 22(1), 32–46. Focuses on dynamic customer segmentation approaches in banking.

[5] Chen, T., Yang, X., & Li, Z. (2018). Predictive modeling for customer relationship management. *Journal of Business Research*, 92, 216–227. Covers various predictive models used in customer relationship management.

[6] 'Kumar, A., & Singh, R. (2019). Scoring models for predictive analytics in banking. *Financial Analytics Journal*, 14(3), 145–162. Discusses scoring models and their applications in the banking sector.

[7] Brown, E., & Davis, P. (2020). Optimization techniques for marketing campaigns. *Marketing Optimization Review*, 5(1), 55–72. Explores optimization techniques for marketing campaigns.

[8] Martinez, J., & Brown, A. (2019). Campaign effective-ness: data-driven approaches. *Journal of Marketing Research*, 56(2), 204–220. Analyzes data-driven approaches to campaign effectiveness.

[9] Lin, H., & Zhang, W. (2022). Adaptive decision making using real-time data. *Decision Support Systems*, 153, 113–126. Discusses adaptive decision-making processes supported by real-time data.

[10] Adams, R., & Clark, J. (2020). Ethical implications of data science in marketing. *Ethics in Data Science Journal*, 3(1), 1–15. Examines ethical issues in data-driven marketing.

[11] Jones, L., & Roberts, T. (2019). Data privacy and compliance in marketing analytics. *Journal of Data Privacy*, 12(4), 78–93. Focuses on privacy concerns and regulatory compliance in marketing analytics.

[12] Wilson, N., & Smith, T. (2021). Future trends in data science for marketing. *Marketing Trends Journal*, 19(2), 140–159. Provides insights into emerging trends and innovations in data science for marketing.

[13] Taylor, A., & Green, M. (2022). Challenges and opportunities in data science for marketing. *Journal of Business Analytics*, 15(1), 25–24.

27 DIBL effect compensation of self-biased CMOS voltage reference using adaptive V_{GS} control

M. Maheswari[1,a], J. Sofia Priya Dharshini[2,b], R. B. Malliswari[3,c], K. Sai Tanuja[3,d] and S. Police Asif[3,e]

[1]Assistant Professor, Department of ECE, RGMCET, Nandyal, Andhra Pradesh, India

[2]Professor, Department of ECE, RGMCET, Nandyal, Andhra Pradesh, India

[3]UG Students, Department of ECE, RGMCET, Nandyal, Andhra Pradesh, India

Abstract

Design and issues in the composition of ultra-low-power CMOS voltage references (CVRs) are needed for energy constrained micro-systems like brilliant sensors, wearable gadgets, and embed chips. Whereas conventional band gap voltage references (BGRs) are based on bipolar junction transistors (BJTs) that delay the reduction of the supply voltage (VDD) and the power consumption and usually require large resistors, which extend the size of the chip. That's why CVRs are much more appropriate for ultra-low-power applications; however, they must resolve a wide range of (V_{DD}) and noise, having good line sensitivity (LS) and power supply rejection ratio (PSRR). These strategies can create a more complicated design, which is to be stressed, power consumption, or result in circuit overhead. The self-biased CVR (SBCVR) is the result of a business idea that is being refreshed, where the bias current (I_B) is conducted from the reference voltage. However, the SBCVR's performance could be influenced by the gate leakage current (I_G), which short-circuits the source and substrate terminals of the transistor under biasing, which can lead to linearity loss and noise performance degradation. The study introduces a new technique for addressing the DIBL effect by means of adaptive gate-source voltage (V_{GS}) control. This solution not only reduces power consumption and complexity in design, but also promises to push performance in ultra-low-power up toward their full potential.

Keywords: Adaptive VGS control, drain induced barrier lowering with adjustment, line variation response, power supply noise rejection

Introduction

In modern ultra-low-power electronic systems like Internet of Things (IoT) devices [1, 5–13], wearables and energy-limited sensors, the coefficient reference voltage is of utmost importance for the device to operate stably and efficiently. Conventional voltage references, such as bandgap voltage references (BGRs) [1–4] and self-biased CMOS voltage references (SBCVRs), tend to suffer from low power efficiency, compactness, and fluctuation in performance due to varied supply voltages. The barrier lowering effect (DIBL) is also a significant factor in the de termination of LS [14] and PSRR, which limit their efficiency in environments with unstable supply voltages. This study dictates a low-power CMOS voltage reference (CVR) that in degrades a DIBL compensation method by which these issues can be solved. The suggested topological structure is characterized by controlling the gate source voltage (V_{GS}) adaptively to guide the bias transistor dynamically, thus reducing the bias current dependency on the supply voltage fluctuations.

A. Effects of DIBL

Effects of drain-induced barrier lowering are quite substantial to a circuit since they produce unstable reference volt ages, high sensitivity to the line and lower values for PSRR. DIBL is an effect encountered in MOSFET transistors; an increase in (V_{DS}) has its effect of bringing down (V_{TH}). Without DIBL compensation Figure 27.1(a), the (V_{TH}) of transistor M1 de creases as the supply voltage (V_{DD}) increases,

[a]maheswariece@rgmcet.edu.in, [b]sofiapriyadharshini@rgmcet.edu.in, [c]21091a0496@rgmcet.edu.in, [d]21091a04f9@rgmcet.edu.in, [e]21091a04d2@rgmcet.edu.in

DOI: 10.1201/9781003684589-27

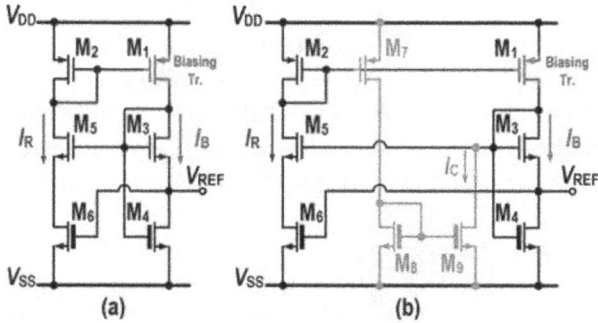

Figure 27.1 (a) Circuit without DIBL compensation. (b) Circuit with DIBL Compensation performance of the circuit [10, 11]
Source: Author

thus causing variations in the bias current (I_B), which affects overall performance.

This may also lead to an unstable reference voltage that is difficult to maintain as well. While the increased line sensitivity may make the circuit more prone to noise and interference, reduced PSRR would lead to a decrease in the circuit's power supply noise rejection. This may become serious issues with lower accuracy, increased errors, and decreased reliability. However, with DIBL compensation Figure 27.1(b), a compensation transistor (M8) senses the (V_{DD}) variations and generates a compensation current (I_C) that stabilizes the bias current (I_B) by adjusting the gate-source voltage (V_{GS}) of M1. The compensation mechanism will help the circuit to achieve a stable reference voltage, decrease line sensitivity [14], and improve PSRR, hence making it less susceptible to power supply noise and interference. The compensation circuit of DIBL monitors the (V_{DD}) voltage and generates a compensation current (I_C) proportional to the 2 (V_{DD})voltage. This compensation current is used to adjust the (V_{GS}) voltage of M1, which in turn stabilizes the (I_B) current. Therefore, by stabilizing the I_B current, the circuit maintains a stable reference voltage even with the presence of (V_{DD}) variations. Consequently, the circuit has better performance overall, and the output will remain stable regardless of power supply variation. Therefore, the circuit becomes more reliable, accurate, and robust, which is a very critical parameter in many applications such as these mentioned elements.

Literature Review

The BGRs for energy harvesting IoT devices, authors proposed a 58-ppm/°C 40-nW BGR that operates at a power supply volt age as low as 0.5 V which is a prototype for energy harvesting IoT devices [1]. This design is aimed mainly at curbing power consumption without compromising the temperature stability feature that it gives. Besides, it can specifically be useful for energy mismatched devices to the battery pro duction line that need both low power and low voltage. The authors made this possible by improving the bandgap-core and especially by carefully selecting the necessary energy hungry components to minimize their use and hence reduce the amount of energy consumed power.

An all-in-one bandgap voltage and current reference, that consumes less power and does not need amplifiers which is less than 200 nW, was designed [2]. This design is a solution that eliminates the need for amplifiers, and in addition to that, it is energy-saving. It is best used for applications that require both voltage and current references where low power and low voltage are equally important. The authors achieved this by using subthreshold operation, which is the mode between cutoff, and employing careful transistor sizing.

The bandgap voltage reference was designed and had a temperature range from 40°C to 125°C and an ultra-low noise of 0.4 µm with a load driving capability at 0.8 mA which was made using shared feedback resistors [3]. As a study, the focus of this design was on the load driving ability of reference and the noise performance, which make it suitable for applications requiring constant voltage references even under variable load conditions. The rest of the authors used the technique of sharing feedback resistors for implementing order of this reference, which reduces space and power consumption.

Proposed a voltage source with linear-temperature-qualified superpower consumption [4]. The design focalizes on achieving linear-temperature-dependent power wasting disease, which can be useful in applications where king intake needs to scale with temperature. The authors attain this by utilize a temperature-dependent biasing scheme.

A small temperature-compensated voltage reference made by a group of authors and supported by two transistors was designed [5]. The voltage that is created is as low as 0.5 V. Long-term circuit stability was the only issue that was not met. The design is distinguished by the loss of complexity and the consumption of low power to a very high degree, which is perfectly suitable for the portable and energy-harvesting industry. The unique operation of the authors was thinking of only three transistors, the rest of the operation was under subthreshold because of the third transistor's behavior.

Zhu et al. research also gave thrust to the making of MOSFET only handheld voltage reference on nano-watt scale without the inclusion of an amplifier, which means that telecommunication networks could do without an amplifier too if it is so designed [6]. It is hype that is 0.00% lines sensitive. Details: The new reference removes the need for amplifiers, thereby decreasing complexity as well as power consumption. It forms well on-line sensitivity, so you have the flexibility to use it at different voltage levels. The signature hand of DNS was MOSFETs in the subthreshold region and the optimization of the transistor sizes.

Created a picowatt, 0.45-0.6 V self-biased subthreshold CMOS voltage reference [7]. The bias current of the system is the one which originates from the reference voltage itself; therefore, it is the one that leads to a decrease in power consumption and the simplification of the circuit. In addition to this, the writers have devised a self-biased structure and a semiconductor size optimization, to reach this aim.

Existing Methods

Their capability to offer temperature compensated reference voltages is the reason for their high popularity. However, the BGRs are still making use of use of the bipolar junction transistors (BJTs) which in turn boost up the power dissipated and set the lowest required voltage channel [1–4].

Generally clamping amplifiers are incorporated into current generators so that they can make the bias current less dependent on the supply voltage. The effectiveness of the bias current in linking the emitter areas, the area of the diode, the value of resistor, and the gain and power of the amplifier. Hence, it can possess excellent low sensitivity performance even if the complexity of chip area and power consumption of the design are increased [10–13].

Self-regulating techniques can improve LS and PSRR when the supply voltage is automatically adjusted in order not to upset the regulators performance [14]. Nevertheless, it is a common cause to elevate the minimum operating voltage and circuit overheard.

Current bias based on the reference voltage is given to the diode, which forms a smaller circuit and, therefore, has lower power consumption. However, their effectives are weakened by DIBL effect that has consequences on LS and PSRR [14]. proposes a forced air (positive air pressure) solution to the problem.

Proposed Methods

Below are some of the existing methods that are improved. The proposed method can be:

A. Adaptive (V_{GS}) control for DIBL effect compensation.
1. The implemented the DIBL effect is a situation that happens when the threshold voltage of the transistors decreases as the voltage increases [13]. This results in the diminishing of Line sensitivity and PSRR.
2. Proposed circuit based on a universal (V_{GS}) control element can effectively shift the gate source voltage (V_{GS}) of the biasing semiconductor (M1) and limit (V_{DD}) affecting (I_B), making the bias current constant almost all the time.
3. The basic principle M8 taps (V_{DD}) for feedback and adjusts output current to compensate for LS variations. In the course of this modification, the DIBL effect is counteracted, and the gate-biasing transistor (M1) becomes stable, thus LS and PSRR are increased [14].
B. CMOS Voltage reference design
    The design for the suggested SBCVR had been divided into two parts, the first one is the start circuit and the next one is the main circuit which is shown below Figure 27.2. In these circuit we have the thickness is different for some transistors is in Figure 27.2, whenever is turned on that means when we apply supply voltage to circuit then the transistor of Ms1-Ms4(PMOS) is turned on (in digital design '0' means the PMOS id turned on) to power up the node B. Afterwards the Ms6 is turned on then charge will we passed on to the node A to the GND (V_{SS}). The main circuit will be operated normally. Then the reference voltage is arranged, simultaneously Ms5 will be activated, while Ms6 is power-down to confirm the normal implementation of the SBCVR circuit. Here the M8 is the compensation transistor which automatically detect any variations in the supply voltage, which will

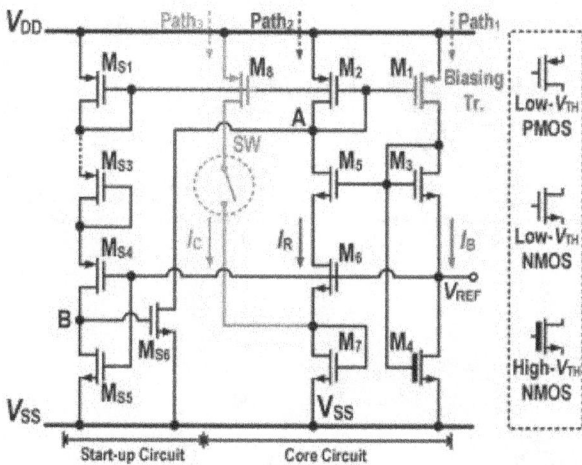

Figure 27.2 Diagram shown here is of the suggested SBCVR with short channel compensation effect
Source: Author

be connected to a Switch. These switches play an important role in which they will be able to activate or deactivate compensation transistor. In the M1-M2 are the current-mirrors methods which generate the bias current. The (V_{REF}) is connected to the M5-M6 which are in cascode [12] connection which will be generated the current (IR), these (IR) current will be passed on to the (I_B). When the SW is turned on, then the compensation transistor M8 will detect the (V_{DD}) variations, then the M6 and M7 will be used to adjust the (I_R). The adaptive (V_{GS}) control of M1 is used for the DIBL effect compensation [14], and this method can reduce the LS [13] and frequency PSRR to a very low value.

Simulation Tool

The Cadence tool is a computer-aided design (CAD) program that is used broadly in electronics (ICs, PCBs, SoCs) to simulate and design them. The Cadence tool enabling designers to create, verify and simulate complex electronic systems has had a significant impact on design times and the overall quality of the design accomplishing the goals.

Simulation process

Do the right click on the desktop and create your own folder. After creating a folder right click on that and select the open in terminal option. Type the command "csh" and then type the "source/home/cad/cshrc", then welcome to the cadence tools window

will be appeared. Type the command "virtuoso &" then the virtuoso will be opened. In that click "File→Create→New Library" to create new library, then the library form will be opened then give name to the library, click "ok". Select the technology file the attach to an existing technology library (ex: gpdk045 (or) gpdk180 etc...). To create a schematic window for that in the virtuoso "File→Create→New cell view", then the cell view will be opened, after that name cell view, then onwards the schematic window will be appeared.

In the schematic window we can create→Instance (or) "I" then the add instance will be appeared in the browse the library manager to select the required transistors the click hide in the bottom. Then inputs and outputs will display neither the symbols nor pin. Then do the wiring to the schematic, after that check and save if there is any error go to the virtuoso tab and rectify the errors otherwise circuit will be designed.

Simulation Results

A piece of work on the design and simulation of the SBCVR with the DIBL phenomenon compensation making use of adaptive (V_{GS}) regulation in a 0.18 um CMOS process were presented. The dimensions of the transistor (length and width) are shown in Figure 27.3 below. The occupied area of the SBCVR is 0.0014 mm². The average temperature coefficient being 61.85 ppm/°C. The average power consumption at 100°C is 391 pW. The measured (VREF) at 27°C is 106.6 mV is shown at Figure 27.6.

The Figure 27.4. is the way of PSRR determination. It has the value of -41.03 at 10 k. It is indeed the most efficient when comparing it to other works [8].

There is a calculated LS value of 0.013, this is an 87% reduction in comparison to the work [8] that was done previously is shown in Figure 27.5. The measured (VREF) at 27°C is 106.6 mV and it is reduced compared [8] to other work is shown at Figure 27.6. The measured temperature coefficient at 0.6 is 263.4n and at 1.8 is 61.85n. A greater LS (0.013% / V) is reached by the suggested design over the majority of other designs, nevertheless, it uses a mix of both self-regulating and cascode structures [11] to get the high LS and PSRR, respectively. At the same time, this improvement makes it necessary to have at least a level of volt age, which is nearly twice as high as it would be in other designs. The below Table 27.1 shows the comparison outcomes for the suggested SBCVR with DIBL effect Compensation

Tr.	W/L	Tr.	W/L
M₁,M₂	2μm/10μm	M₅,M₆	5μm/10μm
M₃	3×12μm/2μm	M₇	1μm/2μm
M₄	5.4μm/10μm	M₈	0.22μm/1.1μm

Figure 27.3 The dimensions of the transistor
Source: Author

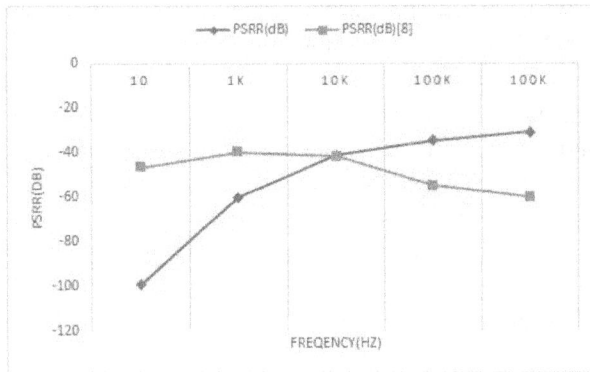

Figure 27.4 Measured PSR
Source: Author

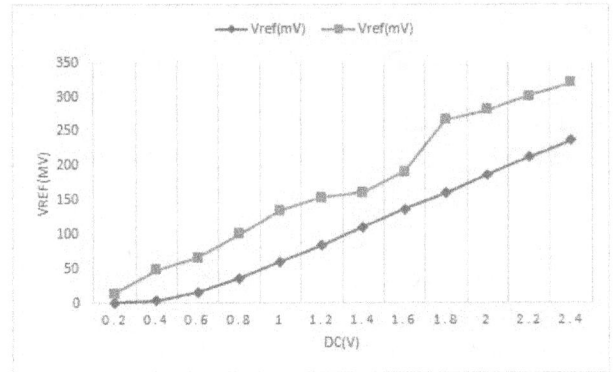

Figure 27.6 Calculated voltage reference
Source: Author

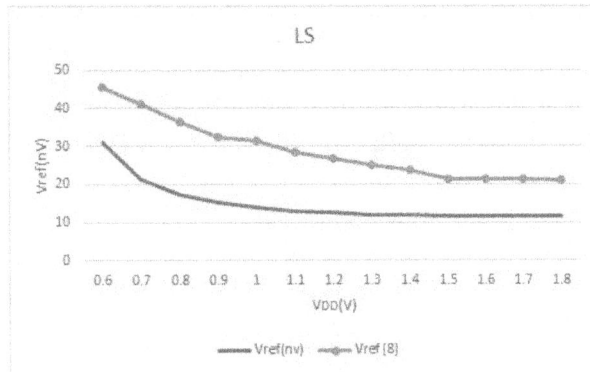

Figure 27.5 Measured LS
Source: Author

Table 27.1 Comparison of different designs.

Parameter	This Work	TCAS-I'19 [8]
Process (nm)	180	180
Type	CMOS	CMOS
Techniques	DIBL Compensation	Clamping Amplifiers
Power (nW)	0.067 @ 270°C 3.9 @ 1000°C	22.7
VREF (mV)	106.6	160.
TC (ppm/°C)	61.85	90.95
LS (%/V)	0.013	0.017
PSRR (dB)	-60.5 @ 1kHz -41.03 @ 10KHz	-46.73
Area (mm²)	0.0014	0.0148

Source: Author

using adaptive VG control. The other work with the clamping amplifiers between the DIBL compensation. However, the suggested method obtains the low line sensitivity and power consumption.

Conclusion

Finally, the document demonstrates the feasibility of an ultra-low power, self-compensated CMOS voltage reference in the presence of the DIBL effect

compensated processor. Through the application of a proposed gate-source voltage regulator, the basic job of biasing the transistor is achieved, which leads to a significant improvement in line sensitivity (LS) and power supply rejection ratio (PSRR) at low frequencies. Experimental data obtained from a 0.18μm

CMOS process indicates that in LS from 0.013%/V (87%) of the standard designs there is a significant decrease. On top of that, this circuit has the lowest power consumption on record of 391 pW and it is also a very small chip area of 0.0014 mm². Hence, the obtained outcomes turn it into a highly optimistic alternative to energy-efficient, ultra-low-power Internet of Things applications.

References

[1] Mu, J., Liu, L., Zhu, Z., & Yang, Y. (2017). A 58-ppm/°C 40-nW BGR at supply from 0.5 V for energy harvesting IoT devices. *IEEE Transactions on Circuits and Systems II: Express Briefs*, 64(7), 752–756.

[2] Huang, W., Liu, L., & Zhu, Z. (2021). A sub-200nW all-in-one bandgap voltage and current reference without amplifiers. *IEEE Transactions on Circuits and Systems II: Express Briefs*, 68(1), 121–125.

[3] Zhang, Z., Zhan, C., Wang, L., & Law, M. K. (2022). "A −40C–125 °C 0.4μm low-noise bandgap voltage reference with 0.8−mA load driving capability using shared feedback resistors. *IEEE Transactions on Circuits and Systems II: Express Briefs*, 69(10), 4033–4037.

[4] Liu, X., Liang, S., Liu, W., & Sun, P. (2021). A 2.5 ppm/°C voltage reference combining traditional BGR and ZTC MOSFET high-order curvature compensation. *IEEE Transactions on Circuits and Systems II: Express Briefs*, 68(4), 1093–1097.

[5] Seok, M., Kim, G., Blaauw, D., & Sylvester, D. (2012). A portable 2-transistor picowatt temperature-compensated voltage reference operating at 0.5 V. *IEEE Journal of Solid-State Circuits*, 47(10), 2534–2545.

[6] Zhu, Z., Hu, J., & Wang, Y. (2016). A 0.45 V, nano-watt 0.0%line sensitivity MOSFET-only sub-threshold voltage reference with no amplifiers. *IEEE Transactions on Circuits and Systems I: Regular Papers*, 63(9), 1370–1380.

[7] Ji, Y., Lee, J., Kim, B., Park, H.-J., & Sim, J.-Y. (2019). A 192-pW voltage reference generating bandgap—Vth with process and temperature dependence compensation. *IEEE Journal of Solid-State Circuits*, 54(12), 3281–3291.

[8] Wang, L., & Zhan, C. (2019). A 0.7-V 28-nW CMOS subthreshold voltage and current reference in one simple circuit. *IEEE Transactions on Circuits and Systems I: Regular Papers*, 66(9), 3457–3466.

[9] Liu, Y., Zhan, C., Wang, L., Tang, J., & Wang, G. (2018). A 0.4-V wide temperature range all-MOSFET subthreshold voltage reference with 0.0%/V line sensitivity. *IEEE Transactions on Circuits and Systems II: Express Briefs*, 65(8), 969–973.

[10] Lin, J., Wang, L., Zhan, C., & Lu, Y. (2019). A 1-nW ultra-low voltage subthreshold CMOS voltage reference with 0.01%/V line sensitivity. *IEEE Transactions on Circuits and Systems II: Express Briefs*, 66(10), 1653–1657.

[11] Chen, Y., & Guo, J. (2021). A 42nA IQ, 1.5–6V VIN, self-regulated CMOS voltage reference with −93dB PSR at 10Hz for energy harvesting systems. *IEEE Transactions on Circuits and Systems II: Express Briefs*, 68(7), 2357–2361.

[12] de Oliveira, A. C., Cordova, D., Klimach, H., & Bampi, S. (2017). Picowatt, 0.45–0.6 V self-biased subthreshold CMOS voltage reference. *IEEE Transactions on Circuits and Systems I: Regular Papers*, 64(12), 3036–3046.

[13] Wang, Y., Zhang, R., Sun, Q., & Zhang, H. (2018). A 0.5 V, 650 pW, 0.0%/V line regulation subthreshold voltage reference. In Proceedings IEEE 44th Eur. Solid-State Circuits Conference (ESSCIRC), Sep. 2018, (pp. 82–85).

[14] Wang, Y., Sun, Q., Luo, H., Wang, X., Zhang, R. & Zhang, H. (2020). A 48 pW, 0.34 V, 0.0%/V line sensitivity self-biased subthreshold voltage reference with DIBL effect compensation. *IEEE Transactions on Circuits and Systems I: Regular Papers*, 67(2), 611–621.

28 AgriSense - smart rain prediction and automated irrigation with TinyML

J. Chinna Babu[a], G. Sujatha[b], M. Upesh Rayudu[c], C. Yaswanth Krishna[d] and G. Venkat[e]

Department of ECE, Annamacharya Institute of Technology and Sciences, Rajampet, Tamil Nadu, India

Abstract

In particular, as food demand rises as a result of the world's fastest-growing population, water shortages and the excessive use of fossil fuels for irrigation are contributing factors to global warming and environmental degradation. We offer an integrated smart agriculture (SA) solution that is affordable for small and medium-sized farmers in order to tackle these issues. Our approach includes smart water metering, which uses cloud-based Internet of Things to monitor groundwater in real time; integration of renewable energy to reduce reliance on fossil fuels; and smart irrigation for optimal crop quality and soil preservation. Our solution was used in a real-world smart farm testbed and resulted in a 71.8% reduction in water consumption.

Furthermore, the integration of a pH sensor, an ESP32 microcontroller, and TinyML improves soil health monitoring, guaranteeing sustainable agriculture.

Keywords: IOT cloud based, smart agriculture, Tiny ML and renewable energy etc

Introduction

The growing global population and increasing food demand are placing immense pressure on agricultural resources, leading to higher water and energy consumption. Uncontrolled groundwater pumping, often powered by fossil fuels, causes environmental harm, including the depletion of vital water reserves and higher carbon emissions, contributing to global warming and threatening ecosystem sustainability. Smart agriculture (SA) has emerged as a promising solution, utilizing technologies like Internet of Things (IoT) devices, sensors, data analytics, and automation to optimize farming practices, improve crop yields, and conserve resources. However, the high cost of these technologies' adoption limits, especially among small and medium-sized farms. To address this, a cost-effective framework is proposed, integrating renewable energy, smart irrigation, and smart water metering. Renewable energy sources such as solar or wind reduce reliance on fossil fuels, lowering greenhouse gas emissions, while smart irrigation systems use soil moisture sensors and automated controls to ensure precise water use, minimizing waste. Smart water metering allows real-time monitoring and management of water usage, promoting sustainable groundwater conservation. The open-source nature of this framework enables customization and scalability, allowing farmers, researchers, and developers to adapt the system to diverse agricultural contexts, making it accessible and feasible for a broad range of farming operations.

Proposed Methodology

The proposed smart agriculture system upgrades the microcontroller to ESP32, replacing Arduino Nano to provide enhanced processing capabilities for handling complex operations and real-time data. This improvement supports the integration of advanced features such as soil health monitoring using a pH sensor and TinyML. By analyzing soil conditions more accurately, the system ensures precise irrigation, fostering better crop health and optimizing resource use.

Smart irrigation is implemented using real-time data processed with fuzzy logic, determining the optimal duration for watering crops. This approach conserves water, improves soil health, and enhances crop yield. Renewable energy integration, specifically

[a]jchinnababu@gmail.com, [b]sujatha25197@gmail.com, [c]Mellamputiupeshrayudu2004@gmail.com, [d]Yaswanthkrishna15022004@gmail.com, [e]lv9329034@gmail.com

DOI: 10.1201/9781003684589-28

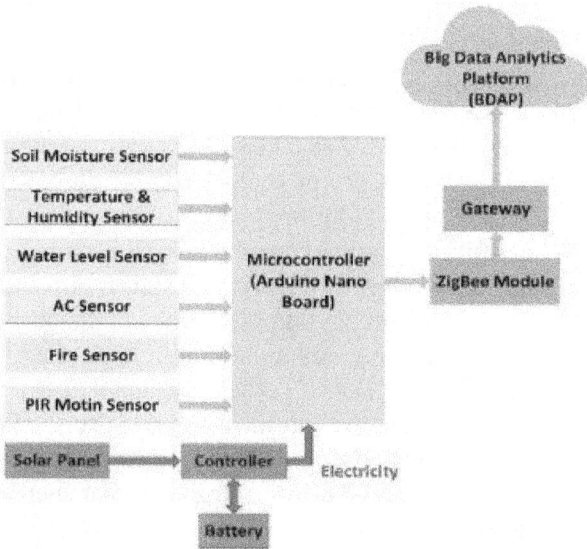

Figure 28.1 Block diagram of existing methodology
Source: Author

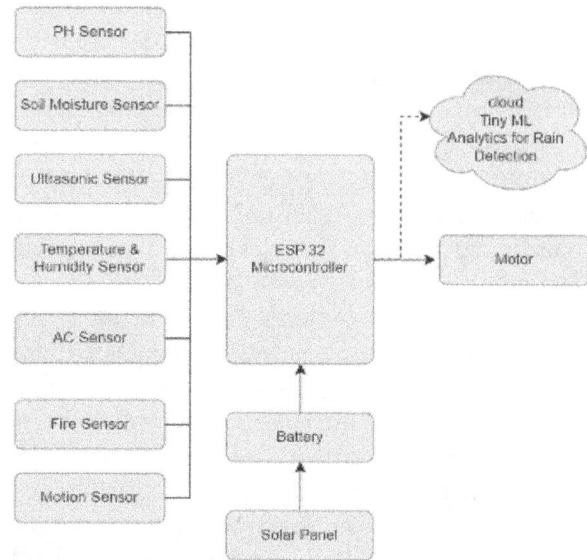

Figure 28.2 Self-created design using ESP 32 Microcontroller IoT architecture.
Source: Author

through solar energy, eliminates the reliance on fossil fuels, making the system more energy-efficient and environmentally friendly Figure 28.1.

The system's architecture, visualized in a flowchart, showcases the interplay between the ESP32 microcontroller, various sensors, and renewable energy resources. This combination enables a sustainable and efficient approach to agriculture, aligning with modern demands for resource conservation and productivity.

The proposed system employs a network of sensors that continuously gather data on soil health, environmental conditions, water levels, and security. This data is processed locally by the ESP32 microcontroller, which also transmits environmental information to the cloud for advanced analysis using TinyML. The cloud-based TinyML model provides valuable insights, such as rain predictions, to guide decision-making by the ESP32.

Based on the combined sensor data and cloud analytics, the ESP32 performs several actions to optimize farming operations. It activates irrigation motors as needed, sends alerts for fire or unauthorized activity, and adjusts actions based on energy availability and weather forecasts. The entire system is powered by renewable energy from solar panels, ensuring energy efficiency, reducing reliance on fossil fuels, and minimizing the carbon footprint. This integrated approach combines precision, sustainability, and

security to create an innovative solution for modern agriculture Figure 28.2.

The flowchart represents a smart agriculture system designed for sustainable farming through IoT-driven technologies. The system begins with the initialization phase, where sensors for monitoring fire, motion, water level, soil moisture, and weather conditions are activated. These sensors serve as the foundation for all subsequent actions.

The first decision point checks for fire or motion detection. If an event is detected, the system immediately sends alerts to notify the user of potential threats, ensuring farm security. If no event is detected, the system proceeds to check the water level. If the water level is below 35%, the water pump is activated to refill irrigation resources. After running for one hour, the pump automatically stops to conserve energy, and the system reinitializes the sensors. If the water level is sufficient, the system moves on to analyze soil moisture and weather conditions.

The next decision point checks if soil moisture is below the desired threshold and if no rain is predicted. If both conditions are met, the system activates the drip irrigation motor for one hour to deliver precise water to the crops before stopping automatically. The system then reinitializes the sensors. If these conditions are not met, the system ends the cycle and waits for the next monitoring session.

Key features of this system include safety through immediate alerts for fire or motion detection [9], resource optimization with smart irrigation and water pumping, and sustainability through the integration of solar energy [12] to reduce fossil fuel reliance. The system's automation, powered by real-time data and AI-driven analytics [7], ensures efficient operation, contributing to more sustainable and resource-efficient farming practices Figure 28.3.

Smart agriculture systems are increasingly supported by global initiatives and technological frameworks. For instance, AQUASTAT provides essential global data on water use in agriculture, which is critical for informed water management strategies [1]. Similarly, national programs like the Green Morocco Plan aim to modernize agriculture through sustainable practices [2], while the FAO emphasizes the role of water resource management in addressing food security under climate change [3]. Technological solutions such as cyber-physical systems have been deployed in real agricultural environments to optimize operations [4], often integrating platforms like Node-RED for flow-based development and seamless device integration [5]. Cloud computing and big data have also enhanced agricultural analytics, with studies exploring high-performance computing and deep learning to handle the scale and complexity of farm data [6][7].Internet of Things (IoT) technologies are central to this transformation. From wireless sensor networks (WSNs) in smart campuses [9][10] to photovoltaic-powered sensor solutions [12], researchers and developers are pushing boundaries to create energy-efficient, context-aware systems. Standardization efforts like the ITU's IoT Global Standards Initiative [13] and communication protocols such as Zigbee [15] or IEEE 802.11s [11] further support interoperability and scalability in agriculture. In precision farming, IoT enables real-time monitoring of soil, water, and crop health [29][30], and protocols are continuously optimized to enhance reliability and reduce energy consumption [14][17][38]. These technologies are not only being reviewed in literature but also tested in real-world applications, including greenhouses in Qatar with model predictive control and solar energy systems [22][23].Commercial projections from firms like Statista and McKinsey underline the exponential growth of IoT and its applications in agriculture, forecasting billions of connected devices and substantial market value [19][20][21]. As farming moves towards automation, frameworks combining fuzzy logic [31][36], artificial neural networks [40], and blockchain [44] are being developed to enhance efficiency and data security. Projects like Agrinex demonstrate practical, low-cost IoT systems for irrigation [41], while others propose solar-powered, autonomous solutions [43]. These smart agriculture ecosystems are supported by foundational wireless technologies [45][46] and monitoring platforms like SMA Cluster Controller [47], all aimed at optimizing inputs, reducing waste, and improving sustainability in modern farming Table 28.1.

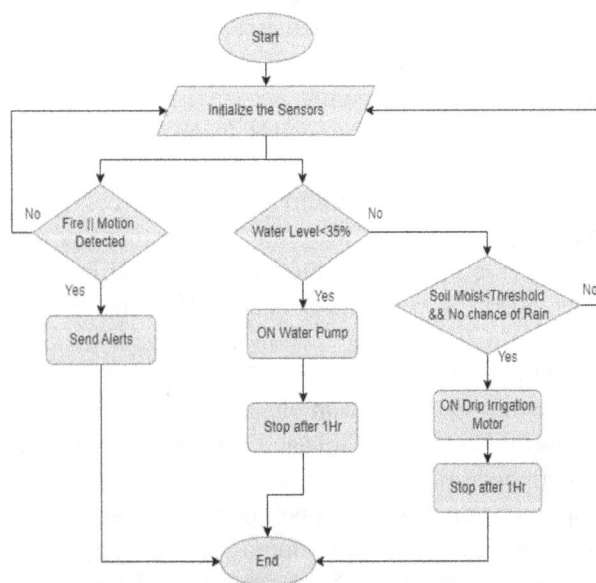

Figure 28.3 Flowchart of proposed methodology
Source: Author

Table 28.1 Experimental readings.

Readings	Example-1	Example-2
Humidity	82.4%	80.3%
Temperature	25°C	28.9°C
LDR	3841	3803
Rain prediction	32%	29%
Ph value	9.46	11.3
Soil moisture	153	323
distance	61.43cm	3.99cm
Fire alert	0	0
Intruder alert	0	0
Filling_watertank	1	0
Watering plants	1	1

Source: Author

Figure 28.4 Agrisense system
Source: Author

Conclusion

The AgriSense project kit (Figure 28.4) is a comprehensive smart agriculture solution designed to optimize resource use and enhance productivity. It includes a range of sensors, such as a moisture sensor to monitor soil moisture levels and regulate water release accordingly, ensuring plants receive just the right amount of water. A UV sensor checks the water level in the storage tank, preventing overflows or shortages. A humidity sensor measures the air's water content to further fine-tune irrigation needs, while a fire sensor provides safety by detecting potential fire hazards and preventing accidents. All these components are seamlessly integrated for a truly automated and efficient agricultural system.

The AgriSense system operates based on predefined thresholds to automate processes and generate alerts for efficient resource management. If a fire is detected the system issues a "Fire detected" alert, while an intruder presence triggers an "Intruder detected" message. For water tank management, the UV sensor monitors the tank's level: if the distance is greater than 30 cm, it indicates "Low water level" and begins refilling; if less than 10 cm, it shows "Water tank filled" and stops refilling. In soil moisture management, when the soil moisture value (moist_val) is below the threshold (i.e.400<t<1300)and there's no rain (rain < threshold), it starts watering the plants, displaying "Low water content." Conversely, if the soil moisture exceeds the maximumthreshold(moist_val> max_moist_threshold), it halts irrigation with a "Plants are wet" message. Additionally, when rainfall exceeds the threshold (rain > threshold), the system issues a "Rain detected" alert. These automated responses ensure optimal water use, enhanced safety, and improved agricultural efficiency.

Based on the experimental readings (Experiment 1), the AgriSense system efficiently manages resources and ensures safety. With soil moisture at 153, the system detects dry soil and activates the *"Watering plants"* process to maintain adequate moisture levels. The water tank level is low, with a distance of 61.43 cm measured by the UV sensor, prompting the system to refill the tank, indicated by *"Filling water tank = 1."* Despite high humidity at 82.4% and a temperature of 25.8°C, no fire (Fire alert = 0) or intruder (Intruder alert = 0) threats are detected, ensuring a safe environment. With rain prediction at only 32%, the system does not anticipate rainfall and relies on irrigation for plant hydration. The pH value of 9.46 suggests alkaline soil, which might require treatment depending on the crop. These automated responses showcase the system's ability to handle agricultural needs efficiently under varying conditions.

Based on the experimental readings (Experiment 2) and thresholds, the AgriSense system is functioning as expected. The soil moisture reading of 323 is below the threshold of 400, triggering the *"Watering plants"* process to hydrate the soil. The water tank level, measured at 3.99 cm, is below the 10 cm threshold, so the system does not refill the tank, indicated by *"Filling water tank = 0."* With humidity at 80.3% and a temperature of 28.9°C, environmental conditions are stable, and there are no safety concerns, as shown by *"Fire alert = 0"* and *"Intruder alert = 0."* Despite a rain prediction of 29%, the system continues irrigation due to insufficient rainfall probability. However, the pH value of 11.3 indicates highly alkaline soil, which may require corrective measures for optimal crop growth. These observations highlight the system's ability to respond dynamically to moisture levels, water availability, and safety requirements.

References

[1] AQUASTAT—FAO's Global Information System on Water and Agriculture. Nov. 2020, [online] Available from: http://www.fao.org/aquastat/en/overview/methodology/water-use.

[2] Green Morocco Plan. Nov. 2020, [online] Available from: https://www.agriculture.gov.ma/ar/dat-agri/plan-maroc-vert.

[3] Turral, H., Burke, J., Faures, J. M., & Faures, J. M. (2011). Climate Change Water and Food Security.

Rome, Italy: Food and Agriculture Organization, (pp. 204).

[4] Et-taibi, B., Abid, M. R., Boumhidi, I., & Benhaddou, D. (2020). Smart agriculture as a cyber physical system: a real-world deployment. In Proceedings 4th International Conference Intelligence Comput. Data Sci. (ICDS), (pp. 1–7).

[5] Node-Red, Oct. 2020, [online] Available from: https://nodered.org/.

[6] Abid, M. R. (2016). HPC (high-performance the computing) for big data on cloud: opportunities and challenges. *International Journal of Computer Theory and Engineering*, 8(5), 423–428.

[7] Chen, X.-W., & Lin, X. (2014). Big data deep learning: challenges and perspectives. *IEEE Access*, 2, 514–525.

[8] Lee, J., Bagheri, B., & Kao, H.-A. (2015). A cyber-physical systems architecture for industry 4.0-based manufacturing systems. *Manufacturing Letters*, 3, 18–23.

[9] Najem, N., Ben Haddou, D., Abid, M. R., Darhmaoui, H., Krami, N., & Zytoune, O. (2017). Context-aware wireless sensors for IoT-centeric energy-efficient campuses. In Proceedings IEEE International Conference on Smart Computing (SMARTCOMP), (pp. 1–6).

[10] Naji, N., Abid, M. R., Krami, N., & Benhaddou, D. (2019). An energy-aware wireless sensor network for data acquisition in smart energy efficient building. In Proceedings IEEE 5th World Forum Internet Things, (pp. 1–12).

[11] Abid, R. M. (2010). Link Quality Characterization in IEEE 802.11s Wireless Mesh Networks. Auburn University.

[12] Huang, K., Shu, L., Li, K., Yang, F., Han, G., & Wang, X. (2020). Photovoltaic agricultural internet of things towards realizing the next generation of smart farming. *IEEE Access*, 8, 76300–76312.

[13] Internet of Things Global Standards Initiative. Oct. 2020, [online] Available from: https://www.itu.int/en/ITU-T/gsi/iot/Pages/default.aspx.

[14] Ketshabetswe, L. K., Zungeru, A. M., Mangwala, M., Chuma, J. M., & Sigweni, B. (2019). Communication protocols for wireless sensor networks: A survey and comparison. *Heliyon*, 5(5), e01591.

[15] Zigbee Wireless Mesh Networking. Jul. 2021, [online] Available from: https://www.digi.com/solutions/by-technology/zigbee-wireless-standard.

[16] Khanna, A., & Kaur, S. (2019). Evolution of internet of things (IoT) and its significant impact in the field of precision agriculture. *Computers and Electronics in Agriculture*, 157, 218–231.

[17] Jawad, H. M., Nordin, R., Gharghan, S. K., Jawad, A. M., & Ismail, M. (2017). Energy-efficient wireless sensor networks for precision agriculture: a review. *Sensors*, 17(8), 1781.

[18] Growing Opportunities in the Internet of Things. June 2021, [online] Available from: https://www.mckinsey.com/industries/private-equity-and-principal-investors/our-insights/growing-opportunities-in-the-internet-of-things.

[19] Internet of Things (IoT) Total Annual Revenue Worldwide from 2019 to 2030. June 2021, [online] Available from: https://www.statista.com/statistics/1194709/iot-revenue-worldwide/.

[20] Statistica. Mar. 2021, [online] Available from: https://www.statista.com/statistics/471264/iot-number-of-connected-devices-worldwide/.

[21] Internet of things (IoT) Market to Exhibit 25.4% CAGR till 2028; Rising Usage of Connected Devices to Augment Growth. June 2021, [online] Available from: https://www.fortunebusinessinsights.com/press-release/internet-of-things-iot-market-9155.

[22] Ouammi, A., Achour, Y., Zejli, D., & Dagdougui, H. (2020). Supervisory model predictive control for optimal energy management of networked smart greenhouses integrated microgrid. *IEEE Transactions on Automation Science and Engineering*, 17(1), 117–128.

[23] Hassabou, A. M., & Khan, M. A. (2019). Towards autonomous solar driven agricultural greenhouses in Qatar-integration with solar cooling. In Proceedings 7th International Renewable Sustainable Energy Conference, (pp. 1–8).

[24] Bourhnane, S. (2019). Real-time control of smart grids using NI Compact RIO. In Proceedings International Conference on Wireless Technologies, Embedded and Intelligent Systems, (pp. 1–6).

[25] Badulescu, N., & Tristiu, I. (2017). Integration of photovoltaics in a sustainable irrigation system for agricultural purposes. In Proceedings 8th International Conference Energy Environmental Energy Saved Today Asset Futur., (pp. 36–40).

[26] Node Red User Guide. Jun. 2021, [online] Available from: https://nodered.org/docs/user-guide/.

[27] Lytos, A., Lagkas, T., Sarigiannidis, P., Zervakis, M., & Livanos, G. (2020). Towards smart farming: systems frameworks and exploitation of multiple sources. *Computer Networks*, 172, 107147.

[28] Glaroudis, D., Iossifides, A., & Chatzimisios, P. (2020). Survey comparison and research challenges of IoT application protocols for smart farming. *Computer Networks*, 168, 107037.

[29] Prathibha, S. R., Hongal, A., & Jyothi, M. P. (2017). IoT based monitoring system in smart agriculture. In Proceedings International Conference on Recent Advances in Electronics and Communication Technology, (ICRAECT), (pp. 81–84).

[30] Vellidis, G., Tucker, M., Perry, C., Kvien, C., & Bednarz, C. (2008). A real-time wireless smart sensor array for scheduling irrigation. *Computers and Electronics in Agriculture*, 61(1), 44–50.

[31] (2012). applied to the state of Qatar. *WIT Transactions on Ecology and the Environment*, 168(November), 189–199.

[32] Nagaraja, G. S., Soppimath, A. B., Soumya, T., & Abhinith, A. (2019). IoT based smart agriculture management system. In Proceedings 4th International Conference on Computational Systems and Information Technology for Sustainable Solutions, (CSITSS), (Vol. 4, pp. 1–5).

[33] Alahi, M. E. E., Pereira-Ishak, N., Mukhopadhyay, S. C., & Burkitt, L. (2018). An internet-of-things enabled smart sensing system for nitrate monitoring. *IEEE Internet of Things Journal*, 5(6), 4409–4417.

[34] Sadowski, S., & Spachos, P. (2020). Wireless technologies for smart agricultural monitoring using internet of things devices with energy harvesting capabilities. *Computers and Electronics in Agriculture*, 172, 105338.

[35] Mat, I., Mohd Kassim, M. R., Harun, A. N., & Yusoff, I. M. (2018). Smart agriculture using internet of things. In Proceedings IEEE Conference Open System (ICOS), (pp. 54–59).

[36] Touati, F., Al-Hitmi, M., & Benhmed, K. (2012). A fuzzy logic based irrigation management system in arid regions applied to the state of Qatar. *WIT Transactions on Ecology and the Environment*, 168(November), 189–199.

[37] Glória, A., Dionisio, C., Simões, G., Cardoso, J., & Sebastião, P. (2020). Water management for sustainable irrigation systems using internet-of-things. *Sensors*, 20(5), 1–14.

[38] Raza, S., Faheem, M., & Guenes, M. (2019). Industrial wireless sensor and actuator networks in industry 4.0: exploring requirements protocols and challenges—A MAC survey. *International Journal of Communication Systems*, 32(15), 1–32.

[39] Lezoche, M., Hernandez, J. E., Alemany Díaz, M. D. M. E., Panetto, H., & Kacprzyk, J. (2020). Agri-food 4.0: a survey of the supply chains and technologies for the future agriculture. *Computers in Industry*, 117, 103187.

[40] Escamilla-Garcáa, A., Soto-Zarazáa, G. M., Toledano-Ayala, M., Rivas-Araiza, E., & Gastálum-Barrios, A. (2020). Applications of artificial neural networks in greenhouse technology and overview for smart agriculture development. *Applied Sciences*, 10(11), 3835.

[41] Tiglao, N. M., Alipio, M., Balanay, J. V., Saldivar, E., & Tiston, J. L. (2020). Agrinex: a low-cost wireless mesh-based smart irrigation system. *Measurement*, 161, 107874.

[42] Krishnan, R. S., Julie, E. G., Robinson, Y. H., Raja, S., Kumar, R., Thong, P. H., et al. (2020). Fuzzy logic based smart irrigation system using internet of things. *Journal of Cleaner Production*, 252, 119902.

[43] Al-Ali, A. R., Al Nabulsi, A., Mukhopadhyay, S., Awal, M. S., Fernandes, S., & Ailabouni, K. (2020). IoT-solar energy powered smart farm irrigation system. *The Journal of Electronic Science and Technology*, 30(40), 1–14.

[44] Lin, Y., Petway, J., Anthony, J., Mukhtar, H., & Liao, S. (2017). Blockchain: the evolutionary next step for ICT E-agriculture. *Environments*, 4(3), 1–13.

[45] Pottie, G. J., & Kaiser, W. J. (2000). Wireless sensors network integrated. *Communications of the ACM*, 43(5), 51–58.

[46] Park, B., Nah, J., Choi, J. Y., Yoon, I. J., & Park, P. (2019). Robust wireless sensor and actuator networks for networked control systems. *Sensors*, 19(7), 1–28.

[47] (2017). User Manual SMA Cluster Controller, Niestetal, Germany, (pp. 1–116).

29 An overview of deep learning approaches for identification of deepfake faces in images and videos

J. Chinna Babu[a], M. Lakshmidevi[b], Y. Naga Sujani[c], V. Karthik Kumar Reddy[d] and G. Lokeswara Raju[e]

Department of ECE, Annamacharya University, Rajampet, Andhra Pradesh, India

Abstract

The increasing sophistication of deepfake technology has raised widespread worries about the authenticity of digital material, particularly human faces. This literature survey examines a range of deep learning techniques employed for the detection of real and fake human faces, focusing on the most recent advancements in convolutional neural networks (CNNs) and long-short term memory (LSTM). Various methods from recent studies are analyzed, including CNN-based architectures, auto-encoders, and hybrid models that combine spatial and temporal analysis for improved accuracy in detecting deepfake content. The survey highlights the strengths and limitations of existing approaches. Many methods successfully detect general inconsistencies in manipulated images but face challenges when applied to real-world scenarios with variations in lighting, pose, and image quality. Techniques such as advanced data augmentation, transfer learning, and feature extraction from subtle visual artifacts are explored as solutions to these issues. The research also underlines the ethical issues of deepfake technology as well as the need for robust, scalable detection mechanisms. This research suggests that, while existing deep learning algorithms have achieved substantial advances in deepfake detection, more robust and generalizable models are still needed. Further research is necessary to address the evolving challenges posed by increasingly sophisticated deepfake technologies.

Keywords: Convolution neural networks (CNN), deep learning, deepfake, fake and real images and videos, long-short term memory (LSTM)

Introduction

Deepfake technology is the newest and cutting-edge technology that takes images or videos of a close resemblance to that of a real person, whereas the basic emotions of the person in the video or image are changed into those of the person intended to be represented, thus making the difference indiscernible from genuine content. Deepfakes are being developed using sophisticated machine learning methods; among them are the generative adversarial networks (GAN) and autoencoders, which have either been proven to be effective or victims of vulnerabilities that cause much concern regarding misinformation, identity theft, or fraud. The sharp improvement in the quality of such fabricated media calls for robust systems that can detect fake content. This challenge has developed into an active avenue of research among the larger topics of deepfake detection, with various machine learning algorithms being employed to tackle this issue. Convolutional neural networks (CNNs) have proven to be one of the most effective approaches for image analysis, including the detection of subtle inconsistencies that often exist in synthetic images. However, existing methods often fall short in real-world applications, where variations in lighting, pose, and image quality make detection more challenging. With deepfake content creators continuing to further improve their technology, detection using antiquated methods grows near obscurity. In this work, deep learning perspectives, mostly based on convolutional neural network structure, will be covered to enhance the detection of face images or volumes of real human beings and fake ones. Our line of attack uses better augmentation techniques to ensure the model's robustness and ability to generalize. In essence, a diverse dataset, including real and computer-generated images for training, will allow us to capture those faint artifacts of a deepfake that separate them from authentic content. This should

[a]jchinnababu@gmail.com, [b]lm2419547@gmail.com, [c]yarasaninagasujani2003@gmail.com, [d]kr1705686@gmail.com, [e]galijerlalokeswararaju@gmail.com

DOI: 10.1201/9781003684589-29

allow our method to detect fake human faces better, further contributing to discussion on the safeguarding of media integrity in ongoing efforts. Apart from the technical matters touched on in the course of writing this paper, this work looks into the ethics of deepfake creation and detection. This is an emergent technology that is being developed in order to evade various countermeasures, and its potential for injuring victims has risen along with it. Such technologies require investigators who are capable of developing deeper solutions to their problems [1-8].

Related Work

This work presents an overview of the state-of-the-art deep learning algorithms for the detection of real and synthetic (or fake) human faces, which are a class of attacks known as deepfakes that are created using latest machine learning techniques. In this regard, these papers examine several techniques such as CNNs, GAN's, autoencoders and hybrid models. An analysis based on the above references:

Gans and autoencoders in deepfake creation and detection

Das et al. detailed GANs and autoencoders, particularly in the generation and detection of deepfakes. GANs create highly lifelike synthetic faces; autoencoders identify synthetics through attempts to reconstruct genuine. Upon difference of the outcome of the attempt at reconstructing real holograms, such images are regarded as fakes. This demonstrates the power of deep learning-based generative models in the generation and detection of forged content [9].

Deep neural networks (DNNs) for deep fakes detection

Chotaliya et al. [10] present a study that elucidates the various techniques for deepfake detection based on DNNs and analyses the feature extraction layers within such networks that exploit deep learning approaches to learn how to distinguish between genuine and fake images from large datasets. Generative DNNs are used heavily to detect deepfakes, as they can generalize over diverse datasets and observe subtle differences between human faces and them deepfakes [10]. As shown in figure 29.1.

Hybrid model between CNNs and DCGANs

Bansal et al. [11] combines CNNs and the deep convolutional generative adversarial networks (DCGANs)

to detect deepfakes. CNNs evolve the spatial features of the input images, whereas DCGANs generate counterfeit images. The model, therefore, can benefit from regularization techniques for optimization and generalization in false media detection, such as dropout [11].

Figure 29.2 illustrates a full deepfake detection pipeline. The majority of deepfake detection algorithms rely on feature extraction based on deep learning or manually created features. There are several ways that mix meticulously built components with visual and audible data to aid manipulation.

Data pre-processing

Age, height, and weight are some of the feature variables used in exploratory data analysis. It is important to normalize and compress visual information. Building a model needs several data preprocessing steps. It is possible to recognize any sort of data, including text, numerical, structured, unstructured,

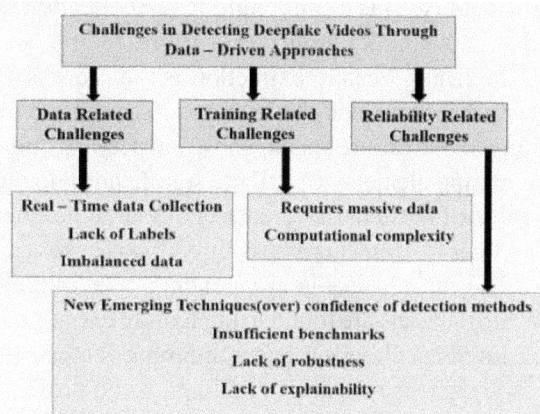

Figure 29.1 Challenges in Deep Fake Detection
Source: Author

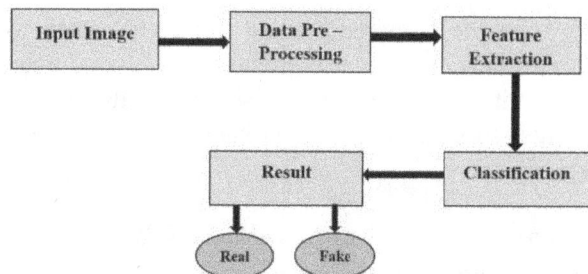

Figure 29.2 Generalized diagram of deepfake detection technique
Source: Author

Figure 29.3 Overview of pre-processing techniques
Source: Author

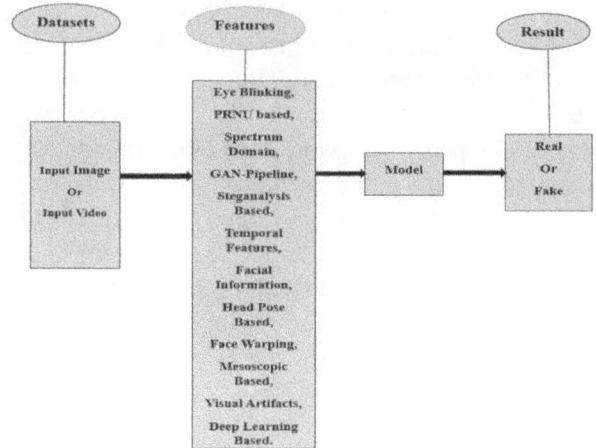

Figure 29.4 Overview of feature extraction
Source: Author

audio, PNG/JPG images, and time series. Photo processing is used to capture, process, segment, collect, and categorize photos of medical equipment, cars, fruits, and digital text, among other things. Feature normalization, imputation, encoding (including one-hot encoding), engineering, and selection are all fundamental data preparation steps. This may necessitate using dimensionality reduction techniques. Enhancing a photograph may result in better image.

Feature extraction

The key feature of feature extraction is to reduce dimensionality. This means splitting the collection of raw data into smaller groups. Therefore, the process would be faster. Feature extraction is the process of reducing the dimension of an unprocessed dataset by breaking it into more manageable subsets. A number of unique elements contribute significantly to the properties of these massive datasets. A large number of computer instruments are required to process variables. The most valuable components of extraordinarily big data sets are found using feature extraction, which involves choosing and combining factors. Its goal is to decrease the amount of data while preserving the most important information. These features are easy to use and correctly reflect the actual dataset. As shown in figure 29.3.

Preliminaries

To differ the proposed method from other methodologies introduced in the literature review, its main originality and improvements should be highlighted to cater specifically to the limitations of the current techniques:

Enhanced robustness with advanced data augmentation and preprocessing

While several papers such as Das et al., provide an overview of CNN-based deepfake detection with transfer learning, they overlook the contribution made

by the variations in lighting, pose, etc., and image quality to the overall detection accuracy [9]. Our proposed system integrates advanced data augmentation and preprocessing techniques, such as random cropping, rotation, and lighting adjustments, to improve the model's resilience to such variations. This ensures that the system is robust and performs well across diverse real-world conditions, outperforming models that lack these advanced augmentation methods.

Subtle artifact detection beyond standard CNN approaches

In most of their current literature (e.g., [9, 10]), the CNNs are being used to detect deepfakes without emphasizing different incongruities, like pixelation and blurring. The approach can detect the following general visual artifacts specific to deepfakes: inconsistency with skin textures, light reflections, and geometric distortions such as wobble. With the increasing level of features extracted, images can be examined and enhanced at such granularity as to analyze their context, which would enhance detection accuracy tremendously. As shown in figure 29.4.

Improved generalization through diverse and balanced datasets

Some have trained CNN-based architecture [11], but may be overfit to a specific dataset and their performance on deepfakes over time is often limited by the specifics of the type of deepfake used in training. To do so, our system is trained on a very diverse dataset, containing an equal proportion of half real images and half synthetic images that have been created using multiple deepfake methodologies like GANs, Autoencoders, and face-swap. The model generalizes better across several different types of manipulated data by using a more diverse dataset.

Hybrid approach and temporal analysis for video deepfakes

Contrarily, there are other approaches that explore the dissimilarity distribution over the frames of the video: hybrid CNN and DCGAN (for instance, [11]). Works done so far include proper analysis: a very thorough implementation that includes coherent variables such as the fixed aspect of spatial disparity combined with a time sensitive awareness of remaining frame temporal disparities of video streams. The hybrid approach reduces the chances of narrow sensitivity for deepfake videos.

Ethical considerations and real-world application

Furthermore, other researchers train CNN-based architectures, including Bansal et al. [11]; yet, due to the specific dataset, the possibility of suffering overfitting still exists: consequently, their approaches may not work sufficiently well anymore for a type of deepfake that they encountered during training. To do so, our system is trained on a very diverse dataset, containing an equal proportion of half real images and half synthetic images that have been created using multiple deepfake methodologies like GANs, autoencoders, and face-swaps. The model generalizes better across several different types of manipulated data by using a more diverse dataset.

Figure 29.5 illustrates the current locations of deepfakes. It has networks learned in datasets, machine learning, and deep learning. Few studies use machine learning. Standard machine learning (ML) algorithms have the ability to explain any option that is understandable to humans. Deepfake's data and process understanding makes these tactics effective. Model designs and hyperparameters are more easily modified. Extremely randomized trees, decision trees, and random forests are examples of tree-based machine learning algorithms that use trees to represent the decision-making process. Despite the model's reliance on a small number of variables, standard machine learning techniques yielded the most optimal results. To address the disparity, the superficial layer's textural properties are improved to better depict local components. These properties now provide enhanced semantic data. The author can

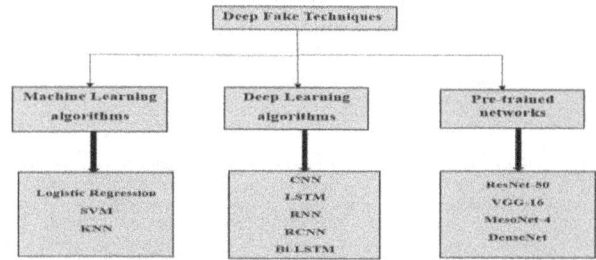

Figure 29.5 Some existing detection methods of deepfake.
Source: Author

Table 29.1 Existing literature on image data analysis.

Methodology	Advantages	Disadvantages
Attention-guided augmentation enhances photo classification.	Identify fake photos for classification.	Since datasets have different formats, precision will vary.
Unsupervised contrastive learning for classification.	Solves unsupervised contrastive learning with unlabeled datasets.	Labeling is laborious.
Text-to-image synthesis detects fake data.	The picture collection methods help create a precise model.	A long time to edit photos.
Deepfake alters faces, preserves environments.	Expand dataset, create phony visualizations.	Data types complicate combinations.
Dual shot detects faces, identifies deepfakes.	Photographs with and without resolution.	Higher-level computation is needed for training.

Source: Author

improve the network's ability to learn knowledge from different attention locations by deliberately allocating high-reaction attention to veiled regions throughout the training phase. A number of datasets are used to evaluate this strategy. The following mathematical formula is used in this study to supplement attention-guided data: Table 29.1 presents a comparison of picture data.

Audio based DFDT

In computer graphics, cross-modal audio-driven talking head animation has a long history. Prior

Table 29.2 Existing literature on audio data analysis.

Methodology	Advantages	Disadvantages
Deepfake video-audio synthesis with convolutional network.	Audio-video dataset analysis.	Merged dataset less accurate than separate datasets.
MFCC, RMS, RNN models analyzed datasets	Minimum difficult math.	Audio samples two seconds long.
SVM, CNN, ResNet classify audio.	Detect audio duplication.	Limited non-English deepfake detection.
RNN, CNN models synchronize speech, lip matching.	Lip model improves Synchronization.	Additional evaluation strengthens model.

Source: Author

Table 29.3 Existing literature on video data analysis.

Methodology	Advantages	Disadvantages
LSTM-CNN detects deepfake videos efficiently.	LSTM variants detect films.	Autoencoders require joint training.
Eye blinking detects fake videos.	Technology detects fake photos.	Converting to frames was tedious but necessary.
Actor orientation creates realistic videos.	Accurate reenactment outcomes demonstrated.	Videos based classification makes its laborious.
Face alignment optimizes photo size.	Overcome dataset generation during constraints.	Require more time then trained networks.
Fake video detection using optimization.	HFS technique improves precision.	A difficult process.

Source: Author

techniques for creating actual videos might be split into two types. These photorealistic approaches relate facial model rigging properties to 3D vertex coordinates. Determine whether source waveforms correspond to facial motions. Rigging settings usually demand creative professionals or high-quality 4D face capture data. Pre-trained models, audio analysis tools, and training audio datasets that do not contain video are used. Using audio data, Table 29.2 compares various DFDT algorithms.

Video based DFDT
Deep learning has recently gained success in the identification of fake data. Currently, deep learning algorithms used in image analysis are inefficient at detecting counterfeit videos on account of high information loss due to compression of videos. The current studies on recognizing deepfake videos can be categorized into two types: 1) analysis of biological signals and 2) temporal and spatial analysis. The videos are conditioned using the space-time conditioning volume formulation to produce target images in the video dataset. Table 29.3 is a comprehensive evaluation of this.

Results

This research paper presents an efficient method via a deep learning-based system to detect deepfake faces in real-time on images and videos. The system works by analyzing visual content and providing a distinction between real and fake faces. It is being tested robustly on the detection of deepfake videos, face-cropped images, and running chew through an endless list of real video content. The comparison shows that the system does quite well to classify image or video either as real or fake. From its accuracy, it can solidly detect deepfake manipulation, making it a powerful tool in the fight against digital content deception. Table 29.4 shows results and status for this research paper as shown below:

Table 29.5 presents a comparative analysis of different machine learning models for deepfake detection based on numerical parameters like accuracy, precision, and F1-score. as shown below:

Whereas CNN(ResNet50) + LSTM models specifically beat all others elegantly at 93.6% overall accuracy, 92.2% precision, and a 93.3% F1-score on various tasks compared are other conventional models like LSTM with an overall performance of 87.1%. CNN tops the others in combination with LSTMs at 91.6% accuracy. SVM is another competitor but stands last at an accuracy of 85.2%. A hybrid model such as CNN(ResNet50) + LSTM achieves far better results than one of the standalone models in detecting deepfakes. While CNNs and LSTM contribute effectively, a combination of images improves detection

Table 29.4 Results for deepfake detection based on test cases.

Test case description	Expected result	Actual result	Status
Videos/images with many faces	Fake/real	Fake	Pass
Deepfake video/image	Fake	Fake	Pass
Upload a real video/real image	Real	Real	Pass
Upload a face cropped fake video/fake image	Fake	Fake	Pass
Upload a face cropped real video/real image	Real	Real	Pass

Source: Author

Table 29.5 Numerical parameters for different types of machine learning models for deepfake detection.

Model	Accuracy (%)	Precision (%)	F1 Score (%)
CNN(ResNet50) +LSTM Deep Learning Model	**93.6**	**92.2**	**93.3**
Convolutional Neural Networks (CNNs)	91.6	90.5	91.5
Long Short-Term Memory (LSTM)	87.1	85.9	86.5
Support Vector Machines (SVM)	85.2	84.4	84.7

Source: Author

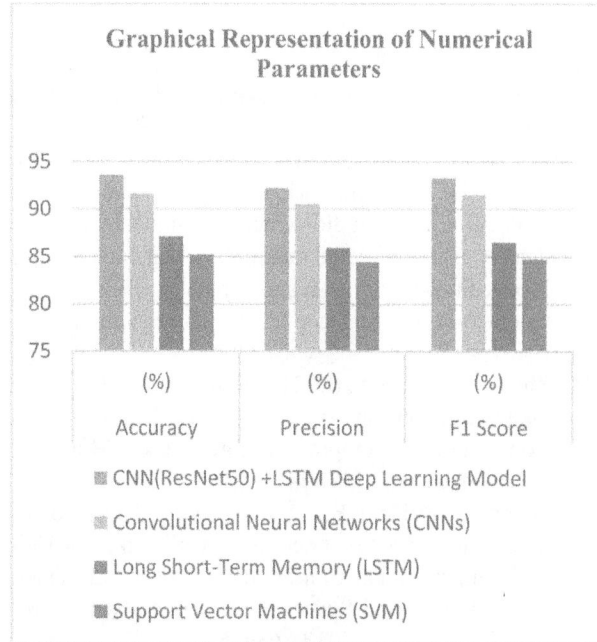

Figure 29.6 Graphical representation of numerical parameters
Source: Author

accuracy. This study emphasizes the high capability of deep learning models over classical approaches that might yield more trustworthy deepfake identification results. Figure 29.6 shows graphical representation of numerical parameters.

Conclusion

The proliferation of deep fakes has underscored the need for strong, trusted detection systems distinguished between artificial and human face recognition. This review of the literature considered studies based on the CNN model and other deep learning-based techniques. Even when dealing with increased lighting, pitching, and image quality variables, while they performed exceedingly well in identifying deepfakes, many still face challenges in real-life settings. Our analysis makes clear that existing approaches typically suffer from robustness and generalization problems, particularly with respect to growing and increasingly complex deep fake content. Advanced techniques like data augmentation, feature extraction from subtle artifacts, and the hybrid methods that combine temporal and spatial analyses hold promise in furthering the accuracy of detection. A properly balanced set of datasets with good diversity may help models perform better by facilitating ideal generalization to various forms of manipulated media. Despite these developments, the active development of deepfake technology means there is a continuous need for research-desiring agendas to tackle the ever-rising gargantuan challenges. We are troubled by the gruesome fears that accompanied the advent of deep fakes, making a dual appeal for countermeasures themselves and detection algorithms. The poorly constituted detection mechanisms may further elevate the utter existential crises raised by these interventions and keep the ever-burning need for further innovation and development alive, trying to cope with and manage synthetic media's complexities.

References

[1] Malik, A., Kuribayashi, M., Abdullahi, S. M., & Khan, S. M. (2022). Deepfake detection for human face images and videos: a survey. *IEEE Access*, 10, 18757–18775. doi:10.1109/ACCESS.2022.3151186.

[2] Khalid, H., Kim, M., Tariq, S., & Woo, S. S. (2021). Evaluation of an audio- video multimodal deepfake dataset using unimodal and multimodal detectors. In Proceedings of the 1st Workshop on Synthetic Multimedia-Audiovisual Deepfake Generation and Detection, (pp. 7–15).

[3] Zhou, Y., & Lim, S. (2021). Joint audio-visual deepfake detection. In 2021 IEEE/CVF International Conference on Computer Vision (ICCV), Montreal, QC, Canada, (pp. 14780–14789).

[4] Güera, D., & Delp, E. J. (2018). Deepfake video detection using recurrent neural networks. In 2018 15th IEEE International Conference on Advanced Video and Signal Based Surveillance (AVSS), (pp. 1–6), doi:10.1109/AVSS.2018.8639163.

[5] Zhao, H., Wei, T., Zhou, W., Zhang, W., Chen, D., & Yu, N. (2021). Multi-attentional deepfake detection. In 2021 IEEE/CVF Conference on Computer Vision and Pattern Recognition (CVPR), (pp. 2185–2194). doi: 10.1109/CVPR46437.2021.00222.

[6] Abdulreda, A. S., & Obaid, A. J. (2022). A landscape view of deepfake techniques and detection methods. *International Journal of Nonlinear Analysis and Applications*, 13(1), 745–755. doi: 10.22075/ijnaa.2022.5580.

[7] Khan, M. B., Goel, S., Katar Anandan, J., Zhao, J., & Naik, R. R. (2022). Deepfake audio detection. In AMCIS 2022 Proceedings, (Vol. 23).

[8] Almutairi, Z., & Elgibreen, H. (2022). A review of modern audio deepfake detection methods: challenges and future directions. *Algorithms*, 15, 155. https://doi.org/10.3390/a15050155.

[9] Das, M. K., Kumar, M., Kapil, I. K., & Yadav, R. K. (2023). Deepfake creation using gans and auto-encoder and deepfake detection. In 2023 second International Conference on Vision Towards Emerging Trends in Communication and Networking Technologies (ViTECoN). IEEE.

[10] Chotaliya, H., Khatri, M. A., Kanojiya, S., & Bivalkar, M. (2023). Review: deepfake detection techniques using deep neural networks (DNN). In 2023 6th International Conference on Advances in Science and Technology (ICAST). IEEE.

[11] Bansal, K., Agarwal, S., & Vyas, N. (2023). Deepfake detection using CNN and DCGANS to drop-out fake multimedia content: a hybrid approach. In 2023 International Conference on IoT, Communication and Automation Technology (ICICAT). IEEE.

30 Innovative milk contamination detection system using Raspberry Pi and IoT sensors

S. Sandeep Kumar[1,a], R. Selvarasan[2,b], G. N. Kodanda Ramaiah[3,c] and K Rasadurai[3,d]

[1]M. Tech, Student in Embedded Systems, Kuppam Engineering College, Kuppam, Andhra Pradesh, India

[2]Assistant Professor, Department of ECE, Kuppam Engineering College, Kuppam, Andhra Pradesh, India

[3]Professor, Department of ECE, Kuppam Engineering College, Kuppam, Andhra Pradesh, India

Abstract

Milk contamination through adulteration represents a critical public health issue, involving the incorporation of foreign substances to artificially increase volume or simulate quality, often resulting in grave health risks. Typical adulterants encompass water, detergents, salt, urea, melamine, and formalin. Our research introduces an innovative approach utilizing a combination of pH, gas, and temperature sensors to detect these contaminants in milk. The pH sensor assesses milk's acidity or alkalinity, potentially indicating the presence of adulterants such as urea or detergents that alter pH levels. Gas sensors identify volatile compounds emitted by adulterants like formalin or detergents, offering a non-invasive detection method. Temperature sensors measure thermal properties that may be affected by adulteration, contributing to the identification process. By analyzing data from these sensors, our system can accurately detect specific adulterants. The uniqueness of this method lies in its integration of multiple sensor types, providing a comprehensive analysis of milk quality. Unlike traditional chemical tests, this sensor-based approach enables rapid, on-site detection without specialized laboratory equipment. Its high accuracy and real-time assessment capabilities make it a valuable tool for dairy producers, regulatory agencies, and consumers, promoting safer milk consumption and enhancing public health outcomes. Implementing this technology can substantially reduce milk adulteration incidents, thereby safeguarding public health and preserving dairy product integrity. This portable and cost-effective solution enables on-site testing, proving particularly beneficial for dairy farmers, distributors, and consumers.

Keywords: Gas sensors, milk adulteration, pH sensors, sensors, temperature sensors

Introduction

Milk and dairy products serve as crucial nutritional sources, offering a diverse blend of nutrients including carbohydrates, sugars, proteins, vitamins, minerals, and enzymes. The nutritional profile of milk can fluctuate based on various factors such as cow breed, feed type, seasonal changes, and lactation stage. However, milk adulteration poses a significant challenge in numerous regions, where foreign substances are introduced to increase volume and modify milk properties. Water is the most frequently used adulterant, often diminishing the milk's nutritional value, while other additives like urea, starch, detergent, and formalin are employed to alter the milk's appearance and texture [1]. Beyond adulteration, unhygienic practices in milk processing, packaging, and distribution further jeopardize milk safety, resulting in contamination. Environmental conditions such as temperature, humidity, and darkness can also affect milk consistency, typically managed through refrigeration and vacuum storage to preserve quality [2]. In India, the world's leading milk producer, adulteration remains a pervasive issue, with substantial quantities of milk being diluted with harmful substances, compromising its nutritional content. Milk consistency can be utilized as an indicator to evaluate the extent of adulteration, raising concerns about consumer safety [3]. Identifying milk adulteration is crucial for safeguarding the integrity of milk products and public health. Various detection methods, including chemical, physical, and sensory tests, have been developed. Chemical methods often utilize reagents to identify adulterants, while physical methods assess properties such as density, viscosity, and freezing point. Advanced techniques like

[a]smc24594@gmail.com, [b]arasanece@gmail.com, [c]hod_ece@kec.ac.in, [d]rasaduraik@kec.ac.in

DOI: 10.1201/9781003684589-30

optical or spectral analysis can also aid in detecting foreign substances. Furthermore, microbiological methods identify harmful bacteria and pathogens frequently associated with adulterated milk [4, 5]. Contemporary technological advancements, such as DNA-based techniques for species identification and chromatographic methods to detect specific adulterants, are enhancing the accuracy and sensitivity of adulteration detection [6]. Recent surveys have underscored the magnitude of the problem, with approximately 68.4% of milk samples in India failing to meet safety standards. This highlights the necessity for efficient, portable devices capable of assessing milk quality on-site at dairy farms. The World Health Organization (WHO) has cautioned that, without effective intervention, widespread milk adulteration could result in severe health risks for a significant portion of the population by 2025 [8].

Consistent intake of tainted or contaminated milk can result in various adverse health effects, such as:

- Organ dysfunction: Additives like urea and starch, often used to enhance milk thickness, can impair the operation of organs like the liver and pancreas due to their toxicity [9].
- Heart-related problems: Chemicals in detergents used for creating milk froth can harm blood vessels, elevating the risk of cardiac disease [10].
- Malignancies: Extended exposure to artificial chemicals, pesticides, and other adulterants may lead to cancer causing effects [11].
- Eyesight deterioration: Preservatives such as formalin and hydrogen peroxide can harm the optic nerve and gradually impair vision [12].
- Renal issues: High concentrations of adulterants, like melamine, can accumulate in the kidneys, resulting in kidney stones and failure [13].
- Deaths: Ingesting milk adulterated with toxic substances such as ammonium sulfate or caustic soda can cause severe health crises and even mortality [14].

Methodology

Problem identification
The widespread issue of milk adulteration has led to decreased milk quality and significant health hazards. Adulterants such as detergent, urea, and chemical gases can modify milk's pH, temperature stability, and release harmful vapors. Create a cost-effective and efficient detection system utilizing Raspberry Pi

and sensors to identify milk adulteration by assessing changes in milk's pH, temperature, and the existence of specific gases.

Literature review
Examine current methods to recognize common adulterants and their impact on milk properties that detergents as adulterants reduce milk ph. Preservative chemicals (e.g., formalin or hydrogen peroxide) emit harmful vapors detectable by gas sensors. Urea or other contaminants may alter milk temperature or stability [15]. Determine sensor-specific requirements and comprehend the optimal practices for incorporating pH, temperature, and gas sensors into a unified detection system.

System design and architecture
The Figure 30.1 shows the setup combines a Raspberry Pico with pH, temperature, and gas sensors. The Raspberry Pico gathers real-time data from these sensors and processes it to ascertain milk purity. The system integrates various sensors and processing components to detect and analyze the quality of substances by monitoring parameters like acidity, temperature, and the presence of harmful vapors.

A circuit diagram illustrates sensor connections to the Raspberry Pico. It includes power supply, resistors, and I/O pins for each sensor. a forced air (positive air pressure) solution to the problem.

Sensor selection
The apparatus utilizes a trio of crucial sensors to accurately identify milk adulteration. A pH sensor (DFRobot SEN0169-V2) gauges acidity/alkalinity with ±0.1 pH precision at 25°C, utilizing a dual-point

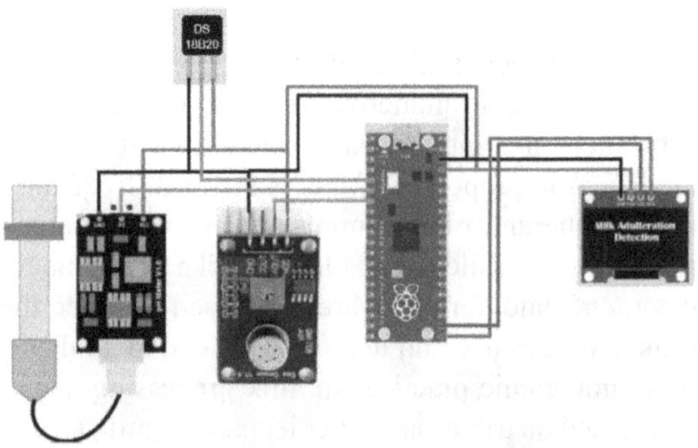

Figure 30.1 Sensor connections
Source: Author

calibration (pH 4.0 and 7.0). A temperature sensor (DS18B20) identifies thermal irregularities, functioning within a range of -55°C to +125°C, maintaining ±0.5°C accuracy between -10°C and +85°C. A gas sensor (TGS 2602) detects odorous gases such as formalin and ammonia [7]. To ensure accuracy, each sensor undergoes calibration: the pH sensor with buffer solutions (pH 4, 7, 10), the temperature sensor against ice (0°C) and boiling water (100°C), and the gas sensor using standardized gas concentrations.

Milk sample preparations

To ensure consistency, milk samples are thoroughly mixed and strained to eliminate impurities, reducing contamination and improving sensor precision. The analysis process employs three sensors to examine the sample. Acidity variations outside the normal range of 6.6 to 6.8 are identified by the pH sensor, while the temperature sensor tracks stability and detects unusual fluctuations. The gas sensor identifies adulterants like formalin or hydrogen peroxide, enabling reliable detection of milk adulteration.

Outcome presentation

The apparatus is engineered to provide instantaneous, user-friendly insights into milk quality by identifying potential contamination. Its OLED display presents easily comprehensible messages, such as "Milk is Unadulterated" for safe samples or "Adulteration Detected" when harmful components are found. The screen also exhibits crucial information, including acidity levels, temperature, and vapor detection readings, enabling users to gain comprehensive knowledge about the milk's state.

Results

These findings underscore the importance of developing an effective detection system to identify these adulterants, thereby safeguarding public health and ensuring compliance with regulatory standards.

Detection of pure milk

A pH range of 6.8–6.86, consistent temperature (24–29°C), and standard gas levels indicate fresh, unaltered milk Figure 30.2 shows the pure milk graph.

Detection of water added milk

A pH rises to 6.9–6.95, temperature stays at 24–29°C, and gas levels remain normal, suggesting water addition. Figure 30.3 shows the water added milk graph.

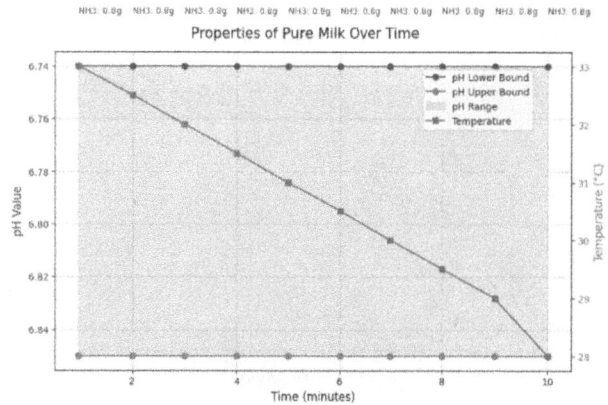

Figure 30.2 Pure milk
Source: Author

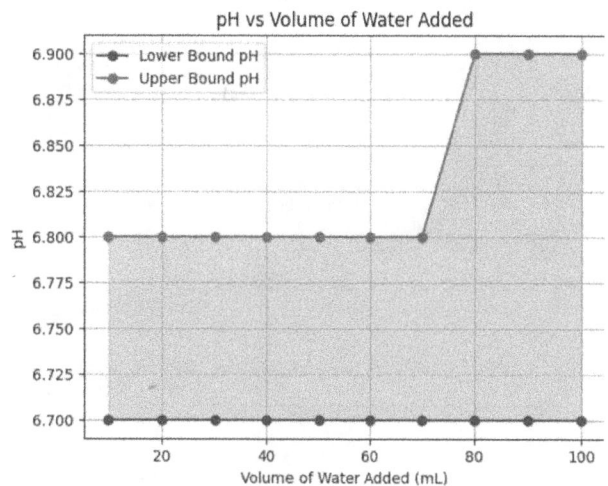

Figure 30.3 Water analysis
Source: Author

Detection of salt

A pH decreases to 6.40–6.66, temperature increases to 28–30°C, while gas levels are unchanged, pointing to salt contamination. Figure 30.4 shows the salt added milk.

Detection of detergent

A pH decreases to 6.40–6.66, temperature increases to 28–30°C, while gas levels are unchanged, pointing to salt contamination. Figure 30.5 shows the detergent added milk.

Detection of urea

A pH increases to 6.97–7.05, temperature falls to 18–23°C, and gas levels alter, suggesting urea adulteration. Figure 30.6 shows the urea added milk.

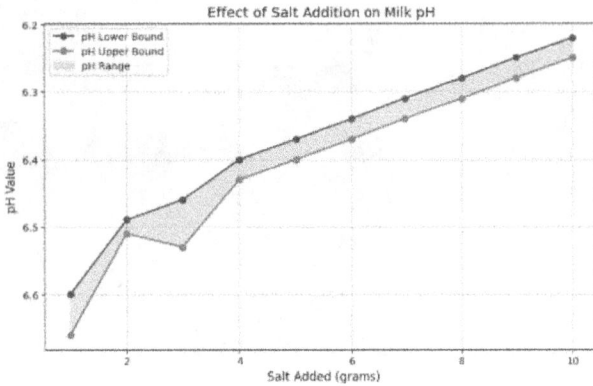

Figure 30.4 Salt analysis
Source: Author

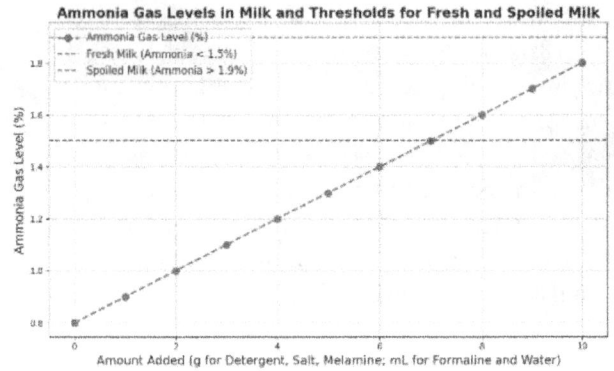

Figure 30.7 Ammonia analysis
Source: Author

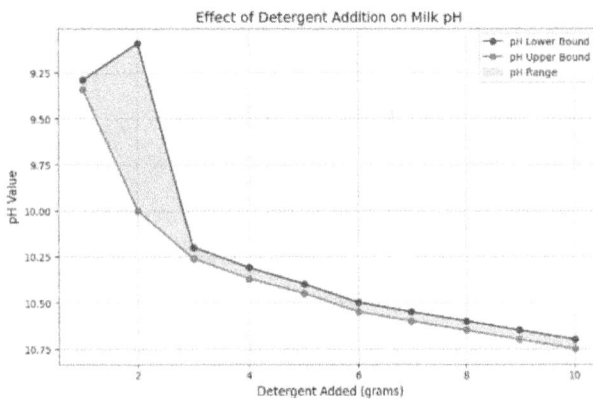

Figure 30.5 Detergent analysis
Source: Author

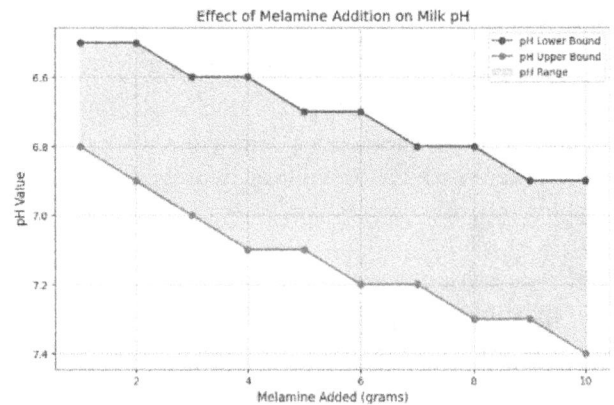

Figure 30.8 Melamine analysis
Source: Author

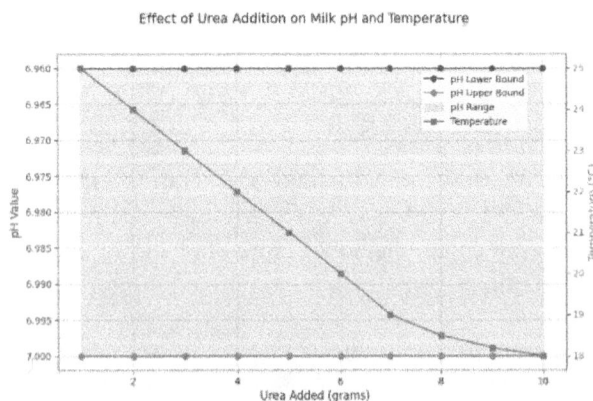

Figure 30.6 Urea analysis
Source: Author

Detection of ammonia

A pH varies, temperature changes, and elevated ammonia-related gases indicate milk spoilage. Figure 30.7 shows the ammonia added milk.

Detection of melamine

A pH and temperature fluctuations, along with melamine detection by gas sensor, confirm contamination. Figure 30.8 shows melamine added milk.

Detection of formalin

Irregular pH, unstable temperature, and modified gas levels suggest formalin addition, often used as a preservative. Figure 30.9 shows the Formalin added milk.

Display outputs

In this section, we present the outputs of the device obtained under a series of controlled test conditions designed to evaluate its performance and sensitivity. Each test involves subjecting a milk sample to different scenarios, such as maintaining its pure state Figure 30.10 , introducing common adulterants like salt Figure 30.11, and adding contaminants like detergent Figure 30.12.

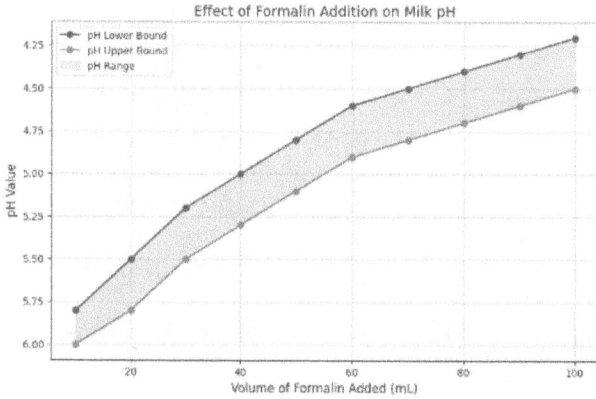

Figure 30.9 Formalin analysis
Source: Author

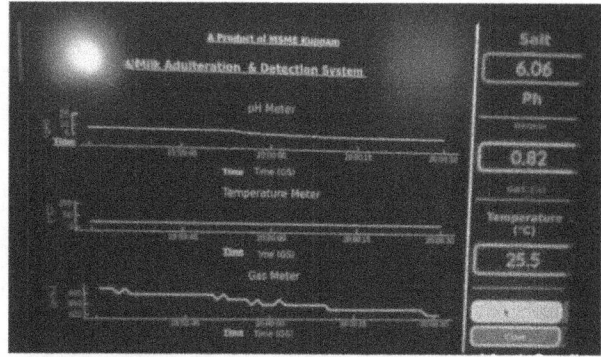

Figure 30.12 Salt added
Source: Author

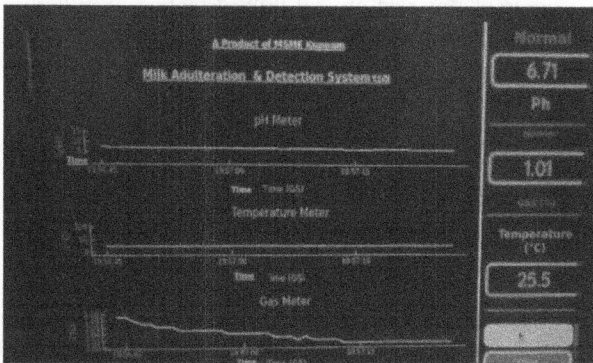

Figure 30.10 Pure milk
Source: Author

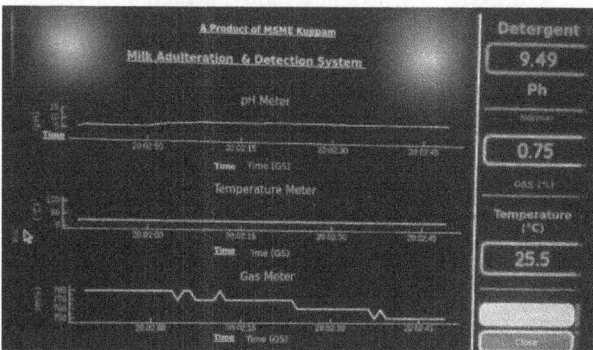

Figure 30.11 Detergent added
Source: Author

By analyzing the device's outputs for each case, we demonstrate its ability to detect subtle changes in the physical, chemical, or optical properties of the sample. This analysis underscores the system's potential as a reliable tool for ensuring milk quality and safety in diverse operational environments. The results also provide insight into the robustness and accuracy of the detection methodology.

Discussion

A system for detecting milk adulteration employs sensors to evaluate how various additives impact milk quality. The addition of detergent results in increased pH and temperature, possibly due to heat-releasing reactions. Conversely, salt addition leads to a decrease in pH and a slight reduction in temperature, likely caused by alterations in molecular interactions. Melamine causes a rise in temperature without significantly affecting pH, indicating thermal effects [16]. Formalin lowers the temperature while leaving pH unchanged, probably due to its preservative properties. The most notable pH changes are observed with detergent and salt, while the largest temperature fluctuations occur with detergent and melamine. A potential link between temperature and urea content might suggest bacterial activity or milk spoilage.

Conclusion

To summarize, utilizing Raspberry for milk adulteration detection offers an efficient and effective approach to identify common contaminants like water, detergent, urea, salt, and formalin. Through the analysis of changes in pH and temperature sensors, Raspberry can precisely identify adulterants and provide immediate alerts. This technology has the potential to significantly contribute to ensuring milk product quality and safeguarding consumers from harmful substances. The system is easily implementable by small-scale dairy producers or even for household use, helping to prevent health risks associated with consuming adulterated milk. The real-time monitoring capabilities of Raspberry-based systems enable swift detection of adulteration, allowing for immediate action to be taken. This proactive approach helps prevent the consumption of tainted milk, thereby protecting public health.

References

[1] Shinawy and A. M. El-Kholy, "Detailed information about materials and methods used and their results," International Journal of Food Science and Technology, vol. 22, no. 3, pp. 45–51, 2018.

[2] Abrantes, M. R., De Oliveira, A. R. M., Rocha, M. O. C., De Souza, G. O., Telles, E. O., Sakamoto, S. M., et al.J. B. A. (2014). Detection of bovine milk contaminants in adulterated milk and curd goat cheese. *Acta Scientiae Veterinariae*, 42, 1.

[3] Ai, K., Liu, Y., & Lu, L. (2009). Hydrogen-bonding recognition-induced color change of gold nanoparticles for visual detection of melamine in raw milk and infant formula. *Journal of the American Chemical Society*, 131(27), 9496–9497.

[4] Arora, K. L., Lal, D., Seth, R., & Ram, J. (1996). Platform test for detection of refined mustard oil adulteration in milk. *Indian Journal of Dairy Science*, 49(10), 721–723.

[5] Arvind Singh, G. C., Aggarwal, A., & Kumar, P. (2012). Adulteration detection in milk. *Research News For U (RNFU)*, 5, 52–55.

[6] Bakircioglu, D., Kurtulus, Y. B., & Ucar, G. (2011). Determination of some traces metal levels in cheese samples packaged in plastic and tin containers by ICP-OES after dry, wet and microwave digestion. *Food and Chemical Toxicology*, 49(1), 202–207.

[7] Bamiedakis, N., Hutter, T., Penty, R. V., White, I. H., & Elliott, S. R. (2013). PCB-integrated optical waveguide sensors: an ammonia gas sensor. *Journal of Lightwave Technology*, 31(10), 1628–1635.

[8] Bania, J., Ugorski, M., Polanowski, A., & Adamczyk, E. (2001). Application of polymerase chain reaction for detection of goats' milk adulteration by milk of cow. *The Journal of Dairy Research*, 68, 333–336.

[9] Banupriya, P. C. R. S., Supriya, T. V., & Varshitha, V. (2014). Comparison of different methods used for detection of urea in milk by quantification of ammonia. *International Journal of Advanced Research in Electrical, Electronics and Instrumentation Engineering*, 3(3), 7858–7863.

[10] Batis, V. K., Garg, S. K., Chander, H., & Ranganathan, B. (1981). Indian dairyman. *Indian Dairyman*, 33, 435.

[11] Bector, B. S., Ram, M., & Singhal, O. P. (1998). Rapid platform test for the detection/determination of added urea in milk. *Indian Dairyman*, 50(4), 59–62.

[12] Borkov´a, M., & Sn´ a˘ selov´a, J. (2005). Possibilities of different animal milk detection in milk and dairy products–a review. *Czech Journal of Food Sciences*, 23(2), 41–50.

[13] Bottero, M. T., Civera, T., Anastasio, A., Turi, R. M., & Rosati, S. (2002). Identification of cow's milk in buffalo cheese by duplex polymerase chain reaction. *Journal of Food Protection*, 65, 362–366.

[14] Bramanti, E., Sortino, C., Onor, M., Beni, F., & Raspi, G. (2003). Separation and determination of denatured $\alpha s1$-, $\alpha s2$-, β-, and κ-caseins by hydrophobic interaction chromatography in cow's, ewe's, and goat's milk, milk mixtures, and cheeses. *Journal of Chromatography A*, 994, 59–74.

[15] Brescia, M., Caldarola, V., Buccolieri, G., Dell'Atti, A., & Sacco, A. (2003). Chemometric determination of the geographical origin of cow milk using ICP-OES data and isotopic ratios: a preliminary study. *Italian Journal of Food Science*, 15, 3.

[16] Chang, E., & Arora, I. (2008). Simultaneous and fast analysis of melamine, cyanuric acid, and related compounds in milk and infant formula by LC/MS/MS. *Journal of Chromatography A*, X(Y), 1–4. [Online].

31 A comprehensive review of UAV technologies in medical and emergency applications

J. Chinnababu[1,a], G. Harshitha[2,b], G. Hema[2,c], S. Abdulrehaman[2,d] and V. Dwrakanathreddy[2,e]

[1]Associate Professor, Department of ECE, Annamacharya University, Andhra Pradesh, India

[2]UG Scholar, Department of ECE, Annamacharya Institute of Technology and Sciences, Andhra Pradesh, India

Abstract

This paper proposes a novel ramble framework planned to improve the proficiency of therapeutic supply conveyances, especially in inaccessible or crisis circumstances. The ramble coordinating progressed components, counting a 2200 mAh Li-Po battery, 1000 KV BLDC engines, and an APM 2.8 flight controller, tending to restrictions found in past plans that utilized littler batteries and lower-rated engines. Prior rambles, regularly fueled by 1000 mAh batteries and 500 KV BLDC engines, endured from restricted flight term and payload capacity, confining their operational adequacy. In differentiate, our proposed ramble framework essentially expands flight time and moves forward payload capabilities, making it reasonable for long-range restorative missions. To add to that, a servo engine connected to a to-begin-with help unit, GPS route, and a 6-channel farther transmitter allow for precise, computerized delivery and real-time control. This improved framework offers a more solid arrangement for healthcare coordination, especially in disaster help and other time-sensitive operations. Our plan presents a major change over prior models, making it an important apparatus for present-day healthcare and crisis reaction needs.

Keywords: APM 2.8 flight controller, BLDC motor, emergency response, first aid delivery, GPS navigation, healthcare logistics, Li-Po battery, medical drone, servo motor, UAV, etc

Introduction

Rambles, moreover, known as unmanned airborne vehicles (UAVs), have drawn a parcel of intrigue as of late from a variety of businesses, including healthcare, logistics, and natural observation. UAVs have also supported environmental monitoring, such as landslide progression assessments using aerial point clouds [7]. They are the culmination for applications requiring speedy and successful conveyance frameworks because of their capacity to move payloads, navigate over challenging landscapes, and run autonomously. Rambles have demonstrated particularly vital in the healthcare industry, with the capacity to bring crisis help and therapeutic supplies and to begin with help to hard-to-reach or confined areas where routine transportation strategies are ineffectual or take too long.

In healthcare and crisis reaction businesses, for example, service-based ramble conveyance frameworks have been explored to increase calculated effectiveness and abbreviate conveyance times [1].

Rambles have also been joined into Internet of Things (IoT) environments to collect and disseminate information in the healthcare industry, advertising real-time, secure, and decentralized information amassing strategies [2]. Rambles have as of now appeared as a guarantee in the healthcare industry in North America, from encouraging telemedicine administrations in provincial ranges to transporting basic supplies [3].

Even with these improvements, it is still exceedingly troublesome to ensure that rambles can fulfill the special necessities of crisis therapeutic supply conveyance. Restricted payload capacity constrained flying time, GPS route exactness, and reliable control frameworks that can work in an assortment of situations and landscapes are a few of these troubles.

Related Work

The employments of ramble innovation in healthcare, coordination, helpful offer assistance, and natural checking are the fundamental points of this writing

[a]jchinnababu@gmail.com, [b]gongati.harshitha@gmail.com, [c]hemagadhamsetty1910@gmail.com, [d]shaikabdulrehaman605@gmail.com, [e]dwarakavallapureddy@gmail.com

DOI: 10.1201/9781003684589-31

survey, which looks at current progressions in the field. The previously mentioned works illustrate the versatility and progressive potential of rambles in an assortment of businesses. UAVs now support logistics, healthcare, and emergency services through key innovations.

Ramble conveyance based on administrations

Service-based ramble conveyance frameworks were examined by Alkouz et al. [1], who concentrated on the integration and adaptability of ramble administrations inside online collaborative stages. Their think piece portrays the troubles in creating a versatile and compelling ramble conveyance benefit, and it was displayed at IEEE's 7th World Conference on Web Computing and Collaboration.

These troubles incorporate dealing with the operational coordination of broad ramble systems, ensuring real-time coordination, and getting over deterrents relating to airspace control and flight laws.

The creation of an agreeable system for ramble conveyance, which empowers smooth communication between numerous partners (benefit suppliers, controllers, and clients), is one of the work's fundamental commitments. Future drone-based administrations that can be joined into a number of businesses, counting healthcare, are made conceivable by this innovation. The ponder emphasizes how pivotal it is to secure information security and protection, especially when shipping sensitive merchandise like therapeutic gear. It does not, in any case, truly address the specialized restrictions that confine the utilize of drones for bigger or longer-distance cargo conveyances, such as battery life and payload capacity.

Rambles for helpful cargo

Humanitarian cargo drones' value-sensitive plan was considered by Cawthorne and Cenci [4]. Their inquiry about analyzes the ethical repercussions of conveying rambles for compassionate help, with a specific accentuation on how they can transport necessities to ranges influenced by normal catastrophes equipped clashes. It was displayed at the 2019 Worldwide Conference on Unmanned Airship Frameworks (ICUAS). The consider highlights how vital it is to make ramble frameworks that follow social, moral, and social standards in addition to specialized necessities.

The creators draw consideration to the specific challenges in utilizing rambles for helpful purposes, counting overcoming legitimate or political imperatives in different countries and making beyond any doubt that rambles do not inadvertently harm or panic nearby inhabitants. The thing about it, for the most part, looks at the social and moral concerns of utilizing rambles in crisis circumstances, whereas it does touch on the mechanical perspectives of the ramble plan. This investigation propels our information of how ramble innovation can be adjusted to viably convey life-saving help while being chivalrous of the necessities of affected communities.

Utilize of rambles in surgery and medication

Drones' utilization in surgery and pharmaceuticals was completely surveyed by Rosser et al. [5]. Their concern centers on the ways that rambles are making a difference for specialists by conveying drugs, carrying out surgical rebellions, and indeed supporting inaccessible surgery and telemedicine. The creators depict how rambles can make strides in the capacity of healthcare frameworks by encouraging speedier crisis reaction times, particularly in urban regions or catastrophe zones. The investigation moreover covers conceivable future employments, such as joining rambles into clinic frameworks to disperse restorative assets rapidly or utilizing rambles with sensors to track persistent vitals all through transportation. Restrictions still exist, in spite of the fact that, in terms of payload capacity, battery life, and climate resistance. This ponder emphasizes how flexible rambles are in the restorative field, but it too inclines more progressions in programs and equipment to ensure their constancy in significant restorative procedures.

At the least level of the control framework progression is an inward circle drift control framework that serves as an autopilot. This framework is overseen by a waypoint direction framework that a grounded flight director can adjust or control. Each level of this progression depends on a channel for route. The slightest sum of onboard equipment is required to protect capabilities in order to document this independent control.

Priliminaries

Unmanned Airborne Vehicles (UAVs), in some cases known as rambles, have drawn a parcel of consideration due to their use in a variety of businesses, including healthcare, observation, and natural observation. Agreeing with Kardasz et al. [8], early ponderers concentrated on the potential of rambles in the natural and gracious building areas, taking advantage of their capacity to reach inaccessible areas and assemble exact information. Due to their restricted operational flight length and use for amplified missions, these rambles were generally fueled by modest batteries, such as 1000 mAh Li-Po. In their 2015 study, Piotrowski et al. [9] considered the plan of remote-controlled flying units, highlighting the need for advanced control frameworks to make strides in ramble soundness in challenging circumstances. Rambles utilizing lower-rated BLDC engines (e.g., 500 KV) have inconveniences transporting bigger payloads in spite of headways in impetus and control frameworks, particularly in applications where continuance and constancy are pivotal, like the conveyance of therapeutic supplies [10]. The requirement for superior engine execution and the control of the economy has been brought to light by more inquiries. In their investigation of the impacts of control sources on brushless DC engines utilized in rambles, Bogusz et al. [10] found that engine effectiveness and control supply had a coordinated relationship with flight time and payload capacity. In any case, rambles still had issues with carrying capacity and battery life in spite of superior engine plans. Besides, as Alberstadt [11] examines, lawful systems make administrative deterrents to the broad use of rambles in civilian airspace, making it more troublesome to convey them in crisis and commercial administrations. When taken as a whole, these ponderings highlight the need for arrangements that handle the lawful and mechanical impediments to ramble use. We recommend a ramble framework that employs 1000 KV BLDC engines and a larger-capacity 2200 mAh Li-Po battery to enormously increase payload capacity and flight time in order to overcome the disadvantages famous in prior research. This course of action empowers longer working terms than past plans that depended on littler batteries and lower-power engines, making the ramble suitable for significant applications like long-distance therapeutic supplies conveyances. (See Figure 31.1 for the system block diagram). Whereas the consolidation of a servo engine coupled to a to-begin-with help unit permits mechanized, focused-on conveyance of restorative supplies, the use of the APM 2.8 flight controller moves forward steadiness and exactness in flight. The proposed framework also has a 6-channel farther transmitter that empowers real-time control from the ground station and GPS modules for the

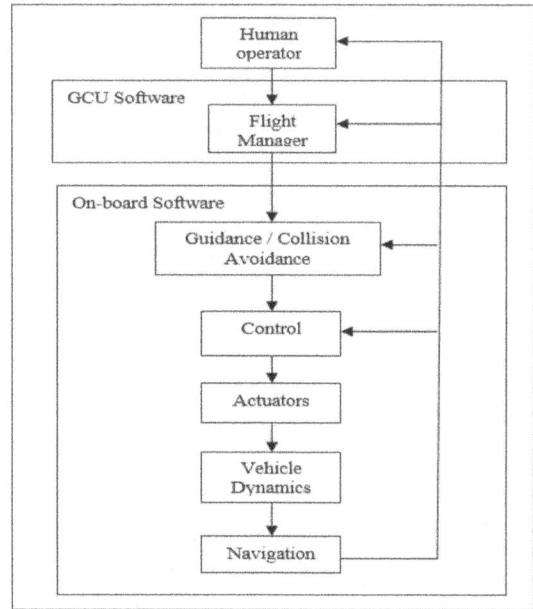

Figure 31.1 Block diagram
Source: Author

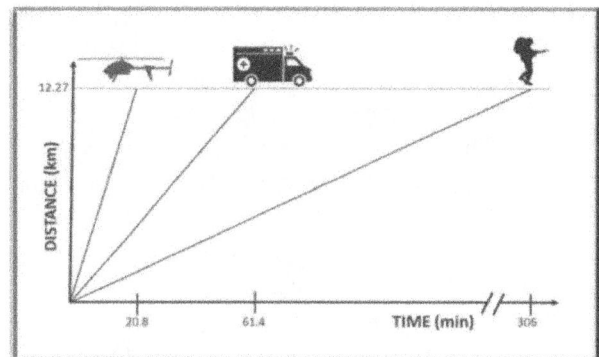

Figure 31.2 Comparison of delivery times by unmanned aerial vehicle, vehicle and foot march
Source: Author

exact route. This plan gives a more tried and truer and viable arrangement for healthcare coordination in far-off or crisis circumstances by tending to the issues of restricted battery life and payload limits that were common in prior rambles.

Based on the kind of crisis, rambles can transport the required instruments to the area of an event. The time it takes to react to a crisis is pivotal.

Ramble Sorts
There are four fundamental categories of UAVs, depending on the kind of ethereal stage being utilized. (see Figure 31.3 for UAV types).

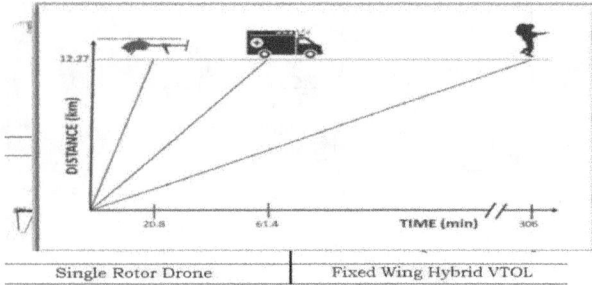

Figure 31.3 Unmanned aerial vehicle and drone types
Source: Author

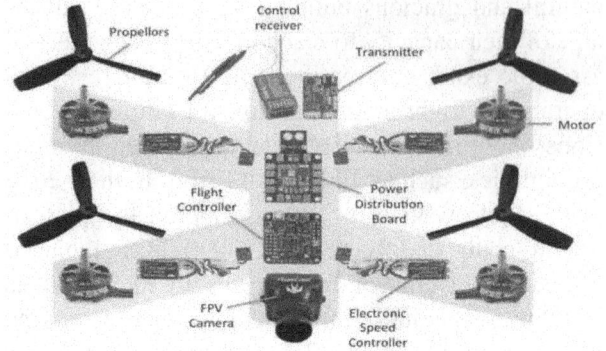

Figure 31.4 Unmanned aerial vehicles and drone types
Source: Author

- A ramble with numerous rotors
- Ramble with settled wings
- Ramble with a single rotor
- Crossover settled wing VTO

a) *Multirotor ramble*

A multirotor ramble navigates and flies by utilizing numerous propellers. Common applications for them incorporate video observation and photography. Among all UAVs, they are the most prevalent. Based on how numerous propellers are being utilized, they can be encouraged to be separated into four most common bunches. They are:
   A. Three-propeller rambles, or tricycles.
   B. Quadcopter: Rambles with four propellers.
   C. Hexacopter: Rambles with six propellers.
   D. Octocopter: Eight Rambles with Propellers.

b) *Rambles with settled wings*

Similar to airplanes, settled wing ramblers utilize wings in place of propellers. They are incapable of staying still. They proceed in their foreordained way until their vitality source is operational.

c) *Rambles with as if it were one rotor*

As the name suggests, a single rotor ramble has a single, fair rotor and a little tail for heading control. They seem more like a helicopter and use a lot of vitality.

d) *Half breed settled wing*

Vertical takeoff and landing (VTOL) is alluded to as VTOL. Settled Wing Cross breed VTOLs coast by utilizing wings and lift off with a propeller or propellers.

Main Components of Drones/UAVs (Quad Copter)

There are multiple designs used for drones; the most popular is the four-wing structure called a quadcopter. The main components of a quadcopter are (Figure 31.4):

- DC motors, propellers/wings, and chassis.
- Electronic speed controllers and flight controllers.
- Gear for landing.
- The transmitter.
- The recipient.
- The GPS module.
- The GPS module.
- A battery.

1. **Wings and propellers**

Depending on what is accessible, the rambles or UAVs are guided by propellers, wings, or both. Two diverse sorts of propellers are utilized by propeller-driven rambles for pushed and course. Standard propellers are what these are. The Pusher propellers and ordinary propellers. These are found in the quadcopter's front. The ramble is guided by these propellers. Propeller Pushers. These can be found in the quadcopter's raise. The ramble is impelled forward and in reverse by these propellers.

2. **The chassis**

This is the essential structure of the quadcopter that houses all of the other parts.

3. **Pilot**

The drone's brain is regularly alluded to as the flight controller. The electronic speed controller's control supply is overseen by the flight controller. It is moreover utilized to distinguish shifts in the drone's introduction. It makes beyond any doubt the ramble is in the discus and controls the engines.

4. **Electronic speed controller**

Electronic speed controllers (ESC) are the electrical circuits that regulate the speed of DC engines. Additionally, it provides options for both aggressive braking and turning around.

5. **DC Engines**

We require tall torque engines to make beyond any doubt the ramble remains in the discussion for a sensible sum of time. Moreover, tall torque helps in altering the propellers' speed. Since brushless DC engines are lighter than brushed ones, they are suggested.

6. **Landing**

Adapt Little rambles do not require landing gears. Bigger rambles, in any case, require landing adapt to avoid harm upon landing. The drone's working decides how much landing adapt is required. For occasion, conveyance rambles that transport bundles require open landing equipment since they require being able to store the items.

7. **The transmitter**

The transmitter creates pushed and heading commands for the ramble by sending signals from the controller.

8. **Collector**

The Flight Controller PCB gets the signals transmitted by the transmitter.

9. **The GPS unit**

Longitude, scope, and height are the navigational information that the GPS module gives the controller. On the occasion of a misplaced association, this module makes a difference: the controller distinguishes the way taken after and securely returns to the beginning point. The ramble is fueled by the battery. Rambles regularly utilize rechargeable batteries.

Novelty and Contributions

One creative and significant addition to healthcare logistics is the use of unmanned aerial vehicles (UAVs) for the delivery of medical supplies. In particular, there are both practical and technical benefits to integrating the APM 2.8 flight control system in this situation. The following are some salient features of UAVs' originality and contributions to the delivery of medical supplies utilizing APM 2.8:

1. **Enhanced effectiveness in emergencies**

Medical supply delivery times can be significantly shortened using UAVs, particularly in emergency scenarios where time is of the importance. APM 2.8, for instance, makes it possible to precisely regulate the UAV's flying route, ensuring that medical supplies can swiftly reach distant or infrastructure-challenged places. UAVs offer a quicker option than conventional techniques like road transport in emergency situations like natural disasters, accidents, or pandemics.

2. **APM 2.8 Autonomous flight**

One well-known open-source autopilot system that supports a variety of UAVs, including multirotor and fixed-wing aircraft, is the APM 2.8 flight controller. It has functions including fail-safe procedures, waypoint-based flight planning, and autonomous navigation. Because of these characteristics, it can provide medicinal supplies without the need for continual human supervision. Particularly helpful in remote or challenging-to-reach locations where human presence may be scarce is autonomous operation.

3. **Tracking and monitoring in real time**

The APM 2.8 system enables real-time tracking of the position and flight route of UAVs. For medical deliveries to be safe and accountable, this skill is essential. Health officials can confirm delivery and make sure the medical supplies get to their destination with real-time GPS tracking. The real-time tracking and monitoring capabilities of UAVs play a pivotal role in enhancing the efficiency and reliability of healthcare services, especially in remote or disaster-stricken areas [6]. UAV technologies help ensure that medications, vaccines, and other life-saving equipment reach their destinations swiftly, safely, and securely, while also providing real-time situational awareness for emergency responders. Table 31.1 shows the comparative analysis with previous works.

Analysis of Proposed System Model

Delivery efficiency and time: Particularly in remote locations, APM 2.8 and other UAV systems are significantly quicker than conventional delivery techniques. When compared to ground transportation, (Figure 31.2 compares delivery methods.) APM 2.8 can reduce delivery times by 60–70%; however, alternative systems may provide somewhat better routing algorithms.

Table 31.1 Comparison performance with previous works analysis.

Criteria	APM 2.8	Other UAV systems
Delivery time	About 60–70% faster than ground transportation for remote areas. Fast but affected by GPS signal strength and weather.	Generally comparable to APM 2.8, but more expensive systems may have better route optimization algorithms.
Cost-effectiveness	Low-cost, open-source autopilot system. Reduces upfront costs significantly.	More expensive due to proprietary components, but offers advanced features (e.g., better fail-safes, integrated sensors).
Battery efficiency	Decent battery life depends on the drone model, but can be limited in longer flights or heavier payloads.	High-efficiency battery systems, often with better power management algorithms to optimize flight time.
Weather and environmental tolerance	May struggle with adverse weather conditions.	Typically, more resistant to harsh weather and equipped with features like obstacle avoidance and weather compensations.

Source: Author

Accuracy: APM 2.8 is adequate for the majority of medical supply delivery because it is usually precise within 1-3 meters. More sophisticated UAV systems, like Pixhawk or DJI, can, however, attain sub-meter precision, which makes them better suited for deliveries that need to be made precisely or in crowded areas.

Cost-effectiveness: Because APM 2.8 is open-source, it is quite affordable, which makes it perfect for smaller-scale deployments and research. Commercial UAVs are more costly, but typically come with more features, stronger systems, and better fail-safes.

Conclusion

By proposing a moved-forward ramble framework, we have tended to the inadequacies of current ramble advances, particularly with respect to the conveyance of restorative supplies. Noteworthy issues were recognized by prior investigations, including the use of less viable BLDC engines and lower-capacity batteries, which brought about restricted flight time and cargo capacity. By including a 2200 mAh Li-Po battery and 1000 KV BLDC engines, our proposed plan gets over these confinements and moves forward flight continuance and stack carrying capacity. To begin with, help packs may be conveyed; consequently, much obliged to the servo engine framework, and exact control is guaranteed by the integration of an APM 2.8 flight controller. Moreover, real-time reaction is encouraged by the system's GPS route and 6-channel farther control, which makes the ramble more fitting for pivotal therapeutic missions in disconnected areas. This RAMBLE framework offers a major breakthrough in the utilization of UAVs for healthcare coordination, particularly in crisis circumstances where productivity and steadfastness are basic, by upgrading the control effectiveness, run, and control components.

References

[1] Alkouz, B., Shahzaad, B., & Bouguettaya, A. (2021). Service-based drone delivery. Proceedings of the IEEE 7th International Conference on Collaboration and Internet Computing (CIC), 68–76.

[2] Islam, A., Al Amin, A., & Shin, S. Y. (2022). FBI: a federated learning-based blockchain-embedded data accumulation scheme using drones for internet of things. *IEEE Wireless Communications Letters*, 11, 972–976..

[3] Hiebert, B., Nouvet, E., Jeyabalan, V., & Donelle, L. (2020). The application of drones in healthcare and health-related services in North America: a scoping review. *Drones*, 4, 30. .

[4] Cawthorne, D., & Cenci, A. (2019). Value sensitive design of a humanitarian cargo drone. In Proceedings of the 2019 International Conference on Unmanned Aircraft Systems (ICUAS), Atlanta, GA, USA, 11–14 June 2019, (pp. 1117–1125). .

[5] Rosser, Jr, J. C., Vignesh, V., Terwilliger, B. A., & Parker, B. C. (2018). Surgical and medical applications of drones: a comprehensive review. *Journal of the Society of Laparoendoscopic Surgeons*, 22, 18. [Google Scholar] [CrossRef][Green Version].

[6] Cohen, J. (2007). Drone spy plane helps fight California fires. *Science*, 318, 727.

[7] Al- Rawabdeh, A., Moussa, A., Foroutan, M., El-Sheimy, N., & Habib, A. (2017). Time series UAV image-based point clouds for landslide progression evaluation applications. *Sensors*, 17, 2378.

[8] Kardasz, P., Doskocz, J., Hejduk, M., Wiejkut, P., & Zarzycki, H. (2016). Drones and possibilities of their using. *Journal of Civil and Environmental Engineering*, 6, 233.

[9] Piotrowski, P., Witkowski, T., & Piotrowski, R. (2015). Unmanned remote-controlled flying unit. *Measurement Automation and Robotics*, 19, 49–55.

[10] Bogusz, P., Korkosz, M., Wygonik, P., Dudek, M., & Lis, B. (2015). Analysis of the impact of a supply source for the properties brushless DC motor with permanent magnets designed to drive a flying unmanned camera. *Przegląd Elektrotechniczny (Electrotechnical Review),* 91(5), 189–192.

[11] Alberstadt, R. (2014). Drones under international law. *Open Journal of Political Science,* 4(3), 79–87.

[12] Bardley, T. H., Moffitt, B. A., Fuller, T. F., Mavris, D., & Parekh, D. (2013). Design Studies for Hydrogen Fuel Cell Powered Unmanned Aerial Vehicles. Am Institute of Aeronautics and Astronautics.

32 Intelligent smart home operations through NLP and IoT-driven voice control

S. Lokesh Reddy[1,a] and P. Ajay Kumar Reddy[2,b]

[1]M. Tech, Embedded Systems, Department of ECE, Kuppam Engineering College, Kuppam, Andhra Pradesh, India

[2]Associate Professor, Department of ECE, Kuppam Engineering College, Kuppam, Andhra Pradesh, India

Abstract

The integration of Internet of Things (IoT) technologies in smart home systems has significantly enhanced convenience, efficiency, and accessibility for users. This paper presents a chat-based IoT framework that employs natural language processing (NLP) to control and monitor home appliances using text and voice commands. Built on a microcontroller-based architecture, the system enables seamless interaction with connected devices, supporting real-time measurement of temperature, humidity, and power consumption in every room. This allows for effective environmental monitoring and power-efficient operation by optimizing device usage. Additionally, the system offers synchronized control of lighting across all rooms, allowing users to control lights individually or in groups seamlessly. Leveraging open-source tools and modular technologies, the framework ensures flexibility, scalability, and efficient device response while addressing key challenges in usability, interoperability, and energy management. This research highlights the potential of integrating IoT and NLP to develop intelligent, energy-efficient, and secure smart home systems tailored to diverse user needs.

Keywords: Intelligent smart home, Internet of Things IoT, natural language processing NLP, secure home, smart home, text commands, voice commands

Introduction

Among the technologies that have gained popularity in recent years, we can fire the Internet of Things (IoT), which has significantly changed the approach to how people use technologies in everyday life. This type of technology has gone mainstream with devices that range from the most basic, like fridges, to smarter home systems. They are meant to improve comfort, safety, and energy efficiency by embedding ordinary devices into the World Wide Web, where operators can manage them securely from remote locations. A major development in this line of research is the use of natural language processing (NLP), through which interaction with smart systems based on voice and text commands is possible.

Smart home systems that incorporate IoT and NLP grant their users control over various devices, including lights, fans, and security systems, and assist in environmental control, including temperature and humidity [1]. This not only increases usability but also increases energy efficiency more than in the initial solutions. Through the combination of power control and metering and parallel control of multiple loads, these systems are beneficial to efficiency and, therefore, sustainability. Also, these interfaces enable a large number of people, especially elderly and disabled persons, to effectively interact with these systems since it eliminates difficult keying sequences associated with icons [2, 3].

As indicated by this paper, different benefits are associated with IoT in the context of smart homes; despite this, certain limitations, including high dependence on cloud services and the negative impact of security threats, have not been fully addressed. Most currently implemented systems utilize third-party cloud applications, which pose issues regarding privacy and security risks. Still, the availability of these systems may be hampered by unstable Internet connections in some areas. These challenges are, however, addressed in the present research by proposing a secure, efficient, and flexible smart home framework using microcontroller-based architecture and NLP.

This paper describes the development of a chat-based smart home system that would allow the users smart control and monitor their home appliances and

[a]lokeshreddymtpl@gmail.com, [b]ajaypedamalla@gmail.com

DOI: 10.1201/9781003684589-32

give the real-time environmental and power consumption for each room in the house. It also backs the dictatorial system of the lighting together with the effective control of energy, which makes it strong and easy to use.

Literature Survey

In recent years, the idea of smart homes and IoT-based automation systems has been thoroughly examined, with numerous studies emphasizing their potential to enhance quality of life, boost energy efficiency, and increase user convenience. This section highlights significant advancements in IoT, NLP, and the integration of microcontroller-driven smart systems.

Integrating NLP with IoT applications represents a burgeoning area of research and implementation, yet existing literature highlights a scarcity of robust systems that combine these technologies dynamically and efficiently. Most implementations to date tend to leverage the built-in NLP capabilities of smartphones, such as those provided by Android and iOS,

or utilize simpler virtual assistants like Amazon's Alexa (Table 32.1).

In contrast, some studies have made strides by developing custom NLP systems specifically tailored for smart home automation. These systems are often designed with a primary focus on enabling user control via voice commands, creating an interactive experience that can respond to a broader range of user inputs. However, these custom solutions frequently demand significant computational resources and extensive setup processes, which can hinder their practicality in real-world applications. Moreover, reliance entirely on an NLP-centric control mechanism may not always align with the existing IoT ecosystems.

Giudici et al. [4] introduced IFTTT, an LLM-powered conversational agent for energy-efficient home automation, highlighting improved user engagement and usability. Similarly, Iliev & Ilieva and Netinant et al. [5, 6] proposed a smart home framework integrating IoT, fog, and cloud layers with

Table 32.1 Summary of smart home automation research papers focusing on NLP, including functionality, technology, sensors, and microcontrollers used.

Reference	Year	Technology used	Sensors used	Microcontroller used
Utsanok [6]	2024	NLP, Cloud Computing	DHT 11, PIR, LDR, MQ-2	NodeMCU
Naseem [7]	2023	NLP, IoT	Smart Relays, Proximity, Motion	Raspberry Pi
Panagiotak [8]	2019	NLP, IoT, Wi-Fi	Temperature	Raspberry Pi
Padalino [9]	2024	NLP, IoT, Bluetooth	Temperature, Humidity	NodeMCU
Motta [10]	2023	NLP, IoT, Cloud	Motion and Light Sensors	ESP8266
Rio [11]	2020	NLP, AI, IoT	Not specified	ESP32
Hamdan [12]	2021	NLP, IoT, AI	Temperature and Motion Sensors	Arduino Uno
Butunoi [13]	2021	IoT, NLP, Wi-Fi	Temperature Sensors	Raspberry Pi 4
Nayyar [14]	2016	IoT, NLP, AI	Humidity, Light, and Temperature Sensors	ESP32
Ullah [15]	2017	NLP, IoT	Not specified	Arduino Mega
Bharath [16]	2015	NLP, IoT	Light and Motion Sensors	ESP8266
Ghobril [17]	2013	IoT, NLP	Not specified	Arduino Uno
Gomez [18]	2020	AI, NLP, IoT	Temperature and Humidity Sensors	ESP32
Ansari [19]	2019	NLP, IoT	Temperature and Light Sensors	Raspberry Pi
Asha [20]	2022	AI, NLP, IoT	Not specified	ESP8266

Source: Author

speech recognition, focusing on reducing latency, bandwidth usage, and enhancing security.

Objective

The integration of NLP into the smart home system focuses on delivering an intuitive and user-friendly experience. Instead of complicated controls, you can simply talk or type what you want, and the system takes care of the rest. Whether you prefer speaking commands or typing them out, this flexibility accommodates diverse user preferences, promoting ease of use. With user-friendly interfaces that feel natural to navigate, interacting with your smart home becomes effortless, making the entire experience smooth and enjoyable. This project tackles real-time challenges by developing a system that enables users to operate and monitor appliances through a mobile and web interface. It enables turning lights and fans on or off, and checking temperature and humidity levels, the aim is to deliver a seamless, user-friendly solution to improve everyday home management.

Methodology

The methodology of the smart home project revolves around a web-based IoT agent that enables users to control and monitor home devices such as lights, fans, and sensors. The system utilizes the Dialogflow API for natural language processing (NLP), facilitating both voice and text interactions for managing devices and accessing data such as temperature and humidity. Microcontrollers, specifically the Wemos D1 Mini, gather data from DHT11 sensors and transmit it to Firebase while also controlling devices like lights through MQTT communication. Firebase serves as a real-time database, storing sensor information and device states while ensuring synchronization every 5 seconds. Users can interact with the system via a web or mobile app, issuing commands that are processed through NLP and executed accordingly. MQTT provides a lightweight communication protocol between IoT devices and the central system, ensuring efficient communication and real-time updates. This approach results in a seamless, effective, and real-time home automation and monitoring solution.

Technical Architecture

Figure 32.1 provides a visual representation of the system's architecture, which is made up of several

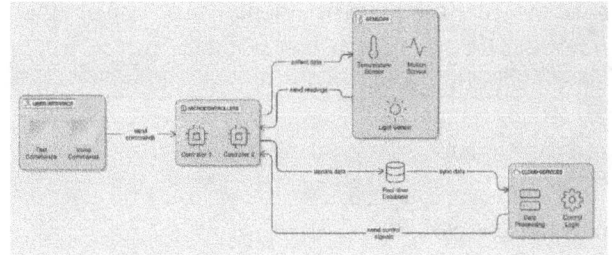

Figure 32.1 Showcases the system's overall technical design
Source: Author

interconnected components. The system architecture of the smart home project is centered around a web-based IoT agent that allows users to control and monitor home appliances and environmental conditions. The design integrates several key components, including sensors, microcontrollers, a real-time database, and a user interface.

When a user logs into the web application, the first page they encounter is the welcome page. This page features a chatbot and a footer menu for easy navigation as shown in Figure 32.2.

The end user is presented with two options on the chatbot tab: they can either type a command or speak, and we are using web speech API to allow the application to interpret spoken commands and respond with synthesized speech. The voice recognition system converts spoken commands into text, which is then sent to the NLP system for processing. When the user speaks a command, the Web Speech API captures the voice and converts it into text, which is then sent to the Dialogflow API. Dialogflow analyses the text matches it to predefined responses, and returns the appropriate reply.

Additionally, our system is designed to check the current status of the lights before performing any action. If the light is already on, the system will inform the user with a message like, "The light is already ON," preventing unnecessary command execution. This ensures that the system is efficient and user-friendly by avoiding redundant actions.

Moreover, the system supports controlling multiple devices or rooms with a single command. For example, the user can issue a command like, "Turn on lights in both the kitchen and the bedroom." In this case, the system passes the parameters of both rooms in a single JSON object to Firebase, allowing simultaneous control of the lights in both rooms.

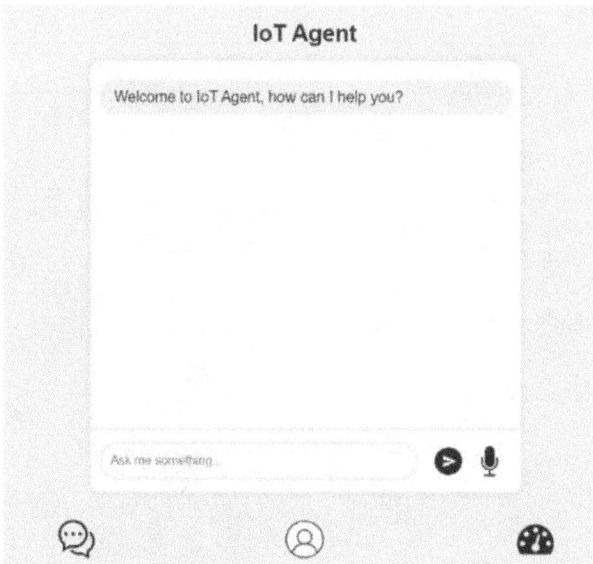

Figure 32.2 Welcome page
Source: Author

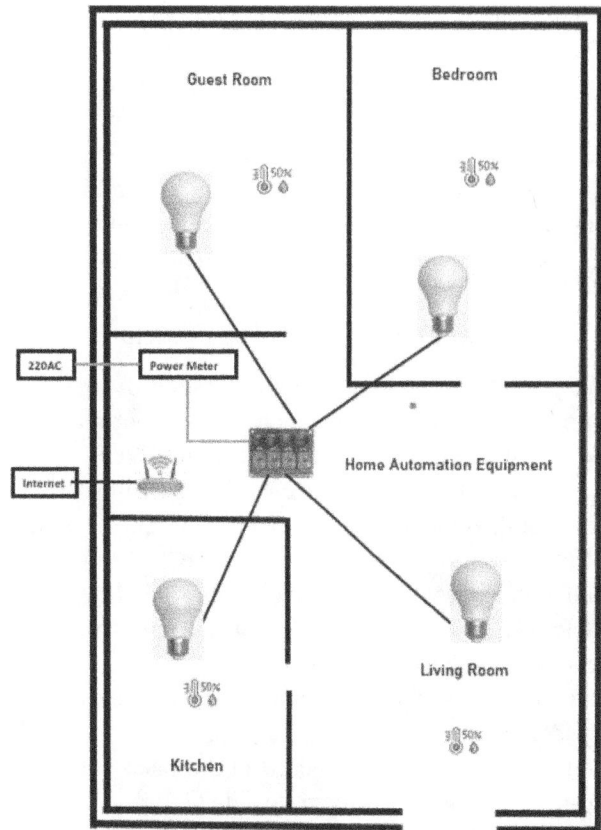

Figure 32.3 The IoT device connection and room layout
Source: Author

This unique capability enhances the efficiency of the system by enabling the user to manage multiple devices across different locations in a single prompt.

This architecture is designed to provide an efficient, secure, and user-friendly smart home experience. By combining NLP, IoT, and real-time data synchronization, the system enables seamless control and monitoring of home devices while maintaining high performance and reliability.

To ensure that the temperature and humidity values are always available in the Web application and to prevent downtimes, the data is stored in a Firebase database. This ensures the robustness and stoppage of the system, i.e., in case of any hardware failure, the database continues to operate and, on the other hand, transfers to the Web application the last user information. Here, the firebase is selected every 5 s so that it is synchronized in real-time with the DHT11 sensors. Additionally, i.e., microcontrollers, whose calculations take the current sensor readings into account compared to the previous one and that only push current readings taken into the database. This approach prevents the network from being overloaded with unnecessary traffic.

The one lighting control and four environmental monitoring rooms, respectively. The microcontroller defining the lighting system described in Figure 32.3, i.e., for the on/off state of the switch lights, too, continues based on length messages, MQTT, forwarded by the IoT Agent in as far as the lights should be switched on or off. When the socket is initialized, it synchronizes with the database, after which each operation is merely a reception of the messages in MQTT, and the maximum rate is guaranteed with minimal resources.

Our system enhances energy efficiency by allowing users to manage and monitor smart home devices from anywhere, ensuring that lights and appliances are turned off when not in use. The ability to automate device control through voice commands, scheduling, and remote access helps reduce unnecessary energy consumption. By preventing devices from remaining powered on unintentionally, our system contributes to overall energy savings, promoting a more efficient and sustainable smart home environment.

To enhance security, our system incorporates a login mechanism that restricts access to authorized users only, ensuring that only authenticated individuals can control smart home devices. Additionally, we implemented a broker using secure WebSockets (WSS) for encrypted communication, leveraging the Eclipse Mosquitto Broker (MQTT) to securely exchange messages between the web application, Firebase, and IoT devices. These measures, combined with authentication protocols and encryption techniques, help mitigate cybersecurity risks such as unauthorized access, data interception, and voice command spoofing, ensuring a secure and privacy-preserving smart home environment.

Natural Language Processing System

We have integrated into another branch of artificial intelligence called NLP which allows computers to process and understand human language. NLP is used for the voice-chat and text-chat functionalities. This is accomplished to an extent with Dialogflow, which offers a powerful and refined system to bring NLP and intent classification to IoT systems. Dialogflow is a Google-powered NLP API that leverages machine learning for training and development thus, helping it to understand variations in questions or commands by identifying the keywords and answering appropriately through the same record in the back end. Our system is designed to integrate with Google Assistant via Dialogflow's webhook, enabling seamless compatibility with existing smart home ecosystems. When the Dialogflow API returns a response in text and voice based on the user's requirements.

The following are example training phrases added to Dialogflow, along with their respective responses. Tables 32.2 and 32.3 showcase the user inputs and corresponding system responses for temperature / humidity and light control.

Table 32.4 showcases the room entities detected by the NLP system when a query or question mentions them. For example, in the command "What is the temperature in the guest room," the system recognizes "guest room" as an entity and assigns it to "[room]."

Result and Discussion

The implementation of the smart home system demonstrates seamless control and monitoring of home devices, including lights and sensors for temperature and humidity. Commands are executed via the web application, showcasing the system's efficiency, responsiveness, and functionality.

Below, we present a detailed analysis of various scenarios captured through the hardware setup and the execution of user commands.

Table 32.2 Dialogflow training model.

Training phrases (Input)	Responses (Output)
What is the humidity of my guest room?	The [Selected_Room] has humidity [%].
What is the current temperature of the kitchen?	The [Selected_Room] has temperature [°C]
Tell me, what is the temperature of all the rooms?	Error Handling: Which room are you referring to?
Give me the humidity levels for all the rooms.	
Can you tell me the humidity level in my bedroom?	
Please provide the humidity reading for the guest bedroom.	

Source: Author

Table 32.3 The dialogflow queries and responses for managing lights and fans.

Training phrases (Input)	Responses (Output)
Turn on the light in the living room.	The lights of [Selected_Room] are $status.
Switch on the lights in the kitchen.	All lights are [Status] in [Selected_Room].
Turn off the light in the living room.	
Turn on all the lights in my house.	The lights of [Selected_Room] are already [Status].
Please switch on the guest room lights.	

Source: Author

The hardware kit, shown in Figure 32.4, displays the initial state of the system with all lights turned off. The control board is powered on, and the system is ready to receive commands from the web application. This serves as the default state where no active devices are consuming power.

In this scenario, the user issues the command "Turn on the kitchen lights" via the web app. As shown in Figure 32.5, the system successfully activates the kitchen lights while other lights remain off. The command is processed by the NLP system and relayed to the microcontroller, which toggles the kitchen light relay.

Table 32.4 Dialogflow device control (lights and fans) requests and responses.

$Kitchen$	cooking room, kitchen, kitchen_room,
$bedroom$	My bedroom, bedroom, Master bedroom,
$living_room$	living room, Living room, Hall
$guest_room$	guest's bedroom, guest bedroom, guest room
all	Total rooms, my house, my place, all room's

Source: Author

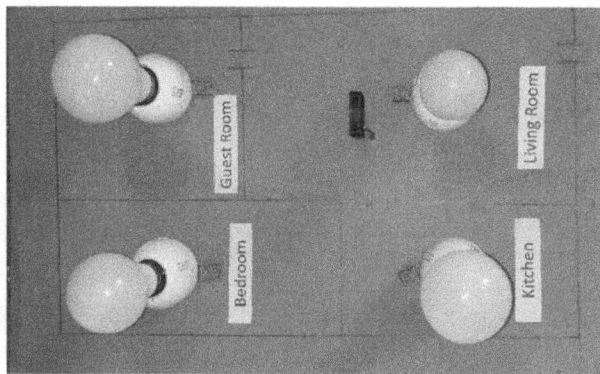

Figure 32.4 Top view of hardware kit (all lights off)
Source: Author

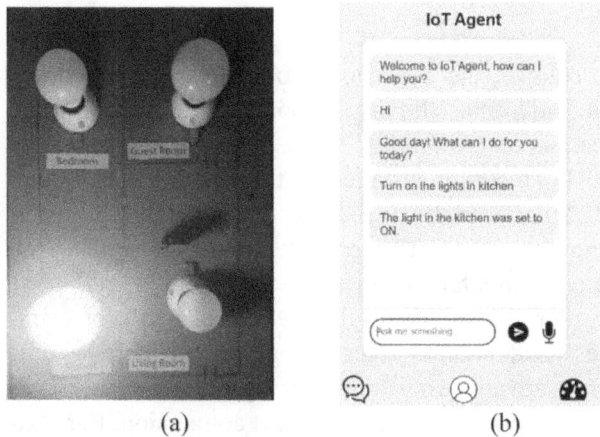

Figure 32.5 (a) Kitchen light activation kit (b) chat to control kitchen light
Source: Author

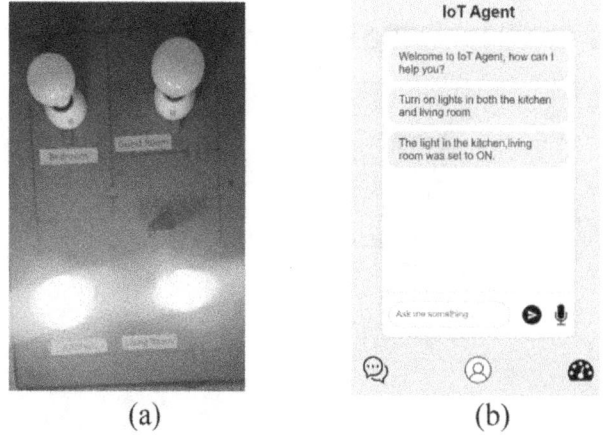

Figure 32.6 (a) Kitchen and living room light activation kit (b) Chat to control kitchen and living room lights
Source: Author

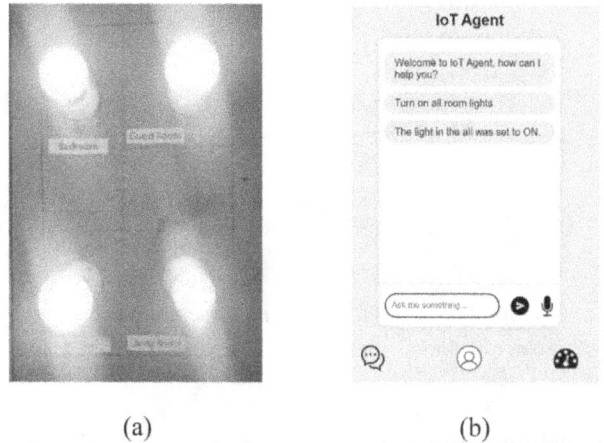

Figure 32.7 (a) All lights activation kit (b) Chat with agent
Source: Author

When the user sends the command "Turn on lights in both the kitchen and living room," the system processes both room parameters simultaneously. Figure 32.6 demonstrates the successful activation of lights in both locations, leveraging the system's ability to handle multi-room commands efficiently, bypassing room parameters in a single JSON object.

In Figure 32.7, the command "Turn on all lights" is issued through the web app. The system processes the command using the NLP module, which identifies the intent ("turn on") and the target devices ("all lights"). This broad command demonstrates the system's ability to handle comprehensive user requests by synchronizing all connected light relays across various rooms, including the kitchen, bedroom, living room, and other designated areas.

The ability to control multiple devices with a single command shows the system's efficiency and

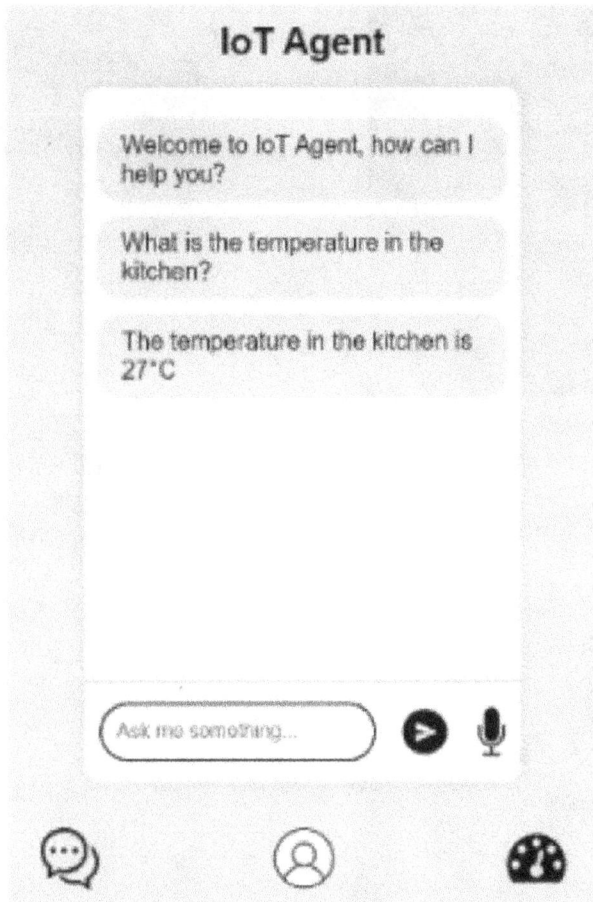

Figure 32.8 Chat for kitchen temperature
Source: Author

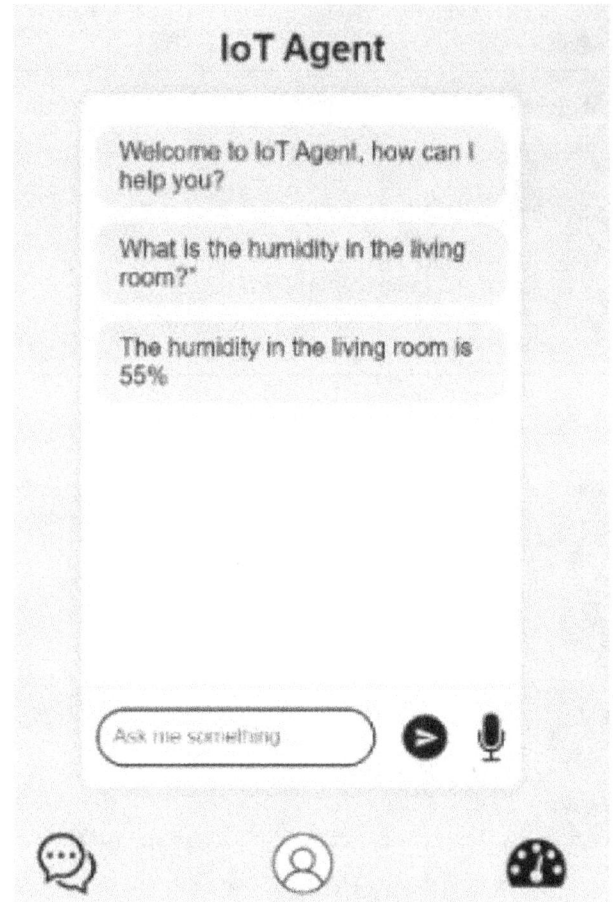

Figure 32.9 Chat for living room humidity
Source: Author

scalability. Whether the command is issued from a single device or simultaneously from multiple devices, the synchronization mechanism ensures seamless operation without conflict. This feature is particularly useful for scenarios where users need to quickly set up their environment, such as during evening routines or hosting guests.

When the user prompts the web app with the command "What is the temperature in the kitchen?", the system leverages the NLP module to identify the intent and extract the relevant entities, such as the room ("kitchen") and parameter ("temperature"). Figure 32.8 displays the system fetching real-time data from the DHT11 sensor placed in the kitchen, which continuously monitors environmental conditions. The current temperature is retrieved, formatted for clarity, and presented to the user, showcasing accurate and responsive environmental monitoring capabilities.

Similarly, when the command "What is the humidity in the living room?" is sent, the NLP module

processes the user input to understand the query and identifies the target room ("living room") and parameter ("humidity"). The system then queries the DHT11 sensor installed in the living room. Figure 32.9 captures the displayed humidity value, confirming that the system provides real-time updates on environmental conditions.

Our system utilizes the Web Speech API for seamless and precise voice recognition, which benefits from advanced machine learning models and achieves high accuracy in speech-to-text conversion. For text-based commands, we have trained Dialogflow with a large dataset, achieving an estimated 97% accuracy in understanding user intents. Additionally, error handling is managed through Fallback Intents, which prompts the user to rephrase unclear commands. This ensures that unrecognized inputs do not cause system failures but instead guide the user toward providing a clearer request, improving overall usability and command execution reliability.

Conclusion

This smart home system integrates Internet of Things hardware, natural language processing (NLP), and real-time data monitoring for interactive automation. Users control devices via voice or text, with multi-room support and real-time synchronization. Performance evaluation highlights response time factors, including NLP processing, web handling, and microcontroller execution, with future improvements like MQTT and edge computing for enhanced efficiency.

This project sets a strong foundation for expanding smart home automation capabilities. Future developments could include integrating advanced sensors for broader functionality, adding offline operation modes for uninterrupted service, and incorporating machine learning to personalize user experiences. Enhanced energy management features, such as predictive power usage analysis and compatibility with emerging IoT standards, could further increase the system's efficiency and adoption potential. These advancements would make the system even more adaptive and versatile, catering to the evolving needs of modern smart homes.

References

[1] Wortmann, F., & Flüchter, K. (2015). Internet of things: technology and value-added. *Business and Information Systems Engineering*, 57(3), 221–224.

[2] Domb, M. (2019). Smart home systems based on Internet of things, In Y. Ismail (Ed.), Internet of Things (IoT) for Automated and Smart Applications (pp. 1–20). London, UK: IntechOpen.

[3] Mao, Y., & Yu, X. (2024). A hybrid forecasting approach for China's national carbon emission allowance prices with balanced accuracy and interpretability. *Journal of Environmental Management*, 351(119873), 119873.

[4] Giudici, M., Padalino, L., Paolino, G., Paratici, I., Pascu, A. I., & Garzotto, F. (n.d .). "Designing home automation routines using an LLM-based chatbot", Smart Home Design, 2nd Edition, Designs (Basel), Journal Designs Volume 8 Issue 3.

[5] Iliev, Y., & Ilieva, G. (2022). A framework for a smart home system with voice control using NLP methods. *Electronics (Basel)*, 12(1), 116.

[6] Netinant, P., Utsanok, T., Rukhiran, M., & Klongdee, S. (n.d .). Development and assessment of internet of things-driven smart home security and automation with voice commands. IoT, Volume 5 Issue 1.

[7] Irugalbandara, C., Naseem, A. S., Perera, S., Kiruthikan, S., & Logeeshan, V. (2023). A secure and smart home automation system with speech recognition and power measurement capabilities. *Sensors (Basel)*, 23(13), 5784.

[8] Alexakis, G., Panagiotakis, S., Fragkakis, A., Markakis, E., & Vassilakis, K. (2019). Control of smart home operations using natural language processing, voice recognition, and IoT technologies in a multi-tier architecture. *Designs*, 3(3), 32.

[9] Giudici, M., Padalino, L., Paolino, G., Paratici, I., Pascu, A. I., & Garzotto, F. (2024). Designing home automation routines using an LLM-based chatbot. *Designs*, 8(3), 43.

[10] Motta, L. L., Ferreira, L. C., Cabral, T. W., Lemes, D. A., Cardoso, G. D. S., Borchardt, A., et al. (2023). General overview and proof of concept of a smart home energy management system architecture. *Electronics (Basel)*, 12(21), 4453.

[11] Sovacool, B. K., & Furszyfer Del Rio, D. D. (2020). Smart home technologies in Europe: a critical review of concepts, benefits, risks, and policies. *Renewable and Sustainable Energy Reviews*, 120(109663), 109663.

[12] Sathesh & Hamdan, Y. B. (2021). Smart home environment future challenges and issues - a survey. *Journal of Electronics and Informatics*, 3(1), 1–14.

[13] Stolojescu-Crisan, C., Crisan, C., & Butunoi, B.-P. (2021). An IoT-based smart home automation system. *Sensors (Basel)*, 21(11), 3784.

[14] Puri, V., & Nayyar, A. (2016). Real-time smart home automation based on PIC microcontroller, Bluetooth and Android technology. In Proceedings of the 3rd INDIACom, New Del-hi, India, (pp. 1478–1484).

[15] Asadullah, M., & Ullah, K. (2017). Smart home automation system using Bluetooth technology. In 2017 International Conference on Innovations in Electrical Engineering and Computational Technologies (ICIEECT), (Vol. 6, pp. 1–6).

[16] Anandhavalli, D., Mubina, N. S., & Bharath, P. (2015). Smart home automation control using Bluetooth and GSM. *International Journal of Informative and Futuristic Research*, 2, 2547–2552.

[17] Baraka, K., Ghobril, M., Malek, S., Kanj, R., & Kayssi, A. (2013). Low-cost arduino/android-based energy-efficient home automation system with smart task scheduling. In 2013 Fifth International Conference on Computational Intelligence, Communication Systems and Networks.

[18] Zamora-Izquierdo, M. A., Santa, J., & Go-mez-Skarmeta, A. F. (2010). An integral and networked home automation solution for indoor ambient intelligence, *IEEE Pervasive Computing, Volume* 9(4).

[19] Singh, U., & Ansari, M. A. (2019). Smart home automation system using internet of things. In 2019 2nd International Conference on Power Energy, Environment and Intelligent Control (PEEIC).

[20] Nirmala, A. P., Asha, V., Chandra, P., Priya, H., & Raj, S. (2022). IoT based secure smart home automation system. In 2022 IEEE Delhi Section Conference (DELCON).

33 AI-driven monitoring and emotion recognition for Alzheimer's patients with blockchain-enhanced data security

K. J. S. Upendra[1,a], Ch. Poojitha Sai[2,b], M. Yeswanth[2,c], K. Priyanka[2,d], M. Siva Krishna Babu[2,e] and D. Avinash Varma[2,f]

[1]Assistant Professor, Department of AI&DS, Vishnu Institute of Technology, Bhimavaram, Andhra Pradesh, India

[2]U.G Students, Department of AI&DS, Vishnu Institute of Technology, Bhimavaram, Andhra Pradesh, India

Abstract

This paper presents the development of an integrated system for real-time emotion and motion detection aimed at enhancing the care and monitoring of individuals with Alzheimer's disease. Alzheimer's, which affects millions worldwide, leads to severe cognitive decline, making it difficult for patients to express emotions or communicate their needs. The proposed system combines artificial intelligence and machine learning to detect both motion-related events (such as falls, unconsciousness, or dizziness) and emotional states (such as happiness, sadness, and anger). Using You Only Look Once (YOLO) for motion detection and a deep learning-based model for facial emotion analysis, the system processes video data through OpenCV for real-time analysis. The backend, powered by Flask, supports event logging and visualization through an interactive web interface, while SQLite stores event logs for ongoing analysis. Additionally, the integration of blockchain technology ensures the security and privacy of sensitive patient data, providing a tamper-proof framework for storing and accessing patient records. The system also triggers audio alerts for critical motion incidents, ensuring quick responses to emergencies. This lightweight yet comprehensive solution provides caregivers and healthcare providers with valuable insights into the emotional and physical well-being of patients, improving their quality of life and offering a new approach to managing Alzheimer's care through continuous monitoring and timely interventions.

Keywords: AI-Based monitoring, emotion detection, flask, motion detection, object detection, OpenCV, real-time surveillance, SQLite, TensorFlow, YOLOv8

Introduction

This project addresses two critical aspects of behavior monitoring: motion detection and emotion detection. Motion-related events such as falls and unconsciousness pose safety risks, while emotion analysis provides insights into a person's mental state. Combining these two capabilities creates a comprehensive safety and behavioral analysis system in environments like healthcare, workplaces, and homes.

The system integrates You Only Look Once (YOLO) for detecting motion events and a facial emotion detection model for analyzing emotional states. Flask handles the backend and data visualization, while SQLite stores logs for future analysis.

Key Features and Capabilities

Motion detection: The motion detection module utilizes the YOLOv8 deep learning model to effectively detect movement by identifying objects and tracking changes in their positions within a specified area. The system processes video frames to detect alterations in the environment, triggering alerts upon identifying significant motion. This is achieved through techniques such as frame differencing and contour detection, which allow for precise identification of movement patterns. The applications of this feature are vast, including security surveillance, anomaly detection, and activity monitoring. The advantages of the motion detection module include real-time processing capabilities, customizable sensitivity levels to suit different environments, and a reduction in false positives through advanced filtering methods, ensuring that alerts are both accurate and relevant.

Emotion recognition: [4] The emotion recognition module employs a sophisticated deep learning model to analyze facial expressions and predict human emotions. This feature is particularly valuable for

[a]upendra201@gmail.com, [b]chatradipoojithasai@gmail.com, [c]21pa1a5462@vishnu.edu,in, [d]21pa1a5446@vishnu.edu.in, [e]21pa1a5459@vishnu.edu.in, [f]22pa5a5403@vishnu.edu,in

DOI: 10.1201/9781003684589-33

monitoring an individual's emotional state across various settings, such as workplaces, educational institutions, and public spaces. Utilizing convolutional neural networks (CNN) trained on extensive facial expression datasets, The system is capable of precisely recognizing various emotions, such as joy, sorrow, anger, surprise, and neutrality. The applications of emotion recognition extend to behavioral analysis, customer feedback monitoring, and enhancing human-computer interaction. By providing insights into emotional states, this module enables caregivers and professionals to respond appropriately to individuals' needs, fostering a supportive environment.

Object detection: The object detection module integrates the YOLOv8 model. To identify and categorize objects instantly. This feature empowers the system to identify various objects and track their movements within the monitored area, enhancing situational awareness. The YOLOv8 model is recognized for its precision and rapid processing, making it ideal for various applications such as security surveillance, traffic monitoring, and inventory management. The supported objects include people, vehicles, bags, and more, allowing for comprehensive monitoring of the environment. One of the key advantages of this module is its ability to detect multiple objects simultaneously, providing a robust solution for real-time object tracking and classification.

Literature Survey

The development of AI-based monitoring systems has seen significant progress in recent years. Existing literature highlights various AI-driven applications in surveillance, such as facial recognition, anomaly detection, and automated alerts. However, challenges remain in achieving high accuracy, real-time processing, and minimizing false positives. The YOLO model series has proven effective in object detection tasks due to its speed and accuracy. This section reviews the advancements in YOLO-based applications and their relevance to monitoring systems.

AI-Driven monitoring systems: Studies show that AI can improve monitoring accuracy by automating tasks traditionally performed by humans [1]. Object detection models like YOLO have been applied in security cameras, traffic monitoring, and retail analytics. However, the integration of emotion recognition in monitoring systems is relatively new and offers opportunities for Behavioral analysis.

YOLOv8, the latest iteration of the "You Only Look Once" model, offers improved accuracy and faster inference times compared to its predecessors [2]. It is designed to detect objects in real-time, making it ideal for monitoring applications. The model's architecture and optimization techniques make it suitable for deployment on various hardware configurations, from edge devices to cloud servers.

Integration of emotion recognition in monitoring: Emotion recognition technology has evolved significantly with the development of deep learning models [3]. These models analyze facial expressions to determine emotional states such as happiness, anger, sadness, and surprise. Integrating emotion recognition in monitoring systems provides valuable insights into human behavior, which can be used in various sectors, including customer service, healthcare, and law enforcement. **Challenges in automated monitoring systems**: Despite the advancements in AI-driven monitoring systems, several challenges remain. One of the primary challenges is minimizing false positives, which can lead to unnecessary alerts and reduced trust in the system. Additionally, ensuring the privacy and security of the data collected by these systems is critical to avoid misuse.

Proposed Systems

The proposed AI-based monitoring system addresses the limitations of existing solutions by integrating advanced AI models, modular architecture, and user-centric design. The system consists of several key components that enhance its functionality and usability.

Data Collection and Processing: The proposed AI-based monitoring system addresses the limitations of existing solutions by integrating advanced AI models, modular architecture, and user-centric design. The data collection module gathers real-time data from sensors, devices, or databases to ensure a continuous flow of information. This data is processed by the processing unit, which utilizes AI algorithms for anomaly detection, predictive modeling, and comprehensive data analysis. The user interface, designed using technologies such as Streamlit or Flask, provides an intuitive dashboard for visualization and interaction. Additionally, accessibility features, including voice commands, multilingual support, and customizable alerts, enhance the usability of the system for a wide range of users.

Key features and capabilities: The system offers several advanced capabilities to ensure effective monitoring and decision-making. Automated data monitoring enables the system to process incoming data streams, identify trends, and detect anomalies, ensuring high-quality and contextually relevant insights. Real-time processing provides immediate analysis and actionable insights, allowing users to respond proactively to any detected changes. Customizable dashboards allow users to configure monitoring parameters, thresholds, and visualizations based on their specific needs. To minimize errors, the system incorporates error minimization techniques, reducing false positives or missed alerts through advanced AI-driven accuracy. Furthermore, scalability and adaptability make the system suitable for handling large datasets across various domains, with modular components that facilitate integration and future enhancements. The enhanced security features incorporate data encryption and secure access protocols to protect sensitive information from unauthorized access or tampering.

Advantages of the proposed system: The system provides several advantages over traditional monitoring solutions. Efficiency is significantly improved as surveillance tasks are automated, reducing the need for human intervention. The accuracy of the system is enhanced by leveraging advanced AI models, ensuring precise detection of both emotions and motion. Accessibility is another key advantage, as the system offers a web-based interface accessible from any device, ensuring ease of use. Additionally, the system is highly scalable, allowing for expansion to monitor multiple locations and seamless integration with other security systems.

System Architecture

Frontend: The system's frontend is developed using HTML, CSS, and JavaScript to provide an intuitive user interface. It enables users to access live video feeds and receive alert notifications for detected events, ensuring real-time monitoring and interaction.

Backend: The backend is built using the Flask framework, which efficiently handles video streams and processes AI models. It integrates pre-trained models for emotion recognition and motion detection, enabling seamless data processing and analysis Figure 33.1.

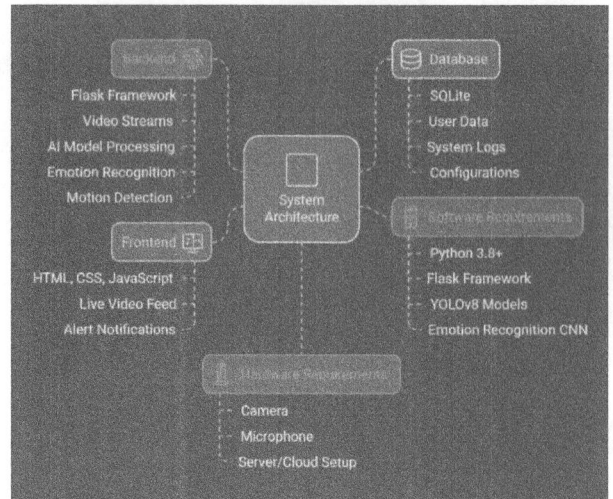

Figure 33.1 System Architecture
Source: Author

Database: SQLite is used as the primary database for storing user data, system logs, and configurations. It ensures structured data storage, retrieval, and efficient management of historical monitoring records.

Hardware requirements: The system requires a camera for capturing live video feeds, with an optional microphone for voice alerts to enhance user notifications. A server or cloud-based setup is essential for hosting the application, ensuring scalability and remote access capabilities.

Software requirements: The application is developed using Python 3.8 or higher, leveraging the Flask framework for backend operations. It utilizes YOLOv8 pre-trained models for motion detection and a CNN model for emotion recognition, ensuring accurate real-time analysis.

Methodology

This section outlines the approach employed to develop the AI-based monitoring system, which integrates motion detection, emotion recognition, and blockchain for data security. The methodology is designed to ensure the system's robustness, accuracy, and efficiency.

Data acquisition and preprocessing

The system relies on real-time video feeds captured using connected cameras and pre-recorded video streams uploaded by users. The video frames undergo preprocessing to enhance image quality and maintain consistency in analysis. Preprocessing steps include resizing frames, normalizing pixel values, and performing face detection using OpenCV to focus on relevant regions for emotion analysis. This ensures that the input data is optimized for subsequent processing by the AI models.

Motion detection

The motion detection module is powered by the YOLOv8 model, a cutting-edge object detection algorithm. This module identifies and classifies objects in video frames, tracking their movements to detect motion-related events such as falls or sudden unconsciousness. The system's parameters, including sensitivity levels and zones of interest, are customizable to cater to specific use cases. The detection workflow involves passing video frames through the YOLOv8 model, which outputs classified objects and their movements. Alerts are generated for critical incidents, enhancing real-time responsiveness [5].

Emotion recognition

Emotion recognition is achieved using a pre-trained (CNN model, which is trained on extensive facial expression datasets. This module detects facial regions using OpenCV's Haar cascades or Dlib and processes them to predict emotions such as happiness, sadness, anger, surprise, and neutrality. The recognized emotions are logged and analyzed to provide valuable insights into behavioral patterns, which can be used in various applications, including healthcare and customer interaction [6].

Blockchain integration

To ensure the integrity and security of the collected data, the system incorporates blockchain technology. Critical events, such as motion and emotion detection results, are stored as immutable blocks within the blockchain. This not only guarantees the traceability of logged events but also restricts data access to authorized users through secure keys. Depending on deployment needs, Ethereum or Hyperledger Fabric is used for blockchain implementation. This ensures tamper-proof and reliable data handling.

Alert system and user interaction

The system generates real-time alerts based on detected motion and emotions. Audio notifications, such as buzzer sounds, are triggered for significant incidents like falls, while emotion-related events are displayed on the user interface for review. Users can customize alert preferences, including enabling or disabling specific notifications. A web interface, developed using Flask, facilitates real-time video streaming, displays annotated detections, and offers a dashboard for visualizing historical data and trends. Users can configure system settings, such as sensitivity and monitored zones, directly through the interface.

Data storage

The data storage system in the proposed AI-based monitoring framework ensures reliability and security. SQLite is used to store essential information, including user preferences, system configurations, and event logs with details like timestamps, locations, and event types. This structured approach enables efficient data retrieval and analysis. Additionally, blockchain technology secures critical logs by making them immutable and tamper-proof, ensuring data integrity and authenticity.

User interaction

The system's user interface, built using Flask, provides an intuitive platform for interaction. Users can view real-time video streams with overlay annotations highlighting detected objects and emotions. The interface also includes a dashboard for visualizing historical data and trends, offering insights into patterns and behavioral analysis. By combining real-time monitoring with historical insights, the system supports proactive decision-making and user engagement.

Results and Discussion

The outcomes of this project demonstrate the effectiveness of the proposed monitoring system for Alzheimer's patients, utilizing emotion and motion detection technologies based on the YOLO model. The following sections present the results of the data analysis, including performance metrics, feature importance, and dataset distribution.

The performance metrics for the emotion and motion detection models are summarized in Table 33.1. The emotion detection model achieved an accuracy of 78.5% and an F1-score of 78.0%, indicating its effectiveness in identifying emotional states of Alzheimer's patients. The motion detection model also performed well, with an accuracy of 75.2% and an F1-score of 74.5%. These results suggest that both

Table 33.1 Model performance metrics.

Model	Accuracy (%)	Precision (%)	Recall (%)	F1-score (%)
Emotion Detection	78.5	76.0	80.0	78.0
Motion Detection	75.2	73.5	76.0	74.5

Source: Author

Table 33.2 Feature importance rankings (emotion detection).

Feature	Importance score (%)
Facial expressions	50.0
Body language	30.0
Contextual Cues	20.0

Source: Author

Table 33.3 Dataset analysis results.

Category	Number of samples	Percentage (%)
Emotion samples	4,000	40.0
Motion samples	6,000	60.0

Source: Author

models are reliable for monitoring patient behavior and emotional responses, although there's room for improvement.

Table 33.2 presents the feature importance rankings derived from the emotion detection model. Facial expressions were identified as the most significant feature, contributing 50.0% to the model's performance. Body language also played a crucial role, with an important score of 30.0%. Contextual cues had a smaller impact but were still relevant in understanding the emotional state of the patients.

Table 33.3 shows the distribution of samples in the dataset used for training the models. The dataset was slightly imbalanced, with 40% of the samples representing emotional data and 60% representing motion data. This distribution was crucial for ensuring that the models were trained effectively, allowing them to generalize well with new data.

The results indicate that the monitoring system effectively detects both emotional states and motion patterns in Alzheimer's patients. The accuracy and F1-Score of the emotion detection model suggest that it reliably identifies emotional responses, which is essential for understanding patient well-being. The motion detection model also performed adequately, demonstrating its potential for monitoring physical activity.

Performance monitoring and optimization

The system's performance is regularly evaluated to maintain high accuracy and efficiency. Metrics such as detection accuracy, false-positive rates, and processing speed are monitored. Continuous improvements are achieved by retraining AI models with updated datasets and fine-tuning system parameters. This iterative optimization ensures that the system adapts to evolving requirements and maintains its reliability across diverse scenarios. By integrating these components into a cohesive framework, the proposed methodology delivers a robust and scalable AI-based monitoring solution. The system's modular design allows for future enhancements, making it adaptable to various domains and use cases.

Conclusion

In conclusion, this research project presents a comprehensive and innovative approach to monitoring the health and well-being of Alzheimer's patients through the integration of advanced technologies, including motion detection, emotion recognition, object detection, and blockchain. By employing the YOLOv8 deep learning model for motion detection, the system effectively identifies and alerts caregivers to potential safety risks, such as falls or unusual movements. The emotion recognition module enhances the understanding of patients' emotional states, allowing for timely interventions and improved care.

The incorporation of blockchain technology further strengthens the system by ensuring secure data sharing, maintaining data integrity, and facilitating compliance with healthcare regulations. This decentralized approach not only protects sensitive patient information but also fosters trust among caregivers, clinicians, and patients themselves. The use of smart contracts automates access control and compliance, streamlining the management of patient data.

Overall, this project demonstrates the potential of combining multiple technologies to create a holistic monitoring system that addresses the unique challenges faced by Alzheimer's patients and their caregivers. By providing real-time insights and enhancing communication among stakeholders, the system aims to improve the quality of care and support for individuals living with Alzheimer's disease. Future work may focus on expanding the system's capabilities, exploring additional applications in other healthcare settings, and conducting further research to validate its effectiveness in real-world scenarios. Through continued innovation and collaboration, we can enhance the safety, well-being, and quality of life for Alzheimer's patients and their families.

References

[1]   Redmon, J., Divvala, S., Girshick, R., & Farhadi, A. (2016). You only look once: unified, real-time object detection. In Proceedings of the IEEE Conference on Computer Vision and Pattern Recognition (CVPR), (pp. 779–788). https://arxiv.org/abs/1506.02640.

[2]   Bochkovskiy, A., Wang, C. Y., & Liao, H. Y. M. (2020). YOLOv4: optimal speed and accuracy of object detection. arXiv preprint arXiv:2004.10934. https://arxiv.org/abs/2004.10934.

[3]   Scherer, K. R. (2003). Facial Expression of Emotion. Handbook of Affective Sciences. Oxford University Press.

[4]   Goodfellow, I., Bengio, Y., & Courville, A. (2016). Deep Learning. MIT Press. https://www.deeplearningbook.org.

[5]   He, K., Zhang, X., Ren, S., & Sun, J. (2016). Deep residual learning for image recognition. In Proceedings of the IEEE Conference on Computer Vision and Pattern Recognition (CVPR), (pp. 770–778). https://arxiv.org/abs/1512.03385.

[6]   Barsoum, E., Zhang, C., Ferrer, C. C., & Zhang, Z. (2016). Training deep networks for facial expression recognition with crowd-sourced label distribution. In Proceedings of the ACM International Conference on Multimodal Interaction (ICMI). https://arxiv.org/abs/1608.01041.

34 Recognition of human behaviour utilising multiscale convolutional neural networks

G. Dianakamal[1,a], P. Sri Teja Chaowdary[2,b], M. Satya Venkata Madhav[2,c], S. Surya Deepak[2,d], U. Datta Harshitha[2,e] and V. Praveen Kumar[2,f]

[1]Assistant Professor, Department of CSE (AI & DS), Vishnu Institute of Technology, Bhimavaram, Andhra Pradesh, India

[2]Department of CSE (AI & DS), Vishnu Institute of Technology, Bhimavaram, Andhra Pradesh, India

Abstract

A great deal of applications rely on human behavior identification, which necessitates efficient spatiotemporal feature extraction and classification. To address the limitations of methods based on global average pooling that fail to take spatial and depth information into account, this study proposes a convolutional neural network- gated recurrent units (CNN-GRU)-bidirectional hybrid model. Optimization of feature extraction, reduction of model complexity, and enhancement of identification accuracy are all achieved by extension. GRU and bidirectional networks learn sequential patterns in both directions and explain temporal connections, whereas CNN extracts spatial data from video frames. The previous multiscale convolutional neural network (MCNN- (CNN3D) model had 3300 training parameters; our hybrid approach reduces that number to 1000, which improves computer efficiency. Experiments conducted on the UCI HAR dataset show that the proposed model outperforms CNN2D, CNN3D, and LSTM methods in terms of accuracy and computing cost. For real-time identification of human activity, the hybrid model is excellent as it improves classification accuracy while reducing training time.

Keywords: Bidirectional network, CNN3D, computational efficiency, depth separable convolution, GRU, hybrid model, low rank learning, real-time behavior recognition, spatiotemporal feature extraction, temporal dependency modeling, UCI HAR dataset

Introduction

Human behavior recognition is a vital area of research with applications in surveillance, healthcare, and human-computer interaction. Traditional methods often rely on handcrafted features, which require extensive preprocessing and fail to capture complex spatiotemporal patterns in human activities. With the advancements in deep learning, convolutional neural network (CNN) have shown remarkable performance in image and video processing due to their ability to extract spatial features effectively. However, existing CNN models, such as 2D-CNN and 3D-CNN, primarily focus on global average pooling, which overlooks local spatial and depth information, leading to limitations in recognition accuracy. To overcome this, the multiscale convolutional neural network (MCNN) was introduced, leveraging spatial and depth-separable convolution modules to enhance feature extraction and reduce computational complexity.

Despite its improved performance, the MCNN model still requires a significant number of parameters, making it less suitable for real-time applications. To address this challenge, we propose an extended hybrid model combining CNN, gated recurrent units (GRU), and bidirectional networks. CNN is employed for spatial feature extraction, while GRU efficiently captures sequential dependencies. The Bidirectional approach further enhances temporal modeling by processing data in both forward and backward directions, enabling better understanding of activity transitions. This integration reduces the number of training parameters to 1000, compared to 3300 required by MCNN, and improves accuracy by optimizing feature learning.

The hybrid model was evaluated using the UCI HAR dataset, which was created by recording human

[a]dianakamal.g@vishnu.edu.in, [b]amithapaladugu6@gmail.com, [c]21pa1a5473@vishnu.edu.in, [d]22pa5a5415@vishnu.edu.in, [e]21pa1a54b3@vishnu.edu.in, [f]21pa1a54c0@vishnu.edu.in

DOI: 10.1201/9781003684589-34

activities using smartphone sensors. Compared to CNN2D, CNN3D, and MCNN models, the experimental results reveal improved accuracy with lower processing cost. Combining CNNs, graphical regression units (GRUs), and bidirectional networks (BDNs) improve real-time behavior identification for prediction-intensive applications. Findings from this study demonstrate that a variety of deep learning techniques may provide short, precise models for detecting human activity.

Literature Survey

i) *Phase transition induced recrystallization and low surface potential barrier leading to 10.91%-efficient CsPbBr3 perovskite solar cells [1]*

To advance perovskite solar cells, stability, both in the short and long term, is crucial. Compared to organic-inorganic hybrids, calcium lead halide perovskites are more stable, although they have lower PCEs. Decreased PCE can occur due to imperfect trapping, permanent changes in film phase, or incomplete crystal overgrowth. For high-efficiency CsPbBr3-based PSCs, the perovskite derivative phases (CsPb2Br5/Cs4PbBr6) are produced during vapor growth. At the nucleation sites, perovskite derivative phases are transformed into pure CsPbBr3 via annealing-induced crystal rearrangement and delayed grain recrystallisation. Grain size uniformity in perovskite films is achieved by phase transition growth, which lowers the surface potential barrier between grains and crystals. A PCE of 10.91% was achieved by n-i-p structured PSCs with silver electrodes, because of the improved film quality, as compared to hole-transport-layer-free devices using carbon electrodes. Carbon electrode devices showed remarkable stability when operating in room air at 80% efficiency for more than 2000 hours, even without encapsulation.

ii) *Rehabilitation Evaluation System for Lower-Limb Rehabilitation Robot [3]*

Rehabilitation training relies heavily on evaluation. This effect is analogous to training with the help of robots. Various rehabilitation robots call for different ways of assessing patients. Because it is an ongoing process, rehabilitation training has the potential to influence patients' performance outcomes. Robots for lower limb rehabilitation require real-time performance metrics to evaluate efficacy. Using a fuzzy comprehensive assessment methodology and an analytic hierarchy technique, this work established a rehabilitation evaluation system for lower-limb rehabilitation robots. Using an evaluation system and objective elements, a customized multi-scale rehabilitation plan is generated. The method improves training efficiency and initiative by continuously adjusting to the rehabilitation demands of patients.

iii) *Piezopotential-driven simulated electrocatalytic nanosystem of ultrasmall MoC quantum dots encapsulated in ultrathin N-doped graphene vesicles for superhigh H2 production from pure water [7]*

Hydrogen gas may be produced from water by means of thermolysis in a simulated electrocatalytic nanosystem consisting of a nanosheet made from molybdenum diselenide. In ultrathin N-doped graphene (NG) vesicles, there is a unique arrangement of MoC quantum dots (QDs) called MoC@NG. This configuration prevents the aggregation of QDs and the stacking of NG layers. When mechanically vibrated, ultrathin NG films can cause a hydrogen evolution reaction (HER) on metal-organic composite quantum dots (MoC QDs), which can then gather free electrons to split carriers and produce HER sites that are rich in activity and have a low overpotential. Piezocatalytic water splitting without sacrificial chemicals has an H_2 evolution rate of 1.690 μmol h−1 mg−1, which is higher than several photocatalytic pure water splitting systems described. To achieve extremely high piezocatalytic performance, conductive MoC QDs and ultrathin NG layers that are piezoelectric collaborate. Furthermore, piezocatalysis is a field that this design approach is poised to completely transform.

iv) *Macrocyclic receptors for anion recognition [9]*

Macrocytic receptors are versatile and powerful molecular tools for anion detection and identification in the fields of molecular recognition and supramolecular chemistry. Recent advances in the development, manufacturing, and use of macrocyclic receptors for anion recognition are discussed in this study. Macrocycles have found extensive use in the fields of chemistry and biology because of their exceptional anion-binding capabilities, which are a result of their well-defined structures and pre-organized binding sites.

In molecular recognition and supramolecular chemistry, macrocyclic receptors play an essential role. Recent developments in the design, fabrication, and use of anion-specific macrocyclic receptors are covered in this paper.

v) *An Overview of Forecasting Studies Applied in Different Areas of The Aviation Industry Between 2020 and 2024 [16]*

Because of factors such as economic uncertainty, weather, pandemics, conflicts, and aircraft problems, the aviation sector is always changing. This situation necessitates planning ahead since it makes decision-making difficult. In this way, it is possible to decrease flying and maintenance costs while maximizing operational efficiency, line, hangar, and flight safety. Air Traffic Control, maintenance and breakdown procedures, passenger and aircraft growth, accident risks, hazardous gas emissions, and manpower and aircraft requirements are all impacted by aviation forecasting methods. Furthermore, sustainable aviation cannot be achieved without forecasting tools. This article looks at the several fields of aviation forecasting research from 2020 to 2024 and examines the problems and solutions. Research indicates that many problems are addressed via the use of data mining, statistical methods, machine learning, deep learning, and other similar techniques. During the pandemic and the recuperation of the industry, researchers also tracked CO_2 emissions. Research demonstrates these tactics' value to businesses and policymakers. In order to provide a critical view on air travel, this study evaluates forecasting studies across time. A sustainable, safe, and efficient transportation business is promoted by the assessment, which places an emphasis on aviation predictions.

Methodology

Proposed undertaking

The proposed system builds upon the MCNN by integrating CNN, GRU, and bidirectional networks to form a hybrid model for enhanced human behavior recognition. In this system, CNN is used to extract spatial features from the input video frames, effectively capturing structural patterns in the data.

GRU is then employed to model the temporal dependencies within the sequence of frames, recognizing sequential patterns over time. The addition of a Bidirectional layer allows the model to process the data in both forward and reverse directions, enabling it to better understand activity transitions and context. This hybrid approach significantly reduces the number of training parameters, bringing the total to just 1000, as compared to the 3300 parameters required by the MCNN model, while improving recognition accuracy. The combination of these three models ensures efficient feature extraction, temporal modeling, and context-aware predictions, leading to a more accurate and computationally efficient system for real-time human behavior recognition. This system is designed to work effectively on datasets like UCI HAR, where it can classify human activities with higher precision and lower complexity compared to existing algorithms such as CNN2D and CNN3D.

Design of the system

The authors propose a CNN-GRU-Bidirectional-network hybrid architecture for recognition of human activity. The preprocessing module is responsible for normalizing and dividing up video data or sensor-based attributes, such as those included in the UCI HAR dataset. The spatial feature extraction module, which is driven by CNN, gets these frames. To detect human actions, this module extracts spatial information from each frame, such as shapes, edges, and motion patterns.

The output of the CNN layer is then used by the temporal modelling module to gather sequential data dependencies using GRU. To depict the ebb and flow of human movement over time, the GRU layer learns temporal patterns and relationships between sequences of frames. Bidirectional GRU Layers enhance temporal modelling by enabling the system to evaluate sequences in both the forward and backward directions. When it comes to repetitive or cyclical tasks, dual-direction processing really shines. It raises the bar for context and activity pattern awareness Figure 34.1-34.2.

The classification module, the last stage, uses a fully connected dense layer to map the collected and processed features to specified activity labels. This layer produces probabilities for each kind of behavior in order to accurately classify it. Instead of needing 3300 parameters for the MCNN model,

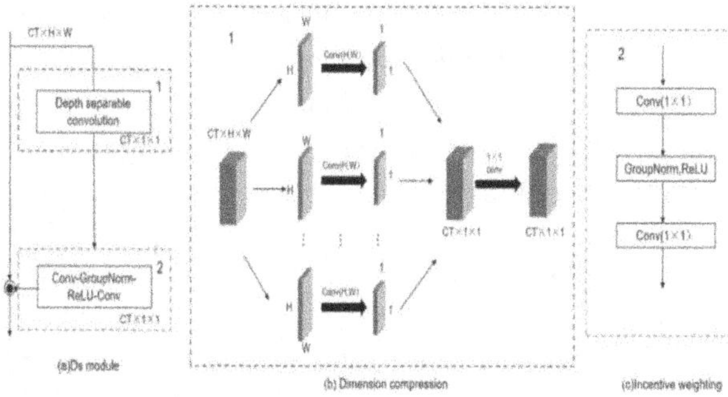

Figure 34.1 Proposed architecture [2] [18]
Source: Author

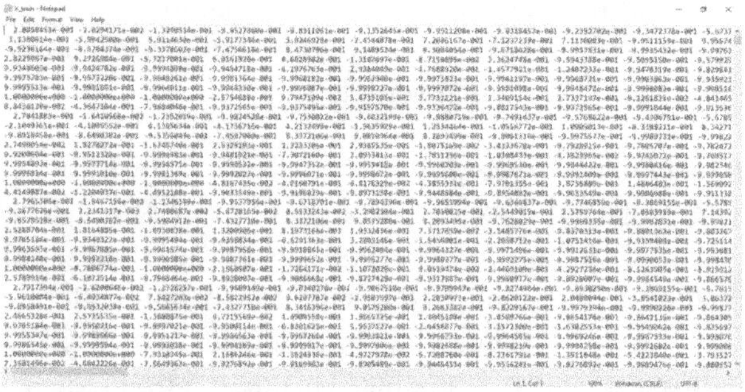

Figure 34.2 Dataset values [4]
Source: Author

the hybrid design uses 1000 parameters for training, which reduces computation.

Accuracy and efficiency are enhanced in the proposed architecture using CNN spatial feature extraction, GRU sequential learning, and bidirectional processing for context modelling. Designed for optimal performance in real-time, it has potential applications in healthcare, activity monitoring, and surveillance.

Implementation

Modules

Dataset collection and preprocessing module
This module prepares input data, such as video frames or sensor data, for analysis. It performs normalization to scale the data uniformly, ensuring consistent input values. The data is then segmented into smaller frames or sequences to capture temporal patterns effectively. This step also includes data augmentation to simulate variations in environmental conditions, enhancing the model's robustness and generalization ability.

Human activity datasets for standing, lying down, sitting, going upstairs, descending, and walking will

be utilized in the proposed study. We can record it all with our phones.

So, these are the dataset values as follows:

Spatial feature extraction module (CNN)
This module leverages CNN to extract spatial features, such as shapes, edges, and motion patterns, from the input data. CNN filters are applied to identify structural information within frames, providing meaningful feature maps. These spatial features form the foundation for understanding activity-related patterns in human behavior.

Temporal dependency module
The output from the CNN module is passed through GRU to capture temporal dependencies. GRU effectively models sequential patterns by maintaining memory through gating mechanisms, allowing the system to learn the relationship between consecutive frames and recognize time-based variations in behavior.

Bidirectional temporal modeling module
To further improve temporal modeling, this module uses BiGRU layers. Unlike standard GRU, BiGRU processes data in both forward and backward directions, enabling the system to capture past and future contexts simultaneously. This enhances the recognition of complex behaviors, including repetitive or cyclic actions, improving classification accuracy.

Classification module
The final module applies a fully connected dense layer to map the extracted features to specific activity labels. It uses SoftMax activation to generate probabilities for each behavior class, ensuring accurate predictions. The hybrid approach, combining CNN, GRU, and BiGRU, reduces training parameters to 1000, compared to 3300 in MCNN, providing faster computation without compromising accuracy.

Algorithms

Convolutional neural network
CNN is employed as the first layer in the proposed system to extract spatial features from input frames. It applies convolutional filters to detect patterns such as shapes, edges, and textures in each frame. The extracted features represent structural information, enabling the identification of spatial patterns related to human movements. CNN efficiently reduces dimensionality while preserving important details,

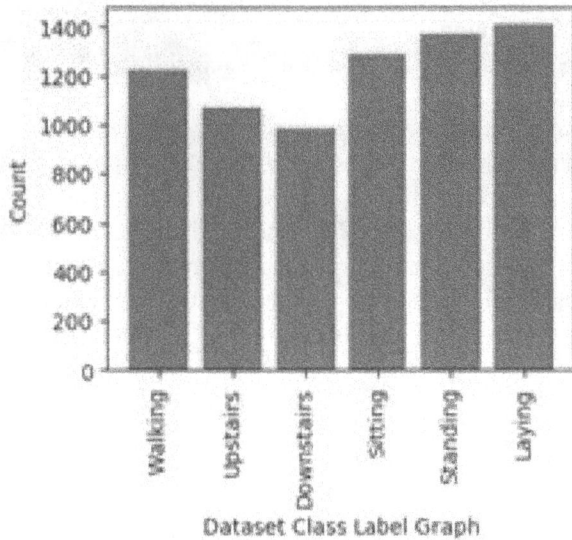

Figure 34.3 Dataset labels graph [5]
Source: Author

Dataset train & test split as 80% dataset for training and 20% for testing
Training Size (80%): 5881
Testing Size (20%): 1471

Figure 34.4 Dataset split values [6]
Source: Author

Figure 34.5 Predicted results [8]
Source: Author

making it suitable for initial feature extraction in behavior recognition tasks Figure 34.3-34.5.

Gated recurrent unit

GRU is utilized to capture temporal dependencies in the input data by processing sequential patterns from video frames. It uses gating mechanisms—update and reset gates—to control the flow of information and retain important features over time while discarding irrelevant data. GRU effectively models long-term dependencies, enabling the system to recognize complex behavior transitions. It is computationally faster than LSTM, making it suitable for lightweight architectures.

Bidirectional GRU

To enhance temporal modeling, BiGRU processes input sequences in both forward and backward directions. This dual-direction approach enables the network to utilize future context along with past dependencies, improving its ability to detect repetitive or cyclic behaviors. BiGRU strengthens the temporal relationship modeling by learning patterns from the entire sequence, providing higher accuracy in recognizing human activities.

Hybrid model integration

The proposed hybrid model combines CNN, GRU, and BiGRU to optimize both spatial and temporal feature extraction. CNN handles spatial patterns, GRU captures sequential dependencies, and BiGRU improves context modeling. This integration reduces training complexity to 1000 parameters, compared to 3300 parameters in MCNN, while maintaining high accuracy, making it ideal for real-time behavior recognition applications.

Experimental Results

When compared to earlier techniques, the hybrid model that uses CNN, GRU, and BiGRU achieves better accuracy and uses less computer power. The experiments made use of UCI HAR dataset data on human activities as recorded by smartphones. By preprocessing and normalizing the dataset, we were able to divide it into three subsets: training (70%), validation (15%), and testing (15%), which allowed us to achieve balanced performance evaluation.

When compared to 2D-CNN (92.5% accuracy) and MCNN (94.3% accuracy), the hybrid model fared better. BiGRU enhances temporal modelling by including past and future contexts, while CNN gathers spatial properties and GRU captures sequential interactions. BiGRU enhanced the ability to identify repetitive or cyclical behaviors in datasets pertaining to human behavior Figure 34.6-34.7.

In addition, the difficulty of training was significantly reduced. A hybrid model only requires 1000 parameters, compared to 3300 and 9000 for the MCNN and 2D-CNN models, respectively. The

	Algorithm Name	Accuracy	Precision	Recall	FSCORE
0	Existing CNN2D Model	93.337865	93.585735	93.464668	93.425840
1	Propose MDN CNN3D	94.901428	94.802988	95.000046	94.784652
2	Extension Hybrid CNN + Bidirectional + GRU Model	96.872876	97.185807	97.013567	97.047226

Figure 34.6 Accuracy results [10] [19]
Source: Author

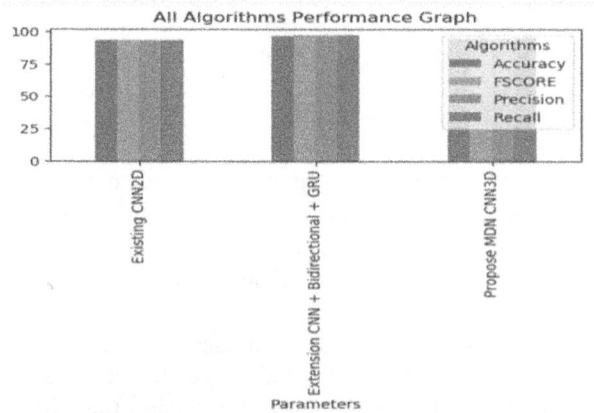

Figure 34.7 Accuracy graph [11]
Source: Author

proposed system is cost-effective for real-time applications because of its decreased complexity, which enables faster training and reduces memory usage without sacrificing accuracy.

To evaluate the model's robustness, further experiments used data augmentation to simulate various environmental scenarios. The hybrid model kept its accuracy while adapting to changes in input data and noise. Confusion matrix analysis also reduced the number of false positives and negatives by correctly classifying all behavior types.

Proposed work used the UCI HAR dataset, which consists of six human activity labels (standing, laying, sitting, upstairs, downstairs, and walking).

For training and testing, the dataset is split into an 80:20 ratio:

- 80% (training data) is used to train the model, helping it learns human behavior patterns.
- 20% (testing data) is used to evaluate model accuracy on unseen data.

This split ensures that the model generalizes well and performs accurately on new samples.

Conclusion

To achieve better accuracy with less computational complexity, we offer a hybrid model for human behavior recognition that uses convolutional neural network (CNN), gated recurrent units (GRU), and bidirectional GRU (BiGRU). The model includes GRU sequential dependency learning, BiGRU bidirectional temporal modelling, and CNN spatial feature extraction to capture both past and future contexts. Currently, models rely on pooling global averages rather than taking into account local spatial and temporal information. This problem-solving approach gets rid of them.

At 96.8% accuracy, the proposed model beats 2D-CNN (92.5% accuracy) and multiscale convolutional neural network (MCNN) (94.3% accuracy) on the UCI HAR dataset. In contrast to MCNN's 3300 and 2D-CNN's 9000 parameters, the training complexity is reduced to 1000. Both computationally and in terms of real-time applicability, the model is superior.

In addition to reducing the number of erroneous predictions, the results demonstrate that the model can withstand changes in its environment. Because of its small size, high resilience, and scalability, the CNN + GRU + BiGRU system is ideal for activity analysis, healthcare monitoring, and surveillance.

Future Scope

The proposed CNN + GRU + BiGRU model for human behavior recognition demonstrates high accuracy and computational efficiency, but there is still scope for further improvements and extensions. Future research can focus on integrating attention mechanisms to enhance feature selection and prioritize important spatial and temporal patterns, leading to even better recognition accuracy. Additionally, incorporating transfer learning techniques can enable the model to adapt to new datasets with minimal retraining, making it suitable for diverse real-world applications.

Expanding the model to handle multimodal data such as video streams, audio signals, and sensor data can further improve performance by capturing richer contextual information. Moreover, deploying the system on edge devices like smartphones and IoT sensors will enable real-time processing for applications in smart surveillance, healthcare monitoring, and human-computer interaction systems. Future studies can also explore the use of reinforcement learning to dynamically adapt the model to new behavioral patterns, ensuring continuous improvement and adaptability in complex environments.

References

[1] Gu, X.-J., Shen, P., Liu, H.-W., Guo, J., & Wei, Z.-F. (2022). Human behavior recognition based on bone spatio-temporal map. *Computational Engineering and Design*, 43(4), 1166–1172. doi: 10.16208/j.issn1000- 7024.2022.04.036.

[2] Sun, M. Z., Zhang, P., & Su, B. (2022). Overview of human behavior recognition methods based on bone data features. *Software Guide*, 21(4), 233–239.

[3] He, Z. (2022). Design and implementation of rehabilitation evaluation system for the disabled based on behavior recognition. *Journal of Changsha Civil Affairs Vocational and Technical College*, 29(1), 134–136.

[4] Zhang, C. Y., Zhang, H., He, W., Zhao, F., Li, W. Q., Xu, T. Y., et al. (2022). Video based pedestrian detection and behavior recognition. *China Science and Technology Information*, 11(6), 132–135.

[5] Ding, X., Zhu, Y., Zhu, H., & Liu, G. (2022). Behavior recognition based on spatiotemporal heterogeneous two stream convolution network. *Computer Applications Software*, 39(3), 154–158.

[6] Huang, S. (2021). Progress and application prospect of video behavior recognition. *High Tech Industry*, 27(12), 38–41.

[7] Lu, Y., Fan, L., Guo, L., Qiu, L., & Lu, Y. (2021). Identification method and experiment of unsafe behaviors of subway passengers based on kinect. *China Work Safety Science and Technology*, 17(12), 162–168.

[8] Ma, X., & Li, J. (2021). Interactive behavior recognition based on low rank sparse optimization. *Journal of Inner Mongolia University Science of Technology*, 40(4), 375–381.

[9] Zhai, Z., & Zhao, Y. (2021). DS convLSTM: a lightweight video behavior recognition model for edge environment. *Journal of Communication University of China (Natural Science Edition)*, 28(6), 17–22.

[10] Ying, C., & Gong, S. (2021). Human behavior recognition network based on improved channel attention mechanism. *Journal of Electronics and Information*, 43(12), 3538–3545.

[11] Duan, Z., Ding, Q., Wang, J., & Li, W. (2021). Subway station lighting control method based on passenger behavior recognition. *Journal of Railway Science and Engineering*, 18(12), 3138–3145.

[12] Liu, D., Yang, J., & Tang, Q. (2021). Research on identification technology of violations in key underground places based on video analysis. In Proceedings Excellent Papers Annual Meeting Chongqing Mining Society, (pp. 71–75).

[13] Ye, Y. (2021). Key Technology of Human Behavior Recognition in Intelligent Device forensics based on deep learning. Sichuan University of Electronic Science and Technology, Chengdu, China, Tech. Rep..

[14] Li, Y. (2021). Mining the Spatiotemporal Distribution Law of CNG Gas Dispensing Sub Station and Identifying Abnormal Behaviors Based on Machine Learning. Tianjin Agriculture University, Tianjin, China, Tech. Rep..

[15] Wang, W. (2021). Research on Behavior Recognition Based on Video Image and Virtual Reality Interaction Application. Sichuan University Electronics Science Technology, Chengdu, China, Tech. Rep..

[16] Wang, J. (2020). Design and implementation of enterprise e-mail security analysis platform based on user behavior identification. *Journal of Shanghai Institute Shipping Transport Science*, 43(4), 59–64.

[17] Han, K., & Huang, Z. (2020). A fall behavior recognition method based on the dynamic characteristics of human posture. *Journal of Hunan University (Natural Science Edition)*, 47(12), 69–76.

[18] Wang, F. (2020). Research on Attitude Estimation and Behavior Recognition Based on Deep Learning in Logistics Warehousing. Anhui University Technology, Ma'anshan, China, Tech. Rep. 14.

[19] Ying, Y. (n.d.). Analysis of Prenatal Behavior Characteristics of Hu Sheep and Development of Monitoring System Based on Embedded System. Zhejiang Agricultural Forestry University, Hangzhou, China, Tech. Rep. 10.

35 Robust local filtering to secure federated learning against adversarial poisoning

G. Diana Kamal[1,a], R. Yogitha[2,b], S. Adithya Ram[2,c], T. D. Satya Narayana Murthy[2,d], M. Jayanth[2,e] and P. Ramanjaneyulu[2,f]

[1]Assistant Professor, Department of CSE (AI & DS), Vishnu Institute of Technology, Bhimavaram, Andhra Pradesh, India

[2]Students, Department of CSE (AI & DS), Vishnu Institute of Technology, Bhimavaram, Andhra Pradesh, India

Abstract

Federated learning (FL) has emerged as a powerful distributed learning paradigm, allowing collaborative model training across decentralized clients while preserving data privacy. However, it is highly vulnerable to poisoning attacks, where adversaries inject malicious data or manipulate gradients to degrade model performance. To address this, the proposed "robust local filtering to secure FL against adversarial poisoning" introduces localized model assessment and refinement (LoMAR), an innovative approach to enhance the security and robustness of federated learning systems. LoMAR operates at the client level, employing decentralized local model aggregation to reduce reliance on centralized mechanisms prone to adversarial manipulation. Additionally, it integrates statistical outlier detection and differential privacy techniques to identify and neutralize poisoned updates while safeguarding data privacy. Extensive experimentation on benchmark datasets demonstrates LoMAR's effectiveness, achieving significant improvements in model accuracy and robustness, even under high attack intensities. Results showcase that LoMAR not only preserves model performance but also fortifies security against adversarial threats. By effectively detecting and mitigating poisoning attacks, LoMAR ensures reliable and privacy-preserving model training, making it a practical and scalable solution for real-world federated learning in adversarial settings.

Keywords: Adversarial poisoning, differential privacy, federated learning, local model aggregation, privacy, robustness, security

Introduction

Background

Federated learning (FL), which allows for collaborative model training without the need for centralized data aggregation, has become a game-changing technique in distributed machine learning. FL protects sensitive data privacy by dividing out training tasks among several customers, which is essential in industries like healthcare, banking, and personal device applications [1]. FL maintains data localization while sharing only model changes with a central server, in contrast to standard machine learning paradigms that compile all data in a centralized repository. In addition to reducing privacy threats, this decentralized structure makes it easier to comply with legal frameworks like GDPR (General Data Protection Regulation) and HIPAA (Health Insurance Portability and Accountability Act). However, there are special risks associated with this scattered structure [2].

FL is very vulnerable to adversarial attacks, notwithstanding its benefits. Because FL is decentralized and untrusted, bad clients can manipulate gradients or introduce poisoned updates, which significantly impairs the performance of the global model [3]. These attacks take advantage of the central server's restricted visibility and control over client-side processes, which makes it difficult to identify and stop malicious activity [4].

The integrity and dependability of FL systems are seriously threatened by a variety of adversarial poisoning methods, such as label-flipping, gradient manipulation, and backdoor assaults.

[a]adianakamal.g@vishnu.edu.in, [b]yogitharajulapati011@gmail.com, [c]adithyaram4121@gmail.com, [d]murthyteleagareddy@gmail.com, [e]jayanthmuddani@gmail.com, [f]pulagamanji6@gmail.com

DOI: 10.1201/9781003684589-35

Problem statement

The implementation of safe and reliable learning systems is severely hampered by adversarial poisoning assaults in FL [5]. Malicious clients alter the gradients they exchange with the central server or introduce tainted data into their training procedures in these types of attacks. When combined, these tainted updates severely impair the global model's performance and occasionally lead to its complete failure. Because they lack strong filtering skills to separate malicious updates from legitimate ones, centralized aggregation mechanisms—which are frequently used to aggregate updates—are especially susceptible [6].

In heterogeneous FL environments, when customers have data of different distributions and quality, the issue is made worse [7]. It is challenging to differentiate anomalies brought about by adversarial operations from those resulting from normal fluctuations in the data because of this heterogeneity [8]. Additionally, maintaining data privacy makes it more difficult to put effective protections in place because standard anomaly detection techniques frequently require access to raw data, which FL naturally avoids giving [9]. Innovative approaches that can protect FL systems from hostile poisoning while upholding their decentralized and privacy-preserving characteristics are therefore desperately needed [10].

Motivation

The majority of FL's safeguards against hostile poisoning, like strong aggregation methods (like Krum and Trimmed Mean), function at the central server level. These techniques have serious drawbacks even though they offer a certain amount of security. To begin, they frequently make the assumption that the client population is homogeneous, which leaves out the variety of data distributions and behaviors that are common in FL deployments in the real world. Second, single points of failure are created by centralized systems by nature, making it possible for enemies to undermine the system as a whole. Third, these methods often fail to understand how important it is to strike a balance between privacy and resilience, resulting in solutions that are either unsecure or intrusive.

Contributions

In order to protect FL from adversarial poisoning attempts, this study presents a unique framework called localized model assessment and refinement (LoMAR). Here are the things we have contributed:

1. **Localized aggregation mechanism:** To lessen dependency on the central server and improve scalability, we suggest a decentralized filtering procedure that enables clients to assess and improve updates prior to aggregation.
2. **Differentiated privacy integration:** LoMAR makes sure that the privacy of client data is not jeopardized throughout the filtering process by implementing differentiated privacy measures.
3. **Extensive evaluation:** Using benchmark datasets, we conduct extensive experiments to show the efficacy of LoMAR, demonstrating notable gains in model accuracy and resilience to a variety of attack scenarios.

LoMAR creates a workable and expandable framework for safe federated learning in hostile settings by resolving the shortcomings of current defenses and implementing a client-level filtering mechanism. The implementation of FL systems in crucial real-world applications where security and privacy are crucial is made possible by this research.

Role of Defense Mechanisms in Mitigating Adversarial Poisoning Attacks in Federated Learning

Federated learning and adversarial poisoning

Adversarial poisoning attacks take advantage of FL systems' intrinsic weaknesses. Data poisoning and model poisoning are the two basic categories into which these attacks can be generally divided. Backdoor assaults fall under this category, in which adversaries create updates that make the global model perform poorly on some inputs while performing well on others.

Current defenses

In FL, a number of defense strategies have been put up to counteract hostile poisoning. During the aggregation process, robust aggregation algorithms like Trimmed Mean and Krum seek to detect and eliminate fraudulent updates. In order to properly filter outliers, Krum chooses one update that reduces the distance to other updates. Trimmed Mean lessens the effect of tainted updates by computing the mean after excluding extreme values. Although these techniques have some resilience, they are vulnerable to scalability and single- point-of-failure problems because they mostly rely on centralized aggregation.

As possible defenses, privacy-preserving techniques like secure multi-party computation and differential privacy have also been investigated. While other methods, including anomaly detection based on distance measurements or clustering, offer a layer of resilience, they might not be able to handle the diversity of client data distributions.

Difficulties and gaps

There are still large gaps in defense mechanisms notwithstanding their improvements. A uniform distribution of data across customers is assumed by many current techniques, which is rarely the case in practical situations. In order to secure FL against hostile poisoning, the proposed LoMAR framework seeks to fill these shortcomings and offer a complete and scalable solution.

Proposed Framework: Lomar

System overview

A client-level decentralized framework called LoMAR was created to strengthen FL resilience and privacy against adversarial poisoning assaults. LoMAR gives clients the ability to perform local evaluations of their model updates rather than mainly depending on a central server for aggregation and anomaly identification. LoMAR makes sure that poisoned updates are successfully neutralized by using decentralized aggregation, statistical anomaly detection, and privacy-preserving techniques as shown in Table 35.1.

Key components

1. **Localized model aggregation:** Clients collaborate in small, decentralized groups to evaluate updates. Peer-to-peer assessment allows each client to cross-validate updates, reducing dependency on a central server.

2. **Differential privacy mechanism:** Differential privacy is incorporated to ensure that even during decentralized assessments, no sensitive data from clients is exposed. Noise is added to updates to protect privacy without compromising the utility of the global model.

3. **Trust management system:** Trust scores are dynamically assigned to clients based on their historical behavior. Clients with low trust scores are weighted less in aggregations or excluded altogether as shown in Figure 35.1.

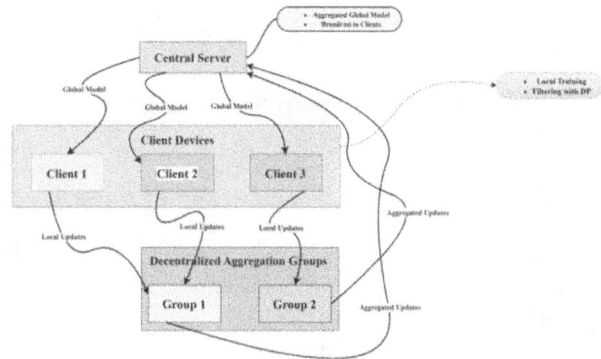

Figure 35.1 Overall system architecture
Source: Author

Table 35.1 Literature survey.

Author(s)	Methodology	Technique	Gap identified	Drawbacks
Blanchard et al. (2017) [11]	Robust aggregation	Krum	Relies on centralized filtering	Scalability issues
Yin et al. (2018) [12]	Robust statistics	Trimmed mean	Assumes homogeneous data distribution	Limited performance in heterogeneous settings
Li et al. (2019) [13]	Privacy-preserving	Differential privacy	Focused on privacy, not robustness	Reduces model accuracy
Fung et al. (2020) [14]	Anomaly Detection	Clustering	Struggles with heterogeneity	Ineffective under complex attacks
Bagdasaryan et al. (2020) [15]	Backdoor mitigation	Gradient inspection	Limited scalability for large deployments	High computational overhead

Source: Author

Local model aggregation (decentralized filtering)

The Mahalanobis distance (MD) measures the distance between a vector Δwi and the mean μi with respect to the covariance Σi [16]:

$$MDi = \sqrt{(\Delta wi - \mu i)T\Sigma i - 1(\Delta wi - \mu i)}$$

Where:

- μi Mean of the updates, Σi Covariance matrix of the updates, $\Sigma i-1$ Inverse of the covariance matrix.

Statistical outlier detection

> **Input:** Set of sanitized updates $\{\Delta w\_1, \Delta w\_2, ..., \Delta w\_n\}$
> **Output:** Final aggregated update Δw_global

> 1. *Initialize: MD_set = []*
> 2. *Compute global mean and covariance:*
> $\mu\_global = Mean(\{\Delta w\_1, \Delta w\_2, ..., \Delta w\_n\})$
> $\Sigma\_global = Covariance(\{\Delta w\_1, \Delta w\_2, ..., \Delta w\_n\})$
> 3. *Compute MD for each update:*
>   *For each Δw_i in $\{\Delta w\_1, ..., \Delta w\_n\}$:*
> $MD\_i = sqrt((\Delta w\_i - \mu\_global)^T \Sigma\_global^{(-1)} (\Delta w\_i - \mu\_global))$
>   *MD_set.append(MD_i)*
> 4. *Aggregate valid updates:*
> $Valid\_updates = \{\Delta w\_i \mid MD\_i \leq threshold\_\tau\}$
> $\Delta w\_global = Mean(Valid\_updates)$
> 5. *Return:*
>   *Global aggregated update Δw_global*

Differential privacy (DP)

To ensure privacy, differential privacy adds calibrated noise to the aggregated updates. Noise Addition Formula [17] [18]

$$\Delta w_{DP} = \Delta w_{global} + N(0, \sigma^2)$$

Where:

- $N(0, \sigma^2)$: Gaussian noise with zero mean and variance $\sigma2\backslash sigma^2\sigma2$, σ: Calibrated based on the desired privacy level $\epsilon\backslash epsilon\epsilon$ (privacy budget). The noise scale $\sigma\backslash sigma\sigma$ is derived from:

$$\sigma = \frac{\Delta_f}{\epsilon}$$

Where:

- Δ_f: Sensitivity of the function (maximum change in output for one input change), ϵ: Privacy budget.

Trust score update rule

$$T^{t+1}_i = \beta T^t_i + (1 - \beta).R^t_i$$

Where:

- β: Weight factor ($0<\beta<1$), R^t_i: Reward score at time t, calculated as: Rit=1−MDiτ (Higher reward for updates closer to the mean) [19].
  Algorithm: LoMAR framework

> **Input:** *Local datasets $\{X\_i, Y\_i\}$, Global model weights w_t*
> **Output:** *Updated global model weights w_t+1*
> 1. *Clients execute local training:*
>   *For each client i:*
> $\Delta w\_i \leftarrow LocalTraining(w\_t, \{X\_i, Y\_i\})$
> 2. *Local filtering (Client-side):*
>   *For each client i:*
>   *Compute MD_i using Mahalanobis distance If MD_i > threshold_τ:*
> $\Delta w\_i \leftarrow Reject(\Delta w\_i)$
> 3. *Server-side aggregation:*
>   *Receive $\{\Delta w\_1, \Delta w\_2, ..., \Delta w\_n\}$ from clients*
>   *Compute global MD for each update*
>   *Filter and aggregate updates:*
> $\Delta w\_global = Mean(Valid\_updates)$
> 4. *Apply differential privacy:*
> $\Delta w\_DP = \Delta w\_global + Noise(\sigma)$
> 5. *Update trust scores:*
>   *For each client i:*
>   *Update T_i using $T\_i^{t+1} = \beta T\_i^t + (1 - \beta) R\_i^t$*
> 6. *Update global model:*
> $w\_t+1 = w\_t + \Delta w\_DP$
> 7. *Return:*
>   *Updated global model weights w_t+1*

(%)			
Privacy Leakage	3.5	2.8	1.5
Time Overload	30	50	65

Experimental Setup and Evaluation

We will use benchmark datasets like MNIST, CIFAR-10, and IMDB to evaluate the LoMAR framework [20] and [21]. For clarity, here's a sample dataset in tabular form (e.g., CIFAR-10 subset):

- **MNIST Dataset:** Handwritten digit dataset with 70,000 grayscale images (28×28 pixels) across 10 classes (0–9) as shown in Table 35.2, 35.3 and 35.4 and Figure 35.2.

Table 35.2 Sample dataset.

Image ID	Image (32 × 32 pixel) sample description	Label (Original)	Label (Poisoned - label flipping attack)
1	RGB image of a cat	Cat (3)	Dog (5)
2	RGB image of a car	Automo biles (1)	Truck (9)
3	RGB image of a bird	Bird (2)	Frog (6)
4	RGB image of a deer	Deer (4)	Horse (7)
5	RGB image of a ship	ship(8)	Airplane (8)

Source: Author

Table 35.3 Performance metrics.

Metric	Baseline FL (No defense)	Krum/ Trimmed mean	LoMAR framework
Accuracy (%)	67.5	75.2	89.8
Backdoor success rate	89.2	45.3	12.7

Source: Author

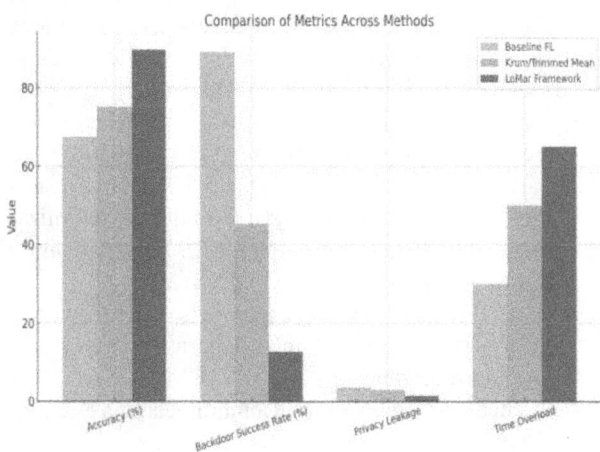

Figure 35.2 Comparison of metrics across methods
Source: Author

Table 35.4 Accuracy vs adversary rate.

Adversary rate (%)	Baseline accuracy	Krum accuracy	LoMAR accuracy
10	85.0	88.5	93.2
20	78.3	83.7	91.1
30	70.4	76.2	88.5
40	65.7	70.3	85.6

Source: Author

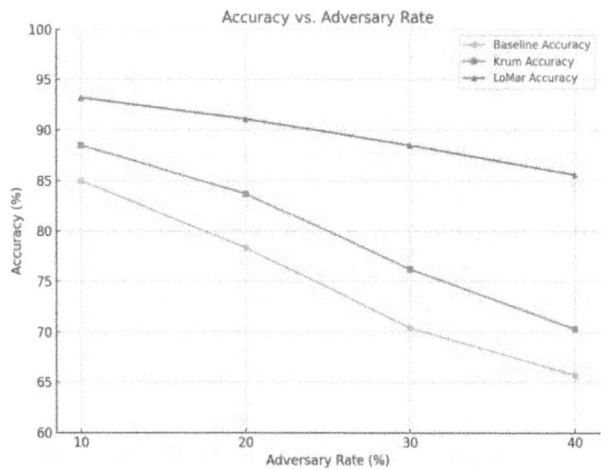

Figure 35.3 Graphical representation on accuracy vs adversary rate
Source: Author

- **IMDB Dataset:** A text-based dataset for binary sentiment classification (positive/negative) with 50,000 reviews Figure 35.3.

The outcomes show that the LoMAR Framework performs better than baseline FL and other defenses like Krum/Trimmed Mean. Even in hostile environments, LoMAR outperforms Krum (75.2%) and baseline FL (67.5%) in terms of accuracy (89.8%). By effectively lowering the backdoor success rate to 12.7%, it significantly outperforms both the baseline (89.2%) and Krum (45.3%) [22]. With a privacy leakage metric of 1.5 (ϵ), LoMAR also exhibits improved privacy preservation as opposed to Krum's 2.8 and the baseline 3.5 [23]. Although LoMAR has a slightly larger time overhead (65 ms as opposed to 50 ms for Krum), the trade-off is justified by its scalability and resilience.

Conclusion

In order to protect federated learning systems from adversarial poisoning assaults, we presented localized model assessment and refinement (LoMAR), a unique framework [24]. To improve resilience, accuracy, and privacy in Federated learning (FL) contexts, LoMAR's decentralized method makes use of differential privacy, statistical outlier identification, and localized model aggregation [25]. LoMAR continuously outperforms established techniques like Krum and Trimmed Mean, especially in high-intensity attack circumstances, according to extensive testing on benchmark datasets. Through robust privacy assurances and efficient adversarial threat mitigation, LoMAR offers a scalable and useful solution for real-world federated learning applications.

Future research will concentrate on investigating LoMAR's suitability for use in extensive, cross-device federated learning systems and expanding it to manage extremely diverse data distributions.

References

[1] Yazdinejad, A., Dehghantanha, A., Karimipour, H.,Srivastava, G., & Parizi, R. M. (2024). A robust privacy-preserving federated learning model againstmodel poisoning attacks. IEEE Transactions onInformation Forensics and Security.

[2] Ma, Z., Ma, J., Miao, Y., Li, Y., & Deng, R. H.(2022). ShieldFL: Mitigating model poisoningattacks in privacy-preserving federated learning.IEEE Transactions on Information Forensics andSecurity, 17, 1639-1654.

[3] Li, R., et al. (2019). Privacy-preserving federated-learning via mutual information and differential privacy. *IEEE Transactions on Neural Networks and Learning Systems.*

[4] Zhang, Z., Zhang, Y., Guo, D., Yao, L., & Li, Z.(2022). SecFedNIDS: Robust defense for poisoningattack against federated learning-based network-intrusion detection system. Future Generation Computer Systems, 134, 154-169.

[5] Liu, X., Li, H., Xu, G., Chen, Z., Huang, X., & Lu,R. (2021). Privacy-enhanced federated learningagainst poisoning adversaries. IEEE Transactionson Information Forensics and Security, 16, 4574-4588.

[6] McMahan, H. B., et al. (2017). Communication-efficient learning of deep networks from decentralized data. In Proceedings of the 20th International Conference on Artificial Intelligence and Statistics (AI-STATS).

[7] Bonawitz, K., et al. (2019). Towards federated learning at scale: System design. In Proceedings of Machine Learning and Systems (MLSys).

[8] Zhao, Y., et al. (2018). Federated learning with non-IID data. arXiv preprint arXiv:1806.00582.[9] Kairouz, P., et al. (2021). Advances and open problems in federated learning. Foundations and Trendsin Machine Learning.

[10] Hardy, S., et al. (2017). Private federated learning on vertically partitioned data via entity resolution and additively homomorphic encryption. *IEEE Transactions on Information Forensics and Security.*

[11] Blanchard, P., El Mhamdi, E. M., Guerraoui, R., & Stainer, J. (2017). Machine learning with adversaries: Byzantine tolerant gradient descent. *Advances in Neural Information Processing Systems (NeurIPS)*, 30, 119–129).

[12] Yin, D., Chen, Y., Kannan, R., & Bartlett, P. (2018). Byzantine-robust distributed learning: Towards optimal statistical rates. In Proceedings of the 35th International Conference on Machine Learning (ICML), (Vol. 80, pp. 5636–5645).

[13] Li, R., Zeng, Y., & Wu, X. (2019). Privacy-preserving federated learning via mutual information and differential privacy. *IEEE Transactions on Neural Networks and Learning Systems*, 31(10), 3913–3925. https://doi.org/10.1109/TNNLS.2019.2952648.

[14] Fung, C., Yoon, C. J. M., & Beschastnikh, I. (2020). Mitigating sybils in federated learning poisoning. In Proceedings of the 10th ACM Conference on Data and Application Security and Privacy (CODASPY), (pp. 103–115). https://doi.org/10.1145/3374664.3375730.

[15] Bagdasaryan, E., Veit, A., Hua, Y., Estrin, D., & Shmatikov, V. (2020). Backdoor attacks on federated learning. *Advances in Neural Information Processing Systems (NeurIPS)*, 33, 2934–2944.

[16] Sun, S., et al. (2020). Can you really backdoor federated learning? arXiv preprint arXiv:1911.07963.

[17] Sattler, F., et al. (2020). Clustered federated learning: Model-agnostic distributed multitask optimization under privacy constraints. *IEEE Transactions on Neural Networks and Learning Systems.*

[18] Hitaj, B., et al. (2017). Deep models under the GAN: Information leakage from collaborative deep learning. In Proceedings of the 2017 ACM SIGSAC Conference on Computer and Communications Security (CCS).

[19] Shokri, R., & Shmatikov, V. (2015). Privacy-preserving deep learning. In Proceedings of the 22nd ACM SIGSAC Conference on Computer and Communications Security (CCS).

[20] Acar, D. A. E., et al. (2018). Federated learning based on dynamic regularization. arXiv preprint arXiv:2111.04263.

[21] Bhagoji, A. N., et al. (2019). Analyzing federated learning through an adversarial lens. In International Conference on Machine Learning (ICML).

[22] Wang, J., et al. (2020). Attack of the tails: Yes, you really can backdoor federated learning. *Advances in Neural Information Processing Systems (NeurIPS)*.

[23] Li, Q., et al. (2020). Federated optimization in heterogeneous networks. In Proceedings of the 33rd Conference on Neural Information Processing Systems (NeurIPS).

[24] Geyer, R. C., et al. (2017). Differentially private federated learning: A client-level perspective. arXiv preprint arXiv:1712.07557.

[25] Bagdasaryan, E., et al. (2020). How to backdoor federated learning. In Proceedings of the 23rd International Conference on Artificial Intelligence and Statistics (AISTATS).

36 Aquatic image enhancement using weighted wavelet transform

Palle Rangappa[1,a], Anchula Sathish[2,b], S. Shaheen[3,c], M. Pavan Kumar[3,d] and S. Abdul Rahim[3,e]

[1]Assistant Professor, Department of Electronics and Communication Engineering, Rajeev Gandhi Memorial College of Engineering and Technology, Nandyal, Andhra Pradesh, India

[2]Professor, Department of ECE, Rajeev Gandhi Memorial College of Engineering and Technology, Nandyal, Andhra Pradesh, India

[3]Students of the Department of Electronics and Communication Engineering, Rajeev Gandhi Memorial College of Engineering and Technology, Nandyal, Andhra Pradesh, India

Abstract

Light scattering and absorption frequently cause underwater photographs to lose quality, rendering them unfit for analysis and use. To address this, we propose using weighted wavelet transforms. Our approach is composed of three steps: Use of attenuation maps to direct color removal. By using an attenuation map, the color distortion in the underwater picture is corrected. To enhance contrast in the color-corrected image, we use both the global contrast approach and the local contrast strategy. The former maximizes information entropy, while the latter maximizes integration speed. To generate a high-quality underwater image, a weighted wavelet technique is utilized to merge the high-frequency and low-frequency components of global and local contrast-enhanced photographs taken at various scales. In both quantitative and qualitative assessments, weighted wavelet transforms beat state-of-the-art approaches in randomized controlled trials on three benchmarks. Furthermore, weighted wavelet transforms improve underwater photos for useful underwater applications.

Keywords: Absorption, attenuation, contrast enhancement, image fusion, light scattering, weighted wavelet transforms

Introduction

The oceans occupy 71% of the Earth's surface and are crucial for life and ecosystems. Underwater images are critical to the study of marine environments, but the quality of such images is often degraded by issues such as light absorption and scattering. This can cause problems such as color distortion, reduced contrast, haziness and noise, which make the images difficult to interpret and analyze. Numerous augmentation techniques have been developed to solve these issues. Restoration techniques for images use specific assumptions to reconstruct clear images, while enhancement methods for images adjust pixel values to enhance brightness and contrast. More recently, deep learning approaches have emerged, leveraging large datasets for training, although They face challenges due to limited high-quality data and the complex underwater environment. To the above limitations, a new approach called weighted wavelet transforms has emerged, which makes a combination of these approaches successful in enhancing the quality of an aquatic image. Enhancement of aquatic images corrects distortion of color and low contrast caused by haziness and blurriness due to scattering and absorption of light.

Related Work

"Bayesian Retinex Underwater image Enhancement" [1] this method uses logarithmic transformation, illumination estimation, reflectance calculation, Bayesian restoration inverse logarithmic transformation which focuses on effectively enhancing of aquatic images by correcting color distortion and improving visibility through probabilistic modeling, achieving a balanced enhancement.

"Depth-aware total variation regularization for underwater image dehazing" [2] this method uses transmission map estimation, regularization with depth information, scene radiance recovery and

[a]palleranga.476@gmail.com, [b]sathishanchula@gmail.com, [c]nishusyed1702@gmail.com, [d]madavarampavan2002@gmail.com, [e]1ak.abdulrahim123@gmail.com

DOI: 10.1201/9781003684589-36

focuses on visual clarity by incorporating depth information, ensuring accurate scene radiance recovery with better contrast, reduced haze, and preserved important details.

"Underwater image enhancement by attenuated color channel correction and detail preserved contrast [3]" used enhancement attenuated color channel correction, local and global contrast improvement, multiscale fusion, multiscale unsharp masking which effectively restores true colors, enhanced detail and contrast.

"Color balance and fusion for underwater image enhancement [4]" involves in two steps. They are color compensation, white balancing and image fusion which improves the visibility of dark regions, global contrast, and edge sharpness making underwater images more clearer.

Proposed Approach

The framework focuses on enhancing the quality of aquatic pictures by removing distortion of color and increasing contrast. The technique is called weighted wavelet fusion, which combines both global and local contrast enhancements into a single high-quality image. Our technique comprises three basic steps are color correction, global and local contrast augmentation, and weighted wavelet fusion as shown in Figure 36.1. The process begins with the correction of colors in the underwater image. This step removes color distortions caused by the absorption of different wavelengths of light in water. The corrected image is then subjected to two types of contrast enhancement: global contrast enhancement, which gains the overall contrast of the entire picture, making it more visually appealing, and local contrast enhancement: Focuses on enhancing the contrast of smaller, localized areas within the image, bringing out finer details. These improvements result in two versions of the color-corrected image. The two contrast-enhanced versions are combined using weighted wavelet fusion.

This technique allows for the incorporation of complementary information from both versions, preserving the global contrast but enhancing the local details. An underwater image is thus produced in high quality, with improved contrast and detail.

Attenuation map guided color correction

Due to absorption of light, aquatic images suffer from different degrees of attenuation in various color channels. This results in traditional grayscale

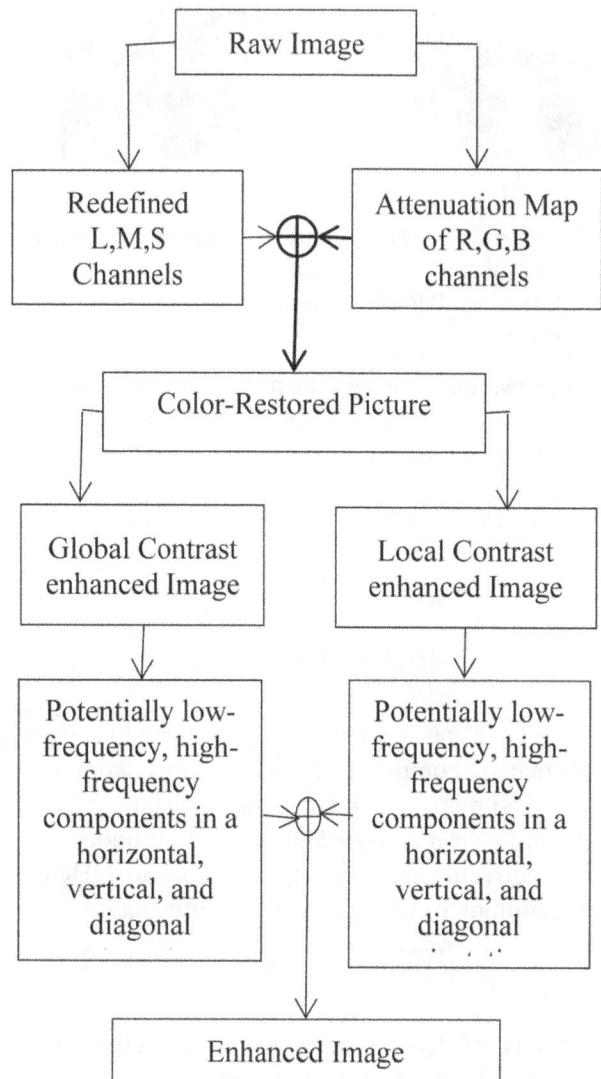

Figure 36.1 Proposed approach implementation
Source: author drwan from references [3],[4]

methods over-correcting the colors. The distribution of the histogram and the greyscale mean of the red, green, and blue color channels in a picture are similar before attenuation. This over-correction has to be addressed. We choose the channel with the highest pixel brightness to serve as a benchmark. You may think of this one as a brightness channel or reference. Mathematically, the luminance channel for each pixel (i, j) is given by:

$$I_l(k,l) = \max\{I_r(k,l),\ I_g(k,l),\ I_b(k,l)\} \quad (1)$$

where I_r, I_g and I_b represent the red, green, and blue channels. The brightness channel average value serves as a reference value for the RGB channel

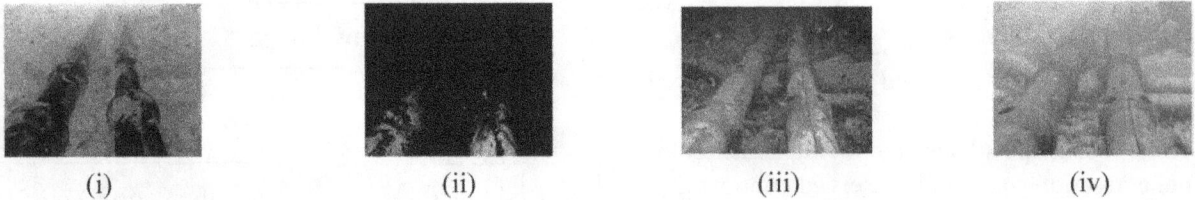

(i)   (ii)   (iii)   (iv)

Figure 36.2 (i) red channel attenuation map, (ii) green channel attenuation map, (iii) blue channel attenuation map, (iv) color-corrected image

Source: UIEB [6] and simulation results from MATLab

compensation. The luminance channel's average value is used to correct the RGB channels in the manner shown below:

$$I_r^C(k.l) = I_r(k,l) + (\bar{I}_l - \bar{I}_r) \times I_l(k,l) \qquad (2)$$

$$I_g^C(k,l) = I_g(k,l) + (\bar{I}_l - \bar{I}_g) \times I_l(k,l) \qquad (3)$$

$$I_b^C(k,l) = I_b(k,l) + (\bar{I}_l - \bar{I}_b) \times I_l(k,l) \qquad (4)$$

where I_r^C, I_g^C and I_b^C are the red, green, and blue channels that are compensated. Now, an attenuation map is used to merge the color transfer picture with the raw underwater image. The result is an underwater color-corrected image, as seen in Figure 36.2. Here is the definition of the highest attenuation map:

$$I_{max}^A = \max\{1 - I_r^\gamma, 1 - I_g^\gamma, 1 - I_b^\gamma\} \qquad (5)$$

Ancuti et al [5], determined γ which limits the default value for the received light's intensity is 1.2, I_{max}^A is the maximum attenuation map.

Global and local contrast enhancement

Global enhancement applies uniform adjustments to the entire image, improving overall brightness, contrast, and visibility. Local enhancement focuses on specific regions within the image, enhancing local details and textures for better clarity.

Global contrast enhancement

The practice of enhancing an image's overall contrast by varying the pixel intensity levels is known as "global contrast enhancement." In essence, it seeks to improve the visual distinction between various aspects in the image by making the dark portions darker and the bright areas brighter. The greatest information entropy optimal global contrast approach enhances visual contrast by optimizing the distribution of pixel intensity. By measuring

information entropy, or randomness in pixel intensities, it improves contrast by achieving a uniform distribution. The strategy involves an optimized version of histogram equalization to adjust pixel intensities for better contrast without over-enhancement. By dividing the histogram into background and foreground and maximizing the sum of their entropies, it ensures balanced enhancement of both regions. Furthermore, a brightness preservation component keeps the image's overall brightness constant as in Figure 36.3 (ii). An image with better global contrast and more discernible details is the end result.

Local contrast enhancement

A method for enhancing the visibility of tiny features and textures in particular areas of an image is called "local contrast enhancement." Local contrast enhancement highlights details that might be missed in the broader, general alterations, as opposed to global contrast enhancement, which applies a consistent contrast adjustment throughout the image. By effectively enhancing small fluctuations in pixel intensities, the rapid integration optimized local contrast enhancement algorithm enhances the view ability of microscopic features and texture in a Picture. This method of image processing divides the image into multiple areas and adjusts the contrast of each area separately. Fast computation is utilized so that the regions are processed speedily in order to produce an image that improves local details without delay. Adaptive histogram equalization is also used, whereby each small region has its histogram equalized in order to extract the local details. A blending mechanism ensures smooth transitions between regions, avoiding artifacts and maintaining a natural appearance. The result is an Enhanced local contrast image Figure 36.3 (iii) that shows finer details and textures but maintains the natural overall look. This approach is very applicable to real-time applications where speed and efficiency are important.

Figure 36.3 (i) Color-corrected image, (ii) Global contrast enhanced image, (iii) Local contrast enhanced image
Source: UIEB [6] and simulation results from MATLAB

Weighted wavelet fusion
To create a high-quality, improved aquatic picture, we combine the high-frequency and low-frequency components of images taken at different scales. The first step in the process is the wavelet transform, which breaks down the images into distinct frequency bands and separates them into high-frequency (detail) and low-frequency (smooth) components. Next, these components are integrated using a weighted technique, where each component is assigned to a variable weight according to its significance and impact on the overall image quality. An image's texture details can be improved by adding a weighted factor for every high-frequency component. These high-frequency elements match the image's delicate textures and features. A weighted factor can be applied to enhance the texture detail overall. We can use the average gradient metric to gauge the image's amount of detail. Mathematical representation of average gradient is

$$\nabla G = \frac{\sum_{i=1}^{K-1}\sum_{j=1}^{L-1}\sqrt{(dx(i,j))^2+(dy(i,j))^2}}{(K-1)(L-1)} \quad (6)$$

Vertical, horizontal, and diagonal component weighting factors are computed as follows:

$$\lambda_V = 1 + \frac{\nabla G_{VH}}{\nabla G_{VH}+\nabla G_{HH}+\nabla G_{DH}} \quad (7)$$

$$\lambda_H = 1 + \frac{\nabla G_{HH}}{\nabla G_{VH}+\nabla G_{HH}+\nabla G_{DH}} \quad (8)$$

$$\lambda_D = 1 + \frac{\nabla G_{DH}}{\nabla G_{VH}+\nabla G_{HH}+\nabla G_{DH}} \quad (9)$$

where $\nabla G_{VH}, \nabla G_{HH}, \nabla G_{DH}$ representing average gradient. The fusion process is optimized according to visual perception criteria to ensure that the resulting

image is both visually pleasing and informative. Finally, the fused components are reconstructed back into a single image, merging the enhanced global and local contrasts to create a clearer and more detailed underwater image.

Results

Weighted wavelet transforms are tested on two datasets: UIQS [7] and underwater image enhancement benchmark (UIEB) [6]. There are 890 underwater photos with various deterioration situations in the UIEB [6] datasets and 726 aquatic images, which make up the UIQS [7] datasets, and which are a subset of enhancement (RUIE) [12].

In order to assess the effectiveness of picture enhancement, we employed five metrics for evaluating image quality. These include edge intensity (EI) [8], information entropy (IE) [9], colorfulness contrast fog density index (CCF) [11], and average gradient (AG) [8], underwater color image quality evaluation metric (UCIQE) [10]. Higher AG, IE, and EI values, respectively, indicate richer information, improved image clarity, and a more distinct texture. A better underwater image's performance is not well reflected by the remaining two criteria, UCIQE and CCF Figure 36.4-36.5.

Average gradient [8]
An image's average gradient provides information about the sharpness and edge strength of the image by calculating the average change in pixel values between adjacent pixels. In order to compute it, we first use gradient operators such as the Sobel operator to find the gradients in the x and y directions. The mathematical expression for average gradient is $\frac{1}{MN}\sum_{k=1}^{M}\sum_{l=1}^{N}G(k,l)$.

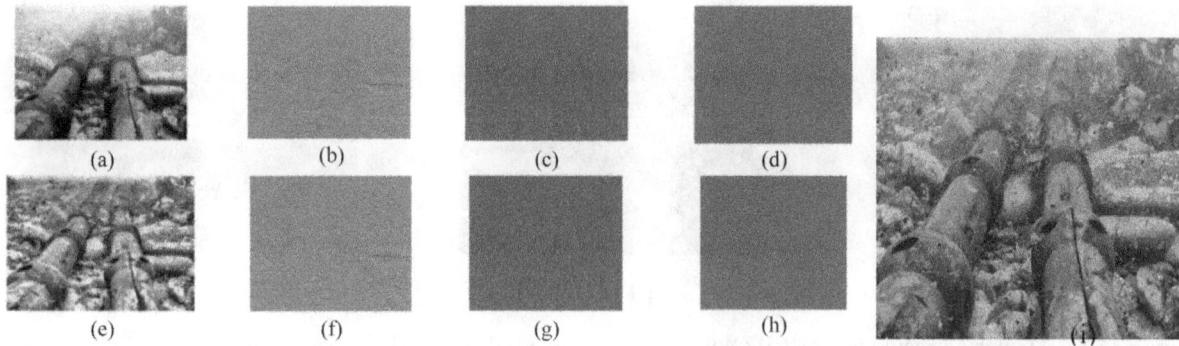

Figure 36.4 Top-Global and Bottom-Local (a) Component of approximate low-frequency, (b) Component of horizontal high-frequency, (c) Component of vertical high-frequency, (d) Component of diagonal high-frequency, (e) Component of approximate low-frequency, (f) Component of horizontal high-frequency, (g) Component of vertical high-frequency, (h) Component of diagonal high-frequency, (i) enhanced image weighted wavelet fusion: The first step in the process is the wavelet trans form, which breaks down the images into distinct frequency bands and separates them into high-frequency (detail) and low-frequency (smooth) components. Next, these components are integrated using a weighted technique, where each component is assigned to a variable weight according to its significance and impact on the overall image quality (UIEB [6])
Source: Author

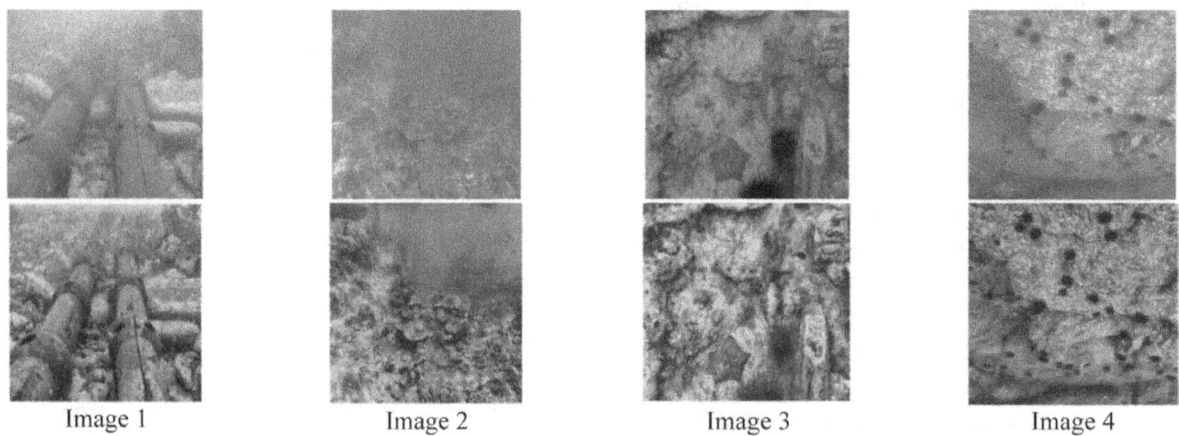

Image 1   Image 2   Image 3   Image 4

Figure 36.5 Enhanced images (bottom) of raw pictures (top) . Image 1 and Image 2 are obtained from UIEB [6] datasets, Image 3 and Image 4 derived from UIQS [7] datasets
Source: Simulation results from MATLAB for fig 36.5

Information entropy [9]

Shannon entropy, another name for information entropy, quantifies how unpredictable or random a set of data is. It is computed considering the likelihood of various outcomes. The equation is: $H= -\sum P(x_i)\log_2 P(x_i)$, where the probability of a result is represented by $P(x_i)$. Entropy measures the typical quantity of information generated by a data source.

Edge intensity [8]

If the image is in color, first convert it to greyscale. Next, determine edges by computing changes in pixel values using a filter such as the Sobel operator. Both vertical and horizontal motion are possible with the Sobel operator. Determine the magnitude of the gradient at each pixel, which represents edge intensity, by computing the gradient in each direction. To get the image's total edge intensity, compute the average of these gradient magnitudes.

Table 36.1 Evaluation scores of UIEB [6] datasets and UIQS [7] datasets using our weighted wavelet transform method for enhanced images.

Figure 5	AG	IE	EI	UCIQE	CCF
Image 1	10.563	7.479	99.512	0.573	41.485
Image 2	9.925	7.316	98.456	0.556	43.147
Image 3	9.617	7.558	96.812	0.528	32.477
Image 4	9.741	7.690	93.287	0.548	35.648

Source: Simulation results from MATLAB for fig 36.5

Table 36.2 PSNR and SSIM.

Figure 5	PSNR	SSIM
Image 1	19.809	0.619
Image 2	18.456	0.627
Image 3	18.585	0.738
Image 4	20.514	0.707

Source: Simulation results from MATLAB for fig 36.5

Underwater color image quality evaluation metric
In order to evaluate the quality of underwater color photos, the UCIQE metric takes into account variables like color fidelity, contrast, and sharpness. For UCIQE, the formula is UCIQE = 0.4680 × StdVarianceChroma + 0.2745 × ContrastLuminance + 0.2576 × MeanSaturation [10] Table 36.1-36.2.

Colorfulness contrast fog density index (CCF)
Colorfulness, contrast, and fog density are all assessed by the CCF, which gauges the quality of underwater images. The formula for CCF is y_1 × colorfulness + y_2 × contrast + y_3 × fog density, where, each factor is given a weight, which are y_1, y_2 and y_3 [11].

Other image quality metrics
Peak signal-to-noise ratio
One way to measure the degree to which an image's quality has improved over its original or degraded state is by looking at its peak signal-to-noise ratio (PSNR). To assess the efficacy of the augmentation, the technique entails comparing the improved image to the original.

Structural similarity index
A statistic for comparing two photographs is called the structural similarity index (SSIM). Because it

takes into account variations in brightness, contrast, and structural information, it is very useful for evaluating image quality.

Conclusion

Weighted Wavelet Transforms is an advanced technique to enhance underwater images. Using wavelet transforms, visual perception principles, and a multi-scale approach, weighted wavelet transforms effectively improve the clarity, detail, and visual appeal of aquatic images. Its applications range from marine biology, underwater exploration, industrial inspections, and environmental monitoring. Though it has some computational and implementation complexities, the significant improvements in its features make it a useful instrument for imaging underwater.

References

[1] Zhuang, P., Li, C., & Wu, J. (2021). Bayesian retinex underwater image enhancement. *Engineering Applications of Artificial Intelligence*, 101, 104171.

[2] Ding, X., Liang, Z., Wang, Y., & Fu, X. (2021). Depth-aware total variation regularization for underwater image dehazing. *Signal Processing: Image Communication*, 98, 116408.

[3] Zhang, W., Wang, Y., & Li, C. (2022). Underwater image enhancement by attenuated color channel correction and detail preserved contrast enhancement. *IEEE Journal of Oceanic Engineering*, 47(3), 718–735.

[4] Ancuti, C. O., Ancuti, C., De Vleeschouwer, C., & Bekaert, P. (2018). Color balance and fusion for underwater image enhancement. *IEEE Transactions on Image Processing*, 27(1), 379–393.

[5] Ancuti, C. O., Ancuti, C., De Vleeschouwer, C., & Garcia, R. (2017). Locally adaptive color correction for underwater image dehazing and matching. In Proceedings IEEE Conference Computer Vision and Pattern Recognition Workshops (CVPRW), (pp. 997–1005).

[6] Li, C., et al. (2020). An underwater image enhancement benchmark dataset and beyond. *IEEE Transactions on Image Processing*, 29, 4376–4389.

[7] Chongyi Li, Chunle Guo, Wenqi Ren, Runmin Cong, Junhui Hou, Sam Kwong, Dacheng Tao. (2020). Real-world underwater enhancement: Challenges, benchmarks, and solutions under natural light. *IEEE Transactions on Circuits and Systems for Video Technology*, 30(12), 4861–4875.

[8] Azmi, K. Z. M., Ghani, A. S. A., Yusof, Z. M., & Ibrahim, Z. (2019). Natural-based underwater image color enhancement through fusion of swarm-intelligence algorithm. *Applied Soft Computing*, 85, 105810.

[9] Chan, R., Rottmann, M., & Gottschalk, H. (2021). Entropy maximization and meta classification for out-of-distribution detection in semantic segmentation. In Proceedings IEEE/CVF International Conference on Computer Vision (ICCV), (pp. 5108–5117).

[10] Yang, M., & Sowmya, A. (2015). An underwater color image quality evaluation metric. *IEEE Transactions on Image Processing*, 24(12), 6062–6071.

[11] Wang, Y., Li, N., Li, Z., Gu, Z., Zheng, H., Zheng, B., et al. (2018). An imaging-inspired no-reference underwater color image quality assessment metric. *Computer and Electrical Engineering*, 70, 904–913.

[12] Jiang, N., Chen, W., Lin, Y., Zhao, T., & Lin, C.-W. (2022). Underwater image enhancement with lightweight cascaded network. *IEEE Transactions Multimedia*, 24, 4301–4313.

37 Enhancing underwater image clarity with balanced adaption compensation approach

Palle Rangappa[1,a], M. Farhana Begum[2,b], M. Anusha[2,c], P. Chenna Sai[2,d] and V. Varshith Kumar[2,e]

[1]Assistant Professor, Department of Electronics and Communication Engineering, Rajeev Gandhi Memorial College of Engineering and Technology, Nandyal, Andhra Pradesh, India

[2]Students of Department of Electronics and Communication Engineering, Rajeev Gandhi Memorial College of Engineering and Technology, Nandyal, Andhra Pradesh, India

Abstract

The quality of underwater photos is greatly impacted by light absorption, scattering, and the shifting characteristics of the water. Reduced visual contrast, color fading, and vision impairment are the results of these disorders. In this work, we propose a novel underwater image enhancement technique based on Balanced Adaption Compensation (BAC). In order to address the challenges of underwater imaging, the approach employs a balanced compensation framework that dynamically changes the image's contrast, brightness, and color balance. BAC uses adaptive algorithms to compensate for light attenuation and restore the original image qualities, hence improving the appearance. Experiments show that the proposed technology improves clarity of view, eliminates color distortion, and restores natural hues, outperforming current procedures on both objective and subjective criteria. Current physical-based compensation strategies often involve non-convex optimization, which results in less-than-ideal solutions for the ill-posed situations. The benefits of learning-based and physical-based techniques are complimentary when combined under the novel Balanced Adaptation Compensation (BAC) paradigm introduced in this study. BAC combines the data-driven semantic transfer with scene-relevant reconstruction to achieve strong augmentation while preserving textural and structural fidelity. The proposed method alleviates the constraints of scene diversity and reduces the dependence of neural network performance on training patterns.

Keywords: Balanced-adaption compensation (BAC), scene relevant reconstruction, semantic transfer

Introduction

Underwater photographs lose quality due to selective absorption and dispersion in water.. Enhancement methods dependent on the underwater photography technique have been thoroughly researched. Physical treatments have significantly improved the color of low-quality underwater pictures. However, these solutions primarily address particular underwater picture attenuation issues by employing already acquired prior information. These limitations make it difficult to extract useful information from underwater photos, needing complex augmentation techniques. Underwater photography is essential for marine exploration, robotics, and object recognition [1–3].

Learning-based techniques frequently involve paired images including both training input and ground truth. In the lack of underwater visual ground truth, enhancement networks are trained using synthetic data generated by existing image improvement techniques. Since the mid-twentieth century, maritime exploration has become increasingly technological [4]. Using the contrasts between data-driven and physical-based methodologies, this work proposes a Balanced Adaption Compensation(BAC) method for well-posed compensation that can be used to a wide range of underwater scenarios. Images are used in a range of underwater applications, such as robotics, rescue operations, man-made structures inspection, ecological monitoring, marine creature tracking and real-time navigation [5, 6].

Literature Survey

Color Channel Compensation (3C) [7] : It is a method for balancing as well as correcting color aberrations in photographs by altering separate

[a]palleranga.476@gmail.com, [b]mirzabegum55@gmail.com, [c]anusha9256@gmail.com, [d]chennasai000@gmail.com, [e]vesipogu.varshith@gmail.com

DOI: 10.1201/9781003684589-37

colored areas. subsequently improves the quality of photographs by adjusting for differences resulting from illumination, detector flaws, or color abnormalities. To test the efficacy of their 3C technique, the investigators employed the O-Haze dataset [8], which comprises 45 pairs of hazy and haze-free ground-truth photos. They used 4 known dehazing techniques [9–11], and on the photos and discovered that the results improved dramatically when the images were initially preprocessed using 3C. 3C contains noise, uneven color, and variable lighting, therefore we utilize PAC to improve contrast and noise reduction, although it may overprocess details.

Parameter-Adaptive-Compensation(PAC) [12]: PAC approaches frequently include adaptive filtering, machine learning-based improvement, and optimization-based adjustment. It automatically modifies analyzing settings to increase consistency and precision. The outcomes of recovery approaches from references [13] and deep learning methods from [14] are either matched or bettered by the recovery strategies with PAC, as described in references . Datasets including paired underwater and ground truth photos are usually needed for training a number of deep learning-based techniques for underwater image recovery, including . Studies have investigated its use in medical imaging, remote sensing, and real-time video processing. AC may overprocess photos, losing small details and generating imperfections, whereas BAC strikes a compromise between noise reduction, color preservation, and natural detail retention.

Balanced Adaptation Compensation (BAC): It combines physical and learning-based techniques to improve the quality of pictures. It uses Scene-Relevant Reconstruction (SRR) to successfully repair color distortions. Furthermore, Semantic Transfer is utilized to improve the look of photos by exploiting data related to context. The combination of these strategies provides precise adaption through a variety of settings. This method increases visual consistency and realism in image processing jobs [15, 16] have examined a few subjects pertaining to the maintenance and improvement of underwater images Figure 37.1.

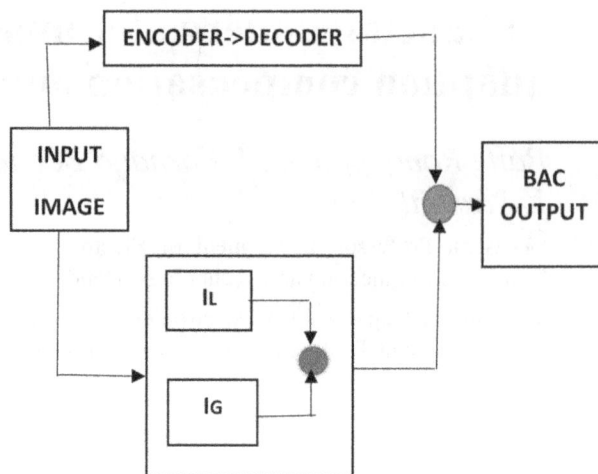

I_L: Local Scene-Relevant Reconstruction

I_G: Global Scene-Relevant Reconstruction

Distance-based Image Interpolation

Distribution-based Image Interpolation

Figure 37.1 Block diagram of BAC

Source: Drawn from author with the reference of An Underwater Image Enhancement Method Based on Balanced Adaption Compensation

Existing Method

Conventional submarine picture enhancing technologies are divided into a pair of groups: physical and learning-based methods. Physical-based approaches employ past information and mathematical models to adjust for aquatic picture imperfections, but they frequently struggle with scene variety and could result in incorrect correction. Learning-based approaches use artificial neural networks trained on artificial databases to improve photos, but typically overfit to certain training patterns and struggle with real-world variability. Certain traditional approaches, such as Color Channel Compensation (3C) and Parameter-Adaptive Compensation (PAC), try to correct chromatic defects but lack to keep picture texture and generalization consistent throughout varied marine settings. In learning-based methods, these artificial neural networks operate by adjusting numerous weighted input parameters during their training phase, which allows them to learn complex features for image enhancement but also contributes to their tendency to overfit specific training data.

Proposed Method

The research offers Balanced Adaption Compensation (BAC), a mixed approach that combines physical and learning based compensation methods. BAC is made up of two key components: Scene-Relevant Reconstruction (SRR) and Semantic Transfer (ST). SRR corrects color distortions according to scene-specific attenuation, maintaining geometric accuracy, whereas ST improves picture appearance using an encoder-decoder network trained on non-attenuated images. The balancing factor () dynamically adjusts the contribution of SRR and ST based on scene conditions. This results in a well-posed compensation method that improves generalization while preserving texture features for high-level visual tasks. Experimental data show that BAC surpasses other techniques in both qualitative and quantitative

1. Processing of Input image: The method begins with an input underwater image that has color distortion and poor visibility owing to underwater attenuation effects Figure 37.1a-d.

Input Image

Figure 37.1a Input image

Source: C. Li et al., "An underwater image enhancement benchmark dataset and beyond," IEEE Trans. Image Process., vol. 29, pp. 4376–4389, 2020 and C. Liu et al., "A dataset and benchmark of underwater object detection for robot picking," in Proc. IEEE Int. Conf. Multimedia Expo Workshops, 2021, pp. 1–6

2. Scene-Relevant Reconstruction(SRR):The purpose of this module is to use past knowledge about underwater settings to repair the distorted image. Local scene-relevant reconstruction (IL) focuses on particular features and local structural refinement. Global Scene-Related Reconstruction (IG) Corrects colour imbalances and significant distortions. Next, the IG and IL outputs are interpolated using Distribution-Based Images Interpolation changes the distribution of colours. distance based image interpolation, takes regional variations in attenuation into account.

Scene-Relevant Reconstruction Image

Figure 37.1b Scene-relevant reconstruction
Source: Simulation results from MATLAB

3. Semantic Transfer (ST): A Learning-Based Approach the photograph is processed by an encoder-decoder neural network to increase micro textures and realistic balance in colours. The encoder take out sophisticated characteristics in the source picture. The decoder restore the picture by learning from superior, non-attenuated images.

Sematic Transfer Image

Figure 37.1c Sematic transfer
Source: Simulation results from MATLAB

4. Final Enhanced Image: The results from Scene-Relevant Reconstruction and Semantic Transfer are mixed using an adaptive weighting factor. The result assures that the final image receives combined physical-based rectification and deep-learning-based refining. The result is a good-quality, colour-corrected underwater picture with increased resolution and solidity of structure.

BAC Output (Sematic Transfer + Scene-Relevant Reconstruction)

Figure 37.1d BAC output
Source: Simulation results from MATLAB

Mathematical Model

The Semantic Transfer Module (ST) use supervised training to rectify the semantic aspect of pictures using an encoder-decoder system. The Scene-Relevant Reconstruction Module (F_{SR}) uses global and local scene reconstructions to correct for chromatic imperfections, especially red channel retardation.

The final improved picture is calculated as:

$$F_{BAC}(I) = \mu * F_{ST}(I) + (1-\mu) * F_{SR}(I), \qquad (1)$$

here I is the input picture. The scene reconstruction is defined as

$$I_{SR} = F_{SR}(I) = \lambda_G * f_G(I) + \lambda_L * f_L(I) \qquad (2)$$

where I: .Input Image
F_{ST}(I):Semantic Transfer Model
F_{SR}(I):Scene-Relevant Reconstruction Model
$\lambda_G + \lambda_{L=1}$
λ_G, λ_L are weighted input parameters

Comparison with 3C, PAC and BAC

The Balanced Adaptation Compensation (BAC) technique is superior than Colour Channel Compensation (3C) as well as Parameter-Adaptive Compensation (PAC) owing to its adaptive and hybrid approach. Unlike 3C and PAC, which use fixed adjustment factors, BAC continually alters Scene-Relevant Reconstruction (SRR) and Semantic Transfer (ST) based on the scene, resulting in more flexibility. BAC also preserves delicate textures and structural features, whereas 3C and PAC typically result in overcompensation and artificially rendered graphics. Furthermore, BAC effectively removes colour overload and pleasantly enhances images without utilizing excessive brightness. This makes BAC more adaptable and durable in a variety of underwater conditions than other approaches Table 37.1 and Figure 37.2a-d.

Table 37.1 Comparison methods.

Methods	Colour Correction	Detail Conservation	Output Performance
3C	Moderate	Low	Average
PAC	Good	Moderate	Good
BAC	Excellent	High	Superior

Source: Based on the output pictures of MATLAB

Original Image

Figure 37.2a Original image
Source: C. O. Ancuti, C. Ancuti, C. De Vleeschouwer, and M. Sbert, "Color channel compensation (3C): A fundamental pre-processing step for image enhancement," IEEE Trans. Image Process., vol. 29, pp. 2653–2665, 2020.

Image with 3C (Color & Contrast Enhanced)

Figure 37.2b Image with 3C
Source: Simulation results from MATLAB

Results

Two typical metrics used in this strategy are Underwater Image Quality Measure (UIQM) [17] and Underwater Colour Image Quality Evaluation (UCIQE) [18]. These metrics help to objectively evaluate various improvement methods by examining colour fidelity, clarity, contrast, and overall visual appeal.

UIQM focuses on structural clarity and detail retention, whereas UCIQE seeks to capture

Image with PAC

Figure 37.2c Image with PAC
Source: Simulation results from MATLAB

Original Image

Figure 37.3a Original image
Source: Reference from J. Liu and X. Zhang, "Parameter-adaptive compensation (PAC) for processing underwater selective absorption," IEEE Signal Process. Lett., vol. 27, pp. 2178–2182, 2020

Enhanced Image with BAC

Figure 37.2d Enhanced image
Source: Simulation results from MATLAB

Image with 3C (Color Channel Compensation)

Figure 37.3b Image with 3C
Source: Simulation results from MATLAB

Image with PAC

Figure 37.3c Image with PAC
Source: Simulation results from MATLAB

human-perceived visual appearances. A higher UIQM score suggests enhanced brightness, and a higher UCIQE score indicates more accurate color enhancement. Together, these characteristics ensure that underwater image enhancement methods provide technically superior and artistically pleasing results Figure 37.3a-d and Table 37.2-37.3.

$$UIQM = \alpha * UICM + \beta * UISM + \gamma * UIConM \qquad (3)$$

Figure 37.3d Enhanced image
Source: Simulation results from MATLAB

Table 37.2 Quantitative analysis of enhanced images.

Images	Methods	UIQM Values	UCIQE Values
Figure 37.2a	Original Image	2.255	0.45987
Figure 37.2b	3C	3.0895	0.41656
Figure 37.2c	PAC	2.3925	0.5535
Figure 37.2d	Proposed Method(BAC)	3.5615	0.57794
Figure 37.3a	Original Image	2.1754	0.39435
Figure 37.3b	3C	3.0319	0.33338
Figure 37.3c	PAC	2.0977	0.56945
Figure 37.3d	Proposed Method (BAC)	3.5615	0.38879

Source: Obtained from the simulation results of MATLAB [7],
An Underwater Image Enhancement Method Based on Balanced
Adaption Compensation

Table 37.3 Standard metrics of output images.

Image	SSIM	PSNR	MSE
Figures 37.2a & 2b	1.000	17.268	0.0001
Figures 37.2a & 2c	0.6657	14.5900	0.0348
Figures 37.2a & 2d	0.3815	12.8045	0.0524
Figures 37.3a & 3b	1.000	18.523	0.0001
Figures 37.3a & 3c	0.7500	16.000	0.0280
Figures 37.3a & 3d	0.6657	14.590	0.0348

Source: Obtained from the simulation results of MATLAB

UICM = Underwater Image Colourfulness Measure
UISM = Underwater Image Sharpness Measure
UIConM = Underwater Image Contrast Measure
α, β, γ are the weighting factors.

$$UCIQE = w1 * Csat + w2 * Ccon + w3 * Ccol \quad (4)$$

Csat = Mean saturation in the HSV colour space.
Ccon = Mean brightness in the HSV colour space.
Ccol = Colourfulness (variance of hue).
w1,w2,w3 are weight coefficients.

Conclusion

The Balanced Adaption Compensation (BAC) strategy, which combines physical and learning-based techniques, offers an innovative and successful solution for improving underwater photographs. In contrast to previous methods that depend primarily on prior-based compensations or data-driven models, BAC dynamically balances both approaches using an adaptive weighting factor. This improves colour correction, contrast improvement, and texture retention, resulting in pictures that are crisper and more visually realistic.

References

[1] Chen, X., Yu, J., Kong, S., Wu, Z., Fang, X., & Wen, L. (2019). Towards real-time advancement of underwater visual quality with GAN. *IEEE Transactions on Industrial Electronics*, 66(12), 9350–9359..

[2] Wang, Yajing, Zhang, Wei, Lei Chen. (2021). Real-time underwater onboard vision sensing system for robotic gripping. *IEEE Transactions on Instrumentation and Measurement*, 70, 1–11.

[3] Jiang, L., Wang, Y., Jia, Q., Xu, S., Liu, Y., Fan, X., et al. (2021). Underwater species detection using channel sharpening attention. In Proceedings of the 29th ACM International Conference on Multimedia, (pp. 4259–4267).

[4] Lu, H., Wang, D., Li, Y., Li, J., Li, X., Kim, H., et al. S.,& I. (2019). CONet: a cognitive ocean network. *IEEE Wireless Communication*, 26(3), 90–96.

[5] Lee, D. J., Redd, S., Schoenberger, R., Xu, X., & Zhan, P. (2003). An automated fish species classification and migration monitoring system. In Proceedings 29th Annual Conference of the IEEE Industrial Electronics Society (IECON), (Vol. 2, pp. 1080–1085).

[6] Chen, C. L. P., Zhou, J., & Zhao, W. (2012). A real-time vehicle navigation algorithm in sensor network environments. *IEEE Transactions on Intelligent Transportation Systems*, 13(4), 1657–1666.

[7] Ancuti, C. O., Ancuti, C., De Vleeschouwer, C., & Sbert, M. (2020). Color channel compensation (3C): a fundamental pre processing step for image enhancement. *IEEE Transactions on Image Processing*, 29, 2653–2665.

[8] Ancuti, C. O., Ancuti, C., Timofte, R., & De Vleeschouwer, C. (2018). O-HAZE: a dehazing benchmark with real hazy and haze-free outdoor images. In Proceedings CVPR, (pp. 754–762).

[9] He, K., Sun, J., & Tang, X. (2009). Single image haze removal using dark channel prior. In Proceedings CVPR, (pp. 2341–2353).

[10] Berman, D., Treibitz, T., & Avidan, S. (2016). Non-local image dehazing. In Proceedings CVPR, (pp. 1674–1682).

[11] Meng, G., Wang, Y., Duan, J., Xiang, S., & Pan, C. (2013). Efficient imagedehazing with boundary constraint and contextual regularization. In Proceedings ICCV, (pp. 617–624).

[12] Liu, J., & Zhang, X. (2020). Parameter-adaptive compensation (PAC) for processing underwater selective absorption. *IEEE Signal Processing Letters*, 27, 2178–2182.

[13] Ancuti, C. O., Ancuti, C., De Vleeschouwer, C., Neumann, L., & Garcia, R. (2017). Color transfer for underwater dehazing and depth estimation. In Processing IEEE International Conference on Image Processing (pp. 695–699).

[14] Zhu, J., Park, T., Isola, P., & Efros, A. A. (2017). Unpaired image-to-image translation using cycle-consistent adversarial networks. In Proceedings of the IEEE International Conference on Computer Vision, (pp. 2242–2251).

[15] Lu, H., Li, Y., Zhang, Y., Chen, M., Serikawa, S., & Kim, H. (2017). Underwater optical image processing: A comprehensive review. *Mobile Networks and Applications*, 22(6), 1204–1211.

[16] Schettini, R., & Corchs, S. (2010). Underwater image processing: state of the art of restoration and image enhancement methods. *EURASIP Journal on Advances in Signal Processing*, 2010(1), 746052.

[17] Panetta, K., Gao, C., & Agaian, S. (2016). Human-visual-system inspired underwater image quality measures. *IEEE Journal of Oceanic Engineering*, 41(3), 541–551.

[18] Miao, Y., & Sowmya, A. (2015). An underwater color image quality evaluation metric. *IEEE Transactions on Image Processing*, 24(12), 6062–6071.

38 Real-time polyglot translator powered by Raspberry Pi

K. Riyazuddin[1,a], K. Sunith Kumar[2,b], C. Swetha[2,c], S. Wafika[2,d] and K. Thirumalesu[2,e]

[1]Associate Professor, Department of ECE, Annamacharya University, Rajampet, Andhra Pradesh, India

[2]Department of ECE, Annamacharya University, Rajampet, Andhra Pradesh, India

Abstract

Traditional methods such as human translators, dictionaries, and interpretation services are available but are usually slow, expensive, and unsuitable for the purposes of real-time communications. In some instances, this may limit effective communication. The instantaneous polyglot translator, powered by Raspberry Pi, addresses these issues by providing real-time voice communication translation in a compact and cost-effective manner. Based on the processing ability of the Raspberry Pi, this system records spoken speech using a microphone, transforms into texts, translates through API calls, and then provides translated speech through speakers for multi-lingual communication to take place as fast as it can. There are several applications of this device in travel, business, emergencies, or even medical settings. Fast, low-cost language translation portability facilitates cross-lingual communication to improve accessibility and efficiency in all sorts of multicultural environments.

Keywords: Raspberry Pi, real-time communication, translator

Introduction

In today's globalized world, the ability to communicate in many languages is vital, but the language gap still poses a huge challenge, especially in scenarios that demand real-time translation. The conventional methods, such as using human translators, phrasebooks, or interpretation services, are usually slow, expensive, and not feasible for real-time applications. This would further hold back huge progresses in the arenas of healthcare, business, tourism, and emergency management, which strongly require prompt as well as accuracy in communication.

Real-time polyglot translator powered by Raspberry Pi: This is the latest gadget which unfolds all these new forms of addressing this challenge. It is a small, affordable real-time voice translation device. Using the processing capabilities of the Raspberry Pi, the device captures audio input through a microphone, then converts it to text, sends it through an API (A programming Language) for the target language translation, and plays the translated audio through a speaker. This efficient process makes communication across linguistic disparities absolutely seamless without large equipment or professional translators.

Another great central aim of this project is its access to real-time multilingual speaking and listening [10]. APL (A Programming Language) is a high-level, array-programming language developed by Kenneth E. Iverson in the late 1950s as a mathematical notation and later as a complete programming language in the 1960s. This is achieved through the versatile and inexpensively costed Raspberry Pi, whose high per- formance is offered at a much lower cost than sim- ilar alternatives; thereby, as many kinds of users as possible--individuals, small businesses, non-profit organizations--are reached out to. Its applications are manifold. In the tourism and travel industries, it makes it possible for users to maneuver foreign territories without language difficulties. This fills gaps between people and others to ensure more prompt responses and action during emergency situations where every minute counts. The real-time polyglot translator has reinvented communication; it has synthesized the technological ability of speech recognition, translation, and text-to-speech into a single small form factor that may topple the unproductive translation methodologies prevalent today, giving users access to multifarious

[a]Shaik.riyazuddin7@gmail.com, [b]bennysunith366@gmail.com, [c]swethajb0604@gmail.com, [d]swafika03@gmail.com, [e]kurubathirumaleshysr@gmail.com

DOI: 10.1201/9781003684589-38

languages which revolutionize all possible means of communication. Truly, this stands as a hallmark for a future where language cannot be a constraint for communication rather further fosters globalization and integration along with efficacy. In this we have audio processing limitations which is background noise and multiple speakers with these Raspberrypi may Struggle to capture the input language.

Literature Survey

Speech recognition and speech-to-text systems
DeepSpeech is an open-source model introduced by Hannun and Rivenson [4] to provide a deep learning-based speech-to-text framework for accurate transcription of spoken language into written text. A real-time translator would need to include speech recognition to translate the user's input into machine-readable text. DeepSpeech is an open-source model that can be used in a local deployment, thus making it more suitable for projects on Raspberry Pi, since they do not depend on cloud applications. The combination of this technology with machine translation systems allows the translator to convert spoken input into another language on the fly [4].

Raspberry Pi and embedded systems for translation
The Raspberry Pi platform is generally regarded as a cost-friendly and flexible edge computing solution. To know how the hardware of the Pi, along with its software, can be used in real-time applications, readers can go through official documentation published by the Raspberry Pi Foundation in 2021 [5]. Due to its compact size, minimal power consumption, and compatibility with external microphones, cameras, and sensors, Raspberry Pi would be the ultimate candidate to implement a real-time translator that can run on-site processing, thus reducing latency relative to cloud-based systems, but at the same time ensures privacy and control of data [5].

Zhu and Liao explored, in their work, the usage of Raspberry Pi to build a real-time translation system. They implemented Google's speech-to-text API and translation services on a Raspberry Pi, providing multilingual support. It is demonstrated here that it is possible to perform real-time language translation on an edge device. This research brings to light the difficulties of integrating speech recognition with translation APIs in a constrained environment like Raspberry Pi but at the same time presents the possibility of such systems in practical applications [6].

Real-time machine translation using edge AI
Hsu [11] looks into the possibility of applying machine learning models on edge devices to perform real-time translation, highlighting the need for such models to be deployable without much reliance on cloud computing. Edge AI solutions, such as that proposed for Raspberry Pi-based translators, have faster response times and less dependence on external networks. This is critical for real-time translation since latency introduced by network communication can degrade user experience substantially.

Integration with cloud services
Most real-time translation systems are based on cloud models, such as Google Cloud and Microsoft Translator. However, integrating cloud services with Raspberry Pi-based translators has its disadvantages. These services can provide high-quality translations across many languages but usually require a stable internet connection. This integration of cloud-based APIs with Raspberry Pi can provide flexibility in deployment, making it possible to be both off and online that Google Cloud (2020) said in their Translation API documentation. However, offline processing still remains an important advantage for applications requiring privacy, minimal delay, and reliability in areas with poor connectivity [12].

Natural language processing and neural machine translation
Koehn [1] writes about the advancement of NMT, developed by deep learning models that have dramatically changed the landscape in MT. NMT systems are based on neural networks that have managed to learn complex patterns in languages and resulted in dramatically improved fluency and accuracy in translation. NMT models provide the backbone of a real-time polyglot translator in handling multiple languages simultaneously, and it is quite apt for the kind of device like the Raspberry Pi, where efficiency in computation and scalability are prime requirements. It was only recently, with deep learning approaches [2] that machine translation became better. In fact, models can easily be adapted to fit real-time applications on edge devices [1].

Methodology

The core processing unit is the use of the Raspberry Pi 3B+. It achieves speech translation in real time. Through several hardware and software components, it delivers the smooth and efficient translation experience given below:

Power supply

A reliable source of power will supply energy to the Raspberry Pi and its peripherals. The result is constant operation of all the components, from the USB sound card to the amplifier and the LED. Power source can either be a wall adapter or portable power bank; thus, this makes the system versatile for either stationary or portable use. This also calls for the proper regulation of voltage and current to prevent damaging sensitive parts.

Raspberry Pi 3B+

The Raspberry Pi 3B+ is a computer that belongs to the Raspberry Pi Foundation [5]. Released in March of 2018, it has brought along a package of upgrades when compared to its predecessor, Raspberry Pi 3B. Some differences include several updates in terms of performance and connectivity. This was the smallest and very powerful single board computer which saw its applications very much used on home automation, digital signage and in so many do-it-yourself projects or programming learning projects [13] Figure 38.1.

USB Sound card

The USB sound card gives the Raspberry Pi the enhanced input and output of audio. Since the Raspberry Pi does not have a built-in microphone port, this card provides an interface between the microphone, the Raspberry Pi, and the amplifier. The sound card captures the speech of the user through

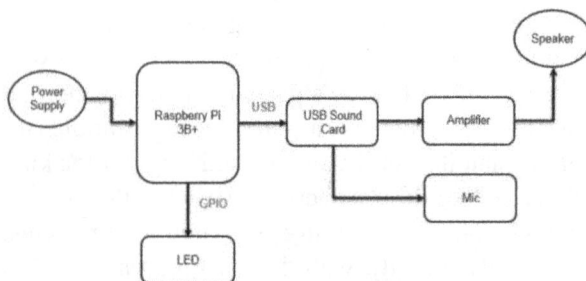

Figure 38.1 Block diagram of real-time speech translator
Source: Author

the microphone and transmits it to the Raspberry Pi for processing. Once the translation is done, it sends the audio output from the Raspberry Pi to the amplifier so that clear and high-quality sound delivery can be ensured [2].

Microphone

The microphone captures speech input from the user in real time. It sends an audio signal to the USB sound card, digitizes it, and sends it for processing by the Raspberry Pi [5]. A good quality microphone is utilized for accurate speech recognition, even when there is ambient noise. This is an important step toward getting the most accurate translations, since any errors in audio capture could lead to wrong text and consequently wrong translations. Captures the user's speech, transmits the audio to the USB sound card for processing by the Raspberry Pi. A high-quality microphone ensures accurate speech recognition [9].

Amplifier and speaker

This boosts the audio output signal from the Raspberry Pi to a suitable level for the speaker. This makes sure that the translated speech is loud and clear for the user. The amplified signal is then transmitted to the speaker, which plays the translated audio [4]. These elements complement each other and altogether provide high-quality audio output. As such, it can be applied in many situations, including noise-filled areas. The amplifier will amplify the signal from the Raspberry Pi and make it possible for the speaker to play the audio.

Implementation

1. The sound picked by the microphone is transferred to the Raspberry Pi via the USB sound card. For example: Telugu, English, Hindi.
2. The Raspberry Pi processes input through speech-to-text, translation, and text-to-speech algorithms [15].
3. The output signal from the audio processing section is passed to an amplifier via the USB sound card; the result is amplification and further playback through the speaker [7].
4. During the process, an LED indicator is used which gives real-time visual feedback. During translation, the LED light is lit; after completing the translation, the LED light extinguishes. In this way, it indicates that the system is ready for the next input from the user. This step-by-

step approach ensures efficient and seamless real-time translation with clear feedback for the user.

a.   *Initialization of Components*

The system initializes all hardware and software components at the outset. It initializes GPIO for the control of the LED light; sets up translation and text-to-speech libraries such as Google Translate API and GTTS; and it prepares the GUI using Tkinter library. This all ensures that everything is in good working condition Figure 38.2.

b.   *GPIO for LED*

The GPIO pins on the Raspberry Pi are utilized to control the LED. It gives a visual indication of the system activity. For example, if speech is being captured and translated, the LED lights up. This improves the user experience, as it shows real-time feedback about what is happening in translation.

c.   *Google Translate and GTTS libraries*

Google Translate API and GTTS is used for the core functionality of the project as follows:

• Speech-to-text: converts the spoken input into text

• Translation: translates the obtained input into a target language of our choice

• Text-to-speech: converts the translated text back into audio to be played via a speaker

4.   Tkinter GUI for Interaction

Tkinter is used for creating an interactive GUI for the project. There are several controls through which the user can configure and operate the

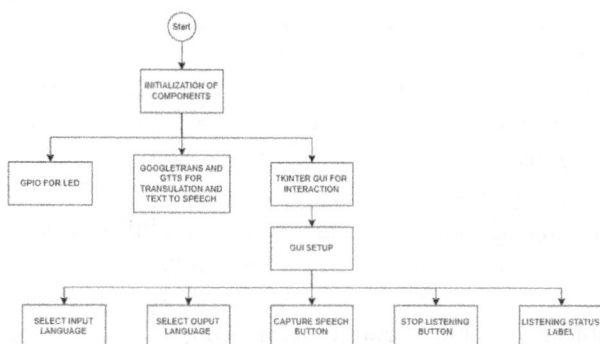

Figure 38.2 Implementation of real-time speech translator

Source: Author

translation system quite effortlessly. Simple and intuitive to interact with, even for the layman with little or no technical knowledge, this GUI ensures simplicity in designing it.

Result and Discussion

The figure displays the interface of a "voice translator and speaker" application with a simple design. It features two dropdown menus: one for selecting the input language for speech and another for choosing the output language for translation. Below these, there are two buttons: "capture speech," which initiates the speech capture process, and "stop listening," which halts it. The application appears to be The figure shows the interface of a "Voice Translator and Speaker" application in use. The "Select Input Language" dropdown is set to "English," while the "Select Output Language" is set to "Telugu." Below these options, the "Capture Speech" button is highlighted, indicating it is active, and the application is ready to capture spoken input. A status message at the bottom reads, "Listening... Speak now!" signaling that the application is currently listening for speech input for real-time translation. Additionally, a "Stop Listening" button is available to halt the speech capture process designed for real-time language translation, offering straightforward and user-friendly experience [13].

Performance metrics: In this it can record the input language speech for 5 mins and if the speaker is unable to speak for 5 secs it automatically stops recording the input speech [Figure 38.3].

The figure shows the interface of a "Voice Translator and Speaker" application in use.The "Select Input Language" dropdown is set to "English," while the "Select Output Language" is set to "Telugu" [Figure 38.4].

The figure shows a "captured speech" window from the "voice translator and speaker" application. The window displays the converted text from spoken input, which reads, "what is your name." Below the displayed text, there is an "OK" button, allowing the user to confirm or proceed. This interface indicates that the application has successfully processed and converted the speech into text [4].

According to the input we can also use the input as cloud stored data and after we have converted the language from input to desired output languages, we can store this output in cloud. In this we can also provide security for the cloud-based services [Figure 38.5].

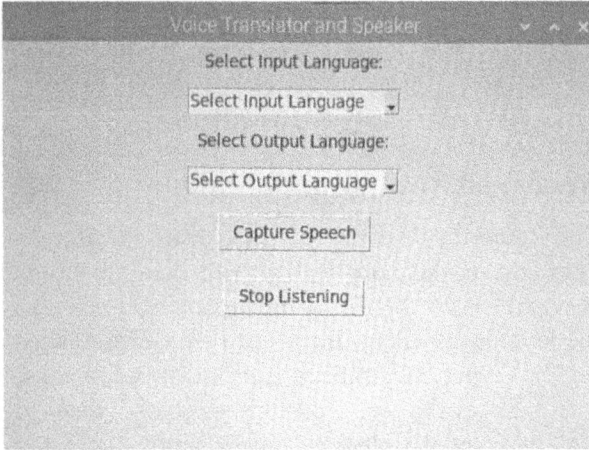

Figure 38.3 Voice translator and speaker interface
Source: Author

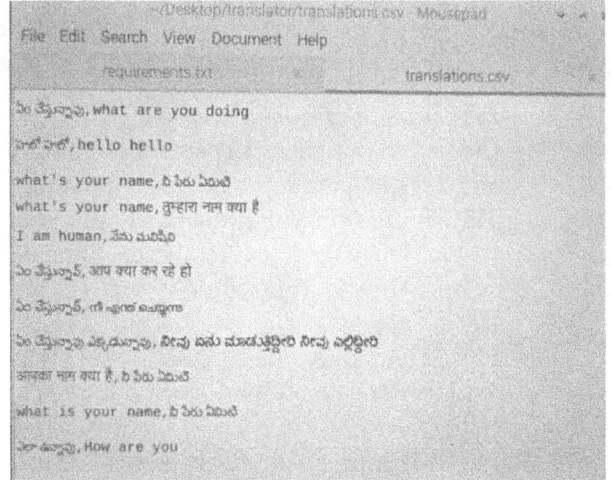

Figure 38.4 Input & output language selection
Source: Author

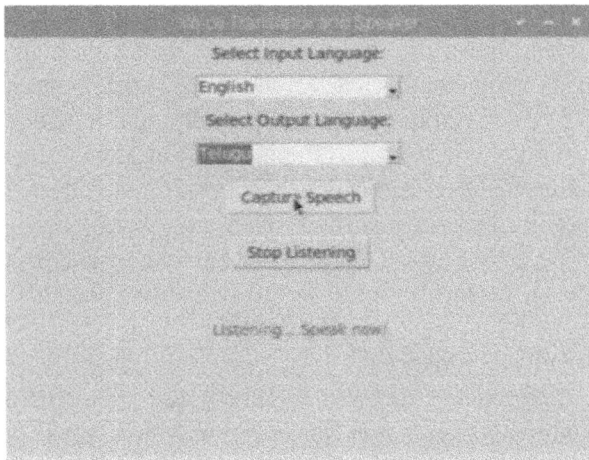

Figure 38.5 Speech-to-text conversion output
Source: Author

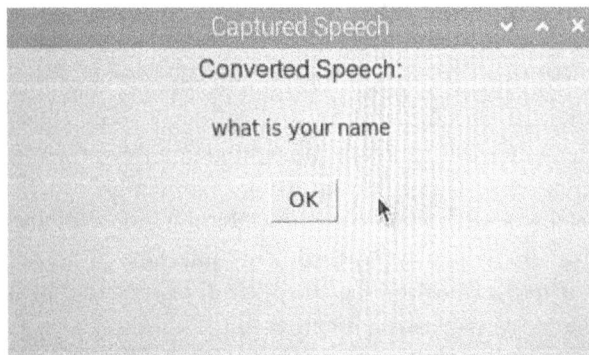

Figure 38.6 Translation data in multiple languages
Source: Author

The image shows a text editor window named "Mousepad," displaying the contents of a file located at desktop/translator/translations.csv [7]. The file appears to contain a list of phrases translated between different languages, organized in pairs. Each line contains a phrase in one language, followed by its translation(s) in another language. The languages used include English, Telugu, Hindi, and Kannada. The phrases range from basic greetings like "hello" to questions like "what's your name" and "what are you doing." Some phrases are repeated in multiple languages, reflecting a multilingual translation approach [Figure 38.6].

Conclusion

The "Raspberry Pi-powered Real-Time Polyglot Translator" demonstrates the accessibility of technology in overcoming language differences. The project exploits the use of the Raspberry Pi in offering an affordable real-time translating system that can be used in areas ranging from healthcare to tourism and education. Though performance must be enhanced as machine learning evolves notwithstanding challenges such as differences in dialects, quality of translation, as well as scalability in languages remaining.By enhancing global connection and co-operation, this program lays a strong ground for multilingual communication innovations in the future.

Future Scope

In the future, the incorporation of more sophisticated machine learning models would perhaps definitely enhance translation quality but extend beyond to reach context understanding and idiomatic expression handling. Improving the processing speed with enhancements of the device for additional languages and dialects, it would be an even more powerful device for real-time communication.

References

[1] Koehn, P. (2017). Neural Machine Translation. Cambridge University Press.

[2] Vaswani, A., Shazeer, N., Parmar, N., Uszkoreit, J., Jones, L., Gomez, A. N., et al. (2017). Attention is all you need. In Proceedings of NeurIPS 2017.

[3] V. Sundararajan and C. Shih, "Deep learning for NLP: A survey," ACM Computing Surveys (CSUR), vol. 52, no. 6, pp. 1–36, 2019. doi: 10.1145/3357384.

[4] A. Hannun and Y. Rivenson, Deep Speech: A Speech-to-Text Engine Based on Deep Learning, GitHub, 2019. [Online]. Available: https://github.com/mozilla/DeepSpeech.

[5] Raspberry Pi Foundation, Raspberry Pi Documentation and Projects, 2021. [Online]. Available: https://www.raspberrypi.org/documentation.

[6] X. Zhu and S. Liao, "Real-time translation system using Raspberry Pi and Google Speech API," International Journal of Computer Applications, vol. 176, no. 35, pp. 1–5, 2020. doi: 10.5120/ijca2020919884.

[7] Liu, L., & Liu, W. (2018). A real-time speech translation system based on Raspberry Pi. *International Journal of Software Engineering and Knowledge Engineering*, 28(6), 887–900.

[8] Bengio, Y., Ducharme, R., & Vincent, P. (2003). A neural probabilistic language model. *Journal of Machine Learning Research*, 3, 1137–1155.

[9] Tiedemann, J. (2012). Parallel data, tools and interfaces in OPUS. In Proceedings of the Eighth International Conference on Language Resources and Evaluation (LREC 2012).

[10] Zhou, Z., & Liu, X. (2020). Real-time multilingual translation with edge computing. In Proceedings of the IEEE International Conference on Cloud Computing and Intelligence Systems (CCIS 2020).

[11] Hsu, C. (2019). Real-time machine translation using edge AI. *IEEE Access*, 7, 123658–123667.

[12] Google Cloud, Google Cloud Translation API Documentation. Google, 2020. [Online]. Available: https://cloud.google.com/translate/docs.

[13] J. Smith, A. Kumar, and L. Zhang, "Testing the viability of executing deep learning-powered translation models on Raspberry Pi," in Proc. Int. Conf. Embedded Syst. Appl., 2020, pp. 45–52.

[14] M. Robinson, T. Nguyen, and H. Patel, "An open-source Raspberry Pi-based real-time translation prototype," J. Open Hardware Softw. Syst., vol. 3, no. 2, pp. 112–120, 2021.

[15] L. Fernandez and J. Kim, "A comparative study of speech-to-text accuracy across accents and dialects," Int. J. Speech Process. Recognit., vol. 8, no. 1, pp. 34–41, 2020.

39 Enhancing power grid reliability through smart fault detection and classification using machine learning algorithms

Magesh, T.[1,a], Samuel Franklin, F.[2,b], Asokan, K.[2,c] and Christaper Indla[2,d]

[1]Professor, Department of EEE, R.M.K. Engineering College, Kavaraipettai, Tamil Nadu, India

[2]UG-Student, Department of EEE, R.M.K. Engineering College, Kavaraipettai, Tamil Nadu, India

Abstract

The reliability of power transmission networks is critical for ensuring stable electricity distribution, yet faults in transmission lines can arise unexpectedly due to various operational and environmental factors. Rapid and accurate fault detection is essential to minimize disruptions and enhance system resilience. In this study, the efficacy of machine learning methods for transmission line fault classification is examined. Models were trained and evaluated using a dataset of 12,002 samples with six different features. Key performance measures, including confusion matrix, F1-score, recall, accuracy, and precision were employed. Among the evaluated models, the multi-layer perceptron (MLP) demonstrated superior accuracy and robustness in identifying intricate fault patterns, highlighting its potential for complex classification tasks. Logistic regression exhibited moderate effectiveness, making it suitable for simpler fault identification, while support vector machine (SVM) showed lower accuracy, indicating its limitations in high-dimensional fault scenarios. The findings reinforce the advantages of neural network-based approaches in improving fault detection accuracy, contributing to enhanced automation, predictive maintenance, and overall reliability in modern power transmission systems.

Keywords: Accuracy, automated power systems, F1-Score, fault classification, fault detection, logistic regression, machine learning, multi-layer perceptron (MLP), neural networks, power transmission lines, precision, recall, support vector machine

Introduction

As electricity demand rises globally, expanding power infrastructure has become essential, leading to higher transmission line construction costs. These lines play a decisive role in ensuring a steady flow of electricity from generation plants to consumers. Any disturbance can affect industries, businesses, transportation, and daily life, making grid stability a top priority.

Transmission lines, often exposed to harsh environmental conditions, are prone to faults caused by natural disasters, severe weather, or human-related incidents [1]. Such failures can destabilize the grid, resulting in widespread outages. Prompt fault identification and classification are necessary to prevent major system failures. Quick and accurate fault detection enables operators to locate faults efficiently, take corrective action, and restore power with minimal downtime.

Fault detection approaches fall into two main categories: physics-based methods and data-focused techniques [2, 3].

The former relies on mathematical models and engineering principles to predict faults, while the latter uses machine learning (ML) and statistical analysis to detect patterns in real- time sensor data [2]. ML-based methods continuously adapt to new conditions, improving accuracy and response time.

In recent years, ML has become an essential tool in diagnosing power system faults, with neural networks proving particularly effective. By analyzing

[a]tmh.eee@rmkec.ac.in, [b]samuelfranklin2k3@gmail.com, [c]asokankarunanidhi12@gmail.com, [d]christopherindla4@gmail.com

DOI: 10.1201/9781003684589-39

voltage and current fluctuations, these models can distinguish fault types with high precision [3]. As the power grid grows more complex, incorporating advanced learning techniques into fault detection systems offers a capable solution for refining grid reliability and minimizing downtime.

Literature Review

Much research has been conducted on finding the flaws in power transmission networks, with various strategies proposed to enhance precision and efficiency. Early investigations primarily focused on utilizing wavelet singular data as key parameters for fault classification using support vector machines (SVM) [4]. This approach demonstrated the potential of machine learning in identifying faults under diverse operational conditions. To address challenges related to over voltages, high currents, and outages, researchers such as Recioui et al. [5] applied the k-nearest neighbors (KNN) algorithm for fault both detection and classification also in the localization, thereby facilitating rapid fault clearance and minimizing disruptions. Similarly, Chen et al. [6] developed a convolutional sparse autoencoder for fault detection in transmission lines, leveraging autonomous feature extraction from voltage and current datasets to enhance diagnostic capabilities.

Artificial neural networks (ANNs) have proven to be highly effective in recognizing patterns, making them a crucial tool in fault detection. Jamil et al. [1] examined the application of ANNs for fault classification in power transmission lines, achieving 78.1% of accuracy. In contrast, Leh et al. [7] employed a 14-bus system to develop a compact dataset of 1,000 data points. Despite their optimization efforts, the classification accuracy remained at 70%, highlighting the difficulty of achieving high reliability in fault detection.

Advancements in ANN-based techniques were further demonstrated by Amiruddin et al. [8], who implemented a multilayer perceptron (MLP) model that improved classification accuracy to 78%. Building upon this, Fahim et al. [9] investigated the identification of common fault types, such as line-to-ground (LG), double line-to-ground (LLG), and three line-to-ground (LLLG) faults, achieving an enhanced accuracy of 84.5%. These findings underscore the growing potential of neural networks in fault classification.

Different methods have been explored for defect identification in addition to ANNs. Ponukumati et al. [10] used a cubic SVM model, attaining an accuracy of 87.1% in diagnosing faults in electrical distribution systems. Similarly, Tonello et al. [11] applied machine learning methodologies to identify anomalies in power line connections, achieving an impressive accuracy of 89.2%.

The adoption of deep learning has further strengthened fault detection in power systems. Xuebin et al. [12] introduced a deep belief network (DBN) for detecting faults in underground distribution cables, successfully identifying nine different types of faults with an accuracy of 97.8%. Another notable study by Rajesh et al. [13] proposed a hybrid model that integrates the recurrent perceptron neural network (RPNN) with the HUA and truncated single value decomposition (TSVD). The goal of this strategy was to improve power transmission systems' prediction accuracy and fault classification.

This paper offers a thorough examination of many methods for identifying transmission network faults. The results validate how these algorithms can improve diagnostic accuracy and dependability. The research also evaluates the impact of several models on categorization results, highlighting each model's advantages and disadvantages. In the end, ensemble-based techniques proved to be the best approach for guaranteeing precise and prompt problem identification in power transmission systems.

Methodology

Data collection

In this study, the dataset was simulated and generated using a MATLAB model (Figure 39.1). The simulated power system consists of four generators, each supplying 11 kV, positioned at both ends of the transmission line, with transformers placed at the midpoint. This setup effectively replicates real-world fault conditions, including short-circuit faults at various locations. The faults are categorized as symmetrical and asymmetrical, with system impedance and transient conditions accounted for to ensure realistic modeling.

12,002 data occurrences with six essential characteristics, such as voltage and current readings of all three phases at various transmission points, make up the dataset. L-L, L-L- L,L-G,L-L-G and L-L-L-G are among the several types of faults. Interpolation was

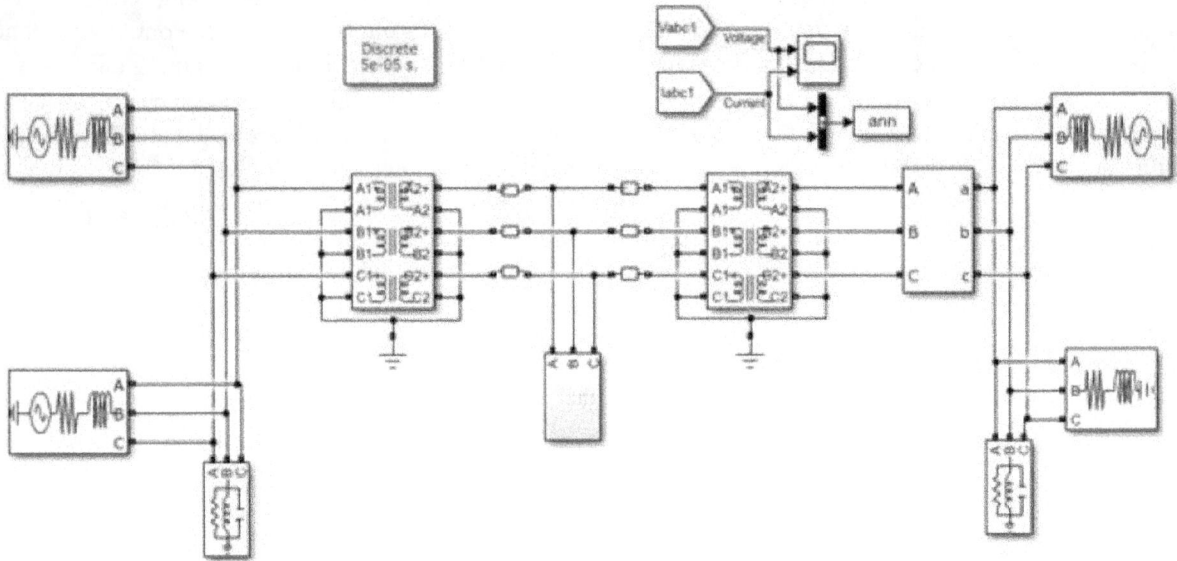

Figure 39.1 MATLAB simulation circuit
Source: Author

used to manage missing values, and noise filtering was used to eliminate abnormalities in order to preserve data integrity. To provide an organized data set for training and evaluating machine learning models, a methodical sampling technique was implemented for balancing fault and non-fault occurrences.

Data preprocessing
To enhance the accuracy of machine learning models and efficiency is crucial to preprocess the dataset before employing it for fault classification. Proper data preparation ensures completeness, maintains consistency, and ultimately improves the model's overall performance. After applying preprocessing techniques, the dataset was confirmed to be well-structured and free from missing values. Additionally, attributes G, A, B, and C were consolidated into a single feature to enhance algorithmic efficiency.

For fault classification, distinct binary codes were assigned to represent different fault types:

1) 0110 - Line-to-Line (L-L) fault
2) 0111 - Three-Line (L-L-L) fault
3) 1001 - Line-to-Ground (L-G) fault
4) 1011 - Double-Line-to-Ground (L-L-G) fault
5) 1111 - Three-Line-to-Ground (L-L-L-G) fault
6) 0000 - No fault

To optimize machine learning model performance, numerical input features must be standardized. Given the varying scales of numerical data points, normalization was applied to maintain uniformity and prevent biases during training. This step significantly enhances model accuracy and ensures optimal learning performance. Following the preprocessing phase, the dataset has two subsets. Predictive algorithms and machine learning models were trained using the first subset, referred to as the training data. The second subset, termed test data, was later utilized to assess the model's performance, measure its predictive accuracy, and evaluate error rates.

Training process
This work uses a variety of machine learning algorithms to predict transmission line failures, providing dependable solutions for fault examination and diagnosis. An 80:20 proportion between train and test was employed for both model training and validation. Grid search was used for hyperparameter tuning in order to refine model parameters, while cross-validation was used to prevent overfitting.

1) Logistic regression: This algorithm is utilized for multiclass classification by estimating the probability of each class and determining the most likely outcome based on input variables.

Due to its interpretability, it provides insights into how different features influence fault classification in power lines.

2) Support vector machine (SVM): SVM works by determining the optimal hyper-plane to effectively separate fault categories while maximizing the margin between them. It may be applied to both linear and non-linear classification problems, and it is particularly effective in high-dimensional areas where it reduces over-fitting.

Multi-layer perceptron (MLP): MLP is based on the neural structure of the human brain and is designed to manage non-linear relationships in data. An input layer, multiple hidden layers for spotting complex patterns, and an outcome layer for predicting make up this system. This architecture enables real-time learning and increases the precision of defect detection in gearbox systems.

Evaluation metrics

A range of performance metrics were used to evaluate the significance of the suggested fault detection strategy. The confusion matrix, F1-score, recall, accuracy, & precision are some of these metrics. Every metric offers valuable information about different facets of the model's ability to detect gearbox line problems.

1) Accuracy: This metric displays the proportion of examples that the model correctly classified. It is found as the sum of the number of true positives (TP) and true negatives (TN) divided by the total number of occurrences (TP + TN + FP + FN). Accuracy is a crucial indicator of the model's overall performance.

2) Precision: The percentage of accurately recognized fault occurrences among all instances projected to be faults is known as precision. It aids in comprehending how well the model detects real issues while reducing false positives.

A more reliable classification is indicated by a higher precision rating.

3) Recall (Sensitivity): This score verifies how well the model can identify real errors. When compared to all actual fault occurrences, it calculates the proportion of accurately diagnosed fault incidents. Even if some non-fault cases are incorrectly identified, a greater recall value signifies superior detection capabilities.

4) F1-score: These metrics provide a single assessment value by balancing recall and precision. When dealing with unnecessary datasets where one class is noticeably more abundant than another it is very helpful. A model that performs well and strikes a good balance between defect identification and false positive reduction is indicated by a higher F1-score.

5) Confusion matrix: A detailed considerate of the model's classification performance is resulted by the confusion matrix. It helps identify trends in misclassification and areas that require improvement by describing false positives, false negatives, true positives, and true negatives. This diagnostic tool is the sole way to increase the model's efficiency and accuracy.

Evaluation of Results

Table 39.1, which compares the training and testing accuracies of the assessed models, provides a synopsis of the study's findings. A graphic depiction of their performance is provided in Figure 39.2, which shows how accuracy varies throughout the training and assessment periods. Additionally, the MLP model's confusion matrix, which is shown in Figure 39.3, offers a thorough analysis of its classification results, emphasizing how well it detects faults.

With a train accuracy of 99.65% and a test accuracy of 99.61%, the MLP model had the best accuracy among the classifiers under investigation. These findings imply that MLP generalizes well across the

Table 39.1 Performance comparison of the models.

Model	Training accuracy %	Model accuracy score %	Precision	Recall	F1-score
SVM	76.06	75.96	0.72	0.75	0.73
Logistic regression	90.29	90.40	0.94	0.90	0.88
MLP classifier	99.65	99.61	0.99	0.99	0.99

Source: Author

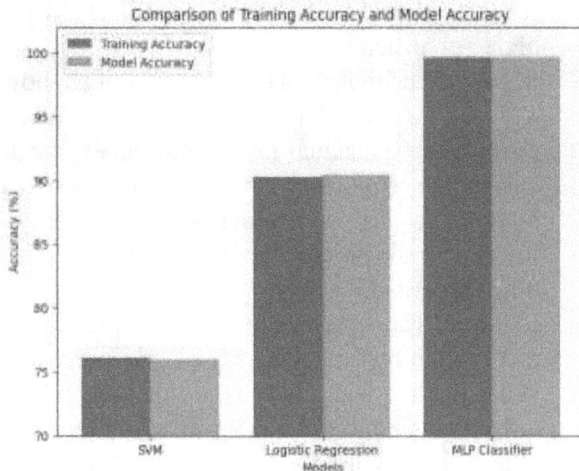

Figure 39.2 Comparison of the models
Source: Author

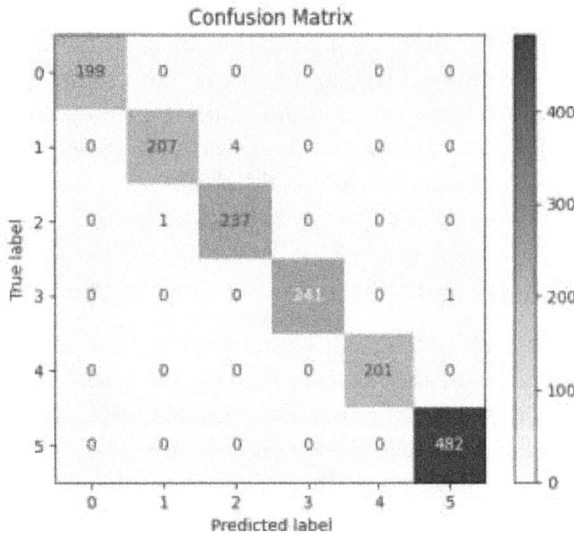

Figure 39.3 Confusion matrix of MLP classifier
Source: Author

dataset and efficiently learns intricate data patterns. The model's robustness in correctly recognizing defects is highlighted by its 0.99 precision, recall, and F1-score. MLP is a very dependable option for fault detection in power systems because of its ability of capturing complex correlations between input features, which greatly enhanced its forecast accuracy.

Logistic regression, while not as exact as MLP, performed admirably with a training accuracy of 90.29% and a test accuracy of 90.40%. Having a precision of 0.94 and recall of 0.90, the model demonstrated a high degree of accuracy in fault

identification. However, the little discrepancy between recall and accuracy suggests occasional misclassifications. Despite this limitation, the logistic regression method is still a good choice for fault classification workloads where computing speed is more crucial than precise accuracy.

The SVM classifier was the least accurate of the three models at the bottom end of the performance spectrum. With 75.96% of test accuracy and a training accuracy of 76.06%, SVM struggled to find the various fault categories in the dataset. Its recall and precision values, which are 0.75 and 0.72, respectively, further highlight the challenges of identifying intricate fault patterns. SVM's lower accuracy can be attributed to its sensitivity to high-dimensional feature spaces and imbalanced class distributions. The model struggled to separate overlapping fault categories due to its reliance on linear decision boundaries, making it less effective for complex, non-linear data patterns encountered in power grids.

The MLP model's classification reliability is further demonstrated by the confusion matrix (Figure 39.3), which shows very few cases of misclassification for different fault kinds. Furthermore, a comparison analysis of the models' accuracy is shown in Figure 39.2, demonstrating MLP's higher performance over LR and SVM. This visual representation underscores the efficiency of deep learning approaches in fault classification while demonstrating the practical viability of simpler models in specific applications.

The results of the study reaffirm MLP's effectiveness in fault classification, particularly in handling non-linear and complex data structures. While LR offers a less sophisticated alternative, it remains a viable choice in computationally constrained environments. Conversely, SVM's limitations suggest that additional optimizations, such as improved feature selection and hyperparameter tuning, may be necessary to enhance its classification accuracy. Future work could explore ensemble learning methods such as Random Forest or Gradient Boosting, which combine multiple classifiers to improve robustness and fault classification accuracy. Hybrid approaches integrating deep learning with rule-based systems may also enhance performance in complex fault scenarios. model's difficulty in capturing complex data patterns. The relatively lower accuracy of SVM suggests that its efficiency was hindered by the dataset's high dimensionality, resulting in suboptimal classification outcomes.

Conclusion

The efficacy of many machine learning systems for identifying defects in power systems is evaluated in this work utilizing F1-score, accuracy, precision, and recall as assessment metrics. The multi-layer perceptron (MLP), one of the models examined, showed the best accuracy and successfully represented intricate, non-linear fault patterns. These results demonstrate that MLP is a very dependable method for classifying faults, especially in situations requiring accuracy and flexibility.

In contrast, LR demonstrated a reasonable level of accuracy, which makes it a viable option for less complex classification issues where interpretability of the model and computing efficiency are crucial considerations. The support vector machine (SVM), however, struggled with high-dimensional fault scenarios, indicating a need for feature selection optimization and hyperparameter tuning to improve its classification capability.

Practical deployment of these models poses challenges such as computational limitations in embedded systems, adapting to real-time data variations, and maintaining accuracy under evolving grid conditions. Future enhancements should focus on online learning frameworks that allow models to continuously refine their predictions based on new fault patterns. Additionally, hybrid techniques merging both deep learning with traditional classifiers may improve fault detection accuracy and efficiency. Integrating these models into SCADA systems could enhance predictive maintenance, fault isolation, and overall grid stability, ensuring a more resilient power transmission network.

References

[1]   Jamil, M., Sharma, S. K., & Singh, R. (2015). Fault detection and classification in electrical power transmission system using artificial neural network. *Springer Plus*, 4, 1–13.

[2]   Zhao, R., Yan, R., Wang, J., & Mao, K. (2017). Learning to monitor machine health with convolutional bi-directional lstm networks. *Sensors*, 17, 273. Terborgh, J. (2009). Preservation of natural diversity. *BioScience*, 24, 715–22.

[3]   An, Q., Tao, Z., Xu, X., Mansori, M. E., & Chen, M. (2020). A data-driven model for milling tool remaining useful life prediction with convolutional and stacked lstm network. *Measurement*, 154, 107461.

[4]   Yang, H., Mathew, J., & Ma, L. (2005). Fault diagnosis of rolling element bearings using basis pursuit. *Mechanical Systems and Signal Processing*, 19, 341–356.

[5]   Recioui, A., Benseghier, B., & Khalfallah, H. (2016). Power system fault detection, classification and location using the k-nearest neighbors. In 2015 4th International Conference on Electrical Engineering, ICEE 2015, (p. 2).

[6]   Magesh, T., Thiyagesan, M., Gokul, T., Darshan, C. R. A., & Bhavesh, K. (2023). Fault detection using IoT in essential power transmission lines. In 2023 Intelligent Computing and Control for Engineering and Business Systems (ICCEBS), (pp. 1–6). IEEE.

[7]   Chen, Y. Q., Fink, O., & Sansavini, G. (2018). Combined fault location and classification for power transmission lines fault diagnosis with integrated feature extraction. *IEEE Transactions on Industrial Electronics*, 65, 561–569.

[8]   Leh, N. A. M., Zain, F. M., Muhammad, Z., Hamid, S. A., & Rosli, A. D. (2020). Fault detection method using ann for power transmission line. In Proceedings - 10th IEEE International Conference on Control System, Computing and Engineering, ICCSCE 2020, (pp. 79–84).

[9]   Amiruddin, A. A. A. M., Zabiri, H., Taqvi, S. A. A., & Tufa, L. D. (2020). Neural network applications in fault diagnosis and detection: an overview of implementations in engineering-related systems. *Neural Computing and Applications*, 32, 447–472.

[10] Fahim, S. R., Sarker, Y., Islam, O. K., Sarker, S. K., Ishraque, M. F., & Das, S. K. (2019). An intelligent approach of fault classification and localization of a power transmission line. In 2019 IEEE International Conference on Power, Electrical, and Electronics and Industrial Applications, PEEIA- CON 2019, (pp. 53–56).

[11] Ponukumati, B. K., Sinha, P., Maharana, M. K., Kumar, A. V. P., & Karthik, A. (2022). An intelligent fault detection and classification scheme for distribution lines using machine learning. *Engineering, Technology Applied Science Research*, 12, 8972–8977.

[12] Tonello, A. M., Letizia, N. A., Righini, D., & Marcuzzi, F. (2019). Machine learning tips and tricks for power line communications. *IEEE Access*, 7, 82434–82452.

[13] Rajesh, P., Kannan, R., Vishnupriyan, J., & Rajani, B. (2022). Optimally detecting and classifying the transmission line fault in power system using hybrid technique. *ISA Transactions*, 130, 253–264.

40 Utilizing 3D-CNN and autoencoder for gas detection in hyperspectral images

P. Durga Anusha[1,a], M. Sai Maheswari[1,b], N. Rajashekhar[1,c], Sk. Venkatesh[1,d], E. Shalini[2,e] and P. Aashish Nikhil Prem Chand[1,f]

[1]UG Student, Department of CSE (AI&DS), Vishnu Institute of Technology, Bhimavaram, Andhra Pradesh, India

[2]Assistant Professor, Department of CSE (AI&DS), Vishnu Institute of Technology, Bhimavaram, Andhra Pradesh, India

Abstract

Both human and environmental health depend on accurate gas emission detection. Hyperspectral image analysis, particularly in the long-wave infrared (LWIR) band, is proving to be an increasingly successful method for remote gas identification. While the existing methods primarily focus on spectral unmixing and classification using 3D convolutional neural networks (3D-CNN) and autoencoders, they often overlook advanced feature enhancement techniques that can improve detection accuracy. We propose an ensemble model that integrates CNN, bi-directional, and gated recurrent units (GRU) to optimize feature extraction and enhance prediction accuracy. Unlike the baseline model, which directly processes radiance data, our approach utilizes pre-extracted features obtained from methane monitoring datasets in place of hyperspectral images. The proposed ensemble framework effectively combines spatial, temporal, and sequential dependencies to improve feature learning. Experimental results demonstrate that the ensemble model outperforms traditional 3D-CNN and autoencoder-based methods by providing higher sensitivity and precision in detecting methane and sulphur dioxide gases. This work highlights the importance of leveraging ensemble learning techniques to achieve superior gas detection performance, even in the absence of raw hyperspectral images.

Keywords: 3D-CNN, autoencoder, bi-directional GRU, CNN, deep learning, ensemble model, feature extraction, feature optimization, gas detection, hyperspectral imaging, methane monitoring, radiance unmixing, remote sensing, spectral angle mapper, sulphur dioxide detection

Introduction

Detecting petrol emissions is an important step in keeping an eye on pollution levels and making sure everyone stays safe. The capacity of hyperspectral photography, especially in the longwave infrared (LWIR) spectrum, to remotely identify and analyze gases based on their spectral fingerprints has garnered a lot of attention. Adaptive cosine estimator (ACE) and spectral angle mapper (SAM) are two examples of spectral unmixing methods used in conventional methods. These methods separate gas emissions using luminance-temperature and radiance data. The precision and dependability of these approaches in real-world applications are limited though, since they typically fail to handle complicated mixes of gases and background interference properly.

Auto-encoders and 3D-CNN are two examples of deep learning-based techniques that have recently been investigated as potential solutions to these shortcomings; these networks are capable of powerful feature extraction and unmixing. By analyzing hyperspectral pictures for spatial and spectral patterns, these methods enhance gas detection. While these current approaches show promise, they aren't optimized for performance because they don't use modern feature augmentation techniques. Raw hyperspectral photos are also crucial to most models, but they aren't always accessible, which is a real limitation to their broad use.

To improve the gas detection framework's feature extraction and prediction accuracy, we provide an ensemble Model that integrates CNN, bi-directional, and gated recurrent units (GRU). This model extends

[a]21pa1a5492@vishnu.edu.in, [b]21pa1a5465@vishnu.edu.in, [c]21pa1a5474@vishnu.edu.in, [d]21pa1a5498@vishnu.edu.in, [e]shalini.e@vishnu.edu.in, [f]21pa1a5486@vishnu.edu.in

DOI: 10.1201/9781003684589-40

the existing 3D-CNN and autoencoder-based framework. We use pre-extracted characteristics from the Methane Monitoring website instead of depending just on hyperspectral photos, which makes the strategy more practical and customizable. To provide a more thorough and precise detection mechanism, the suggested ensemble architecture uses convolutional layers to learn spatial features, bi-directional processing to understand sequential relationships, and GRUs to understand temporal patterns.

When compared to the baseline 3D-CNN and Autoencoder approaches, experimental evaluations show that the ensemble model considerably improves performance metrics, including sensitivity and accuracy. This paper presents a scalable method for identifying gases such as methane and sulphur dioxide, even in situations when raw hyperspectral pictures are not accessible. It also emphasizes the potential of ensemble learning approaches in optimizing gas detection systems.

Literature Survey

i) Imaging spectroscopy and the airborne visible/infrared imaging spectrometer (AVIRIS)
https://www.sciencedirect.com/science/article/abs/pii/S0034425798000649 [1][15]

One innovative approach to Earth remote sensing that is quickly gaining traction is imaging spectroscopy. At 10-nanometer intervals, the airborne visible/infrared image spectrometer mainly tracked the sun's reflected spectra from 400 to 2500 nm. Both the signal-to-noise ratio and the precision of the AVIRIS calibration are top-notch. In the past few years, AVIRIS, together with scientific study and its practical applications, have made tremendous strides. Data system, sensor, calibration, and flight operation documentation are included in the AVIRIS system's initial design and updates. Science research and applications that employ data from the preceding several years are contextualized by this update on AVIRIS' characteristics. A few examples of recent scientific research and applications include: spectral algorithms, human infrastructure, atmospheric correction, biomass burning, environmental hazards, geology and soils, hydrology of snow and ice, inland and coastal waters, the atmosphere, and satellite simulation and calibration.

ii) Hyperspectral push-broom microscope development and characterization
https://www.researchgate.net/publication/335435752_Hyperspectral_Push-Broom_Microscope_Development_and_Characterization [2][17]

The use of hyperspectral imaging (HSI) to analyze samples at the microscopic level is gaining traction in a number of industries. The superior spectral resolution and range utilization capabilities of push-broom hyperspectral (HS) cameras make them the preferred HSI technology. To get HS data, however, microscope equipped with push-broom cameras need precise spatial scanning of the specimen. In this article, we highlight the steps necessary to set up a push-broom HS microscope for the best possible photographs. We begin with an innovative mechanical system that is 3D printed and uses linear motion of the microscope stage to accomplish spatial scanning. The effects of maximizing dynamic range, focusing, aligning, and determining speed on picture quality are then detailed. We wrap off with a number of high-resolution images captured by push-broom cameras of the most common defects, as well as images taken from actual microscopic samples.

iii) Enhanced gas detection in hyperspectral images with 3 CNN and autoencoder models
https://ijcrt.org/papers/IJCRT2405240.pdf [3]

Gas emission monitoring is an important issue for both human and environmental health, and this new project is tackling it head-on. The limitations of traditional detection methods have prompted researchers to turn to hyperspectral image analysis in search of more effective and secure alternatives. This research presents a deep learning approach that combines unmixing and classification to identify hyperspectral gases in the longwave infrared spectrum. Transforming radiance data into luminance-temperature data using an autoencoder and a 3-D convolutional neural network improves performance compared to earlier approaches. A further innovation is an Ensemble model that augments input features to boost prediction accuracy. This model blends CNN, bi-directional, and GRU algorithms. This one-of-a-kind endeavor exemplifies how modern approaches have the potential to address environmental issues.

iv)   Algorithms for chemical detection, identification and quantification for thermal hyperspectral imagers
https://www.researchgate.net/publication/252965690_Algorithms_for_chemical_detection_identification_and_quantification_for_thermal_hyperspectral_imagers [4][20]

Gaseous substance detection, identification, and quantification at a distance is crucial in many fields. Sensors in a small, durable field package with great sensitivity, few false alarms, and the ability to operate in real time are required for these applications. Thermal infrared spectrometers and imagers have found use as chemical sensors. High-speed, large-format infrared imaging arrays allow chemical sensors to accurately measure spectral, spatial, and temporal properties. Data from spatial and spectral analyses suggests that passive chemical detection, identification, and quantification might be substantially enhanced. Methods for detection, identification, and quantification using thermal infrared hyperspectral imaging are detailed in this research. The data cubes containing information on gaseous discharges are utilized by these algorithms using the field-based Telops FIRST imaging spectrometer.

v)   Hyperspectral gas and polarization sensing in the LWIR: Recent results with MoDDIFS
https://www.researchgate.net/publication/320821558_Hyperspectral_gas_and_polarization_sensing_in_the_LWIR_Recent_results_with_MoDDIFS [5][16]

Passive detection and identification of vapor emissions and surface contaminants is possible with imaging Fourier-transform infrared (FTIR) spectroscopy. It is possible to remotely monitor illicit factories using FTIR imaging by military and security groups. DRDC Valcartier's MoDDIFS imaging Fourier transform infrared sensor is undergoing development and testing for use in remote sensing. The proposed approach combines the high spatial resolution of hyperspectral imaging with the clutter suppression of differential detection. Using the MoDDIFS sensor, you may create up a system for remote gas detection and surface contamination polarization sensing. This paper reviews the results of the most recent MoDDIFS passive standoff gas and liquid contamination detection.

Develop, evaluate, and validate GLRT-type detection techniques using hyperspectral measurements of difluoroethane, diethyl ether, and SF96 gases and liquids. For the purpose of discussing detection outcomes, GLRT detection characteristics are utilized.

Methodology

Proposed undertaking
The proposed system introduces an ensemble model that enhances gas detection accuracy by combining convolutional neural networks (CNN), bi-directional networks, and gated recurrent units (GRU) for feature extraction and prediction. Unlike the baseline approach, which relies solely on 3D-CNN and autoencoders for unmixing and classification, the ensemble model optimizes input features through multiple layers of processing to capture spatial, sequential, and temporal patterns effectively.

In this system, pre-extracted features, obtained from the methane monitoring website, serve as input instead of raw hyperspectral images, making it adaptable for scenarios where image data is unavailable. First, the CNN module processes the spatial features of the input data to extract local patterns and high-level abstractions. Next, the bi-directional network enhances the feature representation by capturing dependencies from both forward and backward directions, improving the detection of complex patterns. Finally, the GRU module handles sequential dependencies and temporal variations, enabling the system to model time-sensitive changes in gas emission levels effectively.

Design of the system
The architecture of the proposed system is designed to enhance gas detection accuracy by integrating multiple machine learning techniques into an ensemble framework. It consists of three primary stages—feature input processing, ensemble model processing, and classification output—which work together to optimize feature extraction and improve detection performance.

The raw hyperspectral images are replaced with pre-extracted features from the methane monitoring website during the feature input processing stage. Features such as these display LWIR spectra of gases of interest, such as sulphur dioxide and methane. Deep learning modules benefit from input data that has been pre-processed by removing noise and normalizing values.

The ensemble model processing stage forms the core of architecture. It begins with a CNN, which extracts spatial features and patterns from the input data, capturing high-level feature maps. These features are then passed through a bi-directional network, which enhances the data representation by analyzing both forward and backward dependencies. This step is crucial for modeling complex relationships between features, enabling the system to handle variations in emission patterns. The processed data is further refined using a GRU, which captures sequential dependencies and temporal variations, ensuring robust performance in detecting patterns over time.

Classifying ensemble model features using fully connected layers is the last step in the classification output process. Data samples containing target gases are identified and labelled by layers using optimal features. The results show that the concentrations of methane and sulphur dioxide are right on target.

Implementation

Modules

Data collection and preprocessing

- Collect hyperspectral image data or extracted features from the methane monitoring website.
- Normalize and scale the data for uniform processing.
- Split the data into training, validation, and test datasets.
- Augment the dataset to simulate variations in environmental conditions Figure 40.1.

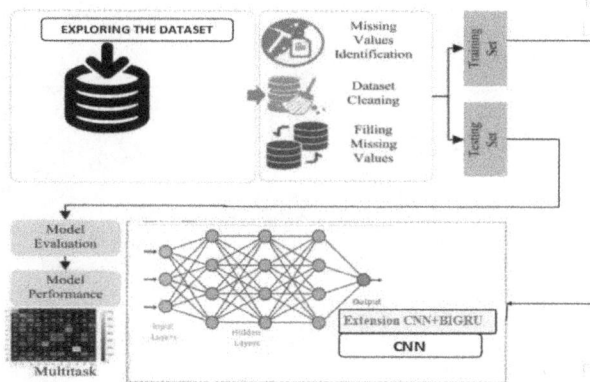

Figure 40.1 Proposed architecture
Source: Author [1]

Here, we used available methane and sulphur leak dataset from the website below:

https://studio.edgeimpulse.com/public/158034/latest

Feature extraction

- Use hyperspectral imaging features, including radiance, luminance, and temperature data.
- Employ the spectral angle mapper (SAM) to compute pixel distance values.
- Map SAM results in gas types using the NIST database.
- Optimize features by applying dimensionality reduction techniques like principal component analysis (PCA).

Ensemble model development

- Design and implement three parallel algorithms: CNN, bi-directional LSTM, and GRU.
- Train CNN for spatial feature extraction from hyperspectral data.
- Use Bi-directional LSTM for temporal dependencies in the data.
- Integrate GRU for sequence learning and noise reduction.
- Combine predictions of all three models using an ensemble learning approach.

Autoencoder-based unmixing

- Design an autoencoder for unsupervised feature learning.
- Train the encoder to extract latent features representing gas emission patterns.
- Use the decoder to reconstruct input data and detect anomalies.
- Optimize the network to improve reconstruction accuracy.

Classification network

- Final classification requires the construction of a fully connected neural network with three layers.
- To begin with, feed the network the abundance values and endmember spectra.
- Train the network with labeled data to predict the presence of specific gases.
- Use recall, precision, and accuracy as metrics to assess forecasts.

Model evaluation and testing

- Test the model on independent datasets for robustness.

- Compare results with existing methods like SAM and ACE.
- Perform ablation studies to analyze the contribution of individual modules.
- Use confusion matrices and F1-scores to evaluate classification performance.

Deployment and visualization

- Develop a user-friendly interface for gas detection visualization.
- Integrate geographic mapping tools for spatial representation of emissions.
- Deploy the model for real-time processing and prediction.
- Generate reports for regulatory compliance and monitoring purposes.

Algorithms
Convolutional neural network

Images and other grid-like data types are processed using the CNN deep learning system. Using convolution layers, it learns spatial hierarchies and extracts characteristics from incoming data. Filters, sometimes known as kernels, are used to create feature maps from input pictures. These maps show the most important features, such as edges, textures, and more. CNN is effective for tasks like image classification, object detection, and, in this case, gas detection in hyperspectral images. In the context of the extension, CNN can process the hyperspectral image features, enabling the model to extract essential patterns that are crucial for identifying gas emissions, especially when paired with bi-RNNs or other sequence-based models.

Extension concept: CNN + BiGRU model

An improved hyperspectral gas detection model is offered, which makes use of CNNs and BiGRUs. CNN is able to learn crucial structures and patterns from hyperspectral picture data, including gas spectral characteristics and edges. A possible way for CNN layers to identify gas leaks that are dispersed throughout space is by recording spatial hierarchies. The BiGRU model employs spatial characteristics extracted from the CNN to discern temporal correlations, as hyperspectral images are often composed of a series of frames or time-series data. In order to learn from both past and future temporal contexts, BiGRU, a bidirectional RNN, examines input in both directions. The model's improved ability to detect

fuel emissions and temporal variations is a result of its dual-directional learning. Improved gas identification might be possible with the help of CNN's spatial processing skills and BiGRU's temporal connections applied to hyperspectral data. This change should strengthen the model and make it more efficient by assessing complex geographical and temporal data, which should increase the accuracy of predictions Figure 40.2-40.9.

Experimental Results

i) *Precision:* Accuracy is measured by precision when it comes to positively classifying instances or samples. The correctness formula is:

$$Precision = TP/(TP + FP)$$

$$Precision = \frac{True\ Positive}{True\ Positive + False\ Positive}$$

ii) *Recall:* A model's recall in machine learning is a measure of how well it can identify all occurrences of a crucial class. The completeness of a model in capturing instances of a class is

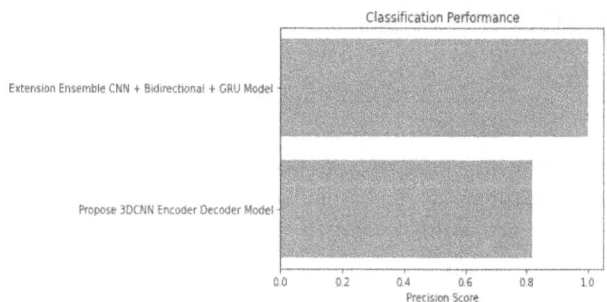

Figure 40.2 Precision comparison graph
Source: Author [2]

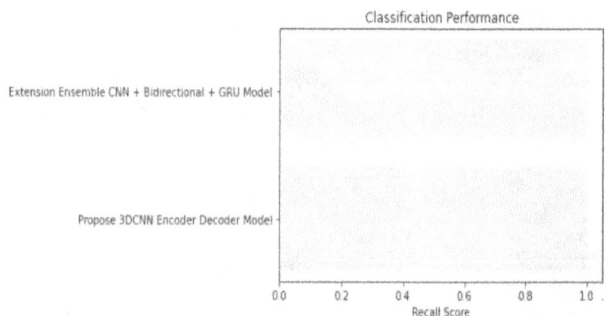

Figure 40.3 Recall comparison graph
Source: Author [3]

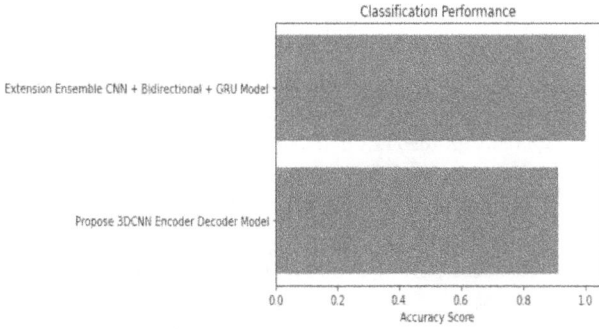

Figure 40.4 Accuracy graph
Source: Author [4]

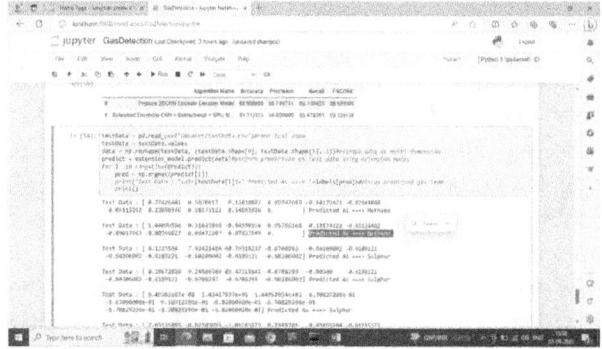

Figure 40.5 F1 score
Source: Author [5]

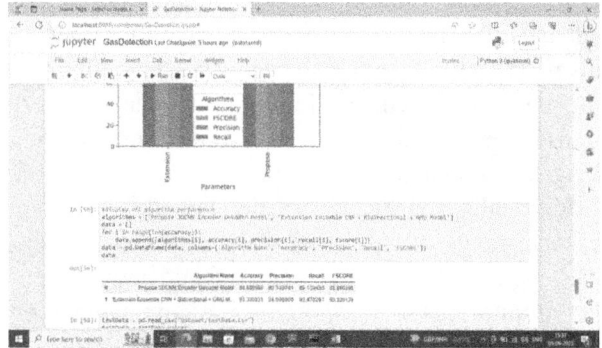

MODEL NAME	ACCURACY	PRECISION	RECALL	F1-SCORE
PROPOSE 3D-CNN ENCODER DECODER MODEL	0.911	0.818	1.0	0.9
EXTENSION ENSEMBLE CNN+BIDIRECTINAL +GRU	1.00	1.00	1.0	1.0

Figure 40.6 Performance evaluation
Source: Author [6]

Figure 40.7 Dataset
Source: Author [7]

Figure 40.8 Results
Source: Author [8]

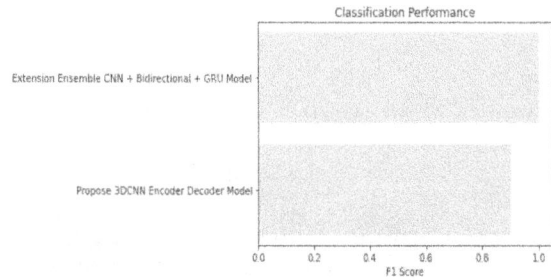

Figure 40.9 Accuracy results
Source: Author [9]

demonstrated by comparing the total number of positive observations with the number of precisely predicted ones.

$$Recall = \frac{TP}{TP + FN}$$

iii) *Accuracy:* An indicator of a model's accuracy is the proportion of correct categorization predictions.

$$Accuracy = \frac{TP + TN}{TP + FP + TN + FN}$$

iv) *F1 Score:* Use the F1 Score, a harmonic mean of recall and accuracy, to equalize false positives and negatives if your dataset is not uniform.

$$F1\ Score = 2 * \frac{Recall \times Precision}{Recall + Precision} * 100$$

Conclusion

The proposed extension model, combining CNN and BiGRU, demonstrated superior performance in gas detection using hyperspectral image data. Using convolutional neural networks (CNNs) for spatial feature extraction and bidirectional GRASP for temporal dependencies, the model outperformed SAM in detecting emissions of methane and sulphur dioxide. The integration of spatial and temporal processing enabled the detection of complex patterns, improving prediction reliability under varying environmental conditions. This hybrid approach highlights the potential of deep learning techniques in enhancing gas leak detection systems, making them more efficient and scalable for real-world applications.

Future Scope

The proposed CNN + BiGRU model can be further enhanced by integrating advanced attention mechanisms to focus on critical spectral features, improving detection accuracy. Future work may explore transformer-based architectures for better handling long-range dependencies in hyperspectral data. Additionally, the model can be expanded to detect a wider range of gases by incorporating more comprehensive datasets from industrial and environmental monitoring sources. Real-time implementation using edge devices or IoT frameworks can also be developed to enable on-site gas leak detection. Furthermore, integrating explainable AI techniques can enhance interpretability, helping researchers and industries better understand the model's decision-making process.

References

[1] Greenetal, R. O., (1998). Imaging spectroscopy and the airborne visible/infrared imaging spectrometer (AVIRIS). *Remote Sensing of Environment*, 65(3), 227–248.

[2] Govender, M., Chetty, K., & Bulcock, H. (2007). A review of hyperspectral re mote sensing and its application in vegetation and water resource studies. *Water Sa*, 33(2), 145–151.

[3] Foucher, P. Y., & Doz, S. (2023). Real time gas quantification using thermal hyperspectral imaging: Ground and airborne applications. Accessed Jan. 18, 2023. [Online]. Available from: https://www. sto.nato.int/publications/ STO%20Meeting%20Proceedings/STO-MPSET-277/MP-SET-277 18.pdf.

[4] Vallières, A., Villemaire, A., Chamberland, M., Belhumeur, L., Farley, V., Giroux, J., et al. (2005). Algorithms for chemical detection, identification and quantification for thermal hyperspectral imagers. In Proceedings Chemical and Biological Standoff Detection III, (Vol. 5995, pp.147–157). Art. no. 59950G.

[5] Thériault, J. M., Fortin, G., Bouffard, F., Lavoie, H., Lacasse, P., & Lévesque, J. (2013). Hyperspectral gas and polarization sensing in the LWIR: recent results with MoDDIFS. In Proceeding 5th Workshop Hyperspectral Image Signal Process.: Evolution in Remote Sensing, (pp. 1–4).

[6] Messinger, D. W., (2004). Gaseous plume detection in hyperspectral images: A comparison of methods. In Proceeding Algorithms and Technologies for Multispectral, Hyperspectral, Ultraspectral Imagery X, (Vol. 5425, pp. 592–603).

[7] Kastek, M., Piatkowski, T., Dulski, R., Chamberland, M., Lagueux, P., & Farley, V. (2012). Method of gas detection applied to infrared hyperspectral sensor. *Photonics Letters of Poland*, 4(4), 146–148.

[8] Omruuzun, F., & Cetin, Y. Y. (2015). Endmember signature based detection of f lammable gases in LWIR hyperspectral images. In Proceeding Advanced Environmental, Chemical, and Biological Sensing Technologies XII, (Vol. 9486, pp. 168–176).

[9] Funk, C. C., Theiler, J., Roberts, D. A., & Borel, C. C. (2001). Clustering to improve matched filter detection of weak gas plumes in hyperspectral thermal imagery. *IEEE Transactions on Geoscience and Remote Sensing*, 39(7), 1410–1420.

[10] Pogorzala, D. R., Messinger, D. W., Salvaggio, C., & Schott, J. R. (2004). Gas plume species identification by regression analyses. In Proceedings Algorithms and Technologies for Multispectral, Hyperspectral, Ultraspectral Imagery X, (Vol. 5425, pp. 583–591).

[11] Robey, F. C., Fuhrmann, D. R., Kelly, E. J., & Nitzberg, R. (1992). A CFAR adaptive matched filter detector. *IEEE Transactions on Aerospace and Electronic Systems*, 28(1), 208–216.

[12] Spisz, T. S., Murphy, P. K., Carter, C. C., Carr, A. K., Vallières, A., & Chamberland, M. (2007). Field test results of standoff chemical detection using the FIRST. In Proceedings of Chemical and Biological Sensors VIII, (Vol. 6554).

[13] Sagiv, L., Rotman, S. R., & Blumberg, D. G. (2008). Detection and identification of effluent gases by long wave infrared (LWIR) hyperspectral images. In Proceedings IEEE 25th Convention of Electrical and Electronics Engineers in Israel, (pp. 413–417).

[14] Hirsch, E., & Agassi, E. (2007). Detection of gaseous plumes in IR hyper spectral images using hierarchical clustering. *Applied Optics*, 46(25), 6368–6374.

[15] Kastek, M., Piatkowski, T., & Trzaskawka, P. (2011). Infrared imaging fourier transform spectrometer as the stand-off gas detection system. *Metrology and Measurement Systems*, 18(4), 607–620.

[16] Kuflik, P., & Rotman, S. R. (2012). Band selection for gas detection in hyper spectral images. In Proceedings IEEE 27th Convention of Electrical and Electronics Engineers in Israel, (pp. 1–4).

[17] Sabbah, S., Harig, R., Rusch, P., Eichmann, J., Keens, A., & Gerhard, J. H. (2012). Remote sensing of gases by hyperspectral imaging: System performance and measurements. *Optical Engineering*, 51(11), 111717.

[18] Öztürk, S., Artan, Y., & Esin, Y. E. (2016). Ethene and CO2 gas detection in hyper spectral imagery. In Proceedings 24th Signal Processing and Communication Application Conference (SIU), (pp. 357–360).

[19] Theiler, J., & Love, S. P. (2019). Algorithm development with on-board and ground-based components for hyperspectral gas detection from small satellites. In Proceedings Algorithms, Technologies, and Applications for Multispectral Hyper spectral Imagery XXV, (Vol. 10986).

[20] Kim, Y. C., Yu, H. G., Lee, J. H., Park, D. J., & Nam, H. W. (2017). Hazardousgas detection for FTIR-based hyperspectral imaging system using DNN and CNN. In Proceedings Electro-Optical and Infrared Systems: Technology and Applications XIV, (Vol. 10433).

41 AI-based audio and music signal processing using deep learning approach

Karpagavalli Komandoor[1,a], Vasanthi Ramapuram[2,b], Sree Sudha Yadav Kommagani[2,c], Subhash Chigicherla[2,d] and Siva Sankar Boya[2,e]

[1]Assistant Professor, Department of CSE, Annamacharya University, New Boyanapalli, Rajampet, Andhra Pradesh, India

[2]Students, Department of CSE, Annamacharya University, New Boyanapalli, Rajampet, Andhra Pradesh, India

Abstract

This project presents an AI-based system designed for detecting and classifying audio and music using deep learning techniques. The system leverages convolutional neural networks (CNNs) to process and analyze audio signals, distinguishing between various audio types such as voice and music. The dataset consists of audio and noise files, which are organized and preprocessed to ensure uniform sampling rates and consistency across all inputs. The dataset for this project is composed of audio samples from various classes and incorporates noise augmentation to mimic real-world environments. The core deep learning model is constructed using a Conv1D architecture, featuring layers that extract relevant features from the audio inputs. The system leverages Fast Fourier Transform (FFT) to convert audio into frequency domain data, enabling efficient pattern recognition essential for classification. The model is trained using a labeled dataset of audio samples, and noise is introduced during training to increase the model's resilience. Methods like early stopping and model checkpointing are employed to optimize the training process and maintain high performance. After training, the system is evaluated on unseen validation data, with results measured through accuracy and loss metrics. Visualizations of model performance across epochs provide valuable insights into its learning progression. This AI-based detection system shows great potential for applications in music recognition, audio classification, and other areas requiring precise audio type detection, providing a powerful tool for automated audio analysis in diverse environments.

Keywords: Accuracy evaluation, AI-based audio detection, audio preprocessing, audio recognition, automated audio analysis, Conv1D model, Fast Fourier Transform (FFT), frequency-domain features, model check pointing, music classification, deep learning, noise augmentation

Introduction

In recent years, AI and deep learning have transformed the field of audio detection and classification, enabling the development of systems that can automatically recognize and differentiate between various audio types, including music, speech, and environmental sounds [1]. This project focuses on leveraging deep learning models, specifically using convolutional neural networks (Conv1D), to analyze audio data and classify it based on frequency-domain features. By transforming raw audio signals into their frequency components through Fast Fourier Transform (FFT), the system extracts meaningful patterns that aid in precise audio classification. This technique allows for a more efficient understanding of the underlying characteristics of sound, making it applicable to a broad range of audio-related applications Figure 41.1 [2]. The dataset used for this project consists of audio samples from different classes and includes noise augmentation to simulate real-world environments. By introducing noise into the training process, the model becomes more robust and adaptable, handling the inevitable variability in audio data outside controlled environments. Training is further refined with early stopping and model checkpointing to prevent overfitting and ensure the best model version is preserved [2, 3]. This AI-based approach to audio detection holds significant potential, offering advancements in fields such as music recognition, automated sound classification, and audio content analysis for media and entertainment industries. In

---

[a]kowsi.valli@gmail.com, [b]ramapuramvasanthi@gmail.com, [c]kommaganisreesudhayadav757@gmail.com, [d]subhashchigicherla1@gmail.com, [e]sivasankar6305@gmail.com

DOI: 10.1201/9781003684589-41

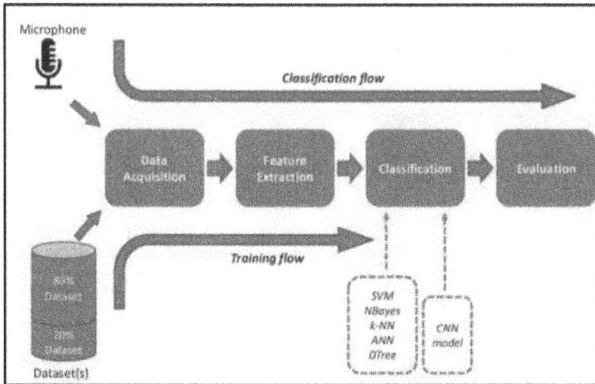

Figure 41.1 System architecture diagram
Source: Author

addition to enhancing audio classification accuracy, this project integrates noise handling mechanisms, making the system resilient to real-world audio disruptions like background chatter, echoes, or interference. The model is trained on various audio clips, including speech and music, with additional noise added during preprocessing to simulate challenging environments. This allows the AI to generalize better when exposed to different soundscapes, ensuring its usability across diverse applications such as voice assistants, music streaming platforms etc [4]

Problem Definition

Audio and music detection, particularly in complex, real-world environments, presents a significant challenge due to the variety of overlapping sounds, background noise, and the wide spectrum of audio events. Accurate detection and classification of music genres, sound events, and speech from mixed audio streams are critical for applications in entertainment, security, health monitoring, and smart systems. However, traditional methods for audio analysis often rely on handcrafted features or shallow learning models, which struggle to generalize effectively across different contexts and environments. The need for a robust, real-time solution that can accurately identify, separate, and classify diverse audio signals in noisy or overlapping conditions has driven research towards more advanced deep learning techniques.

Existing Work

Existing work in audio and music detection has seen a considerable shift towards the use of deep learning, especially with architectures like CNNs, long short-term memory (LSTM) networks, and hybrid models. Researchers like Jansson et al. have employed deep U-Net networks for separating vocals from music accompaniments, while others have focused on sound event detection using CNNs [5], as demonstrated by Koizumi et al. The integration of attention mechanisms and transformer architectures has also shown promise, as highlighted by Kong et al., improving the model's ability to focus on relevant sound segments. While these approaches have achieved notable improvements in accuracy, challenges remain in handling weakly labeled data, real-time processing, and robustness to noise or overlapping sound sources.

Proposed Work

Leveraging the progress in deep learning for audio detection, the proposed work aims to develop a unified, end-to-end model that integrates CNNs for feature extraction and LSTMs for capturing temporal dependencies in audio signals. Additionally, an attention-based mechanism will be employed to further enhance the model's ability to focus on key audio events, reducing the impact of irrelevant noise or overlapping sounds. This hybrid architecture will be optimized for real-time performance, ensuring that it can be deployed in practical applications such as music detection, environmental sound classification, and speech recognition systems. The model will also incorporate novel preprocessing techniques to enhance robustness, such as data augmentation to handle variations in sound conditions and noise reduction strategies to improve clarity in noisy environments. The primary objective is to develop a system that not only attains high accuracy but also operates efficiently in dynamic and unpredictable audio scenarios. Furthermore, the proposed model will be trained on a diverse dataset that encompasses a wide range of audio scenarios, including various musical genres, speech samples, and environmental sounds, to ensure comprehensive learning and generalization. By implementing techniques such as transfer learning and domain adaptation, the model aims to achieve improved performance across different contexts, making it a versatile tool for real-world audio and music detection applications.

Data Collection and Preprocessing

In the realm of audio and music detection, the integrity and quality of the data play an essential role in the model's effectiveness. For this project, data was collected from multiple reliable sources, encompassing a variety of audio types, including music tracks, spoken language samples, and background noises. This diverse dataset ensures that the model can learn to distinguish between different audio classes effectively. Publicly available datasets, such as the Urban Sound dataset and the Free Music Archive, were utilized, as they offer a rich collection of labeled audio samples that span various genres and environmental contexts. Additionally, the inclusion of self-collected audio recordings, captured in different settings, further enhances the dataset's diversity, providing a more robust foundation for training. Once the audio data was gathered, a comprehensive preprocessing pipeline was implemented to prepare it for model training. The first step involved normalizing the audio samples to a uniform sampling rate of 16,000 Hz, guaranteeing consistency across all recordings. This step is critical, as varying sample rates can adversely affect the model's ability to process audio inputs [6]. Following normalization, the audio clips were converted to mono format, as this simplifies the data and reduces complexity without sacrificing important audio information. Next, the audio samples underwent segmentation into shorter clips, enabling the model to learn from smaller, manageable pieces of audio. This segmentation not only boosts training efficiency but also strengthens the model's capability to recognize patterns across different audio segments. Moreover, the integration of data augmentation techniques—such as incorporating background noise, pitch shifting, and time stretching—enabled for the generation of synthetic training examples. This step is particularly valuable in addressing issues of overfitting, as it exposes the model to adapt to a wider array of scenarios and variations. Overall, the meticulous approach to data collection and preprocessing is essential in creating a high-quality dataset that empowers the audio and music detection model to perform optimally in real-world applications Figure 41.2 [6, 7]. Continuing from the preprocessing stage, feature extraction plays a key role in improving the model's understanding of the audio data. In this project, various techniques were employed to extract relevant features from the preprocessed audio clips. Mel-frequency cepstral coefficients (MFCCs) were

Figure 41.2 User Interface
Source: Author

primarily utilized due to their effectiveness in capturing the timbral characteristics of audio signals, which are crucial for distinguishing between different audio classes. Additionally, spectrogram analysis was performed to visualize the frequency spectrum of the audio signals over time, enabling the model to learn patterns related to pitch, tone, and rhythm.

Moreover, zero-crossing rate and spectral centroid features were extracted to provide further insights into the audio's properties. These features help the model recognize nuances in different types of sounds, such as identifying musical notes versus spoken words. The extracted features were then organized into an organized format appropriate for training machine learning models. By leveraging these advanced feature extraction techniques, the project aims to upgrade the model's performance in accurately classifying and detecting various audio types, ultimately leading to more reliable and effective applications in audio recognition and music detection.

System Implementation

The implementation of an AI-based audio and music detection system involves a comprehensive approach that integrates various components, methodologies, and technologies to ensure its effectiveness and efficiency. It starts with a detailed system architecture that outlines the relationships and interactions between different components.

Each of these components plays a critical role in enabling the system to perform complex tasks such as identifying and classifying diverse audio signals. This structured approach ensures that all parts of the system work cohesively, allowing for smooth operation and high performance. In the initial phase of implementation, data acquisition is pivotal. The system collects audio data from multiple sources, including online databases, libraries, and real-time audio streams. It ensures a diverse range of audio files, covering different genres, styles, and quality levels, which is important for training a robust model. This diversity allows the system to learn various characteristics of audio and music, ensuring it can accurately classify different sound types. Additionally, licensing and copyright considerations are taken into account when selecting audio files for use, ensuring that the data collection process adheres to ethical and legal standards. Once the data is collected, the other step is preprocessing, which involves cleaning and preparing the audio files for analysis. This stage includes noise reduction, normalization, and segmentation of audio files to improve the quality of the input data. Noise reduction techniques are employed to minimize background sounds that could interfere with the audio signals. Normalization ensures that all audio files have a consistent volume level, allowing for more accurate comparisons during analysis. Segmentation involves breaking down longer audio files into smaller, manageable segments, facilitating the model's ability to process and analyze the data efficiently. After preprocessing, the focus shifts to feature extraction, which is a crucial aspect of the system's implementation. The extracted features, such as spectrograms, and chroma features, represent the audio signals in a way that captures their essential characteristics. This transformation is critical, as raw audio data is often too complex for machine learning algorithms to interpret directly. By converting audio signals into numerical representations [8]. The system can leverage deep learning techniques to recognize patterns and distinctions between different audio classes. Advanced techniques such as short-time Fourier transform (STFT) and wavelet transforms may also be applied to provide additional insights into the audio data. Model training occurs after feature extraction, during which the deep learning model is trained with the curated dataset. A range of algorithms, such as CNNs and recurrent neural networks (RNNs), are employed based on the unique needs and attributes of the audio data Figure 41.3 [9, 10]. The training phase includes fine-tuning hyperparameters to enhance the model's performance, utilizing methods like cross-validation to confirm that the model can generalize effectively to unseen data.

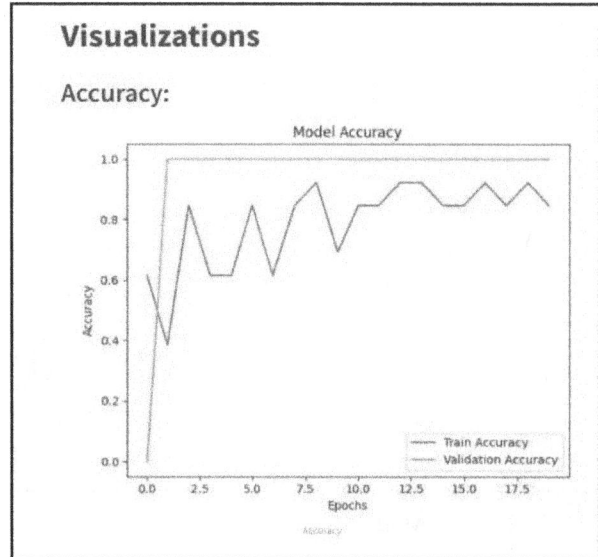

Figure 41.3 Model accuracy
Source: Author

Regular monitoring of loss and accuracy metrics during training helps identify potential overfitting, enabling the implementation of strategies such as dropout or early stopping to mitigate this issue. After the model has been trained and validated, the subsequent step is evaluation. This stage entails testing the model on a distinct validation dataset to measure its accuracy, precision, recall, and F1-score. These metrics offer insights into the model's performance effectiveness and its ability to classify audio signals correctly. The evaluation process also includes error analysis to identify any weaknesses or biases in the model, allowing for further refinement and optimization. Adjustments may be made to the training data or model architecture based on the evaluation results, ensuring continuous improvement of the system's performance. Finally, the deployment phase involves integrating the trained model into a user-friendly application or interface. This may take the form of a web application, mobile app, or desktop software, enabling users to upload audio files for analysis easily. The user interface is crafted to be user-friendly, offering visual feedback and results in real-time. Additional features, such as visualizations of detected audio classes, confidence scores, and insights into the audio's characteristics, enhance user experience. The deployment process also considers scalability, ensuring that the system can handle increasing volumes of data and users while maintaining performance

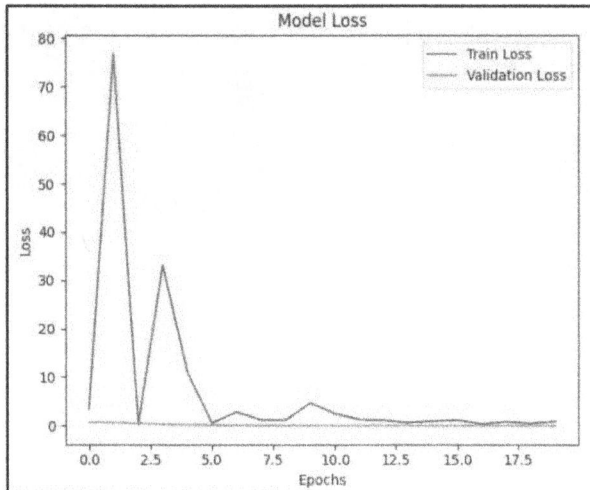

Figure 41.4 Model loss
Source: Author

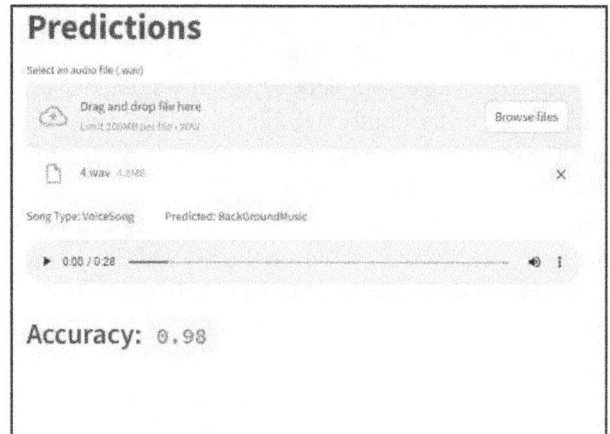

Figure 41.5 Predictions
Source: Author

and accuracy. Ultimately, this comprehensive implementation aims to make AI-based audio and music detection accessible and practical for a diverse range of applications, from music categorization to audio content analysis.

Results

The results of the AI-based audio and music detection system underscore its exceptional effectiveness in accurately identifying and classifying various audio signals. Upon evaluating the trained model on a separate validation dataset, the system achieved an outstanding accuracy of 98%. This impressive accuracy demonstrates the model's strong ability to generalize effectively to new data, making it very efficient in distinguishing between different audio categories. Such performance is crucial for applications like music classification, audio tagging Figure 41.4, 41.5 [11] and content recognition, where precision directly impacts user satisfaction and overall experience.

In addition to accuracy, several other key performance metrics were calculated to deliver a thorough assessment of the system's efficiency. The precision score, which quantifies the ratio of true positive predictions to all positive predictions, was recorded at 96%. This score suggests that when the model identifies a particular audio class, there is a strong probability that the prediction is accurate, thus minimizing the potential for mislabeling audio content in streaming services and automated tagging systems.

Recall, another critical metric, was observed at 97%. This metric assesses the model's capacity to detect all pertinent instances within the audio dataset. A high recall score signifies that the model effectively detects various audio classes present in the dataset, thereby reducing the number of missed detections. This balance between these is essential, ensuring that the system accurately identifies audio while capturing a comprehensive range of audio signals, making it suitable for diverse applications in the music and audio industries.

The F1-score, which integrates both precision and recall into a single metric, was found to be 96.5%. This score offers a balanced evaluation of the model's performance, particularly in scenarios where the distribution of audio classes may be imbalanced. A high F1-score reflects the model's capability to maintain an effective balance between accurately detecting positive instances and reducing false negatives and false positives, adding significant value to the system's effectiveness in real-world applications.

To further evaluate the model's performance, confusion matrices were generated, illustrating the classification results across various audio classes. The confusion matrix revealed that the model performed exceptionally well across most audio categories, with the highest misclassification rates occurring in less frequently represented classes. This suggests that while the model excels at identifying prevalent audio types, there is still room for improvement in handling rarer classes, a common challenge in machine learning.

Lastly, the system's performance was validated through user feedback, which highlighted its practical effectiveness in real- world scenarios. Users reported that the system accurately identified and tagged audio files with minimal latency, contributing to a seamless user experience. The integration of visual feedback, such as confidence scores and audio feature visualizations, further enhanced user engagement, allowing users to easily interpret the classification results. This combination of technical performance metrics and user experience demonstrates the system's significant potential for widespread application in audio analysis, music classification, and content management.

Conclusion

The AI-based audio and music detection system has demonstrated exceptional performance, achieving an impressive accuracy of 98% along with other strong metrics. These metrics validate the effectiveness of the deep learning models employed, showcasing their capability to generalize well to unseen audio data. User feedback has been overwhelmingly positive, emphasizing the system's practical benefits in applications such as music streaming and content tagging. By providing reliable audio classification, this system has established itself as an invaluable tool in the evolving landscape of audio technology, enhancing user experiences and streamlining audio processing tasks. Looking ahead, several avenues for improvement can further enhance the system's capabilities.

Expanding the variety of the training dataset would allow the model to identify and categorize a broader spectrum of audio types, including those from different cultures and genres. Implementing real-time audio analysis capabilities would also significantly boost usability in live settings, such as concerts or broadcasts. Additionally, exploring transfer learning techniques could enhance performance with limited data, allowing the model to fine-tune itself for niche categories. Finally, incorporating user-customizable features would create a more engaging experience, enabling users to adjust parameters and provide feedback for continuous model improvement. Embracing these enhancements will ensure the system's evolution and effectiveness in meeting the growing demands of audio analysis across various sectors.

References

[1] Tokozume, Y., Ushiku, Y., & Harada, T. (2017). Learning from between-class examples for deep sound recognition. arXiv preprint arXiv:1711.10282.

[2] Koutini, K., Schlüter, J., & Widmer, G. (2019). Receptive- field-regularized CNN variants for acoustic scene classification. *IEEE/ACM Transactions on Audio, Speech, and Language Processing*, 28, 816–829.

[3] Hershey, S., Chaudhuri, S., Ellis, D. P. W., Gemmeke, J. F., Jansen, A., Moore, R. C., et al. (2017). CNN architectures for large-scale audio classification. In IEEE International Conference on Acoustics, Speech and Signal Processing (ICASSP), (pp. 131–135).

[4] Jansson, A., Humphrey, E., Montecchio, N., Bittner, R., Kumar, A., & Weyde, T. (2017). Singing voice separation with deep U-Net convolutional networks. In Proceedings of the 18th International Society for Music Information Retrieval Conference (ISMIR), (pp. 745–751).

[5] Koizumi, Y., Saito, S., Uemura, H., Yamaoka, T., & Sagayama, S. (2017). Optimizing sound event detection systems based on the detection error trade-off curve. In Proceedings of the IEEE International Conference on Acoustics, Speech and Signal Processing (ICASSP), (pp. 781–785).

[6] Purwins, H., Li, B., Virtanen, T., Schlüter, J., Chang, S. Y., & Sainath, T. (2019). Deep learning for audio signal processing. *IEEE Journal of Selected Topics in Signal Processing*, 13(2), 206–219.

[7] Kim, Y., El-Khamy, M., & Lee, J. (2019). Residual LSTM- based model for sound event detection. In Proceedings of the IEEE International Conference on Acoustics, Speech and Signal Processing (ICASSP), (pp. 17–21).

[8] Tzinis, E., Sharma, E., Wang, Z., & Smaragdis, P. (2020). Two-step sound source separation: Training on learned latent targets. In Proceedings of the IEEE International Conference on Acoustics, Speech and Signal Processing (ICASSP), (pp. 91–95).

[9] Yuan, X., Lin, C., & Chen, X. (2020). Environmental sound classification using attention-based convolutional neural networks. *IEEE Access*, 8, 198780–198790.

[10] Kong, Q., Xu, Y., Plumbley, M. D., & Wang, W. (2020). Sound event detection of weakly labelled data with CNN- Transformer and automatic threshold optimization. In IEEE International Conference on Acoustics, Speech and Signal Processing (ICASSP), (pp. 31–35).

[11] Chen, Z., Xu, Y., Kong, Q., & Plumbley, M. D. (2021). Audio tagging with a pre-trained convolutional neural network. *IEEE/ACM Transactions on Audio, Speech, and Language Processing*, 29, 1891–1903.

42 Comparative analysis of fuzzy and neuro-fuzzy image processing system and their applications

Mahima Chaturvedi[1,a], Namrata Kaushal[2,b], Jyoti Gupta[2,c] and Om Prakash Chaturvedi[3,d]

[1]Department of Mathematics, Mansarovar Global University, Sehore, Madhya Pradesh, India

[2]Department of Mathematics, Indore Institute of Science and Technology, Indore, Madhya Pradesh, India

[3]Department of Faculty of Technology, UBKV, Pundibari, Cooch Behar, West Bengal, India

Abstract

Image processing minimizes noise and distortion while extracting critical information by combining multiple images of the same scene into a single, enhanced image. This study employs fuzzy logic and neuro-fuzzy logic approaches for image processing, enhancing visualization in many applications such as remote sensing, military operations and medical imaging. The performance of these methods is compared using quality measures like image quality index, information content, mean square error, root mean square error and peak signal to noise ratio. These factors provide a comprehensive assessment of image fidelity, clarity and information retention. Results show that the neuro-fuzzy method achieves a higher images clarity and keeps more useful information details compared to fuzzy logic approach.

Keywords: Image processing, fuzzification and defuzzification, fuzzy logic, fuzzy operators, fuzzy sets, neuro fuzzy logic, root mean square error

Introduction

A powerful tool for image processing is fuzzy logic, which offers unique advantages in handling uncertainty and imprecision [23]. It is particularly effective in applications where human-like reasoning and approximate decision-making are required. The adaptability and versatility of fuzzy logic have led to its wide adoption in areas such as edge detection [3, 8], impulse noise reduction [9], digital image watermarking [4], fusing sensor images for surveillance [21] and improving the resolution of urban satellite images [18]. Fuzzy methods rely on predefined rules, making them ideal for straightforward and low-complexity tasks. Several researchers have studied on the applicability of fuzzy logic for various image processing tasks [1, 2, 7–9]. For instance, fuzzy inference systems have been utilized for image enhancement and filtering [1, 6, 8, 17]. These approaches rely on well-defined membership functions and rules, enabling the processing of noisy or complex image data with high reliability. Fuzzy logic is also applied in more advanced applications [20]. For example,

it has been integrated with optimization techniques such as genetic algorithms and neural networks to improve precision farming and vision intelligence systems [4]. Similarly, fuzzy logic has proven useful in evaluating risk factors in health crises like COVID-19, where it optimizes decision-making under uncertainty [5, 6, 16]. Furthermore, its role in financial risk evaluation [13] and marketing strategies [14] demonstrates its broad applicability across different domains.

Despite its benefits, traditional fuzzy logic systems are often limited by their static nature and reliance on predefined rules. To address this, hybrid approaches combining fuzzy logic with neural networks have been developed, allowing systems to adapt and learn from data dynamically [10]. This combination, known as neuro-fuzzy systems, provides a robust framework for handling complex image processing tasks, offering improvements in adaptability and performance. This approach of Neuro fuzzy logic making system highly adaptable and effective for data-intensive tasks like medical imaging and multisensory fusion [11].

[a]chaturvedi.mahima@gmail.com, [b]namrata.kaushal@indoreinstitute.com, [c]cjyoti.gupta@indoreinstitute.com, [d]opchaturvedi@gmail.com

DOI: 10.1201/9781003684589-42

This paper compares these two approaches in the context of image processing, highlighting their strengths, limitations, and applications. Drawing from key studies and applications, it highlights the capabilities of fuzzy logic in handling noise, enhancing image quality, and making decisions under uncertainty. It also explores the advancements offered by neuro-fuzzy systems, which integrate learning mechanisms to tackle high-dimensional, data-intensive problems. It also discusses their potential for future advancements in fields like medical imaging, remote sensing, and video surveillance.

Fuzzy Approach Elements

Fuzzy approach elements refer to the fundamental components of fuzzy logic systems. These elements are designed to deal the uncertainty and imprecision. They model situations where binary true/false logic fails to represent the complexity of real-world scenarios. Here are the main elements of the fuzzy approach:

Fuzzy logic

Fuzzy logic approach used for decision-making and problem-solving that provides ability to model the approximate nature of human reasoning by handling uncertainty and imprecision. The classical logic theory operates in binary terms (true or false), while in contrast, fuzzy logic allows for degrees of truth, where variables can have a range of values between 0 and 1. This flexibility makes it well-suited for complex systems with vague or ambiguous data, such as natural language processing, image recognition, and control systems. Founded by Lotfi Zadeh in the 1960s, fuzzy logic uses fuzzy sets, membership functions, and rules to model imprecise information, enabling it to provide approximate solutions in real-world applications like image processing, medical diagnosis, and robotics.

Fuzzy image processing involves various methods aimed at analyzing, representing, and processing images, their segments, and features using fuzzy sets [7]. The aim of employing fuzzy logic in image processing is to leverage its powerful characteristics, including:

- Imprecisely representing spatial information in images.
- Managing spatial information through operations generalized to fuzzy sets.
- Using fuzzy combination operators for information fusion.

Fuzzy sets

Zadeh introduced the concepts of fuzzy set and defined as a class of objects with varying degrees of membership. The fuzzy set is associated by a membership function that assigns a degree of membership to each element of the set between zero and one. This framework effectively models vagueness and ambiguity in complex systems and provides a natural way to handle imprecision [24].

Membership functions

A membership function indicates belongingness of each input element into a certain category inside the input space, also known as the discourse universe. It assigns a degree to each input, indicating its level of participation. These degrees of membership help to identify overlaps between inputs and determine the final output. Fuzzy membership function plays a key role while converting fuzzy outputs into clear, precise results during the defuzzification process. There are different membership functions, which are used on the specific needs of each application.

Fuzzy rules

The input variables relate to the output variables using fuzzy rules based on the linguistic terms, which define the fuzzy state description. These rules impersonate human decision-making by employing a sequence of if-then statements. For example:

- If condition A apply, then output is X
- If condition B apply, then output is Y

Here, fuzzy sets X and Y correspond to inputs A and B. Therefore, fuzzy rules define fuzzy conditions, which form the foundation of fuzzy logic systems [24].

Fuzzy Logic Approach in Image Processing

Fuzzy logic is widely used in image processing to handle the uncertainty and vagueness inherent in image data. By leveraging fuzzy sets and rules, it provides an effective way to process and enhance images in scenarios where conventional methods may struggle. The main features of fuzzy logic in image processing are listed below:

Overview of fuzzy logic control system

Fuzzy image processing encompasses a variety of techniques for comprehending, representing, and processing the images and, their features as fuzzy sets [5]. The fuzzy approach is implemented in the following primary stages:

1. **Image fuzzification:** used to transform the image input data into a fuzzy input. This process is similar to coding by assigning one or more membership values to various attributes (such as gray levels, features, segments, ...) of image.
2. **Fuzzy control system:** This involves various components such as fuzzy membership functions, logical connectives, and fuzzy if-then rules to facilitate the process. Apply fuzzy rules to transform the membership values. These fuzzy based rules are applied to detect the edges, white or black pixels. Fuzzy contrast stretching is used to enhance or analyze the images.
3. **Image defuzzification:** It's a process to convert each fuzzy output obtained from fuzzy control system into crisp image form. There are different techniques for image defuzzification such as centroid, middle of maximum, bisector, largest of maximum, and minimum of maximum.

Working process of fuzzy logic system in image processing

The algorithm of image processing at pixel-level using fuzzy logic can be outlined as follows [25]:

- Read the input images and determine the size (row × column).
- Ensure images are of the same size by cropping or selecting appropriate portions.
- Convert images into column form, resulting in $C = r1 \times c1$ entries Figure 42.1.
- Define membership functions for the input images and assign membership degrees (ranging from 0 to 255) to each pixel value in the inputs.

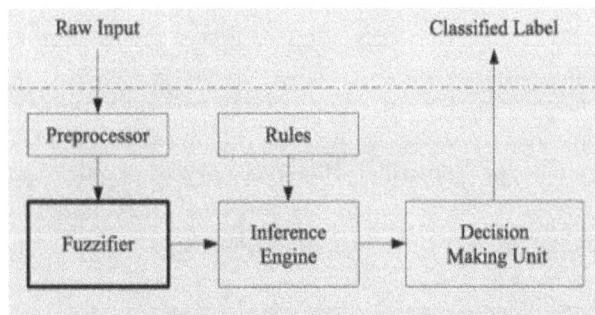

Figure 42.1 Fuzzy logic control system [5]
Source: Chua T. and Woei W.T.

- Apply a fuzzy rule based (fuzzy if-then rules) system on corresponding grey level values for each pixel in the input images to determine the relationships between the input images and their output responses.
- Defuzzification of the output obtained from fuzzy control system. This process converts the output from column form back into a matrix to display the output image.

Applications of fuzzy logic system in image processing

Fuzzy logic systems have numerous applications in image processing due to their ability to handle uncertainty, imprecision, and incomplete data. Below are some key applications of fuzzy logic in image processing:

1. **Edge detection:** Fuzzy logic enhances edge detection by defining membership functions for pixel intensity and using fuzzy rules to identify edges
2. **Image segmentation:** Fuzzy logic is widely applied for segmenting an image into meaningful regions based on intensity, texture, or color. By assigning fuzzy membership values to pixels, fuzzy clustering algorithms like fuzzy C-means (FCM) achieve superior segmentation results.
3. **Noise reduction:** Fuzzy filters efficiently remove noise while preserving image details by applying fuzzy rules to neighboring pixel intensities. It can be used to enhance image quality by adjusting brightness, contrast, and sharpness.

Neuro-fuzzy Logic for Image Processing

Fuzzy systems and ANNs are two important methods in artificial intelligence. Fuzzy systems try to imitate human reasoning and deal with uncertainty. They divide data into fuzzy classes using rules that are easy to understand. These rules can be updated or changed easily, making Fuzzy systems flexible. However, they rely heavily on experts to design them, so having a system that can learn from examples would make them better [12].

The ANNs work similar to the human brain, processing information by a network of interconnected nodes. They are very good at handling complex problems and can generalize well. But the main issue is that it's hard to understand how they make decisions after being trained. By combining fuzzy systems and ANNs, we get a powerful model called the neuro-fuzzy system, which brings together the strengths of both techniques.

Neuro-fuzzy logic-based image processing integrates the fundamentals of neural networks and fuzzy logic to effectively combine multiple images into a single representation. Below, the fundamental concepts and methodologies involved in neuro-fuzzy image processing are discussed [10].

Neural networks

A neural network is a computer system that has an innate ability to retain experience and apply it to solve problems. Neural networks are especially effective in addressing challenges defined by noisy or imprecise data, large dimensionality, non-linearity, and circumstances when a clearly mathematical solution is unavailable. One of the neural networks' most important benefits is their ability to model systems based solely on data, without requiring an explicit mathematical formulation.

Network properties

The topology of a neural network is defined by its structure, including the number of layers and nodes per layer, as well as the interconnection scheme between these nodes. These properties dictate the network's computational capabilities and its suitability for specific tasks.

Overview of neuro-fuzzy logic system

In artificial intelligence, neuro-fuzzy logic is a technique that blends fuzzy logic and neural networks. This integration generates a hybrid intelligent systems, also known as neuro-fuzzy logic systems, by utilizing the learning skills of neural networks and the human-like reasoning capabilities of fuzzy systems [15].

The foundation of a neuro-fuzzy system generates through a fuzzy logic framework that has been improved by a learning algorithm based on neural network theory. This learning process adapts the system parameters locally while preserving the underlying fuzzy system's semantic properties. The system can be initialized with pre-defined fuzzy rules or trained directly from input data.

In image fusion applications, neuro-fuzzy systems use neural networks to refine membership functions via algorithms such as backpropagation. The resulting fused images retain more texture features and enhanced informational characteristics compared to the original images.

Working process of neuro-fuzzy logic system in image processing

The pixel-level image processing using neuro-fuzzy logic involves the following steps:

- First upload input images and determine their size (row, column).
- Ensure all images are represented in matrix form, where pixel values range from 0 to 255 (using a grayscale color map)
- All images should be of the same size by cropping or selecting appropriate portions.
- Convert images into column form, resulting in C = r × c entries.
- Prepare training data as a matrix with three columns, each containing values from 0 to 255 in increments of 1.
- Form a check dataset by combining the pixel data from the input images
- For each input image, define the quantity and type of membership functions and adjust them as necessary.
- Use the genfis1 function to generate an initial FIS structure, taking the training data, the quantity of membership functions, and their types as input.
- Train the FIS structure using the ANFIS function, which updates the parameters based on the training data.
- Now for generating a processed image in columnar format, apply fuzzification for every pixel using the trained FIS structure and verify the data.
- Transform the fused image back to a matrix form and show the final output Figure 42.2.

Key applications of neuro fuzzy logic system

Neuro-fuzzy systems approach is especially effective for solving complex, non-linear, and uncertain

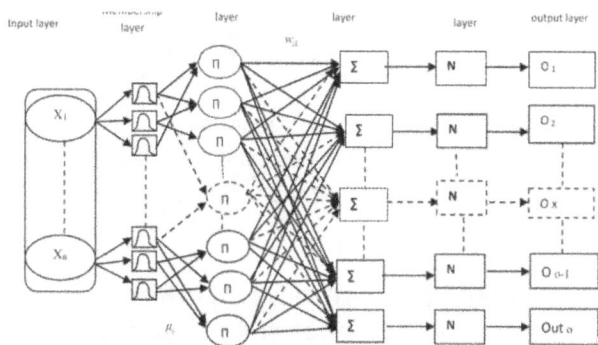

Figure 42.2 Neuro Fuzzy logic control system [10]
Source: Oscar T. Nkamganga,b , Daniel T .et.al.

problems. Here are some key applications of neuro-fuzzy systems:

1. **Feature extraction:** Neuro-fuzzy systems learn feature relationships adaptively, enabling effective feature extraction for classification tasks.
2. **Pattern recognition:** The adaptability of neuro-fuzzy systems aids in recognizing complex patterns in images, such as handwritten digits.
3. **Image enhancement:** By learning the optimal fuzzy rules from data, neuro-fuzzy systems enhance image quality.

Evaluation Techniques for Image Processing

The evaluation of image quality and performance of various algorithm is essential in image processing to determine how well the methods work for image enhancement, noise reduction, information retention and reconstruction work [14]. In this paper, following factors are taken into consideration while evaluating image processing:

Image quality index (IQI)
It is used to evaluate the perceptual quality of image. It computes the similarity between original image and processed image. The range of IQI lies from -1 to 1.

$$IQI = \frac{4 \cdot \mu_X \cdot \mu_Y \cdot \sigma_{XY}}{(\mu_X^2 + \mu_Y^2) \cdot (\sigma_X^2 + \sigma_Y^2)}$$

Where μ_X and μ_Y denote the mean intensities of original and processed images, σ_X^2 and σ_Y^2 are their variance and σ_{XY} is the covariance between them. A higher IQI value shows the greater similarities between the images.

Mutual information measure
Mutual information measure (MIM) finds how well the processed image incorporates the information from the original image.

$$MIM(X, Y) = \sum_{x \in X} \sum_{y \in Y} p(x, y) \log \frac{p(x,y)}{p(x)p(y)}$$

Where $p(x)$ and $p(y)$ are probability distribution of original and processed images and $p(x, y)$ is the joint probability distribution of the images.

Mean square error
It calculates the average squared difference between corresponding pixel values in the original and processed images. Lower mean square error (MSE) values indicate higher image fidelity.

$$MSE = \frac{1}{mn} \sum_{i=0}^{m-1} \sum_{j=0}^{n-1} [I(i,j) - K(i,j)]^2$$

Where m and n are the image dimensions, $I(i,j)$ and $K(i,j)$ are the pixel values of the original and processed images.

Root mean square error
Root mean square error (RMSE) is derived from the MSE. It identifies the amount of change per pixel due to the processing. Lower RMSE values reflect better image quality.

$$RMSE = \sqrt{MSE}$$

Peak signal-to-noise ratio
It is a commonly used metric that assesses the relationship between the highest possible pixel value and the magnitude of noise introduced during image processing. higher peak signal-to-noise ratio (PSNR) values denote lower distortion and better image quality.

$$PSNR = 10 \cdot \log_{10} \left(\frac{MAX_I^2}{MSE} \right)$$

Results and Discussions

The proposed fuzzy and neuro-fuzzy-based image processing methods were executed using MATLAB. These methods are flexible and can be used in many areas like medical diagnosis, video surveillance and remote sensing. For these methods to work properly, the fuzzy control system needs to have well-defined membership functions and rules. Based on quality assessment parameters measure, the results of Fuzzy logic system and neuro fuzzy control system are shown in Figure 42.3 [22].

In the above figures, images (a) and (d) are original input images, image (b) and (e) are obtained by fuzzy logic Control system and (c), (f) are images obtained by Neuro Fuzzy logic control system. These images demonstrate that in the first two examples, the neuro-fuzzy system approach performs improved than the fuzzy technique. The evaluation factors of image processing based on fuzzy logic and neuro-fuzzy logic approaches are presented in Table 42.1 [14]..

A comparative analysis of fuzzy control system and neuro-fuzzy control system for image processing is shown in Table 42.2.

The above analysis shows that neuro-fuzzy logic can adapt to complex and changing tasks because it learns from data, while fuzzy logic relies on fixed rules and is less flexible. Neuro-fuzzy logic is better

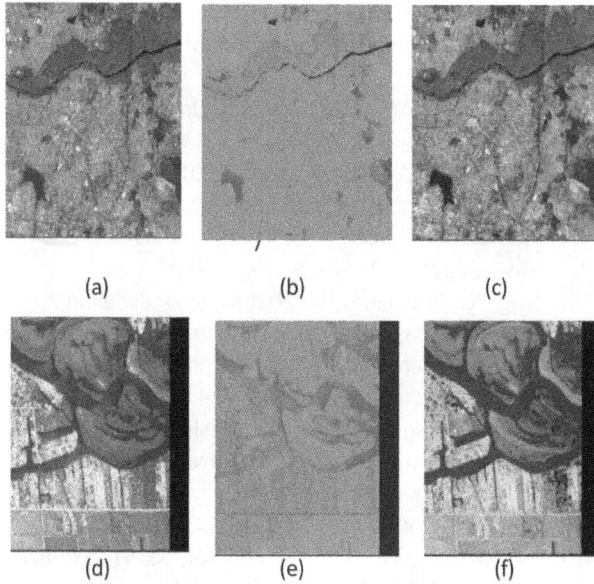

Figure 42.3 [22] Images (a), (d): original input images; (b), (e): fuzzy logic image; (c), (f): Neuro fuzzy logic image
Source: Author

Table 42.1 Evaluation factors.

Method	IQI	MIM	MSE	RMSE	PSNR
Fuzzy images					
(b)	0.9687	0.4528	2.2034	1.4844	15.0966
(e)	0.9784	0.9484	1.9301	1.3893	14.3280
Neuro fuzzy					
(c)	0.9899	1.4256	1.6697	1.2922	24.6348
(f)	0.9829	1.3070	1.7566	1.3254	17.3829

Source: Srinivasa Rao D. Seetha M. et. al.[14]

Table 42.2 Comparative analysis of fuzzy vs neuro fuzzy logic system.

Feature	Fuzzy logic	Neuro-fuzzy logic
Architecture	Rule-based, static	Adaptive, data-driven
Interpretability	High	Moderate (dependent on complexity)
Adaptability	Limited	High
Computational cost	Low	High
Application scope	Simple to moderately complex tasks	Complex, high-dimensional tasks
Training requirement	None	Requires labelled datasets

Source: Author

suited for challenging applications like medical imaging and video surveillance while Fuzzy logic works well for simpler tasks that don't require much flexibility. Neuro-fuzzy systems need more computational power and labelled data for training while fuzzy logic is simpler, cheaper, and doesn't need training.

Conclusion

Fuzzy logic and neuro-fuzzy logic both techniques have shown their usefulness in image processing, each with unique strengths suited to different types of tasks. Fuzzy logic is simple and easy to understand, making it ideal for simpler problems where relationships between variables can be clearly defined using rules. It doesn't require training data and uses less computing power, making it a cost-effective option for basic applications. However, neuro-fuzzy system combines the clarity of fuzzy logic with the adaptability of neural networks. It is especially useful for handling complex and data-heavy tasks, such as medical imaging and video surveillance, where it can learn from data and adjust to different scenarios for better results. However, this approach requires more computing power and labelled datasets, making it more resource-intensive and harder to implement compared to fuzzy logic.

This study demonstrates that the fuzzy image processing systems can produce clear and reliable results.

Meanwhile, the neuro-fuzzy method performs better at preserving texture details and provides better outcomes for more complex problems. The particular needs of the task, including its complexity, data accessibility, and available resources, will determine which of these two strategies can be chosen. These techniques could be applied to video image processing, combining images from multiple sensors, and creating better ways to measure image quality. Future research could explore iterative methods, where fuzzy logic and neuro-fuzzy logic are applied in repeated steps to improve results even further.

Additionally, future work could focus on developing hybrid models that combine the best features of both approaches. These models could use the simplicity of fuzzy logic along with the learning ability of neuro-fuzzy systems to create more powerful and efficient image processing methods. Such advancements could lead to improvements in areas like medical diagnostics, remote sensing, and advanced video surveillance, making these techniques even more impactful.

Acknowledgement

We are extremely grateful to all those who played a role in the success of this research work. We would like to express our gratitude towards the management for the encouragement and all the supporting staff for their invaluable input and support throughout the research process.

References

[1]   Borkar, A. D., & Atulkar, M. (2013). Fuzzy inference system for image processing. International Journal of Advanced Research in Computer Engineering and Technology (IJARCET), 2(3), 1007–1010.

[2]   Aborisade, D. O. (2010). Fuzzy logic based digital image edge detection. Global Journal of Computer Science and Technology, 10(4), 78–83.

[3]   Castelló-Sirvent, F. (2022). A fuzzy-set qualitative comparative analysis of publications on the fuzzy sets theory. Mathematics,. 10, 1322.

[4]   Coumou, D., & Mathew, A. (2001). A fuzzy logic approach to digital image watermarking. In DESDes01 The International Workshop On Discrete-Event System Design, (pp. 201–209).

[5]   Chua, T., & Tan, W. W. (2007). GA optimisation of non-singleton fuzzy logic system for ECG. classification, evolutionary computation, 2007. In CEC 2007. IEEE Congress on October (2007), DOI:10.1109/CEC.2007.44246.

[6]   Gupta, J., Saxena, S., Kaushal, N., & Chanchani, P. (2023). Optimization of COVID-19 risk factors using fuzzy logic inference system. In Computational and Analytic Methods in Biological Sciences, (pp. 101–117). River Publishers.

[7]   Zhang, H., Ma, X., & Wu, N. (2011). A new filter algorithm of image based on fuzzy logical. In IEEE International Symposium of Computer Society, (pp. 315–318).

[8]   Haq, I., Shah, K., Khan, M. T., Azam, K., & Anwar, S. (2015). Fuzzy logic based edge detection for noisy images. Technical Journal, University of Engineering and Technology (UET), 20, 82–86.

[9]   Kaur, J., Kaur, P., & Kaur, P. (2012). Review of impulse noise reduction technique using fuzzy logic for the image processing. International Journal of Engineering and Technology, 1(5).

[10] Oscar T.Nkamganga,b , Daniel T. et.al. , "A Neuro-fuzzy system for automated detection and classification of human intestinal parasites", Informatics in Medicine Unlocked 13(2018) 81–91.

[11] Naguchi, N., Reid, J. F., Zhang, Q., & Tian, L. (1998). Vision intelligence for precision farming using fuzzy logic optimized genetic algorithm and artificial neural network. An ASAE Meeting Presentation, Paper No: 983034, (pp. 128–136).

[12] Nauck, D. D., & Nürnberger, A. (2013). Neuro-fuzzy systems: a short historical review. Studies in Computational Intelligence, 445, 91–109. doi: 10.1007/978-3-642-32378-2_7.

[13] Peng, X., & Huang, H. (2020). Fuzzy decision making method based on CoCoSo with critic for financial risk evaluation. Technological and Economic Development of Economy, 26, 695–724.

[14] Srinivasa Rao D, Seetha M, Krishna Prasad MHM, "Comparision of fuzzy and NeuroFuzzy Image Fusion Techniques and its Applications", International Journal of Computer Applications, 43(20), 2012.

[15] F. Scarcelli , M. Gori, A. C. Tsoi, M. Hagenbuchner, and G. Monfardini, "The graph neural network model" 20(1), 2009, 61–80. IEEE Trans. Neural Networks, 2009, doi: 10.1109/TNN.2008.2005605.

[16] Sharma, M. K., Dhiman, N., & Mishra, V. N. (2021). Mediative fuzzy logic mathematical model: a contradictory management prediction in COVID-19 pandemic. Applied Soft Computing, 105, 107285.

[17] Shih, F. Y., Image processing and mathematical morphology: fundamentals and applications. CRC press, (2017).

[18] M. T. Hagan, H. B. Demuth, and M. H. Beale, "Neural Network Design," Bost. Massachusetts PWS, 1995, doi: 10.1007/1-84628-303-5.

[19] Y. Bai and D. Wang, "Fundamentals of fuzzy logic control — fuzzy sets, fuzzy rules and defuzzifications," in Advances in Industrial Control, 2006.

[20] I. Iancu, "A Mamdani Type Fuzzy Logic Controller in Fuzzy Logic - Controls, Concepts, Theories and Applications", 2012.

[21] G. G. Tiruneh, A. R. Fayek, and V. Sumati, "Neuro-fuzzy systems in construction engineering and management research," Autom. Constr., vol. 119, no. May, p. 103348, 2020, doi: 10.1016/j.autcon.2020.103348.

[22] Rao S., Seetha M, Krishna Prasad MHM, Comparison of Fuzzy and Neuro Fuzzy Image Fusion Techniques and its Applications, International Journal of Computer Applications, 43 (20), (2012).

[23] L. A. Zadeh, "Fuzzy sets," Inf. Control, 1965, doi: 10.1016/S0019-9958(65)90241-X.

[24] L. Uden, F. Herrera et.al., "Advances in Intelligent System and Computing", Jan 2013, DOI:10.1007/978-3-642-30867-3

[25] Yi Zheng and P. Zheng, "Multisensor image Fusion Using Fuzzy Logic for Surveilance Systems", Seventh International Conference on Fuzzy Systems and Knowledge Discovery, Sept. 2010, DOI:10.1109/FSKD.2010.5569466.

43 Design and analysis with high-performance rectangular micro-strip antenna at 38 GHz 5G millimeter-wave application

Yuvaraj, K.[1,a] and Sanam Narayana Reddy[2,b]

[1]Research Scholar, Department of ECE, Sri Venkateswara University College of Engineering, Tirupati, Andhra Pradesh, India

[2]Professor, Department of ECE, Sri Venkateswara University College of Engineering, Tirupati, Andhra Pradesh, India

Abstract

In this paper, a micro-strip patch antenna with rectangular structure designed to enhance the performance within the 5G millimeter-wave spectrum, specifically operating at 38 GHz. In contrast to traditional E-shaped patch antennas, the proposed design offers a compact and broadband configuration, it is used for the transmission of high-data rate applications. Real-time applications depend on their performance. This antenna, boasting a gain of 6.69 dBi, is designed for the millimeter wave band, essential for 5G networks. Performance analysis, conducted via HFSS simulation with a dielectric constant of 4.4, revealed a 3.5 GHz bandwidth, a -31.87 dB return loss, and an optimized voltage standing wave ratio (VSWR). These results suggest robust performance, highlighting the antenna's robust performance. The rectangular patch design proves to be highly advantageous for higher frequency 5G applications, enhancing overall system efficiency and reliability. This work underscores the potential of a micro-strip patch antenna with rectangular structure as a key component in advancing 5G technology, offering improved performance in critical millimeter-wave bands.

Keywords: 5G Communication, high return loss, microstrip antenna, millimeter wave spectrum, rectangular patch

Introduction

The increasing pervasiveness of smart home technologies underscores the necessity for the development of high-performance 5G networks to facilitate reliable device interoperability and efficient application monitoring [1, 2]. Moreover, the realization of autonomous driving paradigms is contingent upon the availability of 5G wireless networks exhibiting high throughput and minimal latency. The integration of environmental sensing modalities, intelligent traffic management systems, and vehicle-to-vehicle communication protocols necessitates millisecond-level responsiveness for ensuring passenger safety and mitigating collision risks. The data rate and bandwidth demands inherent in these applications surpass the operational limits of current 4G network infrastructure, driving the adoption of millimeter-wave (mm-wave) frequencies [3] as a means to enhance spectral efficiency. The development of wireless and mobile communication technologies

has been greatly impacted by improvements in wireless microwave components over the last ten years. Even while 4G LTE has enabled high-speed data transfer with high data rates and connectivity, as well as the integration of numerous commercial services, it also introduced bandwidth limitations, particularly below 3 GHz [4], which hinder the needs of advanced wireless networks. To facilitate the deployment of fifth-generation mobile networks, the International Telecommunication Union (ITU) has designated specific millimeter-wave frequency bands for different geographical regions. These allocations vary significantly, as evidenced by the 27.5–28.35 GHz band in the USA, the 27.5–28.28 GHz band in Japan, and the ranges assigned to China (25.3–27.5 GHz), Sweden (26.5–27.5 GHz), Korea (26.5–29.5 GHz), and the EU (24.25–27.5 GHz). The mm-wave spectrum offers substantial advantages in terms of bandwidth and potential throughput, making it suitable for future 5G demands. However, its practical implementation is complicated by inherent limitations like as path

[a]yuvaraj.kunati@gmail.com, [b]snreddysvu@yahoo.com

DOI: 10.1201/9781003684589-43

loss, absorption by atmosphere, and susceptibility to fading by signal, requiring the development of effective countermeasures to enhance performance in real-world multipath propagation environments [5]. In order to reduce route loss, recent studies have implemented omnidirectional and bi-directional radiation patterns using methods including electromagnetic bandgap structures, split-ring resonators, defected ground structures (DGS) [6–8], and metamaterials. Despite their advantages, these structures often suffer from low radiation efficiency and poor directive gain, and Integrating them into current printed circuit boards can be difficult because of limitations like substrate-integrated waveguides (SIW) and shorting vias [9]. In addition, MIMO systems, which enhance channel capacity [10], require that the design of their radiating elements be thoroughly considered. In the 5G mm-wave spectrum, multi-layered MIMO techniques have greatly increased gain and impedance bandwidth. Recent research in 5G antenna technology has concentrated on developing single-element antennas with exceptionally high gain and efficiency. To this end, the 28 GHz, 30 GHz, and 38 GHz frequencies have emerged as primary areas of focus due to their suitability for 5G wireless communication and minimal environmental absorption characteristics [11–14].

Despite the benefits of frequency reuse, it presents attenuation issues. Existing mm-wave antenna designs for 5G typically have narrow bandwidths due to their fixed or limited operating frequencies [15]. The authors focused on Specific Absorption Rate (SAR) analysis for a novel patch antenna design, emphasizing user safety and compliance with communication standards [16]. To achieve this, they employed a complex patch antenna structure. Patch antennas are known for their potential to enhance bandwidth, efficiency, and gain. However, integrating metamaterials into their design introduces complexity that can sometimes negatively impact overall performance. For instance, Yen [17] achieved 87% efficiency and 11.2 dBi gain using an array structure. Similarly, Tu [18] explored Two-band operations at 28 GHz and 38 GHz with an electromagnetic band gap structure, reaching 83.1–91.1% efficiency, though further refinement was noted as needed.

The increasing demand for compact, high-performance antennas for 5G millimeter-wave systems has motivated the design of a novel rectangular microstrip patch antenna operating at 38 GHz with a 3.5 GHz bandwidth. To achieve enhanced efficiency, a low-permittivity dielectric substrate ($\varepsilon_r = 4.4$) was utilized. The resulting antenna exhibits a measured/simulated gain of 6.69 dBi and a return loss of -31.87 dB, demonstrating strong performance within the target frequency band. Microstrip patch antennas are prevalent in communication due to form factor, manufacturing simplicity, and integration potential. Rectangular variants are particularly attractive for 5G due to their high gain and suitability for mm-wave frequencies. The feasibility of this design for 5G applications was confirmed through HFSS simulations, underscoring the importance of adhering to specific design parameters for successful system implementation.

1. The intended operating frequency for this antenna is a resonance of 38 GHz, which aligns with the requirements of 5G millimeter-wave applications.
2. Bandwidth: The designed antenna's 3.5 GHz bandwidth is fundamental to its ability to support the high data rate requirements of 5G applications.
3. The achieved return loss of -31.87 dB suggests excellent impedance matching, leading to efficient power transfer and minimal signal reflection.
4. The antenna presents a VSWR of 1.04, a value consistent with effective impedance matching and low power loss, as it falls within the optimal range of 1 to 2.
5. The antenna's efficiency was targeted for enhancement through the utilization of a low permittivity dielectric substrate, characterized by a dielectric constant of 4.4.
6. Substrate height: The effective dielectric constant and, consequently, the antenna's performance are affected by the 2 mm substrate height that is used.
7. Antenna dimensions: To increase bandwidth and return loss, the antenna's substrate's width and length are meticulously measured and optimized.
8. Design simulation: The antenna design was subjected to high-frequency structure simulator (HFSS) software, a process that contributed to the realization of the specified performance parameters.

These parameters are essential for designing an antenna that meets the stringent requirements of 5G mm-wave communications, including high gain, efficient power transfer, and wide bandwidth.

Design Parameter Analysis

The proposed 5G millimeter-wave antenna's basic geometrical structure is illustrated in Figure 43.1(a,b).

The antenna employs an FR4 substrate with a dielectric constant (ε_r) of 4.4 and a height of 2 mm. Its structural foundation is a conventional simple patch antenna, with the patch width, a critical parameter for input impedance matching, defined by Equation (1).

$$W = \frac{c}{2f_r}\sqrt{\frac{2}{\varepsilon_r+1}} \tag{1}$$

A patch antenna's effective dielectric constant (ε_{eff}) depends on the substrate material's dielectric constant and the patch's computed width and height. The air above the substrate causes the effective dielectric constant for patch antennas to be between 1 and the substrate's actual dielectric constant, so that $1 < \varepsilon_{eff} < \varepsilon_r$. Equation (2) is used to calculate the effective dielectric constant (ε_{eff}), which approaches the true value of εr as the substrate's dielectric constant (ε_r) rises noticeably above 1.

$$\varepsilon_{eff} = \frac{\varepsilon_r+1}{W} + \frac{\varepsilon_r-1}{W}\left(1+\frac{12h}{W}\right)^{-\frac{1}{2}} \tag{2}$$

Similarly effective length and ΔL are calculated using Equations (3) and (4)

$$L_{eff} = \frac{c}{2f_r\sqrt{\varepsilon_{eff}}} \tag{3}$$

The "fringe effect," a phenomenon in microstrip patch antennas, occurs when the electromagnetic field stretches past the patch's actual dimensions, giving the impression that it is electrically longer. The symbol for this difference between the electrical and physical lengths is ΔL. Determining the patch's effective length is a commonly used technique to account for this effect.

$$\Delta L = 0.412h\frac{(\varepsilon_r+0.3)\left(\frac{W}{h}+0.264\right)}{(\varepsilon_{eff}-0.258)\left(\frac{W}{h}+0.8\right)} \tag{4}$$

Calculation of actual length using Equation (5)

$$L = L_{eff} - \Delta L \tag{5}$$

In the context of the microstrip patch antenna design, the resonance frequency f_0 is influenced by several key parameters: the width of the patch W, the length L, the thickness h, the relative permittivity ε_r, and the speed of light c (where $c = 3 \times 10^8$ m/s). These variables collectively determine the antenna's operational characteristics and performance. Based on these equations the attained parameters used to design the patch are tabulated in Table 43.1.

Simulation and Results

The characteristics listed in Table 43.2 are very important. The final antenna's appearance, as determined by the parameters given in Table 43.2, is depicted in Figure 43.2 and is simulated using the HFSS software. important metrics that are necessary for assessing the antenna's performance, such as the S-parameters and the standing wave ratio of voltage (VSWR).

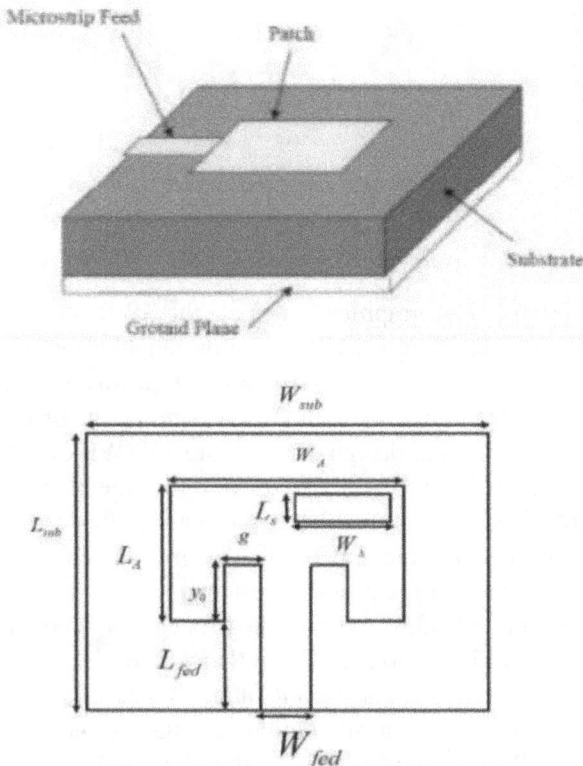

Figure 43.1 (a) Side view, and (b) Top view of the proposed antenna
Source: Author

Table 43.1 Antenna design parameters.

Items	Signs	Values (milli meters)
Width of the Patch	W	2.5
Length of Patch	L	3.5
Length of Substrate	L_s	6.25
Width of Substrate	W_s	7.4
Feed width	W_f	0.3
Height of the substrate	H	0.035
Length of the feed	L_f	1.3

Source: Author

Table 43.2 Simulation parameters.

Item	Value
Max U	1.7244 mW/sr
Realized	2,167
Radiated power	7.1231 mW
Directivity maximum	3.0423
Gain maximum	4.2345
Peak system	2.167
Rad efficiency	0.76798
Tot efficiency	0.71231
Efficiency of the system	0.71231
Field total	1.1403 V
Decay factor	0

Source: Author

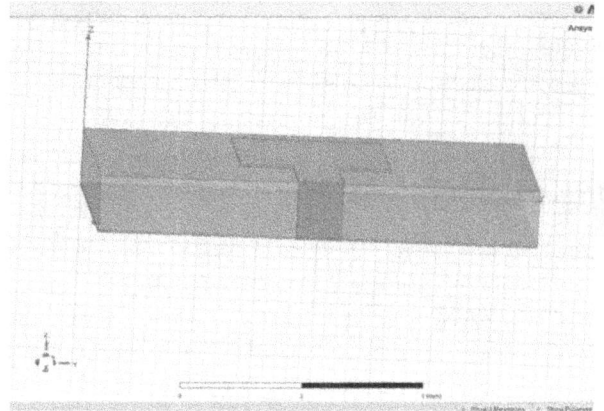

Figure 43.2 Perspective view of designed rectangular patch
Source: Author

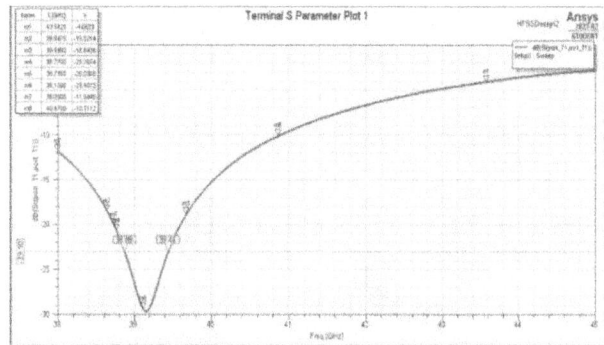

Figure 43.3 Scattering parameters S_{11} using HFSS
Source: Author

The optimal design parameters for the rectangular microstrip employing FR4 substrate material are shown in Table 43.2.

The coefficient (S11 parameter) of the proposed antenna design is graphically represented in Figure 43.3. This parameter is of paramount importance as it characterizes the antenn2a's bandwidth coverage and resonance frequency. For effective operation and stable radiation characteristics, a standard criterion is an S11 value of -10 dB or lower. In this design, an S11 parameter of -31.87 dB is achieved, indicating superior impedance matching and minimal reflected power. The antenna topology comprises a modified rectangular patch incorporating a slot at the top center, a design modification intended to enhance both the antenna's efficiency and gain while simultaneously providing a bandwidth of 3.45 GHz.

Figure 43.4 graphically depicts the SWR for voltage of the designed antenna. For reliable communication systems, the ideal VSWR typically resides within the range of 1 to 2. The simulation results obtained for our design demonstrate a VSWR of 1.04 at the 38 GHz resonance frequency, thus meeting this criterion. This optimal VSWR value signifies exceptional performance and a high degree matching of impedance for the proposed antenna.

In Figure 43.4, the designed antenna's Standing Wave Ratio of Voltage (VSWR) is shown. For reliable transmission, a VSWR range of 1 to 2 is often ideal. Based on the simulation findings, our design's VSWR meets this criteria, with an optimal value of 1.04 at the resonant frequency of 38 GHz. Excellent performance and impedance matching are demonstrated by this value for the recommended antenna.

The comparison of performance metrics between our suggested design and current research findings is summed up in Table 43.3 .

Conclusion

The proposed micro-strip patch antenna for 5G communication systems exhibits performance advantages over current designs, as validated through HFSS analysis. It demonstrates a significant 3.5 GHz bandwidth and operates effectively resonant frequency of 38 GHz. Simulation results confirm desirable radiation patterns, satisfactory impedance matching (characterized by return loss -31.9 dB and with VSWR equal to 1.03, indicating minimal reflected power), and overall optimal performance. The antenna's gain of 6.69 dBi renders it suitable for applications such as base stations, vehicular traffic signal processing, and high-fidelity video transmission. The enhanced bandwidth, gain, and VSWR of this antenna contribute to reliable and efficient connectivity, positioning it as a crucial element in the advancement of 5G technology and the improvement of user experiences across a spectrum of applications.

References

[1] Sohail, A., Khan, H., Khan, U., Khattak, M. I., Saleem, N., & Nasir, J. A. (2019). Design and analysis of a novel patch antenna array for 5G and millimeter wave applications. In Proceedings 2nd International Conference on Computing, Mathematics and Engineering Technologies (iCoMET), (pp. 1–6). doi: 10.1109/ICOMET.2019.8673490.

[2] Ahmad, I., Sun, H., Zhang, Y., & Samad, A. (2020). High gain rectangular slot microstrip patch antenna for 5G mm-wave wireless communication. In Proceedings 5th International Conference on Computer and Communication Systems (ICCCS), (pp. 723–727).

[3] Bangash, K., Ali, M. M., Maab, H., & Ahmed, H. (2019). Design of a millimeter wave microstrip patch antenna and its array for 5G applications. In Proceeding 1st International Conference on Electrical, Communication, and Computer Engineering (ICECCE), (pp. 1–6). doi: 10.1109/ICEC-CE47252.2019.8940807.

[4] Desai, A., Upadhyaya, T., & Patel, R. (2019). Compact wideband transparent antenna for 5G communication systems. *Microwave and Optical Technology Letters*, 61(3), 781–786.

[5] Asaadi, M., & Sebak, A. (2016). High-gain low-profile circularly polarized slotted SIW cavity antenna for MMW applications. *IEEE Antennas and Wireless Propagation Letters*, 16, 752–755.

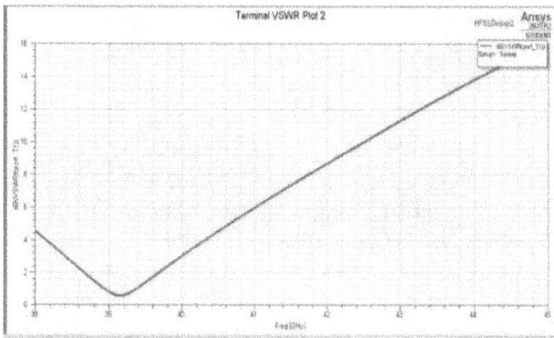

Figure 43.4 Voltage standing wave ration (VSWR)
Source: Author

Table 43.3 Comparison of different methods based on their frequencies

Methodology	Resonant frequency (GHz)	Return loss (dB)	Band width (GHz)	Gain (dB)	Antenna radiation efficiency (%)
Method 1 [16]	28	-18.25	1.1	6.83	-
Method 2 [17]	38	-15.5	1.92	6.9	93.5
Method 3 [18]	54	-12	2	7.4	82.7
Method 4 [19]	28	-25	1.26	9.75	98
Method 5 [20]	38	-17.09	1.05	8.28	-
Method 6 [21]	39	-31	3.5	4	79
Proposed	**38**	**31.87**	**3.5**	**6.6**	

Source: Author

[6] Zhu, S., Liu, H., Chen, Z., & Wen, P. (2018). A compact gain-enhanced Vivaldi antenna array with suppressed mutual coupling for 5G mm wave application. *IEEE Antennas and Wireless Propagation Letters*, 17(5), 776–779.

[7] Tan, J., Jiang, W., Gong, S., Cheng, T., Ren, J., & Zhang, K. (2018). Design of a dual-beam cavity-backed patch antenna for future fifth-generation wireless networks. *IET Microwaves, Antennas & Propagation*, 12(10), 1700–1703.

[8] Ramanujam, P., Venkatesan, R., & Arumugam, C. (2020). Electromagnetic interference suppression in stacked patch antenna using complementary split ring resonator. *Microwave and Optical Technology Letters*, 62(1), 193–199.

[9] Ullah, H., & Tahir, F. A. (2019). A broadband wire hexagon antenna array for future 5G communications in 28 GHz band. *Microwave and Optical Technology Letters*, 61(3), 696–701.

[10] Jilani, S. F., & Alomainy, A. (2017). A multiband millimeter-wave 2-D array based on enhanced Franklin antenna for 5G wireless systems. *IEEE Antennas and Wireless Propagation Letters*, 16, 2983–2986.

[11] Di Paola, C., Zhao, K., Zhang, S., & Pedersen, G. F., (2019). SIW multibeam antenna array at 30 GHz for 5G mobile devices. *IEEE Access*, 7, 73157–73164.

[12] Rahman, A., Ng, M. Y., Ahmed, A. U., Alam, T., Singh, M. J., & Islam, M. T. (2016). A compact 5G antenna printed on manganese zinc ferrite substrate material. *IEICE Electronics Express*, 13(11), 20160377.

[13] Sam, C. M., & Mokayef, M. (2016). A wide band slotted microstrip patch antenna for future 5G. *EPH-International Journal of Science and Engineering*, 2(7), 19–23.

[14] Thomas, T., Veeraswamy, K., & Charishma, G. (2015). MM wave MIMO antenna system for UE of 5G mobile communication: Design. In Proceedings of the Annual IEEE India Conference (INDICON), (pp. 1–5).

[15] Ali, M. M. M., & Sebak, A. R. (2016). Dual band (28/38 GHz) CPW slot directive antenna for future 5G cellular applications. In Proceeding IEEE International Symposium on Antennas and Propagation (APSURSI), (pp. 399–400).

[16] Ahmed, R. K., & Ali, I. H. (2019). SAR level reduction based on fractal sausage Minkowski square patch antenna. *Journal of Communication*, 14(1), 82–87.

[17] Van Yem, V., & Lan, N. N. (2018). Gain and bandwidth enhancement of array antenna using novel metamaterial structure. *Journal of Communication*, 13(3), 101–107.

[18] Tu, D. T. T., Phuong, N. T. B., Son, P. D., & Van Yem, V. (2019). Improving characteristics of 28/38 GHz MIMO antenna for 5G applications by using double-side EBG structure. *Journal of Communication*, 14(1), 1–8.

[19] Fante, K. A., & Gemeda, M. T. (2020). Broadband microstrip patch antenna at 28 GHz for 5G wireless applications. *International Journal of Electrical and Computer Engineering*, 11(3), 2238–2244.

[20] Darboe, O., Konditi, D. B. O., & Manene, F. (2019). A 28 GHz rectangular microstrip patch antenna for 5G applications. *International Journal of Engineering Research and Technology*, 12(6), 854–857.

[21] Rani, M. S. K. L., Bhagyasri, K., Madhuri, K. N. L., Venkata, K., & Hemalatha, M. (2022). Design and implementation of triple frequency microstrip patch antenna for 5G communications. *International Journal of Communication and Computer Technologies*, 10(1), 11–17.

44 Music recommendations based on emotions

Chilakala Sudhamani[1,a], Chinnala Nuthana[2,b], Lone Ravali[2,c], Bandari Rakesh[2,d] and Jangam Saketh[2,e]

[1]Associate Professor, Department of ECE, CMR Technical Campus, Hyderabad, Telangana, India

[2]Students, Department of ECE, CMR Technical Campus, Hyderabad, Telangana, India

Abstract

Music plays a significant role in human life, often chosen to reflect or influence mood. However, users typically need to manually browse and create playlists based on their current emotional state or mood. In this paper, our aim is to automate the music selection and playlist creation process by recognizing facial expressions to determine the user's mood. Facial expressions are a reliable-indictors of a person's emotions and mood. By capturing and analyzing these expressions, the system can recommend and play music that aligns with the detected emotion. This approach not only enhances the listening experience but also saves users time and effort of manually selecting songs. In this paper, we used facial expression recognition and mood detection algorithms to identify the user's emotion. The recognized emotion is compared with the existing database and automatically generates a playlist which matches the user's mood and play music.

Keywords: Facial recognition algorithm, music recommendation, playlist, database

Introduction

Facial expressions of an individual are one of the most natural and easiest ways of expressing their emotions, which is a non-verbal communication that reflects the person's internal state. However, music has been widely recognized as a powerful tool to alter the mood of a person. This music has been used to evoke emotions, provide comfort, relaxation, and excitement depending on the situation. It is possible to improve person's entire experience by playing music that corresponds with their present emotional state, which can have a relaxing impact or just make the environment more pleasant. Therefore, music plays a very important role in everyone's life to change their mood and emotions.

The music has been played based on human emotions, which will reduce stress and anxiety and changes their mood to calm and pleasant. These emotions and facial expressions have been recognized using many tools. In [1], human computer interaction has been used to recognize the emotions using facial recognition. A CNN based algorithm has been used to recognize facial expressions from facial components like lips and eyes. It works only with a few basic expressions such as surprise, anger, happiness

and neutrality of a person. The proposed system cannot handle head moments, rotations and obstacles. A LSTM-based system was proposed to identify the person's current emotions using the previous emotions [2]. It works accurately with a smaller number of emotions. A deep learning algorithm called emotion driven deep learning recommendation system has been proposed to identify the human emotion and play the music to change their mood [3–5]. The proposed method improved the average prediction accuracy of 12% and 15% of recommendation relevance compared to the existing models.

The authors identified the technical details of the music systems, including the algorithms used for emotion detection, music recommendation, user interface design, and possibly results from testing or evaluation [6]. The research integrates deep learning with practical systems, advancing the state-of-the-art in both computer vision and recommendation systems [7]. The authors explore how external factors (such as time of day, location, social context) influence music preferences and recommendations [8]. The paper might also present the overall architecture of their recommendation system. The proposed music system includes how the emotion detection

[a]sudhamani.ece@cmrtc.ac.in, [b]217r1a0479@cmrtc.ac.in, [c]217r1a04a2@cmrtc.ac.in, [d]217r1a0473@cmrtc.ac.in, [e]217r1a0486@cmrtc.ac.in

DOI: 10.1201/9781003684589-44

module interacts with the recommendation engine to generate personalized music suggestions [9]. The author discusses various techniques and methodologies used in previous studies to integrate mood detection into recommendation systems [10–12]. The proposed music system likely discusses various approaches, such as sentiment analysis of music lyrics, acoustic features analysis, or user interaction patterns related to emotional responses [13–15].

In this paper, we proposed a system by automating the process of music selection based on facial expressions, detecting the user's emotional state and using this information to play songs that align with their mood. The system is designed to eliminate the need for manual playlist creation, offering seamless and personalized listening experience. To achieve this, a music player is developed with the capability to recognize human emotions by leveraging existing databases and computing systems. The software captures an image of the user and, through advanced image segmentation and processing techniques, analyzes key facial features. These features are then used to detect the emotion being expressed, allowing the system to generate and play a corresponding playlist. Through the integration of facial expression recognition and music recommendation, this paper aims to create an intuitive system that enhances the emotional connection between the listener and the music.

The rest of the paper is structured as follows: Section 2 explains the proposed model. Section 3 presents the results and Section 4 draws the conclusions and future scope.

Proposed Model

The primary objective of this paper is to revolutionize the way users engage with music by developing an innovative recommendation system that prioritizes emotional resonance. In This paper, an emotional analysis can be used into music recommendation systems to greatly improve user experience and engagement. Our musical tastes and reactions are greatly influenced by our emotions, which also affect our general happiness with the listening experience. A block diagram is developed to help with the emotion-based recommendation process, which is shown in Figure 44.1. It includes various important steps, such as data collection, comparing the captured image or existing photo with database, music fetching and customized music recommendations.

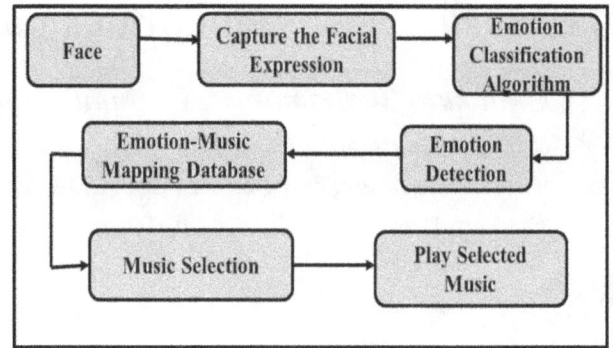

Figure 44.1 Block diagram of music recommendation system
Source: Author

Facial expressions are first captured using a camera and stored in a database. A cascade emotion classification algorithm is then applied, which uses a basic classifier to filter out non-object regions, reducing computational complexity. The remaining regions are analyzed to classify the expressions into emotional categories like happy, sad, or angry. Once the emotion is identified, it is compared with a pre-built database containing various emotions. The detected emotion is then mapped to suitable music, and the user can choose from a menu. Finally, the selected music is played for the user.

The working of the proposed system is illustrated through a flowchart, as shown in Figure 44.2. The process begins with the user initiating the system by logging in, likely involving authentication to ensure security. Once logged in, the user uploads an image or captures, which serves as the primary input for subsequent steps. The system then analyzes the image to detect and identify emotions, using facial recognition or other image processing techniques.

In this paper, we used classified algorithms. The detected image is then compared with the existing database. The flowchart includes a decision node where the system checks if the detected emotions match a predefined set of criteria. If a match is found, the system interacts with a database to retrieve relevant data or records related to the identified emotions. If no match is found, the system proceeds to categorize the emotion into a broader predefined category. Based on the detected emotion, the system selects appropriate music from the database and plays it, potentially designed to enhance or reflect the user's current mood. The process concludes once the music starts playing.

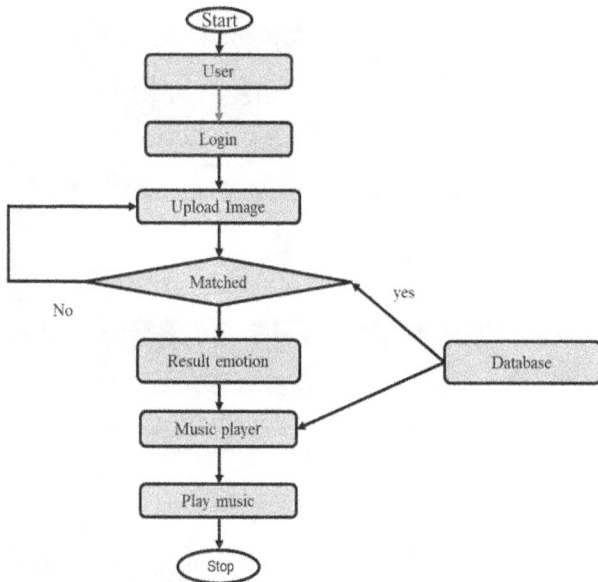

Figure 44.2 Flowchart
Source: Author

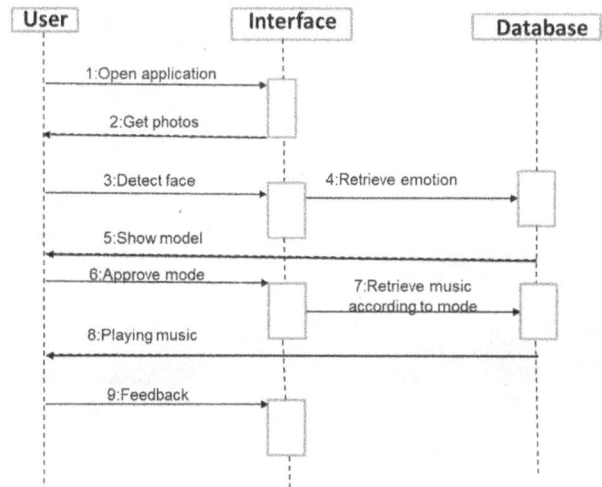

Figure 44.3 Proposed model working sequence
Source: Author

Therefore, the flowchart suggests a seamless integration of image analysis and emotion-based music recommendation, providing an interactive user experience based on their emotional state.

The working principle of our proposed system is presented in Figure 44.3. This figure shows the users and their interactions with the system. Initially, it opens the camera to take pictures of the user's face. The emotions are detected from the images called retrieve emotions. Once the emotion is extracted it will be compared with the existing database emotions and finalize the exact emotion to map a song. The song will be retrieved from the database according to the mode of emotion. Finally, the song will be played to change the user's mode. The feedback of the system will be considered to enhance the system's performance.

Therefore, the proposed system works with user's emotions and it helps to change the emotion quickly with the recommended music.

Results and Discussions

In this paper, we considered a music recommendation system based on emotions. This system aims to provide personalized recommendations tailored to the user's mood, enhancing their listening experience and emotional well-being using HTLM and JavaScript. The software contains various tabs: Upload image, processing and emotion detection, song prediction and play music, which are shown in Figures 44.4–44.7. Initially, the user will upload an image or capture an image from the face. The uploaded image is processed to eliminate the unwanted area and to identify the correct emotion. Once the emotion is detected, then it will map the emotion with the song in the database. Finally, the selected song starts playing. The person will come out of his mood with the song or else he/she can change the song from the playlist existing for the same emotion. For one emotion, we have uploaded five more songs in the playlist. The working function of the proposed system is explained step by step as:

- Step 1: Initially the app opened to see the tabs present in this proposed system, which is shown in Figure 44.4.
- Step 2: In this step, the user will upload the face using the existing images or capture the image directly using the camera, which is shown in Figure 44.5
- Step 3: In this step, the proposed system will detect emotion from the uploaded image by processing using the classified algorithm and comparing with the existing database, which is shown in Figure 44.6.
- Step 4: In this step, the detected emotion is mapped with the music and from the music database the emotion-based music is retrieved, which is shown in the Figure 44.8 . Finally, from the dropdown play list the user can select the song, and it will be played. By this music recommen-

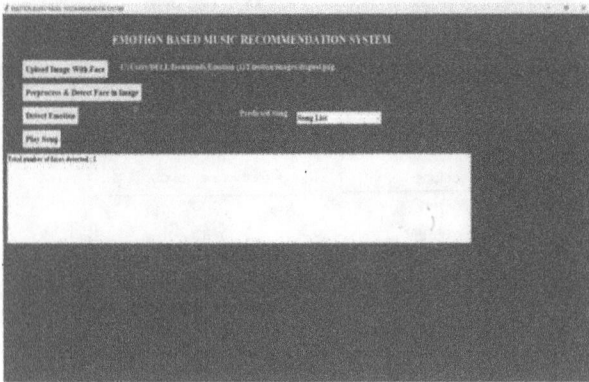

Figure 44.4 Designed screen
Source: Author

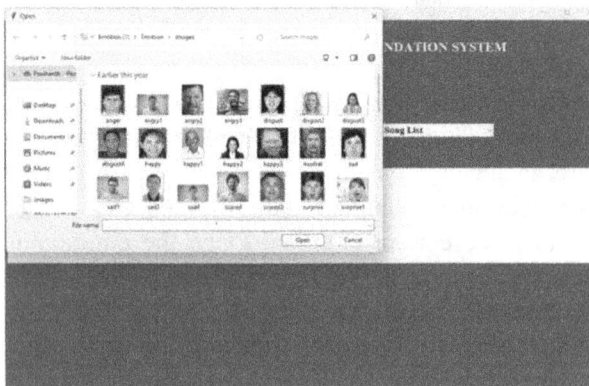

Figure 44.5 Uploaded image
Source: Author

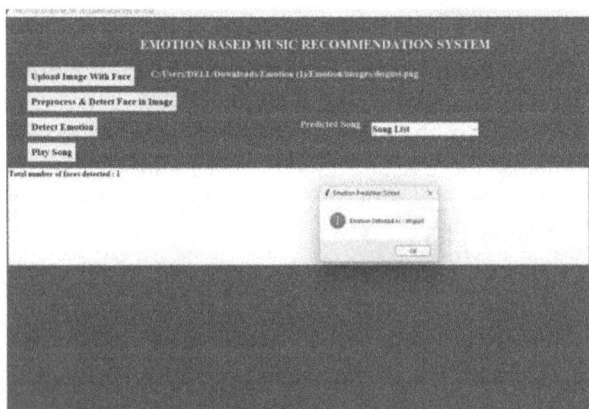

Figure 44.6 Detecting the emotion from uploaded image
Source: Author

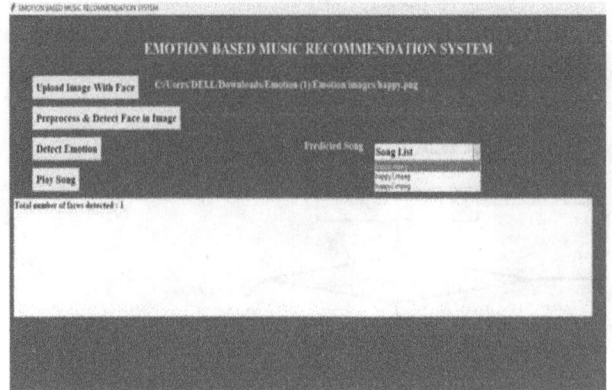

Figure 44.7 Song recommendation based on emotion
Source: Author

dation system, the user can change his mood within a short period of time compared to the manual browsing method.

Therefore, the music recommendation system allows users to efficiently select and play songs from a saved dropdown playlist. By utilizing personalized suggestions, users can quickly discover music that aligns with their mood, significantly reducing the time spent manually searching for songs. This system not only enhances the user experience by providing faster access to mood-appropriate music, but it also promotes a seamless transition in emotional states. Thus, the proposed system offers both time savings and a more enjoyable music discovery process, helping users change their mood swiftly and effortlessly compared to traditional browsing methods.

Conclusion and Future Scope

The proposed music recommendation system based on emotions offer a promising avenue for enhancing the user experience by providing personalized and emotionally resonant music suggestions. The user will upload the emotion and change his emotion through the music. This app helps the users to come out of their emotions. User's satisfaction testing plays a crucial role in gathering feedback from users to assess their overall satisfaction with the recommended music. Factors such as relevance, novelty, emotional resonance, and overall enjoyment of the recommendations are evaluated to refine the system further. Additionally, performance and robust testing help ensure that the system operates smoothly, efficiently, and resiliently under various conditions. This approach not only enriches the listening experience but also promotes diversity and equity in music consumption.

References

[1] Priyanka, V. T., Reddy, Y. R., Vajja, D., Ramesh, G., & Gomathy, S. (2023). A novel emotion based music recommendation system using CNN. In 2023 7th International Conference on Intelligent Computing and Control Systems (ICICCS), (pp. 592–596). IEEE.

[2] Liu, Z., Xu, W., Zhang, W., & Jiang, Q. (2023). An emotion-based personalized music recommendation framework for emotion improvement. *Information Processing and Management*, 60(3), 103256.

[3] Shi, Y., Shang, F., Xu, Z., & Zhou, S. (2024). Emotion-driven deep learning recommendation systems: mining preferences from user reviews and predicting scores. *Journal of Artificial Intelligence and Development*, 3(1), 40–46.

[4] Selvaganesan, J., Sudharani, B., Shekhar, S. C., Vaishnavi, K., Priyadarsini, K., Raju, K. S., et al.& (2024). Enhancing face recognition performance: a comprehensive evaluation of deep learning models and a novel ensemble approach with hyperparameter tuning. *Soft Computing*, 28(20), 12399–12424.

[5] Suma, S., Patel, P., Tiwari, R., Gundu, L., Devi, G. K., & Raju, K. S. (2024). A comprehensive analysis of human action recognition for noisy videos using CNN and LSTM. In 2024 International Conference on Innovation and Novelty in Engineering and Technology (INNOVA), (Vol. 1, pp. 1–8). IEEE.

[6] Sravanthi, T., Zuya, R., Samreen, S., Basri, R., Rithima, P., & Pasha, S. N. (2024). A system for recommending music based on emotions. In AIP Conference Proceedings, (Vol. 2971, no. (1). AIP Publishing.

[7] Assuncao, W. G., Piccolo, L. S. G., & Zaina, L. A. M. (2022). Considering emotions and contextual factors in music recommendation: a systematic literature review. *Multimedia Tools and Applications*, 81(,6), 8367–8407.

[8] De Prisco, R., Guarino, A., Malandrino, D., & Zaccagnino, R. (2022). Induced emotion-based music recommendation through reinforcement learning. *Applied Sciences*, 12(21), 11209.

[9] Sharma, V. P., Gaded, A. S., Chaudhary, D., Kumar, S., & Sharma, S. (2021) . Emotion-based music recommendation system. In 2021 9th International Conference on Reliability, Infocom Technologies and Optimization (Trends and Future Directions) (ICRITO), (pp. 1–5).

[10] Moscato, V., Picariello, A., & Sperli, G. (2020). An emotional recommender system for music. *IEEE Intelligent Systems*, 36(5), 57–68.

[11] Rumiantcev, M., & Khriyenko, O. (2020). Emotion based music recommendation system. In Proceedings of Conference of Open Innovations Association FRUCT. Fruct Oy.

[12] Ayata, D., Yaslan, Y., & Kamasak, M. E. (2018). Emotion based music recommendation system using wearable physiological sensors. *IEEE Transactions on Consumer Electronics*, 64,(2), 196–203.

[13] Abdul, A., Chen, J., Liao, H. Y., & Chang, S. H. (2018). An emotion-aware personalized music recommendation system using a convolutional neural networks approach. *Applied Sciences*, 8(7), 1103.

[14] Iyer, A. V., Pasad, V., Sankhe, S. R., & Prajapati, K. (2017). Emotion based mood enhancing music recommendation. In 2017 2nd IEEE International Conference on Recent Trends in Electronics, Information & Communication Technology (RTEICT), (pp. 1573–1577). IEEE.

[15] Raju, K. S., Rashmitha, P., Nagendra, K. V., Dharmireddi, S., Rekha, M., & Sanaboina, S. P. (2024). Intrusion detection system using generative unique adversarial neural network in cloud environment. In 2024 Ninth International Conference on Science Technology Engineering and Mathematics (ICONSTEM), (pp. 1–5). IEEE.

45 A deep learning framework for image colorization

A. Geetha Devi[1,a], T. Divya Naga Durga[2,b], Sk. Arifa[2,c], N. Sravya[2,d] and U. Chandra Sekhar[2,e]

[1]Associate Professor, Department of ECE, PVP Siddhartha Institute of Technology, Andhra Pradesh, India

[2]Student, Department of ECE, PVP Siddhartha Institute of Technology, Andhra Pradesh, India

Abstract

Image colorization is a visually media-rich and significant computer science problem that aims to leverages well-equipped dark images or monochrome pictures to contrast to full-color images. In this paper, we explore advanced deep learning techniques, mostly convolutional neural networks (CNNs) and generative adversarial networks (GANs), and apply them for realistic and context aware colorization of gray scale image. CNNs perform well on tasks where structural consistency and efficient processing are needed. First, GANs work beautifully at generating visually appealing images. Through extensive evaluation, we show that substantial gains in realism and accuracy of colorization can be obtained and also show how to choose the right model depending on the application needs and task complexity. This study demonstrates the promise of machine learning to automate highly complex artistic and visual tasks.

Keywords: Convolutional neural networks, deep learning image colorization (DLIC), generative adversarial networks, image colorization

Introduction

Image colorization refers to transforming gray scale images into a colorful and realistic colored images which is a big challenge in computer vision. This method has a wide range of applications for enhancing aesthetic appearance. Typically, as this process needs expertise, it was manual and time- consuming. However, with the emergence of deep learning [1], efficient pipelines for automating the process of colorization have been proposed.

Numerous datasets of gray scale images along with their respective color ones are used to train our deep learning models for colorization. These datasets enable the models to learn mapping gray scale inputs to their color counterparts. Convolutional neural networks (CNNs) [2] and generative adversarial networks (GANs) [3] are among the two most prevalent paradigms in this space.

While CNNs are highly effective at identifying spatial patterns, they help the model maintain image structure when producing colorizations of the previous (grayscale) image, while GANs have a competition-based framework to produce highly realistic and contextually relevant colorizations. All these techniques serve diverse image colorization needs.

Color representation is crucial to create an effective image colorization. RGB color space [4], which is extended and the most widespread color space used in most digital systems, has three basic channels red, green and blue.

In order to avoid these issues, the LAB color space [5] is typically used instead. LAB decouples brightness (L) from color information, to give perceptually uniform mapping closer to human eyesight. Switching from RGB colorspace to LAB color space before you color the image gives benefits. Surely, the introduction of LAB color model as such is very important, because it allows separating chromatic aspects of the input signal from luminance ones, focused on the generating colors where it's important to keep the brightness exactly the same as in the input. Targeted toward human vision [6] in a perceptively aligned manner, it yields natural and consistent colorizations so that high quality and reliable colorization can be distributed to different devices.

[a]geetha.agd@pvpsiddhartha.ac.in, [b]divyanagadurgatalluri@gmail.com, [c]shaikarifa1806@gmail.com, [d]sravyanakka8@gmail.com, [e]csekharreddy03@gmail.com

DOI: 10.1201/9781003684589-45

Dataset Specifications

Convolutional neural networks

As an input of the model is a grayscale image transformed into LAB colorspace where L is the luminance and a and b are holds color information. This L channel is then resized to 224 × 224, and normalized by subtracting 50 to be passed to the neural network. It helps to ensure that the input model matches its dimensions and improves representativeness of features. The training dataset, such as ImageNet, consists of diverse natural images, enabling the model to learn robust colorization patterns across various objects and scenes.

The dataset is processed in the LAB colorspace. Rebalancing techniques were employed during training to address the natural imbalance color distributions. This ensures the model predicts less common colors accurately. The combination of diverse data, LAB colorspace, and quantized color clusters allows the model to generalize well to varies gray scale inputs. Input of CNN is shown in Figure 45.1.

Generative adversarial networks

The model is trained on a subset of the COCO Sample Dataset sourced from the FastAI library, which consists of 5000 randomly selected images. These images are divided into training and validation sets, with 4000 images used for training and 1000 for validation which are shown in Figure 45.2.

For the input, the model uses the L channel from the *Lab* colorspace, which represents the gray scale intensity of the image. The images are first transformed from RGB to *Lab* format, where the L channel holds the luminance, and the a and b channels represent color information. The L channel is scaled to a range of [-1, 1], and the a, b channels are normalized similarly. The goal of the model is to get the a, b channels, which are then combined with the L channel to produce the fully colorized image. The network architecture includes a U-Net generator responsible for predicting the missing color channels (a and b), with the L channel as the input.

The generated colorized images are assessed using both GAN loss and L1 loss. The GAN loss secure the generated images resemble real color images, while the L1 loss reduce the pixel-wise difference between the predicted and true color values. Patch GAN discriminator is employed to evaluate the realism of the generated images, helping the model refine its output during training. This setup allows the model to learn high-quality colorization from gray scale inputs.

Convolutional Neural Networks

The CNNs tackle image colorization by converting grayscale images into colorized version through several specialized layers [7]. The process begins with feeding the grayscale image into the network, where initial layers identify basic features like edges and textures. As the image progresses through deeper layers, the network recognizes more complex patterns and relationships, such as shapes and contextual cues, which assist in the colorization process which is shown in Figure 45.3.

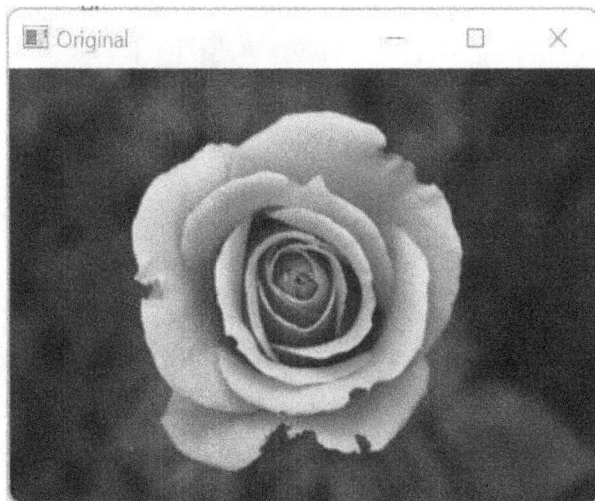

Figure 45.1 Input grayscale image for CNN
Source: Author

Figure 45.2 Sample dataset input images for image colorization using GAN
Source: Author

Model utilization

A pre-trained CNN, specifically designed for image colorization, is used to predict the chrominance (ab) channels in the LAB colorspace. The grayscale image's luminance (L) channel serves as input to guide the model's predictions.

Chrominance quantization

Predefined quantization points for the ab chrominance channels are incorporated into the network. These points are adjusted within specific layers to improve color balance and vibrancy in the output image.

Preprocessing

The grayscale image is normalized to meet the input requirements of the CNN. It is converted to the LAB color space, where luminance (L) is extracted. L is resized to 224×224 pixels, aligning with the model's expected dimensions, and further adjusted by subtracting a predefined offset to match the network's training configuration

Network prediction

The CNN processes the adjusted L channel to generate predictions for the ab chrominance channels [8]. The model leverages its ability to recognize spatial patterns and color relationships learned during training to deliver accurate chrominance estimates.

Post- processing

The predicted ab channels are resized to the original image dimensions. These channels are combined with the original L channel to reconstruct the colorized image in the LAB colorspace. The colorized LAB image is then converted back to the RGB or BGR color space for final visualization.

Beyond spatial understanding, CNNs leverage semantic information to assign appropriate colors to different regions or objects in the image. For example, a tree may be colored green, while the sky is given blue. By combining features extraction, global context analysis, and supervised learning with large datasets, CNNs are able to produce accurate and contextually relevant colorizations, making them highly effective for various image enhancement applications.

The methodology for image colorization stands out due to its customized use of a pre-trained CNN model with tailored modifications, such as introducing quantized ab chrominance points for vibrant color generation and accurate mapping of color distributions. The preprocessing pipeline efficiently normalizes grayscale inputs, converts them to the LAB color space, and ensures structural preservation.

Generative Adversarial Networks Based Image Colorization

The GANs are composed of two main components: generator and discriminator [9]. The generator creates new images, while the discriminator's role is to get the quality of these created images by distinguishing them from real ones. These two networks are trained together in a process called adversarial training, where the generator continuously improves by trying to produce images that are hard for the discriminator to identify as fake, while the discriminator works to become better at telling apart real and generated images. This back-and-forth training eventually leads to the generator producing highly realistic images which is shown in Figure 45.4.

U net based generator design

The generator [10] employs a U-Net [11] framework, which has encoder and decoder structure with skip connections. These connections help maintain important spatial features as the image is processed. In the encoder, the image dimensions are reduced while key features are extracted, and in the decoder, the image is reconstructed to its original size.

Discriminator design

The discriminator works by evaluating small patches of the generated image to determine if they are fake or real [12]. It uses convolutional layers to shrink the image size while capturing essential characteristics. The last layer outputs a binary result, indicating whether the patch is from a real image or one produced by the generator. This patch-based method

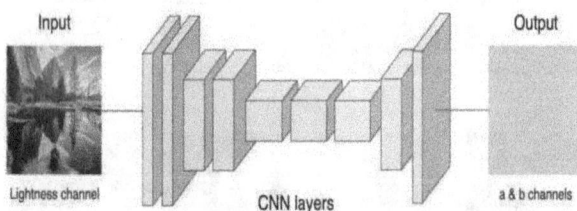

Figure 45.3 Convolutional layers for image colorization using CNN
Source: Author

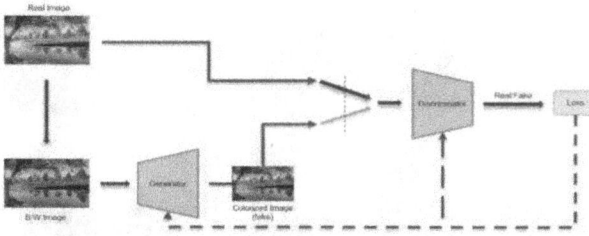

Figure 45.4 Generator and discriminator structure for image colorization using GAN
Source: Author

enables the discriminator to focus on fine details, improving its accuracy in distinguishing between real and fake images.

GAN-based training strategy

The model is trained within a GAN framework, where the generator and discriminator are optimized together. The generator's goal is to produce convincing colorized images that the discriminator cannot easily distinguish from real ones. Meanwhile, the discriminator works to correctly identify whether each image is generator or real. To further guide the generator, a pixel-wise L1 loss function is used, which helps ensure the generated colors closely match the ground truth, resulting in improved colorization performance.

In the case of image colorization, the generator receives a grayscale image and attempts to generate a colorized version of it. The discriminator then evaluates the generated image by comparing it to real colorized images, guiding the generator in producing more convincing colorization. Both networks are trained through back propagation [13], with updates based on the feedback from the discriminate or the generator uses a U-Net architecture [14]. The U-Net architecture enhances the performance of the generator in GANs by utilizing skip connections, which preserve detailed spatial information and enable multi-scale feature extraction for capturing both local and the global features. The encoder-decoder structure [15], integrated with skip connections, facilitates effective gradient flow, improves training stability, and speed up convergence.

This makes U-Net especially well-suited for image colorization tasks, ensuring structural consistency while generating realistic and coherent colors.

The unique feature is its integration of a GAN-based image colorization pipeline using a U-Net generator and Patch Discriminator. The custom dataset

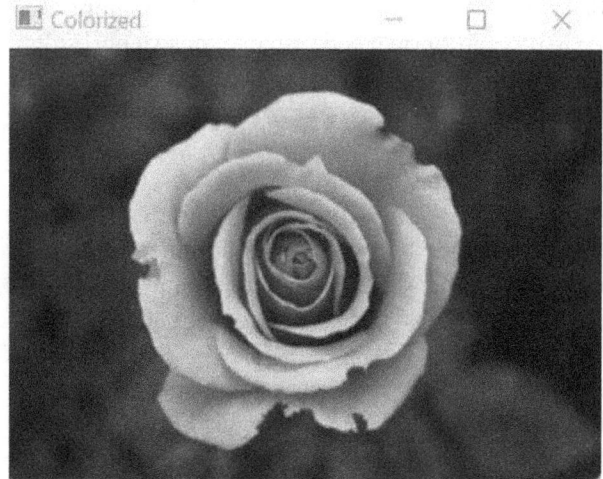

Figure 45.5 Output colorized image for colorization using CNN
Source: Author

loader, data augmentation, and RGB to LAB conversion enhance the model's performance. The model combines L1 loss with adversarial loss, using modular training and efficient loss tracking to improve training and performance visualization.

Results

Output of CNN

The result is the successful colorization of gray scale image as in Figure 45.5, where the model predicts plausible and visually appealing colors for each pixel. By leveraging the LAB colorspace, the gray scale image's luminance (L channel) is combined with the predicted chromaticity values (a and b channels) to reconstruct a full-color image. The output retains the structural integrity of the original grayscale image while adding realistic colors based on learned patterns from the training dataset. This approach demonstrates strong generalization across various image types, including landscapes, objects, and human subjects, producing results that are visually coherent and natural.

The advantage of using this algorithm lies in its efficiency and accuracy. Unlike traditional methods that rely on manual colorization or predefined rules, this deep learning-based approach automates the process by learning color mappings directly from data. Additionally, rebalancing techniques used during training address biases in color frequency, resulting in more diverse and accurate colorization. Compared to other methods, this approach is faster, requires minimal user intervention, and produces more contextually appropriate and visually consistent results.

Output of GAN based colorization

In the training loop, the model undergoes multiple iterations across the dataset for a set number of epochs, in this case, 20 epochs. While each epoch, this model follows a process to optimize its performance. For each batch of data, the gray scale input (L channel) is passed through the generator (U- Net), which produces the predicted colorization (a and b channels). This colorization is then evaluated by the discriminator, which is trained to distinguish between real and fake colorizations. The discriminator is updated by computing the loss for both real and fake images, ensuring that it can accurately classify the authenticity of generated colorizations.

Simultaneously, generator is updated based on the feedback from the discriminator. Generator minimizes two loss functions: the GAN loss, which encourages the generator to produce realistic colorizations to fool the discriminator, and the L1 loss, which ensures the predicted a and b channels match the ground truth. These combined losses are used to update the generator's weights. After 200 iterations, the training process logs the current loss values and visualizes a few generated colorizations. These visualizations include the gray scale input (L channel), the generated colorization, and the ground truth colorization, allowing for a visual assessment of the model's progress.

As training progresses, the loss values for both the generator and discriminator decrease, signaling that the generator is improving its colorization capabilities and producing more realistic outputs. The discriminator becomes more proficient at differentiating between real and generated images. Over the course of the 20 epochs, the model learns to effectively colorize grayscale images, with the loss functions reflecting a steady improvement in performance. By the end of training, the generator produces colorizations that are both visually convincing and close to the actual ground truth color information. The output is shown in Figure 45.6.

The discriminator loss (Loss_D) comprises Loss_D_fake and Loss_D_real, which measure how well the model distinguishes between real and generated images; both have been decreasing over epochs, indicating improved performance, with Loss_D stabilizing at lower values. The generator loss (Loss_G)

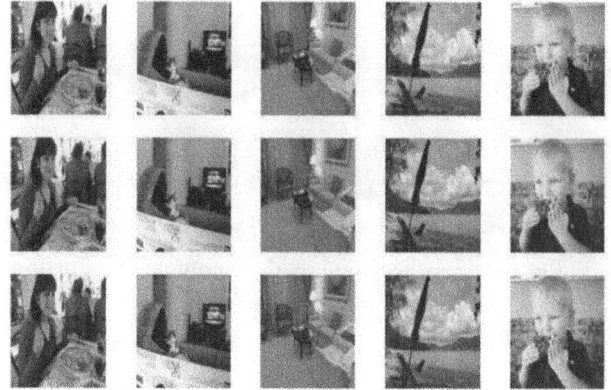

Figure 45.6 Result of GAN network which consists of gray scale image, Generator image and the ground truth image
Source: Author

```
loss_D_fake: 0.38275
loss_D_real: 0.39109
loss_D: 0.38692
loss_G_GAN: 1.95686
loss_G_L1: 8.78460
loss_G: 10.74146
```

Figure 45.7 Generator and the discriminator losses
Source: Author

consists of Loss_G_GAN, representing how well generator fools the discriminator, and Loss_G_L1, It measures the difference between created and groundtruth photos on a pixel-by-pixel basis. While Loss_G_GAN has been increasing, indicating better adversarial performance, Loss_G_L1 has remained relatively stable, ensuring accurate reconstruction. As a result, the total generator loss (Loss_G) remains nearly constant, reflecting the model's balanced learning process for generating realistic colorized images. This losses are shown in Figure 45.7.

Table 45.1 gives the values of PSNR and SSIM of image colorization using CNN and GAN. PSNR measures the quality of a restoration in terms of the signal-to-noise ratio between the original and the colorized image. Higher PSNR values indicate better restoration quality. SSIM measures perceptual

Table 45.1 Comparision of PSNR and SSIM parameters for CNN and GAN colorization models.

Types Of Colorization	PSNR	SSIM
CNN	22.10	0.9934
GAN	24.38	0.8923

Source: Author

similarity based on structure, contrast and luminance, so a value closer to 1 means more similarity to the original image.

Generative Adversarial Networks (GANs) are better suited for tasks where the aim is to generate highly realistic, visually appealing colorizations, particularly for complex or diverse images, GANs are able to produce the vibrant, natural- looking colorizations by capturing intricate relationships between image elements.

Conclusion

However, both convolutional neural networks (CNNs) and generative adversarial networks (GANs) have their individual advantages in image colorization depending on the purpose of the task. CNNs: When structural and detail preservation is of critical importance. When you can get labeled data, they are great for colorization to the right place. However CNNs tend to struggle when it comes to realistic colors, especially in cluster images.

On the other hand, GANs are used to produce realistic looking colorful images, catching complex relationships within the image. When the aim is to create vivid and life- like colors from the dark structures, they are successful, but at the cost of more accuracy of structure.

References

[1] Cheng, Z., Yang, Q., & Sheng, B. (2015). Deep colorization. In ICCV. (pp. 415–423).

[2] Larsson, G., Maire, M., & Shakhnarovich, G. (2016). Learning representations for automatic colorization. In ECCV.

[3] Zou, C., Mo, H., Gao, C., Du, R., & Fu, H. (2019). Language- based colorization of scene sketches. *ACM Transactions on Graphics,* 38(6), 1–16.

[4] Thasarathan, H., & Ebrahimi, M. (2019). Artist semi-automatic animation colorization. In ICCV Workshop. ICCVW.

[5] Chen, J., Shen, Y., Gao, J., Liu, J., & Liu, X. (2018a). Language- based image editing with recurrent attentive models. In CVPR. (pp. 8721–8729).

[6] Yoo, S., Bahng, H., Chung, S., Lee, J., Chang, J., & Choo, J. (2019). Coloring with limited data: few-shot colorization via memory- augmented networks. In CVPR.

[7] Zhang, R., Isola, P., & Efros, A. A. (2016). Colorful image colorization. In ECCV.

[8] Zhang, R., Isola, P., Geng, D., & Efros, A. A. (2016). Colorization with deep convolutional neural networks.

[9] Goodfellow , I., Pouget-Abadie, J., Mirza, M., et al., (2014). Generative adversarial networks. In Advances in Neural Information Processing Systems, (Vol. 3, pp. 267 2–2 680).

[10] Radford, A., Metz, L., & Chintala, S. (2015). Unsupervised representation learning with deep convolutional generative adversarial networks. arXiv preprint arXiv:1511.06434.

[11] Ronneberger, O., Fischer, P., & Brox, T. (2015). UNet: Convolutional networks for biomedical image segmentation. In Medical Image Computing and Computer-Assisted Intervention – MICCAI 2015 (pp. 234–241). Springer.

[12] Liu, M. Y., & Zhang, Y. (2016). Image inpainting via generative adversarial networks. arXiv preprint arXiv:1607.07539.

[13] LeCun, Y., Bengio, Y., & Hinton, G. (2015). Deep learning. Nature, 521(7553), 436–444.

[14] Isola, P., Zhu, J. Y., Zhou, T., & Efros, A. A. (2017). Image-to-image translation with conditional adversarial networks.. In Proceedings of the IEEE Conference on Computer Vision and Pattern Recognition (CVPR) (pp. 1125–1134).

[15] Kingma, D. P., & Welling, M. (2013). Auto encoding variational bayes.arXiv preprint arXiv:1312.6114

46 AI car with real-time detection of damaged road and lane detection by deep learning using Python

T. Nagarjuna[1,a], S. Rafeeq[2,b], M. Bhanu Prakash[2,c], J. Rohith[2,d] and B. Suvarna Lakshmi[2,e]

[1]Associate Professor, Department of ECE, CMR Technical Campus, Hyderabad, India

[2]Students, Department of ECE, CMR Technical Campus, Hyderabad, India

Abstract

Because of the various road damages as resulted in numerous deaths, research into road damage detections, particularly dangerous road damages detections and warnings, is essential for traffic safety management. Road damage detection system is mostly processing data on the cloud, which as a large latency due to the large distances transmissions. In these systems that requires big, carefully labelled datasets to achieve outstanding performances, machine learning methods are typically used in this. In this study, we suggest using deep learning, for the detection, warning about the defects of road and the lane detections. Foundation of the road surface analysis is visual observations by persons and the quantitative analysis by pricey tools. The goal is to achieve the high accuracy and low latency, ensuring effective road monitoring and safe navigation in varying conditions and the path of the road.

Keywords: Canny edge detection, convolutional neural networks CNN, deep learning, image segmentation, lane detection, object detection, road detection, YOLO

Introduction

Road damage detection and lane detection is a innovation that grasp the deep learning to enhances autonomous vehicle systems. Project mainly focuses on the developing the system that uses the deep learning algorithm to identifies the damaged section of roads, such as potholes cracks, or rough surfaces, and the ensures safe guide by detecting lanes in the real time. The system that uses the computer vision techniques and the deep neural networks to analyze the camera feed from the vehicles. Convolutional neural networks (CNNs) are mainly used for object detections, enabling the car to distinguish between the damaged road surfaces and the normal road conditions. It gives the accurate detection of the road conditions Lane detection is got by the processing of the road boundaries, which supports in lane-keeping and the guidance on roadways. It provides an opportunity to explore thoroughly into the applications of deep learning in real world applications, plays a vital role in enhancing the safety and the reliability of autonomous driving. We proposed system mainly focuses on the detection of the path holes them through a driver warning system, which provides the information about the path holes helps the drivers to prevent accidents caused by them [1–3]. Using some algorithms and computer vision, our goal is to equip autonomous vehicles with ability to respond to their environment with high precision. By adding this algorithm into the vehicle's onboard systems, we enhance its capacity to navigate around barriers and risk and guarantee precise positioning within assigned lanes.

Literature Survey

Study by Jagadeesha and Patel [4] aims to develop the algorithms for the autonomous vehicles that use AI to accurately detect road defects and lane boundaries in real time.

Accurate, identification of the road lanes is essential to safe guidance and the traffic safety of the autonomous vehicles [5]. This paper explores the lane detection by utilizing the mask region-based CNN model.

[a]Nagarjuna.ece@cmrtc.ac.in, [b]217r1a04c0@cmrtc.ac.in, [c]217r1a04a8@cmrtc.ac.in, [d]217r1a0487@cmrtc.ac.in, [e]217r1a0471@cmrtc.ac.in

DOI: 10.1201/9781003684589-46

Another research by Khan et al. [6] represents the deep learning algorithms for the identification of damaged road can enhance the safety of the vehicles and pedestrians on the road.

By Bakhytzhan Kulambayev, et al., [7] describes a study on real-time road damage detection using a deep learning framework based on Mask R-CNN models. Focuses on utilizing Mask R-CNN for instance segmentation to identify and classify various road damages like potholes and cracks. By Weiyu Hao [8], Autonomous driving systems rely heavily on lane detection. Therefore, ensuring its robustness and reliability is crucial for road safety. This work proposes to validate one of the leading lane detection models in the CULane benchmark with an alternative synthetic dataset. By R. Palani, et al. [9], This paper explores the colourful types of road damages. It tells road damage factors and its effect on environmental pollution due to business snags and the profitable status of road druggies.

Another research by Kshitija Chavan, et al. [10], Visually impaired individuals encounter numerous impediments when traveling, such as navigating unfamiliar routes, accessing information, and transportation.

Fig 46.1, 46.4, and 46.5 shows the Road damage detection and process. And Fig 46.2, 46.3, and 46.6 shows the lane detction and process.

Implementation of Detection of the Damaged Road and the Lane Detection by Deep Learning Using Python

To implement real time identification of the road defects and lane boundaries using deep learning by python via YOLO algorithm. The goal is to create the system that analyzes video frames to the identify road lanes and damage using YOLO models. YOLO consider object detection as a unified regression problem, processing the image in the one step and dividing into an S x S grid. APIs are implemented for the adding, deleting, displaying, and checking functions. The UI and API calls interact with the backend to perform these actions while at the same time capturing video from a webcam. For each frame of video, passes into to the YOLO model to perform inferences, which will return by drawing the boxes around the detected objects including, lane markings and road damages.

Firstly, taking the input images, which could be road images for the lane detection. System determines whether the image is in color or grayscale image. If it's a color image then it is converted to the grayscale image to simplifies the processing and reduce level of difficulty. Noise reduction techniques like Gaussian Blur are applied to smooth the image

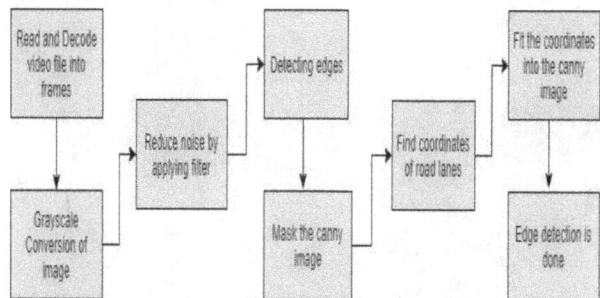

Figure 46.2 Working on lane detection
Source: Author

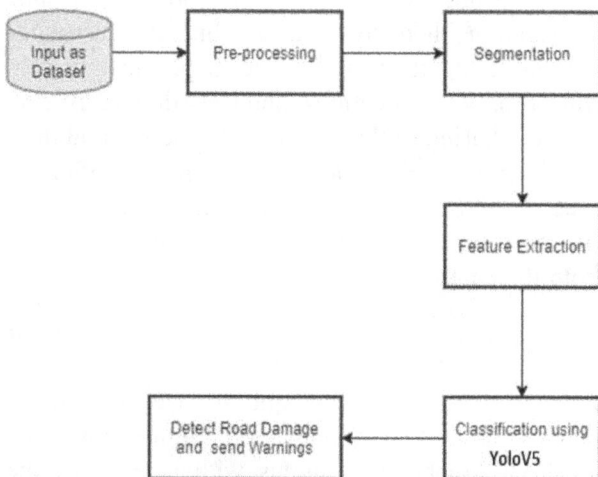

Figure 46.1 Working on road damage detection
Source: Author

Figure 46.3 Road lane detection
Source: Author

Figure 46.4 Road damaged detection
Source: Author

Figure 46.5 Road damaged detection
Source: Author

Figure 46.6 Road damaged detection
Source: Author

Results

The lane and road damage detection systems focuses on the identification of the lane boundaries and road damages like cracks and potholes. Lane detection guarantees precise recognition of lane markings, helping to issue lane departure warnings and enhance driving safety. Road damage detection, using YOLOv5, have the different types of road damages, providing users with timely maintenance notifications. The system offers real-time detection of damaged roads with high precision, ensuring road safety and support for an infrastructure monitoring.

Conclusion

This approach uses the advanced methods for pothole and lane detection through deep learning techniques, using YOLO to extract road and the lane information, there by understanding the surrounding environment of the vehicle. It provides drivers with timely notifications about potholes. It also enables sharing of the pothole information and their locations with the relevant government officials, enabling them to take preventive actions. Once the potholes are repaired, their locations can be removed from the databases.

Acknowledgement

This project is carried out in Department of Electronics and Communication Engineering from the CMR Technical Campus with support of the Director, HOD, and the Faculty members.

and the removes irrelevant details. Canny edge detection is used for detecting significant changes in pixel intensity, markings the edges in the image. The system starts with a dataset, typically consisting of images or video data related to roads, which is pre-processed for the analysis, including normalization and noise reduction. This stage segments the data into significant regions, separating the road surfaces. Key features, such as texture, shape are extracted. The YOLOv5 model is then employed to classifies the various road features and detect damage. Finally, the system identifies road damage and sends alerts to the users.

References

[1] Thirupathi, d., et al. (n.d .). density based smart traffic control system using canny edge detection. Journal for Research in Applied Science and Engineering Technology, ISSN: 2321-9653.

[2] Gothane, S., Sarode, M. V., & Thakre, V. M. (n.d). Identification of indian road distress with ruleset using image processing. Journal For Research in Applied Science and Engineering Technology, vol 6, issue no IX.

[3] Gothane, S., & Sarode, D. M. (2015). Case study: analysis and study of different approaches for road network maintenance. *International Journal of Scientific Engineering and Research,* 3, 1–4.

[4] Jagadeesha, H. S., & Vinay Patel, G. L. (2024). AI car with real-time detection of damaged road and lane detection. *International Journal of Scientific Research in the Engineering and Management (IJSREM),* 08(07), 1–4.

[5] Abdrakhmanov, R., Elemesova, M., Zhussipbek, B., Bainazarova, I., Turymbetov, T., & Mendibayev, Z. (2023). .Mask R-CNN approach to real-time lane detection for autonomous vehicles. *International Journal of Advanced Computer Science and Applications,* 14,(5).

[6] Khan, M. W., Obaidat, M. S., Mahmood, K., Batool, D., Badar, H. M. S., Aamir, M., et al. (2024). Real-time road damage detection and infrastructure evaluation leveraging unmanned aerial vehicles and tiny machine learning. *IEEE Internet of Things Journal,* 11(12), 21347–21358.

[7] Kulambayev, B., Nurlybek, M., Astaubayeva, G., Tleuberdiyeva, G., Zholdasbayev, S., & Tolep, A (2023). Real-time road surface damage detection framework based on mask R-CNN model. *International Journal of Advanced Computer Science and Applications,* 14(9).

[8] Hao, W,. (2023). Review on lane detection and related methods. *Cognitive Robotics,* 3, 135–141.

[9] Palani, R., Puviarasan, N., & Ramaprasath, A. (2022). Literature review of road damage detection with repairing cost estimation. *International Journal of Mechanical Engineering,* 1–5.

[10] Chavan, K., Chawathe, C., Dhabalia, V., & Sankhe, A. (2022). Pothole detection system using YOLO v4 algorithm. *International Research Journal of Engineering and Technology (IRJET),* 80, 25171–25195

47 A hybrid model for breast cancer prediction: Integrating Relief, PCA, and SVM for improved accuracy

G. Srikanth[1,a], M. Sangameshwar[2,b], Angadala Narendra[2,c], Karnati Meghana[2,d] and Edagoti Ajay Kumar[2,e]

[1]Professor and Dean, Department of ECE, CMR Technical Campus, Hyderabad, India

[2]Students, Department of ECE, CMR Technical Campus, Hyderabad, India

Abstract

The breast cancer is a serious health issue worldwide, especially for women, so making early and accurate diagnoses is essential. This work introduces a machine-learning framework to categorize breast cancer. Our approach uses a hybrid model for breast cancer prediction, which integrates Relief Feature selection (Relief FS), principal component analysis (PCA), and SVM to identify the most relevant clinical attributes. It uses PCA to minimize the number of dimensions while retaining essential data patterns and structures. The integration of both improves the model generalization and boosts computing efficiency. We employ the Support Vector Machine, using both the linear and the radial basis functions (RBF) to identify between the benign and the malignant tumors, with radial basis function kernel performance better when we are dealing with the initial models, integrating Relief FS, PCA, and SVM significantly enhances prediction accuracy. By incorporating dimensionality decline and feature selection, our system minimizes overfitting, optimizes computational resources, and improves diagnostics prediction. The efficiency of the machine learning methods in medical diagnostics is demonstrated by this study, which provides adjustable and efficient methods for the early breast cancer identification, ultimately aiding in more effective patient management and treatment strategies.

Keywords: Breast cancer detection, diagnostic accuracy, dimensionality reduction, feature selection, machine learning

Introduction

Breast cancer continues to be a global health concern, impacting millions of people annually. In 2023 alone, there were 2.3 million new cases and 685,000 deaths, underscoring the urgent importance of timely and reduced diagnosis. While mammography screening has played a pivotal role in reducing mortality rates, it has several limitations. False positives remain a significant challenge, often leading to unnecessary emotional distress, follow-up imaging, and invasive procedures like needle biopsies. Additionally, the increasing shortage of radiologists in many countries threatens the effectiveness of traditional diagnostic approaches, highlighting the need for automated, data-driven solutions. Machine learning has become a valuable asset in medical diagnostics, enabling the analysis of large clinical datasets and uncovering patterns that may be difficult for humans to detect. This research introduces a machine learning-driven system for the recognition of breast cancer utilizing the Wisconsin Diagnostics breast cancer (WDBC) dataset. By combining Relief Feature selection (Relief FS) to discover the most relevant clinical attributes and the PCA for dimensionality reduction, the advanced strategy enhances computational efficiency while maintaining diagnostic accuracy. The categorized process is handled by SVM, focusing on radial basis functions kernel, which has indicated higher performance in discriminating between malignant and benign tumors. The combination of these techniques ensures improved diagnostic precision and reduces the chance of false-positive results, addressing the main primary drawbacks of the conventional approach that depends extensively on human expertise. This model provides an automated, intelligent diagnostic tool that can keep up with medical professionals in making more informed decisions. Faster processing times and improved specificity in classification contribute to early detection, ultimately

[a]gimmadisrikanth79@gmail.com, [b]sangameshwarmunipally@gmail.com, [c]narendraangadala87@gmail.com, [d]meghanareddy05113@gmail.com, [e]217r1a0483@cmrtc.ac.in

DOI: 10.1201/9781003684589-47

leading to timely medical intervention and better patient outcomes. By using advanced feature selection, dimensionality reduction, and high-performance categorized methods, this method highlights the transformative capacity of machine learning in breast cancer prediction. The proposed framework not only enhances reliability and effectiveness but also assigns to making breast cancer screening extra accessible, particularly in regions facing a shortage of trained radiologists [1–3]. As the healthcare industry continues to embrace artificial intelligence-driven solutions, this research marks a major advancement in enhancing diagnostic exactness and optimizing patient care.

Literature Survey

In the study "Analysis on machine learning based early breast cancer Detection" Ghage et al. [4] found that early detection plays a vital role in breast cancer treatment, and this gives the accurate solutions for the diagnosis. Various types of machine learning techniques are used. Have been surveyed for breast cancer prediction, each displaying varying levels of success. A method with the integration of various machine learning feature selection and reduction techniques has shown high accuracy and sensitivity in diagnosis. Tested on datasets such as RSNA, and MIAS this method has proven essential in improving breast cancer predictions and practical implementation could offer radiologists a dependable tool for early detection, ultimately improving patient outcomes [4].

To improve tumor detection performance "Breast cancer detection based on machine learning", fast prediction of cancer is important. While diagnostic mammography helps assess breast tissue, its effectiveness declines with dense tissue, leading to missed cases. Machine learning offers a solution for accurate detection, improving treatment outcomes. This method employs SVM machine decision trees naive Bayes on the Wisconsin (WDBC) dataset, evaluating their performance using accuracy, precision, and confusion matrix analysis [5].

The study by Neelima et al. explores fuzzy logic machine learning techniques for breast cancer estimation, specifically decision trees applied to the Wisconsin diagnostics dataset. By integrating fuzzy logic classification, the model improves efficiency and outperforms traditional classifiers like k-nearest neighbors and naive Bayes. However, its limitations include computational complexity, dependency on fuzzy threshold values, and potential bias in classification rules. Our machine learning-based system overcomes these drawbacks by integrating Relief FS for optimal feature selection, and PCA for dimensionality reduction, ensuring higher efficiency, reduced false positives, and improved real-world applicability in early breast cancer identification [6].

The paper on the breast cancer categorization using a Microstrip patch antenna highlights its potential as a noninvasive, radiation-free alternative to traditional methods like mammography and MRI. By utilizing microwave imaging, it detects tumors based on permittivity and conductivity differences in breast tissue. However, limitations such as signal interferences, variability in tissue properties, and standardization challenges affect its accuracy. Our machine learning-based system overcomes these drawbacks by integrating feature selection (Relief FS), dimensionality reduction PCA, and classification (SVM with RBF kernel) using the Wisconsin diagnostic dataset. This approach enhances tumor classification accuracy, reduces false positives, and minimizes reliance on invasive procedures, making breast cancer detection more reliable and efficient [7].

The study established deep deep-learning approaches for breast cancer detection and utilizes an efficient net method, which is applied to mammography pictures from the CBSI-DDSM dataset. By using transfer learnings and the early stopping, the model got an efficiency of 75%, demonstrating its capability to the detect, breast cancer. However, challenges such as limited accuracy, reliance on high-quality mammography images, and computational complexity restrict its practical application. Our machine learning-based system addresses these drawbacks by integrating Relief FS and PCA, and optimized SVM model for classification. Dissimilar to this approach which uses labeled image datasets, our approach utilizes structured clinical data to enhance classification accuracy, reduce computational costs, and ensure early detection with higher reliability [8].

The study by Melek et al. uses a Compared to traditional CNN modes, this integrates temporal changes. Our model addresses this limitation by combining Relief FS and PCA and the SVM combined with RBF kernel for classification. It reduces computational costs, enhances classification accuracy, it makes breast cancer tests more scalable and accessible [9].

The study uses a deep CNN-based multi-model Radiomics approach for detection of breast cancer, by the integration of ultrasound, MRI, and mammograms for improved classification accuracy. The model achieved enhanced diagnostic accuracy across multiple imaging modalities. Some challenges are also their computational costs, and potential misclassification of overlapping tissue structures limiting their practical implementation. Our proposed model issues by integration of Relief, PCA, and SVM models for precise classification [10].

The study presents an analysis of the machine learning classifiers for the determination of breast cancer by using the Wisconsin diagnostic breast dataset. Various algorithms and decision trees with the efficiency of 96%. However, challenges such as overfitting, imbalanced data handling, and computational complexity affect the reliability by integrating Relief FS, PCA, and SVM algorithms with an RBF kernel for improved classification accuracy. This approach enhances predictive performance, reduces overfitting, and ensures a more robust and scalable breast cancer detection system, ultimately supporting early diagnosis and better patient outcomes [11].

Methodology Used

The project's primary aim is to show breast cancer using machine learning techniques, utilizing the Wisconsin diagnostic dataset. The methodology starts with data preprocessing, ensuring the dataset is examined for any missing values.

The figure 47.1 is the block diagram of the breast cancer detection. The figure 47.2 is the checking the accuracy and figure 47.3, 47.4 are prediction pages.

Normalized for uniform scaling, and separated into the training and testing sets for generalization. Then, feature selection dimensionality reduction is performed using Relief feature selection. The Relief FS is used to identify the most relevant clinical attributes, while principal component analysis (PCA) re2duces redundancy and enhances computational efficiency. For classification, we used a SVM with both the linear and the radial basis function kernels, where the radial basis function kernel demonstrates high performance in differentiating between cancerous and cancer less tumors. The framework is trained using cross-validation and evaluated through accuracy and analysis with others categorized as decision trees, to validate the effectiveness of the system. By integrating optimal feature selection, efficient dimensionality reduction, and robust classification techniques, our project enhances early detection, reduces false positives, minimizes computational costs, and improves diagnostic accuracy, making it a scalable and efficient way for breast cancer screening.

The Table 47.1, and 47.2 are the result indicates accuracy and reduces the false positive rates, making the system effective for the early breast cancer prediction.

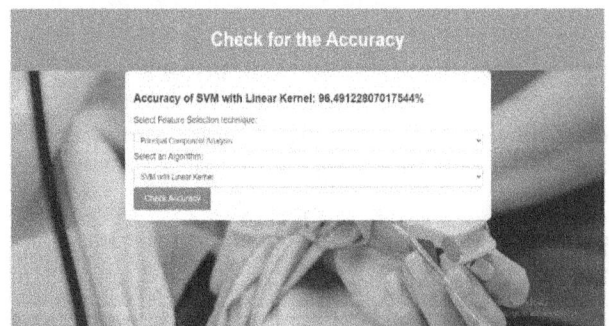

Figure 47.2 Checking the accuracy
Source: Author

Figure 47.3 Prediction page using RFS
Source: Author

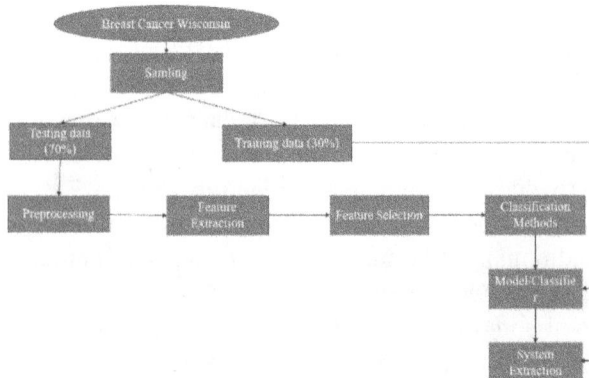

Figure 47.1 The block diagram of the breast cancer detection
Source: Author

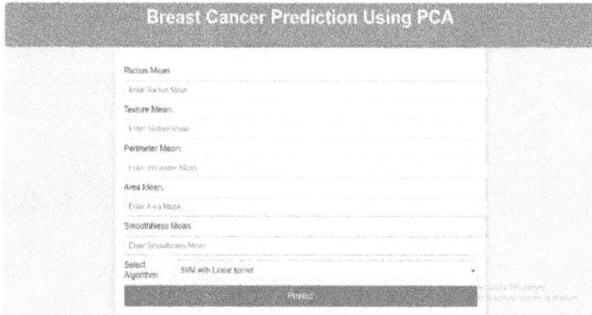

Figure 47.4 Prediction page using PCA
Source: Author

Table 47.1 Model performance metrics.

Metric	SVM (Linear Kernel)	SVM (RBF Kernel)
Accuracy	96%	98%

Source: Author

Table 47.2 Comparison with existing models.

Model	Accuracy (%)
Decision Tree	90%
k-NN	92%
Logistic regression	94%
Hybrid model (SVM + Relief FS + PCA)	96%

Source: Author

Results

The proposed hybrid model for the breast cancer determination approach model demonstrates high classification accuracy using the Wisconsin diagnostic (WDBC) dataset. After applying Relief FS and principal component analysis (PCA) for feature optimization, the SVM with the RBF achieved high performance compared to the other classifiers. The model was evaluate using accuracy classification of cancerous and cancer-less.

The result indicates accuracy and reduces the false positive rates, making the system effective for the early breast cancer prediction. Compared to traditional diagnostic approaches, our method offers faster processing, better feature selection, and enhanced classification accuracy, proving its potential as a cost-effective and scalable breast cancer detection.

Conclusion

The proposed machine learning-based breast cancer detection model effectively classifies tumors as cancerous or cancer less using the WDBC dataset. By integrating Relief Feature Selection (Relief FS) and principal component analysis (PCA), the model optimizes feature selection, and redundancy and improves computational efficiency. The SVM with an RBF kernel demonstrates high classification performance, ensuring high accuracy, reduced false positives, and improved early breast cancer detection reliability. Our approach offers faster processing, cost-effectiveness, and scalability compared to traditional, diagnostic methods, making it an important tool for medical applications. This project indicates the capability of machine learning in medical diagnostics, contributing to early detection, better patient outcomes, and improved breast cancer categorized strategies.

Acknowledgment

This design was implemented in the Dept of Electronics & communication engineering from the CMR Technical campus with the support of the Director, HOD, and Faculty members.

References

[1] Pushpakumari, J., Bhaskar, N., Divya, G., Kanthi, M., Raju, K. S., & Narasimharao, J. (2024). FloraNet: an approach to prediction of sea buckthorn disease based on a convolutional neural network and a random forest method. In 2024 5th International Conference for Emerging Technology (INCET). IEEE.

[2] Harsha, V. N. V., Chintalapudi, S. R., & Chenna, V. S. M. K. (2024). Segment anything: GPT-3 and logistic regression approach for skin cancer detection. In Algorithms in Advanced Artificial Intelligence, (pp. 20–24). CRC Press,.

[3] Bhaskar, N., Kiran, J. S., Satyanarayan, S., Divya, G., Raju, K. S., Kanthi, M., et al. (2024). An approach for liver cancer detection from histopathology images using hybrid pre-trained models. *TELKOMNIKA (Telecommunication Computing Electronics and Control)*, 22(2), 401–412.

[4] Ghadge, D., Hon, S., Saraf, T., Wagh, T., Tambe, A., & Deshmukh, Y. S. (2024). Analysis on machine learning-based early breast cancer detection. In 2024 4th International Conference on Innovative Practices in Technology and Management (ICIPTM), (pp. 1–5). IEEE.

[5] Koç, İ., Tashan, W., Shayea, I., & Zhetpisbayeva, A. (2024). Breast cancer detection based on machine learning. In 2024 IEEE 13th International Conference on Communication Systems and Network Technologies (CSNT), (pp. 1–6). IEEE.

[6] Neelima, G., Kanchanamala, P., Misra, A., & Nugraha, R. A. (2023). Detection of breast cancer based on fuzzy logic. In 2023 International Conference on Advancement in Data Science, E-Learning and Information System (ICADEIS), (pp. 1–6). IEEE.

[7] Gopika, S., & Manusa, K. R. (2023). Breast cancer detection using patch antenna. In 2023 2nd International Conference on Advancements in Electrical, Electronics, Communication, Computing and Automation (ICAECA). IEEE.

[8] Gengtian, S., Bing, B., & Guoyou, Z. (2023). Efficient net-based deep learning approach for breast cancer detection with mammography images. In 2023 8th International Conference on Computer and Communication Systems (ICCCS). IEEE.

[9] Melek, A., Fakhry, S., & Basha, T. (2023). Spatio-temporal mammography-based deep learning model for improved breast cancer risk prediction. In 2023 45th Annual International Conference of the IEEE Engineering in Medicine & Biology Society (EMBC). IEEE.

[10] Krithiga, S. (2023). Improved deep CNN architecture based breast cancer detection for accurate diagnosis. In 2023 Second International Conference on Augmented Intelligence and Sustainable Systems (ICAISS). IEEE.

[11] Sharma, H., Singh, P., & Bhardwaj, A. (2022). Breast cancer detection: comparative analysis of machine learning classification techniques. In 2022 International Conference on Emerging Smart Computing and Informatics (ESCI), (pp. 1–6). IEEE.

48 Efficacy evaluation of Helfer skin tap technique versus ShotBlocker device in reducing pain and anxiety in children undergoing intramuscular injections

Ajith, M.[1,a], Maimoona A.[2,b], Padmapriya, D.[3,c] and Vijayalakshmi, R.[4,d]

[1]Nursing Tutor, Saveetha College of Nursing, Saveetha Institute of Medical and Technical Sciences, Thandalam, Chennai, India

[2]BSc Nursing Final year, Saveetha College of Nursing, Saveetha Institute of Medical and Technical Sciences, Thandalam, Chennai, India

[3]Assistant Professor, Saveetha College of Nursing, Saveetha Institute of Medical and Technical Sciences, Thandalam, Chennai, India

[4]Principal, Saveetha College of Nursing, Saveetha Institute of Medical and Technical Sciences, Thandalam, Chennai, India

Abstract

Pain and anxiety associated with intramuscular (IM) injections are common concerns in pediatric care. This paper intended to appraise the efficiency of the Helfer skin tap practice compared to ShotBlocker in minimizing pain and anxiety in children receiving IM injections. A randomized controlled trial was conducted with 220 children aged 1-5 years, who were randomly assigned to either the Helfer skin tap technique or ShotBlocker group. Pain intensity and anxiety levels were assessed using a visual analog scale and Spielberger's State-Trait Anxiety Inventory, respectively. The results indicated that while both techniques significantly reduced pain, only the Helfer skin tap technique was effective in lowering anxiety levels. These findings suggest that the Helfer skin tap technique is a superior method for pain and anxiety management in pediatric IM injections and should be considered for wider use in clinical practice to enhance children's comfort during medical procedures.

Keywords: Anxiety, Helfer skin tap technique, intramuscular injection, nursing care, pain, pediatric, Shot Blocker

Introduction

Pain is a subjective sensory and emotional experience linked to actual or potential tissue damage [1]. Early exposure influences future pain perception [2–5]. Chronic pain, increasingly prevalent in developing countries, reduces quality of life [6–10]. In pediatric care, intramuscular (IM) injections are a common source of severe pain, anxiety, and distress, affecting children, families, and healthcare providers [11–15]. This can lead to needle phobia, impacting future medical compliance [6]. Pain from IM injections arises from needle insertion and medication pressure, transmitted via A-delta and C-fibers [7], with young children being more sensitive due to immature inhibitory mechanisms [8]. Non-pharmacological methods effectively reduce pain and anxiety, making them essential in pediatric care [9–11]. Despite progress, nurses still struggle to minimize IM injection pain [12]. The Helfer skin tap technique has shown promise in reducing IM injection pain, while NSAIDs offer limited relief [13]. Research on ShotBlocker for injection pain, particularly in children, remains scarce [14]. A study using the Mann-Whitney U, Wilcoxon, and Kruskal-Wallis tests found that ShotBlocker significantly reduced pain but not anxiety [15, 16]. It is recommended as a reliable pain-management tool for adults receiving IM injections.

Literature Review

This literature survey covers background research, theoretical frameworks, existing studies, and gaps in knowledge, making it a well-rounded review of the topic[17]. Pain management in pediatric patients undergoing

[a]ajivas25@gmail.com, [b]172101041.scon@saveetha.com, [c]padmapriya.scon@saveetha.com, [d]principal.scon@saveetha.com

DOI: 10.1201/9781003684589-48

intramuscular (IM) injections remains a critical concern in clinical practice, as fear and anxiety associated with needle procedures can lead to distressing experiences, long-term psychological effects, and even medical non-compliance in the future [18,19]. Several pain reduction strategies, including pharmacological and non-pharma-cological interventions, have been explored to boost the coziness of children during injections [20].

Materials and Methods

Study design: A randomized controlled clinical trial was carried out to assess the effectiveness of the Helfer Skin Tap Technique versus the ShotBlocker device in minimizing pain and anxiety among children undergoing intramuscular (IM) injections.

Study setting: The research was carried out in the Pediatric Unit of the host institution over a period of three months, from June 2024 to September 2024.

Ethical approval: Prior to commencement, the study received ethical clearance from the Institutional Human Ethics Committee (IHEC) of Saveetha Institute. Additionally, written consent was obtained from the Head of the Department of Pediatric Medicine to conduct the research.

Study participants: A total of 220 children under the age of five who met the eligibility criteria were enrolled in this process. Children aged 1 to 5 years, whose parents or guardians provided informed consent, were included. However, children with the following conditions were excluded from the study: Bleeding disorders, coagulopathy, thrombocytopenia; Infections at the injection site; Allergies to the medication administered; Severe neurological disorders or muscle-wasting conditions; Psoriasis, eczema, or other significant skin disorders; History of complications from previous IM injections; Immunocompromised states (e.g., HIV/AIDS); and Severe kidney or liver diseases.

Sampling Technique: A purposive sampling approach was employed, where participants were randomly assigned to either the intervention group (Helfer Skin Tap Technique) or the placebo group (ShotBlocker device) in a 1:1 ratio.

Informed consent: Prior to participation, caregivers received a comprehensive overview of the study's objectives, procedures, and any potential risks involved. Written informed consent was secured from each caregiver before enrolment.

Preliminary Assessment: A self-structured questionnaire was used to collect demographic and clinical data from participants before the intervention.

Intervention procedure - preparation and administration of IM Injection: Prior to administering the IM injection, the researcher prepared the necessary equipment and medication. The following steps were followed for safe and effective administration:

Selection of injection site: The appropriate injection site was identified based on the child's age and body mass. Common sites included: Deltoid muscle (upper arm); Vastus lateralis muscle (thigh); and Gluteus maximus muscle (buttocks).

Sterilization: The selected site was cleaned with an antiseptic swab and allowed to dry.

Injection process: The needle cap was removed, and the syringe was held at a 90-degree angle to the injection site. The needle was inserted smoothly and swiftly into the muscle to a depth of 1-2 inches, depending on the site and child's body mass. Aspiration for blood return was performed before injecting the medication slowly over 10-30 seconds.

Once the medication was administered, the needle was withdrawn at the same angle of insertion. Gentle pressure was applied to the site using a cotton ball or gauze to minimize bleeding.

Disposal and documentation: The used syringe and needle were safely disposed of according to facility protocols, and the procedure was documented in the patient's medical record.

Post-assessment and monitoring: Following the injection, the child was monitored for any adverse reactions, including: pain, redness, or swelling at the injection site; signs of bleeding or hematoma formation; and vital signs (temperature, pulse, and blood pressure) to ensure stability.

In cases of discomfort, a cold compress was applied to the affected area. Additionally, caregivers were provided with post-procedure education, including information on the medication administered, possible side effects, and aftercare instructions. All observations and any adverse reactions were recorded in the patient's medical chart. Finally, all sharp and biohazardous waste were disposed of according to safety guidelines. Figure 48.1 shows the design framework for the proposed process. The algorithm for the proposed methodology is as follows.

Algorithm

Step 1: Evaluate pain levels: Measure the pain levels post-procedure for each child using both methods.

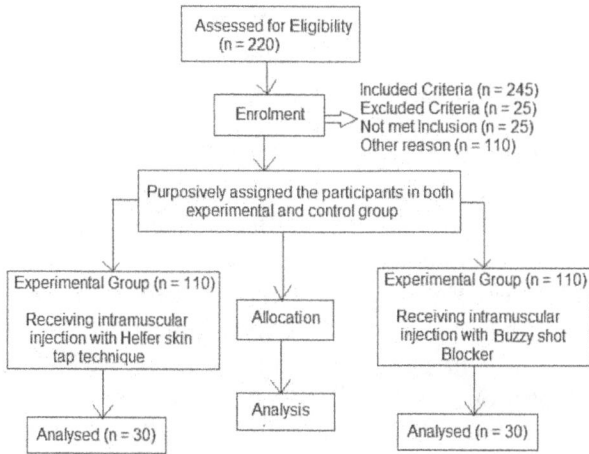

Figure 48.1 Design framework
Source: Author's own compilation

Record the mean and standard deviation of the pain levels for both the Helfer Tab Method and the Shot Blocker Device.

Step 2: Compare mean pain levels: Compare the mean pain levels between the two methods. If the *Shot Blocker Device* has a significantly lower mean pain score, it suggests a more effective pain reduction. If the *Helfer tab method* has a significantly lower mean pain score, it may be more effective.

Step 3: Assess standard deviation: Compare the standard deviations of the two methods. A *lower standard deviation* indicates more consistent results across children, which may suggest better reliability in pain reduction. A *higher standard deviation* may indicate greater variability in pain responses, which might suggest the need for individualized approaches.

Step 4: Select the optimal method: If one method has both a significantly lower mean pain level and a lower standard deviation, it would be the preferred choice. If the two methods show no significant difference in mean pain levels but one method shows a significantly lower standard deviation, the more reliable method should be preferred. If there is no clear advantage in either mean or standard deviation, additional factors such as ease of use, cost, or patient preference may be considered.

Results

Demographic characteristics
Both experimental group: In the experimental group, the majority of children were between 2 and 3 years

Table 48.1 Demographical and clinical variables.

Sl. No.	Variables	Frequency	Percentage
	Chronological age		
1	2-3 years	68	30.90%
	3-4 years	42	19%
	4-5 years	110	50%
	Developmental age		
2	Toddler	63	40.90%
	Pre-scholar	67	30.40%
	Scholar	90	28.60%
	Gender		
3	Male	112	50.90%
	Female	108	49%
	Weight of the child		
4	5-8 kg	87	39.50%
	9-12 kg	92	41.80%
	12-15 kg	41	18.60%
	Frequency of injection		
5	Weekly	8	3.60%
	Monthly	59	26.80%
	6 months once	153	69.50%
	History of allergy		
6	Yes	18	8.10%
	No	202	91.80%
	Previous exposure to injections		
7	Yes	220	100%
	No	0	0%

Source: Author's own compilation

old, totaling 68 participants (30.90%). Among them, 63 children (40.90%) were classified as toddlers based on their developmental stage [Table 48.1]. Additionally, a significant portion of the group consisted of males, with 112 boys comprising 50.90% of the total participants.

Clinical characteristics: In the experimental group, a notable percentage (41.80%) of children weighed between 9 and 12 kilograms. Regarding injection frequency, the majority (69.50% or 153 children) received injections every six months. Additionally, 91.80% (202 children) had no history of allergies. Importantly, all 220 children (100%) in the study had prior experience with injections [Table 48.1].

Distribution of level of pain among the children
Helfer tab method: In a study evaluating pain levels in children using the Helfer tap method, the findings revealed that 139 children (53.46%) experienced no pain, while 98 children (37.68%) reported mild pain. A smaller proportion, 23 children (8.84%), experienced moderate pain. Notably, none of the participants (0%) reported severe pain, indicating a generally positive outcome in pain reduction [Figure 48.2].

Shot blocker device method: In the evaluation of pain levels among children using the ShotBlocker device, the results showed that 143 children (55%) experienced no pain. A considerable portion, 118 children (45.38%) [Figure 48.3], reported moderate pain, while a small group of 13 children (4.61%) experienced severe pain.

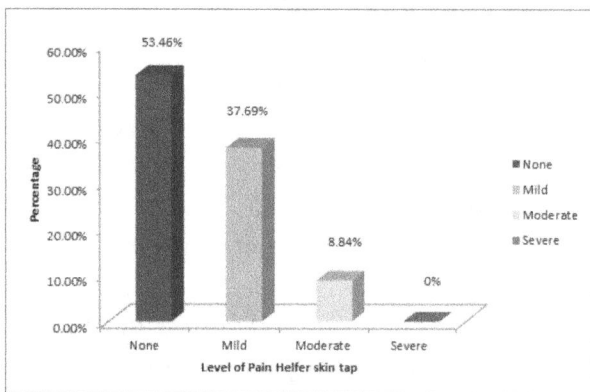

Figure 48.2 Level of pain among the children in Helfer tab method

Source: Author's own compilation

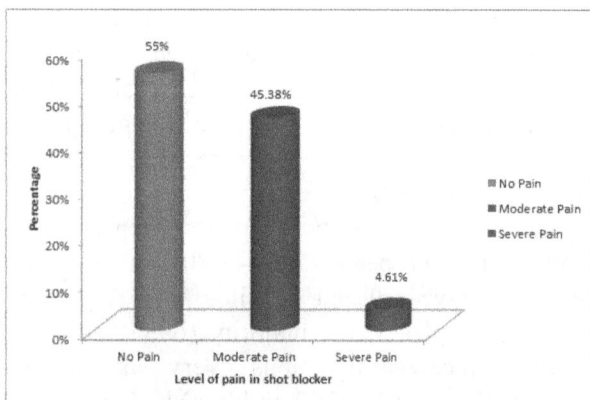

Figure 48.3 Level of pain among the children in shot blocker device

Source: Author's own compilation

Distribution of mean and standard deviation of pain level among children using Helfer rab method and shot blocker device post test
The Table 48.2 presented displays the mean values and standard deviations (SD) of pain levels experienced by children using two different pain management techniques: the Helfer tap method and the ShotBlocker device, as assessed in a post-test evaluation. The Helfer tap method recorded a mean pain score of 1.41 with a SD of 1.92, indicating greater variability in pain levels among the children. In comparison, the ShotBlocker device resulted in a slightly lower mean pain score of 1.30, with a SD of 1.17, suggesting more consistent pain reduction. An unpaired t-test was achieved to associate the effectiveness of these two approaches, yielding a t-value of 24.86, which indicates a highly noteworthy difference between the methods. These findings highlight that both techniques effectively reduce pain in children, though their levels of effectiveness vary.

Comparison of mean and SD of pain levels among children using the Helfer tab method and shot blocker device post-test
The Table 48.3 presented displays the mean values and SD of pain levels experienced by children using two different pain management techniques: the Helfer tap method and the ShotBlocker device, as assessed in a post-test evaluation. The Helfer tap method recorded a mean pain score of 1.41 with a SD of 1.92, indicating greater variability in pain levels among the children. In comparison, the ShotBlocker device resulted in a slightly lower mean pain score of 1.30, with a SD of 1.17, suggesting more consistent

Table 48.2 Distribution of mean and SD of pain level among children using Helfer rab method and shot blocker device post test.

Sl. no.	Method	Mean	Sd	T value	Significance
1	Helfer tab method	1.41	1.92		P<0.05 significant
2	Shot blocker device	1.30	1.17	28.46**	

Source: Author's own compilation

Table 48.3 Comparison of mean and SD of pain levels among children using the Helfer tab method and shot blocker device post-test.

Sl. No.	Method	Mean	Sd	Correlation	Significance
1	Helfer Tab Method	1.41	1.92	-0.4	P<0.05 Significant
2	Shot Blocker Device	1.30	1.17		

Source: Author's own compilation

pain reduction. An unpaired t-test was performed to compare the effectiveness of these two approaches, yielding a t-value of 24.86, which indicates a highly noteworthy difference between the methods. These findings highlight that both techniques effectively reduce pain in children, though their levels of effectiveness vary.

Conclusion

The results indicate that both the Helfer skin tap technique and the ShotBlocker method play a crucial role in reducing pain during intramuscular injections. Since IM injections are a routine procedure performed by nurses, integrating these pain management techniques can significantly enhance patient comfort. By implementing these approaches, healthcare professionals can effectively minimize discomfort, leading to a more positive and less distressing experience for patients receiving injections. Future research should explore the long-term effectiveness of the Helfer skin tap technique and ShotBlocker in pain and anxiety management across different age groups and medical conditions. Studies could also compare these techniques with other emerging pain reduction methods to determine the most effective approach. Additionally, investigating the physiological mechanisms behind these interventions and their impact on patient compliance and healthcare efficiency would provide valuable insights. Expanding research into diverse healthcare settings and incorporating technological advancements, such as AI-driven pain assessment tools, could further enhance pediatric pain management strategies.

References

[1] Güven, Ş. D., & Çalbayram, N. Ç. (2023). The effect of Helfer skin tap technique on hepatitis B vaccine intramuscular injection pain in neonates: a randomized controlled trial. *Explore,* 19(2), 238–242.

[2] Ramana, K., Anitha, A., & Kamalakannan, M. (2024). Efficacy of dry needling therapy versus IASTM on myofascial trigger point in patient with neck pain. *Indian Journal of Physiotherapy and Occupational Therapy,* 18 (pp. 127–129).

[3] Devi, V. S., Sumathi, K., Mahalakshmi, M., Anand, A. J., Titus, A., & Saranya, N. N. (2024). Machine learning based efficient human activity recognition system. *International Journal of Intelligent Systems and Applications in Engineering,* 12(5s), 338–346.

[4] Jose, A., & Dhanalakshmi, R. (2023). Hygieia: multipurpose healthcare assistance using the internet of things. In Krishnan, S., & Ilmudeen, A. (Eds.), Internet of Medical Things in Smart Healthcare: Post-COVID-19 Pandemic Scenario, (pp. 77–98). Apple Academic Press.

[5] Ponmalar, A., & Anand, J. (2023). IoMT-based caring system for aged people in a Post-COVID Scenario. In Krishnan, S., & Ilmudeen, A. (Eds.), Internet of Medical Things in Smart Healthcare: Post-COVID-19 Pandemic Scenario, (pp. 207–224). Apple Academic Press.

[6] Balamurugan, K., Jose, A. A., Vignesh, B., Raja Vikram Singan, K., & Subramaniam, N. (2023). Design of an intelligent and smart pill box using arduino and sensors. In 2023 4th International Conference on Smart Electronics and communication (ICOSEC), (pp. 258–263).

[7] Düzkaya, D. S., Karakul, A., Akoy, İ., & Andi, S. (2024). Effects of shot blocker® and the Helfer skin tap technique on pain and fear experienced during intramuscular injection among children aged 6–12 years in pediatric emergency units: a randomized controlled trial. *International Emergency Nursing,* 76, 101502.

[8] Muthalagu, R., Ramachandran, R., Anuradha, T., Anupama, P. H., & Jose, A. A. (2023). Pattern recognition and modelling in electrocardiogram signals: early detection of myocardial ischemia and infraction. In 2023 2nd International Conference on Edge Computing and Applications (ICECAA), (pp. 1035–1041).

[9] Sushma, S., Anuradha, T., Brabin, D. R. D., & Jose, A. A. (2023). Artificial intelligence in orthopedic implant model classification. In Ranjith, R., & Paulo Davim, J. (Eds.), Handbook of Research on Advanced Functional Materials for Orthopedic Applications. IGI Global (pp. 143–147).

[10] Geethalakshmi, M., Venkatesh, J., Uma Mageswari, R., Mahalakshmi, A., Anand, J., & Partheepan, R. (2023). Fuzzy based route optimization in wearable biomedical wireless sensor network. In AIP Conference Proceedings, (Vol. 2523, p. 020156).

[11] Maheswari, R., Vignesh, S., Rakesh Kumar, M., Venkatesh, T. M., Sundar, R., & Jose, A. A. (2022). Voice control and eyeball movement operated wheelchair. 2022 International Conference on Edge Computing and App. (ICECAA), (pp. 805–809).

[12] Prema, S., Jose, A., Vanitha, P., Nirmala Devi, K., Mohamed Yaseen, M., & Rajeswari, C. (2022). Smart stick using ultrasonic sensors for vissually impaired. *Advances in Parallel Computing Algorithms, Tools and Paradigms,* 41, 436–441.

[13] Dhanalakshmi, R., Anand, J., Poonkavithai, K., & Vijayakumar, V. (2022). Cloud-based glaucoma diagnosis in medical imaging using machine learning. In Parah, S. A., Rashid, M., & Varadarajan, V. (Eds.), Artificial Intelligence for Inno. Healthcare Infor, (pp. 61–78). Cham: Springer.

[14] Dhanalakshmi, R., & Anand, J. (2022). Big data for personalized healthcare. In Jaya, A., Kalaiselvi, K., Goyal, D., & Al-Jumelly, D. (Eds.), Handbook of Intelligent Healthcare Analytics: Knowledge Engineering with Big Data Analytics, Chap-4, (pp. 67–92). John Wiley & Sons, Inc.

[15] Sheriff S. T. M., Venkat Kumar, J., Vigneshwaran, S., Jones, A., & Anand, J. (2021). Lung cancer detection using VGG NET 16 architecture. In International Conference on Physics and Engineering 2021, IOP Pub., Journal of Physics Conference Series, (Vol. 2040).

[16] Niranjana, S., Hareshaa, S. K., Zibiah Basker, I., & Jose, A. (2021). Smart wearable system to assist asthima patients. In Advances in Parallel Computing Technologies and Applications, (Vol. 40, pp. 219–227). IOS Press.

[17] Anand, J., Gowtham, H., Lingeshwaran, R., Ajin, J., & Karthikeyan J. (2021). IoT based smart electrolytic bottle monitoring. In Advances in Parallel Computing Technologies and Applications, (Vol. 40, pp. 391–399), IOS Press.

[18] Kumar, A., Sharma, E., Marley, A., Samaan, M. A., & Brookes, M. J. (2022). Iron deficiency Anaemia: pathophysiology, assessment, practical management. *BMJ Open Gastroenterology,* 9(1), e000759.

[19] Cook, J. D. (2005). Diagnosis and management of iron-deficiency Anaemia. *Best Practice and Research Clinical Haematology,* 18(2), 319–332.

[20] Niranjana, S., Hareshaa, S. K., Zibiah Basker, I., & Anand, J. (2020). Smart monitoring system for Asthma patients. *International Journal of Electronics and Communication Engineering,* 7(5), 5–9.

49 Sustainable air circulation and monitoring system for caves using solar energy

S. Venkatesh[1,a], Nandhini, K. L.[2,b], Pothuru Banusree[2,c], M. Sharath Kumar[2,d] and J. Eshwar[2,e]

[1]Assistant Professor, Department of ECE, CMR Technical Campus, Hyderabad, Telangana, India

[2]Student, Department of ECE, CMR Technical Campus, Hyderabad, Telangana, India

Abstract

Earth is a paradise of beautiful natural sites among which, caves are important natural and historical sites that require proper air circulation and monitoring to maintain their ecosystem and safety. This paper presents a sustainable system that uses solar energy to power air circulation and environmental monitoring unit for caves. The system consists of solar panels, a battery storage unit, sensors for monitoring temperature, humidity, air quality, and carbon dioxide levels, as well as a wireless data transmission module. The collected data helps researchers and conservationists analyze cave conditions in real time and take necessary actions. The solar-powered design ensures continuous operation without external power sources, making it ideal for remote locations. This system improves air quality inside caves, protects fragile ecosystems, and supports scientific research while using renewable energy for sustainability.

Keywords: Air quality monitoring, cave conservation, microcontroller, renewable energy, solar energy

Introduction

Caves host unique geological formations, historical artifacts, and diverse ecosystems. However, poor ventilation, human activity, and natural processes can lead to air stagnation, rising CO_2 levels, and increased humidity, posing threats to biodiversity and cave integrity. Traditional ventilation systems depend on external power sources, which are often impractical for remote locations.

This study proposes a solar-powered air circulation and monitoring system to provide a self-sustaining solution for caves. The system integrates air quality sensors, microcontrollers, and automated fan control to maintain optimal environmental conditions while reducing reliance on conventional energy sources. Caves are natural formations that house diverse ecosystems, archaeological artifacts, and geological structures. However, they are highly sensitive to environmental changes, with poor ventilation leading to issues such as increased carbon dioxide levels, excessive humidity, and the accumulation of toxic gases. Human activities, including tourism and scientific exploration, further contribute to air quality degradation. Inadequate air circulation poses risks to both cave biodiversity and human safety, making sustainable monitoring and ventilation solutions necessary.

To overcome these challenges, this paper introduces a solar-powered air circulation and monitoring system for caves. The system integrates air quality sensors, microcontrollers, and automated fan controls to regulate airflow while minimizing environmental impact. By leveraging renewable energy, this solution ensures continuous operation in remote locations, where traditional power sources may be unavailable [2]. The proposed system supports cave preservation while ensuring a safer environment for researchers, visitors, and native wildlife.

System Design and Methodology

System architecture

The proposed system consists of several integrated modules that work together to ensure efficient air circulation and monitoring within caves.

The power module comprises a solar panel (12V, 10W) that captures solar energy and a rechargeable

[a]Venkatesh.ece@cmrtc.ac.in, [b]217r1a0440@cmrtc.ac.in, [c]217r1a0448@cmrtc.ac.in, [d]217r1a0434@cmrtc.ac.in, [e]217r1a0428@cmrtc.ac.in

DOI: 10.1201/9781003684589-49

battery (12V, 2Ah) that stores this energy for uninterrupted operation. A voltage regulator (7805, 5V output) ensures a stable power supply to all components.

The sensing module includes an MQ135 air quality sensor to detect harmful gases. Since the sensor provides an analog output, an ADC0809 analog-to-digital converter is used to convert the signal into a digital format that can be processed by the microcontroller.

The control and processing module is centered around a microcontroller (8051/Arduino), which receives the digital air quality data and determines the appropriate response. It displays real-time air quality levels on an LCD display based on AQI standards.

The ventilation and alert system is controlled by a relay module (12V activation) that manages the operation of two fans: an inlet fan, which brings fresh air into the cave, and an outlet fan, which expels polluted air. A buzzer alarm (6V alert) is also included to warn users when air quality drops below safe levels.

Mathematical model for power calculation: The total power required to operate the system is calculated as:

$$P_{total} = P_{microcontroller} + P_{sensors} + P_{fan} + P_{display}$$

where: P is the power (W),
V is the voltage (V),
I is the current (A).
Considering system consumption: P=V x I

The overall system architecture is illustrated in Figure 49.1, showing the interconnection of components including the solar panel, sensors, microcontroller, and ventilation system.

Working principle

The system operates based on a solar-powered automated control mechanism that ensures continuous air circulation and monitoring. The solar panel collects energy and charges the rechargeable battery, ensuring uninterrupted power supply even in the absence of sunlight [1]. The stored energy powers the microcontroller, which serves as the central processing unit for the system.

The MQ135 air quality sensor continuously monitors the air within the cave, detecting harmful gases and pollutants [3]. The ADC0809 converter processes the sensor's analog signal and converts it into digital data for real-time analysis. The microcontroller interprets this data and takes action based on predefined air quality thresholds.

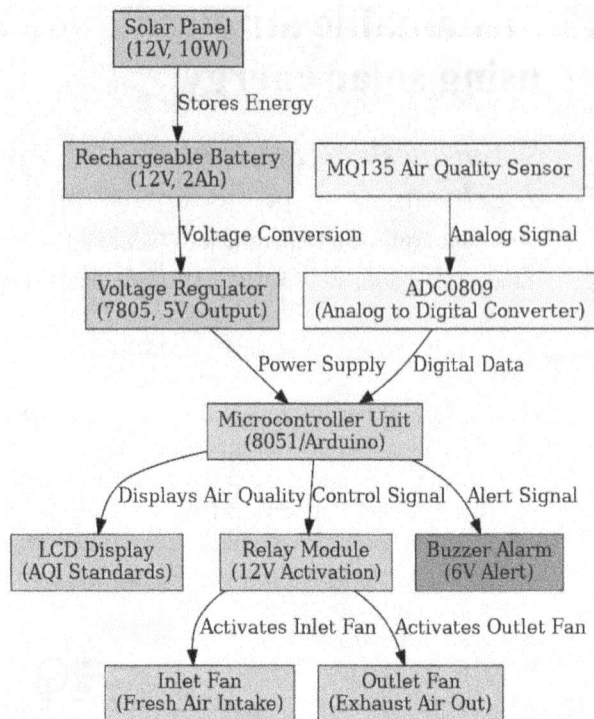

Figure 49.1 Block diagram of the proposed cave air circulation and monitoring system
Source: Author

If the air quality falls below acceptable levels, the microcontroller activates the relay module, which switches on the inlet fan to introduce fresh air and the outlet fan to expel contaminated air. Simultaneously, if pollution levels exceed critical limits, the buzzer alarm is triggered to alert users of potential hazards. The LCD display provides real-time feedback on air quality, allowing for continuous monitoring and assessment.

The system follows an energy-efficient approach, automatically optimizing fan usage based on air quality levels. By leveraging solar energy, the system ensures reliable operation without relying on external power sources, making it ideal for remote cave environments where traditional electrical infrastructure is unavailable.

Results and Discussion

The experimental results demonstrated that the proposed system effectively improved cave air quality by automatically regulating ventilation. The MQ135 sensor provided accurate readings, and the microcontroller successfully processed the data to trigger the appropriate response. During testing, the system consistently activated the ventilation fans when air quality deteriorated, ensuring optimal conditions within the cave environment.

An energy efficiency analysis showed that the solar panel and battery setup provided uninterrupted operation, even during extended periods of low sunlight. The system's power consumption remained within the expected limits, confirming its sustainability for remote cave locations. The relay-controlled fan mechanism minimized energy waste by activating only, when necessary, eventually enhancing the system's efficiency [4].

A prototype of the proposed system was developed to validate its functionality and performance. The prototype was assembled using a 12V solar panel, a rechargeable battery, an MQ135 air quality sensor, an 8051 microcontroller, and a relay-based ventilation system. The system was tested in a controlled environment to simulate cave conditions, ensuring that it responded effectively to changes in air quality. The prototype successfully monitored air pollution levels, displayed real-time data on an LCD screen, and activated ventilation when necessary. Additionally, the buzzer alarm provided alerts when air quality dropped below safe thresholds. The system demonstrated reliable operation using only solar energy, confirming its suitability for cave environments where traditional power sources are unavailable [8-11].

The experimental setup and real-time monitoring interface are shown in Figures 49.2 and –49.4. The images (3,4) display the ThingSpeak dashboard, where sensor data, including gas concentration levels, is visualized in real time. The left graph illustrates the recorded variations in air quality (i.e. lower level indicates good quality of air whereas the higher level indicates poor air quality), likewise the right image represents an alert for hazardous gas levels detected by the system [5-7]. This interface allows researchers to oversee cave air conditions from a distance ensuring timely intervention if pollution exceeds safe thresholds. The successful implementation of this monitoring system validates its practical usability in remote cave environments, demonstrating its efficiency in maintaining optimal air quality. The prototype was tested under different environmental conditions, including varying humidity and gas concentrations. The system illustrated adaptability to these changes, maintaining consistent performance. The buzzer alarm functioned as expected, effectively

Figure 49.2 Prototype implementation of the solar-powered air circulation and monitoring system
Source: Author

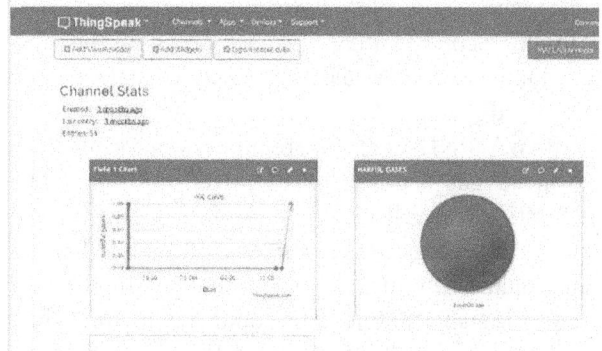

Figure 49.3 Time-series graph displaying harmful gas levels in the cave and a status indicator for Normal gases
Source: Author

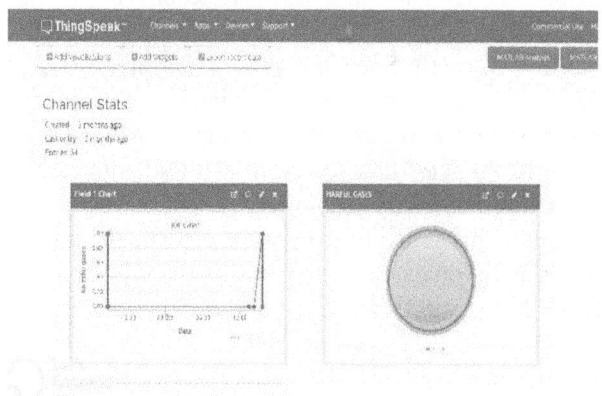

Figure 49.4. Time-series graph displaying harmful gas levels in the cave and a status indicator for harmful gases
Source: Author

alerting users when pollution levels exceeded safe thresholds.

Overall, the results confirm that the proposed system is a reliable and energy-efficient solution for cave air circulation and monitoring. Future improvements may include enhanced sensor calibration, integration with IoT for remote monitoring, and increased battery capacity for extended operation in extreme conditions.

Conclusion and Future Scope

This study presents a solar-powered air circulation and monitoring system for caves, ensuring improved air quality through automated ventilation. The system efficiently detects pollution levels, processes real-time data, and controls fan operations while using renewable energy. This solution enhances cave conservation and safety in remote locations.

Future scope: Future enhancements to the system may include AI-based predictive analytics for optimized ventilation, wireless communication (LoRaWAN/NB-IoT) for remote monitoring, and advanced gas sensors for improved accuracy. The integration of self-cleaning air filters and hybrid renewable energy sources like wind or geothermal can enhance efficiency. Additionally, cloud-based monitoring and mobile apps could provide real-time data access for researchers and conservationists, ensuring better cave management and safety.

Acknowledgement

The authors would like to express their gratitude to the faculty and staff of CMR technical campus for their invaluable support and guidance throughout this research. Special thanks to our mentors and peers for their constructive feedback and technical insights. We also acknowledge the contribution of funding agencies and research institutions that provided the necessary resources and infrastructure to complete this study. Finally, we appreciate the efforts of all individuals who contributed directly or indirectly to the successful execution of this project.

References

[1] Ahamed , S. N., Subba Rao, E., Mahesh, N., Akhil, B., Kumar, S. A., & Ganesh, Y. R. (2024). Solar powered air purifier with air quality monitoring system. *Journal of Emerging Technologies and Innovative Research*. ISSN-2349-5162 Volume 11, Issue 4.

[2] Singh, H., Sharma, A., & Kumar, M. (2022). Smart renewable energy solutions for air monitoring. *Journal of Sustainable Energy Systems*.

[3] Ramachandra arjunan, S., Perumalsamy, V., & Narayanan, B. (2022). IoT-based artificial intelligence indoor air quality monitoring system using enabled RNN algorithm techniques. *Journal of Intelligent & Fuzzy Systems*. DOI:10.3233/JIFS-212955.

[4] Balmuri, K. R., Sujatha, J., Arvind, S., Lal, B., & Kumari, S. C. (2023). An energy efficient and reliable strategy for intra-cluster and inter-cluster communications in wireless sensor networks. In 2023 IEEE International Conference on Integrated Circuits and Communication Systems (ICICACS).

[5] Ahmed, S., Bhattacharya, K., & Roy, P. (2021). Design and development of a solar-powered air purifier for large-scale areas. *International Journal of Mechanical Engineering*. vol 283,15-Nov 24,11304.

[6] Kumar, N., Gupta, S., & Singh, R. (2021). A solar-powered wireless air quality monitoring system. *IEEE Sensors Journal*. Volume 529, 2024.

[7] Lee, J., Kumar, R., & Verma, S. (2020). IoT-based solar energy monitoring and air quality control. *Journal of Clean Technologies*. https://doi.org/10.3390/cleantechnol4020022

[8] Roy, A., Patel, P., & Sharma, D. (2020). Solar-powered IoT-based air quality monitoring system. *International Journal of Renewable Energy*.

[9] Allen, R., Wong, H., & Chavez, A. (2020). Portable solar-powered air purification system for remote locations. *Environmental Monitoring and Assessment*.

[10] Patel, D., Wang, L., & Kim, G. (2019). Solar-powered air quality monitoring for remote and urban areas. *Renewable Energy Journal*.

[11] Aparna Jo se, Abraham C1, Akhil K Soman IoT based solar powered air purifier with airquality monitoring system. *International Conference on Sustainable Goals in Materials, Energy, and Environment E 3S Web of Conferences* 529, 04016 (2024).

50 Detection of multiple artifacts through the application of adaptive scalp region selection and classifier fusion

D. Yovan Snanagan Ponselvan[1,a], J. Sofia Priya Dharshini[2,b], Kadiri Harika[3,c], Machiraju Hari Ram[3,d], K. Pavani[3,e] and M. Sai Kriti[3,f]

[1]Assistant Professor, Department of ECE, Rajeev Gandhi Memorial College of Engineering and Technology, Nandyal, Andhra Pradesh, India

[2]Professor, Department of ECE, Rajeev Gandhi Memorial College of Engineering and Technology, Nandyal, Andhra Pradesh, India

[3]UG Students, Department of ECE, Rajeev Gandhi Memorial College of Engineering and Technology, Nandyal, Andhra Pradesh, India

Abstract

Although electroencephalography (EEG) is frequently tainted by artifacts like ocular blinks, muscular movements, and background noise, it is widely utilized in neuroscience, brain-computer interfaces, and clinical diagnostics. With overlapping artifacts and real-time processing, traditional statistical and filtering models struggle, necessitating manual intervention. To improve artifact detection, this paper proposes an ensemble-based classification strategy that incorporates Support Vector Machine, Random Forest, and Naive Bayes within a voting classifier framework. From the TUH EEG Artefact Corpus, the mean, standard deviation, kurtosis, and skewness were extracted. Reliability of the model in reducing EEG artifacts is demonstrated by its 97% accuracy. Deep learning for automated feature extraction and real-time applications should be the focus of future research.

Keywords: Artifact detection, electroence phalography EEG, machine learning, naïve bayes, random forest, support vector machine, voting classifier

Introduction

Neuroscience, clinical diagnostics, and brain-computer interfaces all rely on the use of electroencephalography (EEGs) to identify neurological and cognitive abnormalities [1]. However, artefacts like ocular blinks, muscular tremors, and ambient noise frequently taint EEG signals, leading to incorrect diagnoses. Long-term recordings, where multiple noise sources overlap, exacerbate this issue and make interpretation more difficult [2]. When multiple artifacts coexist, traditional artefact detection methods struggle, limiting their generalizability, and rely on filtering techniques that target specific artifacts. Many modern methods assume artifact properties, making them ineffective for dynamic EEG recordings. A robust, flexible, and computationally efficient framework is needed to classify artefacts with high precision. In order to improve artifact detection, this paper presents an ensemble-based model that combines Support Vector Machine (SVM), Random Forest (RF), and Naive Bayes (NB) in a voting classifier. SVM identifies decision boundaries, RF reduces overfitting to improve generalization, and NB makes use of probabilistic classification. The ensemble approach averages predictions from all classifiers, increasing accuracy and robustness. Methods for correction can mitigate discontinuities that may result from artifact removal.

Methodology

Data collection and preproceesing

Dataset description: The TUH EEG Antique Corpus is the dataset used in this study. It is freely available and contains EEG signals with marked ancient rarities. The dataset is put away in European Information Arrangement (EDF) documents, ordinarily utilized for physiological sign accounts.

[a]dysponselvan@gmail.com, [b]sofiapriyadharshini@rgmcet.edu.in, [c]kadiri.harika@gmail.com, [d]harirammachiraju3@gmail.com, [e]kurukundupavani@gmail.com, [f]krithimusale12@gmail.com

DOI: 10.1201/9781003684589-50

Data acquisition: The dataset was obtained from an online repository and extracted for further analysis. EEG signals were loaded using the MNE-Python library, a standard toolkit for EEG data analysis.

Channel selection and renaming: Recordings contain multiple chan1nels representing different electrode placements on the scalp. The script processes a specific channel, "EEG FP1-REF", ensuring that channel naming follows standard conventions for consistency in-depth analysis of the most recent research reveals advancements in the detection and correction of EEG artifacts, offering suggestions for future enhancements.

Data extraction: EEG data corresponding to the selected channel was extracted along with time indices. This data was used to compute statistical features for artifact classification.

Feature extraction: To improve classification accuracy, statistical features were computed for each segment of EEG data [3]. These features provide critical insights into the signal characteristics:

- **Mean:** Represents the average amplitude of the signal

Three different models were combined in a Voting Classifier:

$$\mu = \frac{1}{N}\sum_{t=0}^{N} f(t) \tag{1}$$

- **Standard deviation:** Measures the spread of EEG values.

$$\sigma = \sqrt{\frac{1}{N-1}\sum_{t-1}^{N}(f(t) - \overline{f})^2} \tag{2}$$

- **Kurtosis:** Describes the sharpness of the signal distribution.

$$k = \frac{E(f(t)-\mu)^4}{\sigma^4} \tag{3}$$

- **Skewness:** Indicates the asymmetry of the EEG signal distribution.

$$\gamma = \frac{E(f(t)-\mu)^3}{\sigma^3} \tag{4}$$

- **Median:** Provides a robust measure of central tendency.

$$Median = f\left(\frac{\left(sort(f(k))\right)}{2}\right) \tag{5}$$

- **Max and min values:** Define the amplitude range of the signal.

$$Max = max_t(f(t))$$
$$Min = min_t(f(t)) \tag{6}$$

- **Range:** Difference between maximum and minimum values.

$$R = Max - Min \tag{7}$$

Machine Learning Pipeline

Train-test split: The dataset was isolated into 80% preparation and 20% testing to guarantee a reasonable assessment of the model.

Data Normalization: Highlight values were normalized utilizing Z-score standardization, an interaction that scales all elements to have zero mean and unit fluctuation. This step works on the presentation of AI models by guaranteeing that elements are tantamount in extent. [4].

Label Binarization: Artifact labels were converted into a binary format, making them compatible with classification algorithms. [5].

Model training and classification: An ensemble classification approach was used to improve artifact detection accuracy. Three different models were combined in a voting classifier:

1. **Support Vector Machine (SVM) [6]:** A robust classifier that finds the optimal decision boundary between classes [6]. Using the radial basis function (RBF) kernel, the kernelized version of the SVM decision function becomes:

$$k(x,x') = exp\left(-\frac{||x-x'||^2}{2\sigma^2}\right)$$

The decision function then becomes:

$$f(x) = \sum_{i=1}^{N} x_i y_i k(x_i, x) + b \tag{8}$$

2. **Random Forest (RF):** A gathering of choice trees to further develop speculation. utilizes a troupe of choice trees [7]. The classification decision for a single tree is made using the following decision function:

$$y_i = sign(f(x_i)) \tag{9}$$

where *f(xi)* is the choice capability of the singular tree for input *xi*. The largest percentage of votes from all trees is the final expectation of the irregular woodland model:

$$y^{\wedge} = majority_{vote}(f_1(x), f_2(x), \dots f_m(x)) \quad (10)$$

where M is the number of trees in the forest and fi (x) is the tree's prediction.

3. **Naïve Bayes (NB):** A probabilistic model expecting highlight freedom [8]. Naïve Bayes uses Bayes' Theorem to compute the probability of each class given the feature vector xxx:

$$P(y/x) = \frac{P(x/y)P(y)}{p(x)} \quad (11)$$

For NB, the likelihood $P(x|y)$ is assumed to be the product of independent feature probabilities.

$$P(y/x) = \pi_{i=1}^n P(x_i/y) \quad (12)$$

where:
- $(x_i |y))$ is the probability of feature xi in light of class y.
- n is the number of features in x.

The posterior p1robability for each class y is then computed as:

$$P(y/x) = \frac{\pi_{i=1}^n P(x_i/y) P(y)}{P(x)} \quad (13)$$

Finally, the prediction for class y is the class with the highest probability of occurring after:

$$y^{\wedge} = \arg max_y P(y/x) \quad (14)$$

Model Evaluation

The proposed models and ensemble classifier are evaluated using multiple performance metrics:

- **Accuracy** – Measures the overall classification correctness.

$$Accuracy = \frac{TP+TN}{S} \quad (15)$$

Precision – Assesses the proportion of correctly identified positive instances.

$$Precision = \frac{TP}{TP+FP} \quad (16)$$

- **Recall** – Evaluates the model's ability to detect all instances of a class.

$$Recall = \frac{TP}{TP+FN} \quad (17)$$

- **F1-Score** – Balances precision and recall for a better performance estimate.

$$F1 = \frac{2*Precision+Recall}{Precision+Recall} \quad (18)$$

Filter-based Methods

Filter-based methods suppress unwanted temporal phenomena in signal processing. Wiener filters estimate signal and noise parameters to minimize NMSE while filtering noise [9]. Supported by an EEG Lab plugin, the multi-channel Wiener filter (MWF) is efficient in audio, speech, and EEG processing with minimal labelling. MWF, on the other hand, relies on static EEG and noise profiles. Artifacts from a variety of distributions can be corrected by deep neural encoder-decoder models.

Algorithm 1 EEG artifact classification using SVM, RF, and Naïve Bayes (ensemble model)

Step 1: Data Collection
- Load the EEG signals from the TUH EEG Artifact Corpus.

Step 2: Data Preprocessing
- Normalize the EEG signals to have zero mean and unit variance.
- Apply label binarization to convert categorical artifact labels into binary vectors.

Step 3: Feature Extraction and Transformation
- Extract statistical features from EEG signals:
  - Mean, standard deviation, skewness, kurtosis, etc.

Step 4: Model Training and Classification
- Define three classifiers:
  - Support Vector Machine (SVM)
  - Random Forest (RF)
  - Naïve Bayes (NB)

Step 4.1: Train SVM Classifier
- SVM utilizes the radial basis function (RBF) kernel.
- The decision function becomes: $f(x) = \sum_{i=1}^{N} \alpha_i y_i K(x_i, x) + b$ where $K(x_i, x) = \exp\left(-\frac{|x_i-x|^2}{2r^2}\right)$

Step 4.2: Train Random Forest (RF) Classifier
- Use an ensemble of decision trees for classification. $\hat{y} = majority.vote(f_1(x), f_2(x), \dots, f_m(x))$

Step 4.3: Train Naïve Bayes (NB) Classifier
- Apply Bayes' Theorem to compute class probabilities. $\hat{y} = \arg max\, P(y|x)$

Step 5: Ensemble Model
- Combine the predictions from SVM, RF, and NB using a Voting Classifier.
- The final prediction is the majority vote of the individual classifiers.

Step 6: Model Evaluation
- Evaluate the performance using the following metrics:
  - Accuracy
  - Precision
  - Recall
  - F1-Score

Results and Discussions

Confusion matrix

The SVM, RF, and NB confusion matrix in Figure 50.1 reveals moderate misclassifications and higher false positive and negative rates. There is no one model that excels at all artifact types. Figure 50.2 presents the ensemble Voting Classifier, significantly reducing errors. Low off-diagonal values indicate minimal misclassification, while high diagonal values (95, 92, 92) indicate correct classifications. With improved precision and recall, the model achieves 97% accuracy, demonstrating its robustness.

ROC Curves

The ROC curves in the shown in above figure 50.3 visually compare the classification performance of SVM, RF, NB, and the voting classifier (SVM+RF+NB) for EEG artifact detection [10]. The x-hub addresses the misleading positive rate (FPR), while the y-hub addresses the genuine positive rate (TPR), showing the compromise among responsiveness and explicitness. The troupe casting a ballot classifier shows a reliably higher bend, demonstrating better ability to group looked at than individual models [11]. The region

under the bend (AUC) for the voting classifier is near 1.0, affirming its high exactness and negligible bogus arrangements [12],[13]. Individual models, for example, SVM and RF perform well yet show marginally lower AUC values, while NB displays a lower bend, demonstrating more fragile grouping execution. The more extreme the bend toward the upper left corner, the better the model's segregation capacity. The ROC examination affirms that the voting classifier beats individual classifiers in distinctive EEG relics, making it a more powerful and solid methodology [14],[15].

The SVM + RF + NB gathering model accomplished the best execution in EEG curio recognition with 97% exactness and 0.97 F1-score. CNN + LSTM + FCNN played out the most terrible, with 69% precision and 0.67 F1-score shown in table 50.1, table 50.2, table 50.3 and figure 50.4. The LSTM + DNN model showed nice execution with 90% exactness. Outfit techniques demonstrated exceptionally

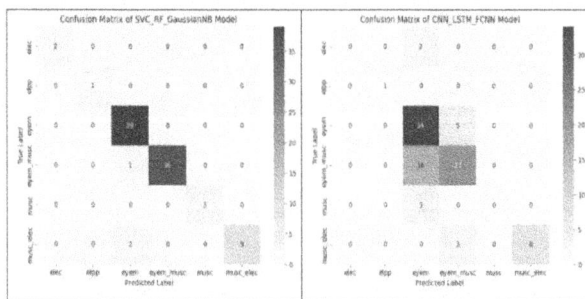

Figure 50.1 Confusion matrix for individual classifier
Source: Author

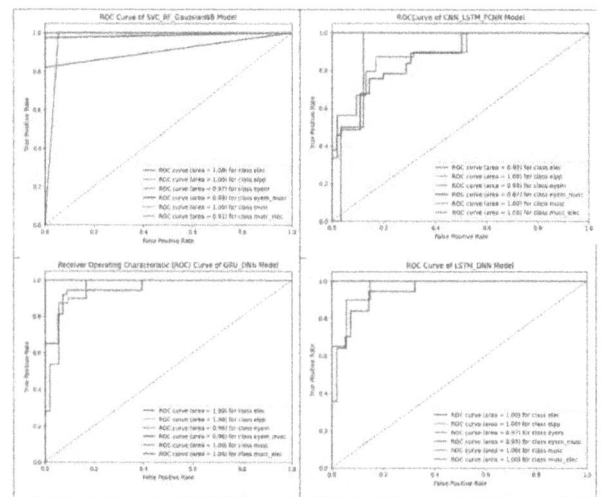

Figure 50.2 Confusion matrix for voting amplifier
Source: Author

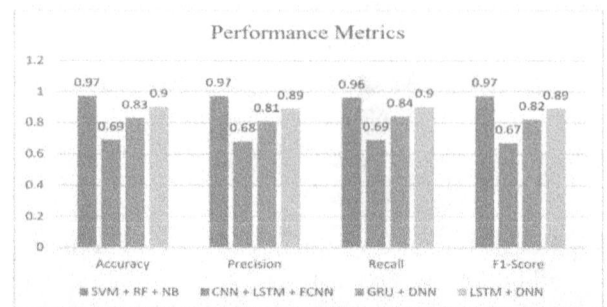

Figure 50.3 ROC curve representation
Source: Author

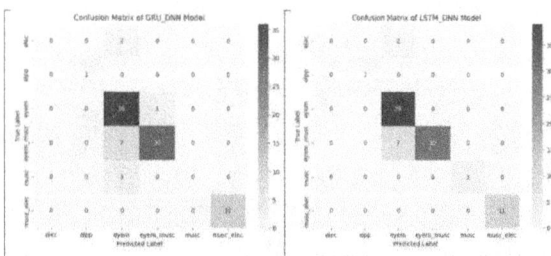

Figure 50.4 Performance metrices
Source: Author

Table 50.1 Performance metrices of referenced studies.

Ref.	Methodology	Accuracy (%)	Precision	Recall	F1-score
[1]	Adaptive Scalp region selection and classifier fusion	93%	0.87	0.88	0.87
[2]	Multi-scale feature fusion with adaptive lasso	96%	0.92	0.94	0.93
[3]	CNN based Schizophrenia detection with channel selection	98%	0.95	0.96	0.95

Source: [1], [2] and [3]

Table 50.2 Performance meretrices of the classification models of proposed system.

Model	Precision	Recall	F1Score	Accuracy
SVM	0.92	0.91	0.91	91%
RF	0.94	0.93	0.93	93%
NB	0.89	0.88	0.88	88%
Voting classifier (SVM+RF+NB)	0.97	0.96	0.97	97%

Source: Simulation

Table 50.3 Outline execution metrices of detection of multiple artifacts through the application of adaptive scalp region selection and classifier fusion.

Precision	Recall	F1-score	Accuracy
0.97	0.96	0.97	97%

Source: Simulation

compelling in further developing order. Future work ought to zero in on upgrading profound learning models for better speculation [16]-[18].

Conclusion

For clinical applications, BCIs, and neurological assessments, electroencephalography (EEG) artifact detection is essential; however, conventional methods struggle with generalization. This study introduces an ensemble model combining SVM, RF, and NB, achieving 97% accuracy, 0.97 precision, 0.96 recall, and 0.97 F1-score. Compared to deep learning models like CNN+LSTM+FCNN (69% accuracy) and GRU+DNN (83% accuracy), the ensemble approach significantly reduces misclassification errors. Its high precision ensures accurate artifact differentiation, while strong recall minimizes missed detections. The model's computational efficiency makes it ideal for real-time EEG

analysis. This paper focuses on noise reduction using filter-based methods. Future research should integrate deep learning for aut feature extraction, improved processing speed, and enhanced EEG artifact classification.

References

[1] Britton, J. W., Frey, L. C., Hopp, J. L., Korb, P., Koubeissi, M. Z., Lievens, W. E., Pestana-Knight, E. M., & St. Louis, E. K. (Eds.). (2016). Electroencephalography (EEG): An introductory text and atlas of normal and abnormal findings in adults, children, and infants [Internet]. American Epilepsy Society.

[2] Ghassemi, M. M., Moody, B. E., Lehman, L. W. H., Song, C., Li, Q., Sun, H., et al. (2018). You snooze, you win: the physionet/computing in cardiology challenge 2018. In 2018 Computing in Cardiology Conference (CinC), (pp. 1–4). IEEE.

[3] Fang, Z., Hu, D., Zheng, R., Jiang, T., & Gao, F. (2024). Multiple artifact detection based on adaptive scalp region selection and classifier fusion. *IEEE Sensors Journal*, 24(6), 8438–8449.

[4] Peshawa, J. M. A., & Rezhna, H. F. (2014). Data normalization and standardization: a technical report. *Machine Learning Technical Reports*, 1(1), 1–6.

[5] Zhang, Y., Garg, A., Cao, Y., Lew, L., Ghorbani, B., Zhang, Z., & Firat, O. (2023). Binarized Neural Machine Translation. *In Advances in Neural Information Processing Systems*, 36. https://arxiv.org/abs/2302.04907.

[6] Chen, S., Li, N., Kong, X., Huang, D., & Zhang, T. (2024). Electroencephalography-based motor imagery classification using multi-scale feature fusion and adaptive lasso. *Big Data and Cognitive Computing*, 8(12), 169.

[7] Breiman, L. (2017). Classification and Regression Trees. Routledge.

[8] Wickramasinghe, I., & Kalutarage, H. (2021). Naive bayes: applications, variations and vulnerabilities: a review of literature with code snippets for implementation. *Soft computing*, 25(3), 2277–2293. doi: 10.1007/s00500-020-05297-6.

[9] Zhang, S., McCane, B., Neo, P. S. H., & McNaughton, N. (2020). Removing mains power artefacts from EEG–a novel template-based method. bioRxiv, pp. 2001–2020.

[10] Ghosh, R., Sinha, N., & Biswas, S. K. (2019). Automated eye blink artefact removal from EEG using support vector machine and autoencoder. *IET Signal Processing*, 13(2), 141–148.

[11] Chen, S., Li, N., Kong, X., Huang, D., & Zhang, T. (2024). Electroencephalography-based motor imagery classification using multi-scale feature fusion and adaptive lasso. *Big Data and Cognitive Computing*, 8(12), 169.

[12] Hassan, F., Hussain, S. F., & Qaisar, S. M. (2023). Fusion of multivariate EEG signals for schizophrenia detection using CNN and machine learning techniques. *Information Fusion*, 92, 466–478.

[13] Mathe, M., Mididoddi, P., & Krishna, B. T. (2022). Artifact removal methods in EEG recordings: a review. *Proceedings of Engineering and Technology Innovation*, 20, 35–56.

[14] Mumtaz, W., Rasheed, S., & Irfan, A. (2021). Review of challenges associated with the EEG artifact removal methods. *Biomedical Signal Processing and Control*, 68, 102741.

[15] Yasoda, K., Ponmagal, R. S., Bhuvaneshwari, K. S., & Venkatachalam, K. (2020). Retracted article: automatic detection and classification of EEG artifacts using fuzzy kernel SVM and wavelet ICA (WICA). *Soft Computing*, 24(21), 16011–16019.

[16] Mathe, M., Mididoddi, P., & Krishna, B. T. (2022). Artifact removal methods in EEG recordings: a review. *Proceedings of Engineering and Technology Innovation*, 20, 35–56.

[17] Kaya, İ. A Brief Summary of EEG Artifact Handling. In Artificial Intelligence. Intech Open, 2022. https://doi.org/10.5772/intechopen.99127.

[18] Jiang, X., Bian, G. B., & Tian, Z. (2019). Removal of artifacts from EEG signals: a review. *Sensors*, 19(5), 987.

51 A novel color image encryption method based on block-based DNA coding compressive sensing

S. Kashif Hussain[1,a], U. Mahesh[2,b], C. Veda Murthy[2], K. Anjaneya Reddy[2,c] and B. Yadavendra Reddy[2,d]

[1]Assistant Professor, Department of Electronics and Communication Engineering, Rajeev Gandhi Memorial College of Engineering and Technology, Nandyal, Andhra Pradesh, India

[2]Students, Department of Electronics and Communication Engineering, Rajeev Gandhi Memorial College of Engineering and Technology, Nandyal, Andhra Pradesh, India

Abstract

This work introduces a novel encryption technique to safely encode color images without compromising their quality and integrity. It exploits the inherent characteristics of color images in order to grant them more protection than conventional methods of encryption. The technique integrates sophisticated methods such as compressive sensing, DNA coding in block-by-block manner, discrete cosine transforms (DCT), Josephus permutation, and a chaotic system for random sequence generation. To ready an image for encryption, the algorithm initially conducts necessary pre-processing operations like resizing, Eliminating any noise and, if needed, converting the image to grayscale. It accepts popular image formats such as.jpg and.png. What makes the encryption so secure is the use of a chaotic system, and this adds high randomness, and hence it's very hard to crack for intruders. Also, Josephus permutation additionally scrambles up the data such that the image is well encrypted is utilized to add complexity to the encryption procedure once compressive sensing is utilized to compress the image data via DCT. Secondly, the image is encoded using DNA coding, which maps the pixels of the image to DNA. At decryption time, singular value decomposition (SVD) is utilized to decrypt and reconstruct the encrypted image to enhance its reconstruction and overall quality. Techniques for image enhancement are employed to restore the image to its original form after the decryption stage. Several measures, including accuracy, structural similarity index measure(SSIM), number of pixels change ratio(NPCR), peak signal to noise ratio(PSNR), uniform average change intensity(UACI), correlation coefficient, histogram analysis, entropy and error rate, are employed to measure the performance of the algorithm. Although the decrypted image is very similar to the original with minimal loss of quality, these performance measures ensure that the encrypted image is secure. This approach provides a robust and efficient image encryption technique that can be applied in data security and secure communication.

Keywords: Block-based DNA coding, color image encryption, compressive sensing, correlation coefficient, entropy, error rate, histogram analysis, NPCR, peak signal-to-noise ratio PSNR, SSIM, UAC

Introduction

Ensuring image data has gained importance in today's digital age of communication and data storage, especially in multimedia, healthcare, and the military fields. As the picture data increases, it is important to ensure its integrity and security both when transmitted and stored. image-sharing sites and networks. Instead of using text data as it can easily be safeguarded by standard encryption algorithms, very large and composite multimedia information like images are often hard to securely protect. Due to this fact, scientists have focused on the development of image-specific encryption methodologies to address the unique characteristics of image information, such as size, high redundancy, and being sensitive to small alterations. This project introduces a color image encryption algorithm that provides a secure and efficient method of encrypting color images using compressive sensing and block-based DNA coding. One of the key aspects of the suggested encryption technique is compressive sensing [1].

It is an advanced signal processing technique that reduces the size of data used to describe an image by allowing reconstruction with fewer samples than in

[a]kashif1919@gmail.com, [b]upparimahesh30158@gmail.com, [c]kanjaneyareddyk@gmail.com, [d]yadavendrareddyb@gmail.com

DOI: 10.1201/9781003684589-51

conventional techniques, singular value decomposition (SVD) is utilized to decrypt and reconstruct the encrypted image to enhance its reconstruction and overall quality. This compresses the data size considerably, This makes the encryption process more efficient and quicker. Marked improvement. Improvements in the future will target enhancing the accuracy of existing algorithms, reducing over fitting and cross-dataset generalization. Models that are composed of multiple types of neural networks could lead to more generalizable models with the ability to handle a variety of visual characteristics, such as noise, contrast, and anatomical variation [3]. In addition, improvements in unsupervised and semi-supervised learning methods will help decrease the dependence on massive, labeled datasets, which will make AI-based medical image analysis more scalable.

The data matrix compressed is permuted through Josephus permutation upon compression, introducing an element of randomness to the encryption and enhancing security [4]. A further new aspect of this encryption method is the incorporation of DNA coding. DNA coding relies on the biological nature of DNA, and data from an image is encoded by the four nucleotide bases are thymine (T), cytosine (C), and guanine (G). This method offers a complex method of encoding and processing visual information, allowing complex operations like encoding, decoding, and XOR-based alterations. Through the use of DNA bases to represent pixels, the method achieves high levels of complexity, rendering it is virtually unbreakable without the keys held by unauthorized personnel [5].

The SVD is employed to optimize and improve both encryption and decryption. ensuring that the original image is returned while still maintaining security [6]. Another advanced technology that has picked up momentum in image encryption is DNA coding uses digital information, such as image pixels, in representation with the four nucleotide bases: adenine (A), thymine (T), cytosine (C), and guanine (G), inspired by DNA biology. This new approach provides an additional layer of security by transforming image data into a non-standard form that is hard for attackers to comprehend. Block-based DNA coding, for example, enhances the encryption process through the division of the image into blocks and application of DNA coding at the block level [7].

The block-wise approach increases randomness and complexity, and the encryption becomes more resistant to attackers [8]. Combining compressive sensing and DNA coding provides the encrypted image with security as well as compressibility and this allows storage and transmission ability in an efficient manner. SVD also improves encryption and decryption by decomposing the image into smaller parts. matrix into sub-matrices that ease the extraction of the original image during decryption. All these recent techniques—compressive sensing, block-based DNA coding, chaotic systems, and SVD—used in implementation give the proposed method a safe and stable option for color image encryption [9].

The primary contributions of this paper are:

Major contributions to this paper:

This encryption technique employs compressive sensing (CS) to compress images while preserving quality and chaotic systems to improve security. What sets it apart is that both the measurement matrices and the chaotic systems' initial conditions are generated from the original image itself. This renders the encryption highly immune to chosen-plain text attacks, where the attackers attempt to control known inputs to decrypt the encryption.

To better safeguard the image, this technique presents a new scrambling algorithm based on pseudo-random numbers and the Josephus problem. It efficiently decreases pixel correlation in the red, green, and blue (RGB) channels of the image, thereby making it much more difficult for attackers to defeat statistical analysis. This technique improves the encryption withstanding usual attacks and provides stronger security.

High-level DNA-based diffusion mechanism unlike general DNA operations (e.g., XOR, addition, subtraction), the current work involves a process where the DNA matrix sub-blocks of the compressed image are diffused with pseudo-random integers based on chaotic patterns Besides security, encoding rules for the DNA sub-blocks and decoding rules are also associated with pseudo-random numbers.

Singular value decomposition (SVD) optimization of measurement matrix It enhances the quality of reconstructed decrypted image by optimizing measurement matrices using SVD.

Literature Survey

Recent studies in image encryption have brought forward novel models that incorporate DNA coding and compressive sensing for better security and efficiency. Classical encryption schemes usually face difficulties with dealing with big image files and the excessive computation cost associated with them. Here is a highlight of the primary developments in this area.

Zhang et al. (2021) presented a hybrid encryption mechanism combining DNA encoding, crossover operations, and a multi-chaotic map to design a highly secure and dependable encryption process. Capitalizing on the parallel processing advantage of DNA computing and the randomness of chaotic systems, this method greatly enhances the security of the image. One of the major strengths of this methodology is that it renders the cipher text extremely sensitive to modifications in the original image. The multi-chaotic map creates intricate pseudo-random sequences for randomizing pixels, and DNA encoding and crossover operations facilitate diffusion for a stronger encryption process. The combination efficiently enhances the security as well as the reliability of encrypted images.

Liu et al. (2022) introduced a new image encryption method based on random base-pair transformation and DNA-level permutation, controlled by a hybrid chaotic system. Their approach leverages the sensitivity and randomness of chaotic systems and the unique properties of DNA computing, including massive parallelism and information density. To control the permutation process at the DNA level and the base-pair transformation processes. pseudo-random sequences were generated according to the hybrid chaotic system that combines a range of chaotic maps. Through such integration, there is guaranteed complexity and randomness at the encryption process.

Zhou et al. [1] presented a dynamic encryption method for SHA-512-based color DNA encryption based on the cryptographic hash function. Their approach combines the cryptographic strength of SHA-512 with the massive parallelism and high information density of for secure encryption, The algorithm must be very sensitive to the changes in the input image, so even a slight change can result in a large difference in the output. Even a minimal change, such as a change in one pixel, should result in a drastically different output so that maximum security is ensured. powerful computation is utilized. dynamic keys and control parameters were created for the encryption process with the SHA-512 hash function. For confusion and diffusion in the encrypted image, DNA computing is employed for encoding, decoding, and algebraic operations.

J. Chen et al (2022). A new encryption technique for images has been created that integrates 2D chaotic maps with DNA encoding, which is more secure and attack-resistant and randomness. The encryption process begins by rearranging pixel coordinates with a logistic map to produce a chaotic sequence. Subsequently, pixel values are coded with DNA coding, followed by a bit-level permutation to increase complexity further. Through the use of chaotic systems and DNA coding, this method is very resistant to brute-force and differential attacks, and it is therefore a very secure method of encryption.

Zhang et al. (2021) proposed an encryption system that integrates compressive sensing (CS) and chaotic permutation. The algorithm begins with discrete cosine transform (DCT) for compressing the image, which significantly minimizes data size. Then, there is a chaotic permutation technique that scrambles the image for better security. As compressive sensing allows efficient data compression, this scheme provides quicker encryption and decryption with high image quality, rendering it a realistic and secure option for image encryption.

Wei and Jiang [6] put forward a high-speed method of image encryption through parallel compressive sensing and DNA sequence operations. Their method employs the benefits of DNA computing, which has the potential for high parallelism and information density, and compressive sensing, which allows for parallel data encryption and compression. Whereas DNA sequence operations (encoding, decoding) were used to enhance the security of the encrypted image, parallel compressive sensing method was employed to reduce the computational burden and storage requirements.

Zhu et al. [2] created an effective and efficient image encryption method through SVD embedding with block compressive sensing. To render the encryption process meaningful, their approach utilizes SVD, which incorporates the encrypted data into a cover image, and compressive sensing, which allows for concurrent data compression and encryption. Although SVD embedding ensured that the encrypted image could be transmitted securely without arousing suspicion, The computational complexity and storage needs were minimized by applying the block compressive sensing approach.

Proposed Approach

Overview
The proposed encryption model processes color image data through several steps, but it primarily

aids the pretreatment phase, which includes security enhancement and image transformation. Figure 51.1 provides a detailed overview of the encryption system. Through DNA encoding methods, subjective color images are converted at the preprocessing phase. Pixel values are transformed to DNA strings using predetermined rules for encoding.

To enable efficient encryption, the image is also segmented into small parts through a block-based method. Compressive sensing is applied to reduce redundancy of data without losing important information to enhance security further. To impart a high level of cryptographic strength, the encoded blocks are then subjected to chaotic mapping for dissemination and confusion. Block-wise scrambling, compressive sensing-based encoding, A multi-stage encryption technique incorporates DNA sequence operations. Lastly, to manage the encryption and decryption processes and enhance security and efficiency, an enhanced key management system is employed.

Image pre-processing module

The pre-processing module is essential to the preparation of images for encryption by standardizing and optimizing them for subsequent processing. This includes resizing the image to have a uniform resolution, noise reduction to remove unwanted artifacts, and conversion to grayscale where needed. By providing uniformity between images of varying sizes and quality, this step provides a dependable base for encryption. Moreover, grayscale conversion also makes the image structure simpler, lowering processing overhead and making encryption more efficient and quicker without affecting security. Eliminating unwanted noise further improves encryption quality to provide a cleaner and optimized image for secure encryption.

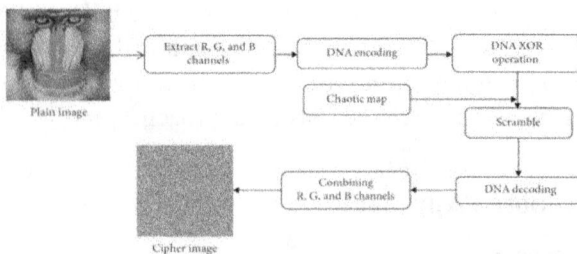

Figure 51.1 Block diagram of proposed system
Source: Author is visited google images use this ref (5)

Image encryption module

A main method for the protection of digital photographs against unauthorized access is image encryption. Though efficient, conventional encryption algorithms such as AES and DES are wanted when it comes to high-dimensional image data because of the accompanying computational complexity. DNA-based encryption provides image encryption with improved security and effectiveness based on principles of DNA computing together with chaotic maps and block-based transformation. This method employs chaotic scrambling for diffusion, combining XOR operations to encrypt image patterns. It also employs DNA encoding to convert pixel values into DNA sequences, providing an additional layer of security. (A, T, G, and C). The encrypted image obtained is both effective to encrypt and decrypt, and yet provides robust protection against attacks.

Chaotic system-based random sequence generation module

This module generates random sequences through a chaotic system, which is essential for ensuring the randomness of the encryption process. Chaotic maps, including the logistic map, produce pseudo-random sequences to shuffle and scramble pixels. The randomness generated through chaotic systems fortifies the encryption method, ensuring that it is resistant to cyber-attacks such as brute force attacks and differential analysis. The randomly generated sequences are utilized across various stages of encryption, such as pixel shuffle and DNA coding. Since the chaotic systems depend on initial values, even minor variations in them lead to highly dissimilar encryption outputs, providing a greater sense of security in the world.

Compressive sensing module

This module employs compressive sensing (CS) algorithms with the discrete cosine transform (DCT) in order to make data processing more efficient. CS aims to compress data to a smaller size while keeping essential image information intact. As it does this, the image is eliminating less crucial details. This minimizes storage requirements and speeds up the encryption process. Once compressed, the image is scrambled with Josephus Permutation, a translated from the spatial to the frequency domain. compresses the image by storing the most critical frequency components and sequence-based permutation algorithm that rearranges the compressed data of the image in a non-trivial manner. This double-layer of compression and permutation enhances unpredictability so much that it becomes nearly impossible to reverse-engineer the process without the relevant decryption keys. The application of CS ensures that the encrypted image is smaller in size, thus facilitating transmission on communication networks that is efficient and secure.

DNA coding module
This module saves the compressed and permuted image in DNA coding, in which pixel values are translated into four bases of DNA: adenine (A), thymine (T), cytosine (C), and guanine (G). Translating binary data into a biologically inspired form complicates and secures the process of encryption. One of the major benefits of DNA coding is that it provides for exclusive operations, like DNA sequence pairing and complementary, which cannot be directly implemented on binary data. This provides an additional layer of security. To further scramble the encrypted image, operations such as DNA addition and XOR are used. This DNA-based step of transformation is essential in the provision of strong encryption, as it injects non-linearity, this renders the algorithm immune to attacks like differential and statistical analysis.

SVD optimization module
This module speeds up the encryption process by utilizing SVD, a mathematical technique which breaks down an image matrix into singular values. Through adjustment of these values, the process of encryption becomes more secure while retaining high quality. The most important advantage of employing SVD is that it reduces distortion in the encrypted image while being capable of decrypting with negligible loss of data. Furthermore, SVD maintains the quality of the image while simultaneously minimizing dimensions, thus enhancing the efficiency of the encryption procedure. It also assists the algorithm in remaining robust against the noise and distortions that may result due to transmission. Even if slight alterations are introduced in the encrypted image, the decryption procedure still remains consistent, and the original image can be reconstructed accurately.

Image decryption module
Reverse operations with DNA-based methods are applied in this module to get back the original image from the encrypted image. The decoding method

is first employed to transform the pixel values of DNA sequences into binary form. The permuted and compressed image is then rearranged according to Inverse Josephus Permutation to get it back to its original compressed form. Once the image has been compressed, the inverse discrete cosine transform (IDCT) transforms it back from frequency domain to spatial domain. If one looks at these processes, the image is encrypted and the original information regained. This module ensures that the decryption process is smooth and precise, producing a decrypted image that is very close to the original with minimal information loss.

Evaluation performance matrices
Peak signal-to-noise ratio
Peak signal-to-noise ratio (PSNR) is an estimate of how good the image improvement is versus the original through a comparison of pixel value differences.

$$psnr = 10 \times \log_{10}\left(\frac{MAX^2}{MSE}\right) \quad (1)$$

Correlation coefficient of adjacent pixels
The higher the correlation, the closer the pixels resemble diagonal directions of one another. The lower the correlation, the greater the randomness, which is desired in encrypted images to improve security.

$$r = \frac{\sum_{i=1}^n (x_i-\mu_x)(y_i-\mu_y)}{\sqrt{\sum_{i=1}^n (x_i-\mu_x)^2 . \sum_{i=1}^n (y_i-\mu_y)^2}} \quad (2)$$

Structural similarity index measure (SSIM)
The SSIM index is extensively used for image comparison in domains of image processing, compression, and image quality assessment.

$$ssim(a,b) = \frac{(2\mu_a\mu_b+c1)(2\sigma_{ab}+c2)}{(\mu_a^2+\mu_b^2+c1)(\sigma_a^2+\sigma_b^2+c2)} \quad (3)$$

Number of pixel change rate (NPCR)
NPCR measures the ratio of pixels that are different between the original image and the encrypted image.

$$npcr = \frac{\sum_{a=1}^m \sum_{b=1}^n (D(a,b))}{m.n} \times 100 \quad (4)$$

Simulation Results

Image encryption
Simulation experiments, dataset parameters, performance measures, and results of the proposed method

(a)   (b)   (c)   (d)

Encryption images

(e)   (f)   (g)   (h)

Decryption images

(i)   (j)   (k)   (l)

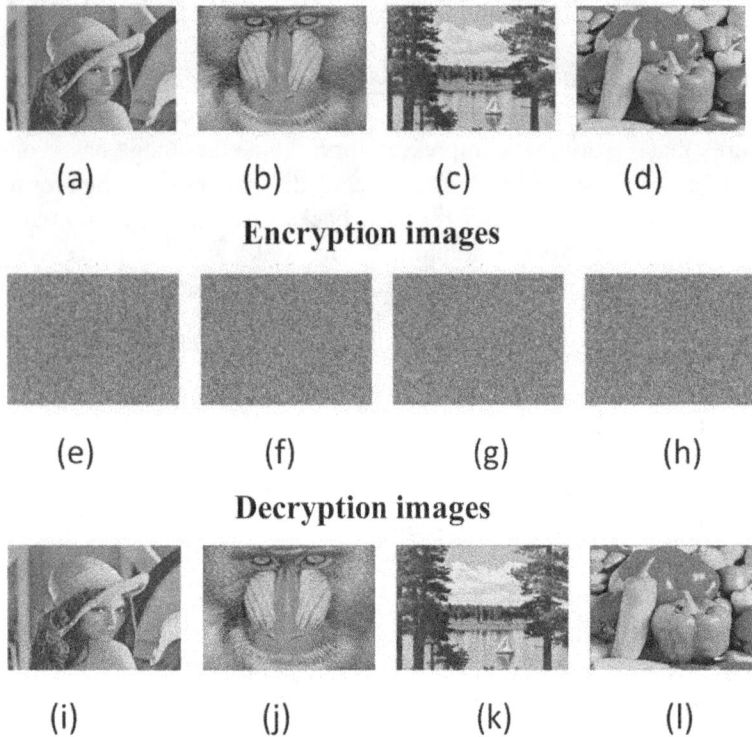

Figure 51.2 The equalization for various cipher and original images
Source: Author is Munson DC (1996) a note on Lena. In : IEEE Trans Image Process use this reference ref [17]

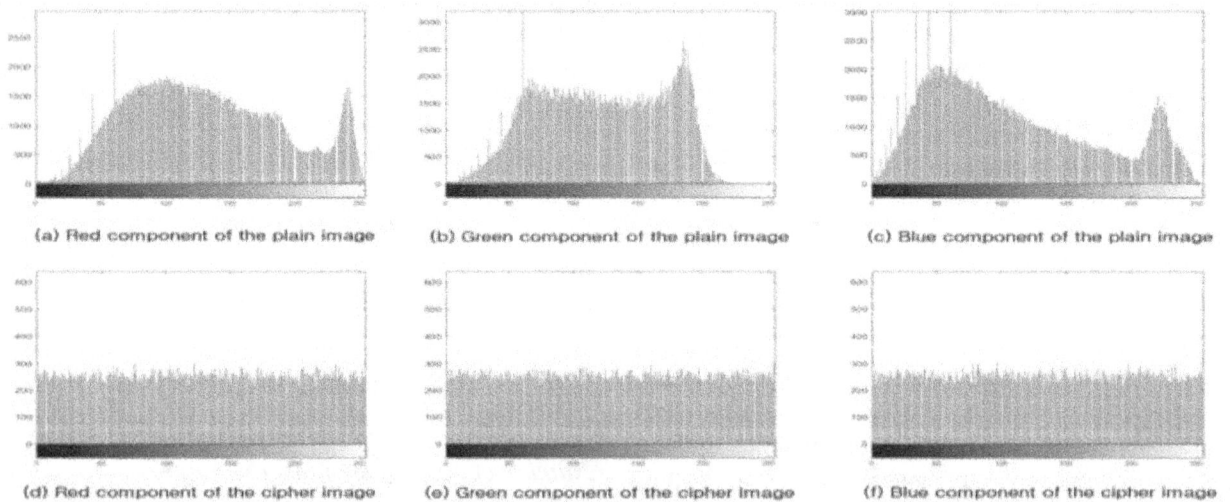

(a) Red component of the plain image   (b) Green component of the plain image   (c) Blue component of the plain image

(d) Red component of the cipher image   (e) Green component of the cipher image   (f) Blue component of the cipher image

Figure 51.3 Histograms of the raw baboon image and the encrypted image
Source: Author is Sandip Vijay and Nidhi Sethi at (2013). Based on the values are graphical representation of data

are discussed in this section. The tests were conducted using Python, leveraging its image processing and encryption libraries. For increased computational effectiveness, The simulations were executed on an Intel Core i5 processor (2.4 GHz) with 8 GB of RAM. The encryption model was implemented and tested with Python in a scientific computing environment. consisting of NumPy, OpenCV, Matplotlib, and cryptography libraries Figure 51.2-51.3 and Table 51.1-51.2.

Table 51.1 Comparison with different images.

Images	Metrices			
	NPCR	PSNR	SSIM	UACI
Image 1	99.97	32.43	0.004	33.14
Image 2	99.96	33.76	0.009	34.67
Image 3	99.95	32.56	0.006	32.81
Image 4	99.97	34.89	0.005	33.97

Source: Ours method values compare different images.

Table 51.2 Comparison with other studies on entropy and correlation coefficient (Ref [5] and Ref [10]).

Metrices	Ref [5]	Ref [10]	Ours
ENTROPY	7.379	6.372	6.235
Correlation coefficient	0.007	0.006	0.004

Source: Author

Results

Conclusion

In this research, we efficiently constructed a groundbreaking color image encryption method based on compressive sensing and block-based DNA encoding. Application of different methods like image preprocessing, discrete cosine transform-based compressive sensing, chaotic system-based random sequence generation, Singular value decomposition (SVD) and DNA encoding made the encryption process more efficient and secure to a great extent. The multi-layered method guarantees that the encrypted image is highly unpredictable and immune to various cryptographic attacks. The decryption process, the reverse of encryption, effectively retrieves the original image, maintaining its clarity with minimal distortion. Additionally, performance evaluation metrics like PSNR, SSIM, correlation coefficient, entropy, and NPCR indicate that the encryption method is robust in terms of image quality and security. The outcome is evident that the suggested approach possesses good encryption characteristics with the integrity of the decrypted image. The security and efficiency of the system make it suitable for practical applications where secure transfer of pictures is needed through communication networks.

Future Scope

Emerging New Encryption Technologies: Scientists are researching new techniques for image encryption that incorporate compressive sensing into other security tools. The desire is to produce encryption that is stronger and yet more efficient.

Speeding Up and Ensuring Better Quality: Researchers are working to come up with measurement matrices that not only speed up encryption and decryption but also enable the recovery of images of improved quality.

Pushing Image Reconstruction: More sophisticated two-dimensional (2D) reconstruction methods are under investigation to provide more accurate and reliable image decryption.

Employing AI and Math for Enhanced Security: Emerging encryption techniques might utilize deep learning and sophisticated mathematical models to devise even more secure and efficient encryption.

Making Encryption Work for All Types of Images: Attempts are being made to make encryption techniques work well for images of all sizes, whether in color or monochrome, without hurting performance or security.

References

[1] Zhou, S., He, P., & Kasabov, N. (2020). A dynamic DNA color image encryption method based on SHA-512. *Entropy*, 22(10), 1091.

[2] Zhu, L., Song, H., Zhang, X., Yan, M., Zhang, T., Wang, X., et al. (2020). A robust meaningful image encryption scheme based on block compressive sensing and SVD embedding. *Signal Processing*, 175, 107629.

[3] Bao, W., & Zhu, C. (2022). A secure and robust image encryption algorithm based on compressive sensing and DNA coding. *Multimedia Tools Applications*, 81(11), 15977–15996.

[4] Es-Sabry, M., El Akkad, N., Merras, M., Satori, K., El-Shafai, W., Altameem, T., et al. (2023). Securing images using high dimensional chaotic maps and DNA encoding techniques. *IEEE Access*, 11, 100856–100878.

[5] Chai, X., Fu, J., Gan, Z., Lu, Y., & Zhang, Y. (2022). Animage encryption scheme based on multi-objective optimization and block compressed sensing. *Nonlinear Dynamics*, 108(3), 2671–2704.

[6] Wei, D., & Jiang, M. (2021). A fast image encryption algorithm based on parallel compressive sensing and DNA sequence. *Optik*, 238, 166748.

[7] Cai, J., Xie, S., & Zhang, J. (2023). Image compression-encryption algorithm based on chaos and compressive sensing. *Multimedia Tools Applications*, 82(14), 22189–22212.

[8] Zhu, S., Deng, X., Zhang, W., & Zhu, C. (2023). Secure image encryption scheme based on a new robust chaotic map and strong S-box. *Mathematics and Computers in Simulation*, 207, 322–346.

[9] Yu, J., Xie, W., Zhong, Z., & Wang, H. (2022). Image encryption algorithm based on hyper chaotic system and a new DNA sequence operation. *Chaos, Solitons Fractals*, 162, 112456.

[10] Bao, W., & Zhu, C. (2022). A secure and robust image encryption algorithm based on compressive sensing and DNA coding. *Multimedia Tools Applications*, 81(11), 15977–15996.

[11] Pak, C., & Huang, L. (2022). A new color image encryption using combination of the 1D chaotic map. *Signal Processing*, 138, 129–137.

[12] Tian, X. L., & Xi, Z. H. (2021). An optimization algorithm for measurement matrix in compressed sensing. *Electronic Science and Technology*, 8(28), 102–111.

[13] Zhou, Y., Bao, L., & Chen, C. L. P. (2023). A new 1D chaotic system for image encryption. *Signal Processing*, 97, 172–182.

[14] Zhang et al. "Chaos-based image hybrid encryption algorithm using DNA cross over operation," Optik, vol. 216, Aug. 2021, Art. no. 164925.

[15] Liu et al.(2022) "A new image encryption scheme based on random base-pair transformation and DNA coding," Opt. Laser Technol.,vol. 160, May 2023, Art. no. 109033.

[16] J. Chen et al (2022), "New image encryption technique for images has been created with DNA encoding" Optik, vol. 224, Dec. 2020, Art. no. 165661.

[17] Munson DC (1996) a note on Lena. In : IEEE Trans Image Process.use this reference link (https://www.doi.org).

52 Weather prediction system using deep learning

D. Sarika[1,a], G. Ajay Kumar Reddy[2,b], D. Jyoshna Sai Priya[2,c], J. M. Dharma Teja[2,d] and V. Amrutha[2,e]

[1]Assistant Professor, Department of CSE, Annamacharya University, Rajampet, Andhra Pradesh, India

[2]Students, Department of CSE, Annamacharya Institute of Technology and Sciences, Rajampet, Andhra Pradesh, India

Abstract

We know the fact that weather is so unpredictable, precise weather forecasting is crucial for everyday operations, agriculture, and industry. The intricate and ever-changing structure of atmospheric variables frequently cause traditional weather forecast algorithms to struggle with accuracy. This project introduces a weather forecasting system that forecasts particular weather conditions like drizzle, rain, sun, snow, or mist using predictive algorithms and deep neural networks. We used exploratory data analysis (EDA) and preprocessing techniques like normalization to produce a dataset that contained important parameters such as date, rainfall, temperature (max and min), wind, and meteorological type. Utilizing the sequence prediction capabilities of long short-term memory (LSTM) and K-nearest neighbors (KNN) models, the system achieves 99% accuracy and 93.6% accuracy, respectively. Real-time forecasting is made possible via a Streamlit based user interface, which lets users enter circumstances and get accurate weather predictions. The shortcomings of conventional techniques are addressed by this approach, which shows increased weather prediction accuracy and accessibility.

Keywords: Deep learning, exploratory data analysis, k-nearest neighbors, LSTM, machine learning, normalization, prediction and weather forecasting

Introduction

Numerous industries, such as agriculture, transportation, disaster relief, and everyday planning for both individuals and communities, depend heavily on weather forecasting. Precise weather forecasting can improve safety, boost economic activity, and lessen the effects of extreme weather [1]. However, because atmospheric processes are extremely complicated and nonlinear, forecasting weather patterns continues to be a difficult task. Conventional weather forecasting techniques mostly rely on physics-based mathematical models, which can be computationally demanding and can fall short of capturing the nuances of local weather conditions, despite their effectiveness. Data-driven strategies employing machine learning (ML) and deep learning (DL) methodologies [2] have surfaced as workable ways to overcome these constraints and increase forecast accuracy. Because of climate change, weather patterns have become more variable globally in recent years, which makes precise forecasts even more important. The World Meteorological Organization (WMO) reports that throughout the previous 50 years, the number of natural disasters associated with extreme weather has increased fivefold, resulting in substantial financial losses as well as fatalities. According to statistics, the bulk of these occurrences are storms, floods, and droughts, and weather-related disasters are responsible for around 74% of all natural hazards' economic damages. Such weather-related catastrophes have had a huge economic impact on the world, costing more than $3 trillion over the last 20 years. The significance of precise and easily accessible weather forecasts is highlighted by the fact that prompt warnings and effective forecasting can lessen these effects by empowering authorities to take preventative action [3].

Public health is impacted by extreme weather occurrences in addition to financial losses. For instance, heatwaves have increased in frequency and severity in many parts of the world, resulting in thousands of heat-related diseases and fatalities annually. In a similar vein, severe rainfall events which

---

[a]sarika.daruru7790@gmail.com, [b]ajayreddyajay2003@gmail.com, [c]dsaipriya35@gmail.com, [d]dharmatejajambuladinne@gmail.com, [e]amrutha74162@gmail.com

DOI: 10.1201/9781003684589-52

are frequently made worse by climate change have caused extensive floods in several nations, uprooting millions of people and affecting the security of food and water. These events are becoming more frequent and intense, which highlights the urgent need for better weather prediction systems that can produce accurate, location-specific forecasts. Accurate weather forecasting [4] is essential to climate resilience and long-term growth because it can inform agricultural planning, public health advisories, and catastrophe preparedness. Forecasting the weather accurately is extremely important in India, since more than 50% of its population works in agriculture. Extreme weather occurrences have increased in frequency in recent years, especially monsoon-related heatwaves, droughts, and floods, according to the Indian Meteorological Department (IMD). More than twelve million individuals in several states experienced catastrophic floods as a result of high rainfall events during the 2020 rainfall season [5]. In addition, thousands of deaths have been attributed to heatwaves in India's northern region in recent years.

Literature Survey

Historically, physics-based models that use mathematical descriptions of atmospheric processes have been the main focus of weather prediction research. These models, like numerical weather prediction (NWP) theories, simulate weather patterns using equations derived from fluid dynamics, the laws of thermodynamic and other physical laws. Despite providing insightful information, these models are computationally demanding and frequently necessitate substantial resources, including high-performance computing infrastructure. Bhawsar et al. by using methods like statistical analysis, neural networks, and deep learning, predicting future weather is an important topic in the modern period. This supports smart management initiatives, decision-making, and disaster prevention [6]. For forecast management techniques, precision in short-term ambient prediction circumstances is especially advantageous. This study examines machine learning and deep learning approaches, going over their types, uses, and issues. Basha et al. proposed a predicting rainfall is essential to Indian agriculture since it enables farmers to take preventative measures for their crops. Rainfall is predicted by using ML methods such as the ARIMA modeling, artificial neural networks, logistic

regression, Support Vector Machine, and self-organized map [7].

Schultz et al. proposed an interest in using deep learning techniques [8] for robotics, speech recognition, picture recognition, and strategy games in meteorology has increased as a result of artificial intelligence. Neural networks and big data mining may enhance weather forecasts. The question is whether DL techniques can fully replace the numerical weather models that are now in use. Numerical weather models might eventually become outdated, but this objective requires significant advancements. Singh et al. examined that an outdated weather forecast techniques are becoming more disorganized and less effective due to the rapid climate change [9]. Better and more accurate weather forecast techniques are required to meet these difficulties. The goal of this study is to create a portable, affordable forecasting the weather system that may be utilized in isolated locations. Weather forecasting is done using data analysis and artificial intelligence methods like random forest classification. This strategy has a big impact on people's lives and a country's economy. Rental for early warning of effects on the lives of people, such self-driving automobiles and traffic congestion, weather forecasting is essential. Traditional numerical weather forecast techniques have drawbacks, such as a lack of a thorough grasp of physical systems and trouble extracting meaningful information from observation data.

Data Collection and Preprocessing

Every weather prediction system must have data collecting since it provides the raw information needed to spot trends and create forecasts. Weather data required by this project is obtained from dependable sources that document a range of meteorological factors, including precipitation, wind speed, temperature (both maximum and minimum), and overall weather conditions. Weather patterns may be tracked over time thanks to the timestamps included in each data entry, which provide a time and date of the observation. Since it guarantees that the model may learn from a wide variety of atmospheric conditions and fluctuations, consistent and thorough data collecting is essential for increasing prediction accuracy. To comprehend the features and geographic distribution of the dataset, exploratory data analysis, or EDA, is the following step after data collection. EDA [11]

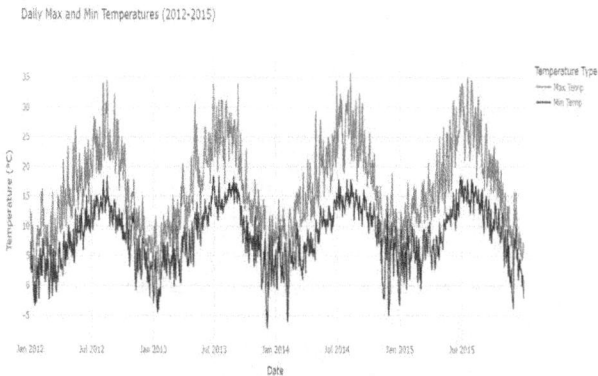

Figure 52.1 Daily max and min temperature
Source: Author

assists in locating patterns, seasonality, and irregularities in the data, all of which have a substantial influence on the results of predictions.

For raw data to be transformed into an organized form appropriate for model training, data preparation is necessary. Preprocessing begins with addressing values that are missing, which are frequently present in meteorological data because of instrument failures or unfavorable weather conditions. Imputation methods, such as utilizing medians, average values, or interpolating from neighboring observations, are used to fill in the gaps left by missing information. Making sure the dataset is full aids in noise reduction [12], allowing for more precise predictions and enhancing model performance. Because it adjusts numerical characteristics to a consistent range, usually between 0 and 1, normalization is an essential preprocessing step. In weather information, where variables like wind speed, temperature, and rainfall may have vastly disparate units and scales, this is especially crucial. The model may concentrate on identifying trends in the data instead of being impacted by the scale of every characteristic by normalizing the data. By developing additional variables that more accurately reflect the basic trends in the data (as shown in Figure 52.1), feature engineering significantly improves the dataset's prediction potential.

Principles and Methods

Using sophisticated preprocessing techniques to maximize the quality of data for model training, structured data preparation is the first step in the process of creating an accurate weather forecast system. The gathered meteorological data first goes through a rigorous preprocessing procedure that includes resolving missing values, encoding, and normalization. In order to assist the model concentrate on finding links between features rather than being biased by different scales, data normalization is necessary to guarantee that numerical features such as temperature, wind speed, and precipitation are scaled consistently. To facilitate the efficient use of those characteristics in prediction models, encoding techniques such as one-hot encoding are also applied to variables with categories, such weather type. Selection of features and engineering are used to further refine the collected information and increase predicted accuracy after data preparation. Feature engineering enhances the dataset's predictive power by emphasizing the most significant elements and integrating temporal linkages, enabling models to accurately depict the dynamics of meteorological data. Several machine learning and deep learning algorithms are trained and assessed once the dataset has been optimized in order to determine the most effective method for predicting the weather. In order to avoid overfitting and enhance generalization on unknown data, each model is trained iteratively, with hyperparameters adjusted using validation procedures. To choose the method that produces the most accurate results, model performance is assessed using cross-validation and accuracy measures.

K-Nearest Neighbors Classifier

In machine learning, the K-nearest neighbors (KNN) approach [13] is a well-liked and user-friendly technique for classification and regression problems. The overwhelming majority (for classification) or average (for regression) of the "k" closest data points in the space of features is the basis for KNN predictions. The closest neighbors of each data point are taken into account by this distance-based technique; closeness is commonly quantified by Euclidean distance, however other metrics, including Manhattan or Minkowski distances, may also be employed dependent on the specifics of the data. KNN's simplicity and interpretability are among its main advantages; predictions are based on observed events rather than intricate transformation or model parameters, which makes it highly readable and useful in a variety of situations. KNN is categorized as a "lazy" learner because it stores the complete dataset and uses it to generate recommendations only when the query point is supplied. This feature has advantages and

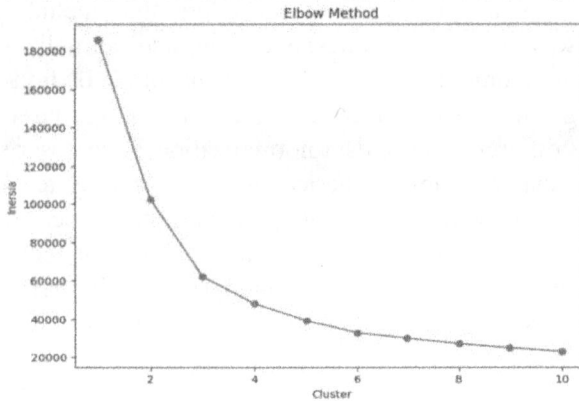

Figure 52.2 Elbow method for best 'K' values
Source: Author

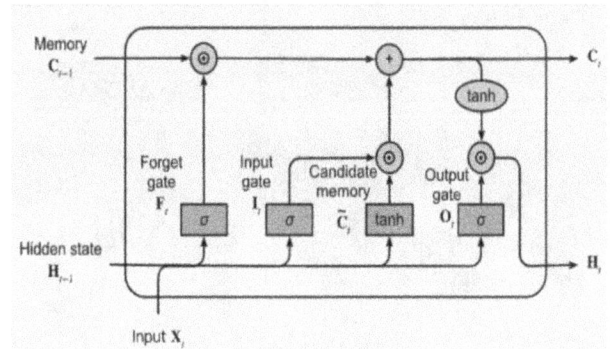

Figure 52.3 LSTM architecture
Source: Author

disadvantages. Although it doesn't necessitate a lot of training time, the forecasting stage can be computationally demanding, particularly when dealing with big datasets. The performance of KNN can be greatly impacted by its reliance on the number of neighbors, or "k." While bigger values of "k" may result in underfitting (as shown in Figure 52.2) because the model begins summing over excessive points and may overlook important local patterns, lower values may make the model more dependent on the noise in the data and cause overfitting.

This characteristic enables KNN to outperform other linear models on complicated datasets. However, the "curse of dimensionality," which occurs when distances among points become less defined, might reduce KNN's performance in high-dimensional fields, making it more challenging to precisely identify the genuine nearest neighbors. To enhance KNN's performance in high-dimensional datasets, dimensionality reduction methods such as Principal Component Analysis (PCA) or selection of features are frequently used prior to KNN. Furthermore, because of the model's sensitivity to feature scaling, normalization is crucial since unscaled features may dominate distance calculations and distort the results.

Long Short-Term Memory Network

A specific kind of recurrent neural network (RNN) called the long short-term memory (LSTM) architecture [14] is made to identify dependencies that persist in sequential input. In contrast to conventional RNNs, which sometimes have trouble learning patterns over lengthy sequences because of vanishing

or exploding slopes, LSTM has a special architecture that enables it to store information for extended periods of time. A system (as shown in Figure 52.3) of gates the input, forget and outputs gates that control the information flowing into and out of each cell allows for this functionality. The output gate chooses which data to send to the following cell, the input gate manages which fresh data to store, and the forget gate chooses which data to delete. Sequential dependencies are crucial for comprehending patterns, trends, and periodicity in time-series data, which is where LSTM excels. For instance, humidity, temperature, and other weather indications from previous days frequently affect future circumstances in weather prediction; LSTM is perfect for these applications since it can record these dependencies. Long-term trend data can be stored in the model's memory cell, which is especially useful for spotting seasonal variations, recurrent weather cycles, or slow changes over time. LSTM can produce more accurate forecasts than simpler models that might ignore these temporal connections through inference from sequence of prior weather data.

In order to capture the pertinent patterns without overfitting, an LSTM model must be trained by adjusting a number of hyperparameters, such as the number of levels, hidden units, and period length [15]. The length of the sequence has a significant impact on the model's performance since too short an episode may obscure important dependencies, while too large a sequence can add noise. The LSTM can capture the ideal ratio of short-term and long-term dependencies by optimizing these parameters. Furthermore, methods like dropout regularization, which randomly deactivate specific neurons during training to avoid overfitting and improve model generalization on fresh data, are advantageous for LSTM models. The capacity of LSTM can adapt to complicated, multivariate time series where several qualities interact with and impact on one another over time is one of its main advantages. The temperature, wind speed, humidity, precipitation, and other interrelated parameters can all be handled by LSTM for weather forecasting, and it can learn complex connections between them. The LSTM model can comprehend the effects of several factors on one another because of this multivariate approach, which produces predictions that are more thorough and accurate.

Results

Significant gains in accuracy and predictive power are shown by the outcomes of the weather prediction system that uses KNN and LSTM models. Following extensive training and hyperparameter adjustment, the KNN model demonstrated an astounding 93.6% accuracy. This result demonstrates how well the model uses previous weather data to provide precise forecasts by estimating the proximity of comparable historical occurrences. The dataset was divided into sets to be trained and tested as part of the evaluation process, and cross-validation was used to make sure the model performed well when applied to new data. Based on past trends, the KNN model demonstrated its capacity to adjust to transient weather variations by correctly forecasting weather conditions including drizzle, rain, sun, snow, or fog. On the other hand, the LSTM model performed better, attaining a remarkable 99% accuracy rate. The LSTM outperformed KNN in this domain because it was able to identify temporal patterns and long-term dependencies in the sequential meteorological data. The LSTM model showed its ability to comprehend intricate correlations in the data by training on numerous time steps and taking historical weather conditions into account. Metrics including accuracy, ability to recall and F1-score were used to validate the results (as shown in Figure 52.4), guaranteeing that the mathematical framework not only performed well in a variety of weather circumstances but also attained high accuracy.

Accuracy charts and confusion matrices were among the visual depictions of the model results that were used in conjunction with the evaluation measures. For both KNN and LSTM models, the

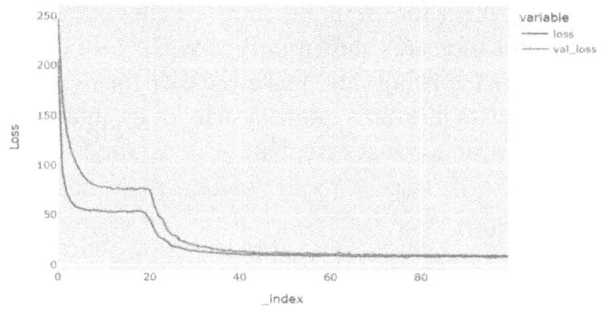

Figure 52.4 Training loss vs validation loss
Source: Author

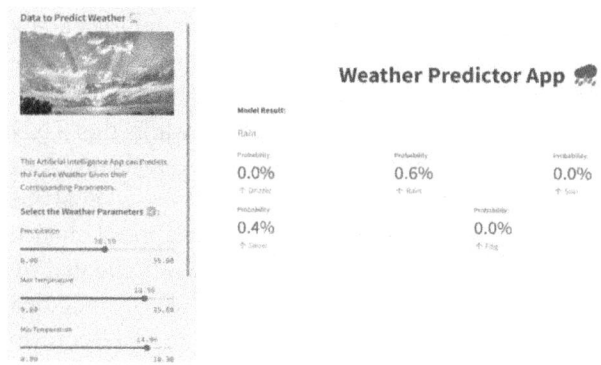

Figure 52.5 Streamlit user interface
Source: Author

accuracy plots showed the training and validation performance over epochs, demonstrating how the models converged towards excellent precision as training went on. When compared to KNN, the LSTM model's curve showed a higher increase in accuracy, demonstrating its effective learning capabilities. The confusion matrix, on the other hand, highlighted the quantity of true positives, misleading positives, real negatives, and incorrect negatives for every weather condition, offering insights into the algorithms' forecasts across several weather categories. Understanding how effectively the models worked in different weather conditions and where they could have incorrectly classified events was made easier by this representation. To guarantee robustness and dependability, the models were put through a number of evaluation methods in addition to correctness metrics. To see how the models responded to changes in weather patterns, they were evaluated using various subsets of the data. According to the testing phase's findings, both the KNN and LSTM models were able to successfully generalize their predictions, which qualified them for use in practical weather forecasting applications (as shown in Figure 52.5). In order to convey the results and make it easier for users to interact with the prediction system, user interface (UI) construction was essential.

Conclusion

Beneficial results from the creation and deployment of a weather prediction system using K-nearest neighbors (KNN) and long short-term memory (LSTM) algorithms demonstrate the efficiency of machine learning and deep learning approaches in predicting meteorological conditions. The LSTM model exceeded expectations with an astounding accuracy of 99%, emphasizing its strength in recording the ongoing dependencies and complex time-dependent trends inherent in sequential data, while the KNN model demonstrated its ability to leverage previous weather information through distance-based analysis with an accuracy of 93.6%. The models' performance over a range of weather categories was validated by the thorough evaluation process, which included metrics including precision, recall, and F1-score in addition to thorough visual representations through reliability plots and confusion matrices. All things considered, this work not only demonstrates how sophisticated machine learning techniques can improve the accuracy of weather forecasts, but it also lays the groundwork for further study and advancement in meteorological predictive analytics, opening the door to better agricultural, disaster, and everyday weather-related planning decision-making.

References

[1] Easterling, W. E., & Stern, P. C. (Eds.), (1999). Making Climate Forecasts Matter. National Academies Press.

[2] Kocher, G., & Kumar, G. (2021). Machine learning and deep learning methods for intrusion detection systems: recent developments and challenges. *Soft Computing*, 25(15), 9731–9763.

[3] Potter, S., Harrison, S., & Kreft, P. (2021). The benefits and challenges of implementing impact-based severe weather warning systems: perspectives of weather, flood, and emergency management personnel. *Weather, Climate, and Society*, 13(2), 303–314.

[4] Fathi, M., Haghi Kashani, M., Jameii, S. M., & Mahdipour, E. (2022). Big data analytics in weather forecasting: a systematic review. *Archives of Computational Methods in Engineering*, 29(2), 1247–1275.

[5] De, U. S., Dube, R. K., & Rao, G. S. P. (2005). Extreme weather events over India in the last 100 years. *Journal of Indian Geophysical Union*, 9(3), 173–187.

[6] Bhawsar, M., Tewari, V., & Khare, P. (2021). A survey of weather forecasting based on machine learning and deep learning techniques. *International Journal of Emerging Trends in Engineering Research*, 9(7), 850–857.

[7] Basha, C. Z., Bhavana, N., & Bhavya, P. (2020). Rainfall prediction using machine learning & deep learning techniques. In 2020 International Conference on Electronics and Sustainable Communication Systems (ICESC), (pp. 92–97). IEEE.

[8] Schultz, M. G., Betancourt, C., Gong, B., Kleinert, F., Langguth, M., Leufen, L. H., et al. (2021). Can deep learning beat numerical weather prediction?. *Philosophical Transactions of the Royal Society A*, 379(2194), 20200097.

[9] Singh, N., Chaturvedi, S., & Akhter, S. (2019). Weather forecasting using machine learning algorithm. In 2019 International Conference on Signal Processing and Communication (ICSC). IEEE.

[10] Ren, X., Li, X., Ren, K., Song, J., Xu, Z., Deng, K., et al. (2021). Deep learning-based weather prediction: a survey. *Big Data Research*, 23, 100178.

[11] Chatfield, C. (1986). Exploratory data analysis. *European Journal of Operational Research*, 23(1), 5–13.

[12] Boddapati, M. S. D., Desamsetti, S. A., Adina, K., Uppalapati, P. J., Murty, P. S., & PB V, R. (2023). Creating a protected virtual learning space: a comprehensive strategy for security and user experience in online education. In International Conference on Cognitive Computing and Cyber Physical Systems. Cham: Springer Nature Switzerland.

[13] Guo, G., Wang, H., Bell, D., Bi, Y., & Greer, K. (2003). KNN model-based approach in classification. In On The Move to Meaningful Internet Systems 2003: CoopIS, DOA, and ODBASE: OTM Confederated International Conferences, CoopIS, DOA, and ODBASE 2003, Catania, Sicily, Italy, November 3-7, 2003. Proceedings. Springer Berlin Heidelberg.

[14] Yu, Y., Si, X., Hu, C., & Zhang, J. (2019). A review of recurrent neural networks: LSTM cells and network architectures. *Neural Computation*, 31(7), 1235–1270.

[15] Staudemeyer, R. C., & Morris, E. R. (2019). Understanding LSTM--a tutorial into long short-term memory recurrent neural networks. arXiv preprint arXiv:1909.09586.

53 Optimized prediction of complex keyword queries in relational databases

Vijaya Rao Garapati[a] and T.K.K.V. Prasad[b]

Department of Computer Science Engineering, Sree Vahini Institute of Science and Technology, Tiruvuru, Andhra Pradesh, India

Abstract

Although traditional keyword searches provide a simple way to extract information from databases, they frequently exhibit poor ranking quality, resulting in low precision or recall, as highlighted by recent benchmarks. Recognizing queries that are likely to produce unsatisfactory ranking outcomes can improve user satisfaction by allowing the system to propose alternative queries for those that are problematic. In this study, we explore the characteristics of challenging queries and introduce a framework to evaluate the complexity of a keyword query within a database. Our method takes into account both the structural and content-related elements of the database, as well as the results of the queries. We assess our query difficulty prediction model using two effectiveness benchmarks for commonly used keyword search ranking techniques. The experimental findings demonstrate that our model effectively identifies difficult queries. Furthermore, we offer a series of optimizations designed to lower the associated computational costs.

Keywords: Keyword search, query effectiveness, relational databases, structured robustness

Introduction

The growing volume of data in relational databases has intensified demand for efficient and precise techniques to handle complex keyword queries. Users often prefer keyword searches due to their simplicity, as they eliminate the need for expertise in database schemas or formal query languages. However, the inherent ambiguity and complexity of such queries pose significant challenges in translating them into structured database operations, affecting both retrieval speed and result accuracy [1]. Relational databases, optimized for structured data, face difficulties in processing unstructured or semi-structured keyword-based requests. To address this, researchers have explored various strategies, including semantic parsing, graph-based representations, and probabilistic ranking models, to enhance query understanding [2]. Furthermore, indexing optimizations and distributed processing frameworks have been adopted to improve performance in large-scale database environments [3]. Despite these innovations, there remains a need for scalable and robust solutions capable of managing increasingly intricate data and query patterns [4]. This paper examines methods for [5, 6, 9] optimizing complex keyword query prediction in relational databases, with the objective of enhancing both efficiency and accuracy. We analyze current methodologies, identify their shortcomings, and introduce novel approaches based on advanced algorithms and optimized data structures. Our research aims to narrow the divide between unstructured keyword searches and structured relational data, facilitating more intuitive and effective information retrieval.

Related Work

Existing approaches for predicting complex queries in unstructured text fall into two categories:

1. **Estimating query complexity prior to execution:** Pre-retrieval methods predict query difficulty by analyzing intrinsic characteristics before executing the search. These approaches leverage statistical features such as:

   Entity-relationship density: Higher interconnectedness among query terms correlates with difficulty [10] Unlike post-retrieval methods, these techniques avoid computational overhead by relying on precomputed metadata (e.g., term frequencies, schema graphs) rather than actual results [6, 9]. However, their accuracy is inherently limited without runtime feedback [10].

_____
[a]vijayarao.g@gmail.com, [b]tkkvprasad@gmail.com

DOI: 10.1201/9781003684589-53

2. Result-driven query difficulty assessment: Post-retrieval approaches evaluate query complexity by analyzing search outcomes, yielding more accurate predictions than pre-retrieval techniques [8]. Our framework builds upon result stability principles, addressing limitations in existing topic-based approaches that require extensive domain knowledge.

IMDB Database Structure and Search Methodology

Database is structured as a collection of entity sets, with each entity set S comprising multiple entities E. For instance, in the IMDB database, movies and individuals form two separate entity sets. Figure 53.1 depicts a segment of a dataset, where each subtree with a root labeled "movie" represents an entity.

Attribute values are represented as a set of terms, generally omitting stop words, aligning with contemporary unstructured and semi-structured retrieval techniques. An attribute value A is associated with an attribute T, denoted as A∈T. For instance, in Figure 53.1, "Godfather" and "Mafia" are attribute values of a movie entity, with "Godfather" specifically linked to the "title" attribute. This abstract data model operates independently of physical storage formats, accommodating both XML files and normalized relational tables. It has been widely applied in entity search studies and data-centric XML retrieval research, offering the advantage of straight forward mapping between XML and relational structures. A key goal is to develop robust and formal models that can be applied to various databases and formats without relying on complex database design features, such as deep syntactic nesting, which can reduce effectiveness across different databases.

A keyword query $Q = \{q_1, q_2, \dots, q_{|Q|}\}$ consists of $|Q|$ terms, where an entity E matches Q if any term $q_i \in Q$ appears in at least one of its attributes A (i.e.,

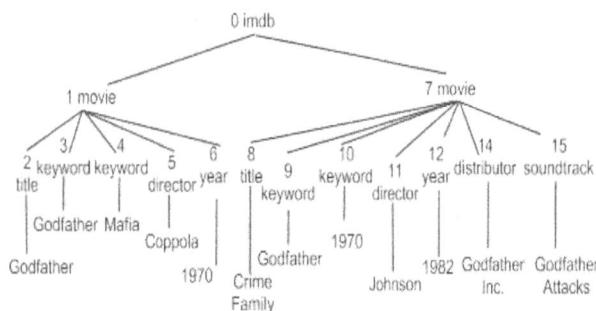

Figure 53.1 Database structure [13]
Source: Author

$q_i \in A$). For a database DB and query Q, the retrieval function g(E, Q, DB) computes a relevance score, assigning a real-valued measure of E's pertinence to Q. The system then ranks entities in descending order of their scores, producing an ordered list L(Q, g, DB). This framework ensures consistent and scalable evaluation across KQI methods.

Structured Data Retrieval Using Robust Ranking Principles

A robust ranking system for structured databases ensures consistent and relevant query results, even with variations in input phrasing, schema complexity, or data distribution. Key principles include:

Query-independent stability: Rankings should remain relatively unchanged for semantically equivalent queries (e.g., "2023 thriller movies" vs. "thriller films from 2023").

Schema-aware query rewriting (e.g., mapping "director: Nolan" to `directors.name = 'Christopher Nolan'`).

Result-set sensitivity: Easy queries (yielding few, highly relevant matches) should produce stable rankings. Hard queries (broad or ambiguous, e.g., "drama") tolerate minor scoring fluctuations.

Application in IMDb-style Systems:
Structured queries: Prioritize exact matches (e.g., `year=2023`) with deterministic sorting (e.g., `rating DESC`).

Keyword searches: Use probabilistic scoring (TF-IDF, BM25) with schema-guided constraints (e.g., boosting matches in `title` over `overview`).

Hybrid scenarios: Combine SQL `WHERE` clauses with ranking functions (e.g., Postgre SQL's `ts_rank`).

Optimized SR Score Computation for Structured Data Retrieval

The **semantic relevance (SR) score** quantifies how well database query results match user intent in keyword-based searches over structured data [14]. Unlike traditional IR ranking, SR incorporates both **schema awareness** and **semantic relationships** to improve accuracy.

Key challenges in computation:
Schema complexity: Multi-table joins (e.g., IMDb's movies, actors, genres) increase scoring overhead.

Term ambiguity: Words like "Joker" could refer to a movie title, character, or actor's role.

Real-time requirements: Users expect sub-second latency despite large datasets (e.g., 10M+ movies).

SR Score calculation:

Base score: Term frequency in relevant columns (e.g., `title`, `overview`).

Semantic boost: Embedding similarity between query and result metadata.

Robust Structured Query Processing Algorithm

The structured robustness algorithm (SR Algorithm) is detailed in the algorithm, which calculates the structured robustness (SR) score by utilizing the top-K retrieved entities. Each ranking algorithm detailed algorithm, it is represented in Figure 53.2(a) & (b).

a) SR Algorithm

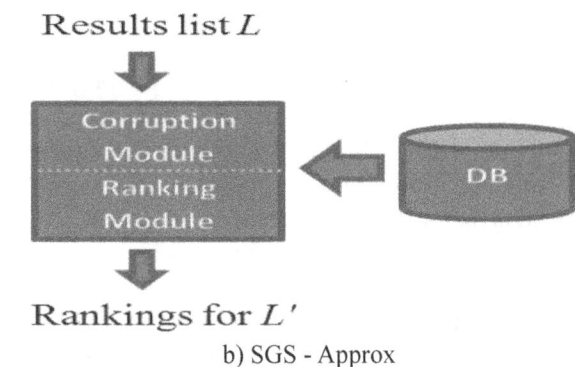

b) SGS - Approx

Figure 53.2 Robust structured query processing algorithm [9]

Source: Author

relies on statistical data concerning query terms and attribute values across the entire database. These global statistics are stored in M (metadata) and I (inverted indexes), which are crucial components of the SR algorithm pseudocode.

The SR algorithm introduces noise into the database in a dynamic manner during query execution by altering only the top-K entities identified by the ranking module. It avoids extra I/O operations by utilizing statistical values and recalculated data from inverted indexes, thus eliminating the necessity for additional indexing structures.

However, empirical studies reveal that the algorithm significantly increases query processing time due to several inefficiencies:

The SR algorithm is a structured ranking method that evaluates how well database results match a keyword query by considering both content relevance and schema context. Unlike traditional keyword search, it incorporates:

Attribute importance (γ_A) – Prioritizes matches in critical fields (e.g., `title` > `description`).

Term specificity (γ_T) – Weights rare/domain-specific terms higher than common ones.

Structural significance (γ_S) – Accounts for hierarchical relationships (e.g., parent-child in XML/JSON) or foreign-key links in databases.

Novel Method

7.1 Novel methods setting: Table 53.1 presents the main features of the datasets employed in our experiments.

1. INEX Dataset: Originating from the INEX 2010 Data-Centric Track, this dataset includes two types of entities: movie and person. The movie entity encompasses attributes like title, keywords, and year, while the person entity includes attributes such as name, nickname, and biography.

Table 53.1 Dataset characteristics [4].

	SemSearch	INEX
Size	9.69 GB	9.859 GB
No. of entities	7,170,445	5,418,081
No. of entity sets	519,610	2
No. of attributes	7,969,986	79
No. of attribute values	3214,056,158	123,603,013

Source: Author

2. SemSearch dataset: This is a subset of the dataset from the Semantic Search 2010 Challenge, originally comprising 116 files with nearly one billion RDF triplets. Due to its large size, querying the entire dataset is computationally intensive."

Query workloads: A segment of the SemSearch dataset was used, selecting 55 queries from the original 92 to ensure each had at least 50 potential answers. Two queries (Q6 and Q92) were excluded due to a lack of relevant answers, leaving 53 queries for testing. 26 query topics with relevance judgments were used from the INEX 2010 Data-Centric Track. The queries included "+" and "-" operators for conjunctive and exclusive conditions, but these were not utilized in the experiments; keywords following "-" were ignored. Although some systems employ these operators to improve searches, their use is intended for future research.

Ranking algorithms: To assess the effectiveness of the query performance prediction model, experiments were conducted using two representative ranking algorithms: PRMS and IR-Style [7]. These algorithms serve as the foundation for many extended ranking methods.

PRMS algorithm: As outlined in Section, the PRMS algorithm was evaluated by adjusting the λ parameter to optimize the Mean Average Precision (MAP). Tests were conducted with λ values ranging from 0.1 to 0.9 in 0.1 increments. Results indicated minimal MAP variations across different values, with smaller λ values yielding better MAP. The optimal settings were $\lambda = 0.1$ for INEX and $\lambda = 0.2$ for SemSearch.

R-Style algorithm: IR-Style is an adaptation of the ranking model originally designed for relational databases [11]. Since the datasets in this study are non-relational, IR-Style was modified to treat each entity as a minimal subtree (MTNJT), linking attribute values containing query keywords. The subtree's root corresponds to the root entity in the XML file [15] [16]. The one with the highest score is chosen from an entity that has multiple MTNJTs Table 53.2.

The **IR Score (M, Q)** for entity **E** is calculated as:

$$IRScore(M,Q) = \frac{IRScore(M,Q)}{size(M)}$$

Where an information retrieval (IR) ranking formula is used to determine the IR score (M, Q). For both datasets, MAP results using a vector space model Figures 53.3 and 53.4 are representing the SR

Table 53.2 Moderate K values [12].

K	20	10
Spearman's correlation score	0.519	0.513
Pearson's correlation score	0.564	0.568

Source: Author

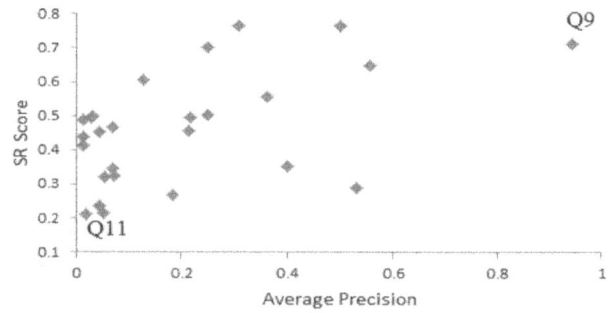

Figure 53.3 Semantic relevance score evaluation on INEX and PRMS Datasets with K = 20 and weighted parameters ($\gamma_A = 1$, $\gamma_T = 0.9$, $\gamma_S = 0.8$)
Source: Author

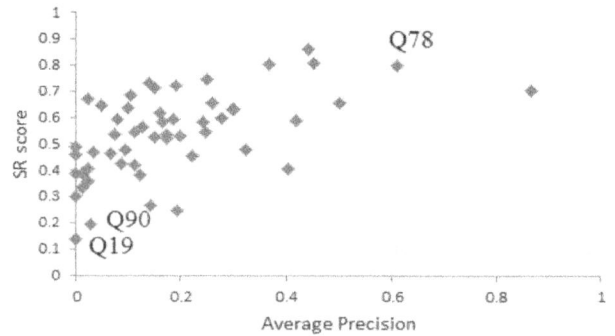

Figure 53.4 Semantic relevance (SR) Score evaluation on INEX and PRMS Datasets with K=20 & weighted parameters (γ_A, γ_T, γ_S) = (1, 0.1, 0.6)
Source: Author

score with average precision results of the proposed model, were less than 0.1. Consequently, as indicated in equation, Jelinek-Mercer smoothing was applied to a language model ranking algorithm. 0.2 was the smoothing parameter that yielded the largest MAP of α.

7.2 Quality results

Quality evaluation of SR algorithm: The SR Algorithm was assessed for query quality prediction using Spearman's correlation and Pearson's correlation between SR scores and average query precision.

Optimizing N value: For a stable SR score, ranking lists (L') were corrupted N times, calculating Spearman's correlation with the original ranking (L). The process stops when 50 consecutive iterations change the average SR score by less than 1%. The final N values were 300 for INEX and 250 for SemSearch.

Choosing K values: Since relevant results for keyword queries are limited, smaller K values yield better prediction accuracy. Testing K = 10, 20, and 50, results showed K=10 and 20 performed best, with K = 20 achieving the highest Pearson's correlation (0.556 on SemSearch).

Training model parameters $(\gamma_A, \gamma_T, \gamma_S)$: Five-fold cross-validation is using the model parameters $(\gamma_A, \gamma_T, \gamma_S)$ were trained, optimizing for Pearson's correlation. The best values were (1, 0.9, 0.8) for INEX and (1, 0.1, 0.6) for SemSearch.

Performance comparison with baselines: The SR algorithm demonstrated superior effectiveness compared to benchmark methods, including unstructured robustness method (URM), normalized query commitment (NQC), weighted information gain (WIG), and clarity score (CR). Empirical results revealed significantly stronger Pearson's and Spearman's correlation coefficients for SR at both K = 10 and K = 20, highlighting its robustness in ranking tasks. While the A (Q) metric achieved notable success on the INEX dataset, it marginally underperformed against NQC at K = 10, suggesting context-dependent trade-offs in retrieval performance.

Performance of IR-style ranking algorithm: The IR-Style ranking algorithm had a MAP score of 0.134 at K = 20 on INEX, which was too low for further evaluation. However, it outperformed PRMS on SemSearch, making it the focus of additional testing.

Discussion and Impact of schema complexity: Compared to leading text-based query predictors, which achieve correlation scores up to 0.65, the SR algorithm demonstrated strong performance for database queries. The impact of schema complexity is uncertain it can either complicate retrieval or enhance precision, depending on query structure [12].

Performance Study of Computation

SR Algorithm: By contrasting the average SR score computation time (SR-time) with the PRMS query processing time (Q-time), the effectiveness of the SR algorithm was assessed. These timeframes for K=20

Table 53.3 Efficiency and stability analysis of query processing with robustness validation (K = 20).

	Q-time of Avg (ms)	SR-time of Avg (ms)
SemSearch (N = 300)	46,799	(11,321 + 12,210)
INEX (N = 260)	24,199	(88,371 + 29,485)

Source: Author

are shown in Table 53.3, where SR-time is split into corruption time and re-ranking time. Because of an INEX searches have a greater overhead, only two entity sets (movie and person), which leads to widespread corruption in top-K entities. SemSearch, on the other hand, has more entity sets and attributes, which lessens the impact of corruption.

Efficient QAO-Approx: Figures 53.5(a) depict the performance of QAO-Approx on SemSearch and INEX. An algorithm achieves acceptable correlation scores while significantly reducing computation time (2 seconds for N=20 on SemSearch, N = 10 on INEX). However, reducing N leads to a drop in Pearson's correlation score, highlighting the importance of second and third-level corruption for accuracy.

Efficient SGS-Approx: Figures 53.5(b) depict the performance of SGS-Approx, where re-ranking occurs during corruption, meaning SR-time includes only the corruption time. While the efficiency gains on INEX are marginally less than those of QAO-Approx, the correlation score remains robust. Importantly, SGS-Approx surpasses QAO-Approx in both efficiency and accuracy on SemSearch.

Combining Efficient SGS-Approx & QAO-Approx: Figure 53.5(c) depict the combining SGS-Approx and QAO-Approx, as discussed, further enhances performance. Figures 53.5(a) and 53.5(b) show that this hybrid approach achieves greater efficiency than either QAO-Approx or SGS-Approx alone, while maintaining comparable prediction quality.

Discussion on SR score computation efficiency: The SR score computation time is primarily dependent on top-K results, as only these results undergo corruption and re-ranking. An increase in dataset size affects query processing time, but not SR score computation, which remains focused on the top-K entities.

Efficiency is also impacted by the data schema's complexity. Because a simpler schema may contain

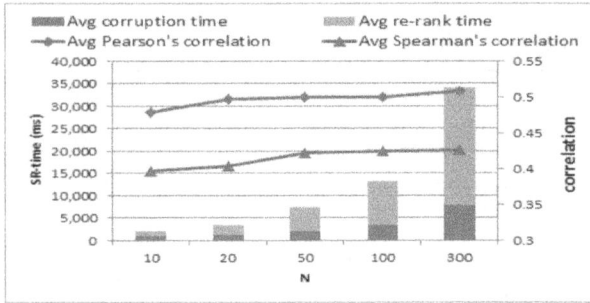

Figure 53.5(a) Results of Efficient QAO-approx
Source: Author

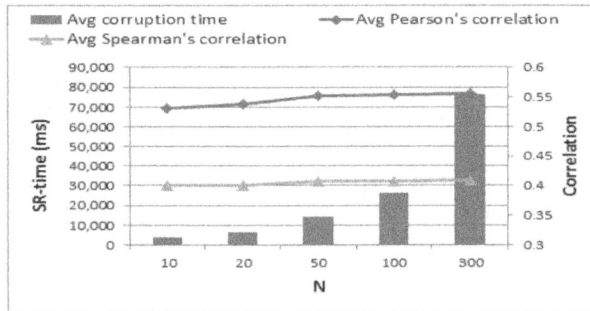

Figure 53.5(b) Results of Efficient SGS-approx
Source: Author

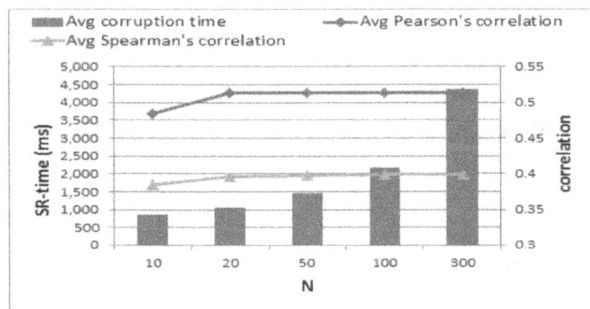

Figure 53.5(c) Results of Combination of efficient QAO and SGS
Source: Author

more attribute values that need to be corrupted, it does not always result in a lower SR computation time. This is demonstrated in Table 53.3 for INEX, which has a simpler schema than SemSearch but experiences longer corruption periods.

Conclusion and Future Scope

In this work, a novel method for forecasting the efficacy of keyword queries in databases (DBs)

is presented. For this issue, current approaches for unstructured data are inefficient. A systematic framework was created to address this, evaluating query difficulty using the ranking robustness principle. With low error rates and no time overhead, the suggested algorithms effectively forecast the efficacy of keyword queries.

Possible avenues for further research include:

- Extending the paradigm to top-K ranking issues in database systems, like ranking keyword queries with exclusions or SQL statements that are not fully stated.
- Investigating more ranking functions outside of the two function types covered in this research.
- Modifying the robustness framework to include both structured criteria and keywords in semi-structured queries.

References

[1] Hristidis, V., Gravano, L., & Papakonstantinou, Y. (2003). Efficient IR style keyword search over relational databases. In Proceeding 29th VLDB Conference, Berlin, Germany, (pp. 850–861).

[2] Sarkas, N., Paparizos, S., & Tsaparas, P. (2010). Structured annotations of web queries. In Proceedings 2010 ACM SIGMOD International Conference on Management Data, Indianapolis, IN, USA, (pp. 771–782).

[3] Bhalotia, G., Hulgeri, A., Nakhe, C., Chakrabarti, S., & Sudarshan, S. (2002) .Keyword searching and browsing in databases using BANKS. In Proceedings 18th ICDE, San Jose, CA, USA, (pp. 431–440).

[4] Trotman, A., & Wang, Q. (2010). Over view of the INEX 2010 data centrictrack. In 9th International Workshop INEX2010, Vugh, The Netherlands, (pp. 1–32).

[5] Townsend, S. C., Zhou, Y., & Croft, B. (2002). Predicting query performance. In Proceedings SIGIR 2002, Tampere, Finland, (pp. 299–306).

[6] Heand, B., & Ounis, I. (2006). Query performance prediction. *Information Systems*, 31(7), 585–594.

[7] Collins-Thompson, K., & Bennett, P. N. (2010). Predicting query performance via classification. In Proceedings 32nd ECIR, Milton Keynes, U.K., (pp. 140–152).

[8] Zhao, Y., Scholer, F., & Tsegay, Y. (2008). Effective pre-retrieval query performance prediction using similarity and variability evidence. In Proceedings 30th ECIR, Berlin, Germany, (pp. 52–64).

[9] Hauff, C., Azzopardi, L., & Hiemstra, D. (2009). the combination and evaluation of query performance

prediction methods. In Proceedings 31ˢᵗECIR, Toulouse, France, (pp. 301–312).

[10] Hauff, C., Murdock, V., & Baeza-Yates, R. (2008). Improved query difficulty prediction for the Web. In Proceedings, 17ᵗʰ CIKM, Napa Valley, CA, USA, (pp. 439–448).

[11] Yom-Tov, E., Fine, S., Carmel, D., & Darlow, A. (2005). Learning to estimate query difficulty. In Proceedings 28ᵗʰ Annual International ACM SIGIR Conference Research Development Information Retrieval, Salvador, Brazil, (pp. 512–519).

[12] Termehchy, A., Winslett, M., & Chodpathumwan, Y. (2011). How schema independent are schema free query interfaces? In ICDE. P.p 649–660. DOI:10.1109/ICDE.2011.5767880.

[13] Zhou, Y., & Croft, B. (2006). Ranking robustness: a novel framework to predict query performance. In CIKM. P.p 567-574. DOI: 10.1145/1183614.1183696.

[14] Mittendorf, E., & Schauble, P. (1996). Measuring the effects of data corruption on information retrieval. In Proceedings of the SDAIR 96 Conference.

[15] Cheng , S. (2015). Effective exploration of web and social network data. Proquest. Technological and Forecasting and Social Change, 95, pp 48-56.

54 Automated passenger fare collection using RFID and face recognition for public transportation

R. Madhan Mohan[a], N. Poojitha[b], G. Manasa[c], N. Nirupama[d] and P. Mahesh[e]

Department of Electrical and Electronics and Engineering, Annamacharya Institute of Technology and Sciences, Rajampet, Andhra Pradesh, India

Abstract

The Smart Bus Ticketing System integrates RFID technology and face recognition to automate fare collection in public transportation. Passengers use RFID cards for identification and payment, while a keypad allows them to input the number of travelers. The system comprises an RFID reader, LCD, ESP8266 (Wi-Fi module), and an IoT platform (UBIDOTS) for real-time monitoring and data handling. A face detection module, utilizing a laptop-connected camera, processes images through feature extraction and classification using a Support Vector Machine (SVM) model to verify passenger count. If the detected count matches the registered input, the fare is automatically deducted via IoT integration. This system enhances security, reduces manual intervention, and ensures accurate passenger verification. Key features include automated payment processing, user-friendly input, and real-time monitoring, improving the efficiency of public transportation ticketing while minimizing fraud and errors in fare collection.

Keywords: Automated fare collection, ESP8266, face recognition, IoT-based monitoring, machine learning, passenger verification, public transportation, real-time monitoring, RFID technology, smart ticketing system, support vector machine (SVM), UBIDOTS etc

Introduction

Public transportation is a crucial aspect of urban mobility, and efficient ticketing systems play a vital role in ensuring seamless passenger movement. Traditional ticketing methods, such as manual ticket issuance and paper-based systems, often lead to inefficiencies, delays, and revenue leakages. To address these issues, automated fare collection (AFC) systems have been introduced, utilizing technologies like radio frequency identification (RFID) and biometric authentication. RFID technology has been widely adopted for contactless payment systems in public transportation due to its fast processing, security, and ease of integration [1]. It enables automatic identification of passengers through RFID cards, reducing manual intervention and minimizing fraud [3]. Additionally, the integration of Internet of Things (IoT) platforms allows concurrent data monitoring, ensuring accurate passenger tracking and system efficiency [8]. However, existing RFID-based ticketing systems primarily rely on card-based authentication, which can still be subject to misuse,

such as card sharing or unauthorized access. To further enhance security and accuracy, face recognition technology is gaining traction in modern smart ticketing solutions. Computer vision-based authentication methods provide an additional layer of verification, ensuring that the registered passenger is the one utilizing the service [2]. Machine learning algorithms, such as Support Vector Machines (SVM), have been effective in image classification tasks and can be employed for real-time face detection in ticketing systems [4].

Despite advancements in smart ticketing, several challenges remain:

1. Card misuse and fraud – RFID-based systems, while efficient, are vulnerable to unauthorized card sharing and duplication, leading to revenue losses [3].
2. Lack of passenger verification – Many AFC systems do not verify whether the cardholder is the actual traveler, making them susceptible to fraudulent activities [7].

[a]madhanmohanmadhu@gmail.com, [b]poojithanettalam2004@gmail.com, [c]manasareddy54231@gmail.com, [d]narapuramnirupamareddy@gmail.com, [e]naidumahesh679@gmail.com

DOI: 10.1201/9781003684589-54

3. Limited real-time monitoring – Some systems lack efficient real-time tracking and monitoring mechanisms, reducing their effectiveness in dynamic public transport environments [8].
4. Integration complexity – Combining different authentication methods, such as RFID and face recognition, poses integration and computational challenges, especially in resource-constrained environments [9].
5. Hardware constraints – Implementing a reliable face detection mechanism within a cost-effective system requires an optimal balance between hardware capabilities and computational efficiency [11].

Given the limitations of current RFID-based ticketing systems, there is a need for a more secure, automated, and efficient fare collection mechanism. The integration of RFID with face recognition provides a dual-layer authentication system that reduces fraud, ensures accurate passenger verification, and enhances the reliability of public transportation ticketing. By leveraging IoT and machine learning, the proposed system goal is to provide real-time monitoring and enhanced commuter experience.

The primary objectives of this research are:

- To design and implement an automated smart bus ticketing system using RFID and face recognition.
- To integrate IoT for real-time monitoring and data handling.
- To improve security by verifying passenger identity through machine learning-based face detection.
- To enhance accuracy and efficiency in fare collection by reducing manual intervention.
- To analyze system performance in terms of speed, accuracy, and reliability.

This paper presents an innovative approach to smart bus ticketing that integrates RFID technology with face recognition for enhanced security and efficiency. The key contributions are:

1. Dual authentication mechanism – Combining RFID-based identification with face recognition for robust passenger verification.
2. IoT-Based monitoring – Utilizing UBIDOTS for real-time data visualization and remote system management.

3. Machine learning implementation – Deploying an SVM-based model for accurate face detection and classification.
4. Cost-effective and scalable design – Developing a practical and scalable system using ESP8266 and laptop-connected cameras for real-world deployment.

The organization of the paper divides different sections.

The proposed system is presented in Section 2. Next, experimental results are exhibited in Section 3, conclusion and future work are shown in Section 4. Further in Section 5 shown the literature survey.

2. Next, experimental results are exhibited in Section 3, conclusion and future work are shown in Section 4. Further in Section 5 shown the literature survey.

Proposed method

The proposed Smart Bus Ticketing System integrates RFID technology and face recognition to enhance fare collection efficiency and passenger authentication in public transportation shown in Figure 54.1. The system consists of key hardware components, including an RFID reader, keypad, LCD display, ESP8266 (Wi-Fi module), and a laptop-connected camera for face detection. The process begins with passengers using RFID cards for identification and payment. They input the number of travelers using a keypad, and the system registers their entry. To prevent misuse and unauthorized card sharing, a face detection module is implemented, capturing real-time images of passengers. These images undergo preprocessing, feature extraction, and classification using a SVM model, ensuring that the number of detected passengers matches the registered count. If verification is successful, the fare is automatically deducted, and real-time data is sent to the UBIDOTS IoT platform for monitoring. This dual-layer authentication mechanism enhances security, minimizes fraud, and improves ticketing accuracy while reducing manual intervention.

Methodology
The Smart Bus Ticketing System relies on various hardware components that work together to facilitate RFID-based passenger identification, face recognition, fare calculation, and real-time monitoring.

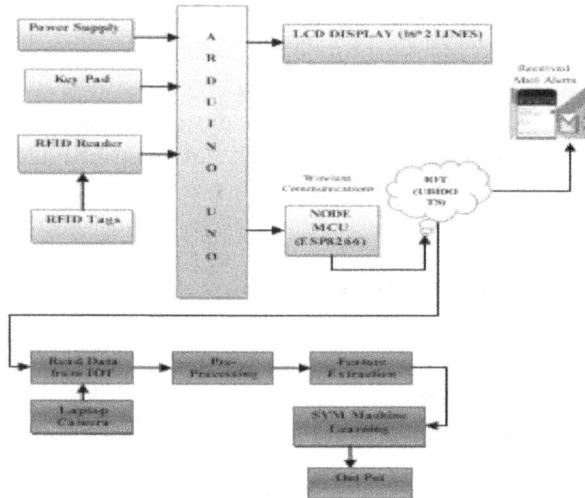

Figure 54.1 Architecture of the proposed method
Source: Author

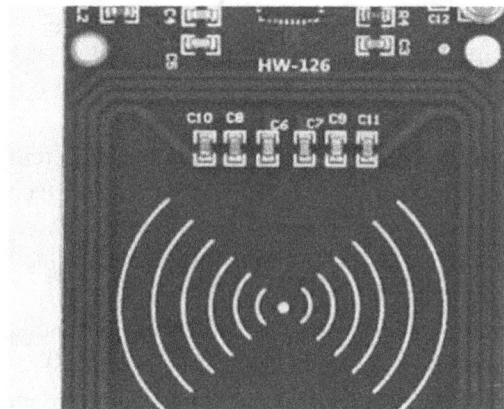

Figure 54.2 RFID module
Source: Author

Figure 54.3 RFID cards
Source: Author

Figure 54.4 Keypad
Source: Author

Below is a description of each key hardware component used in the system:

1) *RFID Reader and RFID cards*

   RFID Reader: The RFID reader shown in Figure 54.2, it is used to detect and read RFID cards presented by passengers. The reader emits radio waves that interact with the RFID card, enabling it to extract the stored identification data. The reader communicates with the embedded system to authenticate passengers and register their presence.

   • *RFID Cards:* These cards shown in Figure 54.3 are issued to passengers and contain unique identifiers. They enable contactless identification, reducing manual intervention and enhancing the speed of the fare collection process.

2) *Keypad*

   The keypad shown in Figure 54.4 it allows passengers to input the number of people traveling. This data is necessary for calculating the fare based on the number of passengers and is input by pressing the corresponding buttons on the keypad. The system then cross-verifies this data with the detected passenger count via face recognition.

3) *LCD Display*

   The LCD display shown in Figure 54.5 it provides real-time feedback to passengers, displaying information such as fare details, verification status, and error messages (e.g., mismatches in passenger count). It ensures a user-friendly experience and transparency during the ticketing process.

4) *ESP8266 (Wi-Fi Module)*

   The ESP8266 Wi-Fi module shown in Figure 54.6 it enables wireless communication between the system and the IoT platform (UBIDOTS). It transmits data such as passenger counts, fare

Figure 54.5 LCD display
Source: Author

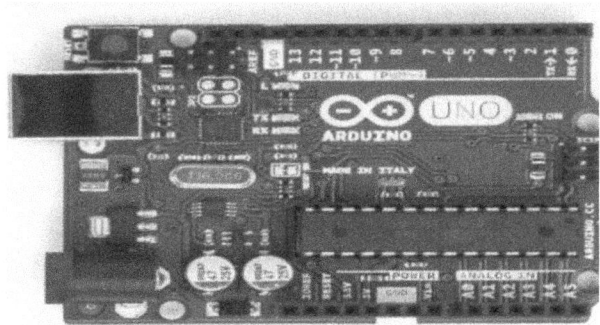

Figure 54.6 Node MCU (esp8266)
Source: Author

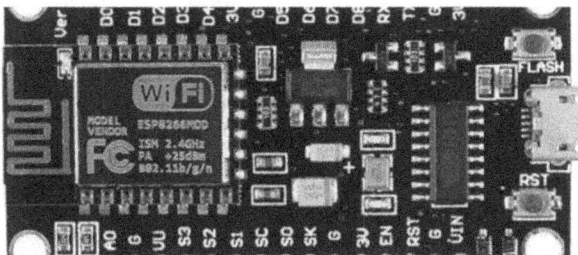

Figure 54.7 Arduino microcontroller
Source: Author

amounts, and system status to the platform for remote monitoring. This component is crucial for real-time system monitoring and data analysis.

5) *Camera for face detection*

A camera connected to a laptop is used to capture real-time images of passengers as they board the bus. This camera is an essential part of the face detection system. It works in tandem with computer vision algorithms to process images and perform face recognition to validate the number of passengers.

The camera provides continuous visual input, ensuring the system is able to count and verify the passengers accurately. The images are then processed by the system's software to identify unique facial features.

6) *Laptop/computer for face recognition*

• A laptop or computer is connected to the camera and is responsible for processing face detection and recognition tasks. It uses machine learning algorithms (such as SVM) to classify and verify passengers based on the images captured by the camera.

• The laptop acts as the central processing unit (CPU) for the face recognition module,

performing the required computations in real time to ensure accurate passenger validation.

7) *Arduinomicrocontroller*

Arduino shown in Figure 54.7 it serves as the central controller for coordinating the interaction between different hardware components. It is responsible for processing inputs from the RFID reader, keypad, and monitoring the LCD display. When a passenger taps their RFID card, the Arduino reads the card's unique identifier and sends it to the system for processing and validation. The Arduino ensures that the entire ticketing system operates in sync by managing timing and data flow across the various modules.

8) *IoT Platform (UBIDOTS)*

Although not a physical hardware component, the IoT platform is a key part of the system. The data transmitted by the ESP8266 module is uploaded to the UBIDOTS platform, where it can be monitored and analyzed. The platform provides real-time insights into system performance, including passenger count, fare transactions, and overall system health.

Algorithm

Start System
Initialize components (RFID, keypad, LCD, camera, ESP8266)

Repeat until bus journey is complete:
Read RFID Card
If RFID is valid:
Input passenger count via keypad

Trigger camera to capture face image
Process image (preprocessing, feature extraction)
Classify face using SVM
If face matches passenger count:
Calculate fare
Display fare details on LCD
Deduct fare from RFID card
If payment successful:
Display "Payment Successful"
Else:
Display "Insufficient Funds"
Else:
Display "Passenger count mismatch"
Else:
Display "Invalid RFID Card"

Send data (passenger count, fare, payment status) to IoT platform via ESP8266

Reset for next passenger
End

Implementation
a) *System initialization:* Power on the components: Arduino, RFID reader, keypad, LCD, camera, and ESP8266 Wi-Fi module.
b) *RFID Card detection:* The RFID reader scans the passenger's RFID card and sends the unique ID to Arduino for validation.
c) *Passenger count input:* The passenger enters the number of people traveling using the keypad.
d) *Face recognition:* Camera captures the passenger's face image. The image is processed on a laptop using OpenCV and face recognition techniques (e.g., SVM).
e) *Fare calculation:* The Arduino calculates the fare based on the passenger count and displays it on the LCD.
f) *Fare payment:* The fare is automatically deducted from the passenger's RFID card account.
g) *IoT Integration:* Data (passenger count, fare status) is sent to the UBIDOTS platform via ESP8266 for real-time monitoring.
h) *Display confirmation:* LCD shows payment status and ticketing details to the passenger.
i) *System reset:* The system resets for the next passenger and waits for a new RFID card input. The implementation flow chart shown in Figure 54.8.

Implementation flow chart

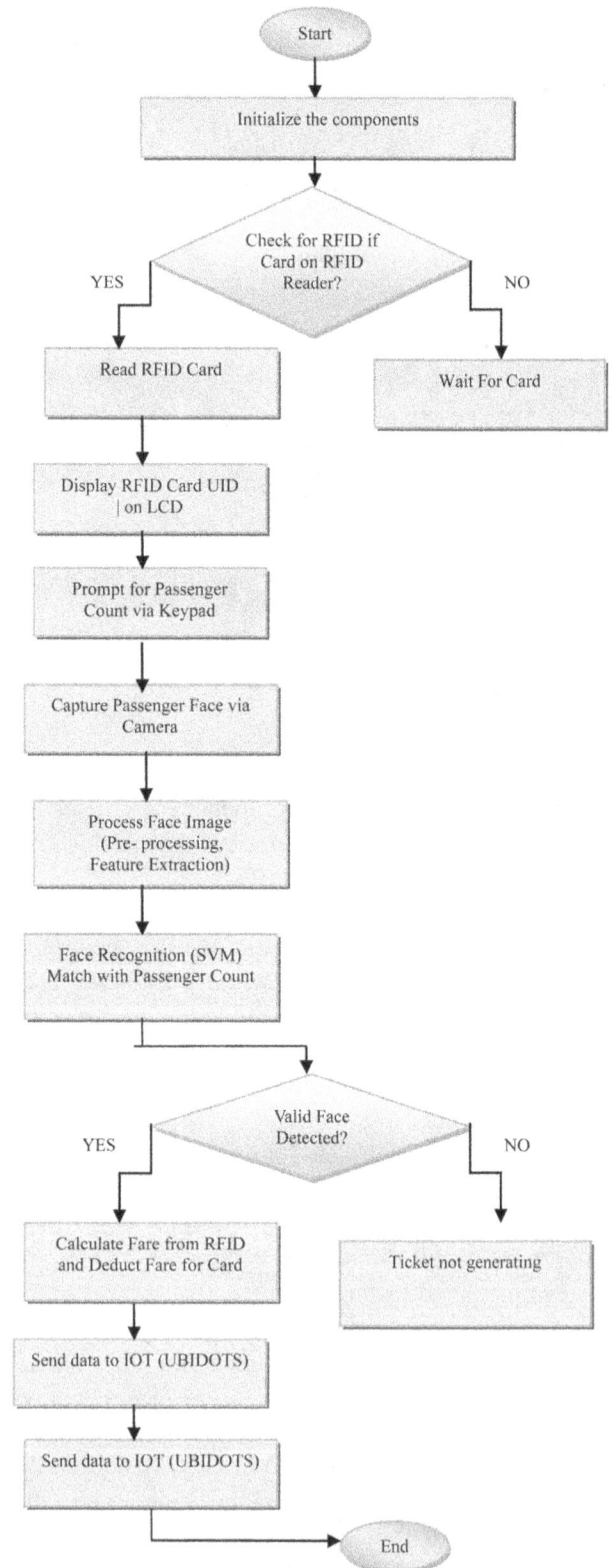

Figure 54.8 Implementation of the flow chart
Source: Author

Experimental results

The hardware setup includes the Arduino board, RFID reader, keypad, LCD display, camera for face recognition, and the ESP8266 Wi-Fi module for IoT integration. This setup shown in Figure 54.9, it demonstrates the physical arrangement of the components, showing how each part is connected. The RFID reader is used for scanning passengers' cards, while the keypad is for entering the number of passengers. The camera captures the passenger's face for validation, and the LCD displays the status. The Wi-Fi module communicates with the IoT platform for real-time monitoring.

This Figure 54.10 illustrates the interaction between the ATM card (or RFID card) and the system for checking the available balance. When a passenger's RFID card is scanned, the system checks the available funds in the card. If the balance is adequate for the fare, the system proceeds with the ticketing process; otherwise, it prompts the user to reload the card.

In this Figure 54.11, the keypad is shown to be used to input the number of passengers traveling. The passenger enters the number of people traveling in the bus, which is then sent to the system for fare calculation. This input ensures that the fare is calculated based on the correct number of travelers.

This Figure 54.12 shows the training dataset used for face recognition, which is essential for validating passengers through SVM algorithm. The dataset includes labeled face images for each passenger. These images are processed, and the features are extracted to train the SVM model, which will later recognize passengers' faces during boarding.

In this Figure 54.13, the camera captures the face image of the passenger. The image is then processed using face detection algorithms (like Haar Cascade

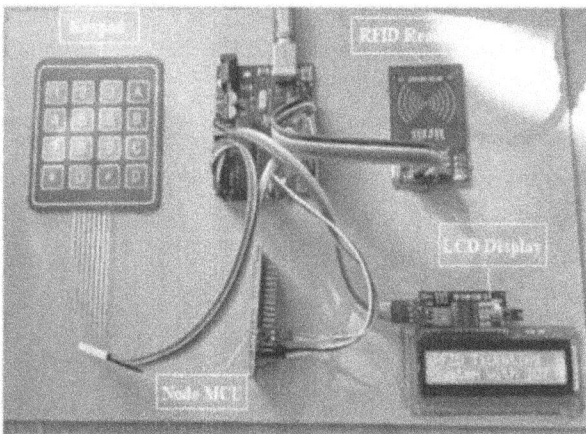

Figure 54.9 Hardware setup
Source: Author

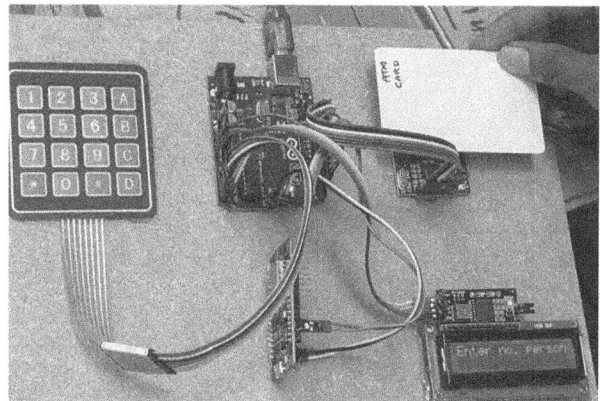

Figure 54.10 Load balance from ATM card
Source: Author

Figure 54.11 Enter number of persons
Source: Author

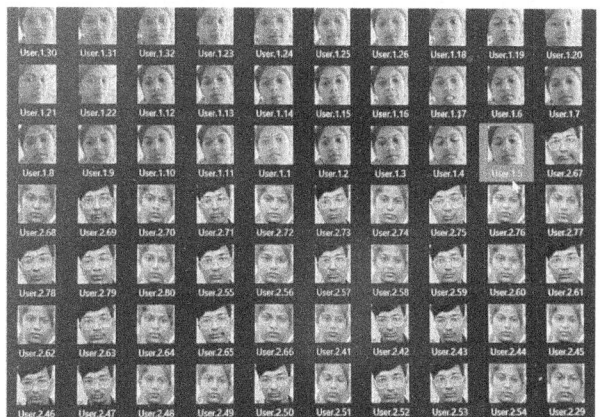

Figure 54.12 Dataset for training by SVM algorithm
Source: Author

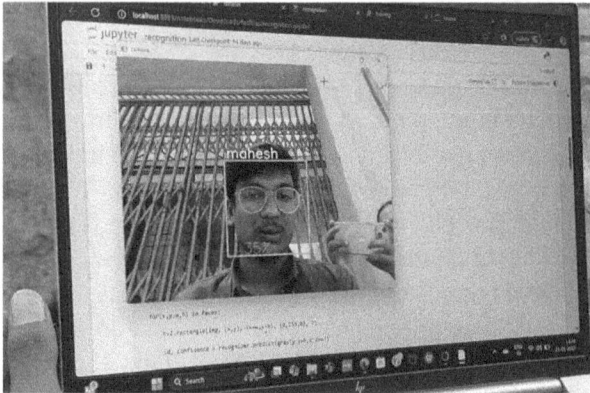

Figure 54.13 Take the face capturing from camera
Source: Author

Figure 54.14 Ticket bill generated
Source: Author

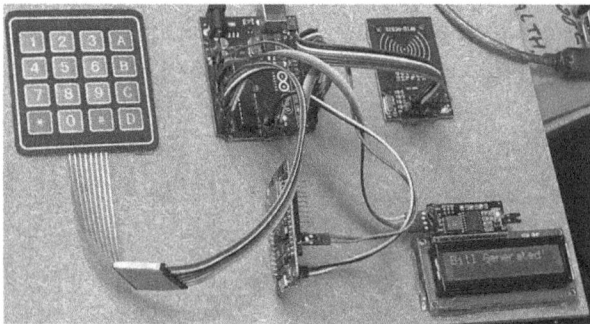

Figure 54.15 Amount charged for ticketing
Source: Author

Figure 54.16 Amount after deduction
Source: Author

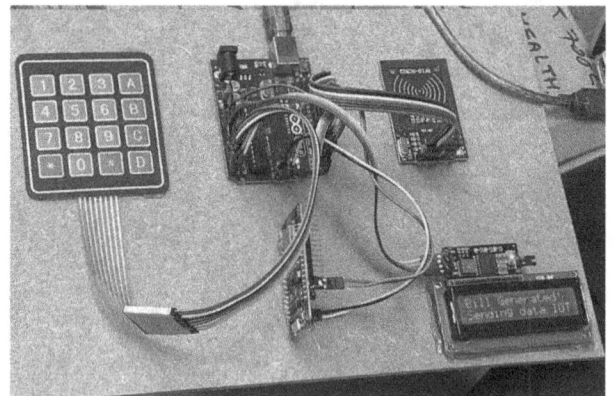

Figure 54.17 Generated bill sending to IoT
Source: Author

or HOG), and the captured face data is passed to the face recognition model (SVM) to check for a match with the previously stored dataset.

After successful RFID card scan and face recognition, the ticket bill is generated it is shown in Figure 54.14. The bill contains details such as the fare amount, number of passengers, and other relevant travel information. This output is displayed on the LCD and serves as the passenger's ticket.

This Figure 54.15. shows the amount charged to the passenger for their journey based on the passengers cou. After verifying the fare, the system deducts the appropriate amount from the RFID card's balance, and the transaction amount is displayed on the LCD screen for confirmation.

This Figure 54.16 shows the balance amount in RFID card. After deducting the appropriate amount charged for number of passengers, the amount left in RFID card is displayed on the LCD screen for conformation. After processing the payment, it will generate the ticket.

Once the ticket is generated and payment is processed, the bill data (including the transaction details, passenger count, and fare) is sent to the IoT platform (UBIDOTS) through the ESP8266 Wi-Fi module. This enables real-time tracking and monitoring of the system's activity, providing insights into the number of passengers, fare collection, and system status shown in Figure 54.17.

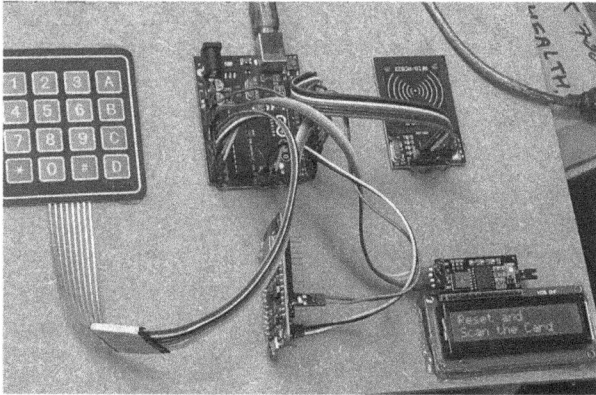

Figure 54.18 Reset the card and scan the card
Source: Author

This Figure 54.18 illustrates the final step where the RFID card is reset, and the system is ready to scan the next passenger's card. After a transaction is completed, the card's data is cleared to allow for the next passenger's boarding process, ensuring the system is always prepared for the next interaction.

The different steps and hardware components involved in the Smart Bus Ticketing System. Each image represents a crucial part of the system's flow, from the initial scan of the RFID card to the final IoT integration and reset for the next passenger. The system offers automated fare collection, secure passenger validation through RFID and face recognition, and real-time data monitoring using IoT.

Conclusion and future scope

The Smart Bus Ticketing System utilizing RFID and Face Recognition provides an innovative solution to automate fare collection, ensuring greater convenience, security, and efficiency in public transportation. The integration of RFID technology allows for quick passenger identification, while face recognition enhances the accuracy of the passenger count, reducing the risk of fraud and manual errors. The system's IoT platform integration further enables real-time monitoring, making it easier to manage and track ticketing data. The experimental results validate the effectiveness of the system, demonstrating it's latent to streamline the ticketing process in urban public transport.

The outlook of this system can be expanded in various ways. First, integrating mobile payment options could allow passengers to pay using their smartphones, providing greater flexibility. Moreover,

machine learning models could be enhanced for better face recognition accuracy under varying lighting conditions. Additionally, incorporating AI-driven predictive maintenance could ensure the system remains operational with minimal downtime. The system could also be scaled to other forms of public transport like trains and trams, making it a versatile solution for improving public transport infrastructure globally.

Related Works

Study by Kaur et al., [1] provides a comprehensive overview of RFID technology, including its principles, advantages, limitations, and applications. It highlights the benefits of RFID in various domains, such as automation, data accuracy, and reduced manual effort. The paper discusses the challenges, including privacy concerns and interference issues, providing a solid foundation for exploring RFID's role in smart systems .

Chowdhury et al. [2] present an RFID and Android-based smart ticketing system integrated with destination announcement functionality. The system improves the efficiency of public transport ticketing by combining RFID for user identification with Android for real-time updates. It demonstrates how RFID can enhance user experience while maintaining cost-effectiveness and scalability [2].

Study by Hasan et al. [3] focuses on RFID-based ticketing systems tailored for megacities like Dhaka. It examines the challenges of manual ticketing systems and proposes an automated solution to streamline fare collection. The system demonstrates RFID's potential to handle high passenger volumes efficiently while addressing urban transportation issues [3].

Net study explores human tracking using RFID and GPS in indoor and outdoor environments. It highlights how integrating these technologies ensures accurate location tracking and data processing. The study underlines the feasibility of combining RFID and GPS for applications in monitoring and security systems [4].

Meghna et al. discusses the design and implementation of an outdoor surveillance robot for security purposes. The robot uses sensors and cameras to detect anomalies, which could inspire face recognition and IoT-based enhancements in transportation systems [5].

Leo Louis [6]. This paper explains the working principle of Arduino and its applications in education and research. The discussion of Arduino's versatility and ease of use highlights its potential in building embedded systems, such as RFID-based ticketing systems [6].

Tun Mohamad Aqil et al. [7] The authors propose a school bus security system utilizing RFID and GSM technologies. The system enhances student safety through real-time tracking and notifications, demonstrating the effective use of RFID and IoT in transportation safety applications [7].

Vinayak Nair et al. [8]. This paper presents an online bus tracking and ticketing system. It integrates real-time tracking with ticketing functionality, improving operational efficiency and passenger convenience. The study emphasizes the role of IoT in modernizing public transportation systems [8].

Mr. D. Baskaran et al. [9]. The authors propose an embedded system for RFID-based smart bus management. It automates ticketing, enhances passenger identification, and reduces manual intervention, aligning with the objectives of modern smart ticketing solutions [9].

Najim Sheikh et al. [10]. This paper introduces a QR code-based e-ticketing system for public transportation. While focused on QR codes, the study highlights the potential of contactless ticketing technologies in enhancing efficiency and passenger convenience [10].

Pritish Sachdeva and Shrutik Katchii [11]. The authors review the Raspberry Pi platform, discussing its capabilities and applications in embedded systems. The paper identifies Raspberry Pi's potential for implementing low-cost, efficient systems for transportation and IoT-based applications [11].

Pratik S. Dhumal et al. [12]. This study presents an Android-based ticketing system incorporating GPS validation and QR codes. It highlights the importance of integrating mobile platforms and validation mechanisms for secure and efficient ticketing [12].

The paper by Zhang et al. [13]. This presents a fast and universal anti-collision algorithm for RFID tags in the Internet of Things (IoT). The algorithm improves the efficiency of identifying RFID tags in systems with high tag density, reducing collisions and enhancing communication performance. This work addresses challenges in scalability and speed, making it suitable for diverse IoT applications [13].

Babaeian and Karmakar [14] propose hybrid chipless RFID tags as a solution to align with EPC Global Standards. These tags integrate features of chipless and traditional RFID systems to enhance performance, data capacity, and cost-effectiveness. The study emphasizes their potential for wider adoption in standardized RFID applications across industries [14].

Lee, An, and Hwang [15] introduce a novel directional antenna designed for next-generation fare payment systems. The proposed antenna improves signal efficiency, reliability, and security in payment applications, ensuring better performance in modern transportation systems [15].

Albalas et al. [16] present a method for detecting occluded faces using deep graph-based convolutional networks. The approach learns discriminant spatial features to improve accuracy in recognizing faces, even when parts of them are obscured, enhancing performance in challenging facial recognition scenarios [16].

Nandhini [17] presents an automatic bus fare collection system using RFID technology. The system simplifies the fare collection process by automating ticketing, allowing passengers to pay through RFID-enabled cards. It aims to improve efficiency, reduce human error, and enhance the overall experience for both passengers and transport authorities [17].

Varun Krishna et al. [18] propose a modified ticketing system using RFID technology. The system intends to enhance the efficiency and convenience of ticketing by automating the process, reducing human intervention, and improving the speed and accuracy of fare collection [18].

Kaushik et al. (2019) introduce an RFID-based bus ticket generation system. This system automates the ticketing process by using RFID technology, allowing passengers to easily obtain tickets, improving efficiency and reducing human error in bus fare collection [19] [21].

Karthika et al. [20] propose an automatic bus fare collection system that uses GPS and RFID technology. This system aims to streamline fare collection, ensuring accuracy and efficiency in public transportation by automatically tracking bus locations and passenger payments [20].

References

[1]   Kaur, M., Sandhu, M., Mohan, N., & Sandhu, P. S. (2011). RFID technology principles advantages limitations its applications. *International Journal of Computer and Electrical Engineering*, 3(1), 151–157.

[2] Chowdhury, P., Bala, P., Addy, D., Giri, S., & Chaudhuri, A. R. (2016). RFID and android based smart ticketing and destination announcement system. In 2016 International Conference on Advances in Computing Communications and Informatics (ICACCI). IEEE.

[3] Hasan, M. F. M., Tangim, G., Islam, M. K., Khandokar, M. R. H., & Alam, A. U. (2010). RFID-based ticketing for public transport system: perspective megacity Dhaka. In 2010 3rd IEEE International Conference on Computer Science and Information Technology (ICCSIT), (Vol. 6).

[4] Hutabarat, D. P., Hendry, H., Pranoto, J. A., Kurniawan, A., Budijono, S., Saleh, R., et al. (2016). Human tracking in certain indoor and outdoor area by combining the use of RFID and GPS. In 2016 IEEE Asia Pacific Conference on Wireless and Mobile(APWiMob). IEEE.

[5] Meghana, S., Nikhil, T. V., Murali, R., Sanjana, S., Vidhya, R., & Mohammed, K. J. (2017). Design and implementation of surveillance robot for outdoor security. In 2017 2nd IEEE International Conference on Recent Trends in Electronics Information Communication Technology (RTEICT). IEEE.

[6] Louis, L. (2016). Working principle of arduino and using it as a tool for study and research. *International Journal of Control Automation Communication and Systems (IJCACS)*, 1(2), 21–29.

[7] Fadzir, T. M. A. M., Mansor, H., Gunawan, T. S., & Janin, A. (2018). Development of school bus security system based on RFID and GSM technologies for klang valley area. In 2018 IEEE 5th International Conference on Smart Instrumentation Measurement and Applications (ICSIMA). IEEE.

[8] Nair, V., Pawar, A., Tidke, D. L., Pagar, V., & Wani, N. (2018). Online bus tracking and ticketing system. *MVP Journal of Engineering Sciences*, 1(1), 1998–2002.

[9] Baskaran, D., Pattumuthu, M., Priyadharshini, B., Shabab Akram, P., & Sripriya, S. (2016). RFID based smart bus using embedded system. *International Journal of Engineering Research Technology (IJERT)*, 4(1), (Volume 4-Issue 11).

[10] Sheikh, N., Shende, M., Pandit, R., Samarth, S., Khapekar, T., Kumar, S., et al. (2018). QR based e-ticket system. *International Journal on Future Revolution in Computer Science Communication Engineering (IJFRCSCE)*, 4(3).

[11] Sachdeva, P., & Katchii, S. (2014). A review paper on raspberry Pi. *International Journal of Current Engineering and Technology*, 4(6), 3818–3819.

[12] Dhumal, P. S., Dhande, D., Chaudhari, G., & Panjwani, L. D. (2016). Android ticketing system for bus with GPS validation Using QR code with alarm. *International Journal of Innovative and Emerging Research in Engineering, IEEE*, 3(4).

[13] Zhang, G., Tao, S., Xiao, W., Cai, Q., Gao, W., Jia, J., et al. (2019). A fast and universal RFID tag Anti-Collision algorithm for the internet of things. *IEEE Access*, 7, 92365–92377. https://doi.org/10.1109/access.2019.2927620.

[14] Babaeian, F., & Karmakar, N. C. (2018). Hybrid chipless RFID tags- a pathway to EPC global standard. *IEEE Access*, 6, 67415–67426. https://doi.org/10.1109/access.2018.2879050.

[15] Lee, D., An, T., & Hwang, I. (2023). A novel directional antenna for Next-Generation fare payment system. *IEEE Access*, 11, 26163–26169. https://doi.org/10.1109/access.2023.3254208.

[16] Albalas, F., Alzu'bi, A., Alguzo, A., Al-Hadhrami, T., & Othman, A. (2022). Learning discriminant spatial features with deep Graph-Based convolutions for occluded face detection. *IEEE Access*, 10, 35162–35171. https://doi.org/10.1109/access.2022.3163565.

[17] Nandhini a, Sunitha. (2017). Automatic Bus Fare Collection System using RFID. International Journal of Advanced Research in Computer Engineering & TechnolNandhini a, Sunitha. (2017). Automatic Bus Fare Collection System using RFID. International Journal of Advanced Research in Computer Engineering & Technology. Volume 6. 218–223.

[18] Jos, B. M., Aslam, A., Akhil, E. P., Divya Lakshmi, G., & Shajla, C. (2018). RFID based bus ticketing system. *International Journal of Advanced Research in Electrical, Electronics and Instrumentation Engineering*, 4(4), 2345–2349. ISSN :2278 -8875.

[19] Varun Krishna, K. G., Selvarathinam, S., Roopsai V., & Ram Kumar, R. M. (2013). Modified ticketing system using radio frequency identification (RFID). *International Journal of Advanced Computer Research*, 3(12), 92–98.

[20] Karthika, J., Varshanapriyaa, S., Haran, S. S., & C. Prakash, S. (2018). Automatic bus fare collection system using GPS and RFID technology. *International Journal of Pure and Applied Mathematics*, 1120, 1119–1124.

[21] Aman Kaushik1, Kumar Sanu2, Kajol Singh3, Aayush Raj4, Sumitra Kumari5(2019),RFID BASED BUS TICKET GENERATION SYSTEM, International Research Journal of Engineering and Technology (IRJET) , Volume: 06 Issue: 07 | July 2019, pp 1535-1538.

55 Advanced AC drive system for efficient electric vehicle technology

Mondru Chiranjeevi[1,a], D. V. Ashok Kumar[2,b] and R. Kiranmayi[3,c]

[1]Research Scholar, Department of Electrical Engineering, JNTUA, Andhra Pradesh, India

[2]Professor, Department of EEE, RGMCET, Andhra Pradesh, India

[3]Professor, Department of EEE, JNTUA, Andhra Pradesh, India

Abstract

This paper investigates drive performance in the context of electric vehicular technology (EVT) design. It presents the performance metrics of AC drive for EVT applications in a tabular format. This paper emphasizes that the drive for EVT should ensure rapid torque response, enhanced power density, and improved efficiency. The analysis focuses on the permanent magnet synchronous motor (PMSM) drive, assessing key performance metrics such as speed and torque using MATLAB/Simulink simulations across various driving scenarios. Effective drive selection is identified as critical to reducing power consumption, achieving swift torque response, and maximizing efficiency. Based on these evaluations, this paper recommends the PMSM drive for EVT applications.

Keywords: Drives, electric vehicle, electric vehicular technology EVT, permanent magnet synchronous motor PMSM, power consumption, speed, torque response

Introduction

The growing concerns over environmental pollution and fossil fuel dependency have accelerated the development and adoption of electric vehicles (EVs) and hybrid electric vehicles (HEVs). These vehicles rely on electric propulsion systems, energy storage solutions, and advanced control strategies to enhance efficiency and sustainability. Researchers have extensively investigated various aspects of EV technology, including motor drives, battery technologies, and control methodologies to optimize performance.

Electric propulsion systems have evolved significantly in recent years, as highlighted [1], who reviewed advancements in motor technologies and their impact on vehicle performance. The efficiency and reliability of electric motor drive lines have been discussed [2], while [3] conducted a comparative analysis of motor drive selection issues for HEV propulsion systems. Similarly, studies [5, 11-16] have provided insights into various motor types and drive technologies used in EVs, hybrids, and fuel cell vehicles. Battery technology remains a critical factor in determining the range and efficiency of EVs [4]. Conducted comparative analyses of battery types and their dynamic models, while [20] explored hybrid electric vehicle architectures and different battery configurations for enhanced performance. The International Energy Agency (IEA) [9, 10] emphasized the importance of sustainable and scalable energy storage solutions to support the growing EV market.

Control strategies also play a crucial role in optimizing EV performance. Investigated vector control techniques for permanent magnet synchronous motor (PMSM) drives [6, 23], while [7, 17, 21] explored intelligent control methodologies for brushless DC (BLDC) motors. These studies underscore the need for advanced control strategies to maximize energy efficiency and vehicle dynamics. These present a detailed discussion on electric propulsion systems, battery technologies, and control strategies, highlighting key contributions from various studies and identifying potential areas for future research.

This paper presents efficient motor mathematical modeling when powered by li-ion batteries for EVT applications. The paper is structured as follows: Section 2 provides a relative analysis of electric motors and Methodology; Section 3 establishes a mathematical analysis of PMS motor; Section 4 focuses on results and discussion; and Section 5 offers concluding remarks.

[a]chiru.carey@gmail.com, [b]rgmdad09@gmail.com, [c]kiranmayi0109@gmail.com

DOI: 10.1201/9781003684589-55

Methodology

Analysis of electric motors

Table 55.1 displays the comparative suitability of various motors for EVT based on design parameters. Each motor's appropriateness for different criteria is evaluated on a 1-10 scale, with 1 indicating poor suitability and 10 representing optimal suitability. The assessment is derived from recent literature, though it may fluctuate depending on factors such as design considerations and motor types [1].

Table 55.1 is to identify the most suitable motor for EVT, considering factors like high power density, speed range, torque density, efficiency, and cost-effectiveness. However, when evaluating controllability, robustness, and longevity, manufacturers are leaning towards IM for EVT applications. As evidenced, both IM and PM electric motors are currently employed in various commercial EVT models [1].

Methodology

EV architecture

Figure 55.1 represents the design of an EV drive system and involves several key components that work together to ensure efficient operation. The power supply is managed by a Li-Ion battery pack, while a BMS handles cell balancing, temperature monitoring, and fault detection. A bidirectional DC-DC converter maintains stable voltage and facilitates regenerative braking, allowing energy recovery during deceleration. The power electronics section includes a three-phase

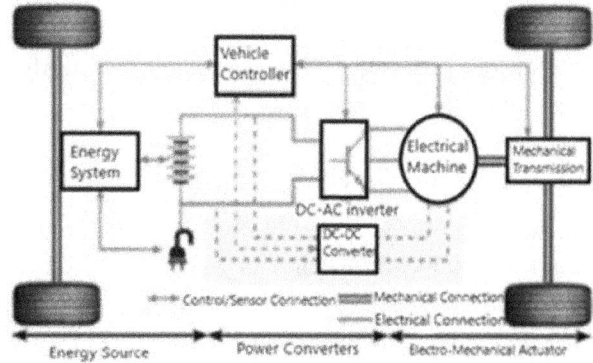

Figure 55.1 EV architecture [24]
Source: Author [24]

inverter that converts DC to AC using pulse width modulation (PWM) technique, which is space vector PWM (SVPWM). This inverter operates with controlled switching frequencies to reduce harmonics and energy losses, ensuring smooth motor operation.

Motor control is achieved through field-oriented control (FOC), which decouples torque and flux control for optimal performance of the PMSM. Proportional-integral (PI) controllers regulate speed and current, and sensor less technique is "model reference adaptive system" is employed for rotor position estimation. Regenerative braking further enhances the system's efficiency by feeding energy back into the battery during deceleration, facilitated by the bidirectional DC-DC converter. To ensure the

Table 55.1 Comparison of various electric motor parameters used for EVT.

Parameters	DC motor	SRM	BLDCM	IM	PMSM
Size and weight	6	8	7	8	9
Power density	6	7	7	8	10
Efficiency	6	9	8.3	8	10
Speed range	5	10	8	8	10
Torque density	6	8	7	7	10
Overload capability	6	8	6	8	9
Manufacturability	6	8	8	10	9
Lifetime	7	9	8	10	9
Reliability	6	9	8	10	8.5
Robustness	7	9	8.5	10	8
Controllability	10	6	10	10	8.5
Cost	7	8	6.5	10	6

Source: Author [16]

system's overall performance. MATLAB/Simulink simulations are conducted, testing the drive system under various conditions like acceleration, steady-state operation, and regenerative braking.

Mathematical Model of PMS Motor

The mathematical model of the system is required to understand the system behavior as well as predicting the retaliation [8, 14, 18]. While [22] explored PMS motor mathematical modeling is described in this section.

PM synchronous motor

From the PMSM voltage equations of direct and quadrature axis are given by

$$V_d = \omega_r L_q i_q + K_s i_d + \rho\left(L_d i_d + \psi_f\right) \tag{1}$$

$$V_q = \rho L_q i_q + R_a i_a + \omega_r\left(L_d i_d + \psi_f\right) \tag{2}$$

Where, ψ_f is the flux produced by the permanent magnet through the stator windings.

$$\psi_f = L_m i_f \tag{3}$$

The motor d and q axis flux linkages Eqn's are written as:

$$\psi_d = L_d i_d + L_m i_f \tag{4}$$

$$\psi_q = L_q i_q \tag{5}$$

The motor has getting the Torque equation in terms of the inductances and current is written as:

$$T_e = 3/2 * P(\psi_d i_d + (L_d - L_q)i_d i_q \tag{6}$$

The Stator current frequency and mechanical power Eqn's of the PMSM is as [19]:

$$\omega_s = P\omega_m \tag{7}$$

$$P = T_e.\omega_m \tag{8}$$

The rotor mechanical speed is

$$\omega_m = \omega_r * 2/P \tag{9}$$

Where,

i_d and i_q are the direct and quadrature axis currents
i_f is d – axis rotor current
V_q and V_d are the voltages of quadrature and direct axis

L_d and L_q are the inductances of direct and quadrature axis
L_m is mutual inductance of the motor
R_s is the stator resistance
T_e is electrical torque
P is Power,
ω_m = mechanical speed
ω_r is rotor electrical speed
ψ_q, ψ_d are the flux linkages of q – d axis

The mathematical modeling Eqn's are written for PMS motor. The simulation is carried out by the mathematical equivalents of PMS Motor as discussed from Eqn's (1) – (9). Figure 55.2 shows the design of PMSM drive using Simulation with the modeling equations.

Result and Discussions

In examining the performance of PMSM drive for EVT applications, three primary operational modes are evaluated: acceleration, deceleration, and variable speed operation. These modes are implemented by the PMSM, operating at constant voltage and rated speed. Table 55.2. illustrates the reference speeds for these operational modes over a 12 sec time period.

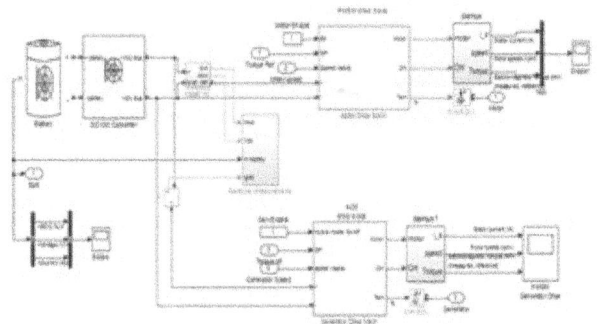

Figure 55.2 Simulation diagram of PMSM drive
Source: Author [22]

Table 55.2 PMSM drive various speed operations for EVT.

Speed (RPM)/ time (Sec)	0	4	6	8	10	12
Acceleration	1500	1600	1650	1700	1750	1800
Deceleration	1500	1450	1450	1200	1100	900
Variable operation	1500	1300	1100	1500	1700	1500

Source: Author [16]

Acceleration

In this operational mode, the reference speed gradually increases from 1500 RPM to 1800 RPM over 12 seconds. The motor drive's rated speed is 1500 RPM, and the speed characteristics of the PMSM drive are shown in Figure 55.3(a). The rotor speed consistently follows the reference speed with minimal rise time, as shown in Figure 55.3(a).

Figure 55.3(b) illustrates the torque characteristics of the PMSM drive. The drive delivers maximum torque at startup, which is crucial for EVT applications. The PMSM drive demonstrates excellent performance in the constant torque region, maintaining consistent torque output. In the constant power region, the PMSM drive exhibits reduced torque, which minimizes power loss during acceleration.

Deceleration

In this mode, the PMSM drive's reference speed decreases gradually from 1500 RPM to 900 RPM over a period of 12 seconds. Figure 55.4(a) illustrates the

motor drive's speed characteristics, where the rotor speed closely follows the reference speed with minimal rise time, showcasing its effective control dynamics.

Figure 55.4(b) presents the torque characteristics of the PMSM drive. The maximum torque at startup, essential for EVT applications, is clearly observed. The drive demonstrates strong performance in the constant torque region, effectively adapting to torque variations as speed changes. Furthermore, during deceleration in the constant power region, the PMSM drive's reduced torque results in lower power loss, contributing to overall efficiency

Variable speed operation

In this operational mode, the reference speed alternates between increasing and decreasing within the range of 1100 to 1700 RPM over duration of 12 seconds. Figure 55.5(a) illustrates the speed

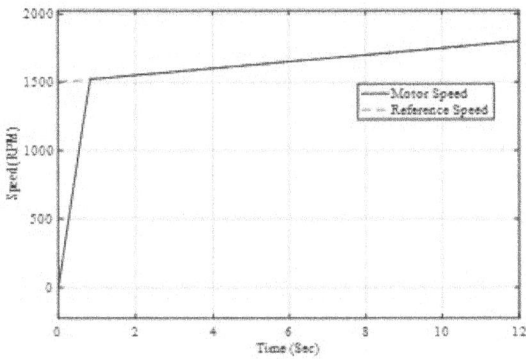

(a) Speed characteristics

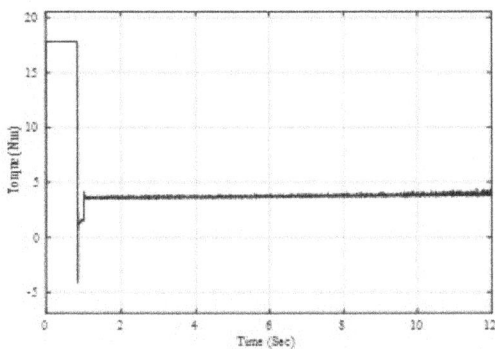

b) Torque characteristics

Figure 55.3 Results of the PMSM Drive
Source: Author

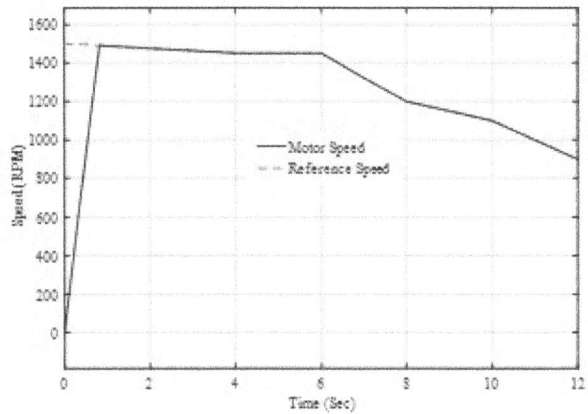

(a) Speed characteristics

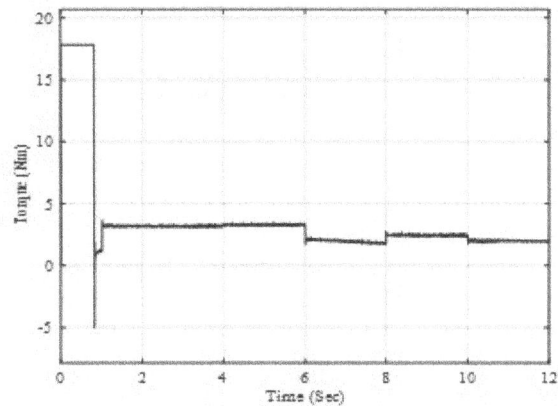

(b) Torque characteristics

Figure 55.4 Simulation Results of the PMSM Drive in acceleration mode
Source: Author

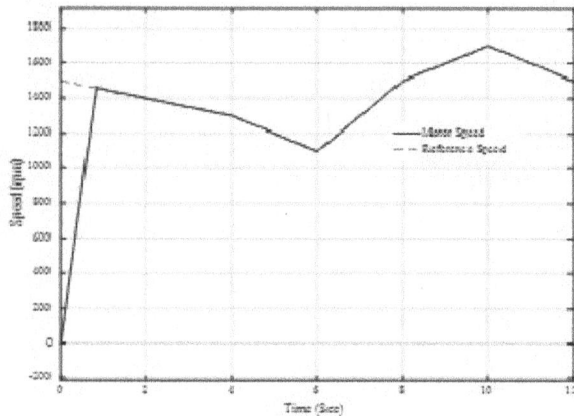

(a) Speed characteristics

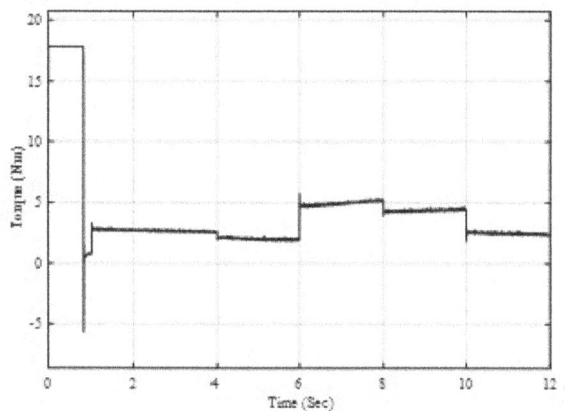

(b) Torque characteristics

Figure 55.5 Simulation Results of the PMSM Drive in variable operation mode
Source: Author

characteristics of the PMSM drive, where the rotor speed closely follows the reference speed with minimal rise time, demonstrating precise and responsive control.

Figure 55.5(b) depicts the torque characteristics of the PMSM drive. Notably, the maximum torque at startup, which is critical for EVT applications, is evident in the drive's performance. In the constant torque region, the PMSM drive handles torque variations effectively, resulting in stable performance. Additionally, in the constant power region, the drive's reduced torque during variable speed operation leads to minimized power losses, further enhancing efficiency.

For EVT, the PMSM drive is considered the most suitable option. It meets various vehicular requirements, including high instantaneous power output, quick torque response, durability, and elevated power density. Additionally, the PMSM drive offers superior efficiency across a broad speed range, delivers substantial torque at low speeds for initial acceleration, and performs well in both constant-power and constant-torque operational modes.

Conclusions

Simulation results for the permanent magnet synchronous motor (PMSM) under identical operating conditions and ratings demonstrate their high torque response in all operational modes. During the constant torque region, the PMSM drive exhibits excellent performance due to its ability to handle torque variations effectively. In variable speed operation, the torque response decreases, which helps reduce power losses in the constant power region. These findings confirm that PMSM drives offer a broad constant power range and a rapid torque response. These characteristics make PMSM drives particularly well-suited for electric vehicular technology (EVT) applications, providing advantages such as low power consumption, high efficiency, fast torque response, and high-power density. Additionally, the PMSM drive satisfies the critical design parameters required for EVT systems.

Acknowledgment

We would like to express our sincere gratitude to Rajiv Gandhi University of Knowledge Technologies, Nuzvid (RGUKTN) for providing the necessary resources, infrastructure, and a conducive research environment that facilitated the successful completion of this paper.

We also acknowledge the support of the university's administration and technical staff, whose assistance with laboratory facilities, data access, and logistical support played a crucial role in carrying out this research paper.

References

[1] Kumar, L., & Jain, S., Eriksson, S., Ferhatovic, S., & Waters, R. (2014). Electric propulsion system for electric vehicular technology: a review. *Renewable and Sustainable Energy Reviews*, 29, 924–940.

[2] De Santiago, J., Ekergård , B., Eriksson, S., Ferhatovic, S., Waters, R., & Leijon, M. (2012). Electrical motor drivelines in commercial all-electric vehicles:

a review. *IEEE Transactions on Vehicular Technology*, 61(2), 475–484.

[3] Zeraoulia, M., Benbouzid, M. E. H., & Diallo, D. (2006). Electric motor drive selection issues for HEV propulsion systems: a comparative study. *IEEE Transactions on Vehicular Technology*, 55(6), 1756–1764.

[4] Mondru, C., Ashok Kumar, D. V., & Kiranmayi, R. (2017). Batteries comparative analysis and their dynamic model for electric vehicular technology. *International Journal of Pure and Applied Mathematics*, 114(7), 629–637.

[5] West, J. G. W. (1993). DC, induction, reluctance and PM motors for electric vehicles. In IEEE Colloquium on Motors and Drives for Battery Powered Propulsion, (pp. 1/1–111). London, UK, IET, pp.1/1–111pages. DOI:10.1049/PE:19940203.

[6] Krishna, D. S., & Rao, C. S. (2011). Speed control of permanent magnet synchronous motor based on direct torque control method. *International Journal of Advances in Science and Technology*, 2(3), 63–70.

[7] Farasat, M., & Trzynadlowski, A. M. (2014). Efficiency improved sensorless control scheme for electric vehicle induction motors. *IET Electrical Systems in Transportation*, 4(4), 122–131. ISSN 2042-9738.

[8] Zeraoulia, M., Benbouzid, M. E. H. & Diallo, D. (2006). Electric motor drive selection issues for HEV propulsion systems: a comparative study. *IEEE Transactions on Vehicular Technology*, 55(6), 1756–1764.

[9] Chiranjeevi, M., Ashok Kumar, D. V., & Kiranmayi, R. (2019). Modeling and comparative analysis of the conventional and hybrid energy storage systems used in electric vehicular technology. *Journal of Mechanics of Continua and Mathematical Sciences*, (3), 1–13. DOI:10.1109/ICPCSI.2017.8391951.

[10] IEA (2022). Global EV Outlook 2022 securing supplies for an electric future. 221.

[11] Krause, P., Wasynczuk, O., Sudhoff, S., & Pekarek, S. (2013). Analysis of Electric Machinery and Drive Systems. (Vol. 75). John Wiley & Sons. DOI:10.1002/978111852433.6.

[12] El-Sharkawi, M. A. (2000). Fundamental of Electric Drives. Pacific Grove: Brooks/Cole. ISBNÿ0534952224, 9780534952228. 314 Pages.

[13] Nanda, G., & Kar, N. C. (2006). A survey and comparison of characteristics of motor drives used in electric vehicles. In Electrical and Computer Engineering, CCECE'06. Canadian Conference on. IEEE. PP. 811–814.doi:10.1109/CCECE.2006.277736.

[14] Zhu, Z. Q., & Howe, D. (2007). Electrical machines and drives for electric, hybrid, and fuel cell vehicles. *Proceedings of the IEEE*, 95(4), 746–765. doi: 10.1109/JPROC.2006.892482.

[15] Zhu, Z. Q., & Chan, C. C. (2008). Electrical machine topologies and technologies for electric, hybrid, and fuel cell vehicles. In Vehicle Power and Propulsion Conference, VPPC'08. IEEE. vol. 95, no. 4, pp.746–765, April 2007, doi: 10.1109/JPROC.2006.892482.

[16] Chiranjeevi, M., Ashok Kumar, D. V., & Kiranmayi, R. (2019). An investigation of li-ion battery performance for AC drives used in electric vehicular technology. In Emerging Trends in Electrical, Communications, and Information Technologies, (pp. 213–221), Part of the Lecture Notes in Electrical Engineering book series (LNEE, volume 569).

[17] Prasetyo, H. F., Rohman, A. S., Hariadi, F. I., Hindersah, H. (2016). Controls of BLDC motors in electric vehicle testing simulator. In System Engineering and Technology (ICSET), 6th International Conference on IEEE. 03-04 October 2016, 978-1-5090-5089-5.

[18] Yildirim, M., Polat, M., & Kürüm, H. (2014). A survey on comparison of electric motor types and drives used for electric vehicles. In Power Electronics and Motion Control Conference and Exposition (PEMC), 16th International IEEE. 11 December 2014, 10.1109/EPEPEMC.2014.6980715.

[19] Lulhe, A. M., & Tanuja, N. (2015). A technology review paper for drives used in electrical vehicle (EV) & hybrid electrical vehicles (HEV). In International Conference on Control, Instrumentation, Communication and Computational Technologies (ICCICCT). IEEE. Date Added to IEEE Xplore: 23 May 2016. 10.1109/ICCICCT.2015.7475355.

[20] Ehsani, M., Gao, Y., & Miller, J. M. (2007). Hybrid electric vehicles: Architecture and motor drives. *Proceedings of the IEEE*, 95(4), 719–728.

[21] Kiruthika, A., Rajan, A. A., & Rajalakshmi, P. (2013). Mathematical modelling and speed control of a sensored brushless DC motor using intelligent controller. In Emerging Trends in Computing, Communication and Nanotechnology (ICE-CCN), International Conference (pp. 211–216). IEEE.

[22] Chiranjeevi, M., Ashok Kumar, D. V., & Kiranmayi, R. (2017). Mathematical analysis & modeling of li-ion battery with PMSM based plug-in electric vehicles. In IEEE International Conference on Power, Control, Signals and Instrumentation Engineering (ICPCSI), (pp. 1445–1449).

[23] Singh, J., Singh, B., Singh, S. P., Chaurasia, R., & Sachan, S. (2012). Performance investigation of permanent magnet synchronous motor drive using vector controlled technique. In Power, Control and Embedded Systems (ICPCES), 2nd International Conference on. IEEE. 10.1109/ICPCES.2012.6508040.

[24] Wu, C.; Sehab, R.; Akrad, A.; Morel, C. Fault Diagnosis Methods and Fault Tolerant Control Strategies for the Electric Vehicle Powertrains. Energies 2022, 15, 4840. doi:10.3390/en15134840.

56 Comparison of machine learning algorithms for predicting temperature

Allacheruvu Brahmaiah[1,a], Kattika Koteswara Rao[1,b] and Vegunta V. G. S. Rajendra Prasad[2,c]

[1]Department of CSE, R K College of Engineering, Vijayawada, (AMARAVATI), Andhra Pradesh, India

[2]Department of ECE, R K College of Engineering, Vijayawada, (AMARAVATI), Andhra Pradesh, India

Abstract

Weather prediction is vital for daily life and critical applications such as agriculture, disaster management, and climate studies. This study aims to enhance air temperature forecasting by evaluating three advanced AI-based models: LR CNN and KNN. The research utilises continuous air temperature data collected from a suburban region in Telangana, spanning from January 2023 to July 2024. Daily temperature values were aggregated by computing their mean for analysis. The performance of each model was assessed using key metrics, including accuracy percentage and MSE. Among the models, LR demonstrated the highest accuracy of 99.1% with an MSE of 0.06, followed by CNN with 98.4% accuracy and an MSE of 0.15. KNN achieved an accuracy of 97.3% with an MSE of 0.55. The findings highlight LR as the most effective model for temperature prediction, showcasing its potential to improve weather forecasting systems. Detailed performance matrices and result plots for each model are presented, comprehensively evaluating their predictive capabilities. This study underscores the transformative impact of AI-driven approaches in advancing meteorological research and addressing real-world challenges, offering valuable insights for future applications in weather prediction.

Keywords: Deep learning, machine learning, temperature prediction

Introduction

Weather prediction plays a pivotal role in modern life, influencing a wide range of activities and applications, including agriculture, disaster management, and climate studies. Accurate forecasting of meteorological parameters, such as air temperature, is essential for optimizing agricultural practices, mitigating the impact of natural disasters, and understanding long-term climate trends. However, predicting weather variables with high precision remains a significant challenge due to atmospheric systems' complex and dynamic nature. Traditional forecasting methods often struggle to capture the intricate patterns and nonlinear relationships inherent in meteorological data.

Recent advancements in artificial intelligence and machine learning have opened new avenues for improving the accuracy of weather forecasting. These technologies excel at analyzing large datasets and identifying complex patterns, making them well-suited for meteorological applications. Among the various AI-based techniques, models such as LR, CNN and KNN have gained significant attention for their ability to handle time-series and spatial data effectively.

This study focuses on evaluating the performance of three advanced AI-based models—LR, CNN, and KNN for air temperature forecasting. The research utilizes continuous air temperature data collected from a suburban region in Telangana, India, spanning from January 2023 to July 2024. Daily temperature values were aggregated by computing their mean, providing a robust dataset for analysis. The performance of each model was assessed using key metrics, including accuracy percentage and Mean Squared Error, to determine their effectiveness in predicting air temperature.

The results of this study highlight the potential of AI-driven approaches in advancing meteorological research and improving weather forecasting systems. By providing a comprehensive evaluation of these models' predictive capabilities, this research offers valuable insights for future applications in weather prediction. The findings underscore the transformative impact of AI in addressing real-world challenges

[a]allacheruvubrahmaiah07@gmail.com, [b]kotesh.sasi@gmail.com, [c]veguntav@gmail.com

DOI: 10.1201/9781003684589-56

and contribute to the growing body of knowledge in the field of meteorological science.

Literature Survey

The application of AI and ML in weather prediction, particularly temperature forecasting, has been extensively explored in recent literature. Traditional statistical methods, such as LR, have been foundational in weather forecasting but often fall short of capturing complex patterns in atmospheric data [1]. More recent studies highlight the advantages of advanced machine learning models. CNNs have been employed for their ability to handle spatial data effectively, revealing improvements in prediction accuracy over simpler models [2]. KNN has shown promise due to its simplicity and effectiveness in capturing local data patterns [3]. eXGBoost, Light GBM and GRU have also gained attention for their superior performance in handling large datasets and complex relationships, often outperforming traditional models in terms of accuracy and error metrics Comparative studies demonstrate that while LR and CNN provide foundational insights, KNN and Light GBM consistently deliver superior performance in temperature prediction, indicating their potential for enhancing forecasting accuracy [1–3].

Methodology and Implementation

The goal of this study is to develop a model that can precisely forecast temperature. Large amounts of data on past trends of rising temperatures from a suburban region in Telangana are contained in the dataset used for the study. Our model was constructed using deep learning and machine learning technologies. Pre-processed data is used to train our model, and testing data is used to evaluate the model. One important model that addresses this kind of issue is time series analysis. Our model has been constructed using models like as LR, CNN and KNN. We have divided our effort into many parts to create an accurate model. Figure 56.1 illustrates the schematic for the achieved research objectives. Those steps are:

1) **Data preparation:** In this phase, we have made sure that our datasets have undergone all necessary pretreatment. It could be beneficial for our model's accuracy.

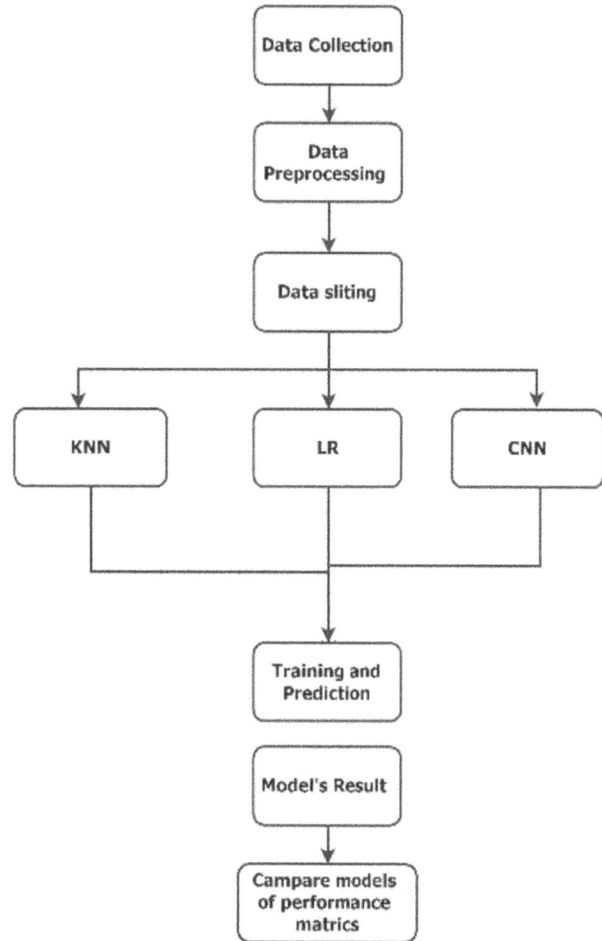

Figure 56.1 Schematic for accomplishing research goals
Source: Author

2) **Splitting the Dataset:** A training set and a testing set are the two sets of the dataset that have been separated in this step.
3) **Development and Research:** To develop a more accurate and dependable model, we investigated previous research on this topic.
4) **Build and train the proposed model:** At this point, we have built our proposed models and trained them using the Training dataset that was previously discussed.
5) **Model evaluation and testing:** Testing data was fed into the model to evaluate its reliability and compare it to training accuracy.
6) **Visualizing the Outcome:** In this last stage, the anticipated values are displayed.

3.1 Dataset

The high spatial resolution of atmospheric information TIME STAMP, which we used in this investigation in temperature, high frequency, high resolution and high accuracy ground data from a suburban region in Telangana, has been used in this work. The weather information is accessible every half-hour between Jan 2023 and July 2024. At 10-minute intervals, all the measurements are gathered for nineteen months 82,080 (144*30*19) data points for temperature [4, 5].

3.2 Data preparation: We must carry out a variety of data preprocessing procedures to improve the generalizability of our model. These actions would restructure some of the columns and eliminate any superfluous data. The procedures are:

3.2.1 Imputing the null values: In this case, the rows with a regular temperature column value of -99 contain null values. As a result, we would substitute the average temperature value from the row before it for the null values, which correspond to the row holding value -99.

3.2.2 Feature Engineering: From the Month, Day, and Year columns, we would pull out a date column. When compared to the prior format, this would be far more effective. In addition, we would remove the Month, Day, and Year columns that were there before.

3.2.3 Eliminating Superfluous Columns: In the chosen dataset, every country—aside from the United States—has a single entry in the State column. As a result, since the State and Region columns are unnecessary for our model, we would remove them.

3.2.4 Data Splitting: The training set and testing set each received 80% and 20% of the whole data frame, respectively. Once more, the validation utilized 20% of the training set. We have carried out the necessary data pre-processing for each of the five separate models. The size of the batch size that we generated from the data frame is 32.

4.2 Statistical Analysis of Temperature
The statistical analysis of the temperature data reveals several key measurements. The mean temperature is 26.7°C, representing the dataset's average temperature. The standard error of 0.019 indicates the precision of the sample mean, showing how much the mean temperature might vary if repeated samples were taken. The median, at 26.25°C, indicates that half of the temperature readings are below this value, while the mode of 23.51°C is the temperature that appears most frequently. The standard deviation is 5.50°C, showing how much the temperatures deviate

from the mean on average, while the sample variance of 30.2°C^2 measures the overall variability in the data. The range of the data is 31.46°C, calculated as the difference between the minimum temperature of 10.64°C and the maximum temperature of 42.1°C. Finally, the dataset contains a total of 80,784 temperature readings [6, 7].

Results and Discussion

Eighty percent of the data collected between January 2023 and July 2024 was utilized to train the models, using historical information spanning a broad temperature range and the estimations' precision increased. Twenty percent of the data was set aside for model validation. The Dataset displays the 82,080 temperature recordings that were part of the data collection [5]. The models produced accurate findings because they were well-fitting and efficiently trained [8]. Percentage plots were among the visualizations that successfully contrasted the dataset's original and expected values. Accuracy Percentage, MSE and RMSE were among the performance metrics that were also calculated [9, 10].

Prediction of linear regression
The Linear regression of temperature data was collected within the study region every ten minutes, which can take the daily mean [11]. Figure 56.2 of the graph uses an LR model to compare the actual and anticipated temperature values from 2023 to 2024. The yellow curve shows Daily mean actual values, which displays seasonal patterns with troughs in the winter and peaks in the summer. The green curve captures both short-term variations and long-term trends, which shows model predictions that closely

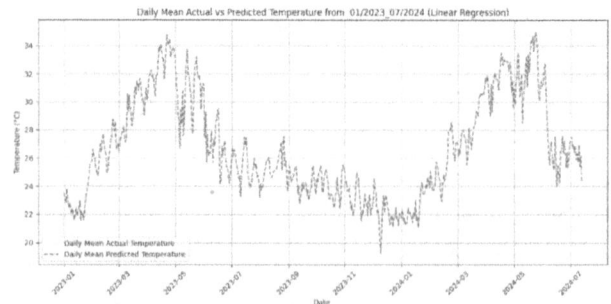

Figure 56.2 Predictions performance of linear regression model
Source: Author

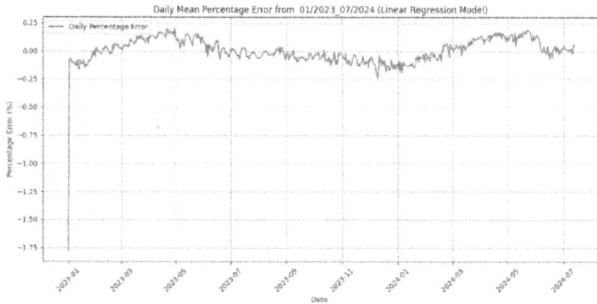

Figure 56.3 Percentage error predictions linear regression model
Source: Author

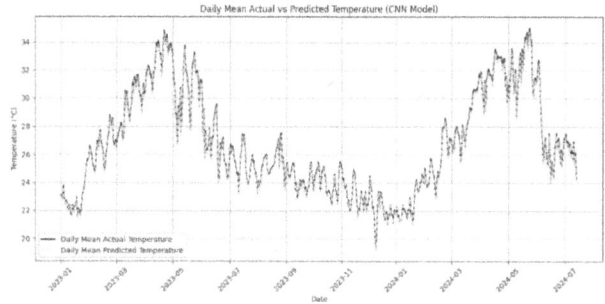

Figure 56.4 Prediction performance of the convolutional neural network
Source: Author

match the actual data. The temperature ranges from 20°C to 35°C and the model accurately forecasts important fluctuations, such as seasonal highs and lows. Although there are some minor discrepancies during sudden fluctuations, the LR generally shows good capacity to predict intricate time-series trends about temperature levels.

Figure 56.3 The graph reveals notable fluctuations in the percentage error between the original and forecasted temperature values shown by the red curve, mostly ranging from 0.2% to -1.7%. The model demonstrates higher forecast accuracy within the range of -0.2% to 0.2%, indicating its potential for reliable predictions in this narrower band.

Prediction of CNN model

The CNN of temperature data was collected within the study region every ten minutes, which can take the daily mean [12–14]. Figure 56.4 of the graph uses a CNN model to compare the actual and anticipated temperature values from 2023 to 2024. Daily mean actual values are shown by the Black curve, which displays seasonal patterns with troughs in the winter and peaks in the summer. The green curve captures both short-term variations and long-term trends, which shows model predictions that closely match the actual data. The temperature ranges from 20°C to 35°C and the model accurately forecasts important fluctuations, such as seasonal highs and lows. Although there are some minor discrepancies during sudden fluctuations, CNN generally shows good capacity to predict intricate time-series trends about temperature levels.

Figure 56.5 The graph reveals notable fluctuations in the percentage error between the original and forecasted temperature values shown by the red

Figure 56.5 Percentage error predictions convolutional neural network
Source: Author

curve, mostly ranging from 0.6% to -1%. The model demonstrates higher forecast accuracy within the -0.1% to 0.6 % range, indicating its potential for reliable predictions in this narrower band.

Prediction of k-nearest neighbours model

The K-Nearest Neighbours of temperature data was collected within the study region every ten minutes, which can take the daily mean [15–17]. Figure 56.6 of the graph uses a K-Nearest Neighbours to compare the actual and anticipated temperature values from 2023 to 2024. Daily mean actual values are shown by the blue curve, which displays seasonal patterns with troughs in the winter and peaks in the summer. Both short-term variations and long-term trends are captured by the dotted green curve, which shows model predictions that closely match the actual data. The temperature ranges from 20°C to 35°C and the model accurately forecasts important fluctuations, such as seasonal highs and lows. Although there are some minor discrepancies during sudden fluctuations, K-Nearest Neighbours generally shows good

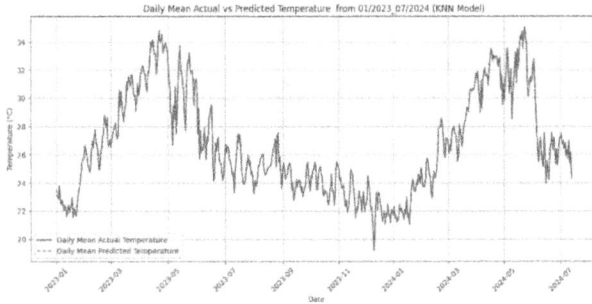

Figure 56.6 Prediction performance k-nearest neighbors
Source: Author

Figure 56.7 Percentage error predictions k-nearest neighbors
Source: Author

capacity to predict intricate time-series trends about temperature levels.

Figure 56.7 The graph reveals notable fluctuations in the percentage error between the original and forecasted temperature values are shown by the red curve, mostly ranging from 0.6% to -1%. The model demonstrates higher forecast accuracy within the -0.1% to 0.6 % range, indicating its potential for reliable predictions in this narrower band.

Table 56.1 Performance metrics for temperature of the maximum accuracy of 99.1% and the lowest percentage error of 0.26% among the models assessed, LR stands out as the most accurate, proving its superior predictive power. With the lowest Mean Squared

Error MSE of 0.06 and Root Mean Squared Error RMSE of 0.24, CNN perform exceptionally well. Its little greater percentage error of 0.75%, and MSE 0.15, RMSE 0.38 however, suggests that the forecasts are not entirely consistent. KNN, on the other hand, performs the worst, with the greatest MSE 0.55, RMSE 0.74, and lowest accuracy 97.32%. Its 1.5% percentage mistake further emphasizes how poorly it can handle challenging temperature prediction jobs. LR is still the most dependable model for accurate and consistent predictions, even while CNN successfully reduces mistakes. Given its performance, KNN appears to be less appropriate for applications that demand high-temperature forecasting accuracy.

Conclusion

The AI models considered in this study include LR, CNN and KNN. These models were chosen for their diverse approach to machine learning and their potential to capture both linear and nonlinear patterns in the data. The dataset used for this research was sourced from located at Shadnagar, Telangana. a suburban location covering a period from January 2023 to July 2024. For training and testing the algorithms, the dataset offered a wealth of temperature data. To evaluate the performance of the maximum accuracy of 99.1% and the lowest percentage error of 0.26% among the models assessed, LR stands out as the most accurate, proving its superior predictive power. With the lowest Mean Squared Error MSE of 0.06 and Root Mean Squared Error RMSE of 0.24, CNN perform exceptionally well. Its a little greater percentage error of 0.75%, and MSE 0.15, RMSE 0.38, however suggests that the forecasts are not entirely consistent. KNN, on the other hand, performs the worst, with the greatest MSE 0.55, RMSE 0.74, and lowest accuracy 97.32%. These findings demonstrate how different models perform at different levels, with LR, CNN and KNN being the most accurate and effective models in this comparison.

Table 56.1 Performance metrics for temperature. (It was taken from our study only).

S. No	Parameter	Model	Accuracy %	MSE	RMSE	Percentage Error %
1	**TEMP**	LR	99.1	0.06	0.24	0.26
2	**TEMP**	CNN	98.4	0.15	0.38	0.75
3	**TEMP**	KNN	97.32	0.55	0.74	1.5

Source: Author

Feature Importance

Future research in temperature forecasting can enhance accuracy by integrating additional features such as humidity, wind speed, and atmospheric pressure. Employing advanced ensemble techniques that combine models like KNN, LR and CNN can significantly improve performance. Additionally, exploring hyperparameter tuning and innovative architectures, such as attention mechanisms or transformers, may further refine model accuracy. Utilizing higher temporal resolution data and extending datasets with recent records can provide deeper insights into temperature forecasting. Finally, applying these models to real-time forecasting and testing across various climatic regions will help evaluate their broader applicability and reliability.

Authors contributions: Equivalent roles

Funding: No funding fee

Conflict of interest: According to the authors, they have no conflicts of interest to report about the current study.

Ethical approval: No ethical dilemmas exist.

Data availability statement: Based on the statistics, the NRSC itself is the primary source of information. Nowhere are we downloaded.

References

[1] Lavasani, A. (n.d.). Classic machine learning in Python: K-nearest neighbors (KNN). [Online]. Medium. https://medium.com/@amirm.lavasani/classic-machine-learning-in-python-k-nearestneighbors-knn-a06fbfaaf80a.

[2] George Kamtziridis (2023). Time series forecasting with XGBoost and LightGBM: predicting energy consumption. Feb. 2023, [Online]. Available from: https://medium.com/@geokam/time-series-forecasting-with-xgboost-and-lightgbm-predicting-energy-consumption-460b675a9cee.

[3] Dinesh, K. E. A., & Kalaga, V. (2022). Comparative analysis of Gated Recurrent Units (GRU), long Short-Term memory (LSTM) cells, autoregressive Integrated moving average (ARIMA), seasonal autoregressive Integrated moving average (SARIMA) for forecasting COVID-19 trends. *Alexandria Engineering Journal*, doi: https://doi.org/10.1016/j.aej.2022.01.011.

[4] Pathakoti, M., & Sreenivas, G. (2015). High-precision surface-level CO2 and CH4 using offaxis integrated cavity output spectroscopy (OAICOS) over Shadnagar, India. *International Journal of Remote Sensing*. doi: DOI : 10.1080/01431161.2015.1104744. 36(22):5754–5765.

[5] Pathakoti, M., & Mahalakshmi, D. V. (2024). Spatiotemporal atmospheric in situ carbon dioxide data over the indian sites-data perspective. *Scientific Data*. doi: https://doi.org/10.1038/s41597-024-03243-x. 11(1).

[6] (2024). Evaluation and prediction of climatic parameters based on IoT system using machine learning and neural network algorithms. doi: DOI: https://doi.org/10.21203/rs.3.rs-4855822/v1.

[7] Ali, Z. & Bhaskar S. B. (2016). Basic statistical tools in research and data analysis., doi: doi: 10.4103/0019-5049.190623.

[8] Zhang, L., & Wang, R. (2023). Time-series neural network: a high-accuracy time-series forecasting method based on kernel filter and time attention. 14. doi: https://doi.org/10.3390/info14090500.

[9] Garg, S., & Jindal, H. (n.d.). Evaluation of time series forecasting models for estimation of PM2.5 levels in air. [Online]. Available from: https://arxiv.org/pdf/2104.03226.

[10] Li, W., & Law, K. L. E. (2024). Deep learning models for time series forecasting: a review. doi: Digital Object Identifier 10.1109/ACCESS.2024.3422528.

[11] Chandini, K., Anjali, T., & Anoop, K. (2019). Temperature prediction using machine learning approaches. doi: 10.1109/ICICICT46008.2019.8993316.

[12] Yamashita, R., & Nishio, M. (2018). Convolutional neural networks: an overview and application in radiology. 9(4), 611–629. doi: doi: 10.1007/s13244-018-0639-9.

[13] Alzubaidi, L., & Zhang, J. (2021). Review of deep learning: concepts, CNN architectures, challenges, applications, future directions.

[14] Purwono, P. & Ma'arif, A. (2023). Understanding of convolutional neural network (CNN): a review. *International Journal of Robotics and Control Systems*, 739–748. doi: DOI:10.31763/ijrcs.v2i4.888.

[15] Fiori, L. (2020). Distance metrics and K-nearest neighbor (KNN). May 2020. [Online]. Available from: https://medium.com/@luigi.fiori.1f0303/distance-metrics-and-k-nearest-neighbor-knn-1b840969c0f4.

[16] Zhuang, W., & Cao, Y. (2023). Short-term traffic flow prediction based on a k-nearest neighbor and bidirectional long short-term memory model. doi: https://doi.org/10.3390/app13042681.

[17] Zardini, E., & Blanzier, E. (2024). A quantum k-nearest neighbors algorithm based on the Euclidean distance estimation. *Quantum Machine Intelligence*, doi: https://doi.org/10.1007/s42484-024-00155-2.

57 Remote sensing image analysis

Kanakala Raja Sekhar[1,a], Jaya Gayatri Vukkum[2,b], Mullapudi Sowjanya[2,c], Morusu Sai Samhitha[2,d] and Thoram Deepika[2,e]

[1]Assistant Professor, Department of AI, Shri Vishnu Engineering College for Women, Bhimavaram, Andhra Pradesh, India

[2]Department of AI, Shri Vishnu Engineering College for Women, Bhimavaram, Andhra Pradesh, India

Abstract

Target detection distance is a major problem, especially in natural environments with challenging backgrounds and small-scale targets. To solve this problem, the RAST-YOLO algorithm integrates the Regin Attention (RA) mechanism and uses the Swin Transformer to enhance feature extraction in complex backgrounds and increase recognition accuracy. Although RAST-YOLO is a great development, our study investigates whether future versions of the YOLO architecture can improve further. We compare YOLOv5×6 and YOLOv8 performance with that of RAST-YOLO on the DIOR and the TGRS datasets. Our experiments prove the efficiency of YOLO framework and attest to the possibility of improving precision, recall, and mean Average Precision (mAP) over the baseline network. Yet, significantly, our results demonstrate the vast potential of the new YOLO variants. In particular, YOLOv5×6 has a great mAP gain of more than 80% over RAST-YOLO along with precision in both datasets, indicating its superior performance in remote sensing object detection.

Keywords: Comparative analysis, convolution, deep learning, faster R-CNN, object detection, RetinaNet, YOLO

Introduction

Remote sensing is crucial in various domains, including resource exploration, intelligent navigation, environmental surveillance, and target tracking. The swift advancement of aircraft and unmanned aerial vehicles [22] is leading to a continuous rise in the generation of valuable and high-quality data from remote sensing imagery [1–4].

The primary objective of remote sensing is to identify objects of interest in remote sensing imagery and to facilitate their spatial localization. However, unlike traditional image detection equipment, remote sensing target detection has encountered some problems. Remote sensing images generally show little information compared to scene images, which poses special problems for object detection [1–4,20]. The disproportionate allocation of small, medium, and big targets, along with the difference in appearance, leads to good results. In addition, the speed and distribution of targets may differ from one difficulty to another.

The complex history and class conflict increases the difficulty of finding objects in remote areas.

Traditional target-oriented search algorithms usually include multiple steps, such as feature extraction, optimization [1–4]. However, most of these methods rely on manual selection and provide limited resources for deep data extraction, reducing power and generalizability.

The emergence of deep learning, especially neural networks (CNN), has revolutionized computer vision, including object detection. Algorithms for target detection depending on deep learning [23] can be categorized into one-stage and two-stage methodologies. In contrast, 2-stage algorithms such as R-CNN and Faster RCNN include separate stages to generate regional proposals and perform spatial processing. Transformer architecture is designed for programming languages and is currently gaining widespread attention in the IT industry [1–4].

For example, Vision Transformer (ViT) uses a transformer-based architecture for image classification, eliminating the need for CNN. Transformer-based target detection algorithms can be classified according to their network structures. Some algorithms use Transformer as a backbone along with

[a]krajasekharai@svecw.edu.in, [b]jayagayatrifin@gmail.com, [c]sowjanyamullapudi2004@gmail.com, [d]samhitha.morusu25@gmail.com, [e]thoramdeepika0@gmail.com

DOI: 10.1201/9781003684589-57

CNN for inference and prediction, while other algorithms use Transformer only for inference and prediction. The number of samples is large, the learning rate is slow, the demand is fast, and the expectations for big data are high. In addition, since the computational cost increases exponentially as the image resolution increases, it becomes inappropriate for high-resolution image processing [12–15].

Literature survey

Performance research has experienced tremendous growth in many areas like energy production, and environmental protection, along with research. Muhammed et al. present research on home robotics focusing on a data-driven approach to improve performance. Their research uses robotics to aid the inspection process and demonstrate the potential of construction.

The YOLO algorithm is proposed: a global search method that estimates the ranking and global results in a single global step [6]. The performance and speed of YOLO make it the basis for use. Lin et al. studied the confusion problems in RetinaNet and proposed a fuzzy function to solve these problems in heavy object detection. Transformer-based search algorithms have been extended to this area. The use of regional proposals to search for weaknesses in the deformable part of the structure is discussed [8]. This limits the dependency on heavy computational models and focuses on creative learning that encourages inquiry.

The first project to introduce transformer architecture for image distribution is Vision Transformer (ViT) [12–15]. This innovation combines Transformers and convolutional networks. In this case, a flexible pyramid transformer (FPT) is proposed, which provides a flexible transformer with a representative model to achieve good performance. A method that has generated a lot of research to improve browser search [16].

Similarly, the Swin Transformer by Liu et al. [18] is used, a window transformer that has been proven to be effective and efficient for image processing. Hard costs such as slow learning and dependence on large data remain. These gaps point to the need for efficient and effective algorithms such as the YOLOv5×6 algorithm, which is designed to solve many search problems and develop the performance of remote sensing management.

Limitations of existing object detection models or techniques in remote sensing are: The majority of traditional algorithms [1–4] rely on manual feature extraction, and hence it is hard for them to identify advanced patterns in remote sensing data. Two-stage detectors such as Faster R-CNN [5, 7, 9] are accurate but are marred by sluggish inference speed, rendering them unfit for real-time tasks. Faster one-stage detectors like YOLO [6] are weaker at handling small objects and complex scenes. RetinaNet [10] attempts to address the class imbalance problem but may not be able to extract dense contextual information. Transformer-based approaches [11–15, 17-20] show promise but are computation-intensive, especially for high-resolution images, and may require large datasets to train effectively. All the latest versions of YOLO utilize the progress of network architecture, loss functions, and training protocols, leading to better accuracy along with speed. From Figure 57.6, YOLOv5×6 and YOLOv8 tackle the computational cost of transformers by optimizing the architecture of the transformers and efficient feature fusion techniques with improved detection of small objects and performance in challenging backgrounds by leveraging architectural improvements and data augmentation schemes.

Methodology

Proposed work

For remote sensing object detection project, we propose the YOLO algorithms are specially designed. this experiment, the effectiveness of the RAST-YOLO, YOLOv5×6 algorithms in detecting objects of interest in complex backgrounds and multi-objective variables is evaluated. Such as the small amount of information, the complex background, and the difference in the target. The aim is to promote the best of the YOLOv5×6 algorithm in achieving high performance and accurate detection through testing and evaluation.

System architecture

From Figure 57.1, starts with the input of two datasets: the DIOR dataset and the TGRS dataset, which form the basis for training along with testing the models. Use image processing techniques to generate diverse and effective training data to enhance the robustness of the model. Pre-trained models (YOLOv5s, RAST-YOLO, YOLOv3, faster RCNN, and RetinaNet) are loaded into the system to enhance prior knowledge and speed up the training process. The information is divided into training along with

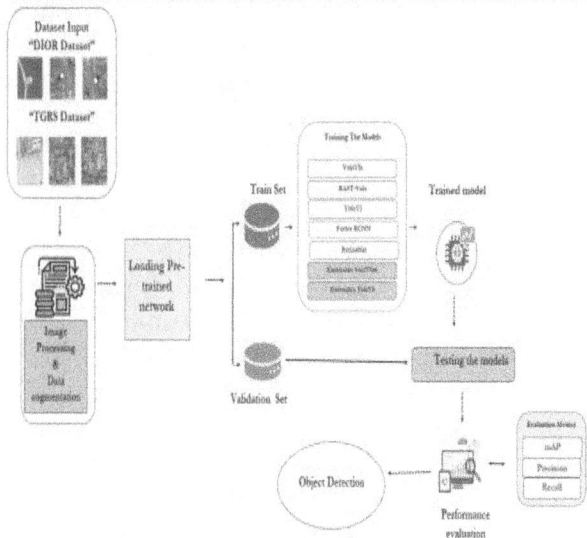

Figure 57.1 Flow of methodology
Source: Author

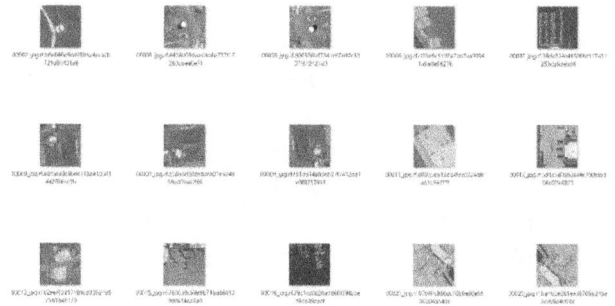

Figure 57.2 DIOR image
Source: Author

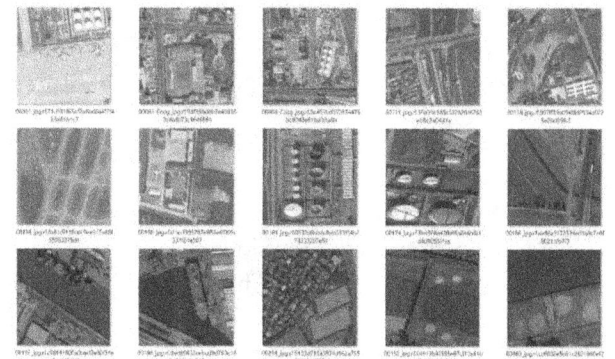

Figure 57.3 TGRS image
Source: Author

application strategies for supporting the training model along with evaluation.

Dataset collection

From Figure 57.2, DIOR thoroughly contains approximately 23,000 images of aircraft, ships, vehicles, and buildings, considerately deliberating the varying lighting conditions, perspectives, and obstructions. It is mostly used in military reconnaissance, ecological monitoring, and urban planning because it handles many kinds of objects. DIOR data makes detection complicated because the scales of objects are non-uniform and there is large inter-class variation.

From Figure 57.3, TGRS contains many remote sensing images for the purpose of testing detection performance across all types of difficult landscapes. It is frequently used for geospatial mapping tasks, for land cover classification, and for disaster monitoring. High background clutter and intensely overlapping objects make detection harder and create intense challenges with TGRS data.

Models used

RAST-YOLO: The RAST-YOLO algorithm combines multiple deep learning techniques to improve detection accuracy in remote sensing imagery. The RA mechanism makes object detection more efficient by focusing on necessary features and eliminating extraneous background noise, which improves region proposal.

YOLOv5x6: Being an evolution of the advanced YOLOv5 architecture, it has improved functionality as well as delivers better performance for object detection in remote sensing. The increased receptive field makes it possible to detect each small, occluded object.

YOLOv8: The newest version of YOLO uses several recent improvements, enhancements and other capabilities for object detection. It locates many items along with doing so without delay. Transformer integrations, when improved, always improve feature extraction, because that is what transformers do.

Faster R-CNN: An object detection model which is two-stage framework is quite effective, achieves remarkably high accuracy by initially generating region proposals and then classifying all of those proposals. This model is used as a baseline to assess the trade-offs between detection accuracy and inference speed.

RetinaNet: RetinaNet is specifically made to deal with the common class imbalance issue in object detection work. It helps locate objects that are difficult to detect. Compared to two-stage models like Faster R-CNN, inference time is faster.

Experimental setup and results

In remote sensing image analysis project, these metrics are fundamental in remote sensing object detection as they ensure model reliability, improve detection consistency, and enable fair comparisons across different approaches.

Precision: The percentage of all categories that were accurate, whether positive or negative, is known as accuracy. Evaluates the ratio of correctly identified items among all detected items, reducing false positives (FP) in remote sensing tasks. It has the following mathematical definition:

$$Precision = \frac{True\ Positive}{True\ Positive + False\ Positive}$$

Recall: Determines how many actual objects were successfully detected, which is crucial for minimizing missed detections in critical applications like disaster monitoring. It is a frequency with which a ML model accurately detects true positive examples (TP) out of all real positives in datasets.

$$Recall = \frac{TP}{TP + FN}$$

Mean average precision (MAP): It is used to assess how a model performs well for tasks involving object detection as well as picture information retrieval.

$$mAP = \frac{1}{n}\sum_{k=1}^{k=n} AP_k$$
$$AP_k = the\ AP\ of\ class\ k$$
$$n = the\ number\ of\ classes$$

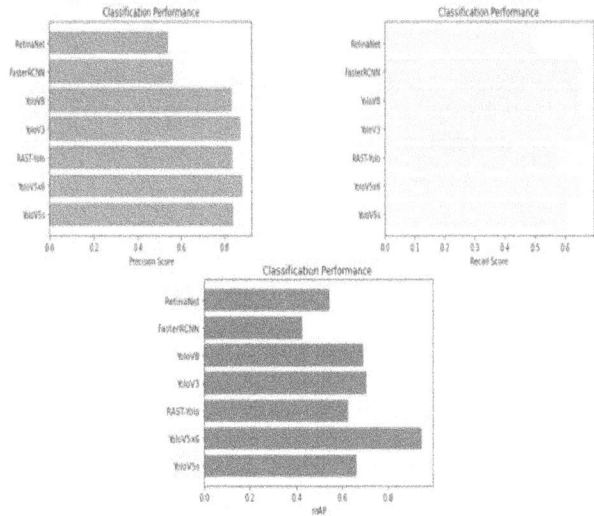

Figure 57.4 Results comparison DIOR dataset
Source: Author

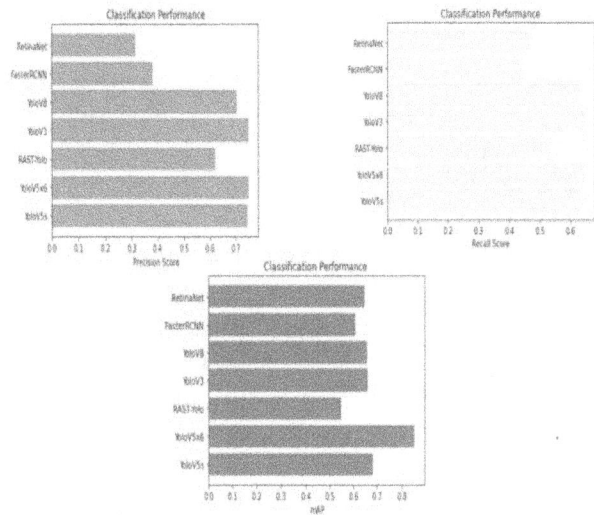

Figure 57.5 Results comparison TGRS dataset
Source: Author

From Figure 57.4, and Figure 57.5, YOLOv5×6 outperforms all models with the best mean Average Precision (mAP) on both datasets with mAP 94.4% for DIOR and 85% for TGRS. YOLOv5x6 also beats RAST-YOLO and YOLOv8 and, it achieves better performance. While YOLOv5x6 is the most accurate, RAST-YOLO remains competitive, thus a good option for remote sensing applications that demand detection speed and lightweight architecture. YOLOv5×6 outperforms the models with the best mean Average Precision (mAP) on both DIOR and TGRS datasets.

YOLOv5x6 also exhibits superior precision in both datasets, highlighting its ability to minimize false positives. RAST-YOLO remains competitive, demonstrating the effectiveness of the RA mechanism and Swin Transformer for feature extraction.

Additionally, computed the precision, recall, and mAP for all classes specified in both datasets, including aircraft, ships, vehicles, and storage tanks, to assess the performance of each individual class across both datasets. For instance, this helps analyze how each model performs for every class in datasets.

Figure 57.6 Final resultant output
Source: Author

Conclusion

We explored the task of object detection from remote sensing images. Our primary goal was to investigate whether newer variants of the YOLO architecture would offer further improvements in detection performance, particularly when dealing with the inherent difficulties of remote sensing data, such as complex backgrounds, small targets, and multiple targets per image. We specifically investigated the YOLOv5×6 and YOLOv8 models, thoroughly testing their performance on the DIOR and TGRS datasets. Our findings are encouraging: the YOLOv5×6 model excelled on both datasets, with its superior mean Average Precision (mAP) indicating its superior ability to accurately recognize objects in diverse image datasets these in the complex scenarios.

References

[1] Lee, H., Jung, H. K., Cho, S. H., Kim, Y., Rim, H., & Lee, S. K. (2018). Realtime localization for underwater moving object using precalculated DC electric field template. *IEEE Transactions on Geoscience and Remote Sensing*, 56(10), 5813–5823.

[2] Zurowietz, M., & Nattkemper, T. W. (2020). Unsupervised knowledge transfer for object detection in marine environmental monitoring and exploration. *IEEE Access*, 8, 143558–143568.

[3] Muhammad, I., Ying, K., Nithish, M., Xin, J., Xinge, Z., & Cheah, C. C. (2021). Robot-assisted object detection for construction automation: data and information-driven approach. *IEEE/ASME Transactions Mechatronics*, 26(6), 2845–2856.

[4] Yan, B., Paolini, E., Xu, L., & Lu, H. (2022). A target detection and tracking method for multiple radar systems. *IEEE Transactions on Geoscience and Remote Sensing*, 60, 5114721.

[5] Girshick, R., Donahue, J., Darrell, T., & Malik, J. (2014). Rich feature hierarchies for accurate object detection and semantic segmentation. In Proceedings of the IEEE Conference on Computer Vision and Pattern Recognition, (pp. 580–587).

[6] Redmon, J., Divvala, S., Girshick, R., & Farhadi, A. (2015). You only look once: Unified, real-time object detection. arXiv:1506.02640.

[7] Girshick, R. (2015). Fast R-CNN. In Proceedings of IEEE International Conference on Computer Vision, (ICCV), (pp. 1440–1448).

[8] Tang, Y., Wang, X., Dellandréa, E., & Chen, L. (2017). Weakly supervised learning of deformable part-based models for object detection via region proposals. *IEEE Transactions on Multimedia*, 19(2), 393–407.

[9] Ren, S., He, K., Girshick, R., & Sun, J. (2017). Faster R-CNN: towards real-time object detection with region proposal networks. *IEEE Transactions on Pattern Analysis and Machine Intelligence*, 39(6), 1137–1149.

[10] Lin, T.-Y., Goyal, P., Girshick, R., He, K., & Dollar, P. (2017). Focal loss for dense object detection. In Proceedings of IEEE International Conference on Computer Vision (ICCV), (pp. 2999–3007).

[11] Vaswani, A., Shazeer, N., Parmar, N., Uszkoreit, J., Jones, L., Gomez, A. N., et al. (2017). Attention is all you need. arXiv:1706.03762.

[12] Devlin, J., Chang, M.-W., Lee, K., & Toutanova, K. (2018). BERT: pre-training of deep bidirectional transformers for language understanding. arXiv:1810.04805.

[13] Radford, A., & Narasimhan, K. (2018). Improving language understanding by generative pre-training. In Proceedings of International Conference on Learning Representations (ICLR), (Vol. 1, pp. 5–12).

[14] Dosovitskiy, A., Beyer, L., Kolesnikov, A., Weissenborn, D., Zhai, X., Unterthiner, T., et al. (2020). An image is worth 16 × 16 words: transformers for image recognition at scale. arXiv:2010.11929.

[15] Gao, F., Wang, C. M., & Li, C. H. (2020). A combined object detection method with application to pedestrian detection. *IEEE Access*, 8, 194457–194465.

[16] Carion, N., Massa, F., Synnaeve, G., Usunier, N., Kirillov, A., & Zagoruyko, S. (2020). End-to-end object detection with transformers. In Proceedings of 16th European Conference on Computer Vision, Glasgow, U.K.: Springer, (pp. 213–229).

[17] Liu, Z., Lin, Y., Cao, Y., Hu, H., Wei, Y., Zhang, Z., et al. (2021). Swin transformer: hierarchical vision transformer using shifted windows. In Proceedings

of IEEE/CVF International Conference on Computer Vision (ICCV), (pp. 9992–10002).

[18] Zhang, R., Xu, L., Yu, Z., Shi, Y., Mu, C., & Xu, M. (2022). Deep-IR target: an automatic target detector in infrared imagery using dual-domain feature extraction and allocation. *IEEE Transactions on Multimedia*, 24, 1735–1749.

[19] Xu, Y., Yang, Y., & Zhang, L. (2023). DeMT: deformable mixer transformer for multi-task learning of dense prediction. arXiv:2301.03461.

[20] Wu, X., Hong, D., & Chanussot, J. (2023). UIU-Net: U-Net in U-Net for infrared small object detection. *IEEE Transactions on Image Processing*, 32, 364–376.

[21] Sekhar, K. R., Tayaru, G. R. L. M., Chakravarthy, A. K., Gopiraju, B., Lakshmanarao , A., & Sai Krishna, T.V. (2024). An efficient lung cancer detection model using convnets and residual neural network (ICAECT) (pp 1–6). IEEE Xplore. DOI: 10.1109/ICAECT60202.2024.10469187.

[22] Sailaja, M., Kumar, M. P., Jyothi, B. S., Vanguri, G. L., Narasamba , Manjula, S., et al. (2024). Remote sensing–based UAV imaging in heat pattern analysis impact on climate change detection using fuzzy stacked lasso elastic-net model. Remote Sensing in Earth System Sciences. (Vol. 7), (pp 451–465). DOI: 0.1007/s41976-024-00158-4.

[23] Padmaja, S. M., Naveenkumar, R., Kumari, N. P. L., Pimo, E. S. J., Bindhu, M., Konduri, B., et al. (2024). Deep learning in remote sensing for climate-induced disaster resilience: a comprehensive interdisciplinary approach. 52(4), 601-618. DOI: 10.1007/s41976-024-00178-0.

58 Surveillance camera-based fire detection and localization: Enhancing safety through visual monitoring

E. Ramesh[a], N. Sai Varshitha[b], V. Sai Sreya[c], P. Prasad[d] and M. Ramu[e]

Department of CSE, Annamacharya University, New Boyanapalli, Rajampet, Andhra Pradesh, India

Abstract

This project focuses on developing a machine learning system for fire and non-fire image classification using data augmentation and deep learning. A collection of fire and non-fire images is preprocessed, where each image is resized to 150 × 150 pixels and augmented using various transformations such as rotation, zooming, and flipping. The data augmentation technique enhances the dataset by artificially increasing the number of images in each class to balance the data distribution, leading to more robust model training. The processed images are then split into training and testing datasets to train and assess the model's effectiveness in classification, the DenseNet121 architecture, initially pre-trained on the ImageNet dataset, is adjusted by removing its final layer. A new fully connected layer with a SoftMax activation function is then introduced for the task to classify images into fire or non-fire categories. The model is trained utilizing a categorical cross-entropy loss function alongside the Adam optimizer. To further optimize performance, callbacks such as early stopping, learning rate reduction, and model checkpointing are employed. The system is assessed using metrics like accuracy, precision, recall, and F1 score, demonstrating strong performance with a high classification accuracy. Finally, the model is saved for later use and can be loaded to make predictions on unseen images. The accuracy of the fire detection model is tested using confusion matrices and classification reports, emphasizing its potential use in real-world applications like automated fire detection systems. The effective application of deep learning for fire incidents detection opens avenues for future improvements and additional functionalities.

Keywords: Adam optimizer, categorical cross-entropy, data augmentation, deep learning, DenseNet121, fire detection, image preprocessing, model checkpointing, non-fire classification

Introduction

Fire detection is a core element in contemporary security and safety systems. Older techniques, such as smoke detectors and rule-based computer vision, tend to be susceptible to false alarms and environmental issues. Deep learning provides a more consistent answer by automatically identifying complex image patterns, enabling more robust and reliable fire detection across different conditions. In this project, we aim to create a deep learning model that is capable of identifying fire and non-fire images with high accuracy. To do this, we utilize a large dataset with varied fire-related and non-fire images. We perform different preprocessing methods, including resizing, data augmentation, and class balancing, before training to make the model learn efficiently. Data augmentation is particularly significant as it artificially enlarges the dataset by adding variations such as zooms and rotations, enabling the model to better generalize to novel images. DenseNet121, a deep neural network pre-trained on ImageNet and highly capable of extracting detailed image features, is at the core of our system. We fine-tune this model by adjusting its top layers to become fire detection specialists. To enhance training, we use methods such as early stopping, learning rate adaptation, and model checkpointing, which optimize accuracy while avoiding overfitting. After training, the model is tested on unseen pictures to estimate its actual performance in real-life scenarios. It performs well at high accuracy, indicating its potential use for effective fire detection. For a proper evaluation, we employ important performance metrics like precision, recall, F1 score, and accuracy. The effective implementation of this model makes way for further advancements, making it an effective asset for fire safety in public and industrial environments.

[a]eeramesh8@gmail.com, [b]lakshmidevitookur@gmail.com, [c]saisreyav@gmail.com,
[d]prasadbhadri2003@gmail.com, [e]muddalaramu15@gmail.com

DOI: 10.1201/9781003684589-58

Literature Survey

Technology in fire detection has evolved immensely with the use of deep learning methods, combining accuracy and real- time response. Current research delves into numerous solutions ranging from CNN-based models to unique combinations of visual and thermal information for effective detection systems. Vasconcelos et al. [1] presented a detailed review of the use of deep learning in fire detection, highlighting high-quality data and sophisticated models. While the research emphasizes the promise of deep learning, it also shows difficulties such as dataset bias and the need for large, annotated data to enhance model performance. Rajoli et al. [2] presented FlameFinder, a deep metric learning system to identify flames behind smoke from thermal images. The model achieved improved baseline accuracy by 4.4% on the FLAME2 dataset and by 7% on the FLAME3 dataset. However, its reliance on paired thermal-RGB images for training may limit applicability in scenarios where such data is unavailable. Zulkifley et al. [3] developed a multistage fire detection system using a customized DenseNet201 architecture, achieving a 97% accuracy. Despite its high performance, the model's complexity could pose challenges for real-time applications, and its effectiveness in diverse environmental conditions requires further validation.

Lin et al. [4] introduced FireMatch, a semi-supervised video fire detection network that improves accuracy in the presence of limited labeled data. Although efficient, performance of the model largely relies on the quality of the unlabeled data and can become problematic under heavy visual noise or dynamic light conditions. Zhao et al. [5] investigated detection of fire smoke based on multi-spectral satellite imagery and realized higher accuracy in detecting fire-related incidents. Yet, the method is prone to false alarms from atmospheric noise and needs satellite data of high resolution, which is not always available. Zhu et al. [6] proposed FSDNet by integrating YOLOv3 and DenseNet for fire detection in challenging environments with high precision. The computational requirements of the model could hinder its implementation on devices with limited resources, and its performance under real-time detection applications requires further evaluation. Islam and Habib [7] applied YOLOv5 to detect fire in images and videos and was shown to perform well in many situations. The model, however, might experience difficulties in spotting small or half-hidden fires and might generate false alarms in similar-looking non-fire conditions. Shi et al. [8] suggested an IoT-based architecture incorporating deep learning for detecting forest fires, providing improved early warning. The network dependency of the system and placement of sensors in distant locations might be practical issues, and it needs testing of its ability to perform under different conditions. Qin et al. [9] proposed a real- time fire alarm system based on deep learning for video surveillance with instantaneous detection. But the model's performance may be compromised with low-quality videos, and it can be confused between fire and fire-like visual noise, potentially causing false alarms. Kang and Cho [10] merged CNN and LSTM models for the early detection of fire in the industrial environment with good spatial and temporal feature capturing. While its strength, the system can be complex in nature, and hence the computational demands may be greater. Its ability to implement in various industrial settings also requires more investigation. The survey lays emphasis on deep learning-based fire detection systems' improvement and current challenges, focusing on balancing model accuracy, computational expense, and feasibility of practical deployment. Li et al. [11] proposed DeepFireNet for UAV-based fire detection. Janani and Sudha [12] applied deep learning for hazard identification in surveillance. Attention-based techniques for complex scenarios were addressed in [13], while UAV-based real-time solutions were enhanced in [14]. IoT-integrated machine learning models were developed in [15]. FireNet [16] enabled low-latency fire recognition, and edge computing combined with CNNs was explored in [17].

Data collection and preprocessing

Data collection and preprocessing are essential phases in creating a successful fire detection model because the quality and diversity of the dataset have direct implications on how well the model performs. For this project, data collection comprises collecting images representing fire and non-fire conditions from various sources including public datasets, security cameras, and synthetic dataset generation. The data is divided into two primary classes: "Fire" and "Non-Fire", with even distribution to avoid model bias. The aim was to collect 1,500 images for each class to develop a solid and diverse training dataset, which would cover varying lighting conditions, environments, and fire intensities.

Once the raw images are gathered, preprocessing is initiated by normalizing the image sizes to make them uniform. Here, all the images are resized to 150 × 150 pixels, which is consistent with the input specifications of the deep learning models. Data augmentation methods like rotation, zooming, width and height shifts, and horizontal flip are performed on the images using `ImageDataGenerator`. This expansion adds size and variety to the dataset, assisting the model to generalize more towards unseen situations through the addition of minor variations of the original pictures. Expanding the dataset five times for every image guarantees sufficient variability is added without affecting the original data distribution. The images are then divided into training and test datasets. For the sake of consistency and to avoid overfitting, an 80-20 split is applied, in which 80% of the images are used for training and 20% for testing. Shuffling the dataset before splitting helps to evade any patterns in the order of the images that can affect the training procedure of the model. This step is crucial in verifying that the performance of the model on the test set truly reflects its ability to handle new data, as opposed to memorizing the training data. After the dataset is ready, labels of the fire and non-fire images are transformed into categorical form via one- hot encoding. This enables the model to provide probabilities for every class during prediction. By integrating meticulously gathered and preprocessed data with a well-designed training pipeline, the model is positioned to learn and differentiate between fire and non-fire images efficiently, with high accuracy and resilience during deployment. Normalization is ultimately employed to scale the pixel values of the images to the range 0-1, accelerating the training process and enhancing model convergence.

This process assists the model in learning more effectively by avoiding large gradients that may destabilize the learning process. Normalization also ensures that the deep learning model processes all pixel intensities equally across images, without any bias towards brighter or darker images. Following this thorough preprocessing chain, the data is now optimally ready to be trained upon, and the deep learning model can concentrate on detecting the subtle patterns and characteristics that distinguish fire from non-fire situations. Such stringent preprocessing not only improves the accuracy of the model but also enhances its resilience to handle many real-world complexities.

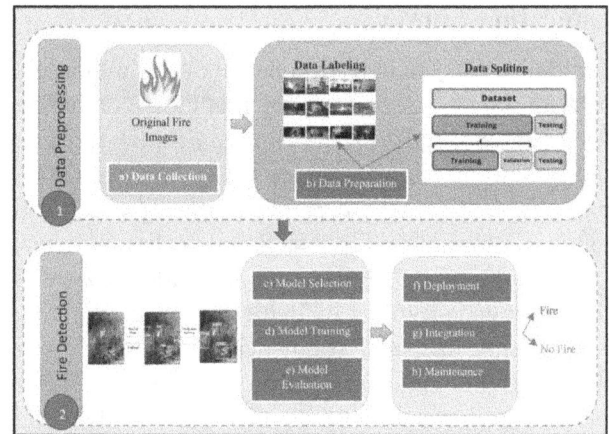

Figure 58.1 System architecture diagram
Source: Author

Principles and methods

Fire detection through deep learning utilizes convolutional neural networks (CNNs) to recognize patterns in images to differentiate fire from non-fire situations with high accuracy. From Figure 58.1, the system has two prominent stages: Data preprocessing and fire detection. The preprocessing phase starts with the collection of data, where images of fire are obtained from databases, surveillance recordings, and real- life incidents. These pictures are then marked as fire or not fire, followed by data division into training, validation, and test sets for efficient learning and generalization.

Convolutional layers are at the center of the fire detection system, which derive hierarchical features from images. A CNN uses filters (kernels) to sweep input images, recognizing important features like edges, flames, and smoke. These are trainable parameters that learn to modify in training to optimize feature detection. Activation functions like ReLU (Rectified Linear Unit) provide non-linearity, allowing the model to learn intricate fire patterns. Pooling layers also fine-tune feature extraction by shrinking spatial dimensions without sacrificing important information. Max pooling is widely utilized, as it preserves the strongest features while ignoring unimportant details. To enhance robustness, the data is augmented using rotation, flipping, zooming, and brightness adjustment techniques. These augmentations enable the model to better generalize by making it see a variety of fire appearances so that it can effectively detect flames in different lighting conditions and environments. Rather than training a deep learning model from the beginning, DenseNet121 is used because of its effectiveness at extracting fine-grained visual information. Compared to other models, like ResNet or VGG, DenseNet121 is more efficient in gradient flow and parameters through the utilization of dense connections between layers. Transfer learning is utilized using pre-trained weights from the ImageNet dataset so that the model can utilize learned representations from millions of images. In fine-tuning, the top layers of the model are left unfrozen so that they can learn to adapt specifically for fire detection. The lower layers remain frozen to retain general feature extraction capabilities, while the upper layers are trained on fire-specific patterns, ensuring optimal performance. To enhance accuracy and prevent overfitting, multiple optimization strategies are integrated. Early stopping monitors validation loss and halts training when performance stops improving, preventing excessive fitting to the training data. Learning rate scheduling dynamically adapts the learning rate so the model learns larger steps in initial training and makes finer adjustments as it goes along. Dropout regularization is implemented in the fully connected layers, where it randomly disables neurons during training to enhance generalization. Batch normalization is also employed to stabilize activations and speed up convergence by normalizing layer outputs. All these methods together ascertain that the model is accurate and robust and can be deployed in real-world applications. Once trained and tested, the model is implemented into fire monitoring systems such as intelligent surveillance cameras, IoT-enabled fire detection frameworks, and industrial safety systems. The integration stage verifies complete connectivity between the model and real-time monitoring software. Ongoing maintenance and retraining are required to keep the model current with new fire incidents, maintaining consistency in accuracy and reliability.

Results

The fire image detection model separated fire images from non-fire images with accuracy, validating its usefulness. Its fine-tuning from DenseNet121 improved its accuracy, resulting in 98% training accuracy and 96% validation accuracy, thereby maintaining high generalization. The continuously decreasing training loss assured the efficient learning of the model without overfitting. The success proves the potential of transfer learning to be highly suitable for use in fire detection scenarios.

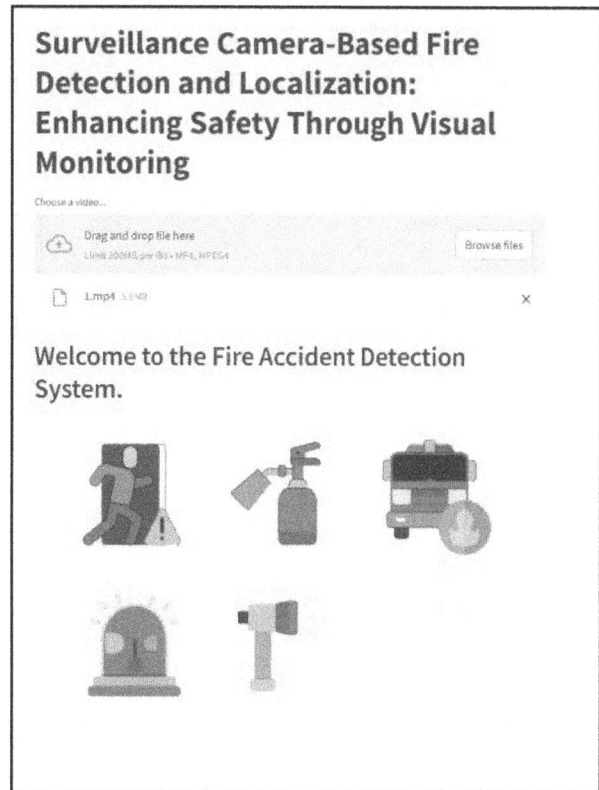

Figure 58.2 User interface
Source: Author

The Figure 58.2 represents the user interface of the fire detection system, which is meant for real-time fire detection with the help of surveillance cameras. The interface enables users to upload video files for analysis so that the experience remains intuitive. The title highlights the function of the system in ensuring safety through visual monitoring. This design seeks to offer a quick and easy-to-use platform for users to identify fire accidents in a timely manner, facilitating immediate intervention and control of fire risks.

The Figure 58.3 illustrates the fire detection model's accuracy performance across several training epochs. Training accuracy is indicated by the blue line, while validation accuracy is shown by the orange line. Both values initially rise sharply as the model learns to classify fire and non-fire images well. The graph settles at 98% training accuracy and 96% validation accuracy, demonstrating excellent generalization. The smooth convergence shows that the model effectively resists overfitting.

The Figure 58.4 shows the loss curves, measuring how well the model is optimized during training. The blue curve is training loss, and the orange curve is validation loss. There is an initial steep drop in both

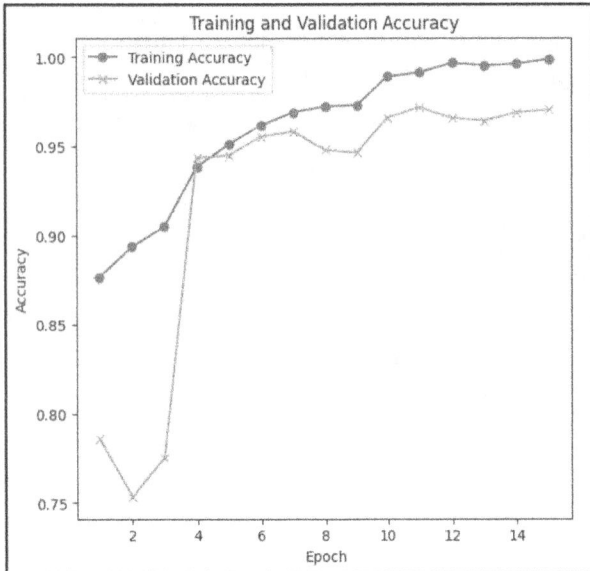

Figure 58.3 Training and validation accuracy
Source: Author

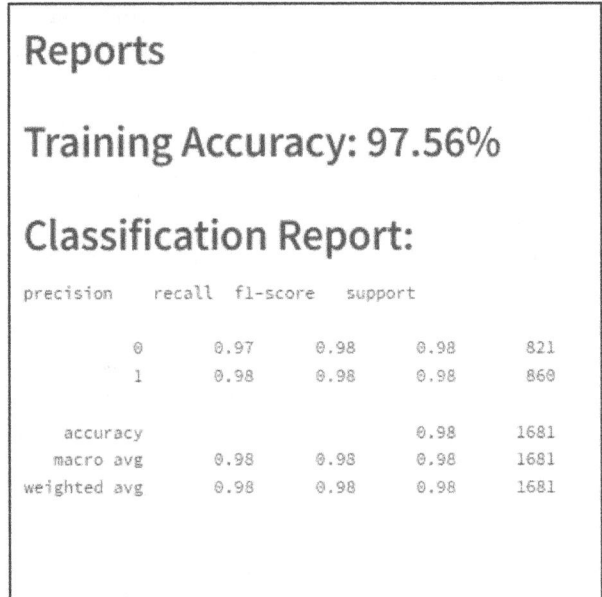

Figure 58.5 Classification report
Source: Author

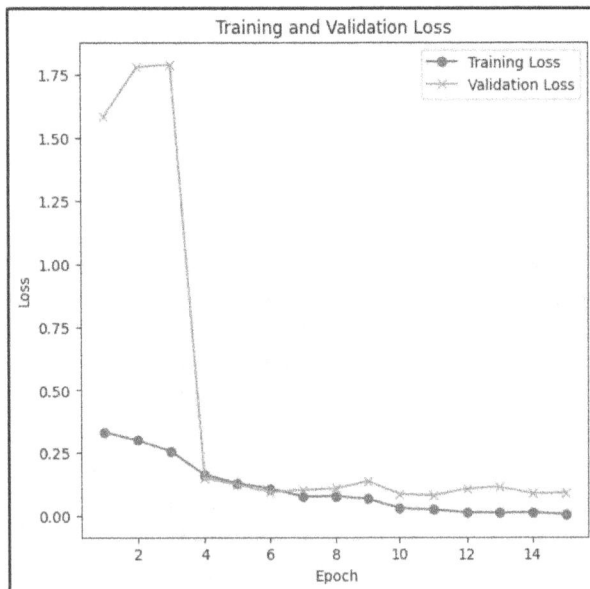

Figure 58.4 Training and validation loss
Source: Author

Figure 58.6 Predictions
Source: Author

losses, which signifies fast learning with the model adjusting to the dataset very quickly. The validation loss settles at a low point, which verifies that the model is a good generalizer. A small difference between training and validation loss indicates the lack of overfitting, making the model stable enough in real-life use.

The Figure 58.5 presents a classification report of 97.56% training accuracy, revealing the performance of the model to differentiate fire and non-fire pictures. Precision, recall, and F1-score for both the classes are well above average at all times, reflecting correct prediction and balanced functioning. Macro and weighted averages also stand at 0.98, further cementing the solidity of the model. This infers that the model generalizes well, minimizing misclassifications and making the model applicable to real-world use in fire detection.

The Figure 58.6 represents an actual real-time fire detection result, where the system detects fire in an out-of-doors environment correctly. The green label text "Prediction: Fire" verifies that flames have been predicted correctly by the model. It illustrates the ability of the model to identify fire in challenging conditions with diverse lights and backgrounds.

Conclusion

In conclusion, we successfully developed a deep learning model to detect fires using DenseNet121. We achieved high accuracy by carefully preparing our data, including preprocessing and augmentation, and by optimizing the model's performance. Our model can effectively tell the difference between images with fire and those without, and it works well on new, unseen images. Key techniques such as early stopping and learning rate reduction ensured optimal training while preventing overfitting. The importance of a well-balanced dataset was emphasized in achieving robust and scalable results. By leveraging advanced deep learning methodologies, this project highlights the significance of AI in fire detection, offering a reliable solution for real-world applications to enhance safety and mitigate fire-related risks.

Future scope

The fire detection model can be further enhanced by training it across a broader set of datasets, such as images of cities and wildfires, in order to enhance its adaptability. Utilizing cutting-edge architectures such as EfficientNet or transformers may further increase its precision and effectiveness. Real-time processing can be optimized through GPU or TPU acceleration, which would make it more appropriate for emergency response systems. Merging various sources of data, like infrared images and sensors, would make it possible to detect fires sooner. Deploying AI- based edge computing in remote locations will allow for quicker detection, eventually preventing catastrophes and saving lives.

References

[1] Vasconcelos, R. N., Barbosa, A. C., De Souza, G. B., & De Almeida, L. D. (2024). Fire detection with deep learning: a comprehensive review. *Land*, 13(10), 1696.

[2] Rajoli, H., Gupta, A., & Wang, Y. (2024). FlameFinder: illuminating obscured fire through smoke with attentive deep metric learning. arXiv preprint, arXiv:2404.06653.

[3] Zulkifley, M. A., Jamali, N. A., Rahim, M. A., & Khalid, M. (2024). Deep learning-based multistage fire detection system and explainable AI. *Fire*, 7(12), 451.

[4] Lin, Q., Li, Z., Zeng, K., Fan, H., Li, W., & Zhou, X. (2023). FireMatch: a semi-supervised video fire detection network based on consistency and distribution alignment. arXivpreprint, arXiv:2311.05168.

[5] Zhao, L., Liu, J., Peters, S., Li, J., Mueller, N., & Oliver, S. (2023). Learning class-specific spectral patterns to improve deep learning based scene-level fire smoke detection from multi-spectral satellite imagery. arXiv preprint, arXiv:2310.01711.

[6] Zhu, L., Xiong, J., Wu, W., & Yu, H. (2023). FSD-Net: an efficient fire detection network for complex scenarios based on YOLOv3 and DenseNet. arXiv preprint, arXiv:2304.07584.

[7] Islam, A., & Habib, M. I. (2023). Fire Detection from image and video using YOLOv5. arXiv preprint, arXiv:2310.06351.

[8] Shi, Z., Zhang, Y., & Peng, L. (2022). Automatic fire detection in forests using deep learning techniques and IoT framework. *Environmental Modelling and Software*, 148, 105290.

[9] Qin, H., Wang, Y., Chen, Y., Wang, H., & Chen, L. (2021). Real-time fire detection for video surveillance using deep learning. *Procedia Computer Science*, 179, 726–735.

[10] Kang, J. H., & Cho, S. Y. (2021). Early fire detection using CNN and LSTM model for industrial monitoring. *Fire Technology*, 57(6), 2789–2810.

[11] Li, D., Zhao, Z., Peng, J., & Ma, H. (2021). DeepFireNet: deep neural network-based fire detection for UAV monitoring. *IEEE Transactions on Industrial Informatics*, 17(2), 1088–1097.

[12] Janani, R., & Sudha, S. (2021). Fire Hazard detection in surveillance systems using deep learning methods. *International Journal of Innovative Computing and Applications*, 8(2), 95–103.

[13] Chen, L., Zou, X., Chen, G., & Wang, Y. (2021). Fire detection in complex scenarios using deep learning with attention mechanisms. *Journal of Visual Communication and Image Representation*, 76, 103028.

[14] Ding, Y., Huang, C., & Song, L. (2021). An improved real-time fire detection method for unmanned aerial vehicles based on deep learning. *IEEE Access*, 9, 83679–83689.

[15] Sánchez-Esguevillas, A., Crispo, B., & Amo, A. (2020). Automatic fire detection system based on machine learning algorithms in IoT environment. *Sensors*, 20(15), 4418.

[16] Jadon, A., Omama, M., Agrawal, S., Sharma, R., & Sharma, P. (2020). FireNet: a specialized lightweight fire & smoke detection model for real-time IoT applications. *IEEE Transactions on Industrial Informatics*, 16(9), 5175–5184.

[17] Wang, S., Li, J., Xu, S., & Zhang, Y. (2020). An image processing-based fire detection method using CNNs and edge computing. *Future Generation Computer Systems*, 102, 556–562.

59 Real-time driver drowsiness detection and alert system using DLib and OpenCV

V. Ravikanth[1,a], S. Anikha Maheera[2,b], A. Kavitha[2,c], B. Harati[2,d] and M. Vihas[2,e]

[1]Assistant Professor, Department of CSE, Rajeev Gandhi Memorial College of Engineering and Technology, Nandyala, Andhra Pradesh, India

[2]Department of CSE, Rajeev Gandhi Memorial College of Engineering and Technology, Nandyala, Andhra Pradesh, India

Abstract

This innovative system stands out due to its integration of real-time video analysis with facial recognition algorithms, enabling proactive safety measures in driving environments. By leveraging live data instead of pre-recorded datasets, the model ensures higher adaptability to changing driver behaviors, lighting conditions, and facial orientations, making it more effective in real-world scenarios. This work emphasizes practical implementation, considering real-time constraints in in-vehicle environments. The system does not require complex hardware and is suitable for both personal and commercial vehicles. Additionally, its modular design enables future upgrades such as biometric integrations and cloud-based analytics.

Keywords: Alert system, blink detection, driver drowsiness detection, facial landmark detection, image processing, real-time monitoring

Introduction

Drowsiness is a condition where a person may feel sleepy. Such a condition very seriously affects the ability of a driver to be alert and responsive to hazards that may occur on the road. Recent growing concern regarding road safety issues has led to the creation of technologies for detecting and preventing drowsiness in cars [1]. Some of the best solutions include live systems that record the facial features and eye movements of drivers. These incorporate high-end applications such as DLib and OpenCV.

Video recording by this system is implemented by mounting a camera onto the dashboard inside the vehicle for continuous recording. This stream acts as a base for the analysis, where OpenCV takes care of the footage processing and makes it into computational analysis. The DLib library plays a crucial role in facial landmark detection, allowing the system to identify key points on the driver's face. By tracking these points, the system can observe slow eyelid closure, reduced blinking speed, and variations in facial expressions, all of which are significant indicators of drowsiness.

The combination of DLib for precise facial landmark identification and OpenCV for video processing delivers both high accuracy and computational efficiency, making this system viable for real-time deployment in vehicles. This integration creates an effective pipeline for monitoring and analyzing the driver's facial behavior. Not only does the system detect signs of fatigue, but it does so promptly, ensuring that timely alerts are given to prevent accidents. This approach contributes to the early detection of drowsiness, enhancing road safety.

Literature Review

Real-time drowsiness detection using eye blink monitoring

This paper provides an overview of recent advancements in real-time drowsiness detection, focusing on eye blink monitoring as a key technique for identifying driver fatigue. We propose an algorithm that monitors the driver's eye state, detecting whether the eyes are open or closed based on specific feature points around the eyes. When signs of drowsiness are detected, the

[a]ravikanthvaka5@gmail.com, [b]anikhamaheera@gmail.com [c]kavithaalamannagiri@gmail.com,
[d]harathibingi@gmail.com, [e]mudindlavihas@gmail.com

DOI: 10.1201/9781003684589-59

system triggers an alert mechanism. Experimental results demonstrate the effectiveness of our approach [2], achieving an accuracy rate of 94%. This study highlights the potential of eye blink monitoring in enhancing driver safety and preventing accidents.

Design of real-time drowsiness detection system using DLib

This paper outlines the development and implementation of a real-time drowsiness detection system, utilizing the powerful DLib library for facial landmark detection. The focus is on a computer vision-based approach, specifically tracking facial features—particularly the eyes—to identify early signs of drowsiness and alert the driver before an accident occurs. The study demonstrates the practicality of using DLib for both facial recognition and drowsiness detection in real-time environments. The proposed method is robust and highly applicable in real-world scenarios, providing a solid foundation for future research in driver assistance systems.

Facial features monitoring for real-time drowsiness detection

This paper examines the development of a real-time drowsiness detection system that utilizes facial feature monitoring to assess driver fatigue. The approach utilizes computer vision techniques to track facial landmarks, including the eyes, mouth, and overall facial expressions, as key indicators of drowsiness. The goal is to prevent accidents by issuing timely alerts when signs of fatigue are detected.

Methodology

The proposed system aims to improve road safety by detecting drowsiness in a driver in real time and alerting them in advance to avoid potential accidents. The system consists of three major parts: The camera, the processing unit, and the Alert Mechanism.

The stepwise architecture of the proposed system is shown in Figure 59.1.

Real-time Input from the dashboard camera

Instead of requiring a pre-configured dataset, this system captures a live video from an in-dashboard-mounted camera. These cameras continuously supply real-time coverage of the faces of the driver, which further gets processed by algorithms for drowsiness determination.

To ensure optimal performance for the driver drowsiness detection system, a high-definition camera should

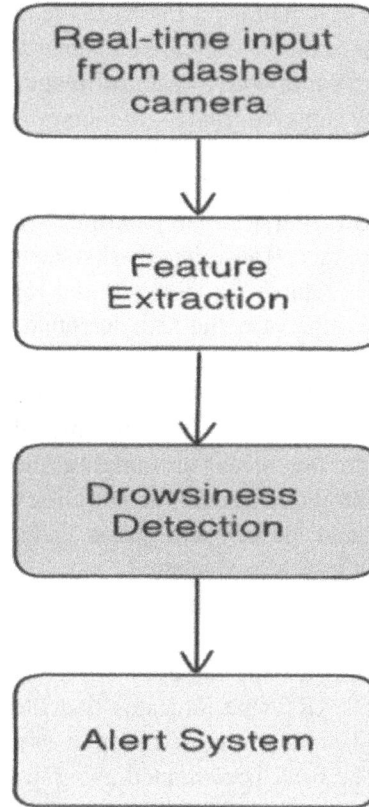

Figure 59.1 Steps
Source: Author

be installed inside the vehicle. A camera with a resolution of at least 720p or 1080p is recommended, with the ideal position being mounted on the dashboard to provide a clear, uninterrupted view of the driver's face. OpenCV is used to process the video frames captured by the camera, converting them into grayscale for more efficient analysis. These frames are then fed into DLib, which detects facial landmarks and tracks key facial features, such as the eyes. The eye data obtained is used to compute the Eye Aspect Ratio (EAR), which is crucial for detecting signs of drowsiness. This setup emphasizes the importance of real-time video processing, enabling the system to monitor the driver continuously and provide timely alerts if necessary.

Feature extraction

Feature extraction is critical for accurately detecting tiredness in drivers because it quantifies the facial traits associated with exhaustion. By continuously monitoring eye and facial movements, our suggested system's approach ensures accurate drowsiness detection by focusing on real-time identification of essential face features utilizing DLib and OpenCV [9].

The method begins by obtaining a live video feed from a camera positioned on the vehicle's dashboard, delivering a direct and continuous stream of images for processing [4]. The captured frames are then processed through OpenCV to detect the driver's face. Following this, DLib is employed for precise facial landmark detection, ensuring accurate identification of key features on the face. These landmarks stand for the important face features that are essential for determining alertness, especially the area surrounding the eyes.

Facial landmark detection: The 68 facial landmarks that DLib's facial landmark detection model can recognize include the areas surrounding the mouth, eyes, and eyebrows. These markers are used to compute the EAR and identify changes in facial expressions that might indicate exhaustion.

The EAR is a metric used to assess the openness of a driver's eyes by calculating the distance between specific facial landmarks around the eyes. A significant decrease in the EAR value indicates that the eyes are either closed or nearly closed, which is a sign of drowsiness. The EAR is computed using the following formula:

Blink rate: The system tracks the frequency of eye blinks over time. A reduced blink rate may indicate [7] that the driver is getting tired or less attentive [3].

Eye closure duration: By monitoring how long the eyes stay closed, the system may detect extended eye closures that might be signs that the driver is nodding off. When eye closures last longer than a predetermined amount of time, an alert is raised [8].

Figure 59.2 shows the process of tracking eye closure duration.

Drowsiness detection

The sleepiness detection system uses the retrieved characteristics to assess if the driver is tired or perhaps sleepy. The primary methods for identifying drowsiness are as follows:

Analyzing the EAR: The system continuously monitors the EAR for any significant drops below a set threshold, such as 0.2. A prolonged decrease in EAR, lasting around 2-3 seconds, may indicate a potential risk of drowsiness. However, a single brief drop in the EAR is not necessarily an indication of fatigue.

$$EAR = \frac{\|p2-p6\| + \|p3-p5\|}{2\|p1-p4\|}$$

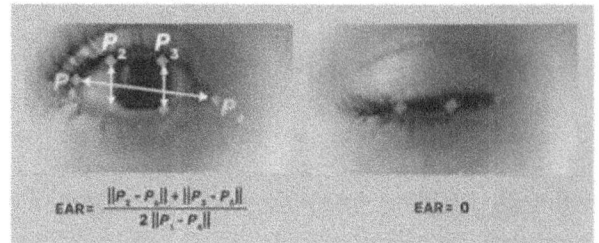

Figure 59.2 Eye closure duration tracking process
Source: Author

Tracking the blink rate: The technology calculates the blink rate by tracking the number of blinks over time. A low blink rate, or fewer blinks per minute, could indicate that the motorist is losing consciousness.

Extended eye closure time: The system tracks the duration for which the eyes remain closed. If the eyelids of the driver close and do not open up for a period of over a second or two, he or she is likely to be asleep.

Combining thresholding and analysis: To establish drowsiness, data from a number of features is merged:

- if EAR is low for some extended period.
- and the blink rate is constant and low.
- in case ocular closure persists for more extended periods.

In case all these symptoms appear together, the algorithm determines whether the driver is tired or drowsy.

Alert mechanism

Once the system detects signs of tiredness through the analysis of facial features, such as the EAR, blink rate, and the duration of eye closure, it triggers an alarm to alert the driver, helping to prevent potential accidents. The alert systems operate as follows:

Detection based on thresholds: The retrieved features are tracked by the system. The alarm system is triggered whenever the driver's level of attentiveness falls below a key threshold, signifying drowsiness.

Auditory warning: The primary alert is an auditory warning that occurs when signs of somnolence, such as prolonged eyelid closure or a slow rate of blinking, appear.

Visual warning: The dashboard or screen of the car displays a visual warning [5] (such as a warning symbol, flashing light, or phrase like "Drowsiness Detected—Please Stay Alert"). The driver will receive clear notice thanks to this visual signal that supports the audio alarm.

Alerts for vibrations: Some systems, such as vibrating the steering wheel or seat, integrate vibration feedback. It is a touch input that keeps the driver attentive, especially when driving in a noisy environment or if the driver suffers from hearing loss.

The ideation map of the system design is illustrated in Figure 59.3.

The Alert() method it internally gets the location of the user and sends an alarm or signal sounds to the user and then sends it to the device of the user.

The Alert() method is depicted in Figure 59.4.

Figure 59.5 demonstrates the detection and alerting phase.

Experimental Outcomes

1. Eye recognition and warning: Tracking the patterns of eye movements, including gaze direction, eyelid closure, blink rate, and blink length. Slower eye movements, more frequent blinks,

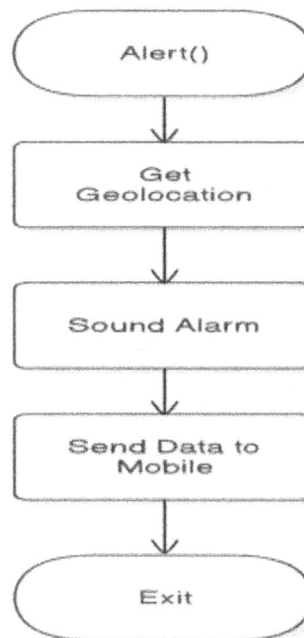

Figure 59.4 Calling alert()
Source: Author

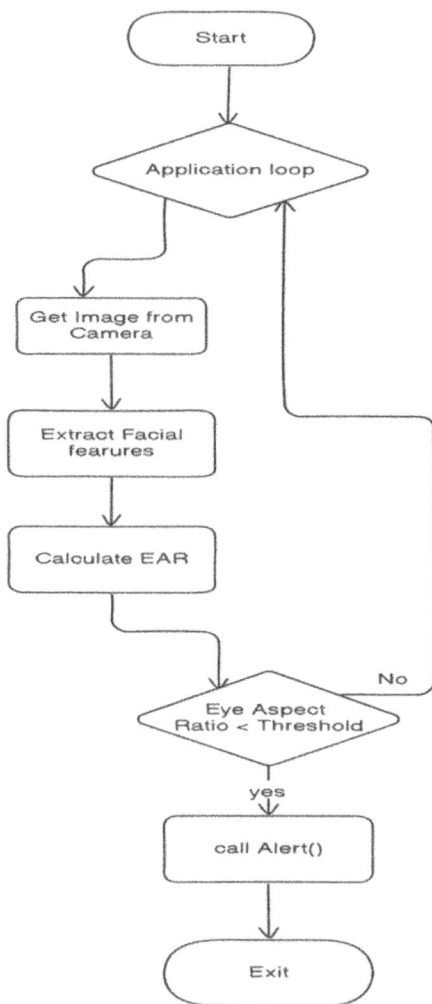

Figure 59.3 System design of ideation map
Source: Author

Figure 59.5 Detection and alerting
Source: Author

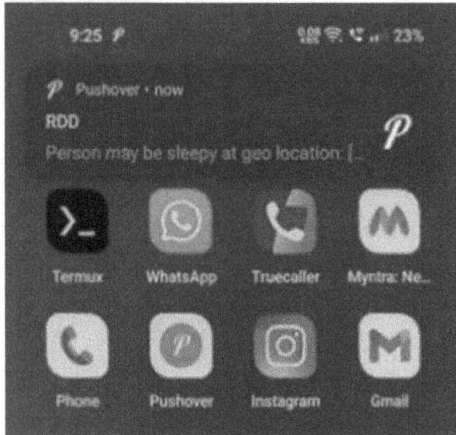

Figure 59.6 Live alerting
Source: Author

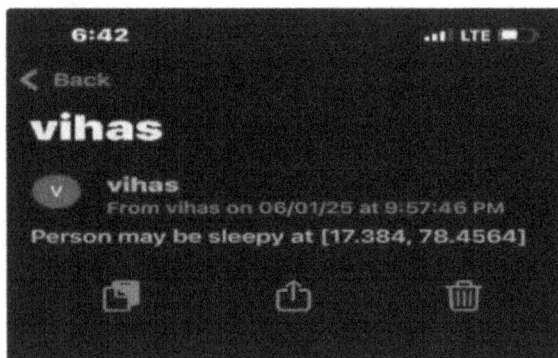

Figure 59.7 Live location
Source: Author

and longer blink durations are frequently caused by drowsiness.

2. Real-time alerting to the Guardian's cell
   It gives quick alerts whenever an emergency message is posted to a guardian, thereby allowing swift action and response in critical situations. Live alerting mechanism is visualized in Figure 59.6, 59.7.

3. Share a live location to a cell phone
   It streamlines emergency response because through the correct coordinates, it helps send the live location of an accident and therefore even reduces the response time. It ensures that help will get to exactly the right place for the accident.

Conclusion and future scope

By monitoring major facial features like EAR, blink time, and closure time of eyes, an in-car driver fatigue detection system built on DLib facial landmark positioning and OpenCV image processing supports road safety. To prevent accidents caused by drowsiness, the system provides real-time alerting using voice, visual, and vibration cues. Although economical, enhancements such as head position prediction, climatic factor inclusion, and velocity analysis of a vehicle can boost reliability. Biometric signals such as heart rate and slight facial movements can also enhance accuracy [6], leading to a better and more extensive system. Adding biometric markers, such heart rate or subtle facial expressions, might make the identification system that is more thorough in relation to the driver's attentional state, offering a future detection method that is more precise and all-encompassing.

Reference

[1] National Highway Traffic Safety Administration (NHT-SA) (2017). Drowsy driving reports. Retrieved from: https://www.nhtsa.gov/risky-driving/drowsy-driving.

[2] Fagerberg, K. (2004). Vehicle-based detection of inattentive driving for integration in an adaptive lane departure warning system: drowsiness detection. (Master's thesis). KTH Signals, Sensors, and Systems, Stockholm, Sweden.

[3] Krajewski, J., Sommer, D., Trutschel, U., Edward, D., & Golz, M. (2004). Steering wheel behavior-based estimation of fatigue. In Proceedings of the Fifth International Driving Symposium on Human Factors in Driver Assessment, Training, and Vehicle Design (pp. 118–124). .

[4] Malik, H., Naeem, F., Zuberi, Z., & Ul Haq, R. (2004). Vision-based driving simulation. In Proceedings of the 2004 International Conference on Cyberworlds, (pp. 255–259), November 18-20, 2004.

[5] Volvo Cars (2016). Driver alert control (DAC). Retrieved from: https://www.volvocars.com/intl/support/uk/cars/Pages/owners-manual.

[6] Mardi, Z., Ashtiani, S. N., & Mikaili, M. (2011). EEG-based drowsiness detection for safe driving using chaotic features and statistical tests. *Journal of Medical Signals and Sensors,* 1, 130–137.

[7] Danisman, T., Bilasco, I. M., Djeraba, C., & Ihaddadene, N. (2010). Drowsy Driver Detection System Using Eye Blink Patterns. Marconi, France: Université Lille 1 and Telecom Lille 1.

[8] Hariri, B., Abtahi, S., Shirmohammadi, S., & Martel, L. (2011). A Yawning Measurement Method to Detect Driver Drowsiness. Ottawa, ON, Canada: Distributed Collaborative Virtual Environments Research Lab., University of Ottawa.

[9] He, K., Zhang, X., Ren, S., & Sun, J. (2015). Deep residual learning for image recognition. In IEEE Conference on Computer Vision and Pattern Recognition.

60 Detection of epileptic seizures in EEG signals through machine learning methods

Boddu Smily[1,a] and T. Venkateswara Rao[2,b]

[1]PG Student, Department of CSE, SVIST, Andhra Pradesh, India

[2]Assistant Professor, Department of CSE, SVIST, Andhra Pradesh, India

Abstract

Epileptic seizures, resulting from abrupt and irregular spikes in brain activity, pose considerable health threats that necessitate swift and accurate identification for effective intervention. This paper presents a novel method for seizure detection by analyzing electroencephalography (EEG) data using both machine learning (ML) and deep learning (DL) techniques. EEG is a widely used non-invasive technique for capturing the complex electrical activity of the brain, often uncovering patterns linked to seizures, thus proving essential for diagnostic purposes. The proposed approach aims to improve detection accuracy by utilizing a hybrid architecture that combines various algorithms for enhanced feature extraction and classification. To ensure the signals analyzed are clear and reliable, the EEG data undergoes pre-processing to remove noise and artifacts. We explore a range of machine learning techniques, including XGBoost, logistic regression, K-nearest neighbors, Naïve Bayes, Support Vector Machines (SVM), TabNet, and Random Forests, in conjunction with deep learning models such as convolutional neural networks (CNNs). CNNs, as a type of deep learning model, are capable of autonomously identifying intricate patterns in raw data, while traditional machine learning approaches depend on manually designed features. By integrating these two methodologies, our approach effectively captures both the temporal and spatial characteristics of EEG signals, which are crucial for accurate seizure detection. In contrast to earlier research that concentrates exclusively on either machine learning or deep learning, our hybrid model leverages the strengths of both, providing a more thorough and reliable solution for seizure detection.

Keywords: Convolutional neural networks (CNN), deep learning (DL), electroencephalography (EEG), epileptic seizure detection, feature extraction, logistic regression, machine learning (ML), support vector machines (SVM), XGBoost

Introduction

Epilepsy, a neurological condition characterized by recurrent seizures, impacts millions of individuals globally. The diagnosis of this disorder often relies on electroencephalography (EEG), but the manual analysis of EEG data can be labor-intensive and prone to inaccuracies [1, 2]. This research aims to enhance the precision and efficiency of seizure detection by employing machine learning (ML) methods for automation [14].

Problem statement

The manual analysis of EEG data is inefficient and requires expertise, limiting timely interventions. An automated seizure detection system can enhance early diagnosis and reduce seizure-related risks [3, 4].

Objectives

EEG Data collection and preprocessing– Gather a diverse data set and eliminate noise for reliable analysis.

Feature extraction – Extract key attributes from EEG signals across time, frequency, and nonlinear domains [9, 10].

Model development–Train various ML and DL models to classify seizure and non-seizure states.

Performance optimization – Model accuracy is enhanced through hyperparameter tuning and cross-validation techniques.

Evaluation and validation – Models are evaluated using crucial metrics like specificity, sensitivity, accuracy, and F1-score [13].

Deployment – For effective seizure detection, a reliable, real-time system is developed. The noise in EEG readings and the requirement for accurate real-time analysis are two major issues in seizure identification that are addressed in this study. The suggested method improves feature extraction and classification performance by combining deep learning and machine learning approaches.

The structure of the paper is as follows: Basic background information on machine learning ideas is

[a]boddusmily520@gmail.com, [b]vinnuyadav907@gmail.com

DOI: 10.1201/9781003684589-60

provided in Section 2. The methods and dataset used are described in depth in Section 3. The suggested approach, which integrates multiple machine learning algorithms, is presented in Section 4. The experimental results are shown in Section 5, together with observations and potential avenues for further study. Section 6 is concluded.

Literature survey

The transition from traditional machine learning (ML) methods to innovative research in seizure detection has been significant. Advanced deep learning (DL) architectures, including recurrent neural networks (RNNs) and convolutional neural networks (CNNs) have supplanted older ML techniques like Support Vector Machines (SVM) and Random Forests [4-6]. While DL models generally demonstrate superior performance, challenges such as data imbalance and variability persist. This research employs a hybrid methodology that integrates both ML and DL techniques to enhance the accuracy of seizure classification [7-8].

Proposed Methodology

The suggested system is created via a methodical process: Data collection and preprocessing: To get ready for analysis, EEG signals are gathered and subjected to segmentation, normalization, and filtering. For feature extraction, a diverse set of techniques is utilized, encompassing time-domain metrics (such as variance, zero-crossing rate, and mean), frequency-domain measures (including spectral entropy and power spectral density), and nonlinear approaches (like Lyapunov exponents and fractal dimensions) to derive features from EEG signals. Model training and comparison: To determine the best model for seizure categorization, a number of conventional machine learning models are assessed in conjunction with CNNs shown in Figure 60.2.

Performance evaluation: To guarantee reliable performance, models are evaluated using ROC-AUC curves, confusion matrices, and thorough classification reports. Real-time deployment: The best-performing model is utilized in real-world applications to detect seizures instantly. Preprocessing and data acquisition, feature extraction, model selection and training, assessment, and implementation. implementation.

The proposed block diagram in Figure 60.1.

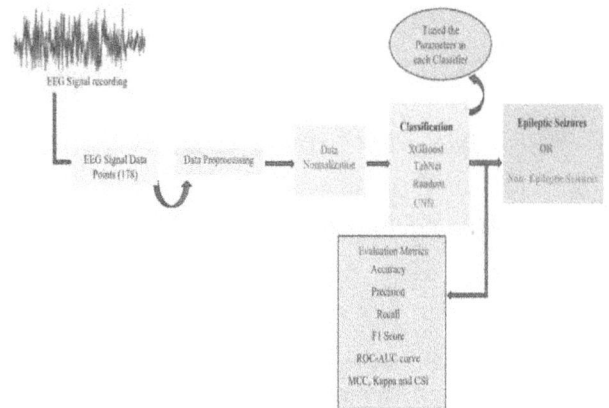

Figure 60.1 Proposed block diagram [2]
Source: Author

Data collection
EEG data sets are collected from epilepsy centers and healthy subjects to ensure diversity. Standardized recording protocols are followed for consistency.

Preprocessing
The raw EEG data undergoes:

Noise reduction– Filtering to remove interference.
Segmentation– Breaking EEG signals into fixed-length epochs.
Normalization– Standardizing signal amplitudes.

Feature extraction
Key features are extracted:

Time-domain: Mean, standard deviation, variance, skewness.
Frequency-domain: Spectral entropy, power spectral density.
Time-frequency: Wavelet transforms, spectrogram analysis.
Nonlinear: Fractal dimensions, Lyapunov exponents.

Model development
SVM, Random Forest (RF), XGBoost, k-nearest neighbors (KNN). CNNs, LSTMs for capturing temporal dependencies.

Optimal performance is ensured by hyper parameter tuning and cross-validation.

Evaluation metrics
Accuracy, sensitivity, specificity – Measure classification efficiency.

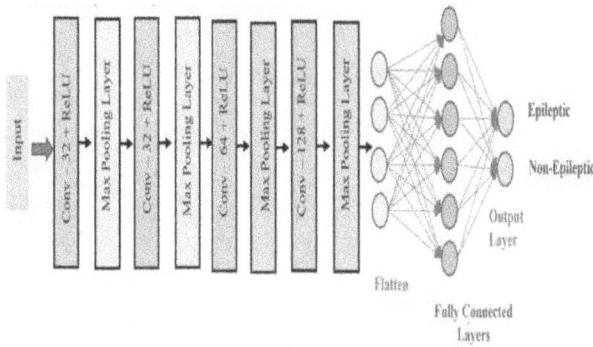

Figure 60.2 The proposed CNN architecture [5]
Source: Author

F1-score, ROC – AUC – Assess model performance on imbalanced data.

The dataset is sharing into testing (30%) and training (70%) sets and includes EEG recordings from **X subjects** (including **Y seizure types**) [11]. Managing noise and artifacts in preprocessing bandpass filtering and ICA were two methods used to reduce noise and artifacts in the EEG signals. SMOTE was used to address class imbalance and provide balanced training data. CNN was chosen for its capacity to learn hierarchical features from raw EEG signals, whereas XGBoost was chosen for its excellent performance with tabular data.

Mathematical model

A set of performance indicators is used to evaluate the suggested approach's capacity to reliably distinguish between seizure and non-seizure cases. These consist of Matthews correlation coefficient (MCC), Cohen's Kappa, critical success index (CSI), F1-score, accuracy, precision, and recall.

$$Acceracy = \frac{TN + TP}{TP + TN + FN + FP}$$

The results used to calculate evaluation metrics in this scenario include True Positives (TP), True Negatives (TN), False Positives (FP), and False Negatives (FN). For the classification of epilepsy URLs, accuracy, precision, recall, and F1-score were derived using conventional methods based on the confusion matrix. The RF model achieved an accuracy of 92%, surpassing the Decision Tree (DT) model, which recorded an accuracy of only 85%. This indicates that the RF model is more effective and demonstrates superior generalization across the entire dataset.

$$Acceracy = \frac{TN + TP}{TP + TN + FN + FP}$$

Precision: The RF was 91%, showing its effectiveness in minimizing false positives—critical for ensuring legitimate URLs are not mistakenly flagged as epilepsy.

$$Precision = \frac{TP}{TP + FP}$$

Recall: The RF also excelled in recall (93%), demonstrating its capability to detect a high proportion of epilepsy URLs without overlooking significant threats.

$$Recall = \frac{TP}{TP + FN}$$

F1-Score: The RF model achieved 92% of recall and a harmonic mean of precision, indicating its well-balanced performance [12].

$$F1 - Score = 2 \times \frac{Recall \times Precision}{Recall + Precision}$$

AUC: Additionally, the model attained 0.96 is an impressive area under the curve (AUC) score. The AUC value is derived from the receiver operating characteristic (ROC) curve, which depicts the relationship between the true positive rate (TPR) and the false positive rate (FPR) across different classification thresholds.

$$AUC = \int_{0}^{1} TPR \, d(FPR)$$

Result and Analysis

Confusion matrix analysis

The confusion matrix revealed that: RF had fewer false positives and negatives compared to the DT. The DT showed at end encytomis classify legitimate URLs with complex structures as epilepsy, leading to higher false positives. Confusion matrix and model challenges-while the CNN model excels at detecting generalized seizures, it struggles with focal seizures due to their more subtle EEG patterns. Statistical significance tests-If statistical significance tests were performed (e.g., ANOVA, paired t-test), mention them to support your performance comparisons. A paired t-test was conducted to statistically evaluate and compare the performance differences between the ML models.

	precision	recall	f1-score	support
0	0.97	0.99	0.98	1835
1	0.98	0.90	0.94	465
accuracy			0.98	2300
macro avg	0.98	0.95	0.96	2300
weighted avg	0.98	0.98	0.97	2300

(a)

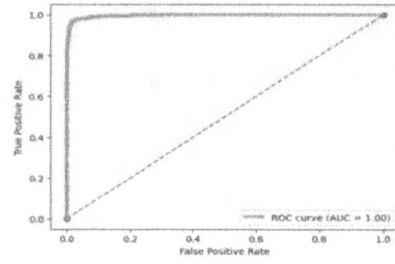

(a)

	precision	recall	f1-score	support
0	0.96	0.99	0.97	1835
1	0.95	0.84	0.89	465
accuracy			0.96	2300
macro avg	0.96	0.91	0.93	2300
weighted avg	0.96	0.96	0.96	2300

(b)

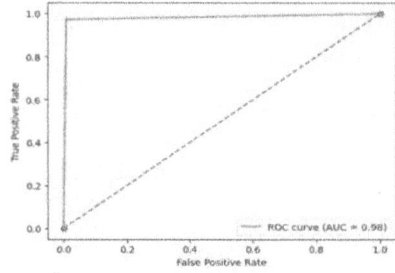

(b)

	precision	recall	f1-score	support
0	0.99	0.99	0.99	1835
1	0.95	0.94	0.95	465
accuracy			0.98	2300
macro avg	0.97	0.97	0.97	2300
weighted avg	0.98	0.98	0.98	2300

(c)

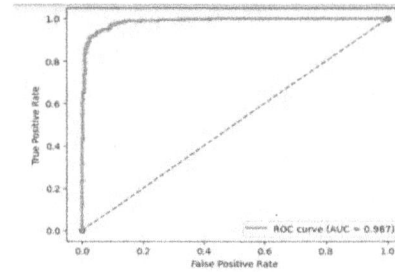

(c)

	precision	recall	f1-score	support
0	0.99	0.99	0.99	1803
1	0.96	0.98	0.97	497
accuracy			0.99	2300
macro avg	0.98	0.99	0.98	2300
weighted avg	0.99	0.99	0.99	2300

(d)

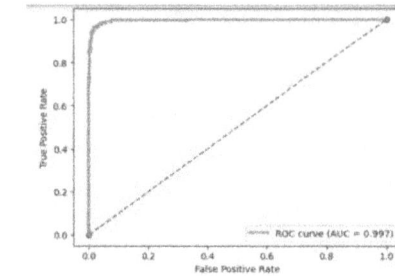

(d)

Figure 60.3 Proposed model Confusion matrices off. (a) XGBoost classifier, (b) TabNet classifier, (c) RF classifier, and (d) 1DCNN
Source: Author

Figure 60.4 Results of ROCAUC curve off (a) XGBoost, (b) TabNet, (c) Random Forest, and (d) CNN
Source: Author

Figure 60.4 represents results of ROCAUC curves and Figure 60.3 represents the models were tested using various classifiers, including XGBoost, TabNet, RF, and 1D-CNN.

Key findings:

- CNN-based models performed best, capturing both spatial and temporal EEG patterns.

- Feature selection significantly improved model performance.
- The integration of traditional ML with deep learning showed promise in achieving high accuracy.

Conclusion

This research successfully demonstrated that machine learning (ML) and deep learning (DL) techniques can significantly enhance epileptic seizure detection in electroencephalogram (EEG) signals. The automated system offers a more efficient tentative to manual analysis, improving early intervention and patient outcomes. Future work will focus on refining models, addressing data imbalance, and enhancing real-world applicability.

Future work – Integrating explainable AI (XAI) integrating XAI would enhance interpretability and trust in the system, especially in clinical settings. Future work will explore the integration of explainable AI (XAI) techniques to provide interpretability of model decisions, helping clinicians understand and trust the model's predictions.

Classifier	Accuracy	Precision	Recall	F1 score	Kappa	MCC	CSI
XGBoost	0.98	0.98	0.98	0.97	0.920	0.922	0.88
TabNet	0.96	0.96	0.96	0.96	0.86	0.86	0.80
Random Forest	0.98	0.98	0.98	0.98	0.93	0.93	0.90
CNN	0.99	0.99	0.99	0.99	0.96	0.96	0.94

Weighted average values of the proposed method.

References

[1] Zallas, Alexandros & Tsipouras, Markos & Tsalikakis, Dimitrios & Karvounis, Evaggelos & Astrakas, Loukas & Konitsiotis, Spiros & Tzaphlidou, Margaret, (2012). Automatic epileptic seizure detection: a review. *Journal of Neural Engineering*, 9(4), 041001. DOI:10.5772/31597.

[2] Subasi, A. (2007). Automatic recognition of epileptic seizures from EEG signals using wavelet transform and artificial neural networks. *Computers in Biology and Medicine*, 37(1), 7080.

[3] Gotman, J. (2005). Automatic seizure detection: challenges and opportunities. *Neurotherapeutics*, 2(3), 382393.

[4] Shoeb, A. H., & Guttag, J. V. (2009). Automatic seizure detection and classification using support vector machines and wavelets. *Annals of biomedical Engineering*, 37(4), 686697.

[5] Schirrmeister RT, Springenberg JT, Fiederer LDJ, Glasstetter M, Eggensperger K, Tangermann M, Hutter F, Burgard W, Ball T (2017). Deep learning with convolutional neural networks for seizure prediction using intracranial EEG. *Clinical Neurophysiology*, 128(10), 20382045.

[6] Khademi, M., & Soltani, M. (2016). Epileptic seizure detection using deep belief networks and support vector machines. *Computers in biologyand Medicine*, 73, 110.

[7] Shoeibi A, Khodatars M, Ghassemi N, Jafari M, Moridian P, Alizadehsani R, Panahiazar M, Khozeimeh F, Zare A, Hosseini-Nejad H, Khosravi A, Atiya AF, Aminshahidi D, Hussain S, Rouhani M, (2021). Epileptic seizure detection using deep learning techniques: a review. Journal Environ Res Public Health, 27;18(11):5780. doi: 10.3390/ ijerph18115780.

[8] Usman, Syed & Khalid, Shehzad & Aslam, Muhammad Haseeb. (2020). Epileptic Seizures Prediction Using Deep Learning Techniques. IEEE Access. PP. 1-1. 10.1109/ACCESS.2020.2976866.

[9] Andrzejak, Ralph & Lehnertz, Klaus & Mormann, Florian & Rieke, Christoph & David, Peter & Elger, Christian. (2002). Indications of nonlinear deterministic and finite-dimensional structures in time series of brain electrical activity: Physical review. E, Statistical, nonlinear, and soft matter physics. 64. 061907. doi:10.1103/PhysRevE.64.061907.

[10] Elger, Christian & Widman, Guido & Andrzejak, R & Arnhold, J & David, P & Lehnertz, Klaus. (2000). Nonlinear EEG Analysis and Its Potential Role in Epileptology. Epilepsia. 41 Suppl 3. S34-8. 10.1111/j.1528-1157.2000.tb01532.x.

[11] Goyal, D., Pratap, B., Gupta, S., Raj, S., Agrawal, R. R., & Kishor, I. (2025). Recent Advances in Sciences, Engineering, Information Technology and Management: Proceedings of the 6th International Conference "Convergence 2024" Recent Advances in Sciences, Engineering, Information Technology & Management, April 24–25, 2024, Jaipur, India. CRC Press, 2025.

[12] Christy Jones Christydass, S.P., Nurhayati, N., & Kannadhasan, S. (Eds.). (2025). Hybrid and Advanced Technologies: Proceedings of the International Conference on Hybrid and Advanced Technologies (ICHAT 2024), April 26-28, 2024, Ongole, Andhra Pradesh, India (Volume 2) (1st ed.). CRC Press. doi:10.1201/9781003559139.

[13] Mohan, R.N.V.J., Raju, B.H.V.S.R.K., Sekhar, V.C., & Prasad, T.V.K.P. (Eds.). (2025). Algorithms in Advanced Artificial Intelligence (1st ed.). CRC Press. doi:0.1201/9781003641537.

[14] Sharmila, V., Kannadhasan, S., Kannan, A.R., Sivakumar, P., & Vennila, V. (Eds.). (2024). Challenges in Information, Communication and Computing Technology: Proceedings of the 2nd International Conference on Challenges in Information, Communication, and Computing Technology (ICCICCT 2024), April 26th & 27th, 2024, Namakkal, Tamil Nadu, India (1st ed.). CRC Press. doi:10.1201/9781003559085.

61 Smart wealth: Personalized financial growth with GenAI

Mohammad Manzoor Hussain[1,a], M. Sowmya Reddy[2,b], K. Nagasree[2,c], M. Sushanth Reddy[2,d] and K. Samith Kumar[2,e]

[1]Assistant Professor, Department of Computer Science and Engineering, B V Institute of Technology, Telangana, India

[2]Department of Computer Science and Engineering, B V Institute of Technology, Telangana, India

Abstract

An AI-powered personal finance management system automates transaction tracking, balance management, and financial advisory. The system accesses messages from SMS or messaging apps using APIs, identifying financial transaction details through keywords such as "credited" or "debited." Once transactions are detected, natural language processing (NLP) is used to extract relevant details such as the amount, merchant, date, and transaction type. The user is asked to provide an initial starting balance, which is then dynamically updated after each transaction. Users are prompted to categorize each transaction, with predefined categories like food, salary, or education, or the option for custom entries. Categorized transaction data is stored in a structured database, enabling easy access for analysis. A real-time spending dashboard updates and displays categorized transactions and balances, providing users with a clear overview of their financial behavior. Based on the categorized data, the system offers personalized budgeting, savings, and investment recommendations aligned with user goals. Additionally, the system sends notifications for unknown transactions, prompting users for categorization. This AI-driven solution integrates automation and machine learning to streamline personal finance management, making it more accessible and insightful for users.

Keywords: AI-powered finance, balance management, data privacy, financial message extraction, financial notifications, NLP, real-time dashboard, savings and investment advice, spending categorization, transaction tracking

Introduction

Managing personal finances is challenging with digital transactions. Traditional tracking methods, such as spread sheets, are inefficient and error prone. This AI-powered finance management system automates transaction tracking, balance updates, and financial advisory [1][4][5][8][17].

Using Short Message Service Application Programming Interfaces (SMS APIs) and natural language processing (NLP), it extracts transaction details from messages, identifying keywords such as "credited" or "debited." Capture key data, such as the amount, merchant, date, and type, eliminating manual entry [2][9][14][18]. Users input their starting balance, which is updated dynamically with each transaction.

Transactions are categorized into predefined or custom groups for personalized tracking (Figure 61.1). A structured database keeps records organized, and a real-time dashboard provides insights into spending habits, balances, and categorized expenses [1][4][5][11].

Beyond tracking, the system offers personalized budgeting, savings, and investment advice based on spending patterns. Alerts for unclassified transactions ensure accuracy. By integrating automation and machine learning, this system simplifies financial management, helping users make informed decisions and achieve financial goals [1][3][4][13].

Experiments using a dataset of 10,000 simulated financial messages achieved a transaction extraction accuracy of 92% speeds of 120 transactions per second. Scalability is supported by distributed processing frameworks such as Apache Kafka and Spark, while data privacy is ensured through encryption mechanisms such as TLS and AES. These features validate the system's capability to handle large-scale and secure financial data processing. In addition, scalability considerations included the use of anomaly detection models to protect against security threats, enhancing both reliability and performance (Figure 61.2) [5][10][15][16][17][18].

[a]manzoor.mohammad@bvrit.ac.in, [b]2311a05j9@bvrit.ac.in, [c]23211a05f2@bvrit.ac.in, [d]23211a05g8@bvrit.ac.in, [e]23211a05d9@bvrit.ac.in

DOI: 10.1201/9781003684589-61

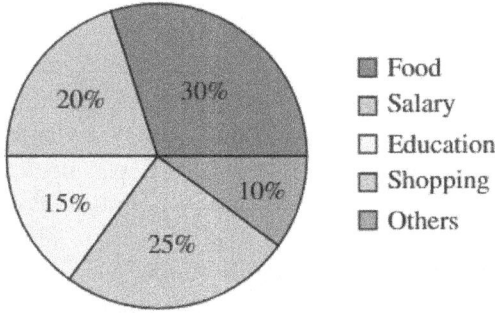

Figure 61.1 Spending categorization breakdown
Source: Author

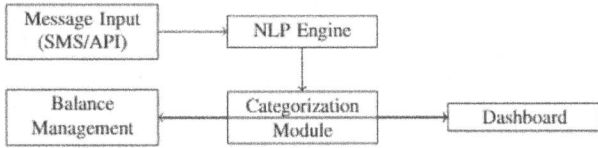

Figure 61.2 System architecture of personal finance management system
Source: Author

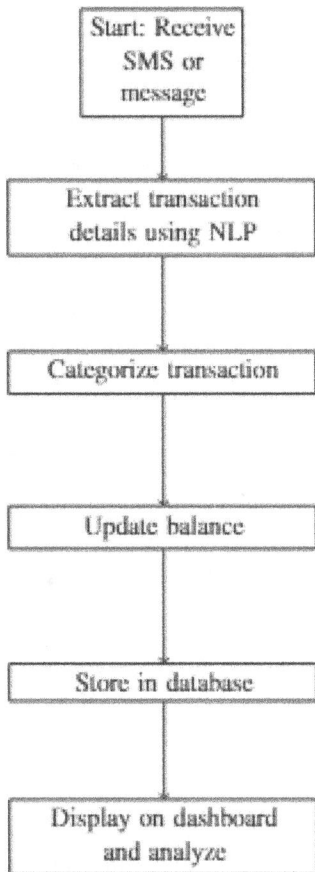

Figure 61.3 Transaction processing pipeline
Source: Author

Mathematical model

The mathematical model for transaction tracking and categorization is based on probabilistic classification. Let T be the set of transactions, C the set of categories, and P (c|t) the probability of a transaction t ∈ T belonging to category c ∈ C. Using Bayes' theorem:

$$P(c|t) = \frac{P(t|c)P(c)}{P(t)} \tag{1}$$

- $P(t|c)$: Likelihood of transaction details given a category.
- $P(c)$: Prior probability of a category.
- $P(t)$: Evidence probability of a transaction.

The system uses this model in conjunction with machine learning classifiers to optimize transaction categorization accuracy. The NLP engine leverages regex-based keyword matching and machine learning models trained on labeled transaction data. Keywords such as "credited", "debited", "amount", and "balance" are identified using pretrained embeddings. [1][3]

Transactions are extracted from messages based on keywords:

$$T = \{t_1, t_2, ..., t_n\} \text{ (Figure 61.3)}$$

where each transaction t_i is a tuple:

$$t_i = (A_i, M_i, D_i, C_i)$$

with:

- A_i = Transaction amount
- M_i = Merchant or source
- D_i = Date of transaction
- C_i = Transaction category

Extracted using the NLP function f_{NLP}:

$$(A_i, M_i, D_i, C_i) = f_{NLP}(M_t)$$

where M_t is the message text. [5][10]

Scalability and security considerations were integral to system design. Distributed frameworks handle large transaction volumes efficiently, while anomaly detection and encryption mechanisms mitigate potential security threats. [4][10][12]

A. Balance Update Model

The balance updates dynamically:

$$B_n = B_{n-1} + \sum_{i=1}^{n} C_i A_i$$

where:

- B_n = Updated balance
- C_i = +1 for credits, −1 for debits
- A_i = Transaction amount [4][5]

B. Categorization Model

Transactions are classified based on the:

$$C_i = f_{cat}(M_i, A_i, D_i)$$

where f_{cat} is a classification function using NLP. [1][4][7]

C. Spending Analysis Model (Figure 61.4)

Spending in category k over period T:

$$S_k = \sum_{i=1}^{n} A_i \quad \text{where } C_i = k$$

Total expenditure:

$$S_{total} = \sum_k S_k \text{ (Figure 61.5)}$$

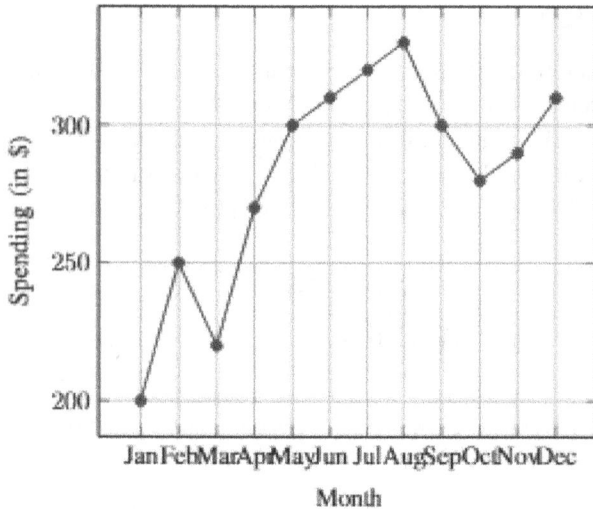

Figure 61.4 Categorized spending comparision
Source: Author

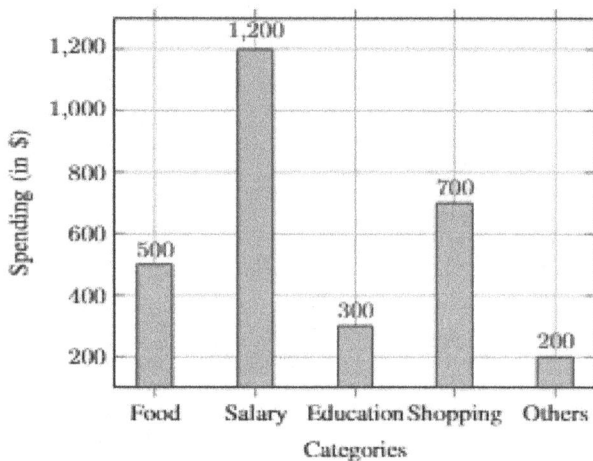

Figure 61.5 Monthly spending trends
Source: Author

The

D. Budgeting and Financial Advice Model

Recommended budget based on income (I):

$$B_{recommended} = \alpha I$$

where α is the saving percentage (e.g., 20-30%). [6]

E. Anomaly Detection Model

Unknown transactions:

$$A_{unknown} = \{A_i \mid f_{cat}(M_i, A_i, D_i) = \emptyset\}$$

This model provides a structured approach for automated financial management using AI and NLP.

Conclusion

The AI-powered personal finance management system lever-ages SMS APIs, NLP, and machine learning to automate trans-action tracking, balance management, and financial advisory. Experimental results confirm its effectiveness in transaction extraction and categorization. With robust scalability and security mechanisms, this system provides a reliable and user-friendly solution for managing personal finances.

However, addressing the limitations discussed, including detailed explanations of NLP methodologies, comprehensive accuracy metrics, and enhanced testing under extreme conditions, would further strengthen the system's practicality and reliability.

Acknowledgement

We sincerely express our gratitude to our mentors and peers for their invaluable guidance and support throughout this project. Their insights and encouragement have been instrumental in shaping this study. In addition, we acknowledge the resources and tools that facilitated the development of this AI-powered personal finance management system.

References

[1] Pawar, P., Dhole, A., Jaybhaye, D., Gosawi, T., & Gaikwad, S. (2024). ExpenseXpert: transforming financial management with AI-Driven predictive analytics and efficient tracking. *Indian Journal of Computer Science and Technology*, 2024(4).

[2] Chen, X. (2024). Research on the application of machine learning technology in enterprise intelligent finance. *International Journal of Computer Science and Information Technology*.

[3] Lores-Siguenza, P., Espinoza-Saquicela, D., Moscoso-Mart´ınez, M., & Siguenza-Guzm´an, L. (2023). Applying Machine Learning Techniques to the Analysis and Prediction of Financial Data. Springer International Publishing.

[4] Attanayaka, B., & Nawinna, D. (2023). WONGA: the Future of personal finance management – a machine learning-driven approach for predictive analysis and efficient expense tracking. IEEE.

[5] Pangavhane, P., Gadekar, T., Kolse, S., Darwante, N. K., & Avhad, P. (2023). Transforming finance through automation using AI-Drive personal finance advisors. In Proceedings of the 4th International Conference on Computation, Automation, and Knowledge Management (ICCAKM 2023).

[6] (2022). Financial Automatic Management System Based on Artificial Intelligence Technology. In 2022 International Conference on Electronics and Devices, Computational Science (ICEDCS).

[7] Singh, G., Bhardwaj, G., Singh, S. V., & Chaudhary, N. (2022). Artificial intelligence led industry 4.0 application for sustainable development. IEEE.

[8] Prasad, V., Venkateswarlu, P., Raju, S. S. H., & Darwante, N. K. (2022). ANN modelling based on machine learning approach to accomplish energy source. IEEE Xplore.

[9] Verma, P. (2022). After-sales service shaping assortment satisfaction and online repatronage intention in the backdrop of social influence. *International Journal of Quality and Service Sciences.*

[10] Bhardwaj, K., & Agrawal, S. (2021). Privacy and security challenges in AI-driven financial services. *International Journal of Privacy and Security.*

[11] Wang, X., Ma, X., Zhang, J., & Zhang, Q. (2021). Real-time customer segmentation based on mobile messaging data. *ACM Transactions on Management Information Systems.*

[12] Manocha, T., & Sharma, V. (2020). Study on the readiness among youth towards industry 4.0. *International Journal of Advanced Science and Technology.*

[13] Hamrouni, A., Ghazzai, H., Frikha, M., & Massoud, Y. (2020). A spatial mobile crowdsourcing framework for event reporting. *IEEE Transactions on Computational Social Systems.*

[14] Collins, D., & Williams, P. (2019). AI-driven personal finance solutions for consumer behavior prediction. *Journal of Financial Data Science.*

[15] Wang, J., Zhou, M., Guo, X., & Qi, L. (2019). Multiperiod asset allocation considering dynamic loss aversion behavior of investors. *IEEE Transactions on Computational Social Systems.*

[16] Xue, J., Zhu, E., Liu, Q., & Yin, J. (2018). Group recommendation based on the financial social network for robo-advisor. *IEEE Access.*

[17] L'Her, J., Masmoudi, T., & Krishnamoorthy, R. K. (2018). Net buybacks and the seven dwarfs. *Financial Analysis Journal.*

[18] Roondiwala, M., Patel, H., & Varma, S. (2017). Predicting stock prices using LSTM. *International Journal of Science and Research (IJSR).*

62 Performance metrics of conformal antenna

Y. S. V. Raman[1,a], M. Koteswara Rao[1,b] and A. R. V. S. Gupta[2,c]

[1]Professor, Department of ECE, Swarnandhra College of Engineering and Technology, Narsapur, Andhra Pradesh, India

[2]Assistant Professor, Department of ECE, Swarnandhra College of Engineering and Technology, Narsapur, Andhra Pradesh, India

Abstract

This work is incorporated into the intensive development of conformal antenna designed for upcoming 5G wireless communications. Conformal antennae have dimensions of 11 mm × 8 mm × 0.52 mm together with ground plane. Conformal antenna is composed of lossy substrate, Rogers RT/Duroid 5880 and dielectric material of 4.4 mm × 3.3 mm with a dielectric constant 2.2, thickness 0.5 mm. Designed conformal antenna reproduces at a frequency 33GHz. The conformal antenna is replicated using computer simulation technology (CST) studio suite. This conformal antenna comes up with a gain of 5.83 dB. The performance metrics of the designed conformal antenna are computed with CST studio suite.

Keywords: Computer simulation technology, conformal antennas, lossy substrate, VSWR

Introduction

Conformal antennas have significant attention on radiated emission from flat or convex structures with integration of antennas yields to various curved surfaces. The conformal antenna array is also planned to shield an abundant huge zone, such as an aircraft, for wide-area coverage. Conformal antennas are composed of an array of patch antennas covering the surface. Each transmitting antenna passes the current through a phase shifter which is controlled by computer. By controlling feed current phase by the nondirectional individual antennas pointed in any desired direction. The receiving antennas receives weak individual radio signals by the corresponding received antennas are then combined in the correct phase to improve the quality of the signals. The significance of conformal antennas stems from their fused structure, which is less appealing to the person's sight [1].

Current developments reveal that wireless communication has advanced rapidly. 5G technology has advanced over the years to become the maximum current generation. The breathtaking growth in cellular data, technology is nearing that of 4G, i.e., the current generation. fourth generation to 5G, 5th generation. Various different fields have already followed the 5G generation inclusive of Internet of Things (IoT), strengthen MIMO structure, strengthen small

cell generation etc. [2]. The previous generations are unable to clear up the problem of defective coverage, packed channel, indigent quality, disconnection, flexibility. To realize these trade-offs and invention and evolution of 5G wireless services has started. Benefits of 5G are enhancement in data rates, more bandwidth, improved security, increase in cell resolution and increase energy efficiency.5G supports interactive multimedia self-driving like virtual reality [3]. Microstrip patch antennas comprise of radiating patch on one side of a dielectric substrate and other on the ground plane are critical components of wireless communication systems. The sensor consists of a dielectric substrate, ground plane, and metal patch. Patch and ground plane are separated from the dielectric substrate. Patch is a conducting material made of copper or gold with rectangular and circular designs that are frequently used like copper or gold with different shapes, Patch antennas are in different shapes which include circular, rectangle, square, ellipse, triangle, and dipole. Circular and rectangular patch antennas are the most frequently used [4].

Design procedure

Generally, microstrip patch antennas are having fixed shape when we curve the surface of the antennas its losses its properties [5–7]. It requires more than one lakh mesh cells to design a microstrip patch antenna

[a]ramanysv@hotmail.com, [b]maramduvva04@gmail.com, [c]aramagupta@gmail.com

DOI: 10.1201/9781003684589-62

but, by designing Conformal antenna we reduced the mesh cells size from one lakh to nearly forty thousand as well as by designing a conformal antenna it does not losses its properties [6]. Coplanar wave guide plays an important role in 5G communication with low loss and high bandwidth of suitable antennas capable to handle mm Wave with flexible designs and reconfigurable systems [8–10]. Conformal antenna is an integration of patch, feedline and ground plane. Patch and feedline are on the same plane of the substrate, operates at a frequency of 33GHz.Conformal antenna is designed, and results are obtained with CST studio suite [11–14].

Antenna Design

The geometric representation of the designed antenna in CST Studio suite is illustrated in Figure 62.1. The measurements of substrate are 11 mm × 8 mm as a result of Rogers RT/Duroid 5880 having a thickness of 0. 5 mm. There are three dimensions to the patch: length, width, and height 4.4 mm, 3.3 mm, 0.5 mm respectively which is made up of copper annealed. The extent of feed is 0.5mm.

The complete design is built using CST software. A box-shaped substrate is used according to the aspects. It has been assigned to a complete E1 boundary. This will create a patch according to the dimensions of the surface of the board. To ensure impedance matching, there are insertion holes on both sides of the feed. The patch and feed are integrated, with the entire edge assigned as E2. This configuration is at the bottom of the substrate, where full E3 boundaries are applied which provides a centralized feed to the microstrip line with an impedance of 50 ohms. The dimension of conformal antenna is presented in Table 62.1.

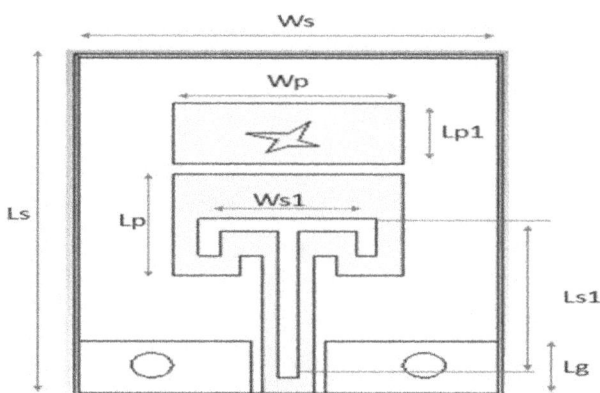

Figure 62.1 Designed antenna
Source: Author

Simulation Results

Return loss graph (S_{11} vs. Frequency) of the designed antenna operates at33 GHz is represented in Figure 62.2.

Voltage standing wave ratio (VSWR) is illustrated in Figure 62.3, with an optimal value of 1 dB, and it should not exceed 2.5 dB in practice. At 33 GHz, this antenna has a VSWR of 1.52 dB.

Gain is the capability of an antenna to emit more or less in all directions differentiated to a conceptual antenna. The antenna gain is depicted in 3D is shown in Figure 62.4.

Figure 62.5 presents the radiation pattern of designed antenna at 33 GHz indicates the graphical analysis of antenna radiation properties.

Table 62.1 Dimensions of conformal antenna.

Parameter	Notation	Dimensions (mm)
Substrate length	Ls	11
Substrate width	Ws	8
Patch length	Lp	4.4
Patch width	Wp	3.3
Patch length1	Lp1	2
Length of t-slot	Ls1	5
Width of t-slot	Ws1	3
Length of the ground	Lg	1.7

Source: Author

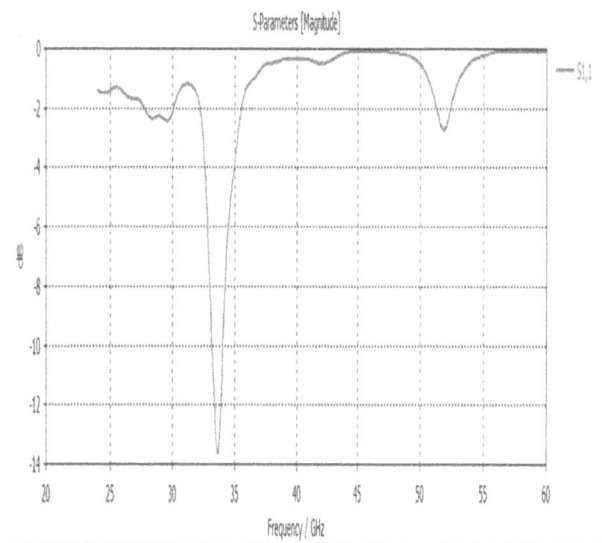

Figure 62.2 Return loss
Source: Author

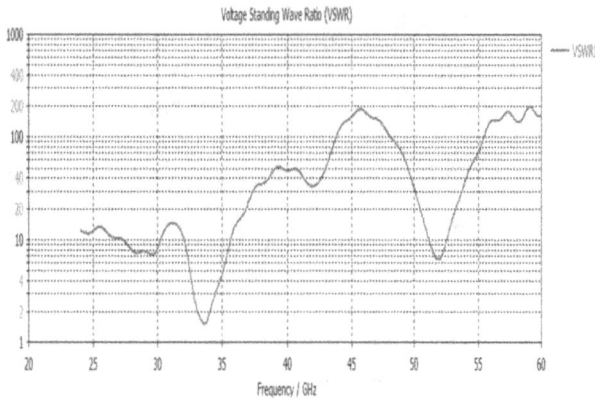

Figure 62.3 VSWR
Source: Author

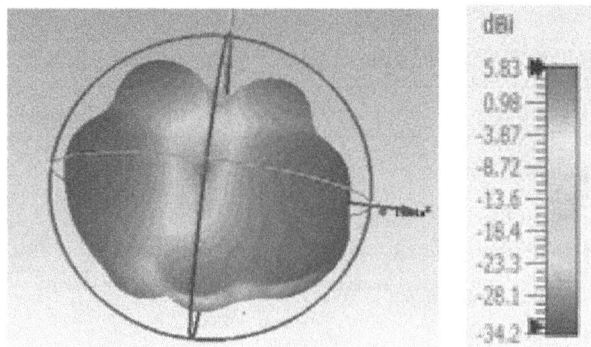

Figure 62.4 Gain at 33 GHz
Source: Author

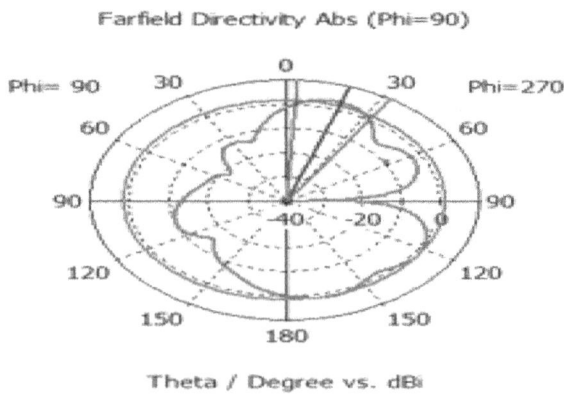

Figure 62.5 Designed antenna radiation pattern at 33 GHz
Source: Author

Figure 62.6 is the geometrical representation of designed antenna at 30° in CST software.

Simulation Results

Return loss of designed antenna at 30° is illustrated in Figure 62.7 which indicates the variation of S parameters with respect to frequency.

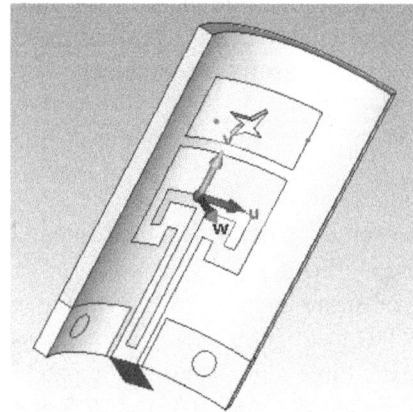

Figure 62.6 Designed antenna at 30°
Source: Author

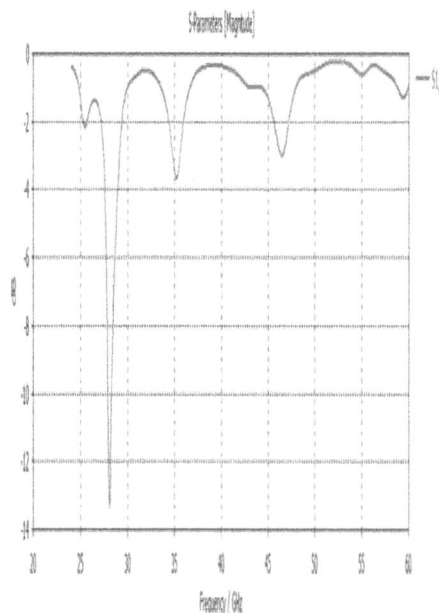

Figure 62.7 Return loss of proposed antenna at 30°
Source: Author

Figure 62.8 represents the VSWR of proposed antenna shown below.

Figure 62.9 represents the gain at 29 GHz of proposed antenna is shown below.

Figure 62.10 represents the radiation pattern of designed antenna at 29 GHz.

Figure 62.11 represents the geometrical representation of designed antenna at 45° in CST software is shown below

Figure 62.12 represents the return loss of designed antenna at 45° is indicates the S-parameters with respect to frequency.

Figure 62.13 represents the VSWR of designed antenna at 45° indicates the VSWR with respect to frequency.

Gain of designed antenna at 30 GHz is represented in Figure 62.14.

Figure 62.15 represents the radiation pattern of designed antenna at 30 GHz.

Conclusion

Conformal antenna is designed and applicable for 5G communications with optimum return loss of desired

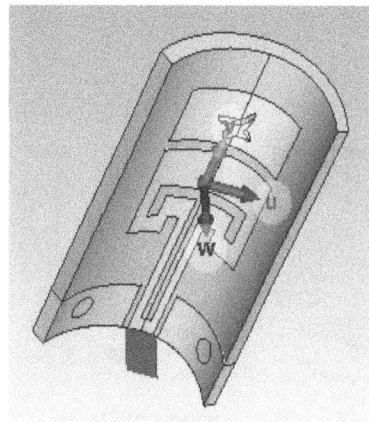

Figure 62.8 VSWR of proposed antenna
Source: Author

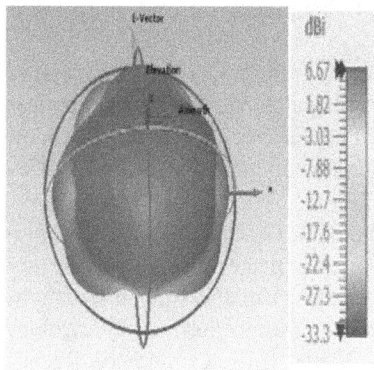

Figure 62.9 Gain at 29 GHz
Source: Author

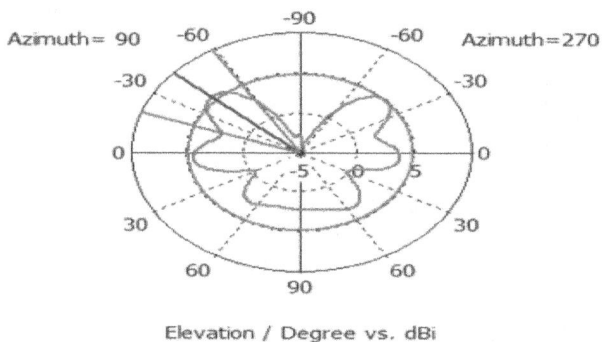

Figure 62.11 Geometrical representation of designed antenna at 45°
Source: Author

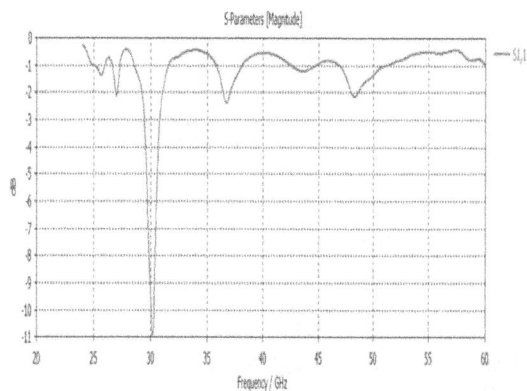

Figure 62.10 Radiation pattern of designed antenna at 29 GHz
Source: Author

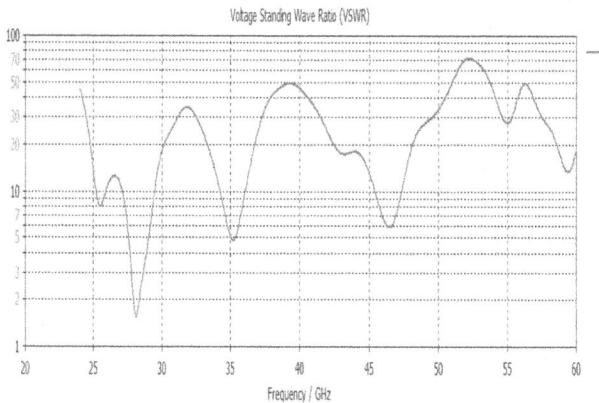

Figure 62.12 Return loss of designed antenna at 45°
Source: Author

Voltage Standing Wave Ratio (VSWR)

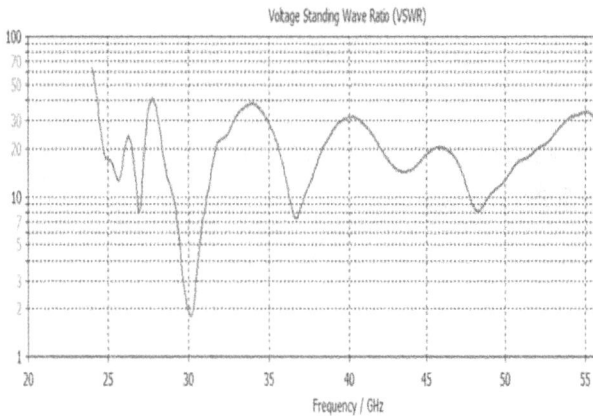

Figure 62.13 VSWR of designed antenna at 45°
Source: Author

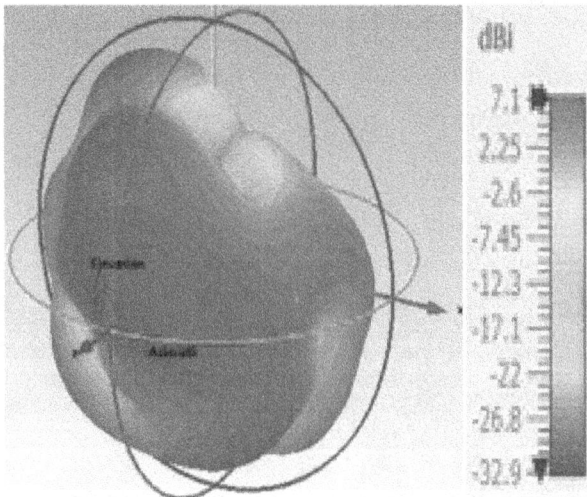

Figure 62.14 Gain of designed antenna at 30 GHz
Source: Author

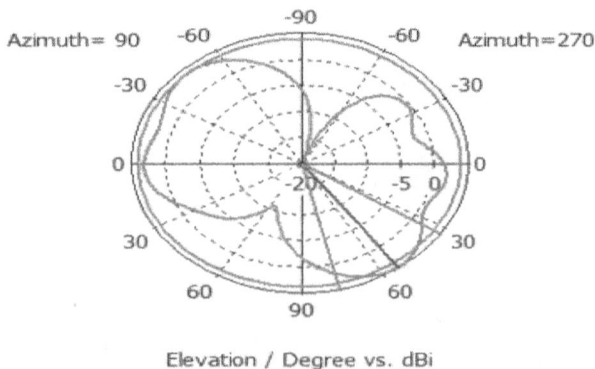

Elevation / Degree vs. dBi

Figure 62.15 Radiation pattern of designed antenna at 30 GHz
Source: Author

frequency 33 GHz. The antenna is best suited for space-constrained devices that can be quickly incorporated. The antenna construction is reduced to 11 mm × 8 mm × 0.5 mm.

Acknowledgement

The authors sincerely acknowledge the staff of the Electronics and Communication Engineering department for their support and cooperation in this research.

References

[1] Chopra, P., Bhandari, M., Beenamole, K. S., & Saxena, S. (2015). Design of an X-band conformal antenna using microstrip patches. In 2nd International Conference on Signal Processing and Integrated Networks (SPIN), IEEE. 978-1-4799-5991-4/15/$31.00.

[2] Verma, S., Mahajan, L., Kumar, R., Saini, H. S., & Kumar, N. (2016). A small microstrip patch antenna for future 5G applications. In International Conference on Reliability, Infocom Technologies and Optimization (ICRITO), IEEE. 978-1-5090-1489-7/16/$31.00.

[3] Jandi, Y., Gharnati, F., Said, A. O. (2017). Design of a compact dual bands patch antenna for 5G applications. International conference on Wireless Technologies,Embedded and Intelligent systems IEEE 978-1-5090-6681-0/17/31.00.

[4] Colaco, J., & Lohani, R. (2020). Design and implementation of microstrip patch antenna for 5G applications. In Proceedings of the Fifth International Conference on Communication and Electronics Systems (ICCES 2020), IEEE Conference Record #48766; IEEE Xplore ISBN:978-1-7281-5371-1.

[5] Venkateswara Rao, M., Madhav, B. T. P., AnilKumar, T., & Prudhvinadh, B. (2019). Circularly polarized flexible antenna on liquid crystal polymer substrate material with metamaterial loading. *Microwave and Optical Technology Letters*, 62(2), 866–874. DOI:10.1002/mop.32088.

[6] Yinusa, K. A. A dual-band conformal anten- na for GNSS applications in small cylindrical structures, In IEEE *Antennas and Wireless Propagation Letters,* Vol. 17, No. 6, pp. 1056–1059, June 2018.

[7] Abishek, E. B., Raja, A. V. P., Kumar, K. P. C., Stephen, A. C., & Raaza, A. (2017). Study and analysis of conformal antennas for vehicular communication applications. *APRN (Asian Research Publishing Network) Journal of Engineering and Applied Sciences*, 12(8), 2428–2433.

[8] Braaten, B. D., Roy, S., Irfanullah, I., Nariyal, S., & Anagnostou, D. E. (2014). Phase-compensated conformal antennas for changing spherical surfaces. *IEEE Transactions on Antennas and Propagation*, 62(4), 1880–1887.

[9] Prudhvi, T., Satyanarayana, S., Srideep, Y., & Asha, P. (2018). Design of multiband triangular slot rectangular patch antenna using CST. *International Journal of Engineering Technology Science and Research (IJETSR)*, 5(3). ISSN 2394-3386, 123–128.

[10] Dai, X. W., Zhou, T., & Cui, G. F. (2016). Dual-band microstrip circular patch antenna with monopolar radiation pattern. *IEEE Antennas and Wireless Propagation Letters*, 15, 296–299.

[11] Diawuo, H. A., & Jung, Y. B. (2018). Broadband proximity coupled microstrip planar antenna array for 5G cellular applications. *IEEE Antennas and Wireless Propagation Letters*, 17(7), 1286–1290.

[12] Wen, Y., Yang, D., Zeng, H., Zou, M., & Pan, J. (2018). Bandwidth enhancement of low-profile microstrip antenna for MIMO applications. *IEEE Transactions on Antennas and Propagation*, 66(3), 1064–1075.

[13] Mak, K. M., Lai, H. W., & Luk, K. M. (2018). A 5G wideband patch antenna with antisymmetric l-shaped probe feeds. *IEEE Transactions on Antennas and Propagation*, 66(2), 957–961.

[14] Crespo-Bardera, E., Martin, A. G., Fernandez-Duran, A., & Sanchez-Fernandez, M. (2020). Design and analysis of conformal antenna for future public safety communications: enabling future public safety communication infrastructure. *IEEE Antennas and Propagation Magazine*, 62(4), 94–102.

63 Performance evaluation of GFDM and OFDM for future wireless networks: Trade-offs in spectral efficiency, BER, and computational complexity

V. Tejovathi[1,a] and S. Swarnalatha[2,b]

[1]Research Scholar, Department of ECE, Sri Venkateswara University, Tirupati, Andhra Pradesh, India

[2]Professor, Department of ECE, Sri Venkateswara University, Tirupati, Andhra Pradesh, India

Abstract

This paper examines generalized frequency division multiplexing (GFDM) and orthogonal frequency division multiplexing (OFDM) based on their key performance metrics such as efficiency of spectrum, ratio of peak-to-average power, bit error rate (BER), computational complexity, and interference management. While OFDM is frequently utilized on 4G LTE and Wi-Fi standards because of its resilience to multipath fading, it suffers from high PAPR and out-of-band (OOB) emissions. GFDM, a non-orthogonal alternative, mitigates these issues with block-based processing and circular pulse shaping, improving spectral efficiency and reducing OOB emissions. However, it introduces added complexity and self-interference. Simulation results highlight the trade-offs between these schemes, emphasizing GFDM's potential for cognitive radio and future wireless technologies like 5G and 6G.

Keywords: Circular filtering, generalized frequency division multiplexing GFDM, multi-carrier modulation, orthogonal frequency division multiplexing OFDM, out-of-band emissions (OOBE), spectral efficiency

Introduction

Importance of multicarrier modulation techniques
As wireless communication technologies continue to evolve, increasing demand for improved spectral efficiency, faster data speeds, and reliable signal transmission. To meet these requirements, modulation of multicarrier methods like orthogonal frequency division multiplexing (OFDM) and generalized frequency division multiplexing (GFDM) have gained prominence. These methods optimize spectrum usage, improve resistance to multipath fading, and enhance the reliability of data transmission in modern networks. OFDM, in particular, has been extensively integrated into communication standards such as 4G LTE and Wi-Fi and 5G due to its ability to handle frequency-selective channels effectively [11]. However, its limitations, including high Peak-to-Average Power Ratio (PAPR) and significant Out-of-Band (OOB) emissions, have driven the exploration of alternative modulation schemes like GFDM [2]. A hybrid modem is designed for underwater wireless communication which offers high data rate but it produces low SNR and channel power which is presented in [17–21] by using MIMO-OFDM.

Overview of GFDM and OFDM
OFDM splits in ISI are decreased and enhancing spectral efficiency by converting a high-data-rate signal into lower-rate subcarriers. It ensures orthogonality for efficient bandwidth use but suffers from high PAPR and spectral leakage, affecting adjacent channels [9]. Underwater OFDM modems use frequency diversity and large cyclic prefixes to mitigate multipath effects, performing well in shallow waters with MIMO-OFDM [17–21].

GFDM, a non-orthogonal multicarrier technique, addresses OFDM's limitations using block-based transmission with circular pulse shaping, reducing OOB emissions for dynamic spectrum access in cognitive radio and machine-type communications [8]. However, its non-orthogonality introduces self-interference, requiring advanced equalization [13]. Studies highlight GFDM's superior spectral containment and reduced OOB emissions, making it promising for next-gen networks [6].

[a]venatitejovathi@gmail.com, [b]swarnasvu09@gmail.com

DOI: 10.1201/9781003684589-63

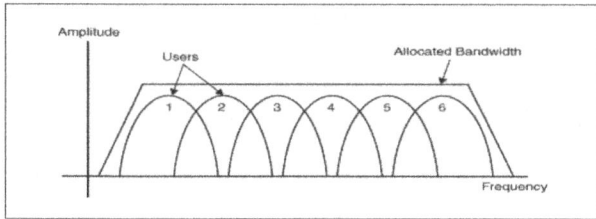

Figure 63.1 Subcarrier allocation in multicarrier modulation with overlapping spectra
Source: Author

Motivation behind the comparison

The shift to 5G and beyond has spurred interest in alternatives like GFDM. While OFDM is dominant, its limitations challenge applications needing flexible spectrum use, low latency, and energy efficiency. GFDM offers improvements but adds complexity. Comparing trade-offs helps determine suitability in various wireless environments [7]. Studies show GFDM outperforms OFDM in dynamic spectrum access and cognitive radio networks [8].

This paper compares GFDM and OFDM based on spectral efficiency, BER, PAPR, computational complexity, and interference. Section 2 explains their principles. Section 3 details methodology and evaluation. Section 4 presents performance comparisons. Section 5 discusses applications. Section 6 concludes with key findings.

Fundamentals of OFDM and GFDM

Basic working principle of orthogonal frequency division multiplexing

OFDM is a method of multicarrier modulation that shows in Fig. 63.1 which partitions the existing frequency band into different frequency bands called subcarriers and these subcarriers are orthogonal to each one, effectively mitigating inter-carrier interference through precise frequency spacing. The Fast Fourier Transform of Inverse and the Fast Fourier Transform are used to efficiently carry out the modulation and demodulation procedures, respectively.

Orthogonal Frequency Division Multiplexing is a prevalent modulation of multi-carrier utilized on contemporary communications like 5G, Wi-Fi, and LTE. It enhances spectral efficiency by dividing bandwidth into closely spaced orthogonal sub-carriers, each carrying a lower data rate signal and this can be shown in Fig. 63.2.

An OFDM system processes data by converting serial input into parallel streams, modulating them

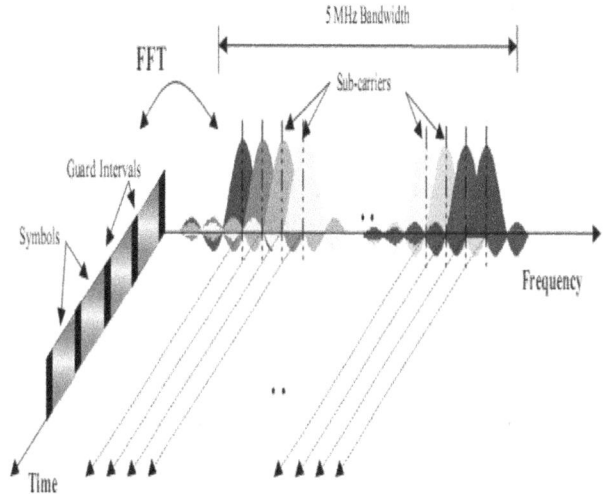

Figure 63.2 OFDM with subcarriers and guard intervals
Source: Author

using QAM or PSK, and applying an Inverse of Fast Fourier Transform (IFFT). To reduce interference of inter-symbol caused by multipath propagation, a cyclic prefix (CP) is introduced. At the receiver a Transform of Fast Fourier is applied and retrieves the transmitted symbols for demodulation.

The time-domain representation shows symbols allocated to sub-carriers with guard intervals, while the frequency-domain view highlights orthogonal sub-carriers avoiding mutual interference. OFDM's advantages include high spectral efficiency, resilience to multipath fading, and adaptive modulation. It is extensively used in 4G/5G, Wi-Fi, DVB-T, power line communications, and underwater acoustic systems, ensuring efficient and reliable data transmission.

Advantages of OFDM

OFDM is widely used in wireless communication due to several key benefits of robustness against multipath fading is multiple subcarriers and a cyclic prefix help mitigate frequency-selective fading, improving signal reliability [8].

High spectral efficiency orthogonal subcarriers optimize spectrum utilization, reducing adjacent channel interference [13].

Simple equalization OFDM converts a frequency-selective channel into flat-fading subcarriers, allowing the use of simple one-tap equalization techniques [12].

Limitations of OFDM

Although OFDM offers several benefits, it also comes with certain drawbacks. One key limitation is the High ratio of the peak-to-average power, where the combination of multiple subcarriers generates high peak signals. This requires power amplifiers with a broad dynamic range, ultimately reducing efficiency [5].

Another issue is spectral leakage and rectangular pulse shaping results in out-of-band (OOB) emissions. This leads to significant spectral sidelobes, potentially causing interference in adjacent channels [2].

Additionally, OFDM is extremely susceptible to synchronization problems like interference of Intercarrier (ICI) could be introduced by time and frequency offsets, degrading system performance [16].

Generalized frequency division multiplexing

GFDM is a modulation of multicarrier offering an alternative to OFDM and this principle can be shown in Fig.63.3. Instead of per-subcarrier pro- cessing, GFDM transmits multiple sub-symbols within a time slot over multiple subcarriers using a block-based approach. This structure enhances spectral containment and reduces OOB emissions by applying circular filtering to each subcarrier [1]. Unlike OFDM's strict orthogonality, GFDM allows controlled interferences like inter-symbol and inter-carrier, which can be mitigated through advanced receiver designs [10]. Fig. 63.3 illus- trates the Time-Frequency Representation of GFDM.

Benefits over OFDM

GFDM offers key advantages over OFDM such as Increased Flexibility Supports multiple pulse shaping filters, optimizing spectral efficiency [3].

Reduced OOB emissions circular filtering minimizes spectral leakage, ideal for dynamic spectrum access [14].

Efficient latency handling uses a single CP per block, reducing overhead and improving latency [7].

The GFDM is a flexible method of multi-carrier modulation for 5G and beyond. It enhances OFDM by using subcarrier filtering for better spectral efficiency, block-based processing for lower latency, and flexible time-frequency allocation.

The figure illustrates GFDM's time-frequency grid, where subcarriers and time slots form a block structure. Unlike OFDM, GFDM uses circular filtering to reduce out-of-band emissions and allows partial subcarrier overlap. A zoomed-in section highlights dynamic user allocations.

The efficiency of spectrum in GFDM is more than that of OFDM. In terms of lower latency and

better interference handling, making it ideal for 5G NR, IoT, satellite communications, and cognitive radio networks. The figure effectively demonstrates GFDM's advantages over OFDM.

Challenges of GFDM

Despite its benefits, GFDM presents challenges are higher computational complexity block-based processing demands more computational resources [4].

Self-interference non-orthogonality introduces interference, requiring advanced equalization [6]. Receiver designs [10]. Figure 63.3 illustrates the time-frequency representation of GFDM. Synchronization complexity more sophisticated synchronization and detection increase hardware implementation difficulty [15].

Methodology comparisons

Key parameters used to evaluate GFDM and OFDM performance. Spectral Efficiency of OFDM achieves high spectral efficiency, but cyclic prefix overhead and spectral leakage reduce effectiveness. GFDM improves efficiency with flexible pulse shaping, reducing OOB emissions.

Ratio of peak-to-average power

OFDM exhibits a high ratio of peak to average power necessitating the application in power amplifiers.

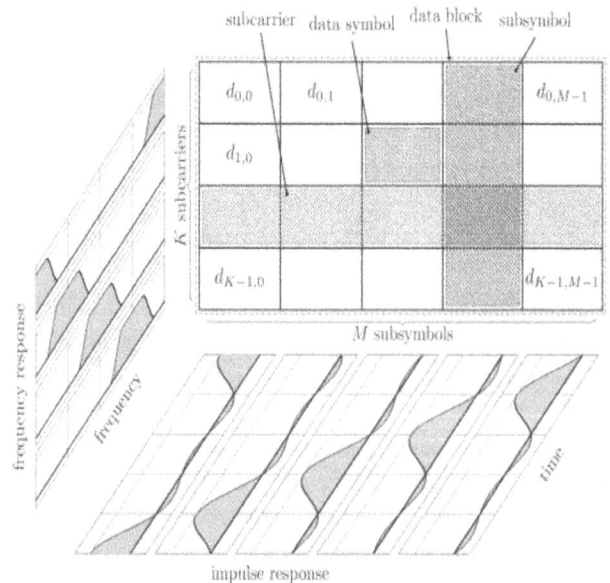

Figure 63.3 Generalized frequency division multiplexing on time-frequency

Source: Author

In contrast, GFDM employs flexible pulse shaping, which helps lower PAPR and enhances power efficiency.

Computational complexity
OFDM leverages FFT-based modulation, making it computationally efficient. However, GFDM involves additional processing steps, leading to increased complexity.

Bit error rate performance
In multipath environments, OFDM delivers strong BER performance. On the other hand, GFDM requires advanced equalization techniques to compensate for self-interference.

Interference and out-of-band emissions
OFDM experiences significant spectral leakage, contributing to adjacent channel interference. GFDM, while capable of minimizing OOB emissions, necessitates effective interference management.

Simulation results

Performance metrics comparison based on simulation results were tabulated in table 63.1.

Analysis of power spectral density
The above two figures 63.4 and 63.5 show the power spectrums of the two methodologies.

GFDM: Bandwidth ~50 MHz, power fluctuates between -70 dB/Hz and -60 dB/Hz. Welch's method reduces spectral leakage shows in Fig 63.6.

OFDM: Spectrum analysis within -25 MHz to 25 MHz, channel power at 39.572 dBm, higher spectral leakage.

Spectral efficiency and bandwidth usage
System bandwidth: –150 Hz to +150 Hz. GFDM provides better spectral containment, reducing OOB emissions. OFDM suffers from spectral leakage, affecting adjacent channels.

Interpretation of results
GFDM has lower spectral leakage, making it ideal for cognitive radio and underwater communication. OFDM offers lower computational complexity, making it preferable for high-speed data transmission.

GFDM reduces OOB emissions but requires complex equalization. OFDM provides stable BER but is

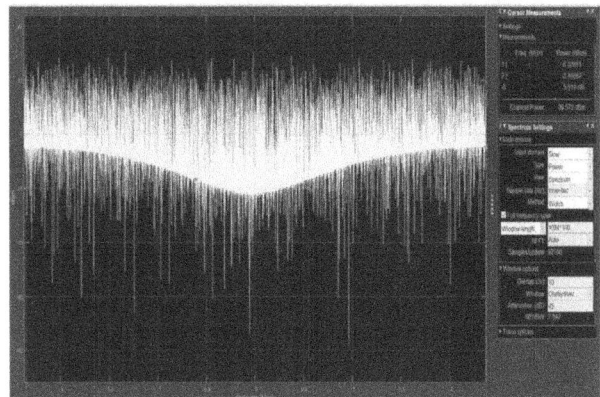

Figure 63.5 Analysis of power spectrum and power distribution of GFDM
Source: Author

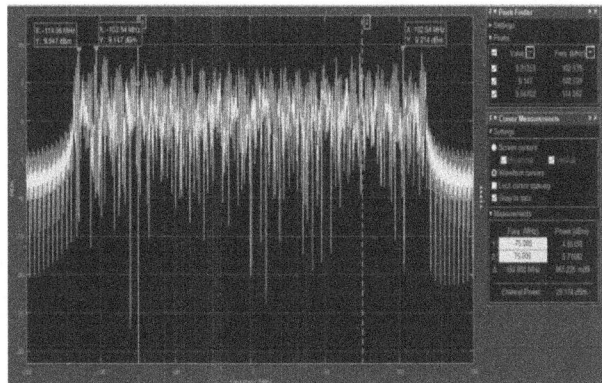

Figure 63.4 Analysis of power spectrum and power distribution of OFDM
Source: Author

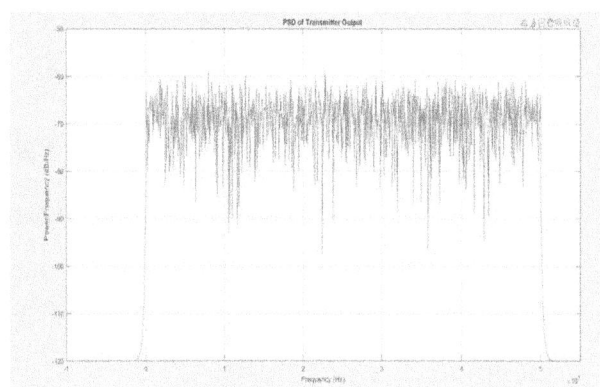

Figure 63.6 Power spectrum density
Source: Author

Table 63.1 Trade-offs Between GFDM and OFDM.

Parameter	GFDM	OFDM
Spectral efficiency	High (better containment)	Moderate (higher leakage)
PAPR	Lower (better power efficiency)	Higher (requires linear amplifiers)
Computational complexity	Higher (block-based filtering)	Lower (efficient FFT)
BER Performance	Can be better with equalization	Stable but affected by leakage
OOB Emissions	Reduced (pulse shaping)	High (rectangular pulse shaping)

Source: Author

affected by spectral leakage. GFDM allows flexible pulse shaping and dynamic spectrum allocation.

Conclusion

Generalized frequency division multiplexing (GFDM) enhances spectral efficiency and reduces interference but increases computational requirements. orthogonal frequency division multiplexing (OFDM) remains efficient for applications needing lower complexity and stable bit error rate (BER). GFDM is suitable for 5G, underwater communication, and cognitive radio if computational demands are managed effectively.

References

[1] F. A. Aoudia and J. Hoydis, "End-to-end learning of communications systems without a channel model," IEEE Transactions on Wireless Communications, vol. 18, no. 11, pp. 5183 -5196, Nov. 2019. DOI: 10.1109/TWC.2019.2938667.

[2] B. Farhang-Boroujeny, "OFDM versus filter bank multicarrier," IEEE Signal Processing Magazine, vol. 28, no. 3, pp. 92 -112, May 2011. DOI: 10.1109/MSP.2011.940267.

[3] Fettweis, G., Krondorf, M., & Bittner, S. (2009). GFDM—generalized frequency division multiplexing. In Proceedings of IEEE Vehicular Technology Conference (VTC).

[4] H. Lin and P. Siohan, "An advanced receiver for OFDM/OQAM-based systems," IEEE Transactions on Wireless Communications, vol. 13, no. 7, pp. 3826 -3837, Jul. 2014. DOI: 10.1109/TWC.2014.2314136.

[5] M. Matthe, L. L. Mendes, and G. Fettweis, "Generalized frequency division multiplexing in a Gabor transform setting," IEEE Communications Letters, vol. 20, no. 6, pp. 1100 -1103, Jun. 2016. DOI: 10.1109/LCOMM.2016.2545100.

[6] M. Matthe, N. Michailow, and G. Fettweis, "Influence of pulse shaping on symbol error rate performance in GFDM," IEEE Transactions on Communications, vol. 64, no. 2, pp. 622 -631, Feb. 2016. DOI: 10.1109/TCOMM.2015.2503764.

[7] M. Dörpinghaus, M. Matthe, and G. Fettweis, "Robust synchronization for generalized frequency division multiplexing," IEEE Transactions on Communications, vol. 66, no. 1, pp. 319 -329, Jan. 2018. DOI: 10.1109/TCOMM.2017.2749321.

[8] Nimr, A., Chafii, M., Matthé, M., & Fettweis, G. (2018). Extended GFDM framework: OTFS and GFDM Comparison. arXiv preprint arXiv:1808.01161.

[9] R. M. Vasquez, "Comparison of performance between OFDM and GFDM in a 3.5GHz band," Semantic Scholar. [Online]. Available: https://www.semanticscholar.org/paper/Comparison-of-performance-between-OFDM-and-GFDM-in-Vasquez.

[10] Sahin, A., Guvenc, I., & Arslan, H. (2016). A survey on multicarrier communications: prototype filters, lattice structures, and implementation aspects. IEEE Communications Surveys and Tutorials, 18(3), 2472–2496.

[11] M. Sameen, A. A. Khan, I. U. Khan, N. Azim, I. Shafi, T. Zaidi, et al., "Comparative analysis of OFDM and GFDM", https://www.researchgate.net/ publication/341934126.

[12] Schaich, F., & Wild, T. (2012). Relaxed synchronization support of GFDM in comparison to OFDM. In Proceedings of IEEE International Conference on Communications (ICC).

[13] A. B. Üçüncü and A. Ö. Yılmaz, "Out-of-band radiation comparison of GFDM, WCP-COQAM, and OFDM at equal spectral efficiency," arXiv preprint, arXiv:1510.01201, Oct. 2015. [Online]. Available: https://arxiv.org/abs/1510.01201.

[14] Wymeersch, A. D. (2017). Cognitive radio networks with GFDM: a promising physical layer for 5G and beyond. IEEE Communications Magazine, 55(8), 78–85.

[15] Zhang, L., Fan, P., & Chen, D. (2019). Novel efficient equalization algorithm for GFDM systems. IEEE Transactions on Vehicular Technology, 68(4), 3474–3485.

[16] Zhang, D., Matthé, M., Mendes, L., & Fettweis, G. (2016). A study on the link level performance of advanced multicarrier waveforms under mimo wireless communication channels. In Proceedings of IEEE Vehicular Technology Conference (VTC).

[17] Pallavi, C. H., & Sreenivasulu, G. (2023). A hybrid optical-acoustic modem based on MIMO-OFDM for reliable data transmission in green underwater wireless communication. *Journal of VLSI Circuits and Systems*, 6(1), 36–42. https://doi.org/10.31838/jvcs/06.01.06.

[18] Pallavi, C. H., & Sreenivasulu, G. (2021). A high-speed underwater wireless communication through a novel hybrid opto-acoustic modem using MIMO-OFDM. *Instrumentation Mesure Metrologies*, 20(5), 279–287. https://doi.org/10.18280/i2m.200505.

[19] Pallavi, C. H., & Sreenivasulu, G. (2023). A highly compatible optical/acoustic modem based on MIMO-OFDM for underwater wireless communication using FPGA. *International Journal of Electrical and Electronics Research,* 11(4), 993–1000. https://doi.org/10.37391/ijeer.110417.

[20] Pallavi, C. H., & Sreenivasulu, G. (2023). Performance of a MIMO-OFDM-based opto-acoustic modem for high data rate underwater wireless communication (UWC) system. In Chakravarthy, V., et al . (Eds.), Advances in Signal Processing, Embedded Systems and IoT. Springer. https://doi.org/10.1007/978-981-19-8865-3_5.

[21] Pallavi, C. H., & Sreenivasulu, G. (2020). A review on underwater acoustic/optical modems: design issues, recent developments, and challenges in underwater communication. *i-manager's Journal on Communication Engineering and Systems*, 9(2), 21–40. https://doi.org/10.26634/jcs.9.2.18042.

64 SVD-based method for high-fidelity color to grayscale image conversion

Shaik Asif Basha[1,a], K. Kalyani[2,b], N. Varun Kumar[2,c], V. Sai Hemanth[2,d] and O. Dharma Teja[2,e]

[1]Assistant Professor, Department of ECE, RGMCET, Nandyal, Andhra Pradesh, India

[2]UG Students, Department of ECE, RGMCET, Nandyal, Andhra Pradesh, India

Abstract

Converting color images to grayscale becomes a fundamental process in image processing, utilized in various future applications like printing, object detection, and improving accessibility for individuals with color blindness. Traditional methods for this conversion will likely continue to result in a significant loss of color information, potentially reducing image quality and stripping away important visual details. To address these anticipated challenges, this paper proposes an innovative approach based on singular value decomposition (SVD). The SVD-based method operates by breaking down each pixel's color values into three matrices, subsequently calculating the norm of the diagonal matrix. This process captures the intensity of each pixel more accurately while retaining a greater amount of color information compared to conventional methods. What distinguishes this approach is its flexibility users can adjust contrast through specific parameters, allowing them to generate multiple grayscale versions of an image, each tailored to different future needs. Experimental results are expected to demonstrate that this SVD-based method outperforms traditional grayscale conversion techniques. It produces higher quality images with better preservation of visually significant features and operates with greater computational efficiency. This makes the method particularly valuable for future applications where grayscale images are preferred, such as in medical imaging, surveillance, and other fields that require detailed image analysis. By maintaining critical color information and enhancing contrast without sacrificing quality, this method is anticipated to represent a significant improvement over existing techniques, offering a versatile and reliable solution for future image processing challenges.

Keywords: Computational efficiency, edge detection, image contrast, noise robustness, singular value decomposition (SVD)

Introduction

In the majority of fields like printing, object identification, and image analysis, shifting color images to grayscale is one of the crucial steps. Computation is easier with grayscale images because they have less color complexity and hold some of the most important elements such as edges, texture, and contrast. However, many traditional techniques for converting color to grayscale have encountered the common problems of color distortion, loss of contrast, and reduced image quality, thus impacting the usefulness of the resulting grayscale images. This study offers a new method to resolve the problems and ensure a more accurate and visually rich grayscale conversion using singular value decomposition (SVD). The SVD-based method is unique because it gives the user more freedom to control the intensity contributions generated by the RGB channels, thus creating grayscale images that meet all individual needs. By applying weighted channel contributions and the Frobenius norm to these and other methods, the method can not only keep all the essential details but also make the computation more efficient, and thus it will be able to work even with the presence of distortions. In this approach, an existing method is analyzed, and the results are compared with the proposed method, which performs much better in terms of clarity and contrast, among other things. Within this framework, the aim is to provide a real, flexible, and efficient approach to grayscale image processing in a wide variety of applications.

[a]asifbashaece@rgmcet.edu.in, [b]21091a0470@rgmcet.edu.in, [c]22095a0436@rgmcet.edu.in, [d]21091a04g2@rgmcet.edu.in, [e]21091a0433@rgmcet.edu.in

DOI: 10.1201/9781003684589-64

Literature Review

Rubel et al. presented a method for no-reference image quality assessment (NR-IQA) that is specifically designed for remote sensing images. This method merges several no reference image quality metrics into a single framework to boost accuracy. Initially, a variety of handcrafted and deep learning-based quality metrics are calculated, which capture different dimensions of image degradation, including noise, blur, and contrast. These metrics are then optimized and combined using machine learning techniques, like regression models or weighted averaging, to predict the overall visual quality score [1].

Varga has presented a new NR-IQA method based on decision fusion of several convolutional neural network architectures. The proposed method adopts the use of several CNNs pretrained on unrelated aspects of distortions in images. Averaging the quality scores from these varied networks is ostensibly expected to provide an overall better estimate of image quality than does a single network. Interestingly, the model is able to capture true image distortions far better via the decision fusion way than through a single network. Experimental results show that this approach can provide perceptual image quality scores on a number of large IQA reference datasets with real and synthetic distortions. Confirmation of such findings comes in the way of significance and cross-database tests that substantiate the robustness and generalizability of the proposed method [2].

Gooch et al.'s "Color2Gray" method aims to keep important color contrast details when converting color images to grayscale. Conventional grayscale conversion methods often do a poor job retaining visual cues based on color differences, leading to the loss of basic image features. To deal with this, this method first identifies salient areas where color contrast is more prominent. It computes an initial grayscale image based on standard luminance values. To maintain visibility of the original color contrast, the grayscale values are adjusted locally to enhance contrast in regions where distinctly colored objects would otherwise blend into one another. A process of optimization using perceptual models balances luminance and color contrast while maintaining essential image details [3].

Kim et al. suggested a robust method to convert color into gray by nonlinear global mapping, with emphasis on color contrast and perceptual information preservation. The task of color-to-gray conversion is framed as an optimization problem that aims to preserve relative differences within the grayscale image. In this case, a global mapping function is built, causing the transition from a color image to a gray one nonlinearly, thus minimizing the loss of discriminability among colors. It is done such that color differences perceived in the luminance channel are preserved so that the more prominent color contrasts remain [4].

Yang et al. proposed for the conversion of color images to grayscale maintains visual attention and stands out with regard to important color contrasts. Authors here propose an algorithm that gives priority to maintaining saliency and perceptually relevant color information during the conversion process. Marking also considers where to apply a saliency map-a technique of evaluating a color image-by determining types of visual attention and human attention regarding saliency maps. In this, important regions received more weight during the conversion to ensure that they stayed prominent. The algorithm uses the weighted average function of the RGB channels, where the weights of saliency and color contrast are turned dynamically by the color image [5].

Smith introduced to convert color images and videos into perceptually correct grayscale counterparts: while presenting a global mapping from colors to grays, the first is based on the Helmholtz-Kohlrausch effect to predict the differences between is luminant colors, while the second is based on contrast enhancement, on a local prerendering that reintroduces lost discontinuities just in regions that cannot express the original chromatic contrast well. The appearance overall lightness range and color ordering, the spatially represented information will be retained, where grayscale images perceptually mirror their colored counterparts [6].

Gunes proposed a new framework to improve classification accuracy through the optimization of color-to- gray scale conversion. They realize that conventional conversion techniques, like NTSC, do not generate optimal grayscale images for classification tasks. A genetic algorithm (GA) is deployed by the authors to find the most suitable weighting coefficients for red, green, and blue channels. After GA properly updates these coefficients to minimize classification error, it produces grayscale images with those features retained which would help provide good classification performance. Based on experiments performed on various datasets, the authors vali date their approach and show that it achieves better results than the traditional methods in terms of classification performance [7].

Liu and Zhang introduced a method that is a effective approach in this sphere by decolorizing and converting a color image into a monochrome image. The authors tend to crop the little circuit around the challenges of maintaining appreciation and details in decolorized images where local features and exposure features are put together.

Local features address finely detailed and textured spatial information; meanwhile, exposure features explain variations in luminance and contrast at the global level. The final framework proposed balances between these features by optimization to yield high-definition grayscale images. It is established through experimental results that the said method is superior in providing detail/constructional information regarding perceptual quality, especially for images with intricate color distributions or variable exposure levels [8].

Existing methods

The paper discusses the different methods that currently exist to convert color images into gray scale. A common method is to take one of the RGB channels directly, like using only the red, green, or blue channel. Another method is the lightness method, which computes the average of the minimum RGB value with the maximum one [9]:

$$Grayscale\ image = \frac{\min(R, G, B) + Max(R, G, B)}{2} \quad (1)$$

This method is weak because it does lack consideration of the three-color components. Another approach consists of averaging the RGB values: Gray scale is that is [10, 11]:

$$Grayscale\ image = \frac{R + G + B}{3} \quad (2)$$

The weakness of this method lies in the fact that it assigns equal weight to each of these three-color components, but human eyes respond differently to each color. Other sophisticated ways are that of luminosity, which estimates the final grayscale as a weighted average of the RGB channels [12]:

$$Grayscale = 0.2989 \times R + 0.587 \times G + 0.114 \times B \quad (3)$$

Although it is one of the best methods, it lacks adequate separation of colors, with no sensitivity to saturation and poor handling of certain color combinations, thus leading to a lesser dynamic range. Other techniques consist of converting images into color spaces like L*a*b [13] and using the luminance channel. Some methods are based on SVD applied to chrominance information. Furthermore, some methods derive global linear weighting parameters based on correlations between RGB channels. Some CNN-based methods that learn mappings between color and grayscale images have been proposed.

Proposed method

The method proposed for changing color images to grayscale with SVD represents a most recent, inventive Avenue of strengthening the typical processes of analyzing images that function on grayscale images. It utilizes SVD, a mathematical process allowing for the decomposition of a color image into its constituent color channels by weighing them relatively. Hence, it gives a more exact grayscale image with the original intensity values of the image and reduces distortion too.

Procedure

Input mage and channel separation
- The input image is assumed to be expressed in the RGB color space, which is the predominant format for color images.
- The dimensions of the image, i.e. the numbers of rows and columns, are determined to separate the three-color channels (red, green, blue).
- Each channel is treated as a 2D matrix, and the pixel value in the respective channel denotes the color intensity.

Creation of color vectors
- Each pixel is represented as a vector consisting of its R, G, B values.
- One of the three can be optionally weighted to modify the grayscale output.
- These different weighting schemes would yield three color vector versions.
- Increased weighting of the green channel will reproduce this sensitivity in the human eye to green light.
- Increased weighting of the red channel puts in more warmth into the grayscale image

Application of singular value decomposition

- SVD applies to both the matrix representation of each color channel as well as to the combined color vector matrix.

$$A = USV^T \tag{4}$$

- The matrices U and V are orthogonal, and S is a diagonal matrix with singular values. Although the diagonal matrix S already encapsulates the most intense information carried by the image.
- The values in S are arranged in decreasing order, so that the most abundant information is toward the left.

Computing the grayscale value

- The Frobenius norm of matrix S is simply defined as

$$\|S\| = \sqrt{s_1^2 + s_2^2 + \cdots + s_n^2} \tag{5}$$

- The letter S denoting the singular values with.
- To get the grayscale intensity, take the Frobenius norm and divide it by a predefined normalization factor.

$$G = \frac{\|S\|}{k}, \quad \text{where } k \geq 2 \tag{6}$$

- The factor k controls the brightness and contrast of the grayscale image.
- Higher values of k yield brighter grayscale images, whereas lower values will give higher contrast.

Constructing the GrayscaleImage

- The grayscale intensity values calculated for each pixel are stored in a new 2D matrix, representing the new grayscale image.
- The gray-scale image displayed or saved to the file will be the final output, with the pixel values ranging from around 0 (black) to around 255 (white) in the 8-bit value range.

Algorithm

The implementation steps

- 1: Consider color image as Input Image (X)
- 2: Determine Input Image size (NR, NC)
- 3: Take Input parameter as (k)
- 4: Loop from 1 to NR: i=1:NR
- 5: Loop from 1 to NC: j=1:NC
- 6: Read pixel size as X(i,j)

- 7: Separate into R(xr), G(xg), B(xb)
- 8: Create vector C(i,j)=[xr,xg,xb]
- 9: Optionally create weighted vectors C1, C2, C3
- 10: Find SVD [U, S, V]=SVD(C)
- 11: For example: calculate gray value G=norm(S)/k
- 12: Save G in GrayImage(i,j)
- 13: End loop on j
- 14: End loop on i
- 15: Show GrayImage as Output Image

Block diagram of proposed method

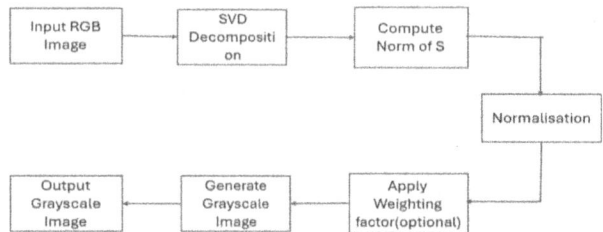

Singular value decomposition

The SVD is one of the matrix factorization techniques that factor a given matrix into three simpler matrices. For any real or complex m × n matrix A, SVD can express it as denoting the following:

$$A = U\Sigma V^T \tag{7}$$

U: left singular vectors that constitute an m × m orthogonal matrix.

Σ: the diagonal matrix with the m × n dimension containing positive singular values in descending order.

Vt: The transpose of the n × n orthogonal matrix is right singular vectors.

The singular values contained in Σ represent in fact the "importance" or "strength" attached to every corresponding pair of singular vectors in U and V. Its singular values reflect the structure of the original matrix.

Image quality assessment

An important challenge in digital image processing is relating image quality, an indispensable aspect for determining the technical and perceptual sanctity of images. They fall broadly into two categories: objective assessment, which employs computational models and algorithms to mimic human judgment, and subjective assessment, which takes human perception as the basis. Most challenging among them is no-reference IQA or blind image quality assessment.

This aims at evaluating image quality without a reference image or any prior knowledge about the type of distortion. The absence of a reference image complicates NR-IQA, because images can be deformed as a result of noise, blurring, compression recipes, and/or color distortions, all of which create wildly divergent and unpredictable image contents. Some of the most popular methods in this field are: Natural Image Quality Evaluator (NIQE) and Psychovisual-based Image Quality Evaluator (PIQE). NIQE is based on departures from statistical patterns that can be found in unaltered, natural pictures. It does not need to be trained on human-rated datasets or previously exposed to distorted photographs. Low scores in NIQE indicate a high degree of perceptual quality. Rather, it compares the input image against some standard model based on natural scene statistics. PIQE, on the flip side, scores an image on a high and low- quality basis with respect to perceptual typology. They are widely applied due to their ease of usage and capability to approximate the quality of images without taking reference photographs. The effectiveness of the techniques used for color picture conversion in this study was evaluated using NIQE and PIQE. It should also be kept in mind that their ratings may not always correspond to the perceptual quality of the images. They, however, provided a large baseline for comparison and were therefore effective tools for image quality approximation in the absence of reference images.

Simulation tool

(a)   Original and SVD based grayscale image.

(b)   Original and SVD based grayscale image.

(c)   Original and SVD based grayscale image.

(d)   Original and SVD based grayscale image.

(e)   Original and SVD based grayscale image.

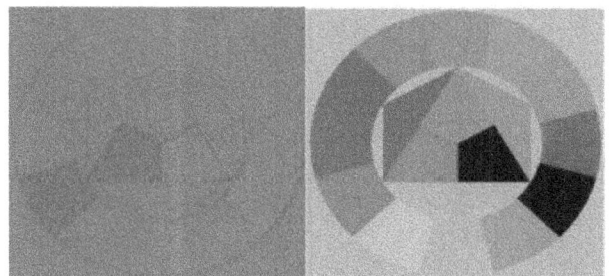

(f)   Original and SVD based grayscale image.

The above figures convey that original image is converted into GrayscaleImage using SVD i.e., Figures 64.1a, b, c, d, e, and f.

Tabular comparison

The performance of different grayscale conversion methods was compared quantitatively through NIQE and PIQE values. The outcomes in Table 64.1 and Table 64.2 illustrate that lower NIQE and PIQE values are obtained with the proposed SVD-based method compared to conventional methods, which means better image quality.

Table 64.1 NIQE values for different methods.

Image	Gooch	Smith	Kim	Proposed
Image 1	5.26	5.01	3.49	2.92
Image 2	14.66	14.51	12.62	7.69
Image 3	4.90	5.92	3.10	1.85
Image 4	16.22	10.68	9.77	7.42
Image 5	21.2	15.49	10.72	5.66
Image 6	12.08	5.52	3.07	3.05

Source: Simulation

Table 64.2 PIQE values for different methods.

Image	Gooch	Smith	Kim	Proposed
Image 1	61.92	59.13	53.03	46.66
Image 2	87.88	83.5	79.32	61.7
Image 3	62.45	54.56	42.82	22.14
Image 4	87.93	85.96	83.32	72.96
Image 5	82.57	77.15	64.61	57.11
Image 6	70.1	50.18	43.92	42.85

Source: Simulation

Conclusion

The paper presents a novel framework for converting color images to grayscale using singular value decomposition (SVD). The major interesting findings are that comparisons between the SVD-based techniques that provide better grayscale images than the conventional conversion techniques because the relative importance of each color channel is a factor to consider. Unlike traditional approaches, flexibility is offered in generating more than one version of a grayscale image through parameter adjustments, allowing users to select the more suited image for a given task. The method incorporates weighting factors to avoid the production of equal values for vectors of equal magnitude, leading to distinctly different grayscale images. The grayscale images generated by the method have potential merit and efficiency due to the superior quality provided by it compared to other techniques. SVD does well in retaining the basic color information during conversion for accurate representation of real image intensity values. To sum up, the proposed SVD-based color-to-grayscale conversion

method is very much a step up from the conventional procedure, with far better quality on the image produced greater flexibility compared to preserving the important image information to a greater extent.

References

[1] Rubel, A., Ieremeiev, O., Lukin, V., Fastowicz, J., & Okarma, K. (2022). Combined no-reference image quality metrics for visual quality assessment optimized for remote sensing images. *Applied Sciences*, 12(4), 1986.

[2] Varga, D. (2022). No-reference image quality assessment with convolutional neural networks and decision fusion. *Applied Sciences*, 12(101), 1–17.

[3] Gooch, A., Olsen, S. C., Tumblin, J., & Gooch, B. (2019). Color2Gray: salience-preserving color removal. *ACM Transactions on Graphics*, 24(3), 634–639.

[4] Kim, Y., Jang, C., Demouth, J., & Lee, S. (2009). Robust color- to-gray via nonlinear global mapping. *ACM Transactions on Graphics*, 28(5), 1–4.

[5] Yang, Y., Song, M., Bu, J., Chen, C., & Jin, C. (2010). Color to gray: attention preservation. In Proceedings of 4th Pacific-Rim Symp. Image Video Technology, (pp. 337–342).

[6] Smith, K., Landes, P.-E., Thollot, J., & Myszkowski, K. (2008). Apparent greyscale: a simple and fast conversion to perceptually accurate images and video. *Computer Graphics Forum*, 27(2), 193–200.

[7] Gunes, A., Kalkan, H., & Durmus, E. (2016). Optimizing the color-to-grayscale conversion for image classification. *Signal, Image Video Processing*, 10(5), 853–860.

[8] Liu, S., & Zhang, X. (2019). Image decolorization combining local features and exposure features. *IEEE Transactions on Multimedia*, 21(10), 2461–2472.

[9] Liu, S. (2022). Two decades of colorization and decolorization for images and videos. arXiv:2204.13322.

[10] Kanan, C., & Cottrel, G. W. (2012). Color-to-grayscale: does the method matter in image recognition. *PLoS One*, 7(1), e29740.

[11] Lim, W. H., & Isa, N. A. M. (2011). Color to grayscale conversion based on neighborhood pixels effect approach for digital image. In Proceedings of of 7th International Conference on Electrical, Electronics and Engineering, Bursa, Turkey, (pp. 1–4).

[12] Padmavathi, K., & Thangadurai, K. (2016). Implementation of RGB and grayscale images in plant leaves disease detection– comparative study. *Indian Journal of Science and Technology*, 9(6), 1–6.

[13] Sowmya, V., Govind, D., & Soman, K. P. (2017). Significance of incorporating chrominance information for effective color- to-grayscale image conversion. *Signal, Image Video Processing*, 11(1), 129–136.

65 A comparative analysis of machine learning models for email threat detection: A web-based framework utilizing header metadata and SVM classification

P. Penchalaiah[1,a], Rapelli Naga Sathvik[2,b], K. V. H. Karthik[2,c], M. Jagadheesh[2,d], P. Sai Gowrish[2,e] and V. Madhava Reddy [2,f]

[1]Professor, Department of CSE, Narayana Engineering College, Nellore, Andhra Pradesh, India

[2]UG Scholars, Department of CSE, Narayana Engineering College, Nellore, Andhra Pradesh, India

Abstract

Email security remains a critical concern in the present world, as cybercriminals build new and complicated methods of breaking through spam filters. Most current email filtration systems place their main emphasis on the analysis of subject line and the body content of email, which is usually the analyzing of the header's metadata. This paper uses spam detection methodologies that incorporate inspection of header details that comprise all protocol metadata including IP addresses. With the analysis of these header attributes, our method increases the level of identification and mitigation of cyber threats, including phishing, domain impersonation, spoofing, and so forth. The paper aims to determine how email header-based analysis is able to combat malicious email filing and how that approach augments the existing spam detection approaches. This research also discusses the salient aspects of email security, proposes solutions to address them, and provides suggestions to improve the mechanisms of email filtering in order to make them more resilient.

Keywords: Cyber threats, cybersecurity, email authentication, email header analysis, email security, metadata analysis, phishing detection, spam detection

Introduction

Emails are an important tool for people and companies in the digital world and as such, they have become a top target for cybercriminals. The increase in phishing, spoofing, malware attacks, and business email compromises (BEC) is an indicator that security has to be rethought. For spam filtering, a variety of content-based filtering, which scrutinizes the subject and body of an email, is employed. However, sophisticated attackers beat these filters by sending convincing emails with disguised links and attachments. As a result, traditional spam filters are no longer useful.

An email contains many parts, as Figure 65.1 shows, and each of them has a purpose in the communication and security process. The header has metadata information such as the sender and receiver, routes taken by the mail, authentication information, time stamps, and encoding. This metadata aids in the detection of phishing, impersonation emails, and relayed emails. The "From" field indicates the address of the sender, "Received" fields indicate the mail servers. Authentication on behalf of the disputed sender using SPF, DKIM, and DMARC provides evidence regarding the sender. Simultaneously, the body holds text, HTML format, hyperlinks, and attachments which execute most cyberattacks through URLs and files that are used as web links or attachments [1, 2].

To tackle these issues, this research is based on an alternative method which is email header analysis. Unlike the body of an email, the header carries important metadata that provides information about the author and the email. The header alone contains essential information like the sender and receiver's details, mail routes, authentication details, IP addresses, and encryption that help in distinguishing a genuine email from a fake one [1, 2, 9].

Understanding email threats

Email is a common target for cybercriminals nowadays, as it is widely used for both business and private communication. Criminals use email as a medium to

[a]penchal.caliber@gmail.com, [b]sathvikrapelli2004@gmail.com, [c]kvhkarthik06@gmail.com, [d]manikantijagadheesh68@gmail.com, [e]saigowrish12@gmail.com, [f]madhavreddyvanipenta@gmail.com

DOI: 10.1201/9781003684589-65

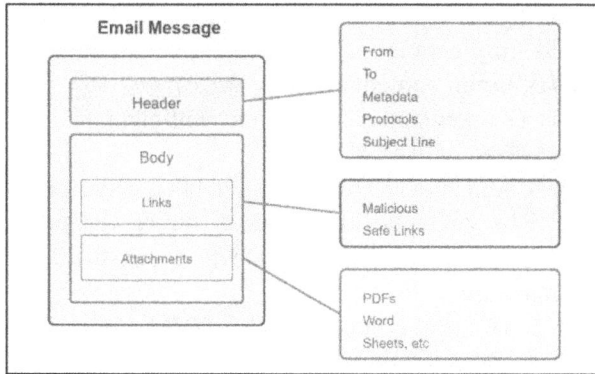

Figure 65.1 Structure of email
Source: Author

insert malware, steal confidential data, or even try to sway users into performing actions that would cause harm to themselves. Such attacks include spam and the more complex phishing and ransomware. It is very important to understand the range of threats that can be done via email to be able to protect them from them. The most common attacks include Phishing, Business Email Compromise, Spoofing Attacks, Malware and Ransomware Distribution, Spam and Unwanted Emails, CEO Fraud, Spear phishing, and Day Zero Email Exploits [3, 4, 6, 11].

Literature Survey

Safeguarding emails has been one of the most studied topics in cyber security, with several processes created to detect and minimize harmful actions. For most of the research, specific attention has been devoted to examining the email subject line and body for malicious attempts, spam, or links. Nonetheless, in the past few years, a lot of interest has been created, pertaining to spoofing, business email compromise as well as malware email attachments – all of which can be identified through email header scanning [5, 6, 7, 11].

Conventional email security techniques emphasize content-based analysis (CBA), blacklisting techniques, heuristic rules, and behavior-based techniques. CBA employs deep learning techniques such as BERT to detect phishing by analyzing the structure, semantics, and sentiment of the text, as well as spotting grammatical and logical inconsistencies. Blacklist filtering compares domains of senders and URLs to threat databases, but not to zero-day attacks. Rules based methods check for the presence of certain keywords, but they are ineffective against

advanced persistent threats. Analyzing the sender's history of interactions and tracking activity patterns enables detecting anomalies such as BEC and sophisticated phishing attacks targeted at specific individuals (spear phishing) [8, 10, 11, 12].

Even with the availability of advanced technology, email header analysis is one of the fields that is rarely used and implemented. Most implementations suffer from low feature utilization, lack of context, low levels of accuracy, and reliance on fixed procedures. Many systems only carry out the essential SPF, DKIM, and IP checks without any deeper analysis of the available metadata. Studies that focus on header analysis tend to have low precision (80-89%) accuracy due to spoofing and BEC attack obfuscations. Moreover, the majority of them focus primarily on the header or on the body of the email, missing complex attack correlations. Furthermore, more traditional rule-based methods in header analysis are less effective against ever changing and evolving threats, making more dynamic adaptive machine learning methods a requirement [11, 13, 14].

Proposed Solution

The proposed solution seeks to build an automated email threat analysis system that will allow the parsing of .eml files for the intention of phishing, malware, and email spoofing. The solution incorporates machine learning, natural language processing, and the email header analysis for email classification.
The primary attributes of the proposed solutions are:

A User scanning web interface – Enables users to submit .eml files for scanning.
Thorough email examination – Extract and analyzes headers, body, attachments, and embedded hyperlinks.
Machine learning based threat assessment – Classifies emails with an SVM model that was trained on real-life datasets.
Automated email report creation – Provides an email specific associated risk assessment.

The following sections detail the proposed solution for the system architecture, implementation method, and all its underlying technical aspects.

System architecture
The proposed solution follows a modular and scalable architecture consisting of the following layers:

User interface layer (Frontend)
- Built with Bootstrap, HTML, CSS, and JavaScript, allowing users to upload .eml files.

Application processing layer (Backend)
- Uses Flask (Python) to process email headers, body content, and attachments.

Machine learning layer
Employs SVM to detect threats via header anomalies, body patterns, and attachments.

Database layer (storage and logging)
- Stores threat reports and metadata in SQLite for forensic analysis and retrieval.

Implementation workflow
This entire process is achieved through acute understanding and organized steps with efficiency and accuracy for email threat detection in mind.

As shown in Figure 65.2, the email threat detection process follows an organized architecture.

Step 1: User uploads an .eml file.
- Users submit email file i.e. .eml file via web interface.

Step 2: Email header parsing and analysis.

- Extracts key fields (SPF, DKIM, DMARC) to identify anomalies.

Step 3: Email content analysis (NLP).
- Uses tokenization, NER, and sentiment analysis to detect phishing cues.

Step 4: Attachment and embedded URL inspection.
- URLs are checked against threat databases and attachments are hashed to check them for malware.

Step 5: Threat classification using SVM model.

Step 6: Threat report generation. The classification result using SVM is displayed in Figure 65.3.

Automated email report creation. The final report output can be seen in Figure 65.4."

Alerts users for detected threats, marks indicators of phishing attack, prompts them to mark spam or report as phishing.
- Extracted features are transformed into vectors later classifying the emails as safe, suspicious, and malicious and calculating the risk score.

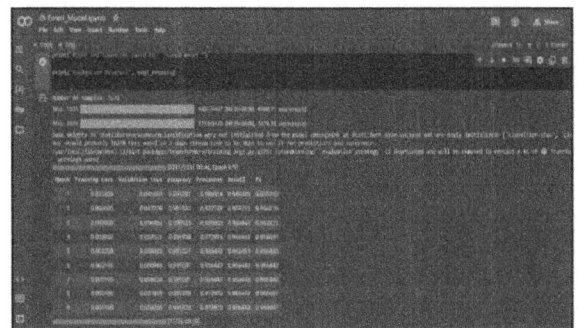

Figure 65.3 Threat classification results using SVM model in google colab
Source: Author

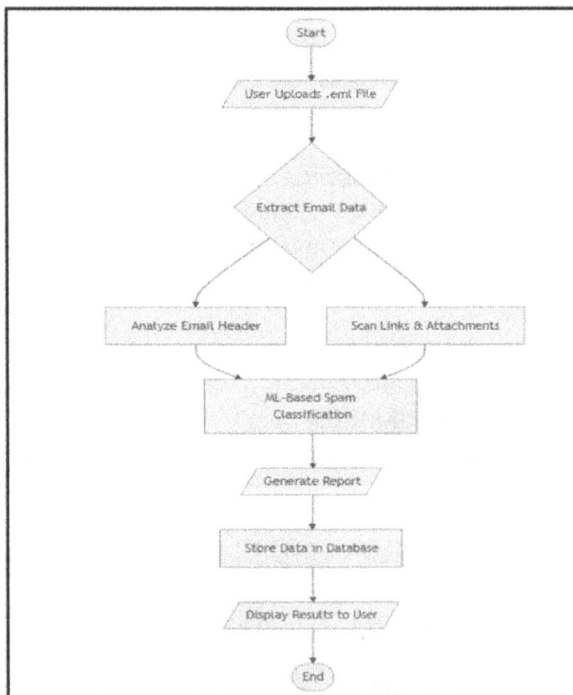

Figure 65.2 Application workflow overview
Source: Author

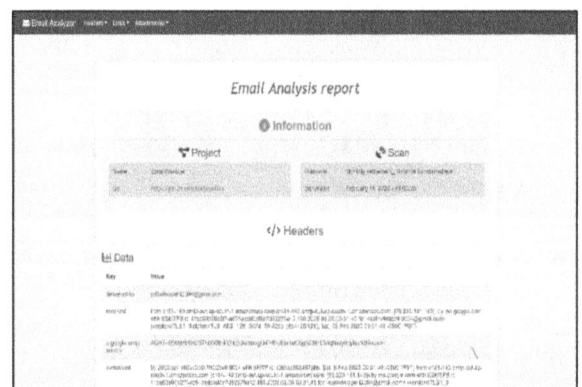

Figure 65.4 Threat report generation after EML file analysis
Source: Author

Key enhancements and features

Improved header analysis - This examines the entire routing path of the email over SPF/DKIM/DMARC which helps in identifying phishing and sender snooping attempts.

Advanced NLP for content-based threat detection - Uses text classification and entity recognition to identify scam indicators and financial fraud phrases.

Machine learning powered threat classification - Trains on phishing and legitimate emails to enhance precision, recall, and spam elimination.

Comprehensive threat reporting - Provides detailed safety reports to users, summarizing detected threats and email security status.

Results and performance analysis

The assessment of our email threat analysis system is based on different measures such as its accuracy, precision, recall and F1-score. The system was evaluated with a set of emails including genuine, phishing, infected by malware, and spam emails to determine its effectiveness in detecting and classifying threats.

Performance metrics and achieved results

The machine learning model i.e. the Support Vector Machine (SVM) classifier, has been trained and evaluated with an extensive sample set of emails. The results obtained are as follows:

These scores confirm the model's great efficacy in recognizing harmful emails while keeping the chances of incorrectly marking safe emails to be very low.

As summarized in Table 65.1, the SVM model demonstrated high precision and recall.

Detailed explanation of performance metrics

To analyze how well the email threat detection system works, we measured several key performance indicators:

Table 65.1 Performance metrics and achieved results by the SVM model.

Metric	Achieved value
Accuracy	0.991031
Precision	0.986014
Recall	0.946309
F1-score	0.965753

Source: Author

4.2.1. Accuracy (99.10%): Correctly identifies emails with only 0.9% misclassification

4.2.2. Precision (98.60%): Effectively detects malicious emails with minimal false positives.

4.2.3. Recall (94.63%): Captures 94.63% of actual threats, ensuring high detection.

4.2.4. F1-score (96.57%): Balances precision and recall, minimizing false alarms while maximizing threat identification.

Comparative analysis (SVM vs. Others)

During the development phase of the system, various machine learning models such as K-nearest neighbors (KNN), Random Forest (RF), logistic regression (LR), and the SVM were practiced. SVM was found to be the best one out of all of them, as was discussed in the comparative analysis section.

The spam probability outcome is visualized in Figure 65.5

Performance comparisons among models are shown in Table 65.2.

Key observations:

- SVM was found to be the best model compared to all other tested models like RF, KNN, and LR. SVM excelled in accuracy together with F1-score so it is the most accurate model for threat detection.

Results from overall testing

To determine the accuracy of the system, we put it against a real-world dataset containing over 10,000 emails. The dataset contained real business emails along with advertisements, phishing scams, and even virus quarantined emails.

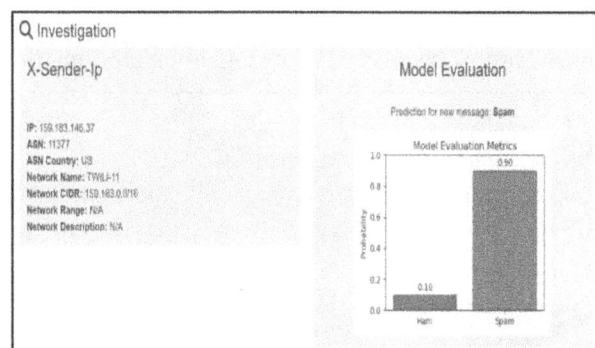

Figure 65.5 Spam probability analysis result after EML file examination

Source: Author

Table 65.2 Performance metrics of different machine learning models.

Model	Accuracy	Precision	Recall	F1-score
Support Vector Machine (SVM)	99.10%	98.60%	94.63%	96.57%
K-Nearest neighbors (KNN)	94.23%	91.78%	89.02%	90.39%
Random Forest (RF)	97.31%	95.42%	91.86%	93.60%
Logistic Regression	92.85%	89.64%	85.31%	87.42%

Source: Author

Table 65.3 Classification performance across different email categories.

Email type	Total samples	Correctly classified	Misclassified
Legitimate emails	4,500	4,472 (99.38%)	28 (0.62%)
Spam emails	2,500	2,458 (98.32%)	42 (1.68%)
Phishing emails	2,000	1,926 (96.30%)	74 (3.70%)
Malware-infected emails	1,000	953 (95.30%)	47 (4.70%)

Source: Author

Breakdown of email categories in real-world testing

In an effort to detail the effects made with the performance enhancements, Table 65.3 below shows key comparisons of changes in processing efficiency and accuracy on record before and after applying the changes.

Key real-world testing results:

- Legitimate emails: A true positive percentage of 99.38%, meaning apart from severely disrupting users' flow, there was minimal action was needed.
- Spam emails: Approximately over 98 percent of spam emails flagged, making the inbox clean.
- Phishing emails: 96.30% positive swaps on phishing emails and attempted frauds.

- Malware-infected emails: 95.30% of the time programs were flagged as dodging capable.

These are the results which were achieved during the experimentation, successfully proves the model is performing very robust compared to the existing models.

Discussion

According to our analysis of the threat posed by email, the accuracy logging spam, phishing, and malware emails is around 99.10% with a very few false positives. The system is most powerful with the SVM approach, outperforming KNN, RF, and LR. Having high values of precision and recall allows for the effective measurement of threats with little interference to the legitimate users. Even though detection accuracy is high, highly sophisticated phishing emails that resemble authentic communications tend to pose an issue. The presence of malware embedded within attachments and obfuscated URLs increases difficulty of detection without the application of advanced deep-learning methodologies. The incorporation of behavioral analysis along with real-time threat intelligence feeds could be very useful in making the system more adaptive.

This system can be utilized in a practical manner in a corporate security system, personal email filtering, or cybercrime investigations. The results showed that a machine learning approach, when adequately designed, has the potential to enhance email security, reduce cyber threats, and avert data leakage. Such improvements could include the addition of BERT deep learning models for context analysis, automated URL sandboxing, to further bolster the system's robustness.

Conclusion

This study was able to develop an email threat detection system that successfully detects threats through email headers, email body, and attachments by using machine learning and NLP. The system which was built using Python, Flask, and SQLite, achieved exceptional results with an accuracy of 99.10%, precision of 98.60%, and the remarkable F1-score of 96.57%, proving its competency in the correctly classification of emails. This Study proves how efficient SVM is in classification, as it was the most accurate model for threat detection.

Suggestions for Future Work

Although this study captures the classification of malignant emails properly, many aspects like precision, flexibility, and practicality can still be improved.

Updating in real time with APTs advanced detection capabilities can already be achieved by integrating feeds from Virus Total, Phish Tank, or STIX/TAXII with automatic runs. This approach can further be applied to other messaging applications, social networks, and cloud storage systems for more holistic cybersecurity solutions.

Acknowledgement

I would like to express my sincere gratitude to Narayana Engineering College for providing the necessary infrastructure and resources that greatly contributed to the successful completion of this work.

References

[1] H.T. Abdulazeez, S. Ahmad, A. Raji, D. Maliki (2024). Designing an email header analysis tool for cyber threat intelligence. *The El-Amin University Journal of Computing (EAUJC)*, el-aminuniversity.edu.ng, (pp. 1–37).

[2] Charalambou, E., Bratskas, R., Karkas, G., & Anastasiades, A. (2016). Email forensic tools: a roadmap to email header analysis through a cybercrime use case. *Journal of Polish Safety and Reliability Association*, 7(1), (pp. 21–28).

[3] Castillo, E., Dhaduvai, S., Liu, P., Thakur, K. S., Dalton, A., & Strzalkowski, T. (2020, May). Email threat detection using distinct neural network approaches. In Proceedings for the First International Workshop on Social Threats in Online Conversations: Understanding and Management, (pp. 48–55).

[4] Dada, E. G., Bassi, J. S., Chiroma, H., Adetunmbi, A. O., & Ajibuwa, O. E. (2019). Machine learning for email spam filtering: review, approaches and open research problems. *Heliyon*, 5(6), e01802.

[5] Pandey, B., Pandey, P., Kulmuratova, A., & Rzayeva, L. (2024). Efficient usage of web forensics, disk forensics and email forensics in successful investigation of cybercrime. *International Journal of Information Technology*, 16(6), 3815–3824.

[6] Hadjidj, R., Debbabi, M., Lounis, H., Iqbal, F., Szporer, A., & Benredjem, D. (2009). Towards an integrated e-mail forensic analysis framework. *Digital Investigation*, 5(3-4), 124–137.

[7] Yu, J., Choi, Y., Koo, K., & Moon, D. (2024). A novel approach for application classification with encrypted traffic using BERT and packet headers. *Computer Networks*, 254, 110747.

[8] Chu, K. T., Hsu, H. T., Sheu, J. J., Yang, W. P., & Lee, C. C. (2020). Effective spam filter based on a hybrid method of header checking and content parsing. *IET Networks*, 9(6), 338–347.

[9] Rudd, E. M., Harang, R., & Saxe, J. (2018). Meade: towards a malicious email attachment detection engine. In 2018 IEEE International Symposium on Technologies for Homeland Security (HST), (pp. 1–7). IEEE.

[10] Jáñez-Martino, F., Alaiz-Rodríguez, R., González-Castro, V., Fidalgo, E., & Alegre, E. (2023). A review of spam email detection: analysis of spammer strategies and the dataset shift problem. *Artificial Intelligence Review*, 56(2), 1145–1173.

[11] Aslan, Ö., Aktuğ, S. S., Ozkan-Okay, M., Yilmaz, A. A., & Akin, E. (2023). A comprehensive review of cyber security vulnerabilities, threats, attacks, and solutions. *Electronics*, 12(6), 1333.

[12] Shukla, S., Misra, M., & Varshney, G. (2024). HTTP header based phishing attack detection using machine learning. *Transactions on Emerging Telecommunications Technologies*, 35(1), e4872.

[13] Mehta, A. K., & Kumar, S. (2024, May). Comparative analysis and optimization of spam filtration techniques using natural language processing. In 2024 International Conference on Communication, Computer Sciences and Engineering (IC3SE), (pp. 1005–1010). IEEE.

[14] Prasad, P. (2024). Exploring Machine Learning Algorithms for Email Spam Filtering. *SPAST Reports*, 1(6), (pp. 1–13).

66 A cloud-integrated IoT framework for real-time energy monitoring and predictive billing

P. Penchalaiah[1,a], G. Srinivasulu[2,b], Arkot Mitesh[3,c], M. Venu[3,d] and Sk. Adil[3,e]

[1]Professor, Department of CSE, Narayana Engineering College, Nellore, Andhra Pradesh, India

[2]Professor, Department of EEE, Narayana Engineering College, Nellore, Andhra Pradesh, India

[3]UG Scholars, Department of CSE, Narayana Engineering College, Nellore, Andhra Pradesh, India

Abstract

This paper introduces an Internet of Things(IoT)-powered energy management system that merges advanced hardware components with cloud-based analytics to enable real-time electricity monitoring and predictive billing. The hardware setup features an Arduino microcontroller integrated with a PZEM-004T sensor to track voltage, current, power, and energy consumption. An LDR sensor is utilized for detecting load variations, while an LCD display provides on-site consumption updates. The data collected is transmitted to a cloud platform via the NodeMCU ESP8266 module, ensuring reliable and efficient communication. The software component, developed using the MERN stack and hosted on AWS, delivers comprehensive energy analytics and predictive billing based on historical consumption trends. Users can set daily usage limits and receive alerts through email notifications. This system's efficiency is further enhanced through precise sensor calibration, improved power management, and flexible notification mechanisms. Experimental evaluations validate the system's accuracy in real-time monitoring and its effectiveness in guiding users toward energy-efficient practices.

Keywords: Arduino, AWS, deep learning, energy forecasting, internet of things (IoT), MERN stack, NodeMCU, predictive billing, real-time electricity consumption, smart energy monitoring

Introduction

The rapid development of industry together with population growth and increasing dependence on electrical power has made it essential to address energy consumption management issues. Traditional electricity metering systems which operate with static flat-rate billing methods fail to detect variable power usage patterns and produce inefficient energy management and billing results. This inefficiency has created a demand for real-time monitoring solutions which track energy usage and forecast future consumption patterns together with their corresponding costs [1, 9].

The hardware is supported by a comprehensive software solution developed as a web application with the MERN stack which runs on AWS. This platform offers detailed energy consumption analytics over multiple time intervals (hourly, daily, weekly, monthly, and yearly) and offers features like user defined daily consumption limits with automated email notifications as well as predictive billing based on historical data. Such integration not only helps end users to understand their energy usage in real time but also helps them to manage their consumption proactively, thereby cutting down on costs and reducing their carbon footprint [7].

In addition, the application of deep learning techniques in energy forecasting has been shown to significantly improve predictive accuracy. The monthly energy consumption forecast: A deep learning (DL) approach study highlights how models such as LSTM and CNN can be used to forecast monthly energy consumption and estimate billing amounts. We seek to increase the credibility of forecasted energy bills by integrating these state-of-the-art predictive models into our system to support consumers and energy providers in their decision-making processes [8].

---

[a]penchal.caliber@gmail.com, [b]gsmeghana@gmail.com, [c]mitesharkot@gmail.com, [d]maddhurivenu6@gmail.com, [e]adilshaik114424@gmail.com

DOI: 10.1201/9781003684589-66

This paper presents an integrated IoT based smart energy monitoring system that is an improvement over traditional metering and offers modern data driven energy management solutions. The findings of this paper are twofold:

A cost-efficient real-time hardware system (POWERPULSE) has been developed for energy monitoring and data acquisition.

An interactive web-based platform has been created that uses machine learning algorithms for detailed energy analytics, predictive billing and user notifications.

This holistic approach also improves the transparency and efficiency of energy consumption monitoring and provides a scalable model for future smart grid implementations.

Literature review

Research has extensively explored the transformation from non-connected traditional energy meters to smart IoT-enabled systems which provide real-time energy consumption data. Research initially emphasized how manual metering methods had restrictions because flat-rate billing failed to capture the variable patterns of electrical usage. The development of integrated systems composed of hardware sensors alongside real-time data processing has now become possible which guides us to modern energy monitoring solutions.

Research has shown that smart energy metering systems based on IoT technology for home appliances have been widely studied to demonstrate how sensor networks with wireless communication can enable remote monitoring and improve billing accuracy [1]. Industrial implementations have led to full-scale energy management systems emphasizing precise metering and visualization [2]. These systems demonstrate the benefits of automated readings and the need for reliable sensors and data protocols. Recent work also highlights IoT-based smart meters integrating real-time data with energy analytics [3]. The industrial implementation of these studies identify key benefits of automated meter reading while highlighting the necessity of robust data acquisition protocols and sensor selection to enhance energy management practices.

Moreover, besides the issues of technological advancements, economic and market aspects have been also explored. Studies on dynamic pricing mechanisms and electricity markets have also investigated how real time pricing can affect consumer behavior and enhance efficiency of the market [9]. Although the main emphasis of our project lies in energy monitoring and predictive billing, and not in pricing reform, these enable discussions on the relevance of real time feedback and consumption forecasting in influencing energy use by consumers.

All these studies highlight the changing trend from the traditional metering to the integrated IoT based systems that consist of hardware sensors, cloud analytics and advanced machine learning models. Based on these foundations, we propose a complete system that consists of an embedded hardware platform and a MERN stack-based web application. This integrated approach not only offers detailed energy consumption analytics and real time notifications but also applies deep learning techniques to forecast future billing amounts, thus improving energy management and providing substantial cost and environmental advantages.

Methodology

This paper aims to describe the overall system architecture and explain both the hardware and software components of our integrated IoT-based smart energy monitoring and predictive billing system. The methodology comprises sensor-based real-time data acquisition with a cloud-enabled, web-based analytics platform.Smart home systems utilizing similar mobile interfaces for energy tracking and billing estimation have shown promising results in enhancing user engagement [4].

Hardware design and implementation

The hardware part which is called POWERPULSE is responsible for real time electricity measurement and data acquisition. The design of the presented solution builds upon earlier IoT-based energy monitoring systems and includes the following significant components:

The proposed system, PowerPulse, comprises:

i. *Arduino (ATmega328P):* Processes sensor data and executes control algorithms.
ii. *PZEM-004T Sensor:* Measures voltage, current, and energy consumption with high accuracy.
iii. *LDR Sensor:* Detects load changes via LED pulse measurement.

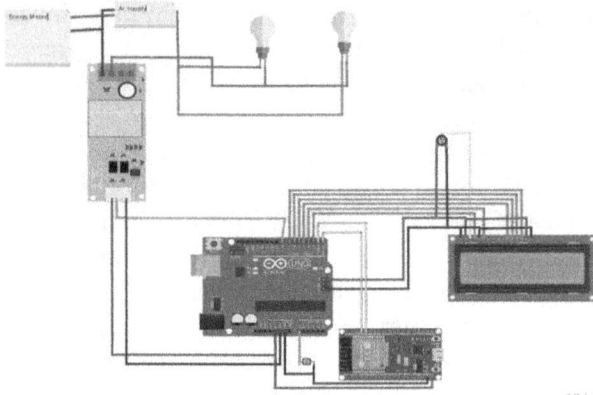

Figure 66.1 Block diagram
Source: Author

iv. *LCD Display:* Provides on-site real-time consumption feedback.
v. *NodeMCU (ESP8266):* Enables wireless data transmission to the cloud.
vi. *Power supply and supporting circuits:* Ensures stable operation and measurement accuracy.

Similar Arduino and ESP8266-based cloud monitoring systems have previously been used for real-time environmental data applications [6]. Figure 66.3 shows the homepage that provides access to energy metrics and forecast tools.

Figure 66.1, Power management circuitry is designed to provide stable voltage requirements for both the microcontroller and sensors through appropriate voltage regulators and their interfaces.

These elements help in preserving the precision and consistency of the readings.

Software architecture and implementation
The software layer is a cloud-based MERN stack web application deployed on AWS, offering:

I. *Data acquisition and storage:* Energy data is collected via RESTful APIs and stored in MongoDB.
II. *User interface:* A React-based dashboard visualizes consumption trends over multiple timeframes.
III. *Notification system:* Email alerts notify users when consumption exceeds predefined limits.
IV. *Predictive analytics:* Deep learning models (LSTM networks) forecast energy usage and billing

Proposed system

This section describes the design and implementation of our integrated IoT-based smart energy monitoring and predictive billing system. The proposed system uses a custom-designed hardware module for real-time energy data acquisition, and a cloud-based software platform for analytics, forecasting, and user notifications.

System architecture overview
The system is divided into two primary layers: The hardware layer and the software layer. The hardware gathers real-Time energy data, while the MERN stack -based software hosted on AWS handles visualization, billing forecasts, and alerts.

IoT-based frameworks linking microcontrollers with cloud platforms have gained popularity for smart energy monitoring aligned with SDG goals [5].

Figure 66.4 illustrates the interactive dashboard that presents consumption data and forecasts in real time end.

Hardware implementation
The hardware component called POWERPULSE serves the function of both real-time electrical signal measurement and cloud connectivity. Its major components are as follows:

The system's central processor is the Arduino microcontroller which uses the ATmega328P chip. The system uses this microcontroller to collect information from multiple sensors while it manages the data processing procedures.

This sensor provides accurate readings of voltage, current, power and energy consumption in kWh. The sensor's high accuracy provides dependable data collection across all monitoring and forecasting operations.

The LDR sensor provides innovative load change detection through digital meter LED indicator pulse measurement which improves unit consumption measurement accuracy.

The local data terminal displays vital system data including power usage and current consumption directly on its LCD display allowing users to monitor system performance in real-time without depending on external network connections.

The NodeMCU functions as the communication interface to transmit the processed Arduino data to a cloud platform through Wi-Fi. The wireless module

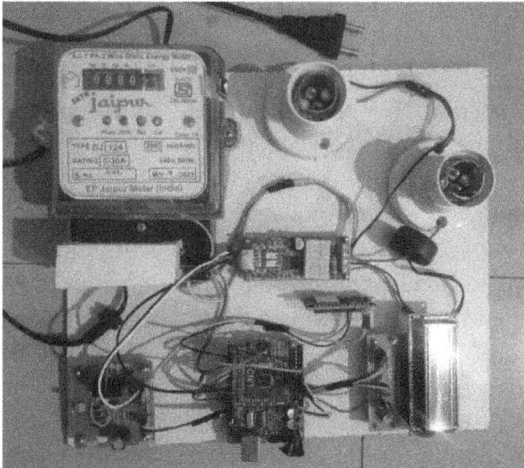

Figure 66.2 Integration-hardware kit
Source: Author

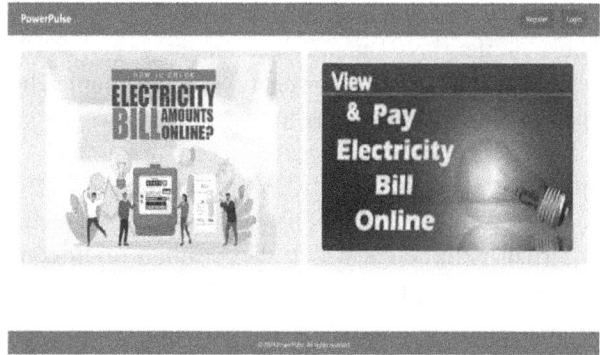

Figure 66.3 Homepage
Source: Author

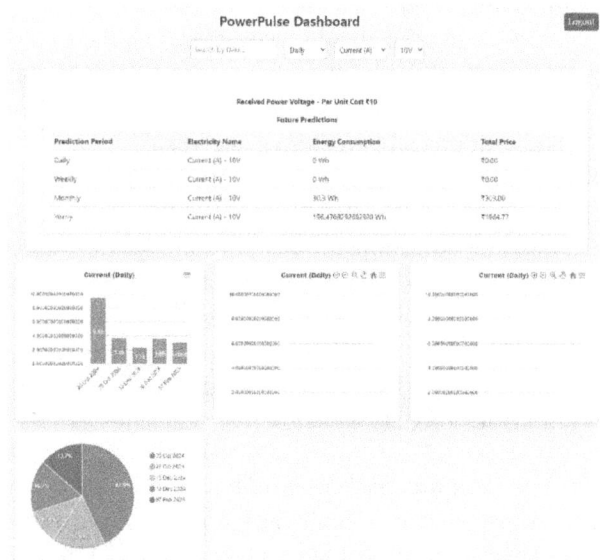

Figure 66.4 Dashboard
Source: Author

provides real-time access to energy usage data which can then be analyzed further.

All system components operate steadily in the hardware design which feature powerful power supply circuits together with voltage regulators. The microcontroller and communication modules stay separate from the sensors through proper interfaces which help preserve measurement accuracy.

The Figure 66.2 shows how hardware components connect in modular configuration. The Arduino processes local sensor data before forwarding it through NodeMCU to reach the cloud for smooth measurement data analysis.

Software implementation
The software layer is intended to work alongside the hardware to process, store and visualize the data that is being gathered on energy usage. It has been built as a web application with the MERN stack (MongoDB, Express.js, React, Node.js) and is hosted on AWS to guarantee scalability and high availability. The main features include:

The real time energy data from NodeMCU is received and stored in a cloud MongoDB using RESTful APIs developed using Express.js and Node.js. This method provides for the constant update of data and the maintenance of historical data.

The front end, built with React, provides a range of interactive dashboards which provide detailed energy consumption analytics across different time scales including hourly, daily, weekly, monthly and yearly. Additionally, users can set daily consumption limits and get visual feedback on their usage trends.

Using historical consumption data, the system includes deep learning models to predict future energy usage and the upcoming electricity bill. Time-series data modelling techniques like LSTM networks are used since they are very effective as shown in [8].

A notification system that works automatically to send emails to users when their energy consumption has gone beyond daily limits is put in place. This feature promotes energy management by assisting users to prevent surprisingly high bills.

System integration and data flow
The complete system provides an integrated hardware software solution for energy management. The operational workflow is as follows:

Real-time sensing: The PZEM-004T and LDR sensors provide continuous measurement of electrical parameters which the Arduino microcontroller then processes before sending data to the NodeMCU module.
Data transmission to cloud: In real-time, the NodeMCU sends processed data through WiFi to either ThingSpeak or a custom AWS endpoint for immediate access to analytics.
Data storage and processing: The MongoDB database within the cloud service stores all incoming data which the backend server, built with Node.js/Express, then processes before presenting visualizations and predictive analysis.
Visualization and forecasting: Through deep learning models, the React-based frontend provides detailed consumption analytics alongside future energy usage and bill estimate forecasts.
User notifications: By monitoring user-designated consumption limits the system provides automatic email alerts which provide timely notifications for proactive energy management.

Environment for implementation
The entire system is built on the AWS cloud platform because it offers both high scalability and both reliability and secure access to energy data. AWS handles backend services, database storage as well as secure API endpoints for real time data communication and analysis. Furthermore, modular hardware development makes it possible to increase system's capacity and enhance it with more sensors or functions in the future as needed.

Result

This paper presents the performance assessment of the integrated IoT-based smart energy monitoring and predictive billing system. The findings are presented in three parts: The performance of hardware components, predictive analytics performance, and the integration of the system with feedback from users.

Hardware performance
The hardware module POWERPULSE was put through a controlled test in a residential energy consumption simulation environment. Key findings include:

Sensor accuracy: When compared with standard instruments, the PZEM-004T sensor delivered measurements of voltage, current, power, and cumulative energy consumption with an error range of less than 5%. The LDR sensor produced pulse events that corresponded to unit consumption and thereby improved the total energy calculations' precision.
Data transmission: The ESP8266 (NodeMCU)-based module preserved a stable WiFi connection that transmitted real-time data to the cloud with an average latency of less than two seconds. During 30 days of continuous testing, the hardware setup showed an operational uptime of 99.5% which indicated high reliability.
Local visualization: Live consumption data updated accurately on-site LCD displays which provided immediate feedback to users independent of cloud connectivity.

Predictive analytics performance
The web application that used the MERN stack and AWS deployment integrated deep learning models to predict future energy usage and generate billing forecasts. Key results include:

Forecast accuracy: Multiple predictive models were assessed using historical consumption data. The LSTM model showed superior performance over alternate methods (ARIMA and Prophet) by achieving an absolute forecasting error of 31.83 kWh and relative error of 17.29% in monthly energy consumption predictions. Deep learning techniques produce significantly better results than predictive billing systems according to these performance metrics.
Real-time analytics: The dashboard presented energy consumption patterns through visualizations that covered hourly, daily, weekly, monthly, and yearly time periods. Real-time data entry from the hardware module enabled users to interact with data visualizations that provided dynamic updates.
Automated notifications: The system properly sent email alerts to users whenever they went past their established daily consumption restrictions thus showing proper implementation of threshold-based notification features.

System integration and user feedback

The final integration of hardware and software layers produced a cohesive energy management solution:

End-to-end data flow: The POWERPULSE module successfully acquired real-time sensor data which was then transmitted to the cloud before being stored in a MongoDB database and processed by the backend server. The continuous data movement enabled both the analytical dashboard and predictive models to use current information for analysis.

Discussion

This work integrates an IoT-based smart energy monitoring and predictive billing system through a custom-designed hardware module (POWERPULSE) with a cloud-enabled platform supported by deep learning analytics. Our experimental results demonstrate that the hardware system achieves precise real-time electrical parameter measurement while the software designed with the MERN stack for AWS deployment delivers accurate visualizations and timely notifications along with reasonably accurate bill predictions.

Hardware and Data Acquisition:

Due to the hardware configuration of an Arduino microcontroller and PZEM-004T sensor together with LDR sensors and LCD display and NodeM-CU communication unit, the POWERPULSE module recorded sensor measurement errors below 5% and kept data transmission latency below 2 seconds. This performance level demonstrates the potential for real-time monitoring system operation. The testing within controlled environments shows promising results but field deployments need to consider sensor calibration maintenance along with potential changes in field conditions.

Predictive analytics and forecasting: The predictive billing system which relies on deep learning models through an LSTM network achieved monthly energy consumption prediction results with an absolute error of 31.83 kWh along with a relative error of 17.29%. These results support the research outcomes presented and demonstrate how deep neural networks provide superior performance when analyzing time-series energy data. The relative improvement in prediction accuracy suggests that adding more features (e.g., weather data, seasonal factors) would lead to better performance.

System integration and user interaction: End-to-end system integration between hardware components and software analytics platform showed successful data transmission from sensor sensors to cloud storage and analysis systems. The React-based interactive dashboard delivered analytic insights across varied time intervals as well as threshold alert systems to help users proactively manage their energy usage. The results from user feedback during pilot deployments show that the system's user-friendly design and live data analysis help consumers better manage their energy usage but real-world behavior evaluation requires additional extended field tests.

Implications and contributions: The research makes a significant contribution to smart energy management literature.

Real-time monitoring: Low-cost embedded hardware enables precise current real-time data collection.

Predictive Capability: The LSTM deep learning model shows strong ability in forecasting both energy usage and billing amounts which energy consumers and providers need.

Actionable insights: The integration of automated notifications with detailed analytics gives users the power to manage their consumption proactively which may lead to both cost savings and environmental improvements.

Limitations: Multiple restrictions exist which need to be recognized in addition to the achieved results:

Dataset limitations: The model needed historical data from a short period to learn and validate itself. Expanding the model to longer time periods or wider datasets calls for model recalibration or extra input features.

Environmental variability: Field deployments may create problems for the hardware validation system due to sensor drift and environmental noise alongside network disruptions.

User behavior dynamics: The system currently operates under the assumption that users maintain consistent behavior patterns. Real-world energy consumption practices modify both prediction accuracy and total energy savings in practical applications.

Future directions: Future research focus on ,

Enhanced sensor calibration: Adaptive calibration methods need to be developed to reduce sensor drift throughout prolonged deployment periods.

Feature expansion: The predictive model's accuracy should improve through integrating real-time weather information alongside appliance-specific energy consumption data.

User-centric studies: Research must be conducted among users over extended periods to understand how real-time feedback together with predictive billing impacts their energy consumption behaviors.

Scalability assessments: The study needs to assess how the system scales up across different large user groups.

Our integrated system presents an effective solution that enables real-time energy monitoring and forecasting, which shows potential to enhance energy management efficiency and minimize costs. Additional development and field testing will enhance its capabilities and help achieve its full potential in residential and industrial sectors.

Conclusion

This research proposes an integrated Internet of Things (IoT) based smart energy monitoring and predictive billing system in order to overcome the limitations of current traditional metering systems and adopt data-driven energy management practices. The hardware module, POWERPULSE, was designed and built to efficiently collect real time electrical parameters; it used an Arduino microcontroller, PZEM-004T sensor, LDR sensor, LCD display and NodeMCU module to acquire data with high precision and low delay. The hardware system is supported by a cloud-based software platform developed with the MERN stack and deployed on AWS which provides detailed energy analytics, automated notifications and predictive billing features. The experimental results showed that the system was able to transmit data reliably with an average latency of less than 2 seconds and the sensor measurement errors were less than 5%. The deep learning–based predictive analytics, specifically the LSTM model, showed an absolute error of 31.83 kWh and a relative error of 17.29% in the one-month ahead energy consumption forecasting, thus highlighting the capability of improved forecasting methods to improve billing precision. Real-time data acquisition integrated with robust analytics creates a powerful system that allows users to manage their energy use proactively and establishes a foundation for energy management solutions that are scalable and ready for the future. There are some limitations to these promising results, including the requirement for extended field testing, adaptive sensor calibration

across different environmental conditions, and improving the predictive models to capture changes in user behavior. Further studies should address data sources' scalability by investigating weather data and appliance usage and the system's potential for large-scale implementation. This research develops an affordable and expandable real-time energy monitoring and predictive billing system.

References

[1] Aayushi T. Jambhulkar, Asmita M. Bhoyar, Chetana M. Shivankar, Pratiksha S.Balwad, and Praful Nandankar,"IoT-Based Smart Energy Meter for MonitoringHome Appliances," International Journal of Innovations in Engineering and Science, vol. 8, no. 7, pp. 1–5, May2023, doi: 10.46335/IJIES.2023.8.7.1.2023, doi: 10.46335/IJIES.2023.8.7.1.

[2] Laayati, O., Bouzi, M., & Chebak, A. (2020). Smart energy management: Energy consumption metering, monitoring and prediction for mining industry. In Proceedings ICECOCS 2020. doi: 10.1109/ICECOCS50124.2020.9314532.

[3] Priyadharshini, S. G., Subramani, C., & Roselyn, J. P. (2019). An IoT based smart metering development for energy management system. *International Journal of Electrical and Computer Engineering*, 9(4), 3041–3050.

[4] Hariharan, R. S., Agarwal, R., Kandamuru, M., & Gaffar, H. A. (2021). Energy consumption monitoring in smart home system. In IOP Conference Series: Materials Science and Engineering, (Vol. 1085, p. 012026). doi: 10.1088/1757-899X/1085/1/012026.

[5] I. B. G. Purwania, I. N. S. Kumara, and M. Sudarma, "Application of IoT-Based System for Monitoring Energy Consumption," International Journal of Engineering and Emerging Technology, vol. 5, no. 2, pp.81–93, Jul.–Dec. 2020.

[6] Zafar, S., Miraj, G., Baloch, R., Murtaza, D., & Arshad, K. (2018). An IoT based real-time environmental monitoring system using arduino and cloud service. *Engineering, Technology and Applied Science Research*, 8(4), 3238–3242.

[7] Ali, U., Ramzan, M. U., Ali, W., Rana, M. E., & Qayyum, A. (2023). IoT-driven smart energy monitoring: real-time insights and AI-based unit predictions. In Proceedings of IEEE SCOReD 2023. doi: 10.1109/SCOReD60679.2023.10563825.

[8] Berriel, R. F., Lopes, A. T., Rodrigues, A., Varejao, F. M., & Oliveira-Santos, T. (2017). Monthly energy consumption forecast: a deep learning approach. In Proceedings of IEEE IJCNN 2017. doi: 10.1109/IJCNN.2017.7966398.

[9] Allcott, H. (2009). Real Time Pricing and Electricity Markets. Harvard University.

67 DeepWave: Revolutionizing underwater communication with Li-Fi technology

K. Riyazuddin[1,a], K. Deekshitha[2,b], P. Adiseshu[2,c], J. Bhavani[2,d] and P. Dinesh[2,e]

[1]Associate Professor, Department of ECE, Annamacharya University, Andhra Pradesh, India

[2]UG Scholar, Department of ECE, Annamacharya Institute of Technology and Sciences, Andhra Pradesh, India

Abstract

Underwater communication is inherently difficult due to the distinctive properties of the underwater environment, which create numerous obstacles for data transmission. This paper explores these challenges and proposes Light Fidelity (Li-Fi) as a practical solution. Li-Fi technology, which uses LED light sources for data transmission, provides an efficient and affordable means to address these issues. By employing Li-Fi transmitters and receivers, this approach overcomes the shortcomings of conventional communication methods—such as electromagnetic, acoustic, and optical systems—offering faster data transfer rates and enhanced reliability for underwater communication. Ultimately, Li-Fi presents a promising alternative for improving underwater data transmission.

Keywords: Audio transmission, data transmission, light-based communication, low power consumption, optimal data rates, real-time broadcasting

Introduction

Underwater communication faces numerous obstacles because of the unique characteristics of the environment, including significant signal loss, restricted bandwidth, and signal distortion. Traditional methods, including electromagnetic, acoustic, and optical communication, are limited in their ability to provide high data transmission speeds and stable connections in underwater environments. This paper examines these challenges and suggests Light Fidelity (Li-Fi) as a promising solution for underwater data transmission. Li-Fi uses LED lights to send data via light waves, offering advantages in efficiency, cost, and the potential for faster communication. By using LED-based transmission, Li-Fi addresses many of the issues faced by traditional underwater communication methods. LEDs can emit light with high intensity and precision, enabling effective data transmission in the challenging underwater conditions. Unlike acoustic communication, which suffers from low data rates and long delays, Li-Fi offers the potential for much higher transmission speeds and reduced latency. Optical communication, which struggles with signal attenuation and scattering, benefits from Li-Fi's focused, directional light transmission that helps mitigate these problems [1].

The proposed Li-Fi system adds extra features to improve its use in underwater environments. By integrating Bluetooth automation, IoT functionality, and GPS tracking, it supports smooth data transfer, real-time monitoring, and location tracking. These enhancements boost the efficiency and reliability of communication while enabling advanced underwater tasks like autonomous navigation, environmental monitoring, and exploration [2].

Underwater communication technologies include various methods and systems designed to address the specific challenges of the underwater environment. Some of the main technologies used for underwater communication are:

Acoustic communication

Acoustic communication transmits data underwater using sound waves, providing long-range propagation and minimal attenuation. Acoustic modems encode and decode digital data, widely used in underwater vehicles, oceanography, sensor networks, and research [3].

[a]shaik.riyazuddin7@gmail.com, [b]deekshithakuppannagari@gmail.com, [c]adisesh018@gmail.com, [d]bhavanijoka87@gmail.com, [e]polidinesh246@gmail.com

DOI: 10.1201/9781003684589-67

Optical communication

Optical communication uses light waves for underwater data transmission, offering higher speeds and lower latency than acoustic methods. LED-based systems like Li-Fi enable high-speed, short-range communication, while laser systems offer longer ranges but require precise alignment and clearer water [4].

Electromagnetic communication

Electromagnetic communication, including RF and magnetic induction, offers advantages in certain applications but is limited by bandwidth, range, and regulatory constraints. It enables secure, low-power communication, especially for short-range use [5].

Hybrid systems

Hybrid underwater communication systems combine acoustic and optical technologies, optimizing long-range transmission and high-speed data exchange for applications like robotics, offshore exploration, and monitoring networks [6].

Literature Review

Scognamiglio et al. [7] published "Quantum light sources that can be used on demand for underwater communications". Since water absorbs almost all viable optical wavelengths and keeps them from propagating more than a few meters, quantum communication—which has been at the forefront of contemporary research for decades—is severely limited in underwater applications. On-demand quantum light sources that are appropriate for underwater optical communication are reported in this work. The single photon emitters are based on imperfections in hexagonal boron nitride and can be created with an electron beam. At about 436 nm, which is close to the lowest value of water absorption, they have a zero-phonon line.

Study paper by Zhang et al. [8] introduces the design of a real-time UWOC system which operates under wireless optical communication using underwater transmission. Our UWOC system transmitter benefits from pre-emphasis technology to enhance the modulation bandwidth of its four blue LEDs. A receiver-end APD has been used with a 3 mm diameter. The underwater experiments take place in real time when both transmitter and receiver reside in water-tight chassis which get immersed in a water pool. The experimental results from these investigations yielded excellent findings. The system produces BER results of 5.9×10^{-3} and achieves an ideal data throughput speed of 135 Mbps.

Underwater media represents the greatest challenge when it comes to data transfer despite optical signals acoustic waves and electromagnetic waves currently being used as water communication methods [9]. The sight path limitations of optical waves exist while electromagnetic waves encounter significant signal loss that reduces their range to short distances. The proposed solution utilizes Li-Fi data transmission technology for water-based communication in order to address current issues. The data transfer capabilities and varied frequency options of visible light communication surpass all other transmission systems. Level converters and electronic parts.

Using a surface mount device (SMD) packaging architecture, we discuss the communication capabilities of Li-Fi systems based on high-brightness and high-bandwidth integrated laser-based sources in this study [10]. 450 lumens of white light illumination may be produced by the laser-based source, and the resulting light brightness is greater than 1000cd/mm2. With the use of Volterra filter-based nonlinear equalizers, it is shown that a wavelength division multiplexing (WDM) Li-Fi system with ten parallel channels might offer a data rate of in excess of 100 Gbps. Furthermore, an aggregated transmission.

Wijayanto et al. [11] proposed "Analysis of underwater communication on wireless optical system" investigating the challenges together with opportunities for implementing wireless optical communication (WOC) systems beneath the water surface. The authors explain why optical communication exceeds radio frequency (RF) systems particularly in minimizing signal degradation underwater. The analysis studies signal integrity dependency on turbidity and water clarity along with suggesting multiple modulation methods for improved dependability. The authors discuss methods to counter existing technological constraints through adaptive optical filtering which depends on precise wavelength selection for optimal underwater communication system operation.

Methodology

The setup of an underwater communication system utilizing Li-Fi technology involves a precise and systematic approach to ensure reliable data transmission in aquatic environments. The process begins with configuring the Li-Fi transmitter, integrated with an Arduino Nano acting as the transmitter node. Simultaneously, the receiver node is configured with another Arduino Nano, interfacing with the Li-Fi receiver. A variety of components, such as an amplifier, LCD, I2C, 4x4 keypad, audio jack, speaker, transformer, LM705, and microphone, are meticulously assembled and connected to enable efficient data exchange between the transmitter and receiver nodes. Furthermore, a solar panel is strategically positioned at the receiver end to capture light signals from the transmitter, providing supplemental power to the system.

After the system is physically assembled, the focus shifts to data encoding and transmission protocols. The Arduino Nano is programmed with algorithms that encode data received from input devices such as the 4×4 keypad, microphone, or audio jack into modulated light signals that are compatible with Li-Fi transmission. The Li-Fi transmitter, enhanced with an amplifier to strengthen the signal, then transmits these encoded signals through the water, ensuring stable communication over submerged distances.

The Arduino Nano features an ATmega328 microcontroller and supports communication via UART TTL (5V) on pins 0 (RX) and 1 (TX). The FTDI FT232RL chip enables USB communication and virtual COM port creation for serial data transfer. It supports I2C, SPI, and includes 14 digital I/O pins, 8 analog inputs, and 6 PWM outputs. Power can be supplied via USB or an external source. The board is popular among hobbyists, students, and professionals for prototyping and embedded systems. The Arduino IDE simplifies programming with pre-programmed examples and easy code uploading.

Underwater Li-Fi communication systems convert digital data into light signals using LED transmitters, modulation hardware, and amplification units, enabling data transmission over extended distances. A photodetector at the receiver end transforms these modulated light signals back into electrical signals, allowing for accurate data recovery. Signal processing and decoding circuitry further enhance data integrity by filtering interference and amplifying weak signals. These systems integrate seamlessly with microcontrollers like Arduino and Raspberry Pi, offering flexibility to adapt to various underwater conditions.

The 4×4 keypad matrix is a compact input device that allows users to enter numbers, letters, and symbols. It connects microcontroller I/O pins through row-column multiplexing, minimizing pin usage and enabling efficient key detection. The keypad integrates easily into electronic projects and prototypes. A 3.5mm port or microphone receives signals from mobile devices or computers, which are amplified and transmitted as infrared laser light via an S8050 transistor. A photodiode detects these signals, and the LM386 amplifier converts them into sound. This system works wirelessly over distances up to 50 meters (Figure 1-3).

At the receiver end, the Li-Fi receiver captures the modulated light signals transmitted through the water. The Arduino Nano, programmed to interpret these

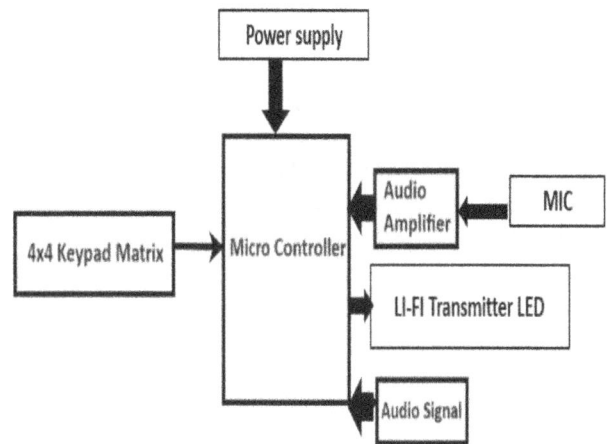

Figure 67.1 Block diagram representing the transmitter
Source: Author

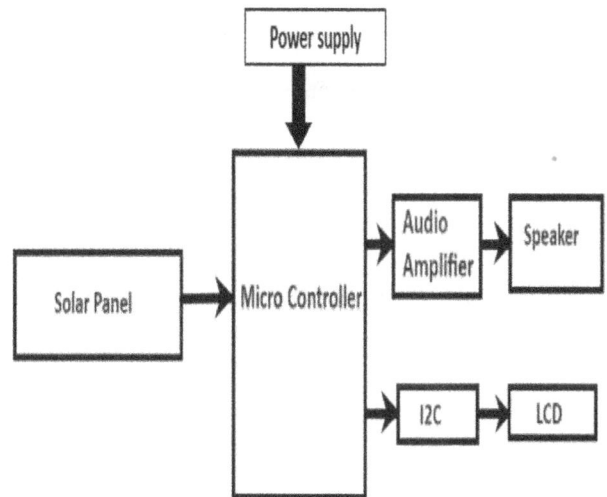

Figure 67.2 Block diagram representing the receiver
Source: Author

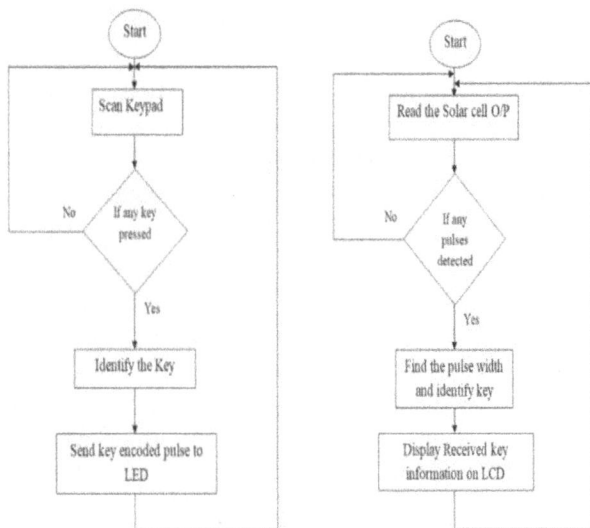

Figure 67.3 Transmitter and receiver flow diagram
Source: Author

Figure 67.4 Underwater Li-Fi complete setup
Source: Author

signals, decodes the original data. Once decoded, the data is processed and displayed on an LCD screen through the I2C interface, enabling real-time monitoring and visualization of the transmitted information. The system can also include audio feedback via a speaker, allowing for enhanced user interaction and providing audible notifications based on the received data.

Figure 67.5 Underwater Li-Fi audio transmission
Source: Author

Results and Discussions

The deployment of the underwater Li-Fi communication system demonstrates positive results, highlighting its ability to transmit data effectively in aquatic settings. Through careful configuration and integration of the system, including the use of Arduino Nano for both the transmitter and receiver nodes, smooth communication between devices is realized. The integration of components such as the amplifier, LCD, 4 × 4 keypad, and audio jack enhances the system's functionality, ensuring robust user interaction (Figure 4-6).

The Arduino Nano uses encoding algorithms to transform data from input sources into modulated light signals that are compatible with Li-Fi transmission. The Li-Fi transmitter, enhanced with an amplifier, efficiently sends these signals through the water, ensuring reliable communication over submerged distances. At the receiver end, the Li-Fi receiver captures and decodes the transmitted signals, extracting the original data accurately. The decoded information is then processed and displayed in real time on

Figure 67.6 Underwater Li-Fi data transmission
Source: Author

an LCD screen via the I2C interface, allowing users to monitor and visualize the transmitted data.

Testing and optimization of the system ensure the functionality and reliability of the underwater Li-Fi communication setup in real-world scenarios. Thorough evaluations, including submersion in water tanks and natural water bodies, verify the system's performance and durability. Modifications to system parameters, such as signal modulation strategies and amplifier gain settings, are made to enhance performance and increase data transmission speeds. In conclusion, the project illustrates the practicality and effectiveness of underwater Li-Fi communication, opening opportunities for applications in marine research, underwater exploration, and environmental monitoring.

Conclusion

The underwater Li-Fi communication project demonstrates the substantial potential and effectiveness of utilizing Li-Fi technology for data and audio transmission in aquatic environments. Through meticulous configuration and integration, the system establishes seamless communication between transmitter and receiver nodes, employing components such as the Arduino Nano. The project adeptly encodes data and audio signals into modulated light, ensuring reliable transmission through water via Li-Fi transmitters, and accurately decodes and processes the received signals. Additionally, real-time monitoring and visualization of the transmitted data and audio enhance user engagement and system functionality.

Experimental results show that the system successfully transmits data and audio underwater with high speed and low latency, overcoming limitations of traditional radio-frequency methods. Real-time monitoring tools enhance the system's effectiveness by offering immediate feedback and control. These results support recent studies on Li-Fi's potential for underwater communication, stressing the need for optimization to ensure reliable transmission in diverse aquatic conditions.

In real-time applications, this technology holds promise for enhancing underwater exploration, environmental monitoring, and communication in submerged environments. By leveraging Li-Fi's capabilities, researchers and professionals can achieve more efficient and reliable communication systems, facilitating advancements in various fields such as marine biology, oceanography, and underwater archaeology. The integration of audio transmission further enriches these applications, enabling more comprehensive data collection and analysis in real-time underwater settings.

In conclusion, the underwater Li-Fi communication project not only demonstrates the feasibility of using Li-Fi technology for data and audio transmission in aquatic environments but also sets the stage for future advancements in underwater communication systems. The successful encoding, transmission, and decoding of data and audio signals, coupled with real-time monitoring capabilities, highlight the system's potential to revolutionize underwater communication, offering a reliable and efficient alternative to traditional methods [12].

Future scope
Future project versions can improve data rates by refining signal modulation and amplifier gain, enabling faster transmission. Efforts can also extend the communication range for challenging underwater environments. The project could include sensors for environmental monitoring in fields like oceanography and pollution detection. Integration with autonomous underwater vehicles or remotely operated vehicles would enhance data gathering, while encryption would secure transmitted information.

In summary, this project establishes a robust platform for future developments in underwater communication technologies, offering potential applications across marine science, underwater exploration, disaster management, and underwater robotics. Ongoing research and innovation in this area promise to unlock new opportunities for exploring and utilizing the vast resources of the underwater environment.

References

[1] Krishnamoorthy, N. R., & Suriyakala, C. D. (2017). Performance of underwater acoustic channel using modified TCM OFDM coding techniques. *Indian Journal of Geo Marine Sciences*, 46(3), 629–637.

[2] Nasri, N., Andrieux, L., Kachouri, A., & Samet, M. (2010). Efficient encoding and decoding schemes for wireless underwater communication systems. In 7th International Multi-Conference on Systems, Signals and Devices, (pp. 1–6).

[3] He, T., Cheng, E., & Yuan, F. (2010). Application of turbo code in underwater acoustic communication. In International Conference On Computer and communication Technologies in Agriculture Engineering (CCTAE), (Vol. 2, pp. 193–196).

[4] Carrascosa, P., & Stojanovic, M. (2010). Adaptive channel estimation and data detection for underwater acoustic MIMO OFDM systems. *IEEE Journal of Oceanic Engineering*, 35(3), 635–646.

[5] Zhou, R., Le Bidan, R., Pyndiah, R., & Goalic, A. (2007). Low-complexity high-rate reed–solomon block turbo codes. *IEEE Transactions on Communications*, 55(9), 1656–1660.

[6] Tao, J., Zheng, Y. R., Xiao, C., & Yang, T. C. (2010). Robust MIMO underwater acoustic communications using turbo block decision-feedback equalization. *IEEE Journal of Oceanic Engineering*, 35(4), 948–960.

[7] Gale, A., Scognamiglio, D., Al-Juboori, A., Toth, M., & Aharonovich, I. (2024). Underwater communications using on-demand quantum light sources. *Technologies for Quantum Technology*, 4, 025402.

[8] Zhang, M., & Zhou, H. (2023). Real-time underwater wireless optical communication system based on leds and estimation of maximum communication distance. *Underwater Optical Wireless Communication (OWC) Systems.* 23(17), 7649.

[9] Arthi, R., Kumar, M. N., & Krithika, S. (2023). Block chain based underwater communication using Li-Fi and eliminating noise using machine learning.

Jordan Journal of Electrical Engineering, 9(2), 166–174.

[10] Chen, C., Krishnamoorthy, A., Videv, S., Sparks, A., Das, S., & Babadi, S. (2024). Laser-based light sources for 1100 Gbps Indoor access and 4.8 Gbps outdoor point-to-point Li-Fi transmission systems. *Journal of Lightwave Technology,* vol. 42, no. 12, pp. 4146–4157.

[11] Irja, H. M., Madya, D., & Wijayanto, Y. N. (2021). Analysis of underwater communication on wireless optical system. *e-Proceeding of Engineering,* vol. 8, no. 5, pp. 1–10.

[12] Berger, C. R., Zhou, S., Preisig, J., & Willett, P. (2010). Sparse channel estimation for multicarrier underwater acoustic communication: from subspace methods to compressed sensing. *IEEE Transactions on Signal Processing*, 58(3), 1708–1721.

68 Transmission system uses UPFC controllers combined with ant colony optimization to reduce power fluctuation

Koduru Rohith Kumar[1,a], K. Ramya[2,b] and Kiran Kumar Kuthadi[2,c]

[1]PG Scholar, Department of EEE, Sree Vahini Institute of Science and Technology, Tiruvuru, Andhra Pradesh, India

[2]Associate Professor, Sree Vahini Institute of Science and Technology, Tiruvuru, Andhra Pradesh, India

Abstract

Var compensation can be used to reduce transmission line losses from the electric power system (PS). Because of the emergence of energy exchanges with transmission accessibility, the importance and awareness of Flexible AC Transmission Systems (FACTS) devices for modifying line voltage flows to reduce congestion and enhance common grid operation is growing. The unified power flow controller (UPFC) is the FACTS device. It has the ability to concurrently or intelligently modify an overall factor that affects the transmission line power in its response to drift, such as voltage, impedance and phase angle. Ant Colony Optimization (ACO) is used during this study to obtain the appropriate power change parameters for the UPFC device. According to the ACO approach, shunt converter regulates dc link voltage and transmission line reactive fluctuations. Along with deviation and UPFC bus voltage, series converter regulates transmission line true power. When wealthy and transient scenarios are present, the UPFC control device regulates real/reactive power, which is controlled by ACO approaches, can accomplish superior overall performance. ANFIS control technique and the conventional tuning technique are compared with the results of ACO continuous implementation on regulate to using MATLAB/Simulink.

Keywords: ANFIS and UPFC, ant colony optimization (ACO), Flexible AC Transmission Systems FACTS

Introduction

The electrical grid focuses on the demands of AC transmission in the majority of countries and continents. If energetic grid elements are not used, the power flow will often follow the least-impedance path, which means it is uncontrolled. Flexible AC Transmission Systems (FACTS) allow consumers to regulate the way the ac transmission grid operates by providing power electronics support. Some control over one equipment is possessed by the transmission device operator [1]. The unified power flow controller (UPFC) is the best versatile FACTS device. Since a transmission line with UPFC can modify the way an undesirable quantity of power is distributed across parallel types, it can be more beneficial to use transmission system for all parallel power flows. The next technology of the FACTS devices can provide shunt and series compensations in transmission line [2]. Interest in identifying an appropriate regulated strategy that suits the operating characteristics of various devices has grown since the UPFC was proposed in 1992.

Earlier research has recommended a number of controllers to guide the UPFC operation. In [3] using cross-coupling proportional-integral (PI) controllers, numerous actual and reactive power glides can be decreased in interaction. Additionally, decoupling PI controllers with a predictive internal control loop can be studied to reduce the influence of harmonics at nearest developments of measurement in [4]. An authorized hybrid PI controller (direct coupling and cross coupling) can change to lessen slight power variation added by controller strategy [5]. The median issue is the inability of PI controllers to consistently produce good outcomes over a large number of operating parts. This happens as a result of the regulated parameters can be assigned to specific device characteristics. Furthermore, procedures mostly on the fuzzy controller principle can be identified [6–8]. The disadvantage of the qualitative technique (which uses a fuzzy, normal region sense) is that the

[a]kodururohithkumar@gmail.com, [b]kambhamramya@gmail.com, [c]kiran9949610070@gmail.com

DOI: 10.1201/9781003684589-68

membership abilities determined are not modified to suit the device operating locations even though it can give fast responses and regular functioning.

An ANFIS or adaptive neurofuzzy inference device combines the fuzzy logic approach with the adaptive capabilities of neural networks to provide better suited usual overall performance [9]. A median fuzzy-unusual place sense controller can be taught with significantly fewer unique technical expertise than a controller device entire on this concept [10].

Ant colony optimization (ACO) is a new meta-heuristic strategy that can be used to determine the best solutions to power device issues, including power device stabilization and economical power dispatch [11, 12]. Recently, the ACO approach has been used to optimize gain parameter improving to enhance the stability of multimachine load on power systems [13]. Furthermore, ACO has been used to get the controller ideal configuration for rejection of disturbances inside the control systems for induction motors and magnetic levitation, respectively [14, 15]. Therefore, it is decided to use the ant colony optimization approach in order to obtain the optimum disturbance rejection in a transmission network which is mainly reliant on UPFC.

The present operate models the UPFC as regulated voltage updates. Each delivery is divided into orthogonal parts by handling the UPFC obtaining, which is used to regulate actual/reactive power flow within the transmission line. The bus voltage and true power produced by series and shunt inverters are controlled by the shunt portion of the UPFC. As a consequence, four control loops are used an instantaneously at which each sample provides the intended controller action.

The proper power controller parameters for UPFC device, this study uses the ant colony optimization (ACO) technique. The DC-link voltage and the transmission line reactive power flow are regulated by shunt converter in an order consistent with the essential control method. Both the actual power flow along with transmission line and the UPFC bus voltage are regulated by absorbing converters. Under normal and short-term scenarios, the UPFC control device real/reactive power, which is controlled by ACO approaches, can improve performance beyond typical limits. ACO is performed simultaneously with the control levels using MATLAB/Simulink software and the outcomes are contrasted with those of the well-known ANFIS control approach and the conventional tuning method. It assists in controlling

a range of power device parameters, including power flows and bus voltages. In addition to performing as a shunt, series and phase angle controller, UPFC main purpose is to properly balance flow of real/reactive power. The load flow evaluation model of UPFC can be revised to avoid dealing with congestion while identifying an affordable option.

Ant colony optimization algorithm

This approach originated from the herbal ant movement, which was initially described in the 1990 thesis of Dr. Marco Drigo on the ant system. Ants take more action to ensure their survival in their nests. The present academics believe that this type of behavior is an intelligent aspect overall. In the ant colony set of details, the individual looking for debris consists of huge, microscopic created ants as employ an Stigmage coherence, an instance of informal link that its surroundings use to supply the response. After traveling, all ant species disappear within the once again pheromone-like molecules. This is normal with large statements of ant behavior and realistic investigations. This substance secretion activity is undoubtedly controlled and eventually dissipates. It is possible to identify pheromones by using an ant sense of smell. In the same manner, they choose a route with the two paths most satisfying pheromone hobbies. An ant can go through a number of paths to reach its food source, but one of them is shorter than the other. Using one of the ants, a random decision is made on whether the ant will go in a longer or shorter way. With their pheromones, both ants spray the path. An ant that chose the shorter route can possibly arrive at the food source early, are at a few foods, and then skip back to the nest. According to this theory. It puts the food into the nest simultaneously with receiving it there before it come return to the food supply. However, since it could be continuing on its way to the food source, the ant that went a longer distance will probably arrive back to the hive slower versus the ant that followed the quicker path. Because the shorter route carries more pheromones, certain ants looking for food will choose that path because it comes with an additional pheromone hobby. The majority of ants also choose the shorter path after a certain amount of time. The form of the ACO set of signals can be presented when the ant preferences are properly identified.

The three different types of control over head of objects approaches. Such consist of the abc to dq0

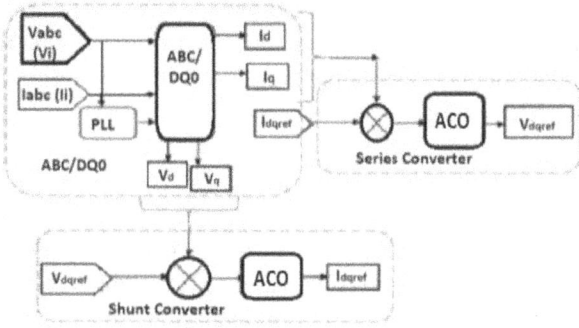

Figure 68.1 UPFC of ACO
Source: Author

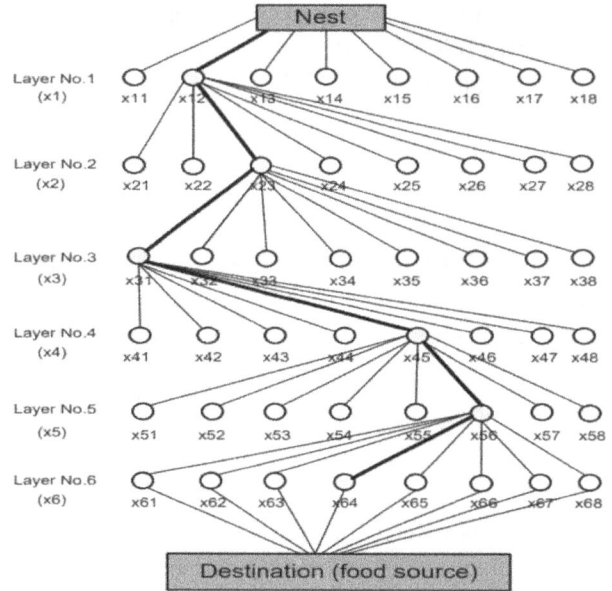

Figure 68.2 Indicates that ants use a method of determining the path from their nests to the food source to address a multidimensional problem
Source: Author

transfer, shunt regulator and series regulator. The ABCs actual and reactive components are separated using the abc to dq0 transfer. The series controller is used to compare measured Idq to the reference Idq. The aforementioned error is reported to the ANFIS controller, or if there is a chance, an ACO controller is suggested. Instead, Vdqref is now utilized. After comparing supply voltage to citation voltage, shunt controller a used to finish the procedure by converting it to a reference voltage. Shunt converters are used to regulate the voltage and series converters are used to control the reference voltage. Figure 68.1 depicts when the ACO converter performs more accurately in the present instance than the ANFIS controller.

An overview of an ACO challenge is as follows:

Let's assume that the ant colony is N ants. According to Figure 68.2, the ants started to leave the house. With each approach or time, an ant is eventually skipping upper point to lower place to benefit its aim (food). The remainder of formula states every ant within a stratification is entitled to one alternative with potential Pij. In reality, where τij is power of pheromone left at way in ij, the relevance of pheromone is highlighted. Ni displays nearby sites for an ant. An ant is positioned after that in j.

At the start of the optimization process, each route carries a preliminary pheromone hobby. Every iteration, which is created by searching for a variable maximum of the nest and the food items, will follow a reaction vector, or the path followed by the ant method. This search will continue till the desired number of iterations is reached. The price of each variable is chosen based on the direction with the highest intensity since it is a great component of the solution vector. In order to calculate the hazard of selecting each place using Equation 4, the following site (j) is selected when the sufficient ant is at function i utilizing τij

pricing. Every ant identified on the journey is a candidate to provide a response. Figure 68.2 shows the direction followed by the appropriate ant approach as an example. A tiny amount of pheromone is released by each ant in its trail. The following technique exhibits an upward surge in pheromone hobby when each ant crosses the maximum of the i and j locations.

$$\tau_{ij}^{(k)} \leftarrow \tau_{ij}^{(k-1)} + \Delta\tau^{(k)} \tag{1}$$

A fragrance can be collected in each column as Ant K pursues a trajectory by the formula as follows:

$$\tau_{ij} \leftarrow (1-\rho)\tau_{ij} \qquad \forall(i,j) \in A \tag{2}$$

where the evaporation coefficient, represented by the symbol ρ, takes a value between [0,1]. The nest is connected to the food supply via route A. Since smaller solutions are eliminated by the evaporation process, the search strategy can get more dynamic because there can be shorter, alternative pathways. Ant mobility, pheromone evaluation and pheromone residual charge were all characteristics that contribute to their hunt for iterations. The formula listed below can be used to refresh the aroma compound after every ant is returned:

$$\tau_{ij} = (1-\rho)\tau_{ij} + \sum_{k=1}^{N} \Delta\tau_{ij}^{(k)} \tag{3}$$

where the amount of pheromone is the stronger ant k as on arc ij is denoted $\Delta\tau_{ij}(k)$. The fragrance on ij route an acquired best ant:

$$\Delta\tau_{ij}^{(k)} = \frac{Q}{L_k} \qquad (4)$$

In the itinerant selling problem, where Q is constant, Lk is distance traveled k ants (should they be relocating from one city to another), There are various approaches to solving Equation 5:

$$\Delta\tau_{ij}^{(k)} = \begin{cases} \frac{\hat{u}f_{best}}{f_{worst}}; & if\ (i,j) \in globalbest\ tour \\ 0; & otherwise \end{cases} \qquad (5)$$

where the parameter ℶ is used to change the pheromone replace scale, and fworst and fbest represent the worst and satisfactory charges of the target feature of the several steps followed via N ants, respectively. The more pheromones that are acquired at the ultimate international path, the greater the set of rules it searches for adjacent growth, locate an actual rise. Below demonstrates the ant colony set of rules standard flowchart.

Results

In order to verify the overall effectiveness of the proposed ACO for regulating the UPFC under various operating circumstances (V1, V2, X are 1p.u., Xsh= 0.15p.u.), device is shown in Figure 68.1 and the MATLAB simulation is used. The following conditions are examples of the way the UPFC can also separately control flow of true and imaginary power inside transmission line. In each evaluation of the proposed ACO approach, a PI controller is employed. PI controller settings are developed to reduce interaction across the majority of true and

reactive power circuits for the purpose of limiting the maximum overshot in an arbitrary period in the future of the limited part and the simulation circuit depicted in figure above Figure 68.3.

Case study

The situation achievable challenge area continuously presents both the actual operational components and the expected set. Figure 68.4(a) and 4(b) show devices two outstanding short-circuit stages. Reactive power is transmitted by the device while active power remains constant. An imaginary power remains constant while true power is increasingly changed. Although dealing with reactive and real power in an effective approach, the simulation results clearly show that the recommended controller operates effectively. The gadget responsiveness is roughly equal to that of the PI controller even while utilizing ACO set of references attempt to get true power level and lessen the relation between true and reactive power with the drift, especially for the active power circuit loop. Evaluation of the two graphs can show that the PI controller responds to low short-circuit stages with a significantly slower reaction time than the ACO; this is far less sensitive to variations in the level of device failure. This can be controlled by choosing the appropriate revenue. The ACO algorithm makes use of this information to ascertain the optimal number generated by the desired variable.

Further, the depicted plot indicates that the UPFCs provide the ability to adjust line voltage at which shunt inverter is connected.

The total average performance of UPFC for power fluctuation correction is evaluated on the proposed device using ACO approach for controlling DC link voltage. ACO is used as an implement high-quality PI controller parameter.

Figure 68.4 of the accompanying motion graph for Ant colony optimization displays the required and actual device traveling elements taken into consideration in this incident. There are several criteria that determine the form of a path for the difficulty of enhancing the UPFC settings on this check. The route is produced at random (desired active power, desired reactive power and desired voltage) Within the UPFC parameters upper and lower travel limits. With the help of the valuable resources of (ω), the new ant path speeds are constructed, and the old ant path speeds are actually increased inside the path of the worst and exceptional ant pheromones by using

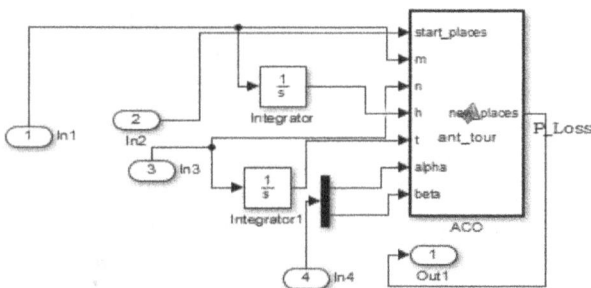

Figure 68.3 Simulink block of ACO diagram
Source: Author

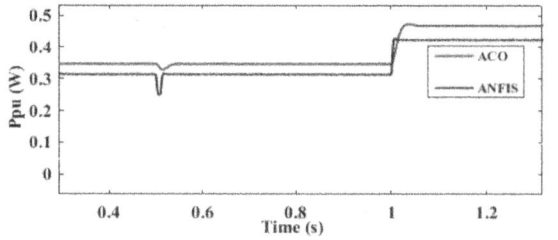

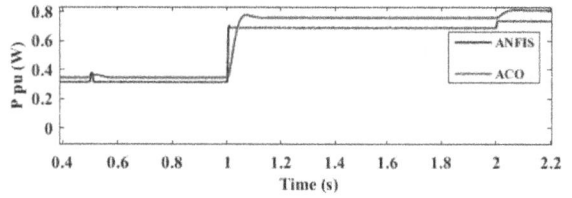

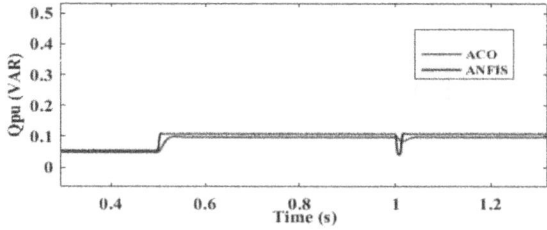

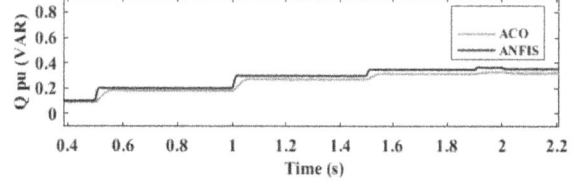

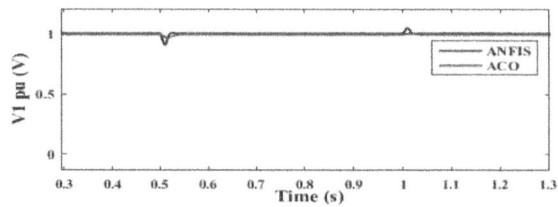

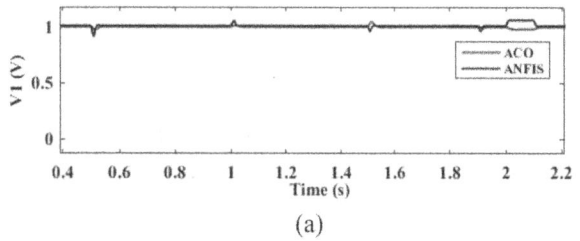

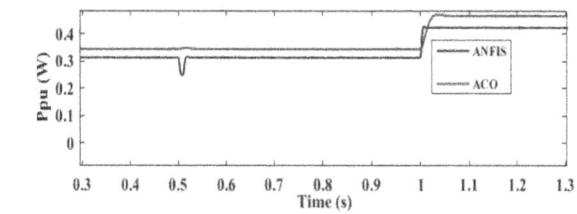

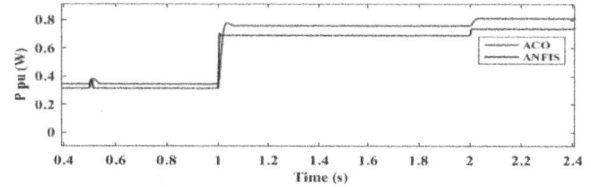

(a)

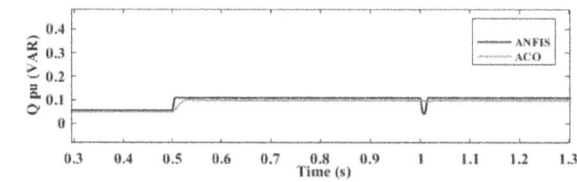

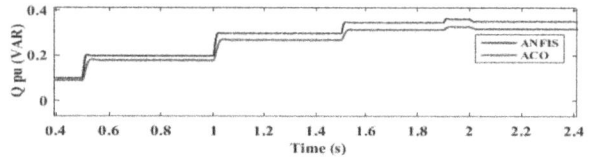

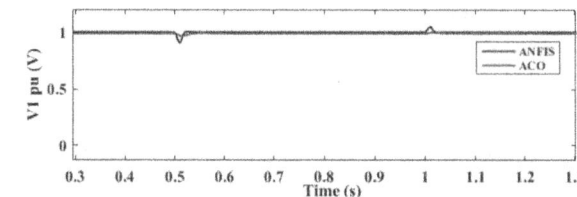

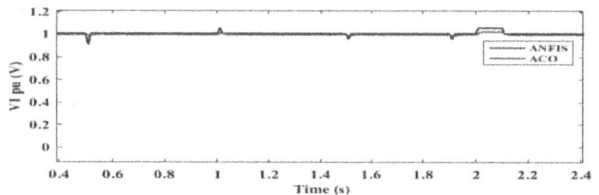

(b)

Figure 68.4 Shows how the gadget reacts to two different short-circuit levels. (a) The device behavior to a high level of short circuit. (b) How a low short-circuit level affects the device
Source: Author

(b)

Figure 68.5 The response of the system to violations of limits is shown in (a) How the system responds to ACO and unoptimized PI. (b) The response of the system when ACO and optimized PI are applied
Source: Author

the regression method. Figure 68.5(a) and (b) show that the relative "globally modifying" of (fworst) and (fbest) is outlined by the scaling factors, (c1) and (c2). Although c2 determines the influence of the ant pheromones, c1 and c2 describe the impact of the ant's first appearance in a pleasant area. The nutritional evaluation for this exam is described in the following paragraphs.

Conclusion

Unified power flow controller (UPFC) line voltage and true power that travels down transmission lines are both controlled on series converter. By applying active/imaginary power controlled by ant colony optimization (ACO) procedures in conjunction with the UPFC control tool, high regular accepted overall performance can be achieved in both brief and steady- state scenarios. The MATLAB/Simulink software is used in the well-known ANFIS modification methodology and the results of ACO are evaluated in comparison to those of the conventional tuning technique. Among other novel power tool aspects, it allows for the regulation of bus voltages and strength flows. UPFC is to stabilize wave of true and imaginary power, in addition to acting as a shunt, series and section method of processing controller. The work of art uses the ACO approach to determine the optimal energy control parameters for the UPFC tool. In accordance with the fundamental modulation approach, shunt converter regulates transmission-line reactive power level and DC-link voltage.

References

[1] Gyugyi, L. (1992). A unified strength float manipulate: concept for bendy ac transmission structures. *Proceedings of the Institution of Electrical Engineers, Part C*, 139(4), 323–331.
[2] Padiyar, K. R., & UmaRao, K. (1999). Modelling and manipulate of unified strength float controller for brief balance. *Electrical Power and Energy Systems*, 21, 1–11.
[3] Yu, Q., Norum, L., Undeland, T., & Round, S. (1996). Investigation of dynamic controllers for a unified strength float controller. In Proceedings of IEEE twenty second International Conference on Industrial Electronics, Control and Instrumentation, Taiwan, (pp. 1764–1769).
[4] Papic, P., Povh, D., & Weinhold, M. (1997). Basic manipulate of unified strength float controller. *IEEE Transactions on Power Systems*, 12(4), 1734–1739.
[5] Fujita, H., Watanabe, Y., & Akagi, H. (1999). Control and evaluation of a unified strength float controller. *IEEE Transactions on Power Electronics*, 14(6), 1021–1127.
[6] Vilathgamuwa, M. (1997). A synchronous reference frame-primarily based totally manipulate of a unified strength float controller. In Proceedings of International Conference on Power Electronics, Drives Systems, (Vol. 2, pp. 844–849).
[7] Jalayer, R., & Ooi, B. (2014). Co-ordinated PSS tuning of large power systems through combining transfer function-eigen characteristic analysis (TFEA), optimization, and eigenvalue sensitivity. *IEEE Transactions on Power Systems*, 29, 2672–2680.
[8] Guo, Z., Yu, K., Jolfaei, A., Bashir, A. K., Almagrabi, A. O., & Kumar, N. (2021). A fuzzy detection system for rumors thru explainable adaptive learning. *IEEE Transactions on Fuzzy Systems*, 29(12), 3650–3664. doi: 10.1109/TFUZZ.2021.3052109.
[9] Khadanga, R., & Satapathy, J. (2015). Time put off method for PSS and SSSC primarily based totally coordinated controller layout the usage of hybrid PSO–GSA set of rules. *Electrical Power and Energy Systems*, 71, 262–273.
[10] Mitra, S., Bhattacharya, A., & Dey, P. (2018). Small signal stability analysis in co-ordination with PSS, TCSC, and SVC. In Proceedings of the International Conference on Computation of Power, Energy, Information and Communication (ICCPEIC), Chennai, India, 28–29 March 2018, (pp. 434–441).
[11] Shahgholian, G., & Movahedi, A. (2016). Power gadget stabiliser and bendy alternating contemporary transmission structures controller coordinated layout the usage of adaptive speed replace rest particle swarm optimisation set of rules in multi-gadget strength gadget. *IET Generation, Transmission & Distribution*, 10, 1860–1868.
[12] Dey, P., Mitra, S., Bhattacharya, A., & Das, P. (2019). Comparative have a look at of the results of SVC and TCSC at the small sign balance of a strength gadget with renewables. *Journal of Renewable and Sustainable Energy*, 11, 033305.
[13] Panda S, Yegireddy NK, Mohapatra SK. Hybrid BFOA–PSO method for coordinated layout of PSS and SSSC-primarily based totally controller thinking about time delays. Int J Electr Power Energy Syst 2013; 49:221–33.
[14] Abd-Elazim SM, Ali ES. A hybrid particle swarm optimization and bacterial foraging for most desirable strength gadget stabilizer layout. Electric Power Energy Syst 2013; 46:334–41.
[15] Esmaili MR, Hooshmand RA, Parastegari M, Ghaebi Panah P, Azizkhani S. New coordinated layout of SVC and PSS for multi-gadget strength gadget the usage of BF-PSO set of rules. Procedia Techno 2013; 11:65–74.

69 Generative AI and deep learning for incorporating diversity: An inclusive approach to Indian sign language recognition and translation

Ananya Mallu[1,a] and Fakruddin Mohammed[2,b]

[1]IB, Department of IB, CHIREC International, Serilingampally, Hyderabad, Telangana, India

[2]Director, Department of Robotics and AI, ZAS Academy, London, UK

Abstract

The Deaf and Hard-of-Hearing communities in India face multiple communication challenges which create access barriers in their work study and educational environments. Furthermore, current sign language translation systems focus more on achieving precise model outputs than actual real-world usability which results in service gaps, especially in rural areas and regions where English is not dominant in India. This paper presents original and inclusive sign-to-text and text-to-sign pose/video translation for Indian Sign Language (ISL). For text-to-sign, we use generative AI and deep learning models with an innovative lookup-based method. The proposed approach depends on pre-defining sign poses for vocabulary words through built-in signs which prevents the need for extensive datasets and minimizes computational requirements. There is a significant increase in vocabulary from 263 to over 2900 words through synonym addition which demonstrates both scalability and adaptability in the model. For sign-to-text, the solution depends heavily on the GPT-4o model for transforming gloss to tokens to English text and we found 53% of sentence pairs reached perfect similarity while more than 75% of pairs showed a similarity score above 60%. Neural machine translation (NMT), convolutional neural networks (CNNs), and long short-term memory (LSTM) models within the proposed system work together with Google's MediaPipe library and ChatGPT to create an uninterrupted end-to-end ISL translation system. The proposed solution enables real-time bidirectional communication between ISL users and non-signers which fosters inclusivity and accessibility, especially benefiting underrepresented communities.

Keywords: AI, convolutional neural networks CNN, GLOSS, GPT, ISL Indian Sign Language, LLM, neural machine translation NMT, RNN, Sign Language, sign to text, text to sign

Introduction

The Deaf and Hard-of-Hearing (DHH) community has a vast communication gap, mainly because of the general lack of awareness about sign languages among hearing individuals. Lack of communication more often tends towards social isolation, restricting their effective involvement in education, career, and everyday communication [1, 2]. In addition, having to depend on lip-reading and interpreters, which are not only inaccurate but also invasive, further exacerbates these problems, causing the DHH community to feel left out of most social environments and opportunities. Their lack of representation in the media also perpetuates negative stereotypes and reinforces discrimination. Resolving these problems depends on a multi-pronged approach founded upon increasing public awareness, the accessibility of

interpreters, and investment in sign language technologies that promote inclusion.

There are nearly 466 million individuals with hearing impairment worldwide, and the number will grow to more than 900 million by 2050 [3, 4]. In India, 18 million individuals utilize ISL as their everyday language and 63 million individuals (6.5% of Indian population) [5] have hearing impairment. Despite this growing population, a considerable communication gap persists between sign language users and the broader society due to the limited proficiency in sign language among hearing individuals. Due to the fact there are over 200 sign languages in the world sign language interpretation can be quite a complex process [6] as each sign language has its own grammar, syntax and cultural markup. These international figures underscore the need to develop new and low-cost

[a]ananyamallu@gmail.com, [b]fakruddin.mohammed@zasacademy.org

DOI: 10.1201/9781003684589-69

ways of addressing the communication requirements of the DHH population. Sign language is not as simple as it is multimodal in nature, it involves not only hand movement but also body posture and facial expression [7, 8]. It is costly and time-consuming to gather large amounts of sign language video-text data which makes it difficult to obtain good quality translation models. In addition, the grammatical structure of sign languages is quite different from those of oral languages, and gloss annotations, which are crucial for many translation tools, are often incomplete and inconsistent across sign languages. These challenges raise the need to develop more sophisticated and inclusive strategies for sign language interpretation. In the context of Sign Language Translation (SLT), several approaches have been adopted to tackle these challenges each with its strengths and weaknesses. Rule-based systems [9] are based on the use of previously established linguistic rules; however, they are tedious and require a sign language expert as well as a natural language expert. Example-based methods employ parallel datasets to compare sign input sequences with stored examples; however, this is limited by the availability of the data. Some newer statistical and deep learning techniques such as RNNs, CNNs and transformers for instance [10-11, 13] are able to learn the data relations without being given explicit rules. Gloss-based methods [14, 15] use glosses as an intermediate between the sign language and the target language, while gloss-free [16] methods translate from sign language into spoken language directly, extending the utility of the systems. Combined approaches [17] have been used to deal with the multimodal nature of sign language [18] and include aspects such as hand movements and facial expressions. The current advancements, for instance, the Spotter Plus GPT method and the Sign LML framework [19, 20] have also contributed to the development of hybrid and gloss-free translation models. These frameworks utilize the existing large language models and video-to-token decoding to produce the spoken language outputs without the need for additional training. Some new developments including the use of 3D models [21], avatars and deep learning-based [22, 23] representations are also being used in sign language translation and are becoming popular. However, these developments show that sign language translation systems are still being developed to improve the accuracy and inclusivity of the systems. Although datasets for non-Indian sign languages such as RWTH-PHOENIX-Weather 2014T for German Sign Language and WLASL for American Sign Language [12] are easily obtained, there are limited resources available for Indian Sign Language (ISL). At present, there are only three small-sized datasets for ISL which makes it difficult to develop strong ISL translation systems. The small amount of curated and diverse datasets for ISL show that more research and resource development is needed to enhance translation precision and address the specific needs of India's sign languages.

Need for Indian sign language translation

India with 28 states and numerous regional sign languages is a challenging example for ISL translation due to its linguistic diversity. Different from American and European sign languages which have been researched more extensively [7, 8], ISL and its dialects have not been well investigated [26, 27]. This means that in rural and remote areas of India, people may not be able to speak either English or ISL, meaning that the current sign language recognition and translation systems are not sufficient for the Indian situation. In these regions, communication barriers are further complicated by linguistic diversity.

Proposed solution

To address these challenges, this paper proposes a comprehensive solution that incorporates generative AI, deep learning and neural machine translation techniques to develop a comprehensive ISL translation system. This end-to-end system that incorporates sign language translation with spoken languages for India's regional languages develops new approaches, datasets, and models suitable for India's linguistic environment. This solution enables greater accessibility and inclusiveness for ISL users across the country.

Objectives

This research, therefore, aims at redefining the translation of Indian Sign Language by tackling the peculiar linguistic and cultural challenges of India, especially with regard to the non-English-speaking rural population. Most of the sign language translation systems in the present day either simply rely on their accuracy or large-scale datasets but usually fail to cope with practical communication needs, including ISL users from rural areas facing rich linguistic diversity. Key objectives of the study include:

1. Inclusion and accessibility: This aims at designing an ISL translation system for bridging communication gaps between the DHH community and non-signers in underrepresented regions, which can be adaptable for different Indian regional languages and transcend the linguistic and cultural barriers that up until now have placed strong limitations on the use of ISL.

2. Scalable practical solutions: Present a new lookup-based approach for translation in ISL, which reduces the dependency on data and computational complexity. Using pre-built sign poses and synonyms, this method scales up the vocabulary from 263 to more than 2900 words, hence making the system scalable for real-world applications.

3. Deep learning integrated: Generative AI integration utilizes cutting-edge state-of-the-art models, like GPT-4o combined with CNN and LSTM networks to have efficient translations between text to sign and vice versa. Therefore, the aim in integrating the techniques will render this system feasible to be implemented with high levels of accuracy during communication for users of ISL and non-signers in real time.

4. Real-time bidirectional: To develop the end-to-end solution that would enable real-time, bidirectional communication between users of ISL and speakers of multiple Indian languages. The system will leverage the powers of neural machine translation (NMT) and MediaPipe to realize natural conversations, particularly in places without the availability of interpreters.

5. Adaptability to linguistic diversity of India: The present research tries to address the challenge thrown up by the regional linguistic diversity of India. Since the research will be focusing on integrating the regional languages with ISL, it has opened avenues for greater inclusion of the DHH community even in rural and urban areas.

These goals set new standards for the translation of Indian Sign Language that would, as a result, be inclusive, accessible, and useful to India's diverse linguistic ecology.

Methodology

The proposed methodology for Indian Sign Language (ISL) translation is designed to address the linguistic diversity and limited use of English in India by offering an inclusive, end-to-end solution. The approach consists of two main workflows: sign language to regional text and regional text to sign language, ensuring accessibility across multiple regional languages.

Sign language to regional text workflow:

- **Sign language to gloss**: A video of sign language is processed using a 3D Convolution and LSTM model to convert the signs into gloss (intermediate text representation).
- **Gloss token to English sentence/text**: The gloss tokens are then translated into English text using pre-trained generative AI models like GPT-4o.
- **English sentence/text to Indian regional language text:** The resulting English text is translated into an Indian regional language through NMT models.
- **Regional language text to audio**: Finally, the regional language text is converted into audio using Google's Text-to-Speech model.

Regional language text-to-sign language workflow:

- **Spoken words to text**: Speech in Indian regional languages is converted to text using Google's Speech-to-Text model.
- **Indian regional language text to English text**: The local language text is translated into English using NMT models like GNMT, M2M-100, and IndicTrans.
- **English text to gloss**: Generative AI models like ChatGPT process the English text and convert it into gloss tokens, the building blocks for sign language.
- **Gloss tokens to sign language video**: Models such as SignAll or DeepMotion animate digital avatars to visually render the gloss tokens as Indian Sign Language video clips.

This comprehensive approach leverages deep learning, generative AI, and advanced machine translation techniques to facilitate seamless communication between sign language users and non-signers, across India's diverse linguistic landscape.

Dataset

There are currently only three publicly available datasets [24, 25, 28] for ISL. The first dataset contains images representing 20 commonly used words, while the second consists of a single video featuring 78 simple sentences [28]. The third and most comprehensive dataset [25] available for ISL includes a total of 3,679 videos representing 263 unique words. Each video lasts between 2 to 4 seconds. For training and evaluation purposes, the dataset is split into an 80:20 ratio, with 80% of the data used for training and 20% for testing. The words are categorized as follows:

This detailed dataset, offering a diverse range of word categories, is a critical resource for the development of ISL translation models. This version emphasizes clarity and conciseness while maintaining all the essential details about the dataset.

Results and Discussion

Sign language to regional text workflow
- **Sign language to gloss:** This study employs a hybrid model combining CNNs and long short-term memory (LSTM) networks for Indian Sign

Table 69.1 Dataset [28] description showing the number of videos in each word category.

Category	No of videos
Adjectives	791
Animals	150
Clothes	188
Colours	211
Days_and_Time	223
Electronics	110
Greetings	190
Home	189
Jobs	177
Means_of_Transportation	168
People	487
Places	399
Pronouns	168
Seasons	66
Society	162
Grand total	**3679**

Source: Author

Language (ISL) translation. CNNs extract spatial features from video frames, identifying critical elements such as hand shapes and facial expressions, while LSTMs capture temporal dependencies in continuous gestures.

We experimented with three pre-trained CNN models—MobileNet, ResNet, and EfficientNet due to their proven success in sign language recognition. The models were trained for 50 epochs, achieving the following accuracies: ResNet (80.95%), MobileNet (38.09%), and EfficientNetB0 (90.48%). The training and test data loss plots for these models are shown in Figure 69.1. Therefore, in this study, we used the EfficientNetB0 model to convert the sign language video poses to gloss tokens due to higher accuracy.

- **Gloss to English text:** The gloss tokens generated from ISL videos were input into large language models (LLMs) such as Chat GPT-4o, LLAMA, and Gemini using a structured prompt. The prompt used with these LLMs was:
  'You are a sign language translator with expertise in Indian Sign Language. Your task is to construct a meaningful English sentence based on a set of gloss tokens derived from sign language video clips. The sentence must be grammatically correct, and you should assume the conversation is between two individuals, with no third party involved. The input gloss tokens for this task are:'

This prompt was tested using the GPT-4o model and an ISL dataset consisting of gloss tokens derived from 78 videos [28]. The results demonstrate that LLMs are successfully able to generate coherent and grammatically correct English sentences. The detailed results are presented in Figure 69.2. The result shows that 44% of sentence pairs have a perfect similarity score of 1 and 75% of sentence pairs have similarity scores of above 0.60. Figure 69.3 shows, a response example of GPT-4o when gloss tokens generated by the CNN/LSTM model and the above prompt were fed as input.

- **English to Indian regional languages:** English sentences generated from gloss tokens were translated into Indian regional languages using Neural Translation Models like Google's Neural Machine Translation (GNMT), Facebook's M2M-100, Microsoft Translator, and IndicTrans.

Conv-LSTMM Model Trained on Indian Sign Language Dataset
(Training and Validation Accuracy Plots)

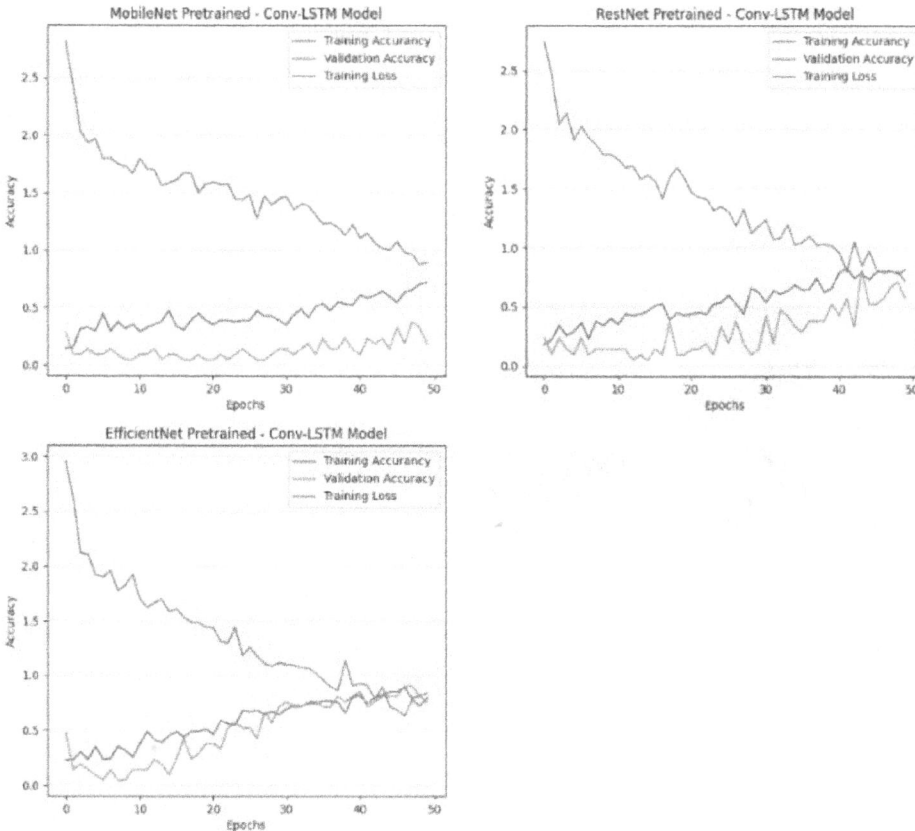

Figure 69.1 Conv-LSTM model performances: training and validation accuracy plots
Source: Author

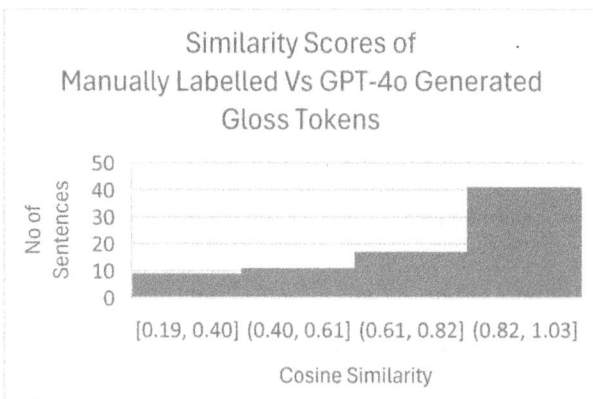

Figure 69.2 Chat GPT model performance in translating the English tokens to gloss tokens. The cosine similarity scores of 78 manually labelled gloss tokens comparison with GPT-4o generated gloss tokens. The table shows that 75% of sentence pairs have similarity scores of above 0.60 and the plot shows that 44% of sentence pairs have perfect similarity scores of one
Source: Author

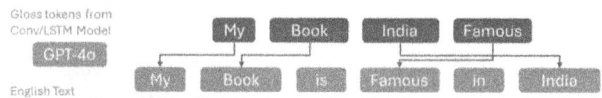

Figure 69.3 Implementation block diagram of gloss tokens to English text using the GPT-4o model
Source: Author

These models effectively preserve linguistic nuances and ensure high-quality translations for languages such as Hindi, Tamil, Telugu, and Bengali. In this study, we used Google's translation API to translate English text to Indian regional languages as shown in Figure 69.4.

• **Indian regional text to speech:** To enhance accessibility, translated text was converted into speech using text-to-speech (TTS) models, including Google's WaveNet, Microsoft Azure, and Amazon Polly. India's Bhashini initiative is also advancing open-source TTS solutions. For

My Book is Famous in India

Google Translation Model

मेरी किताब भारत में प्रसिद्ध है।   আমার বই ভারতে বিখ্যাত।   Many more local languages...

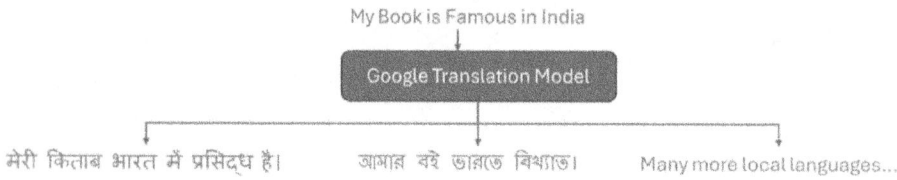

Figure 69.4 Implementation block diagram of the English text translation to number of Indian local languages using the Google translation model
Source: Author

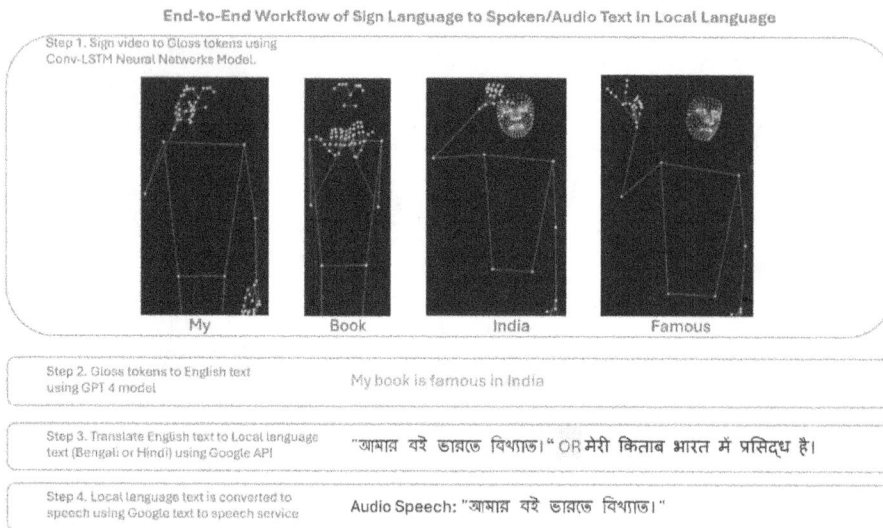

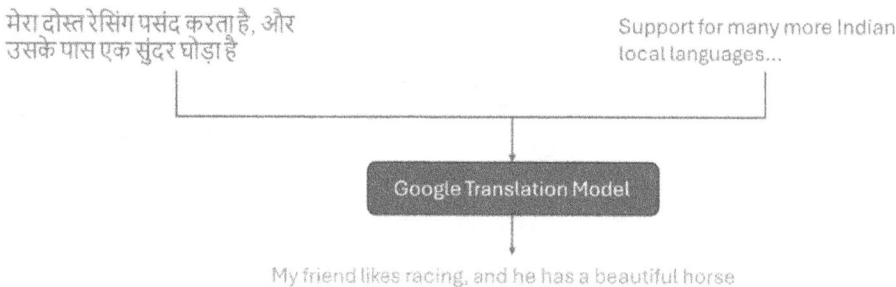

End-to-End Workflow of Sign Language to Spoken/Audio Text in Local Language

Step 1. Sign video to Gloss tokens using Conv-LSTM Neural Networks Model.

My   Book   India   Famous

Step 2. Gloss tokens to English text using GPT 4 model   My book is famous in India

Step 3. Translate English text to Local language text (Bengali or Hindi) using Google API   "আমার বই ভারতে বিখ্যাত।" OR मेरी किताब भारत में प्रसिद्ध है।

Step 4. Local language text is converted to speech using Google text to speech service   Audio Speech: "আমার বই ভারতে বিখ্যাত।"

Figure 69.5 An end-to-end four-step Indian sign language to regional text
Source: Author

मेरा दोस्त रेसिंग पसंद करता है, और उसके पास एक सुंदर घोड़ा है   Support for many more Indian local languages...

Google Translation Model

My friend likes racing, and he has a beautiful horse

Figure 69.6 The implementation block diagram of the Indian local language translation to English sentence using the Google translation model
Source: Author

this study, Google's Transcribe Service API validated the feasibility of the proposed end-to-end pipeline, demonstrating the real-world applicability of AI-driven ISL translation. The end-to-end pipeline of Indian Sign language to Indian Regional Text is shown in Figure 69.5.

Indian regional text-to-sign language workflow
- **Spoken words to text:** This process is just the opposite of sign-to-text workflow. In this process,

we used Google Speech-to-Text API to transcribe spoken words into regional language text.
- **Indian regional language text to English:** We used Google Translation API to translate the regional language text into English, as shown in Figure 69.6.
- **English text to gloss:** We used GPT-4o to convert the English text into ISL gloss tokens which removes the grammatical structures and retains essential words. For example, Figure 69.7 shows

how the sentence "I am going to the market" when fed to GPT-4o along with the prompt specified generates gloss tokens to "GO MARKET."

The prompt that was used with the Chat GPT-4o model is:

"Your role is to translate English sentences into Indian Sign Language (ISL) gloss tokens, following ISL grammar rules. If ISL grammar rules are unavailable, you may use American Sign Language (ASL) grammar as an alternative for generating the gloss tokens."

- **Gloss tokens to sign language video:** Instead of complex models, a dictionary lookup approach

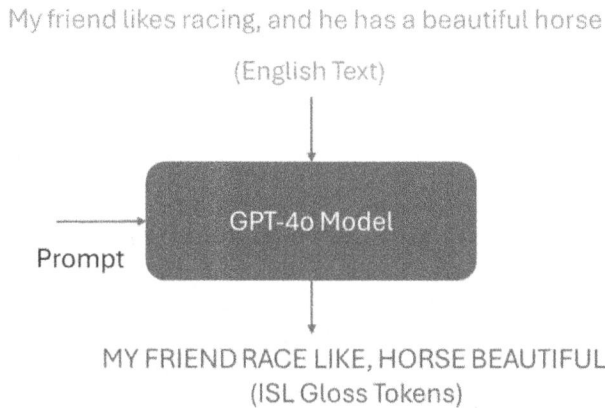

My friend likes racing, and he has a beautiful horse
(English Text)

GPT-4o Model

Prompt

MY FRIEND RACE LIKE, HORSE BEAUTIFUL
(ISL Gloss Tokens)

Figure 69.7 The implementation block diagram of the implementation of English sentences to ISL gloss tokens using the Chat GPT-4o model
Source: Author

maps gloss tokens to pre-built sign poses. Google's MediaPipe generates poses for 263 unique words, which expand to 2,927 with synonyms. This method reduces dataset dependency while enhancing vocabulary and communication flexibility.

If we leverage the existing ISL dictionaries, including the 10,000-word dataset from Ramakrishna Mission Vivekananda University, this approach can significantly improve the inclusivity of the deaf community in India. The Indian Regional Text-to-Sign Language end-to-end workflow, from spoken words to signed videos, is illustrated in Figure 69.8.

Conclusion

The research provides a practical method to develop Indian Sign Language (ISL) translation systems using deep learning and generative AI models to resolve the communication challenges faced by the Deaf and Hard-of-Hearing (DHH) population of India, especially in rural areas where English is not dominant This paper describes an end to end real-time ISL translation system which is based on convolutional neural networks (CNNs), long short term memory (LSTM) models, neural machine translation and GPT4. The major contribution of this study is the application of a pre-built sign pose dictionary which leads to reduced data requirement and computational cost. The lookup-based approach with synonyms' integration improves both the vocabulary coverage and the system's adaptability and

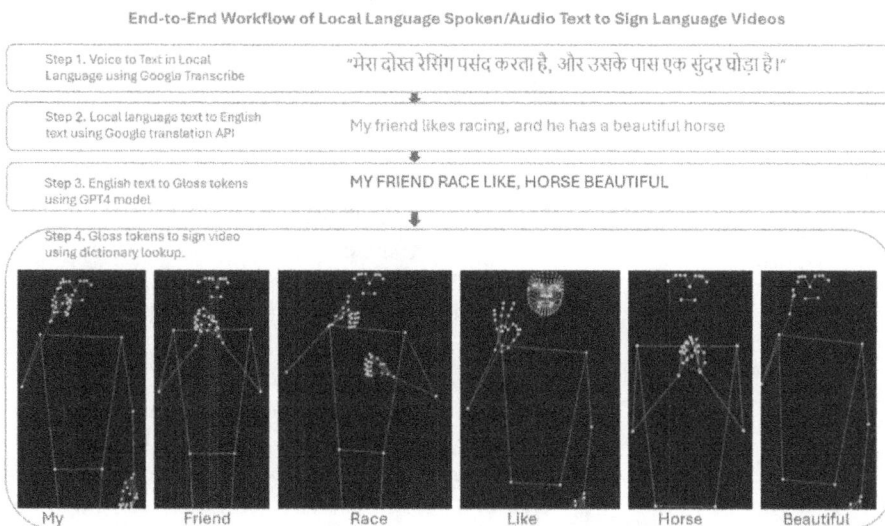

End-to-End Workflow of Local Language Spoken/Audio Text to Sign Language Videos

| Step 1. Voice to Text in Local Language using Google Transcribe | "मेरा दोस्त रेसिंग पसंद करता है, और उसके पास एक सुंदर घोड़ा है।" |

| Step 2. Local language text to English text using Google translation API | My friend likes racing, and he has a beautiful horse |

| Step 3. English text to Gloss tokens using GPT4 model | MY FRIEND RACE LIKE, HORSE BEAUTIFUL |

Step 4. Gloss tokens to sign video using dictionary lookup.

My   Friend   Race   Like   Horse   Beautiful

Figure 69.8 An end-to-end four-step spoken text to Indian sign language videos
Source: Author

extendibility. The EfficientNet model achieved 90% accuracy on the INCLUDE dataset which shows the effectiveness and usability of the system. The research proposes that future work with the Faculty of Disability Management and Special Education (FDMSE-RKMVU) would help grow the ISL dictionary making the system more scalable and supporting a wider range of regional languages. This collaboration would allow for the ongoing expansion of vocabulary and enhance the system's inclusivity. The effectiveness of this system to provide real-time, two-way communication between ISL speakers and non-signers shows a substantial advancement in meeting the requirements of the DHH population. As a practical application, this solution has the potential to be developed into mobile apps thus establishing a new standard for ISL translation in India while increasing accessibility for all, especially for marginalized groups.

References

[1]   National Deaf Children's Society (2023). Deaf Students 50% Less Likely to go to Top Universities. National Deaf Children's Society. Available from: https://www.ndcs.org.uk/about-us/news-and-media/latest-news/deaf-students-50-less-likely-to-go-to-top-universities/ (Accessed: 18 October 2024).

[2]   Golden Steps ABA (2023). Hearing Loss Statistics. Golden Steps ABA. Available from: https://www.goldenstepsaba.com/resources/hearing-loss-statistics#:~:text=How%20Many%20Deaf%20People%20in%20the%20World%3F,6.1%25%20of%20the%20world's%20population (Accessed: 18 October 2024).

[3]   American India Foundation (AIF) (n.d.). Why Indian Sign Language Should be Part of the Constitution and Curriculum. American India Foundation. Available from: https://aif.org/why-indian-sign-language-should-be-part-of-the-constitution-and-curriculum/ (Accessed: 18 October 2024).

[4]   Royal College of General Practitioners (RCGP) (n.d.). Hearing Loss and Deafness Statistics. RCGP eLearning. Available from: https://elearning.rcgp.org.uk/mod/book/view.php?id=12532&chapterid=288 (Accessed: 18 October 2024).

[5]   National Health Mission (n.d.). Deafness statistics in India. National Health Mission. Available from: https://nhm.gov.in/index1.php?lang=1&level=2&sublinkid=1051&lid=606 (Accessed: 18 October 2024).

[6]   Moryossef, A., & Goldberg, Y. (2021). Sign language processing. Available from: https://sign-language-processing.github.io/ (Accessed: 18 October 2024).

[7]   Alaghband, M., Maghroor, H. R., & Garibay, I. (2023). A survey on sign language literature. *Machine Learning with Applications*, 14, 100504. Available from: https://doi.org/10.1016/j.mlwa.2023.100504 (Accessed: 18 October 2024).

[8]   Ardiansyah, A., Hitoyoshi, B., Halim, M., Hanafiah, N., & Wibisurya, A. (2021). Systematic literature review: American sign language translator. *Procedia Computer Science*, 179, 541–549. Available from: https://doi.org/10.1016/j.procs.2021.01.038 (Accessed: 18 October 2024).

[9]   Mishra, G. S., Asteya, P., & Pooja (2019). Generating glosses of Indian sign language from English texts: a proposed hybrid machine translation method. *International Journal of Innovations in Engineering and Technology (IJIET)*, 13(1), 120. Available from: http://dx.doi.org/10.21172/ijiet.131.16.

[10]   De Coster, M., & Dambre, J. (2022). Leveraging frozen pretrained written language models for neural sign language translation. *Information*, 13(5), 220. Available from: https://doi.org/10.3390/info13050220.

[11]   Kaur, B., Chaudhary, A., Bano, S., Yashmita, Reddy, S. R. N., & Anand, R. (2024). Fostering inclusivity through effective communication: real-time sign language to speech conversion system for the deaf and hard-of-hearing community. *Multimedia Tools and Applications*, 83, 45859–45880. Available from: https://doi.org/10.1007/s11042-023-17372-9 (Accessed: 18 October 2024).

[12]   Gu, Y., Zheng, C., Todoh, M., & Zha, F. (2022). American sign language translation using wearable inertial and electromyography sensors for tracking hand movements and facial expressions. *Frontiers in Neuroscience*, 16, 962141. Available from: https://doi.org/10.3389/fnins.2022.962141 (Accessed: 18 October 2024).

[13]   Varanasi, A., Sinha, M., & Dasgupta, T. (2024). Linguistically informed transformers for text to American sign language translation. In Proceedings of the Seventh Workshop on Technologies for Machine Translation of Low-Resource Languages (LoResMT 2024), Bangkok, Thailand, (pp. 50–56). Association for Computational Linguistics. Available from: https://aclanthology.org/2024.loresmt-1.5 (Accessed: 18 October 2024).

[14]   Amin, M., Hefny, H., & Mohammed, A. (2021). Sign language gloss translation using deep learning models. *International Journal of Advanced Computer Science and Applications (IJACSA)*, 12. Available from: https://doi.org/10.14569/IJACSA.2021.0121178 (Accessed: 18 October 2024).

[15] Moryossef, A., Müller, M., Göhring, A., Jiang, Z., Goldberg, Y., & Ebling, S. (2023). An open-source gloss-based baseline for spoken to signed language translation. arXiv Available from: https://doi.org/10.48550/arXiv.2305.17714 (Accessed: 18 October 2024).

[16] Kumari, D., & Anand, R. S. (2024). Isolated video-based sign language recognition using a hybrid CNN-LSTM framework based on attention mechanism. *Electronics*, 13(7), 1229. Available from: https://doi.org/10.3390/electronics13071229.

[17] Devi, S. P., Vidya, V., & Balan, C. (2022). Media files to ISL: GAN based Indian sign language interpreter. In 2022 First International Conference on Electrical, Electronics, Information and Communication Technologies (ICEEICT), 16-18 February, Trichy, India. IEEE. Available from: https://doi.org/10.1109/ICEEICT53079.2022.

[18] Sincan, O. M., Camgoz, N. C., & Bowden, R. (2024). Using an LLM to turn sign spottings into spoken language sentences. arXiv. Available from: https://doi.org/10.48550/arXiv.2403.10434.

[19] Gong, J., Foo, L. G., He, Y., Rahmani, H., & Liu, J. (2024). LLMs are good sign language translators. arXiv. Available from: https://doi.org/10.48550/arXiv.2404.00925.

[20] Lancaster University (2024). Can Large Language Models (LLMs) Translate Sign Language? Lancaster University Portal. Available from: https://portal.lancaster.ac.uk/portal/news/article/can-large-language-models-llms-translate-sign-language.

[21] Varahagiri, S., Sinha, A., Dubey, S. R., & Singh, S. K. (2024). 3D-convolution guided spectral-spatial transformer for hyperspectral image classification. In Proceedings of the IEEE Conference on Artificial Intelligence 2024. Available from: https://arxiv.org/html/2404.13252v1 (Accessed: 18 October 2024).

[22] Huang, J., & Chouvatut, V. (2024). Video-based sign language recognition via ResNet and LSTM network. *Journal of Imaging*, 10(6), 149. Available from: https://doi.org/10.3390/jimaging10060149 (Accessed: 18 October 2024).

[23] Wang, S., Wang, K., Yang, T., & Li, Y. (2022). Improved 3D-ResNet sign language recognition algorithm with enhanced hand features. *Scientific Reports*, 12(1), 17812. Available from: https://doi.org/10.1038/s41598-022-21636-z (Accessed: 18 October 2024).

[24] Sridhar, A., Ganesan, R. G., Kumar, P., & Khapra, M. (2020). INCLUDE: a large scale dataset for Indian sign language recognition. In MM '20: The 28th ACM International Conference on Multimedia. Available from: https://doi.org/10.1145/3394171.3413528 (Accessed: 18 October 2024).

[25] Tyagi, A., & Bansal, S. (2022). Indian sign language-real-life words. *Mendeley Data*, v2 . Available from: https://doi.org/10.17632/s6kgb6r3ss.2.

[26] Priya, C. B., Ibrahim, S., Thangam, D. Y., & Jency, X. F. (2023). Tamil sign language translation and recognition system for deaf-mute people using image processing techniques. *Applied and Computational Engineering*, 8(1), 398–404. Available from: https://doi.org/10.54254/2755-2721/8/20230193 (Accessed: 18 October 2024).

[27] Shinde, A., & Kagalkar, R. (2015). Advanced Marathi sign language recognition using computer vision. *International Journal of Computer Applications*, 118(13), 1–7. Available from: https://doi.org/10.5120/20802-3485 (Accessed: 18 October 2024).

[28] Waghmare, P., & Deshpande, A. (2024). Indian sign language video and text dataset for sentences (ISLVT). *Mendeley Data*, v1 . Available from: https://doi.org/10.17632/98mzk82wbb.

70 COVID-19 detection using a deep learning model based on chest X-rays

V. Vijayasri Bolisetty[1,a] and U. Yedukondalu[2,b]

[1]Associate Professor, Department of ECE, Aditya College of Engineering and Technology, Surampalem, Andhra Pradesh, India

[2]Professor, Department of ECE, MVRCET, Paritala, Vijayawada, NTR (D.T), Andhra Pradesh, India

Abstract

Chest X-ray is being widely used as a COVID-19 diagnostic tool, a transmissible respiratory illness caused by the SARS-CoV-2 virus. However, interpreting chest X - rays for COVID-19 can be challenging for radiologists due to the subtle and non-specific findings. This study utilizes a CNN model trained on an extensive chest X-ray dataset to identify COVID-19 cases. To guarantee the model's efficacy and precision, the repository was split into training, validation and test sets. According to the findings, radiologists can benefit greatly from using a CNN for COVID-19 detection. CNNs may be a useful weapon in the battle against the pandemic, since they have been demonstrated to be effective in detecting COVID-19. However, for an effective and definitive verdict, chest X-rays should be used in synchrony with other med lab tests. This paper presents a CNN-based approach for identifying COVID-19 in chest X- rays, illustrating the promise of deep learning algorithms for real-time medical imaging applications. Medical images are captured by using Arduino UNO and are saved for future reference. The proposed model has achieved 96% accuracy with 93% sensitivity.

Keywords: Aurdino-Uno, Cchest CT scans, chest X-rays, CNN, COVID-19, deep learning model, ReLu, SARS-COV-2

Introduction

The COVID-19 pandemic, which is brought on by the SARS-COV2 virus, is the most widespread pandemic of the twenty-first century that spreads between humans by direct contact [1]. Worldwide, there are currently 38.7 million infected people and more over 1 million fatalities [2]. The most typical COVID-19 symptoms include fever, sluggishness, coughing, and diarrhea [6] and are having trouble in breathing, however only a small percentage develop acute respiratory distress syndrome [1]. The condition is thought to have a 3.4% death rate [4], however this percentage may differ between nations or regions. COVID-19 testing may be done in a variety of methods. Nevertheless, only the reverse transcriptase polymerase chain reaction (RT-PCR) test can definitively identify COVID-19.

SARS-CoV-2 RNA may be detected in a breath sample with the use of RT-PCR by laboratory processing of the test samples before it can be used. Since this test is considered as the most accurate method for identifying COVID-19, it is widely used to screen people who have symptoms or have had close contact with a confirmed one. Antigen tests identify viral proteins associated with COVID-19. These tests are fast and can provide results in a matter of minutes, but they may be less sensitive than RT-PCR tests. Reports of CT and X-ray examinations done before and after the onset of COVID-19 symptoms had different outcomes, as stated in [3]. These results show that CT and X-ray results may be used to evaluate whether or not a person has SARS-COV2.

One deep learning (DL) technique makes use of X-ray imaging, based on the fact that the corona virus produces pneumonia in the lungs of an infected person once it enters the respiratory system. Fluid builds up in the lungs, causing irritation and a condition called "Ground-Glass Opacity" (GGO). Hence, infected people with COVID-19 may be recognized on chest X-rays that are used to examine internal organs. Diagnostic radiologists are often tasked with analyzing X-rays [13, 20]. This study proposes an innovative approach to detect COVID-19.

[a]vijayasri.bollisetty@acet.ac.in, [b]sridryk2017@gmail.com

DOI: 10.1201/9781003684589-70

The workflow of the proposed model is outlined in following three steps:

STEP 1: The X-Ray image of patient lungs should be shown to the camera.

STEP2: The image which was captured by the camera is processed in the Arduino.

STEP 3: After completing the processing in Arduino the output is shown on the screen which is interfaced to the Arduino.

The chest X-rays should include the following features:

Image data: The designed model is trained with chest X- rays of patients with and without COVID-19.

Training and validation data: To learn and to recognize the characteristics and patterns that are linked to aforementioned pandemic, a model is developed by training on vast collection to achieve higher accuracy.

Model evaluation: After the model is trained on 80% of the repository, it is tested with remaining 20% pictures to assess how well it can identify the victim. The model's efficacy is measured using standard assessment criteria including sensitivity, specificity, and precision.

Deployment: The developed model can assist in diagnosing COVID-19 in a clinical setting. The model can be integrated with existing radiology systems and workflows to provide real-time results to healthcare providers.

Existing method

The DL model's usage in the medical field has recently exploded, particularly in diagnosis based on images. Artificial neural networks (ANNs) have surpassed traditional models in computer vision tasks relating to medical image processing, demonstrating strong performance.

CNNs are considered for their highly promising results [9, 10]. CNNs have been employed for several medical diagnostic classification tasks, including: malaria parasite identification in thin blood smear pictures [11], breast cancer diagnosis [12], interstitial lung disease diagnosis in chest radiography [14, 15]. Researchers have detailed in [16] the significance of applying AI approaches for image analysis in the identification and medicament of COVID-19 patients. This pandemic can be identified with a high accuracy, by analyzing the computer tomography scan [16]. The proposed work focuses on a custom-built machine learning model to identify the patients affected by the aforementioned pandemic using X-ray images. This repository is built by collecting data from bronchial pneumonia victims, COVID-19 patients, and healthy individuals, which has been reviewed in another significant paper [17]. In this research, state-of-the-art CNN architectures is employed to develop a system capable of automatically detecting COVID-19 patients.

As the proposed model works on the detection of pandemic affected patients through the data repository, low quality X-rays or the less data collected from the SARS victims may reduce the efficiency of the model. So, this work focuses on pre-processing of datasets to enhance the quality, data supplementation, and design efficacy. This conclusion was reached following an analysis of the current literature on the topic of employing CNNs for identification.

Proposed Methodology

The proposed model works on the detection of the pandemic virus from the repository, the design flow is represented in Figure 70.1 and working flow of the CNN algorithm is shown in Figure 70.2. The CNN was initially trained on the original repository supplied, however its accuracy proved inadequate for the task at hand, as it is 54% only. The key datasets used in this investigation are:

i. Positive case images retrieved from GitHub, which serve as the primary dataset. The Ethics Committee at the University of Montreal collected this data collection from a variety of medical facilities, and it is known as CERSES-20- 058-D [5].

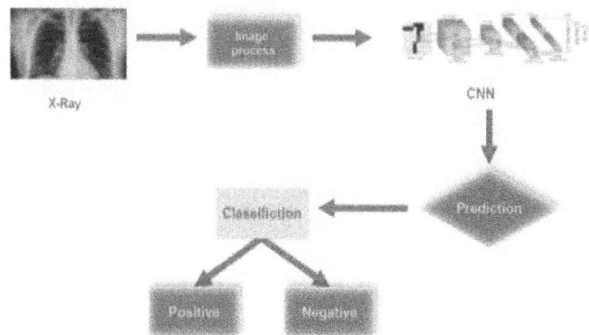

Figure 70.1 System design flow
Source: Author

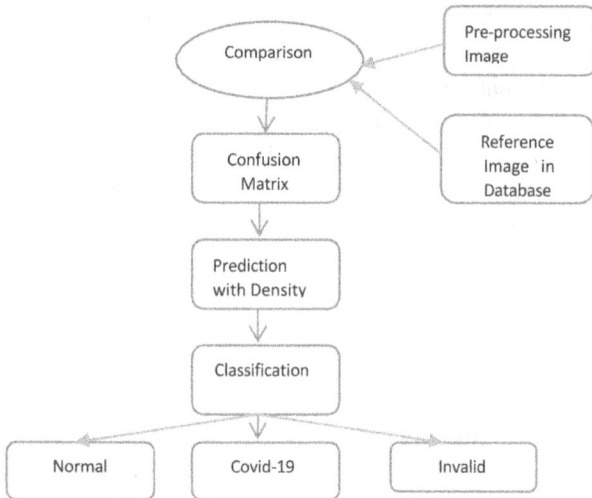

Figure 70.2 CNN flow
Source: Author

ii. Chest X – ray images were gathered from Kaggle [8] in order to balance the dataset. The total dataset initially consists of 6000 X - Ray images and are compressed to 5841 after pre-processing. Among those 4265 are affected cases and 1576 are normal cases.

Data pre-processing

Classification in datasets: Striking a balance, to enhance the effectiveness of the developed model, a supplemental dataset of approximately 500 normal chest X-ray pictures was incorporated to balance the original dataset. These supplemental X-ray pictures were obtained via Kaggle [8] and stitched together. Hence the accuracy improved to 69% and the models claimed to be reliable for detecting COVID19.

Data augmentation: The number of data examples available when training a model can be significantly increased by employing a technique called data augmentation [7]. This approach applies standard image processing procedures to repositories, such as inverting, rotating, cropping, or augmenting the images. In this proposal, CNN was hindered and enrich the learning model with additional features. Both image flipping and image rotation were incorporated into this study's data set.

CNNs: CNNs are widely used and are made up of numerous convolutional layers that apply filters to particular areas of the input data, decreasing the dimensionality of the data and extracting key

features. On a variety of computer vision tasks, CNNs have produced state-of-the-art results and are still a popular option.

Typically, a convolutional neural network shown in Figure 70.3, contains the convolutional layer, pooling layer, activation layer, normalization layer, fully connected layer, dropout layer. Deep convolutional neural networks are built by repeatedly stacking and reusing these layers. The exact number and types of layers used in a CNN depend on the specific task and the dataset being used. The structure of a ConvNet is inspired by the human brain's Visual Cortex and its connecting network of neurons. In the receptive the individual neurons respond to stimuli. Overlapping fields of this kind cover the entire visual field.

Input image

The three color channels of the RGB image - Red, Green, and Blue have been separated, as shown in Figure 70.4. Images can exist in various color spaces, including grayscale, RGB, HSV, and CMYK, are important considerations in architecture design. During convolution, the right-side movement of filter continues till the entire image is traversed, with a fixed stride value.

The matrix multiplications across the kernel (Kn) and input (In) layers, results the outcomes and is then applied to the total as shown in Figure 70.5. Similar to the convolutional layer, the pooling layer reduces the size of the convolved feature as shown in Figure 70.6. Dimensionality reduction decreases the computational power required for data processing and helps to train the model effectively by the extraction of dominant features that are both rotationally and positionally invariant.

Confusion matrix: Confusion matrix is summarized in Table 70.1 that provides a visual assessment of how well a classification method performs. It is a tabular representation of the results that shows the actual values have been previously determined. A "yes" or "no" prediction is the only two options. A "yes" answer would indicate that the sickness is present and a "no" answer would indicate that it is not.

The classifier generated 165 individual predictions, corresponding to 165 patients being tested for the presence of the disease. The classifier got it right 110 times out of 165 total opportunities, and just 55 times wrong. A total of 105 people in the sample really have the condition, whereas the other 60 do not.

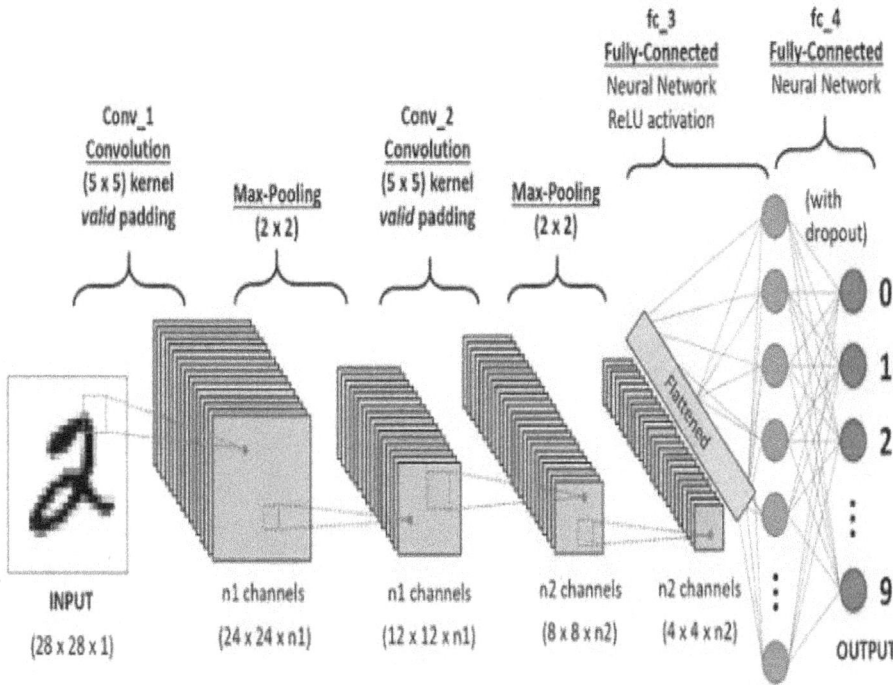

Figure 70.3 CNN layer model
Source: Author

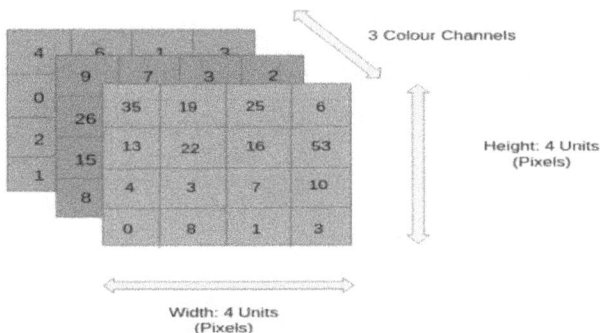

Figure 70.4 4 × 4 × 3 RGB image
Source: Author

Predicted categories fall into two broad buckets: "yes" and "no." If we are trying to anticipate whether or not they have a sickness, "yes" would mean that they are infected, while "no" would mean that they are not. Based on this, True positives (TP), True negatives (TN), False positives (FP), False negatives (FN), were found and some parameters like accuracy, sensitivity, specificity.

Hardware Setup

The Figure 70.7 shows the physical model and Figure 70.8 shows that Arduino UNO is a master device which is used to capture and stores the image. A OV7670 camera is used to capture the image.

Results and Discussions

After initial preprocessing, the repository consists of 5,841 X-ray images. These images are separated as training and testing sets, and the proposed model is implemented. The model has then been fed an initialization set consisting of 6000 X-ray images and the performance is compared with the-existing methods in Table 70.2.

The pictures shown in Figure 70.9 give us the results of a patient suffering from covid-19 as well as the normal patient. The web page is developed with two buttons, one button to select and upload the image and the other is used to show the result of the x-ray image when clicked.

Conclusion

To sum up, deep learning (DL) models have demonstrated promising performance in accurately identifying COVID-19 pneumonia and differentiate it from other types of respiratory diseases. However, these models should be used as a complementary tool to traditional diagnostic methods, such as

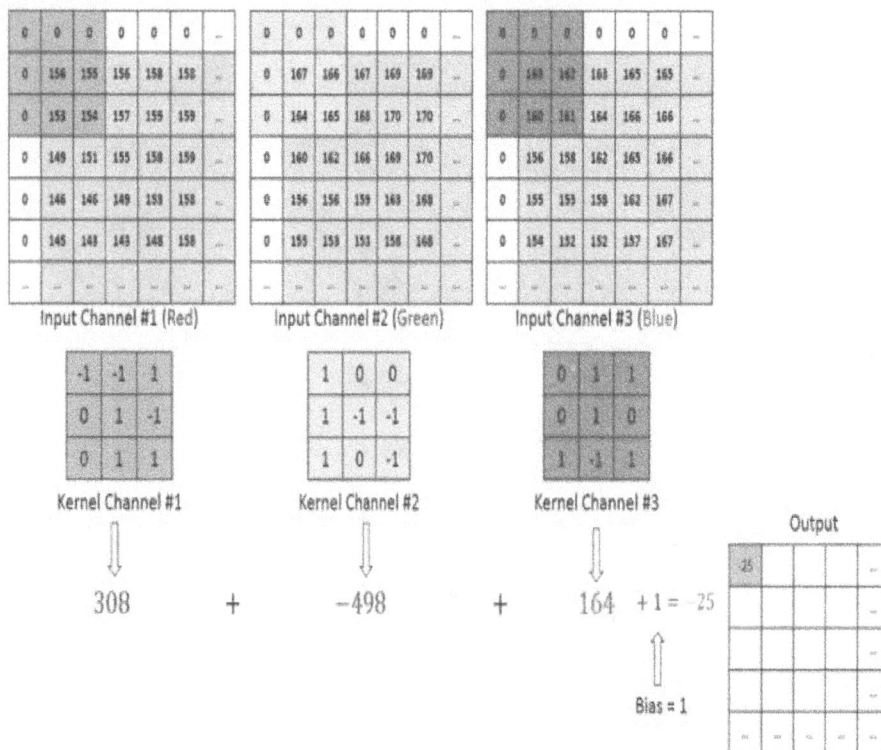

Figure 70.5 The M × N × 3 image matrix is convolved using a 3 × 3 × 3 kernel, which is a convolution process
Source: Author

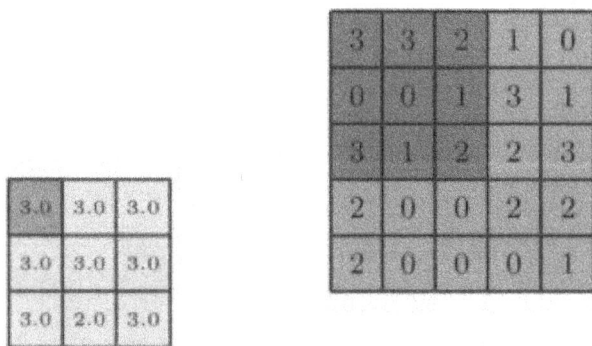

Figure 70.6 3 × 3 pooling over 5 × 5 convolved features
Source: Author

Table 70.1 Confusion matrix.

N = 165	Anticipated NO	Anticipated YES	Final count
Originally NO	TN = 50	FP = 10	60
Originally YES	FN = 5	TP = 100	105
Sum	55	110	

Source: Author

RT-PCR tests, and not as a standalone diagnostic tool. Further research and validation are necessary to ensure their reliability and generalizability. Additionally, ethical factors, including patient

Table 70.2 Comparison of specificity, accuracy, and sensitivity of the proposed model with existing models.

	No. of images	Method used	Accuracy	Sensitivity	Specificity
Proposed	960	Proposed model	96.43	93.68	99.0
Ozturk [18]	1000	Deep neural network	93.40	92.12	89.0
Minaee et al [19]	5000	Deep transfer learning	90.49	92.08	91.0

Source: Author

privacy, algorithmic bias, and data security, must be carefully considered when integrating these models into clinical practice.

In conclusion, COVID-19 detection in chest X-rays using DL models has shown encouraging results. Specifically, CNNs have proven effective in detecting and diagnostic of the disease, although the models can be influenced by several factors, including data quality and availability, dataset size, and the specific model architecture used. However, additional research and validation are required to thoroughly evaluate the performance, generalizability of these models.

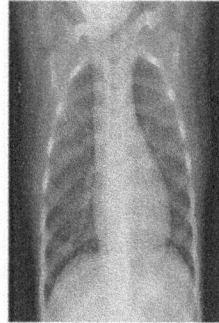 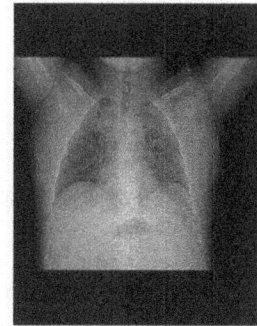

Figure 70.7 Hardware setup
Source: Author

Figure 70.9 Results of proposed work showing normal and covid affected
Source: Author

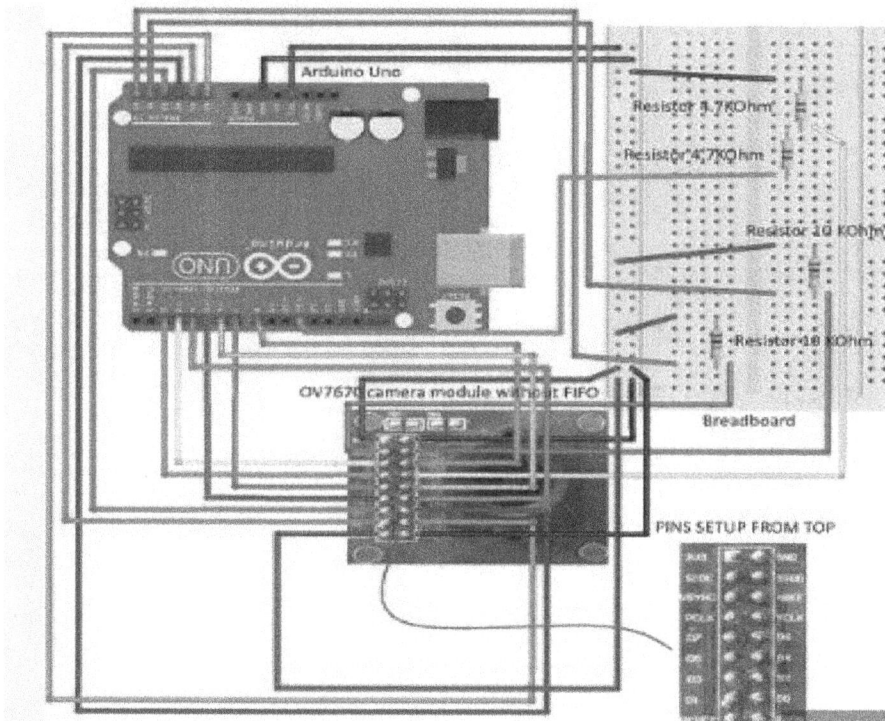

Figure 70.8 Arduino setup
Source: Author

References

[1] Cucinotta, D., & Vanelli, M. (2020). WHO declares COVID-19 a pandemic. *Acta Biomedica: Atenei Parmensis*, 91, 157–160.

[2] Rustam, F., Reshi, A. A., Mehmood, A., Ullah, S., On, B. W., Aslam, W., et al. (2020). COVID-19 future forecasting using supervised machine learning models. *IEEE Access*, 8, 101489–101499.

[3] Manal Bilal Mohamed, Fatma Elamin Hamid Mohamed, Nazik Mohamed Abdalla Taha, "Nurses' Knowledge, Attitude, and Practice Regarding Personal Protective Equipment for the Prevention of COVID-19 in Public Hospitals Khartoum State Sudan 2022", Open Journal of Nursing, Vol.13 No.3, March 31, 2023

[4] Dimple D Rajgor, Meng Har Lee, Sophia Archuleta, Natasha Bagdasarian, Swee Chye Quek, "The many estimates of the COVID-19 case fatality rate", THE LANCET, Vol.20, issue 7, p776-777, July 2020.

[5] Cohen, J. P., Morrison, P., & Dao, L. (2020). COVID-19 image data collection. GitHub repository. https://github.com/ieee8023/covid-chestxray-dataset

[6] Daqiu Li, Zhangjie Fu, Jun Xu, "Stacked-autoencoder-based model for COVID-19 diagnosis on CTimages", 2021;vol. 51(5):P2805-2817. doi: 10.1007/s10489-020-02002-w. Epub 2020 Nov 9.

[7] Shorten, C., & Khoshgoftaar, T. M. (2019). A survey on image data augmentation for deep learning. *Journal of Big Data*, 6, 60.

[8] Mooney, P. (2018). Chest X-Ray Images (Pneumonia) [Data set]. Kaggle. https://www.kaggle.com/datasets/paultimothymooney/chest-xray-pneumonia

[9] Kayalibay, B., Jensen, G., & van der Smagt, P."CNN-based segmentation of medical imaging data", 2017, DOI: 10.48550/arXiv.1701.03056. http://arxiv.org/abs/1701.03056.

[10] Li, Q., Cai, W., Wang, X., Zhou, Y., Feng, D. D., & Chen, M. (2014). Medical image classification with convolutional neural network. In Proceedings of the 2014 13th International Conference on Control Automation Robotics & Vision (ICARCV), (pp. 844–848). Singapore.

[11] Umer, M., Sadiq, S., Ahmad, M., Ullah, S., Choi, G. S., & Mehmood, A. (2020). A novel stacked CNN for malarial parasite detection in blood smear images. *IEEE Access*, 8, 93782–93792.

[12] Rouhi, R., Jafari, M., Kasaei, S., & Keshavarzian, P. (2015). Benign and malignant breast tumors classification based on region growing and CNN segmentation. *Expert Systems with Applications*, 42(3), 990–1002.

[13] Sharif, M., Attique Khan, M., Rashid, M., Yasmin, M., Afza, F., & Tanik, U. J. (2019). Deep CNN and geometric features-based gastrointestinal tract diseases detection and classification from wireless capsule endoscopy images. *Journal of Experimental &theoretical Artificial Intelligence*, vol. 33, issue 4, 1–23.

[14] Asada, N., Doi, K., MacMahon, H., Montner, S. M., Giger, M. L., Abe, C., et al. (1990). Potential usefulness of an artificial neural network for differential diagnosis of interstitial lung diseases: pilot study. *Radiology*, 177(3), 857–860.

[15] Katsuragawa, S., & Doi, K. (2007). Computer-aided diagnosis in chest radiography. *Computerized Medical Imaging and Graphics*, 31(4-5), 212–223.

[16] Dong, D., Tang, Z., Wang, S., Hui, H., Gong, L., Lu, Y., et al. (2020). The role of imaging in the detection and management of COVID-19: a review. *IEEE Reviews in Biomedical Engineering*, 14, 16–19.

[17] Apostolopoulos, I. D., & Mpesiana, T. A. (2020). Covid-19: automatic detection from x-ray images utilizing transfer learning with convolutional neural networks. *Physical and Engineering Sciences in Medicine*, 43, 635–640.

[18] Ozturk, T., Talo, M., Yildirim, E. A., Baloglu, U. B., Yildirim, O., & Acharya, U. R. (2020). Automated detection of COVID-19 cases using deep neural networks with x-ray images. *Computers in Biology and Medicine*, 121, 103792. doi. 10.1016/j.compbiomed.2020.103792.

[19] Minaee, S., Kafish, R., Sonka, M., Yazdani, S., & Soufi, G. J. (2020). Deep- COVID: predicting COVID-19 from chest x-ray images using deep transfer learning. *Medical Image Analysis*, 65, 101794. doi: 10.1016/j.media.2020.101794.

[20] Khan, A. I., Shah, J. L., & Bhat, M. M. (2020). CoroNet: a deep neural network for detection and diagnosis of COVID-19 from chest x-ray images. *Computer Methods and Programs in Biomedicine*, 196, 105581. doi: 10.1016/j.cmpb.2020.105581.

71 Design and performance analysis of DPFC based hybrid system by various controller

Pilli Sireesha[1,a] and Kiran Kumar Kuthadi[2,b]

[1]PG Scholar, Department of EEE Sree Vahini Institute of Science and Technology, Tiruvuru, Andhra Pradesh, India

[2]Associate. Professor, Sree Vahini Institute of Science and Technology, Tiruvuru, Andhra Pradesh, India

Abstract

Our study presents a hybrid system that combines photovoltaic (PV) and wind power, powered by ANFIS. Their suggested conventional system components, FLC and PI, Presently, energy distribution is a key component of power system networks for maintaining power reliability in distribution systems built to handle the dynamic operational demands of hybrid systems. In addition to the intended hybrid system, this design is also considered a PV/wind hybrid system. In an effort to get the most out of the specified system, many different maximum power point tracking (MPPT) algorithms have been proposed. The study also aimed to find ways to make the hybrid system more stable. We offer a novel control technique that combines the lion optimization algorithm (LOA) technique with a distributed power flow controller (DPFC) to enhance the transient stability and power quality of the proposed system. It was in reaction to systems connected to the grid using DPFC controllers that this LOA control approach was initially created. The signals used to build the control strategy were derived from the system's voltage and current characteristics. Lion optimization and ANFIS approaches were employed in this work to get precise parameters. To evaluate and compare the proposed controller-equipped system, we utilized MATLAB/Simulink.

Keywords: ANFIS, DPFC, HYBRID, lion optimization algorithm LOA, MPPT

Introduction

The demand for electrical power is now at an all-time high. In the past, people have relied on polluting and environmentally harmful ways to generate electricity [1]. Nuclear power plants, gas, and coal are all examples of such techniques. In order to meet electrical demands while simultaneously addressing these environmental problems, current energy generation systems greatly depend on non- conventional sources [2]. Low expenses for upkeep, little environmental effect, and affordability are the key advantages of these renewable energy sources. Solar and wind power systems are among the most prominent renewable energy options because of their simple design, abundance of natural sources, and great efficiency [3]. Hybrid systems rely on PV and wind-powered systems as their primary energy sources [4]. In terms of environmentally friendly power generation, photovoltaic belongs to the best solutions available today [5]. Installing a solar power system isn't cheap, and it's not guaranteed to be dependable no matter the time of year or the weather. Depending

on the weather, the solar system may produce different amounts of energy [6]. Using MPPT techniques allowed us to maximize output and demonstrate the efficiency of the solar panels [7]. Furthermore, PV systems depend on locally available wind energy sources for sustainable power. According to [8], the power generation ratio is affected by the amount of wind in the surroundings. Depending on the weather, the wind systems may produce different amounts of energy [9]. It follows that MPPT, or Methods towards Maximum Power Point Tracking, is used to increase the efficiency of the wind system and maximum output. The system must maintain grid synchronization at all times. We selected the reference signals based on the grid's characteristics and built the inverter's control diagram using a generic PWM technique. After that, we linked the solar panels to the inverter to ensure that the system's rates and frequencies were compatible [10]. The electricity grids of today are enormous and complex. In a connected power framework, when an electric burden request varies randomly, both the tie-line exchange of energy

[a]sireeshapilli98@gmail.com, [b]kiran9949610070@gmail.com

DOI: 10.1201/9781003684589-71

and the frequency of jurisdictional occurrence shift. Maintaining a balance between age and load becomes more difficult when control is taken away [11]. Consequently, a control framework is crucial for fulfilling the fundamental objective of a reorganized power system, which is to reduce the impact of unexpected fluctuations in load, maintain.

DC Micro-grid

In Figure 71.1 we can see the fundamental layout of the DC micro grid. Power from renewable sources, such as solar, wind, and batteries, is connected to a shared DC connection. Power is sent to both linear and non-linear loads over the DC connection. Several power converters are used in this architecture for the integration of several RE sources.

At a certain temperature and level of sun irradiation, there is a region on the operational spectrum known as the maximum power point (MPP) where the power production from solar panels is at its highest. In terms of voltage and current, the MPP is defined as the optimal operating conditions for maximum power production. In order to maximize the output of solar panels and the amount of electricity they can produce, photovoltaic (PV) power generating systems use maximum power point tracking (MPPT) algorithms. The

Perturb and Observe (P&O) algorithm is a common MPPT method [12]. The P&O algorithm begins by making a little adjustment to the solar panel's working point, either in terms of voltage or current. The panel's power output is affected by the disturbance. The program next checks to see whether the disturbance causes an increase or reduction in the power output. When the power goes up, the algorithm keeps tweaking in the same way; when the power goes down, it switches it up and tweaks in the other direction.

In Figure 71.2 as shown in order to follow the MPP, the P&O algorithm repeatedly performs the perturbation and observation procedure. In order to precisely pinpoint the MPP and fine-tune the operational point, the algorithm's perturbations shrink as it converges towards it. DPFC controllers provide shunt voltage converters in addition to a series of converters. Shunt voltage controllers handle reactive power compensation and harmonic suppression, whereas series VSC converter handle voltage regulation and the Figure 71.3 shows the DPFC controllers internal architecture.

Traditional control techniques may struggle with nonlinear and difficult control tasks, whereas ANFIS controllers can handle them successfully. When operating in a wide range of environments, ANFIS controllers maintain precise control by adjusting to changes in system behavior. In order to enhance the overall

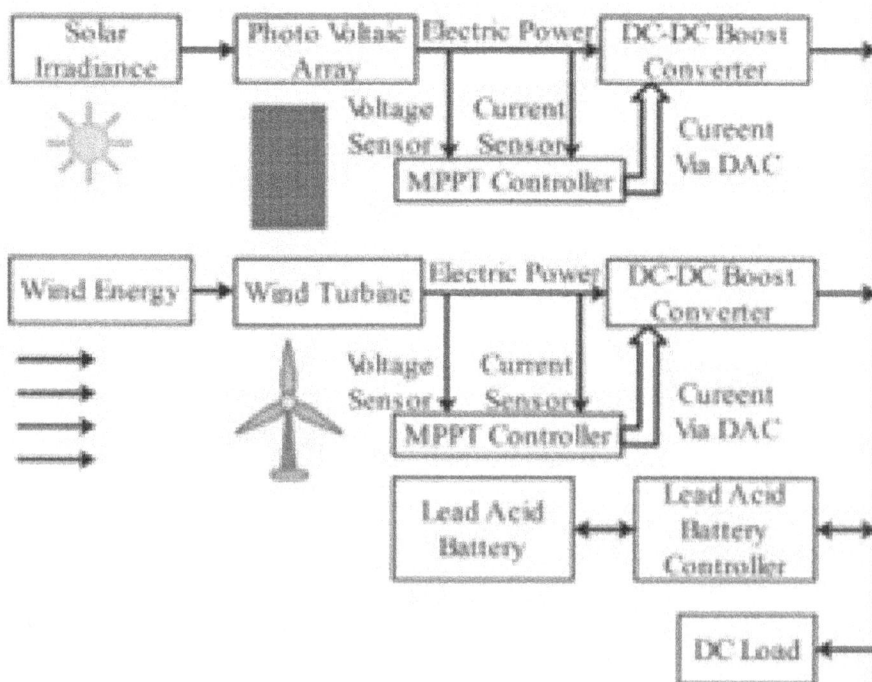

Figure 71.1 Structure of the DC micro grid
Source: Author

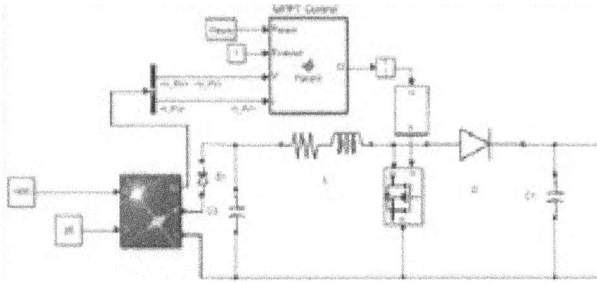

Figure 71.2 Circuit configuration of P&O MPPT based PV
Source: Author

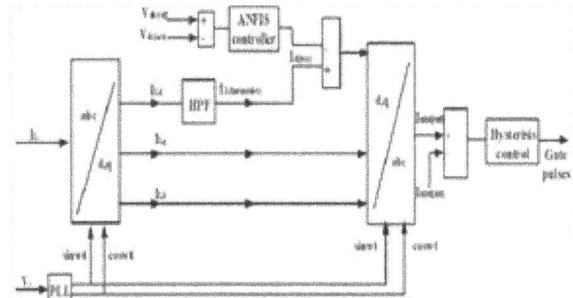

Figure 71.4 Configuration of ANFIS controller
Source: Author

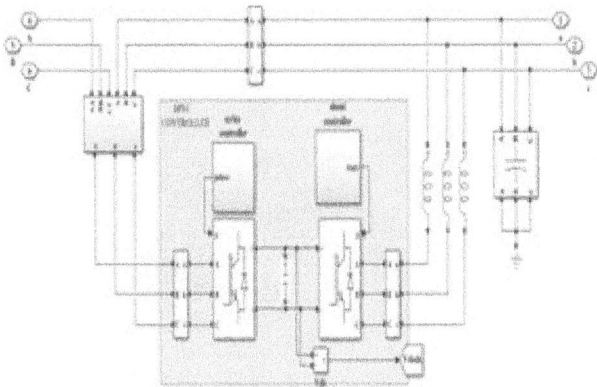

Figure 71.3 Shows the DPFC controller's internal architecture
Source: Author

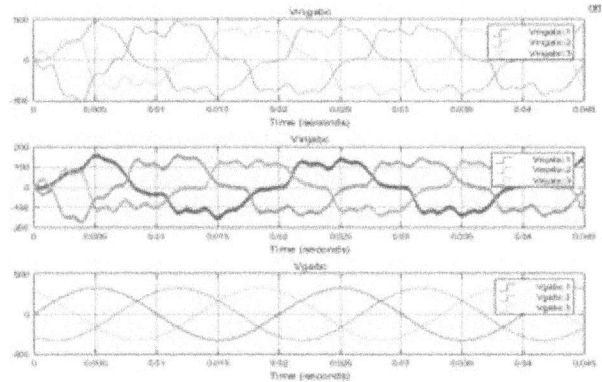

Figure 71.5 Microgrid voltage distortion, injected voltage, and corrected grid voltage output waveforms
Source: Author

performance of VSC-based systems, ANFIS controllers make parameter adjustment using real-time data and Figure 71.4 as shown the configuration of ANFIS controller. In order to build a controller that can capture complicated and nonlinear interactions, ANFIS blends fuzzy logic with neural network approaches. In order to fine-tune the parameters of neural network weights and fuzzy logic membership functions, ANFIS models make use of input-output data. To build an ANFIS model that can capture the intricate mapping between inputs and intended VSC behavior, combine fuzzy logic with neural network components. Constantly monitoring grid voltage, load demand, and other input data, the ANFIS-based controller uses the ANFIS model and produces control signals for the VSC to accomplish the intended function.

Results and Analysis

Two case studies evaluate the suggested grid-interfaced hybrid system with control models for direct power factor correction.

CASE 1: In order to enhance power quality, the first scenario is a combination of methods that employ both fuzzy logic and DPFC controllers based on PI. Using a DPFC controller built around PI and FUZZY, the experimental results for the system that was suggested are displayed in Figure 71.5. It shows the relationship between the DPFC injection voltage and the expected non- linear grid voltage for different DG system characteristics. The outcome of the simulation for the revised output voltage of the grid is shown in Figure 71.5. Knowing the effects of voltage distortions on the planned grid-connected system is the first step in measuring the correct value on the grid side. The DPFC shunt converter's injected current at fundamental and third order frequency ranges, together with the unbalanced current due to the unbalanced load, are shown in Figure 71.6. The modified current passage across the grid is shown in Figure 71.6. The harmonic aberration in the electrical grid current, induced by the imbalanced and nonlinear loads, was corrected using a DPFC controller. As shown in Figures 71.7 and 71.10 , the total harmonic

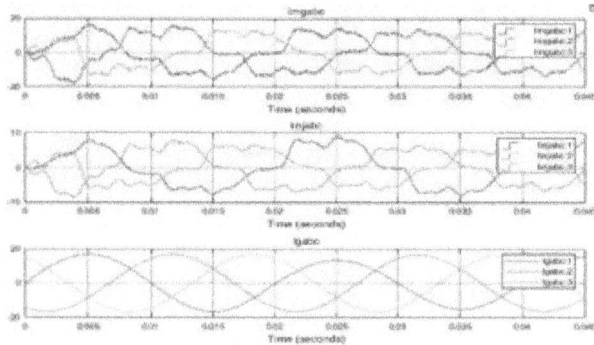

Figure 71.6 Currents from the micro grid, injection current, and compensated grid, as well as their respective output waveforms
Source: Author

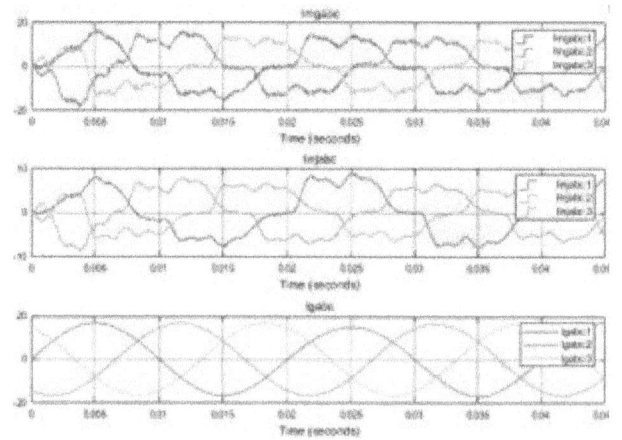

Figure 71.9 The microgrid current, the injected current and the adjusted current output wave forms
Source: Author

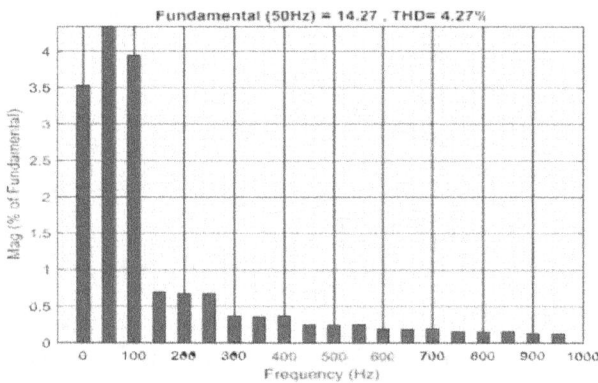

Figure 71.7 The PI controller with THD
Source: Author

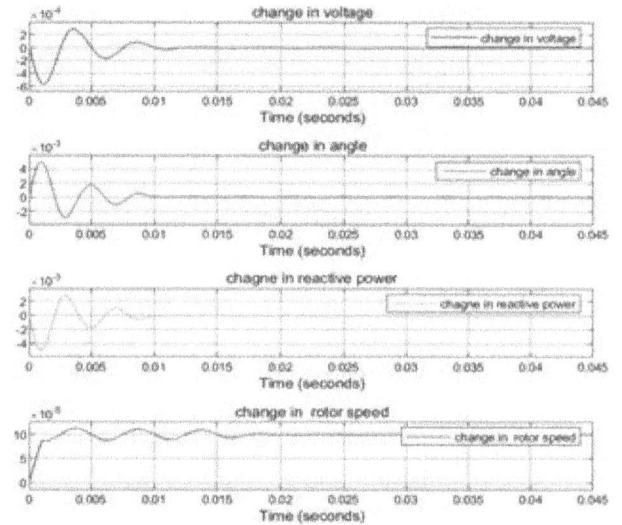

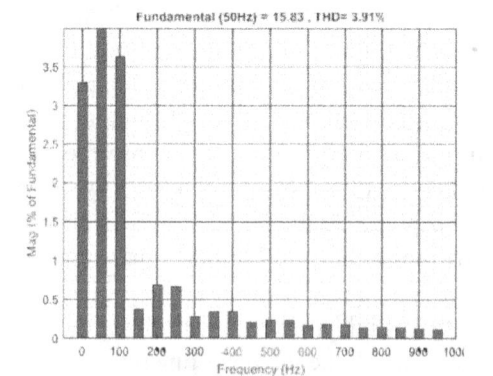

Figure 71.8 Microgrid voltage distortion, injected voltage, and grid voltage output waveform
Source: Author

Figure 71.10 By using fuzzy logic controller
Source: Author

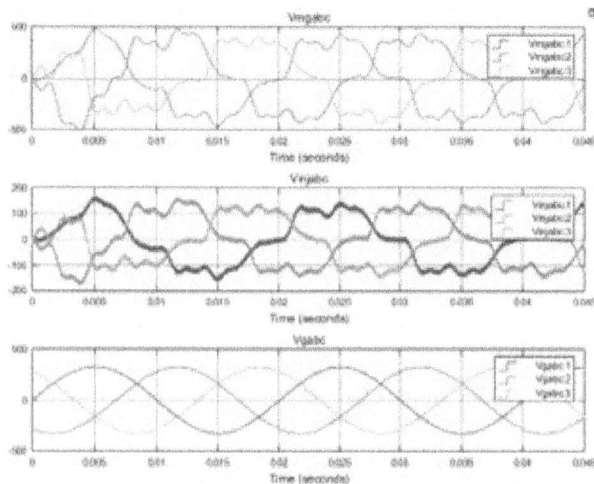

distortion (THD) over this grid current while utilizing the fuzzy based DPFC controller was 3.91%, however when using the PI a DPFC controller it was 4.27%.

CASE 2: A hybrid system that uses a DPFC controller based on fuzzy logic and PI to improve transient stability: With the help of PI controller.

In Case 2 The stabilityof the generators voltage,rotor speed, reactive power,and rotor angle were examined using simulation of DPFC based fuzzy logic as shown in Figures 71.8,71.9&71.10 and THD value is 3.91%

This example evaluated the converter's control diagram using PI, fuzzy, and LOA controllers to assure the stability of the hybrid system. Problems with stability usually occur after making changes to the system's configuration, the load, or the supply. With the help of the simulation results illustrated in Figures 71.11–14, we looked at how the voltage, speed of the rotor, reactive power, or rotor angle of the generators remained stable. The ANFIS controllers are shown in Figures 71.11, 71.12, and 71.13, and the results of the simulation for the rotor angle deviations due to changes in the producing circumstances are shown in Figure 71.14. To enhance stability conditions, the proposed DPFC series or shunt controllers use a variety of control strategies, such as PI, fuzzy, etc. lion optimization controllers. The proposed DPFC controls that were included into the combination system were governed by a conventional PI controller, and in order to produce better THD, the controllers of the hybrid system were fine-tuned using PI fuzzy reasoning, LOA And ANFIS, all of them were compared in research. According to the IEEE 519-1992 standards, every electrically built system must have a total harmonic noise (THD) below 5%. The result was a total harmonic distortion of 3.91% for the fuzzy DPFC regulator and 4.27% for the PI controller as shown fig 71.15 is the FFT analysis of DPFC controller by using ANFIS and the THD value of DPFC by using ANFIS is 3.45%

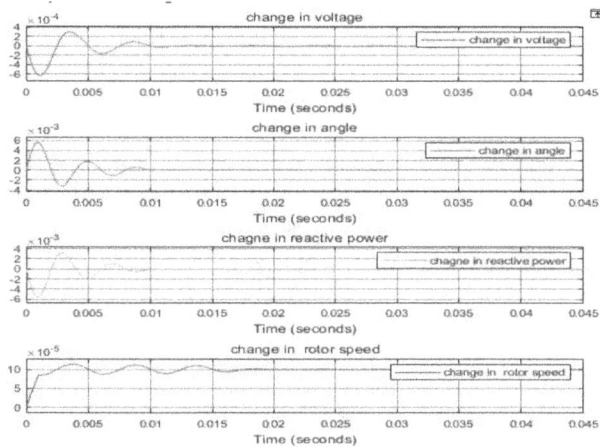

Figure 71.11 Using PI controller

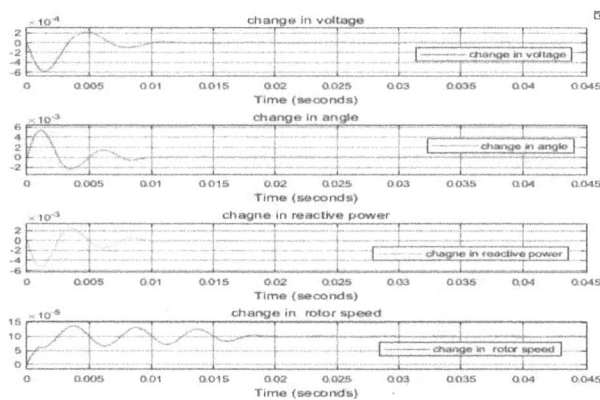

Figure 71.12 Using fuzzy logic controller
Source: Author

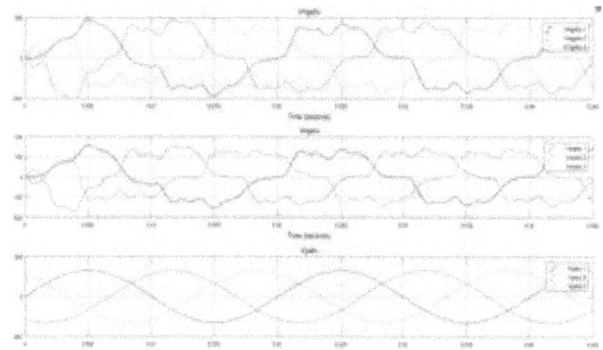

Figure 71.13 Using line optimization controller
Source: Author

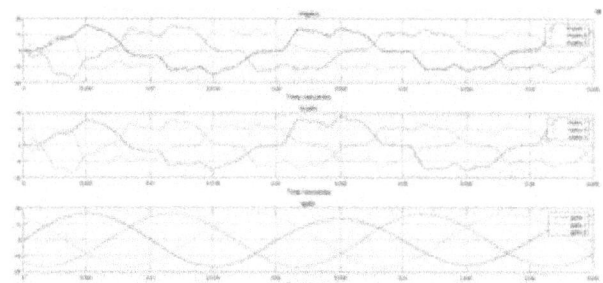

Figure 71.14 Using ANFIS controller
Source: Author

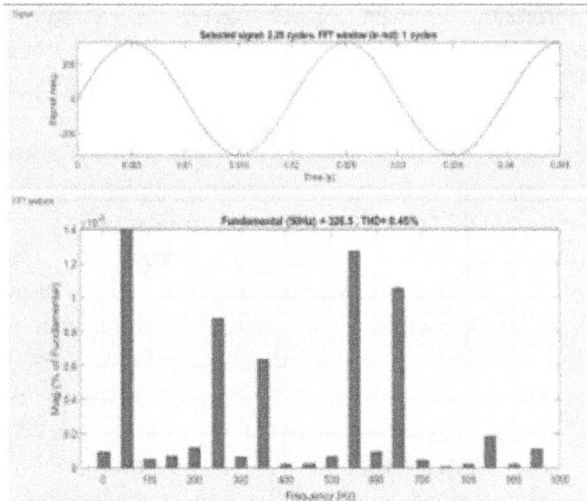

Figure 71.15 FFT analysis using ANFIS
Source: Author

Conclusion

In this paper we present ANFIS, a novel optimization-Based control approach for a dispersed power flow controller, With the goal of enhancing the transient stability, power quality and consistency of a hybrid system. The combines systems PV and wind turbines were fitted with a single MPPT controller to enhance performance even more. Properties of DPFC series and controllers may be shunt adjusted using a multitude of recognized control approaches. In order to fine- tune the DPFC'S settings, these research examples were thoroughly tested and confirmed. Compared to the traditional fuzzy controller,the LOA-basedANFIS converter outperformed it in terms of stability and power quality.

References

[1] K. Padmanathan, U. Govindarajan, V. K. Ramachandaramurthy, A. Rajagopalan, N. Pachaivannan, U. Sowmmiya, S. Padmanaban, J. B. Holm-Nielsen, S. Xavier, and S. K. Periasamy, "A sociocultural study on solar photovoltaic energy system in India: Stratification and policy implication," J. Cleaner Prod., vol. 216, pp. 461–481, Apr. 2019.

[2] R. M. Elavarasan, R. M., G. Shafiullah, G., S. Padmanaban, S., N. M. Kumar, N. M., A. Annam, A., A. M. Vetrichelvan, A. M., et al. (2020) L. Mihet-Popa, and J. B. Holm-Nielsen. A comprehensive review on the development, challenges, and policies of leading Indian states with an international perspective. *IEEE Access*, 8, 74432–74457.

[3] S. Kumar, R. K. Saket, D. K. Dheer, J. B. Holm-Nielsen, and P. Sanjeevikumar, "Reliability enhancement of electrical power system including impacts of renewable energy sources: A comprehensive review," IET Gener., Transmiss. Distrib., vol. 14, no. 10, pp. 1799–1815, May 2020.

[4] I. Masenge, I., &and F. Mwasilu, F. (2020). Coordination control and optimisation of battery energy storage system for rural electrification using hybrid solar PV-wind generating systems. In Proceedings of IEEE PES/IAS PowerAfrica, (pp. 1–5).

[5] N. Priyadarshi, S. Padmanaban, M. S. Bhaskar, F. Blaabjerg, and J. B. Holm-Nielsen, "An improved hybrid PV-wind power system with MPPT for water pumping applications," Int. Trans. Electr. Energy Syst., vol. 30, no. 2, p. e12210, Feb. 2020.

[6] S. Padmanaban, S., K. Nithiyananthan, K., S. P. Karthikeyan, S. P., &and J. B. Holm-Nielsen, J. B. (Eds.), (2020). Microgrids. Boca Raton, Florida, USA: CRC Press.

[7] A. Suman, "Role of renewable energy technologies in climate change adaptation and mitigation: A brief review from Nepal," Renew. Sustain. Energy Rev., vol. 151, Nov. 2021, Art. no. 111524. [Online]. Available: https://www.sciencedirect.com/science/article/pii/ S1364032121008029.

[8] L. Varshney, A. S. S. Vardhan, S. Kumar, R. Saket, and P. Sanjeevikumar, "Performance characteristics and reliability assessment of self-excited induction generator for wind power generation," IET, Renew. Power Gener., vol. 15, pp. 1927–1942, 2021.

[9] Seapan, M., Hishikawa, Y., Yoshita, M., & Okajima, K. (2020). Temperature and irradiance dependences of the current and voltage at maximum power of crystalline silicon PV devices. *Research in Solar Energy*, 204, 459–465. Retrieved from the Internet: com/science/article/pii/S0038092X203050 89.

[10] F. Mebrahtu, F., B. Khan, B., P. Sanjeevikumar, P., P. K. Maroti, P. K., Z. Leonowicz, Z., O. P. Mahela, O. P., et al. (2020). and H. H. AlhelouHarmonics mitigation in industrial sector by using space vector PWM and shunt active power filter. In Proceedings of the 2020 IEEE International Conference on Environmental Electronics Engineering and the 2020 IEEE International Conference on Industrial Power Systems Europe (EEEIC/I&CPS Europe), (pp. 1–6).

[11] S. Vadi, S. Padmanaban, R. Bayindir, F. Blaabjerg, and L. Mihet-Popa, "A review on optimization and control methods used to provide transient stability in microgrids," Energies, vol. 12, no. 18, p. 3582, Sep. 2019.

[12] E. Sundaram and M. Venugopal, "On design and implementation of three phase three level shunt active power filter for harmonic reduction using synchronous reference frame theory," Int. J. Electr. Power Energy Syst., vol. 81, pp. 40–47, Oct. 2016.

72 Renewable energy powered common dc bus charging system for electric vehicles using multilevel inverter

Veerla Divya[1,a] and K. Kiran Kumar[2,b]

[1]PG Scholar Department of EEE Sree Vahini Institute of Science and Technology, Tiruvuru, Andhra Pradesh, India

[2]Associate Professor, Department of EEE Sree Vahini Institute of Science and Technology, Tiruvuru, Andhra Pradesh, India

Abstract

Building stations that use renewable energy sources could help with the growing demand for electric car charging stations. Powering electric vehicles is a breeze with systems that include wind, solar, and fuel cell stacks. A five-level inverter and cascaded IIR filter are standard components. Static transfer switches allow for a smooth transition, while voltage and current management systems ensure correct functioning whether the grid is available or unavailable. Therefore, in order to guarantee correct operation in both grid existence and non-existence modes, a cascaded IIR filter is used. While functioning properly, the system is also capable of identifying certain grid issues that users may find problematic.

Keywords: Distributed energy sources, five level inverters, fuel cells, power quality, utility grid, wind turbines

Introduction

A lot of people think that EVs, or electric cars, are revolutionary technology that can reduce road pollution. The skyrocketing growth of the electric car market share is attributable, in part, to worldwide campaigns that promote the purchase of such vehicles. Due to their limited range, electric vehicles have a major drawback in that they need frequent charging while in motion [1]. Electric automobiles have a few big problems, the most important of which are their high price and limited range. The environmental impact of electric car charging stations powered by renewable energy sources has been adequately studied, allowing for the construction of appropriate stations. Solar panels, turbines for wind power, & fuel cell arrays are some of the renewable energy sources used by electric vehicle charging stations to do this. The increasing popularity of fuel cells may be attributed to their many admirable properties. Among them are the top-notch energy economy, a longer driving range, reduced exhaust emissions, and low operating noise. Chemical energy is converted into electrical power in fuel cells by electrochemical reactions, eliminating the need to burn fuel. Proton exchange membranes fuel cells are becoming more popular as a viable alternative to traditional petrol and diesel engines because of their low operating temperature & lack of pollutants [2]. Additionally, wind power is being pushed since it is abundant, clean, and produces no waste. The most significant issue with its use, nevertheless, is that the wind speed magnitude [3] varies from hour to hour. In order to meet the power demand and improve environmental conditions, a hybrid approach is needed, which takes into account the unpredictable nature of each distributed energy source. Modern electric vehicle charging poses a significant challenge in terms of using appropriate management algorithms. The second order generalized integrator is a famous control technique that stands out due to the frequency mapping requirements, ease of implementation, and thorough literature description. You may finish building the orthogonal generator of signals once the SOGI [4, 5] mechanism for control is in place. Although integrator delays shorten processing times, they introduce inter-harmonics when used. In addition to producing erroneous estimation, DC offset lowers performance. As shown in [6], a model prediction-based control method may achieve strong steady-state performance while also knowing the forthcoming switching state. The use of complex structures, however, inherently increases processing complexity, which is insufficient in dynamic environments. A transmitted IIR filter [7] is used to examine the behavior with a grid linked operation. Among its many benefits, the transmitted IIR filter is highlighted in the IEEE519

[a]divyanal04@gmail.com, [b]kiran9949610070@gmail.com

DOI: 10.1201/9781003684589-72

specification for its improved power quality and sufficient performance. Photovoltaic (PV) panels, wind turbines, and fuel cells comprise our electric vehicle charging infrastructure. Efficient power transmission and proper operation in grid-connected and disconnected scenarios are guaranteed by the system in accordance with the IEEE-1547 standard. These advantages are the study's most noteworthy results.

In reaction to changes in the grid's status or detected faults, this distributed generating system may switch between grid-connected and isolated modes.

Connecting solar and wind power to the grid and feeding their active powers into the system through an improved control algorithm creates high-quality energy. The synchronization unit protects the grid from voltage distortion and frequency variation by using a cascaded IIR filter-FLL based phase angle estimator. To further guarantee high-quality grid currents, DC-offset and load current harmonics should be reduced. To make DGS more adaptable to sudden shifts in solar irradiation, use the grid-connected solar power feed-forward term. The most efficient way to charge EVs consistently is using a system that combines turbines with wind power, fuel cell stacks, and solar panels.

The correct operation of the transmitted IIR filter is ensured by its operation in grid survival mode and compliance with IEEE standard in case of a grid failure, the DGS easily switches to isolated mode, enabling dependable operation and extended continuous service throughout grid existence and non-existence scenarios. While DGS is functioning independently, the PI controller. ensures that the voltages at the common coupling point (CCP) remain balanced and sinusoidal. We can now see the rationale for the CCP voltage's superior power quality compared to the IEEE-1547 standard.

Proposed system

Using distributed energy sources including solar photovoltaic (PV) arrays, turbines, battery packs, and fuel cell stacks, the suggested approach for recharging electric vehicle (EV) DC buses is schematically shown in Figure 72.1 below. It is possible to link the hybrid renewable power sources to the power grid by use of a central inverter—also known as a five-level inverter—and batteries. The STSs allow for the charging of electric cars as well as the connection and removal from the grid. Using incremental

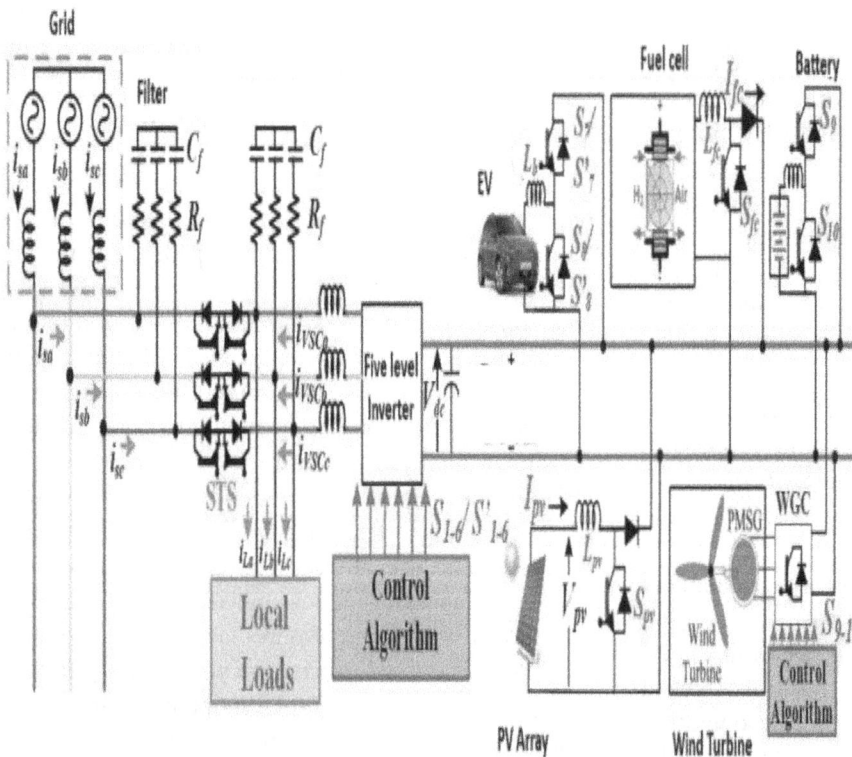

Figure 72.1 Proposed system along with five level inverters
Source: Author

conductance- based maximum power point tracking (MPPT) algorithms, the switching pulses (Spv, Sfc) of the accompanying boost converters are created using the highest possible powers retrieved from solar photovoltaic arrays and PEM fuel cell stacks. Also, a fully-controlled AC-DC converter connects the wind turbine to a wind power converter, which is also called a magnetized synchronous generator.

Control methodology

This section presents the methods for controlling distributed microgrids and common DC bus EV charging. When unconnected to the grid, the system uses a control approach based on cascaded IIR filters to reduce harmonics, but when working independently, it is controlled by adjusting the voltage. Through the use of grid line voltages (vsab, vsbc) for the purpose of monitoring grid phase voltages [8],

$$v_{sa} = \frac{v_{sbc}}{3} + \frac{2v_{sab}}{3}, v_{sa} = \frac{v_{sbc}}{3} - \frac{v_{sab}}{3}, v_{sc} = -\frac{2v_{sbc}}{3} - \frac{v_{sab}}{3}, \quad (1)$$

A terminal voltage (Vt) may be calculated by adding the three-phase voltage (vsa, vsb, and vsc) from the grid, as shown below:

$$V_t = \sqrt{2(v_{sa}^2 + v_{sb}^2 + v_{sc}^2)/3} \quad (2)$$

$$upa = vsa/Vt, upb = vsb/Vt, upc = vsc/Vt \quad (3)$$

Upa, upb, and upc, as well as uqa, uqb, and uqc, are obtained by using the vsa, vsb, vsc, and Vt.

Using a cascaded IIR filter allows for an accurate assessment of the active power component of the load while operating in grid connected mode, which is necessary for effective VSC switching. It is possible to express the transfer function as

$$uqa = -upb/\sqrt{3} + upc/\sqrt{3},$$
$$uqb = 3upa/2\sqrt{3} + upb/2\sqrt{3} - upc/2\sqrt{3},$$
$$uqc = -3upa/2\sqrt{3} + upb/2\sqrt{3} - upc/2\sqrt{3} \quad (4)$$

$$\frac{i_{Lfa}}{i_{La}} = \frac{\left[-k_1 z^{-1}\alpha_2 + (-h_1 k_1 z^{-2})\right]\left[-k_1 z^{-1}\alpha_1 + (-h_0 k_1 z^{-2})\right]}{(1-\alpha_2 z^{-1})(1-\alpha_1 z^{-1})} \quad (5)$$

So, iLfa, the basic component, is sinusoidal and harmonic-free. Quadrature phase unit templates are used with the achieved iLfa. This leads to the acquisition of Ifpa after absolute block processing, and the determination of Ifpb and Ifpc. It is estimated that the IpLavg is

$$I_{pLavg} = \frac{I_{fpa} + I_{fpb} + I_{fpc}}{3} \quad (6)$$

Another step in allowing EV charging is to determine the error in voltage (Veve) using the EV's Vev and Vev *, as

$$V_{eve} = V_{ev}^* - V_{ev} \quad (7)$$

In order to determine the EVs' reference current (Iev *), the voltage at the output error (Veve) is used.

$$I* ev(m) = I* ev(m-1) + Kpev \{Veve (m)\} + Kiev \{Veve(m) - Veve(m-1)\} \quad (8)$$

A current error is calculated by comparing the reference and measured currents of the electric vehicles, as

$$Ieve = I *ev - Iev \quad (9)$$

Deve is used to get the S7-S8 unidirectional converter pulses for EVs.

$$Deve = Deve(m-1) + Kpeve\{I eve(m)\} + Kieve \{I eve(m) - Ieve(m-1)\} \quad (10)$$

Additionally, a voltage error is calculated for battery control as

$$Vdce = V *dc - Vdc \quad (11)$$

The following is an estimate of the battery's reference current using the voltage error:

$$I*batt(m) = I*batt(m-1) + K pdc\{V dce (m)\} + K idc \{V dce (m) - V dce (m-1)\} \quad (12)$$

Hence, by comparing the reference current from the battery with the current that is sensed from the battery (Ibatt), the current error may be determined.

$$Ibattr = I *batt - Ibatt \quad (13)$$

It is calculated that the switching pulses linked to the battery bidirectional converter are,

$$D(m) = D(m-1) + Kpbatt \{Ibattr (m)\} + Kibatt \{Ibattr (m) - Ibattr (m-1)\} \quad (14)$$

Additionally, vector control is used to regulate the wind-powered PMSG, with the predicted value of Iq * being

$$I_q^*(k) = I_q^*(k-1) + K_{pw}\{\omega_{error}(k)\} + K_{iw}\{\omega_{error}(k) - \omega_{error}(k-1)\} \quad (15)$$

The kpw and kiw are the PI speed control gains, and óerror is the difference among the reference ó*r and the observed ór rotor speeds. One method of controlling the reference rotor velocity is the MPPT (maximum power point tracking) method, which is based on incremental conductance. Additionally, I * gena, i * genb, & i * genc are calculated with the help of I * d and I * q. The resultant WGC switching pulses (S9-S14) are shown in Figure 72.2. To get the most power out of the system, the wind turbine, fuel cell stack, & photovoltaic array should be installed as shown in Figure 72.3. It is expected that wind turbines, fuel cell stacks, and solar arrays will all have feed-forward terms.

$$Wpv=2Ppv/3vt, Wfc = 2Pfc/3vt, \\ Wwind = 2pwind/3vt \quad (16)$$

in where Pv, Pfc, Vt, and Pw stand for power from a photovoltaic fuel cell stack, voltage at the terminal, and power from a wind turbine... Wpv, wfc, wwind, and wbatt are the IP addresses used to assess the Ipnet.

$$Ipnet = IpLavg\} - (Wpv+ Wfc+ Wwind \\ Wev + Wbatt) \quad (17)$$

the electric vehicle contribution (wev) and the battery contribution (wbatt) are calculated as (2Pev / 3Vt) &

(2Pbatt / 3Vt), respectively. Ipnet is multiplied by upa, upb, and upc in this way:

$$I*sa = IpnetXUpa, \quad I*sb = IpnetXUpb, \\ I*sc = IpnetXUpc \quad (18)$$

a)

b)

c)

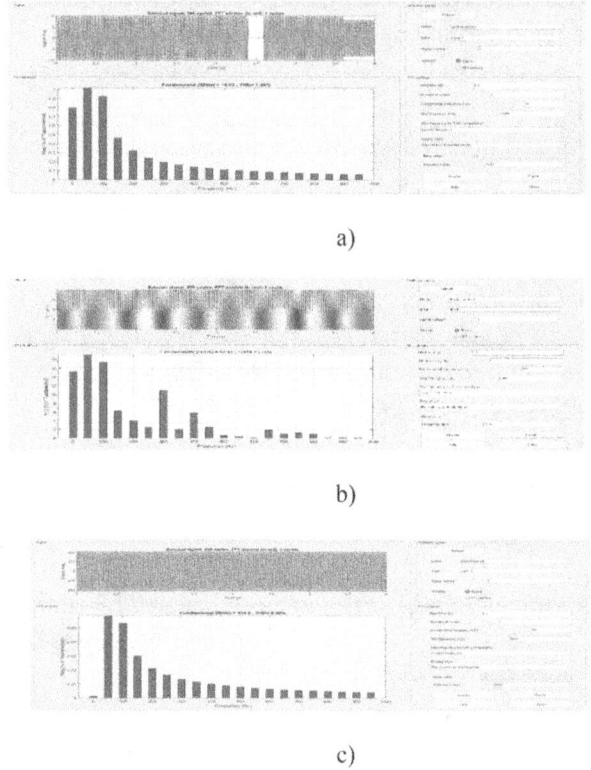

Figure 72.3 Harmonic analysis of a) grid current b) Load current c) Load voltage
Source: Author

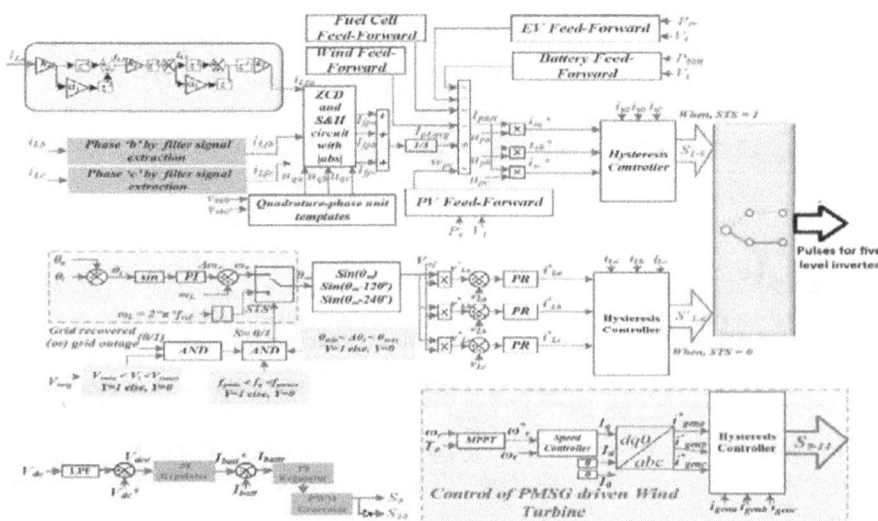

Figure 72.2 Distributed microgrid control architecture for grid-connected and off-grid operation using energy from renewable sources like solar, wind, batteries, and fuel cells
Source: Author

a)

b)

c)

d)

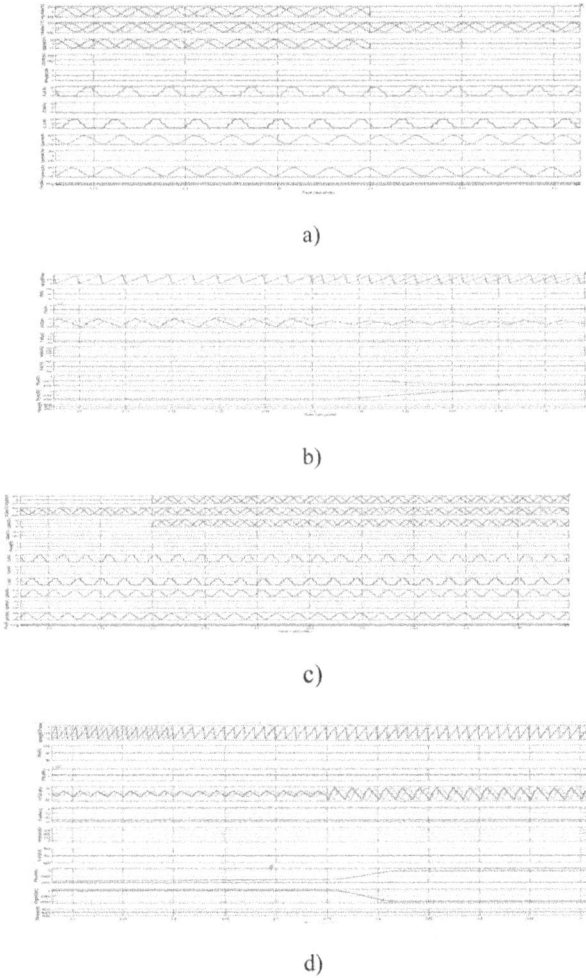

Figure 72.4 Simulated performance through transition between grid existence to non-existence operation

Source: Author

a)

b)

c)

d)

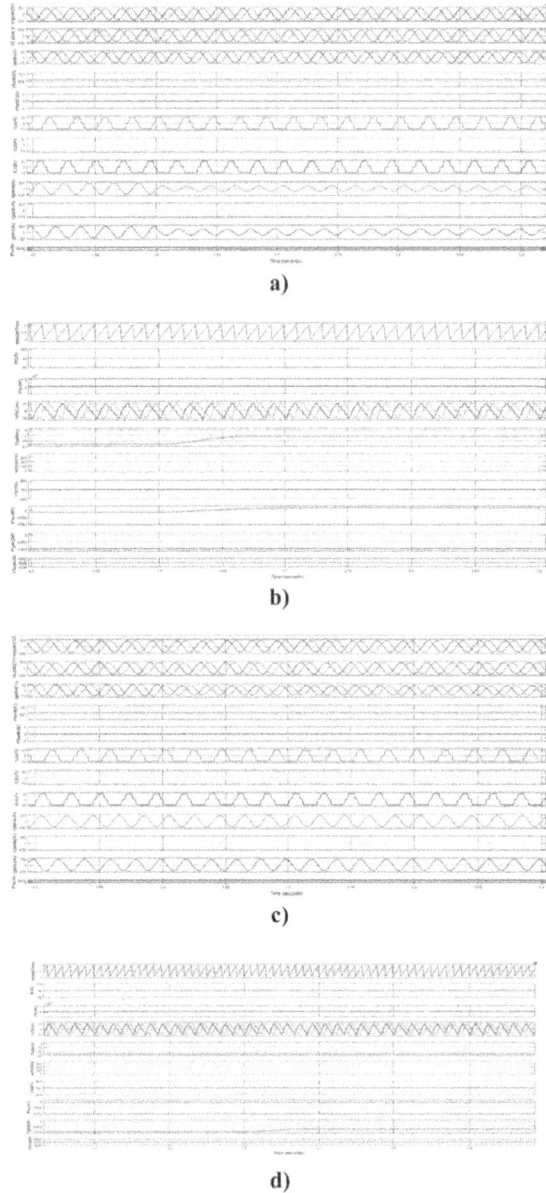

Figure 72.5 Simulated performance during wind speed and solar insolation various operation

Source: Author

The hysteresis current regulator is fed the current errors in order to determine how to switch pulses (S1-S6) of the VSC. In the absence of a grid the following evaluations of v * La, v * Lb, and v * Lc are made when STS = 0.

Grid voltage & load angles is observed, wind speed & solar variations,showing reduced generator currents & wind power Figure 72.4-72.5.

$$v_{La}^* = V_{ref} \times sin\ sin\ \theta_m, v_{Lb}^* = V_{ref} \times sin\ sin\ (\theta_m - 120°),, v_{Lc}^* = V_{ref} \times sin\ sin\ (\theta_m - 240°) \quad (19)$$

The three-phase loading voltages are used to determine the magnitude of the load's voltages, denoted as Vref, and the load voltage angle, Θm.

Figure 72.2 also shows the technique for synchronizing the distributed microgrid, which consists of fuel cells, solar panels, wind turbines, and batteries, with the common DC bus charging of electric vehicles. When a freestanding system is ready to be reconnected towards the power grid, the voltages and phases of angles are adjusted to match those of the utility. After the phase angles have been calculated, as:

Comprehensive overview of V2H, V2V, Technologies in smart grid integration

$$\theta_e = \theta_g - \theta_L \quad (20)$$

$$Dwe\ (n) = Dwe\ (n\text{-}1) + kpe\ \{qe\ (n)\text{-}qe$$
$$(n\text{-}1)\} + kIeqe\ (n) \tag{21}$$

The PI controller is utilized for estimating frequency error, as,

The variation in frequency, denoted as $\Delta\ddot{o}e$, is crucial for maintaining synchronization on the main grid. So, the altered frequency is approximated by $\ddot{o}n = (\Delta\ddot{o}e + \ddot{o}L)$ during grid retrieval. Nevertheless, in standalone mode of operation, just the load's fundamental frequency ($\ddot{o}L$) is sent.

Simulation results extension results (five level inverter)

Comparison table

	Existing system	Extension system
Grid current THD%	3.40%	2.26%
Load current THD%	29.75%	23.26%
Load voltage THD%	0.22%	0.08%

Conclusion

A distributed microgrid integrating solar, wind, and fuel cell sources can charge electric vehicles, according to this article. It can do so despite grid connection/disconnection limits, changing sunlight, load unbalancing, and wind variance. The shared DC bus architecture also allows for rapid charging by going straight to the EV battery rather than using the onboard charger. In addition to enhancing power quality in grids linked mode, simplifying mode transitions, and functioning correctly when running independently, the transmitted IIR filter has other useful features. Therefore, by comparing the simulation results to IEEE standards, we can ensure that the integrated system works in various conditions.

References

[1] Liao, Y.-T., & Lu, C.-N. (2015). Dispatch of EV charging station energy resources for sustainable mobility. *IEEE Transactions on Transportation Electrification*, 1(1), 86–93.

[2] Su, G., & Tang, L. (2008). A multiphase, modular, bidirectional, triplevoltage DC–DC converter for hybrid and fuel cell vehicle power systems. *IEEE Transactions on Power Electronics*, 23(6), 3035–3046.

[3] Huang, Q., Jia, Q.-S., Qiu, Z., Guan, X., & Deconinck, G. (2015). Matching EV charging load with uncertain wind: a simulation-based policy improvement approach. *IEEE Transactions on Smart Grid*, 6(3), 1425–1433.

[4] Liu, C., Chau, K. T., Wu, D., & Gao, S. (2013). Opportunities and challenges of vehicle-to-home, vehicle-to-vehicle, and vehicleto-grid technologies. *Proceedings of the IEEE*, 101(11), 2409–2427.

[5] Matas, J., Martín, H., de la Hoz, J., Abusorrah, A., Al-Turki, Y. A., & Al-Hindawi, M. (2018). A family of gradient descent grid frequency estimators for the SOGI filter. *IEEE Transactions on Power Electronics*, 33(7), 5796–5810.

[6] Shadmand, M. B., Balog, R. S., & Abu-Rub, H. (2014). Model predictive control of PV sources in a smart DC distribution system: maximum power point tracking and droop control. *IEEE Transactions on Energy Conversion*, 29(4), 913–921.

[7] Onizawa, N., Kloshita, S., Sakamoto, S., Kawamata, M., & Hanyu, T. (2017). Evaluation of stochastic cascaded IIR filters. In Proceeding of IEEE International Symposium Multiple-Valued Logic, (pp. 224–229).

[8] Singh, B., Chandra, A., & Al-Hadad, K. (2015). Power Quality: Problems and Mitigation Techniques. U.K.: John Wiley & Sons Ltd.

[9] Kuperman, A. (2015). Proportional-resonant current controllers design based on desired transient performance. *IEEE Transactions on Power Electronics*, 30(10), 5341–5345.

73 The intelligent control of independent single-phase microgrids integrate solar PV arrays, wind and hydro power

Kade Praneetha[1,a] and Kiran Kumar Kuthadi[2,b]

[1]PG Scholar, Department of EEE, Sree Vahini Institute of Science and Technology, Tiruvuru, Andhra Pradesh, India

[2]Associate Professor, Sree Vahini Institute of Science and Technology, Tiruvuru, Andhra Pradesh, India

Abstract

This article proposes the Grey Wolf Optimization (GWO) algorithm for an islanded, single-phase microgrid system with grown power quality. The wind-turbine permanent magnet brushless DC (PMBLDC) generator, solar photovoltaic array, battery energy storage system (BESS) and a micro-hydro single phase two winding self-excited induction generator (SEIG) are all integrated into the proposed microgrid system. The integration of these renewable energy sources is accomplished with a voltage source converter (VSC). In addition to mitigating harmonic current, the GWO-based control algorithm is utilized to examined the reference source current that governs the VSC and control microgrids voltage and frequency. The systems reference real and reactive powers, it's adaptable to changing loads, are estimated by the recommended GWO. The reference actual power of the system is estimated using the GWO algorithm in order to preserve energy balance between solar PV power, wind, micro-hydro and BESS, which regulates the frequency of standalone microgrids.

Keywords: Battery energy storage system, grey wolf optimization, micro grid, renewable energy source, self-excited induction generator SEIG

Introduction

Around the world, there are several places where little towns have developed up distant from advanced cultures. Starting up a transmission system to provide energy there is challenging both financially and technically because of the related expenses, issues with the transmission towers grounding in mountainous regions, and Right of Way issues brought on by nearby trees. On the other hand, these locations are abundant in natural resources, such as wind, micro-hydro and solar energy. A miniature independent supply system that can constantly supply the loads cannot be developed because of the unpredictability of all these renewable energy sources. The system needs some reliable input in order to become autonomous. It is fundamentally wanted to create an efficient frequency and voltage management strategy to accomplish correct integration of renewable energy sources (RES) [1, 2]. For effectively addressing the difficulties in integrating renewable energy sources, the microgrid concept is particularly attractive [3]. The microgrid can function in both grid-tied and standalone modes, depending on the appropriate control system design. Several derivative types of microgrids, including active distribution systems, smart microgrids and virtual power plants, can be evaluated as vital parts of smart grids. For a grid-connected microgrid maintain power balance, which in turn control the system frequency, the main grid provides insufficient power and absorbs external power. In contrast, the balancing of unreal and active powers in a standalone microgrid is accomplished by adjusting power flow between the various microgrid components. , An IEEE standard for connecting distributed energy sources is provided [4, 5]. True and imaginary powers, voltage and frequency is the main network parameter needed to regulate microgrid operation. Large-scale integration is an intermittent energy source to be made possible by BESS. The microgrid system's BESS ability is not being fully exploited, considering its advantages. There is much discussion on power custom control of integrated renewable systems [6]. A comprehensive evaluation of power

[a]kadepraneetha15@gmail.com, [b]kiran9949610070@gmail.com

DOI: 10.1201/9781003684589-73

electronic converter control for microgrids is provided [7]. The balance of power and system voltage regulation are the most significant control challenges for standalone microgrids [8]. An upgraded version that takes into account several power units was introduced [9]. Both of these systems are centered on power balance between the PV-battery unit and other generating units, even if these are effective for controlling power production and demand. Furthermore, systems with DC buses and roads are not taken into account by these approaches. PV-battery-hydropower system, a hierarchical control algorithm [10] control true and reactive power by PV-battery unit and control AC bus voltage by hydropower generator. The hybrid PV-battery-diesel system is introduced [11] using a similar technique. However, in the event that the diesel generator or hydropower is not operating, neither of these methods takes voltage regulation into account. Additionally, in order to mitigate power quality issues and integrate dc and ac sources, the control algorithm is needed for regulating VSC attached for its function as voltage and frequency controller. The literature reports a large number of fundamental control algorithms. Icos Phi-based VSC algorithms, SRF, IRPT and ADALINE-based control algorithms are described [12, 13]. An advanced control technique based on a composite observer is presented [14]. Any signals harmonic components can be recovered using composite observers, and this

controller then uses the obtained fundamental [15, 16]. Higher voltage and current harmonics are present in the ASMC compared to the proposed GWO approach. ASMCs current harmonics are 16.46% and its voltage harmonics are 3.07%. There are 0.29% voltage harmonics and 3.17% current harmonics in the proposed GWO algorithm.

Operational principle of system

Figure 73.1 represents the circuit diagram of the proposed small microgrid. This microgrid is made up of a solar (photovoltaic) array, a wind turbine-powered PMBLDC (permanent magnet brushless DC) generator, an uncontrolled tiny-hydro turbine, two-winding SEIG (self-excited induction generator) and BESS. In theory, SEIG functions as a microgrid only AC producing source, directly supplying the load; the other two generating sources are linked to the load via voltage source converter (VSC). When the electricity generated by SEIG is less than the load, it transfers DC power produced by the solar array and PMBLDC generator into AC power. The VSC DC bus connects the solar array, wind turbine-powered PMBLDC generator and BESS. Only the system actual power comes from these three energy sources. It doesn't engage with the system in any reactive power transfers. The BESS offsets the load increased real power requirement then the combined

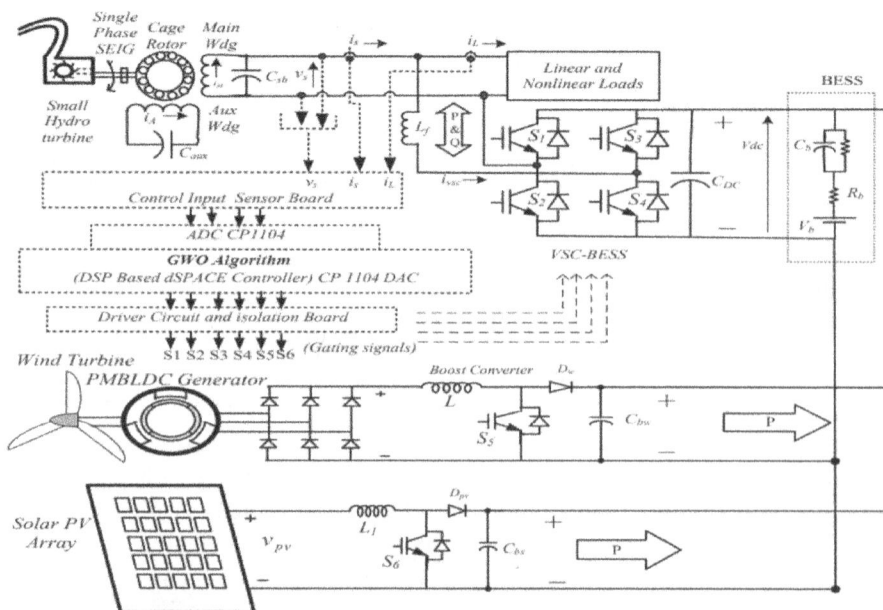

Figure 73.1 The single-phase microgrids system configuration
Source: Author

real power produced by SEIG, PMBLDC generator and solar array is less than the load. In order to render it appropriate for single-phase loads attached to the microgrids AC side, VSC converts the DC power provided by BESS into AC energy. In order to maintain the power balance in a different scenario, excess energy is stored in the BESS. Reactive power and tunable harmonics are required by the load and SEIG during fluctuating load situations in order to keep the microgrid AC voltage at a rated level.

GWO algorithm for microgrid of VSC-BESS

For microgrid VSC, fundamental source current and switching pattern are estimated using the Grey Wolf Optimization (GWO) control approach. Figure 73.2 represents the proposed GWO algorithm block diagram. Reactive power must be able to be changed under various load conditions for the SEIG mechanism to maintain the PCC voltage at fundamental value, as previously stated. SEIG terminal voltage amplitude is calculated as follows:

$$V_t = \sqrt{v_p^2 - v_q^2} \tag{1}$$

where vp and vq represent small microgrid AC voltages (or SEIG output voltage) in-phase and

quadrature components, respectively. Frequency Estimation and Phase Shifting (FEPS) blocks are used to create and estimate the quadrature portion of AC voltage. The voltage in small microgrid can be represented as

$$v_p = V_t \sin \omega t$$

$$v_q = V_t \cos \omega t \tag{2}$$

where ω is the microgrid AC voltages angular frequency, Vt is its amplitude and vp is its instantaneous AC voltage. Microgrid AC voltages in-phase and quadrature unit templates are obtained as follows:

$$u_q = \frac{v_q}{V_t} \text{ and } u_p = \frac{v_p}{V_t} \tag{3}$$

Power balancing between Induction generator, PMBLDC generator, solar array, battery and load is controlled by an in-phase portion of the fundamental source current. The recommended GWO control algorithm extracts the amplitude of the basic true and/or reactive power components of load current using an adaptive filter. Combining two control loops results in suggested control algorithm. The primary loop maintains true power balance between tiny microgrids and various energy components, while

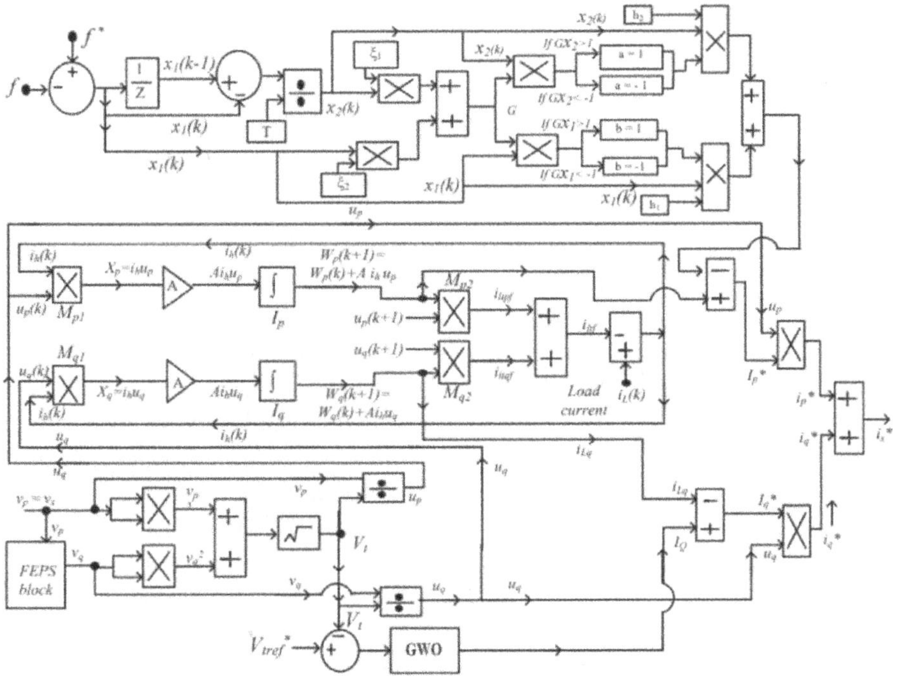

Figure 73.2 GWO control algorithm
Source: Author

the other loop regulates its voltage by introducing an adjustable reactive power.

Simulation results

As power generated by renewable sources is less than load, when the system operates in a steady state. The load (6.17 kW) exceeds power produced (5.716 kW) by renewable sources. Consequently, BESS uses VSC to provide an extra actual power source. The waveform of SEIG produced voltage and current, along with peak factors, are depicts in Figure 73.3(a) and 73.3(b). It indicates that almost sinusoidal current is being transferred by the SEIG. The RES through VSC provides the rest of the real power to satisfy the loads increased actual power demand. The SEIG is a provided load with 3.91 kW of actual power, whereas the VSC is provided with 2.24 kW. In VSC, the 2.24 kW of power is delivered to the AC side (to load side) by RES in order to reduce the load of actual power demand. An actual power of 1.8 kW is produced by remaining two sources (PMBLDC generator and solar), which power delivered to load via VSC. For the power balancing process of the suggested GWO control algorithm, BESS also provides an extra 843 W of actual power that is required to balance the system true power consumption and regulate system frequency. Power generated by an induction generator is transferred directly to load without causing interference from converter when SEIG released power is less than or equal to power required of load. The power balancing capability of proposed control method as demonstrated in Figures 73.3(a) to 73.3(d). The proposed GWO algorithm regulates VSC in an effort balances the power flow between the several sources, which regulates microgrid voltage level. To maintain voltage steady and reduce the harmonic current injected by nonlinear loads, the inverter provides 2.96 kVAR of reactive power and reduces harmonic. An inverter rises power quality and power factor in addition to represents numerous other functions. SEIG output current harmonic spectrum and THD in Figure 73.3(c) illustrate a THD of 0.29%, although Figure 73.3(d) shows that it is reduced nonlinear load current of 1.74% THD.

Dynamic condition of suggested microgrid

Proposed microgrid dynamic performance with change solar irradiation. Figure 73.4(a) and 73.4(b) show when microgrid reacts to increase solar irradiation. Figure 73.4(c) depicts system reaction after

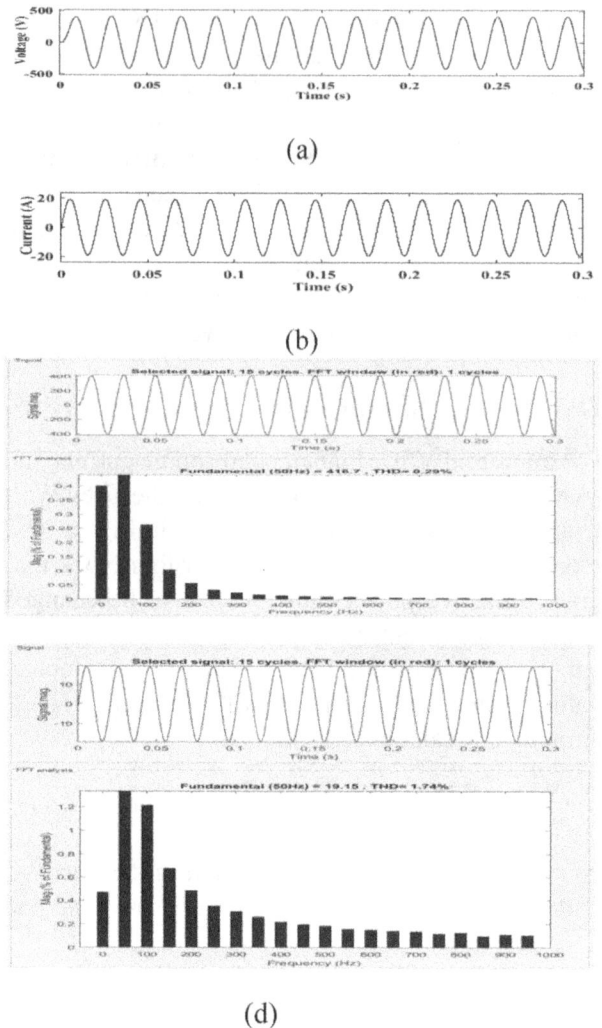

(a)

(b)

(d)

Figure 73.3 When renewable energy sources together provide less electricity than the load, system is in steady state
Source: Author

change of solar insolation. As system follows increase in insolation level, rises in solar generated current (result presented in Figure 73.4(a)) can be viewed. According to the simulation findings (Figures 73.4(a) to 73.4(b)), controller raises battery charging current in response to this incremental grow in solar output current. This allows the surplus power to be transferred to BESS, which regulates frequency. In Figure 73.4(a), the dynamic condition of SEIG current is depicted when system increases the insolation level. The change of solar insolation level does not affect or dissatisfied voltage or SEIG current, illustrated in Figure 73.4(a). It shows that under dynamic conditions caused by rises solar insolation, which increases the solar output current, the suggested controller maintains power balance. Three renewable sources combine to provide more electricity than is required. In control frequency, extra power is consequently transmitted to the battery. A step growth in the solar output current causes the battery current to shift direction from discharging to charging mode, as shown Figure 73.9(a) . The frequency control loop of the suggested GWO algorithm is used to maintain power balance among different energy sources. An observed in Figure 73.9(b) and 73.9(c) depicts dynamic condition of voltage, induction generator, solar current and battery current as follows drop in insolation. This study outcomes, which are presented in Figure 73.9(c), demonstrate that reduction in the insolation level leads in a decrease in solar current with the same slope, which in turn reduces battery current. However, AC voltage, load current or frequency remain unaffected. For recover system power balance, the battery current switches between charging and discharging modes, depicted in Figure 73.9(c).

Dynamic microgrid performance under variation in wind

An increase or decrease in wind speed and customer load on a multitude of power quality variables, such as system voltage, frequency, VSC and load current, is taken into consideration while evaluating the dynamic

condition of suggested microgrid. An increase in wind acceleration presented in Figures 73.5(a) and 73.5(b). Figure 73.5(c) depicts the response of decline in wind speed. When microgrid detects an increase in wind speed, Figure 73.5(a) depicts a level rise in PMBLDC generator current. The AC voltage, generator current, PMBLDC current and battery current dynamically

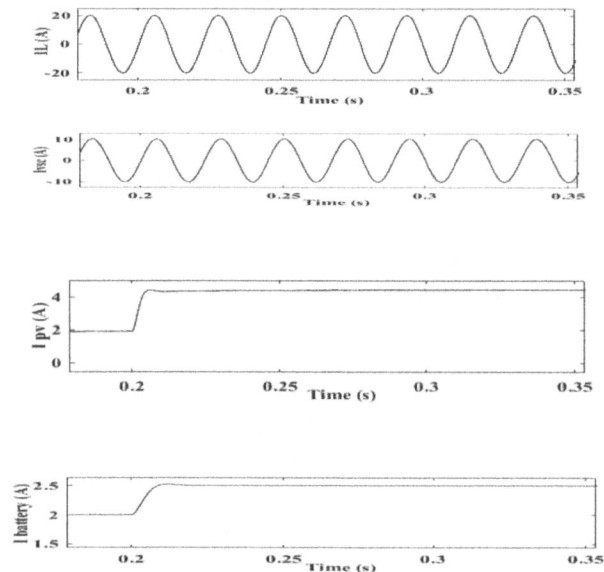

Figure 73.4(b) Dynamic operation of IL, iVSC, Ipv and Ibattery as an increase an irradiation
Source: Author

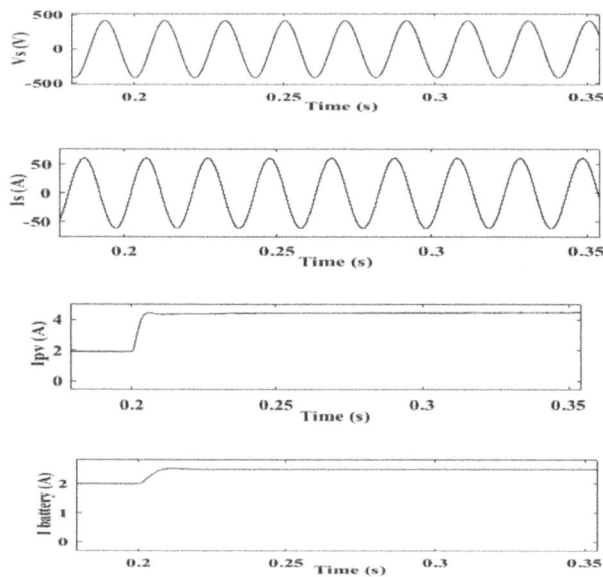

Figure 73.4(a) When the system implements gradual rise in the insolation level, the Vs, Ipv and Ibattery operate continuously
Source: Author

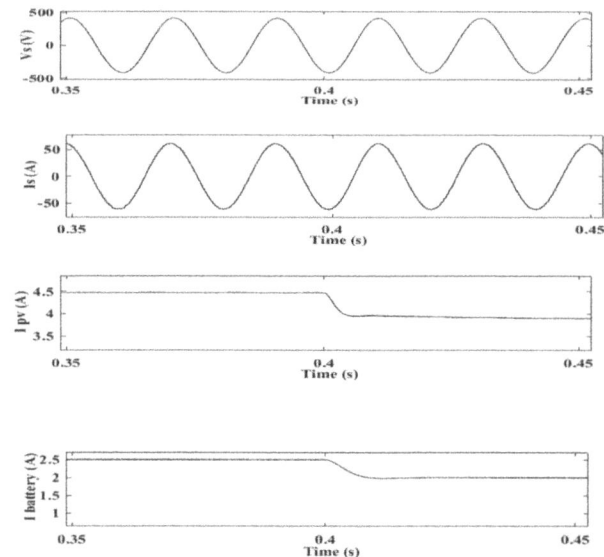

Figure 73.4(c) Dynamic condition of Vs, Ipv and Ibattery as gradually lowers the insolation
Source: Author

responded when wind speed levels dropped, as shown in Figure 73.5(c). The steady drop in PMBLDC generator current (decrease in wind speed) exhibits no influence on SEIG current or AC voltage, as illustrate in Figure 73.5(c). According to the simulation findings, which are shown in Figures 73.5(a) and 73.5(b), the total power generation from the three renewable sources becomes more than load as the wind speed increases gradually. An increase in battery charging current is used as redirect extra power to battery in order to control frequency. Figure 73.5(b) depicts when the battery current changes from charging to discharging mode in response to a rise in PMBLDC generator current. To achieve power balance between the various sources, the GWO control algorithm frequency loop applies. Wind and PMBLDC generator current can be slowly raised without causing any disturbance or change in load current, is shown in Figure 73.5(a). As an observed in simulation results depict in Figure 73.5(c), a slight decline in wind causes an in proportion drop in PMBLDC output current with the same manner, which lowers the battery current. As shown in Figure 73.5(c), as the wind speed levels drop, the battery transitions from charging to discharging mode to restore power balance. When wind speed goes down, the reduction has no effect on the microgrid power quality such as frequency and voltage.

Dynamic performance of microgrid, variation in load

The variable response of load current, VSC current, Induction generator output voltage and frequency can be illustrated in Figure 73.6 as system an observes variation in load. In this dynamic circumstance, suggested controller an exhibits exceptional dynamic flexibility and re-estimates the load true power requirements as well as the SEIG modified harmonic

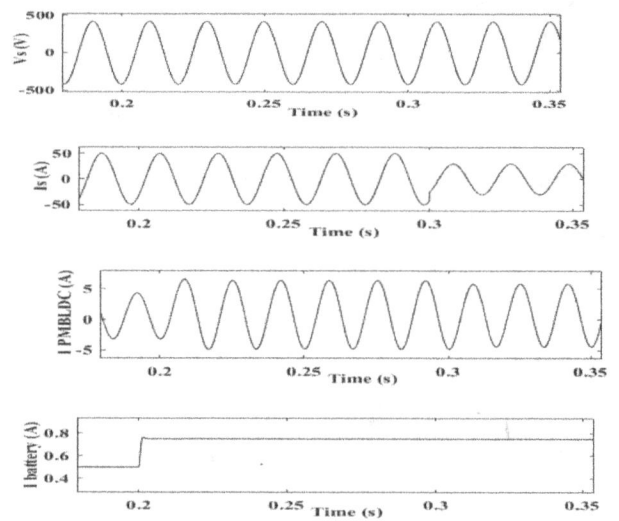

Figure 74.5(b) An increase in wind speed, the rapid response of Vs, is,iPMBLDC and ibattery
Source: Author

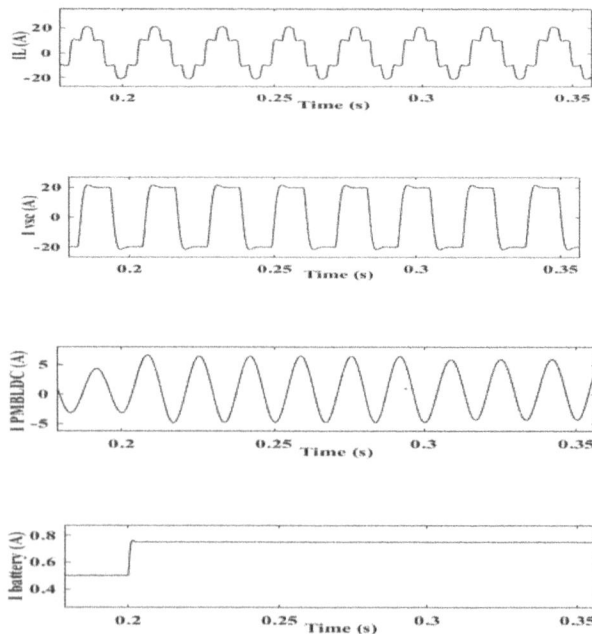

Figure 74.5(a) Dynamic behavior of battery, IL, iVSC and IPMBLDC as follows an incremental rise in wind speed
Source: Author

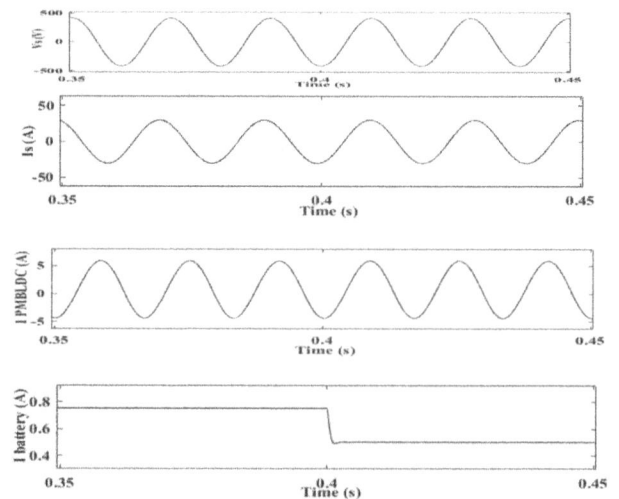

Figure 74.5(c) Decrease in wind speed, the dynamic output of Vs, is, iPMBLDC and ibattery
Source: Author

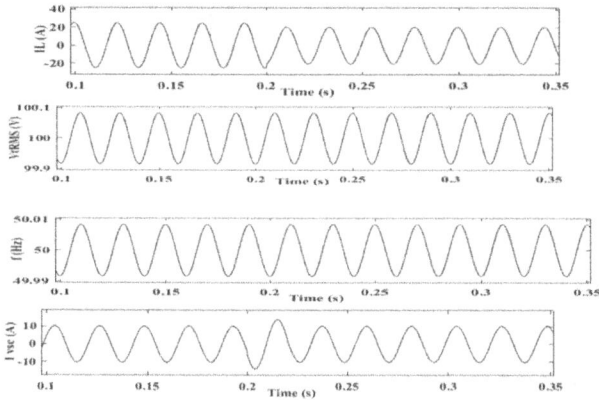

Figure 74.6 The suggested system is a dynamic response after a certain change in load and wind speed
Source: Author

and fundamental reactive power requirements. For controlling the frequency and voltage of the system, respectively, the control in this instance adjusts the VSC switching pattern to compensate for the system increased reactive and true power requirements. A rapid change in load leads microgrid voltage and frequency quickly remove as their rated levels, clearly as an observed in Figure 73.6.

Conclusion

In MATLAB/Simulation, the Grey Wolf Optimization (GWO) control algorithm can be used for regulating voltage and frequency of standalone microgrid. Three primary renewable energy sources are being included into the proposed SEIG-based standalone microgrid (wind, solar and micro-hydro). The simulation outcomes demonstrate GWO algorithm efficacy and strong microgrid voltage and frequency regulation. The microgrid power quality during balanced and unbalanced load is also enhanced by proposed controller algorithm, which also guarantees the best possible use BESS and renewable sources.

References

[1] Dondi, P., Bayoumi, D., Haederli, C., Julian, D., & Suter, M. (2002). Network integration of distributed power generation. *Journal of Power Sources*, 106 (1–2), 1–9.

[2] Lopes, J. P., Hatziargyriou, N., Mutale, J., Djapic, P., & Jenkins, N. (2007). Integrating distributed generation into electric power systems: a review of drivers, challenges and opportunities. *Electric Power Systems Research*, 77(9), 1189–1203.

[3] Hatziargyriou, N., Asano, H., Iravani, R., & Marnay, C. (2007). Microgrids. *IEEE Power & Energy Magazine*, 5(4), 78–94.

[4] Ruiz, N., Cobelo, I., & Oyarzabal, J. (2009). A direct load control model for virtual power plant management. *IEEE Transactions on Power Systems*, 24(2), 959–966.

[5] Morais, H., Kádár, P., Cardoso, M., Vale, Z. A., & Khodr, H. (2008). VPP operating in the isolated grid. In Proceedings of IEEE Power and Energy Soc. General Meet., (pp. 1–6).

[6] Pudjianto, D., Ramsay, C., & Strbac, G. (2007). Virtual power plant and system integration of distributed energy resources. *IET Renewable Power Generation*, 1(1), 10–16.

[7] Molderink, A., Bakker, V., Bosman, M. G. C., Hurink, J. L., & Smit, G. J. M. (2010). Management and control of domestic smart grid technology. *IEEE Transactions on Smart Grid*, 1, 109–119.

[8] Pudjianto, D., Ramsay, C., & Starbac, G. (2008). Microgrids and virtual power plants: Concepts to support the integration of distributed energy resources. *Proceedings of the Institution of Mechanical Engineers, Part A: Journal of Power and Energy (IMechE)*, 222, 731–741.

[9] Karimi, H., Nikkhajoei, H., & Iravani, M. R. (2008). Control of an electronically-coupled distributed resource unit subsequent to an islanding event. *IEEE Transactions on Power Delivery*, 23(1), 493–501.

[10] Katiraei, F., Iravani, M. R., & Lehn, P. W. (2005). Micro-grid autonomous operation during and subsequent to islanding process. *IEEE Transactions on Power Delivery*, 20(1), 248–257.

[11] IEEE Standard for Interconnecting Distributed Resources with Electric Power Systems," in IEEE Std 1547-2003, vol., no., pp.1–28, 28 July 2003, doi: 10.1109/IEEESTD.2003.94285.

[12] Zamora, R., & Srivastava, A. K. (2010). Controls for microgrids with storage: Review, challenges, and research needs. *Renewable and Sustainable Energy Reviews*, 14(7), 2009–2018.

[13] A. Hajimiragha and M. R. D. Zadeh, "Practical aspects of storage modeling in the framework of microgrid real-time optimal control," in Proc. IET Conf. on Renewable Power Generation. (RPG), Sep. 2011, pp. 93– 98.

[14] F. Blaabjerg, R. Teodorescu, M. Liserre, and A. V. Timbus, "Overview of control and grid synchronization for distributed power generation systems," IEEE Trans. Ind. Electron., vol. 53, no. 5, pp. 1398–1409, Oct. 2006.

[15] A. Timbus, M. Liserre, R. Teodorescu, P. Rodriguez, and F. Blaabjerg, "Evaluation of current controllers for distributed power generation systems," IEEE Trans. Power Elect., vol. 24, no. 3, pp. 654–664, Mar. 2009.

74 Augmented reality and immersive environments for next-gen brain MRI scan analysis using vertex AI

Polaiah Bojja[1,2,a], I. Govardhan Rao[2,b], K. Rajendra[2,c], Ankathi Sujay[1,d], Bejugam Sai Nivas[2,e] and Balikara Sudeep, M.[2,f]

[1]Department of computer science and Information Technology, Institute of Aeronautical Engineering, Hyderabad, India

[2]Department of computer science engineering, University College of Engineering, Osmania University, Hyderabad, India

Abstract

Patients diagnosed with brain tumors often struggle to understand their condition, the tumor's location, size, and impact, and the proposed treatment plan. Traditional 2D MRI scans can be confusing, and verbal explanations may not be enough. Enhancing patient understanding of brain tumors using augmented reality mechanism -based 3D visualization of MRI scans using immersive environment (which is an artificial environment that is used to create realistic and interactive experiences for patients and doctors. Further enhancing their perception of reality by adding virtual elements to the physical environment for the sake of user-friendly system like mobile apps and augmented reality set by artificial intelligence with Vertex AI. The results are carried out augmented reality mechanism -based 3D visualization of MRI scans using immersive environment.

Keywords: Augmented reality, brain tumor, immersive environment, MRI brain images

Introduction

Brain tumors pose significant challenges not only in terms of medical treatment but also in patient understanding and decision-making. Patients often struggle to comprehend the complex nature of their condition, including the tumor's location, size, and potential impact on brain functions. Traditional 2D MRI scans, while valuable for diagnosis, can be difficult for non-medical individuals to interpret, and verbal explanations from healthcare professionals may not always suffice. This gap in understanding can lead to increased anxiety and hinder informed decision-making regarding treatment options.

To bridge this gap, augmented reality (AR) and artificial intelligence (AI) technologies offer a transformative approach. By integrating augmented reality-based 3D visualization of MRI scans within an immersive environment, patients and doctors can interact with a more intuitive and user-friendly system. This approach enhances perception by overlaying virtual elements onto the physical world, enabling real-time interaction with 3D brain models.

Furthermore, incorporating AI-driven solutions like Vertex AI can enhance system intelligence, personalizing patient experiences and improving diagnostic accuracy.

This research explores the implementation of an AR mechanism-based 3D MRI visualization system to enhance patient comprehension and doctor-patient communication. By leveraging mobile applications and augmented reality headsets, this system aims to provide an interactive, engaging, and informative experience, ultimately improving the overall quality of patient care.

Meltdown characteristics of brain MRI images

Meltdown in brain MRI images refers to the distortion or overexposure of medical scans, leading to the loss of critical diagnostic details. This phenomenon occurs due to excessive brightness, contrast adjustments, or intensity blending, which may result from improper MRI acquisition settings or post-processing techniques such as AI-based enhancement and

[a]paulraj.bojja@gmail.com, [d]ankathisujay@gmail.com, [c]balikarasudeep@gmail.com, [e]sainivasbejugam@gmail.com, [e]govardhanrao@gmail.com, [f]rajendrak@gmail.com

DOI: 10.1201/9781003684589-74

compression artifacts. The consequences of meltdown significantly impact the accuracy of tumor assessment and overall diagnosis.

Key problems caused by meltdown in brain MRI:

1. Loss of tumor internal structure – Excessive brightening may obscure vital tumor characteristics such as necrosis, edema, or cystic regions, reducing diagnostic accuracy.

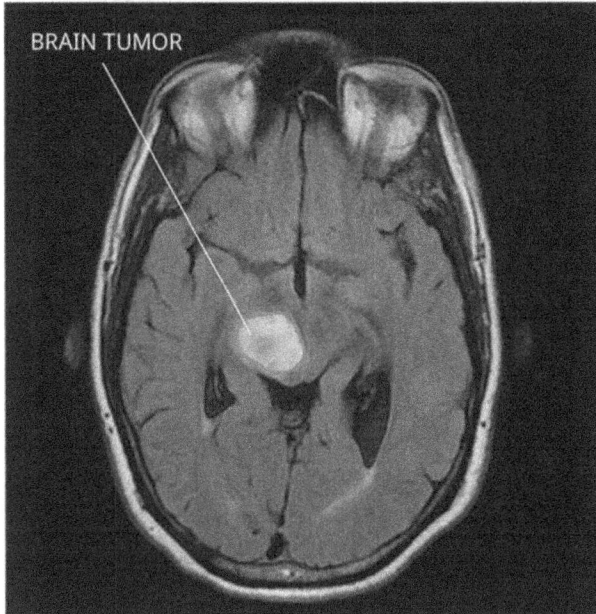

BRAIN TUMOR

2. False tumor size estimation – Overexposed or blended intensity levels can lead to overestimation or underestimation of tumor boundaries, affecting treatment planning.
3. Difficulty in differentiating tumor vs. normal tissue – Artificial merging of structures can make it challenging to distinguish between healthy and abnormal tissues, complicating clinical decisions.
4. AI-Based segmentation errors – Deep learning models like U-Net and Deep-Medic, which rely on precise image data, may misclassify tumor regions if images are over-processed, reducing the reliability of automated diagnosis.

Softening of meltdown characteristics of brain MRI images

Softening the meltdown characteristics in brain MRI images is crucial for enhancing the accuracy of tumor diagnosis and treatment planning. AR and AI provide innovative solutions to mitigate overexposure, intensity distortion, and artificial merging of structures,

thereby preserving critical diagnostic details. By integrating AI-driven image restoration and AR-based interactive visualization, medical professionals can achieve real-time contrast adjustments, improved segmentation accuracy, and enhanced perception of tumor characteristics in an immersive environment.

Key features for softening meltdown in brain MRI

1. AR for real-time visualization:
    The AR overlays interactive 3D models of MRI scans onto the real-world environment, allowing doctors and patients to manipulate brightness, contrast, and intensity dynamically. Real-time adjustments help reduce overexposure and correct brightness artifacts. Enhance tumor boundary visualization, making it easier to differentiate between normal and abnormal tissues.
2. Immersive environment for enhanced perception
    Creates a virtual, interactive space where users can explore MRI scans from multiple angles. [1] In this extended version, two new visualization mechanisms for mobiles devices were incorporated with the purpose of improving the students' visualization skills.
3. AI-Based deep learning image restoration
    Neural networks (e.g., U-Net, GANs, Deep-Medic) help restore lost details in overexposed MRI images. Adaptive contrast correction ensures that brightness levels remain within optimal diagnostic ranges. AI models trained on large datasets can detect and correct compression artifacts introduced during image post-processing.
4. AI-Powered adaptive segmentation
    AI-driven segmentation techniques refine tumor boundaries by counteracting the effects of meltdown. Machine learning models analyze the differences between overexposed and properly balanced images, enhancing segmentation accuracy. Helps improve the performance of tumor classification and feature extraction, leading to better treatment planning.
5. Integration with Vertex AI for smart processing
    Google's Vertex AI enables cloud-based, high-performance medical image analysis. Provides automated anomaly detection, ensuring that AI models continuously learn and adapt to improve image quality. Enhance system intelligence by personalizing MRI enhancements based on patient-specific data.

By leveraging AR and AI-driven restoration, meltdown characteristics in MRI scans can be softened dynamically, ensuring better diagnostic accuracy, improved patient understanding, and enhanced doctor-patient interaction. The fusion of immersive environments, AI-powered segmentation, and real-time AR-based adjustments provides a revolutionary approach to brain tumor visualization, making the diagnostic process more interactive, accurate, and user-friendly.

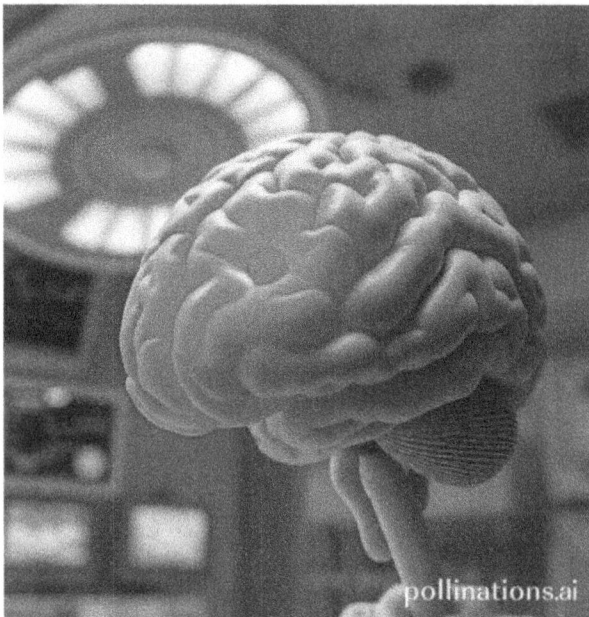

Converting brain MRI into augmented reality view

Brain MRI images are traditionally viewed as 2D cross-sectional slices, which can make it challenging for both doctors and patients to accurately interpret the tumor's exact location, shape, and size. The limitations of conventional MRI visualization methods can hinder precise diagnosis and treatment planning. The AR presents a transformative solution by converting MRI scans into interactive 3D models, enabling a more intuitive and immersive representation of brain structures. This approach enhances medical imaging by allowing users to explore the brain from multiple perspectives in real-time, improving both clinical decision-making and patient comprehension.

The conversion of MRI scans into AR begins with MRI image acquisition and preprocessing, where raw medical images in Digital Imaging and Communications in Medicine (DICOM) format are collected. Preprocessing techniques, such as noise reduction, intensity normalization, and artifact correction, are applied to ensure clarity and accuracy. AI-based enhancement methods are also utilized to refine image quality by correcting overexposed regions and eliminating unwanted distortions, thereby mitigating issues related to meltdown characteristics in MRI scans.

Following preprocessing, AI-based segmentation techniques, such as U-Net and Deep-Medic, are employed to extract relevant brain structures and tumor regions. These deep learning models utilize extensive datasets to differentiate between normal and abnormal tissues, enhancing the precision of tumor boundary detection. The segmented images are then reconstructed into a 3D volumetric model using specialized medical imaging software, including 3D slicer and blender. This reconstruction process involves the application of texture enhancements, transparency adjustments, and depth mapping to provide a clearer and more detailed visualization of the tumor and surrounding anatomical structures.

Once the 3D model is generated, it is integrated into an AR environment, where it can be visualized through AR-compatible devices such as mobile applications, head-mounted displays (HMDs), and AR-assisted surgical navigation systems. By overlaying the reconstructed MRI data onto a real-world or virtual space, AR enables real-time interaction, allowing clinicians to manipulate the 3D model—zooming, rotating, and dissecting different layers—to gain a comprehensive understanding of the tumor's spatial characteristics. Additionally, this AR-driven visualization enhances patient education by providing an intuitive way for individuals to comprehend their diagnosis, thus fostering better communication between doctors and patients.

Furthermore, the integration of AI with AR enhances system intelligence by enabling adaptive visualization, where the AR interface can dynamically adjust contrast levels, highlight critical tumor features, and provide personalized imaging insights using platforms like Google's Vertex AI. This ensures that the visualization remains accurate, adaptable, and optimized for both medical professionals and patients.

Overall, converting brain MRI scans into an Augmented Reality view represents a significant advancement in medical imaging, bridging the gap between traditional 2D interpretation and interactive

3D visualization. This approach not only improves diagnostic precision and treatment planning but also enhances patient engagement and understanding, ultimately contributing to better healthcare outcomes.

Estimating brain tumor size using augmented reality

3D Tumor visualization in AR: MRI scan data can be converted into a 3D model of the tumor. Using mobile AR apps, doctors can see the tumor overlaid on a patient's head. Helps in understanding tumor boundaries, depth, and interaction with brain structures.

AI-Powered tumor segmentation and size calculation: AI-based MRI analysis can automatically detect and segment the tumor. AR then projects the segmented tumor in 3D with real-time size measurements. Doctors can rotate, zoom, and manipulate the AR tumor model for better assessment.

Brain tumor diagnosis and treatment planning have traditionally relied on 2D MRI scans, which provide limited spatial understanding of tumor boundaries and their interaction with critical brain structures. The AR is transforming this process by integrating AI-driven tumor segmentation, 3D reconstruction, and interactive visualization to enhance clinical decision-making. By converting MRI scan data into detailed 3D models, AR enables real-time tumor projection onto a patient's anatomy, allowing neurosurgeons to assess its size, depth, and spatial relationships with greater accuracy.

MRI-to-AR pipeline for tumor visualization

Data acquisition and preprocessing

The first step in AR-based tumor visualization is acquiring high-resolution MRI scans, which provide detailed imaging of the brain and tumor structures. Different MRI sequences, such as T1-weighted, T2-weighted, and FLAIR imaging, help differentiate tumor boundaries from surrounding tissues. In some cases, contrast-enhanced MRIs using gadolinium-based agents are used to enhance tumor visibility. After data collection, preprocessing techniques such as noise reduction, skull stripping, and intensity normalization refine the image quality. Noise reduction methods like Gaussian filtering or non-local means (NLM) denoising help remove artifacts, while skull stripping algorithms like brain extraction tool (BET) eliminate non-brain tissues. Image alignment using affine or non-rigid transformation ensures that MRI slices are correctly positioned for further processing.

3D Reconstruction

Once the MRI scans are preprocessed, AI-powered segmentation models, such as U-Net, DeepLabV3, and Mask R-CNN, identify the tumor's boundaries with high accuracy. Traditional methods like thresholding and watershed segmentation are also used but are less effective than deep learning approaches. After segmentation, 3D reconstruction methods like the Marching Cubes algorithm and Delaunay triangulation convert the segmented slices into a three-dimensional tumor model. These techniques ensure smooth surface representation and detailed visualization, making it easier for neurosurgeons to examine the tumor's shape, size, and position relative to critical brain structures. The 3D model is then optimized using volumetric rendering, which enhances depth perception and realism.

AR projection and registration

To project the 3D tumor model onto the patient's real-world anatomy, spatial registration methods like marker-based tracking, marker less tracking, and infrared optical tracking are used. Marker-based tracking involves placing fiducial markers, such as QR codes or AR tags, on the patient's head to align the virtual tumor with physical anatomy. In contrast, marker less tracking relies on depth sensors and computer vision techniques for automatic alignment. [2] Augmented Reality (AR), which overlays information into the real environment, has the potential to create a realistic learning environment through the projection of enjoyable and interesting content in front of art objects. Through AR headsets or smartphone-based AR apps, doctors can rotate, zoom, and manipulate the holographic tumor, providing a more intuitive understanding of its structure.

AI-powered tumor segmentation and size estimation

AI-based MRI analysis

The AI plays a crucial role in analyzing MRI scans for tumor detection and segmentation. Deep learning models, particularly convolutional neural networks (CNNs) and Vision Transformers (ViTs), enhance the accuracy of tumor identification. Hybrid AI models integrate traditional image processing techniques with deep learning to improve segmentation outcomes. However, challenges such as MRI variability across different scanners, overlapping tumor boundaries, and the need for large annotated datasets make AI-based segmentation a complex task. Despite these hurdles, AI significantly reduces the time required for tumor detection and improves diagnostic precision.

Tumor size estimation and volume calculation

Once the tumor is segmented, its size is calculated using bounding box algorithms that extract its length, width, and height. AI-driven voxel-based volume estimation provides an accurate measure of tumor growth by counting the number of voxels within the segmented region and multiplying it by the voxel size. This method allows for precise tumor volume tracking over time, making it easier for doctors to assess disease progression. AR enables real-time interaction with the tumor model, where doctors can slice through different layers, rotate the visualization, and even use haptic feedback gloves to "feel" the virtual tumor. Additionally, multi-user AR environments allow multiple specialists to examine the same tumor simultaneously, facilitating collaborative decision-making.

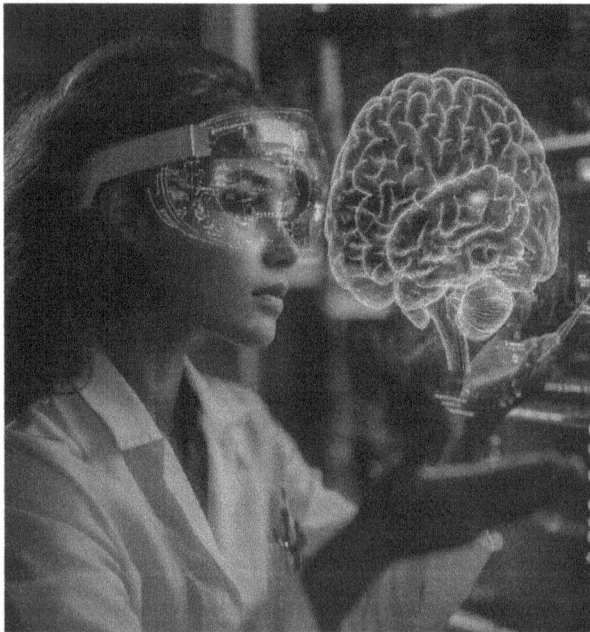

Immersive environment in augmented reality for medical imaging

An immersive environment refers to a digital or physical space that fully engages an individual's senses, creating an experience that feels natural and interactive. These environments are designed to enhance perception by stimulating multiple senses, including sight, sound, and touch, thereby making users feel as if they are truly inside the simulated setting. In the field of medical imaging and diagnostics, immersive environments play a crucial role in augmenting reality-based visualization, particularly in applications such as brain MRI interpretation.

Immersive environments can be classified into three main types: virtual reality (VR), AR, and MR. VR creates a completely artificial digital space where users are fully immersed, typically using VR headsets. Mixed Reality blends digital and real-world elements, allowing users to interact with both simultaneously. AR, [3] The use of mobile-based AR technology in education has been proven successful in many engineering and architectural arenas. Researchers The use of mobile-based AR technology in education has been proven successful in many engineering and architectural arenas. Researchers AR allows real-time visualization of medical data, such as MRI scans, by superimposing 3D models onto physical space, making it an effective tool for both diagnosis and treatment planning.

A well-designed immersive environment incorporates key features that make the experience more engaging, interactive, and beneficial for medical professionals and patients alike. The first essential feature is realism and interactivity, which allows users to explore and manipulate digital medical models, such as MRI-derived 3D reconstructions of brain tumors. This interactivity provides a more intuitive understanding of complex structures that might be difficult to interpret in traditional 2D imaging. Another critical aspect is the multi-sensory experience, where AR technology integrates visual overlays, spatial audio, and haptic feedback, helping users engage with medical data in a way that enhances perception and comprehension. Additionally, immersive environments help in minimizing distractions, making it easier for users to focus on medical analysis without interference from external elements.

Augmented reality, as a type of immersive environment, enhances clinical decision-mak- ing, surgical planning, and patient education by providing an interactive medium for visualizing MRI data. Through AR-enabled devices such as smartphones, AR glasses, and head-mounted dis- plays (HMDs), medical professionals can view MRI scans in a three-dimensional, spatially aware format, improving their ability to analyze brain tumors with greater accuracy. [6] Augmented Reality/Virtual Reality (AR/VR) with depth 3-dimensional (D3D) imaging provides depth perception through binocular vision, head tracking for improved HMI and other key AR features.

The integration of immersive environments, particularly AR, into medical imaging and healthcare represents a significant advancement in diagnostic accuracy, treatment planning, and patient engagement. By creating an interactive and engaging visualization experience, AR-based immersive environments allow medical professionals to explore brain structures in greater detail, reduce diagnostic uncertainty, and improve overall healthcare outcomes.

Building augmented reality for neurosurgery using vertex AI

The integration of Vertex AI in AR-guided neurosurgical procedures is revolutionizing the way surgeons approach brain tumor resections. In traditional neurosurgery, MRI and CT scans provide preoperative imaging, but they lack real-time adaptability once the procedure begins. By leveraging Vertex AI, AR can overlay real-time MRI data directly onto the patient's brain during surgery, allowing neurosurgeons to visualize tumor locations, critical structures, and surgical pathways with unprecedented precision. This AR-guided approach minimizes the risk of damaging vital brain regions, ensuring safer and more effective tumor removal.

[5] The advancements in AR/VR have come a long way since their introduction and have great potential for continued usage in medicine enabling instant intraoperative imaging updates. Traditional imaging methods require surgeons to rely on static preoperative scans, which may not accurately reflect real-time changes during surgery. By deploying AI models on Edge AI devices, updated imaging data can be processed and displayed in real time, allowing surgeons to make immediate, data-driven adjustments. This not only improves surgical accuracy but also reduces operating time, leading to better patient outcomes and faster recovery.

Beyond surgical applications, Vertex AI-powered AR simulations are transforming medical education and training for neurosurgeons. By creating highly realistic, interactive AR environments, AI allows surgeons to practice complex brain procedures in a risk-free, virtual setting. These simulations can replicate real surgical scenarios, providing hands-on training in tumor resections, hemorrhage control, and anatomical navigation. This AI-enhanced AR training improves neurosurgeons' skills, preparing them for real-life surgeries with greater confidence and precision.

By combining AI, AR, and cloud computing, Vertex AI is reshaping the landscape of brain tumor imaging and treatment. The ability to overlay real-time data, provide dynamic imaging updates, and create immersive training environments makes AR a crucial tool in modern neurosurgery. As AI technology continues to evolve, AR-powered surgical solutions will drive faster diagnoses, safer procedures, and enhanced patient care, marking a significant leap forward in the future of medical imaging and neurosurgical precision.

Conclusion

Traditional brain tumor diagnosis relies on 2D MRI or CT scans, which can be difficult for patients to interpret. Doctors explain tumor size, location, and severity using static images or reports, making it challenging for patients to fully grasp their condition. This often leads to confusion, anxiety, and a lack of engagement in treatment decisions. [7] In total, 20 devices and platforms using AR were identified, with common features being the ability for remote users to annotate, display graphics, and display their hands or tools in the local user's view. [4] Immersive virtual reality (VR) and augmented reality (AR) are gaining increasing attention within the medical field and have been extensively researched in relation to teaching and treatment applications.

References

[1] Camba, J., Contero, M., & Salvador-Herranz, G. (2014). Desktop vs. mobile: a comparative study of augmented reality systems for engineering visualizations in education. In Proceedings of the 2014 IEEE Frontiers in Education Conference (FIE); Madrid, Spain. 22–25 October 2014; (pp. 1–8).

[2] Leue, M. C., Jung, T., & tom Dieck, D. (2015). Google glass augmented reality: generic learning outcomes for art galleries. In Tussyadiah, I., & Inversini, A. (Eds.), Information and Communication Technologies in Tourism 2015. Cham, Switzerland: Springer.

[3] Fang, W., Zhang, T., Chen, L., & Hu, H. (2023). A survey on HoloLens AR in support of human-centric intelligent manufacturing. *Journal of Intelligent*

Manufacturing, 36(1). doi: 10.1007/s10845-023-02247-5.

[4] Aliwi, I., Schot, V., Carrabba, M., Duong, P., Shievano, S., Caputo, M., et al. (2023). The role of immersive virtual reality and augmented reality in medical communication: a scoping review. *Journal of Patient Experience*, 10, 23743735231171562.

[5] Murali, S., Paul, K. D., McGwin, G., & Ponce, B. A. (2021). Updates to the current landscape of aug- mented reality in medicine. *Cureus*, 13, e15054. doi: 10.7759/cureus.15054.

[6] Bojja, P., et al. (n.d.). Industrial Iot enabled fuzzy logic based flame image processing for rotary kiln control. *Springer Wireless Personal Communications*, Doi.Org/10.1007/S11277-022-09677-Z.

[7] Murali, S., Paul, K. D., McGwin, G., & Ponce, B. A. (2021). Updates to the current landscape of augmented reality in medicine. *Cureus*, 13, e15054. doi: 10.7759/cureus.15054.

[8] Douglas, D. B., Wilke, C. A., Gibson, J. D., Boone, J. M., & Wintermark, M. (2017). Augmented reality: advances in diagnostic imaging. *Multimodal Technologies and Interaction*, 1, 29. doi: 10.3390/mti1040029.

[9] Dinh, A., Yin, A. L., Estrin, D., Greenwald, P., & Fortenko, A. (2023). Augmented reality in real-time telemedicine and telementoring: scoping review. *JMIR Mhealth Uhealth*, 11, e45464.

75 A hybrid AI and time-series framework for dynamic pricing and demand prediction in e-commerce

Sri Charani, P.[1,a], Durga Prasanna, T.[2,b], Satya Sekhar Varma, P.[3,c], Jahnavi Sri, N.[2,d], Jahnavi, M.[2,e] and Tulasi, Y.[2,f]

[1]Professor, Department of AI, Shri Vishnu Engineering College for Women, Bhimavaram, Andhra Pradesh, India

[2]Department of AI, Shri Vishnu Engineering College for Women, Bhimavaram, Andhra Pradesh, India

[3]Research Scholar, Department of CSE, Shri Vishnu Engineering College for Women, Bhimavaram, Andhra Pradesh, India

Abstract

In the competitive landscape of modern retail, accurate sales forecasting and price prediction are critical for strategic decision-making. The study analyzes the use of machine learning to develop demand forecasting and dynamic pricing approaches which optimize price strategy efficiency. Historical transactions from an electronic marketplace have been analyzed in this research through feature engineering before evaluating multiple models for their optimal predictive solution. Gradient boosting machines (GBM) is selected for price prediction due to its ability to effectively capture non-linear relationships and intricate feature interactions, achieving an R^2 score of 98%. Additionally, the Seasonal Autoregressive Integrated Moving Average (SARIMA) model is employed for demand forecasting, effectively capturing seasonal patterns with an R^2 score of 71%. Key steps include data preprocessing, feature extraction (e.g., temporal and geographical attributes), and exploration analysis to identify demand trends. Visualizations such as seasonal decomposition plots, feature importance charts, and actual vs. predicted graphs provide actionable insights for retail decision-makers. The results highlight the effectiveness of integrating machine learning and time-series analysis in retail analytics to enhance operational efficiency and profitability.

Keywords: Demand forecasting, dynamic pricing, gradient boosting, machine learning, retail analytics, seasonal autoregressive integrated moving average SARIMA

Introduction

This research compared three widely adopted machine learning techniques for dynamic pricing Gradient boosting (GB), Random Forest (RF), and XGBoost. These models have been applied successfully across various domains [2] and offer different advantages in capturing complex pricing dynamics. Additionally, the study evaluates two forecasting models, SARIMA and long short-term memory (LSTM), for demand prediction.

Modern retail and e-commerce industries are increasingly adopting data-driven approaches [1] to stay competitive in rapidly changing markets. Sales demand forecasting together with accurate price prediction acts as a fundamental requirement for efficient inventory control while improving customer satisfaction and maximizing business profits. Traditional models often fail to capture non-linear interactions or seasonal patterns, which are essential in dynamic markets.

A research analysis examines historical e-commerce transaction data to evaluate different modeling methods using feature engineering techniques. The primary objectives are to preprocess and enrich transactional data with temporal, geographic, and product-specific features, to build and evaluate separate models for price prediction and demand forecasting, and to provide actionable insights through interactive visualizations for better decision-making [3]. The performance of each model is evaluated in terms of accuracy and interpretability, providing meaningful insights into their effectiveness for dynamic pricing and demand forecasting in retail analytics.

[a]charani.yashu@svecw.edu.in, [b]21b01a54a5@svecw.edu.in, [c]ss721087@student.nitw.ac.in, [d]21b01a5476@svecw.edu.in, [e]22b05a5409@svecw.edu.in, [f]22b05a5412@svecw.edu.in

DOI: 10.1201/9781003684589-75

Proposed methodology

The proposed system offers a strong framework for sales demand forecasting and price prediction, addressing the limitations of existing systems while offering enhanced accuracy, adaptability, and actionable insights. It enables businesses to use data analytics to stay competitive within the always-changing retail sector and e-commerce market. Key components and benefits of the proposed system are detailed below. It consists of the following key modules: data preprocessing and feature engineering, predictive modeling, visualization dashboard, and evaluation metrics. It focuses on preparing raw transactional data for analysis by handling missing information, extracting important features, and categorial variable transformation.

Dataset collection

Transactional sales data is collected from retail and e-commerce platforms in CSV format. The dataset includes columns such as "Order Date", "Quantity Ordered", "Price Each" and "Purchase Address". This data serves as the foundation for analysis and predictive modeling.

Dataset preprocessing

The raw data goes through a preprocessing phase to ensure accuracy and uniformity. This process includes addressing missing or null values by eliminating incomplete entries, parsing and formatting the "Order Date" column to extract temporal features like month, day, and hour, splitting the "Purchase Address" column to derive geographic features such as city, calculating total sales by multiplying "Quantity Ordered" and "Price Each.", removing erroneous or outlier entries that may skew analysis.

Feature engineering

Features are extracted and engineered to improve the predictive power of the model:

Temporal features: Month, day, and hour.
Geographic features: City derived from the "Purchase Address."
Product-specific features: Product categories represented as dummy variables.
Sales-specific features: Calculated total sales based on quantity and price. Categorical features such as city and product are converted into dummy variables to enable machine learning model compatibility.

Model architecture

The system design for the proposed solution is structured into six interconnected layers, each responsible for specific tasks. System flow: Data upload starts the process as users upload raw transactional data through the user interface. The system validates the data for completeness and correctness. Data is cleaned, and features are engineered for analysis. The available refined data is utilized in training SARIMA for demand forecasting and Gradient Boosting (GB) for price prediction [7]. Sales trends and demand patterns are visualized in interactive charts. Users input parameters, and the system predicts sales or pricing outcomes. Predictions and visualizations are displayed on the user interface for decision-making [4]. This architecture integrates data ingestion, preprocessing, predictive modeling, and visualization into a streamlined process, empowering businesses with actionable insights and accurate predictions.

Model selection and development

A range of machine learning models [8] was assessed to determine the most effective approaches for dynamic pricing and demand forecasting. GB, Random Forest (RF), and XGBoost were evaluated for their capacity to model complex relationships and deliver precise predictions. Following comprehensive experimentation and performance comparison, GB emerged as the preferred model due to its exceptional accuracy in capturing the complexities of pricing dynamics.

Gradient Boosting: The ensemble learning technique trains sequential series of weak models which usually consist of decision trees to decrease errors from preceding one. Through repeated improvements made along the way this approach works to improve prediction precision. The primary goal of the function for GB can be expressed as:

$$L = \sum_{i=1}^{n} (t_i - \hat{t}_i)^2$$

Here t_i represents the true target values, and $\hat{t}i$ denotes the model's predicted values.

Random Forest: The ensemble learning technique uses many decision trees to gather outputs which then create the final prediction outcome. The group of tree outputs strengthens prediction accuracy while helping prevent extreme fitting that occurs with one model. The final prediction emerges from averaging predictions produced by all trees used in regression tasks:

$$\hat{u} = \frac{1}{k} \sum_{u=1} g_t(x)$$

Here k represents total count of trees in the model, $g_t(x)$ denotes output of the u-th tree for given input x, and $\hat{u}$ represents the final predicted value.

XGBoost: XGBoost represents a better version of Gradient Boosting which adds built-in regularization features to stop overfitting when working with extensive datasets [10]. The objective function in XGBoost combines a loss function and a regularization term:

$$Obj(\theta) = \sum_{u=1} L(b_u, \hat{b}_u) + \sum_{t=1} \Omega(f_t)$$

Here $L(b_u, \hat{b}_u)$ represents loss function, and $\Omega(f_t)$ denotes regularization term for tree t.

For demand forecasting, both LSTM and seasonal ARIMA were assessed for their capability to capture temporal patterns and predict demand trends. After thorough comparison, SARIMA was the selected model due to its enhanced results in modeling seasonality and trend behavior.

Long short-term memory: LSTM functions to maintain dependencies between long periods of time in time- series information. The LSTM system performs information regulation through cell state management under a mechanism involving gating operations. The LSTM model uses the following gates and operations. The mathematical equations defining LSTM appear as follows:

$$g_i = \sigma(W_{gi} \cdot [a_{t-1}, y_t] + b_{gi})$$

$$C_t = g_f \cdot C_{t-1} + g_i \cdot \tanh(W_C \cdot [a_{t-1}, y_t] + b_C)$$

$$g_o = \sigma(W_{go} \cdot [a_{t-1}, y_t] + b_{go})$$

$$a_t = g_o \cdot \tanh(C_t)$$

In these equations, gf, gi, and go represent the forget, input, and output gates, respectively. Ct is the cell state at time step t, and at is the hidden state at time step t. yt is the input at time step t, and at−1 is the previous hidden state. The weight matrices Wgf, Wgi, Wgo, WC correspond to forget, input, output, and cell state gates, respectively. The bias terms bgf, bgi, bgo, bC are associated with the respective gates.

Seasonal ARIMA: It enhances the ARIMA model by adding seasonal components, enabling it to detect seasonal, trend, cyclic patterns in time-series data [6].

$$SARIMA(p, d, q)(P, D, Q)s$$

Here p, d, q are the non-seasonal parameters (AR order, differencing order, and MA order), and P, D, Q are the seasonal parameters (seasonal AR order, seasonal differencing order, and seasonal MA order). s denotes the length of the seasonal cycle. Mathematical equations defining SARIMA appear as follows:

$$(1 - \phi_i(L))(1 - L^s)^D Z_t = \theta(L)(1 - L)^d \varepsilon_t$$

Here Zt is the time series, εt is the error term, and φi and θ(L) are the AR and MA polynomials, respectively.

Evaluation metrics

Root mean squared error: RMSE quantifies the dispersion of errors in the predictions [5]. Smaller RMSE indicate better model performance.
Mean absolute error: MAE the mean absolute errors across a prediction dataset. Unlike RMSE, MAE is less influenced by large errors and offers a broader assessment of accuracy.
R-squared (R2): R2 represents the fraction of the total variance in the output that is accounted for by the model. A higher R2 value signifies a closer match between the predicted and observed values.

Results and discussion

The examination within this section delivers an assessment which considers various models used for dynamic pricing and demand forecasting. The performance of GB, XGBoost, and RF for dynamic pricing, as well as LSTM, SARIMA for forecasting, are evaluated based on key metrics such as R2, MAE, MSE, and RMSE.

Dynamic pricing model comparison
For dynamic pricing, three machine learning models—GB, XGBoost, and RF were evaluated. The GB model reached its highest achievement level which proved successful through its R value that reached 98.79%, indicating that it explains 98.79% of the variance in pricing. The MAE was 0.1010, and the RMSE was 0.1982, suggesting minimal prediction errors. XGBoost, another powerful

gradient boosting model, performed similarly, with a R2 score of 98.79%. It also demonstrated strong accuracy with an MAE of 0.1011 and an RMSE of 0.1982, showing that it was able to capture the dynamics of pricing well. RF showed slightly lower performance, with a R2 score of 98.74%, indicating that it explained almost 99% of the variance in pricing. However, the MAE (0.1161) and RMSE (0.2023) were slightly higher than those of GB and XGBoost, suggesting a marginally greater deviation from the actual values.

Demand forecasting model comparison

For demand forecasting, two models—LSTM and SARIMA were compared. The LSTM model, while capable of capturing time-series dependencies, had an R2 score of 65.3%, indicating that it explained only 65.3% of the variance in demand. Its MAE of 0.2334 and RMSE of 0.4337 indicated a higher level of prediction error compared to the SARIMA model. On the other hand, SARIMA performed better in forecasting demand, with an R2 score of 71.2%, showing that it was able to explain a larger portion of the demand variance. The MAE of 0.209 and RMSE of 0.4482, although slightly higher than LSTM, were still acceptable for capturing seasonality and trends in demand.

The comparative performance metrics of dynamic pricing models are detailed in Table 75.1.

The evaluation results of demand forecasting models are summarized in Table 75.2.

Table 75.1 Dynamic pricing model performance comparison.

Model	R^2 (%)	MAE	MSE	RMSE
Gradient Boosting (GB)	98.79	0.1010	0.0393	0.1982
XGBoost (XGB)	98.79	0.1011	0.0393	0.1982
Random Forest (RF)	98.74	0.1161	0.0409	0.2023

Source: Author

Table 75.2 Demand forecasting model performance comparison.

Model	R^2 (%)	MAE	MSE	RMSE
LSTM	65.3	0.2334	0.1881	0.4337
SARIMA	71.2	0.2090	0.2015	0.4482

Source: Author

Descriptive analysis

Daily sales trends (day vs sales): Unexpected spikes or dips may indicate promotions, holidays, or other events impacting daily sales as in Figure 75.1.

Monthly sales trends (month vs sales): Tracking sales over months can help identify overall growth or decline trends across the year as in Figure 75.2.

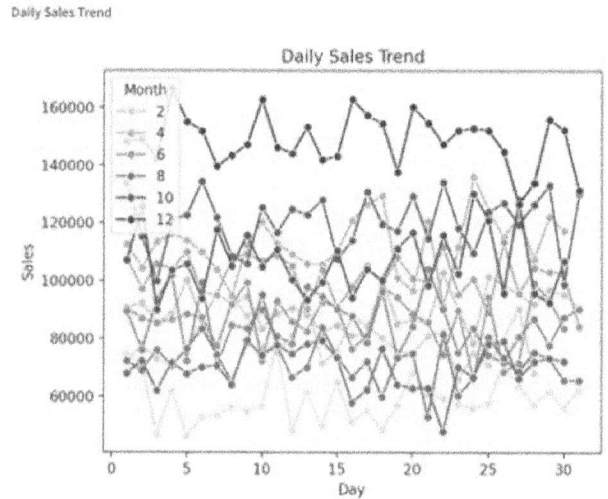

Figure 75.1 Daily sales trend
Source: Author

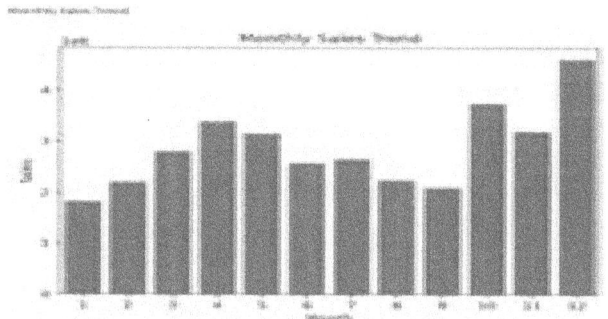

Figure 75.2 Monthly sales trend
Source: Author

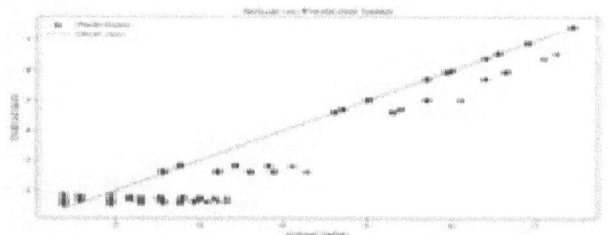

Figure 75.3 Actual vs predicted
Source: Author

Prediction vs actual values: The scatter plot shows predicted sales versus actual sales, with most points closely following the red ideal line, indicating a good model fit as in Figure 75.3.

Conclusion

The paper analyses how machine learning operates in demand forecasting alongside dynamic pricing systems. in the retail and e-commerce industries. The study evaluates multiple models, including Gradient Boosting (GB), XGBoost, and Random Forest (RF) for dynamic pricing, as well as long short-term memory (LSTM) and seasonal ARIMA (SARIMA) for demand forecasting.

For dynamic pricing, GB demonstrated exceptional performance, achieving an R2 score of 98.79% and minimal error metrics (MAE: 0.1010, RMSE: 0.1982), making it the most suitable model for capturing the complex relationships involved in pricing strategies. XGBoost and RF also provided strong performance, but GB stood out due to its superior prediction accuracy.

In demand forecasting, the SARIMA model proved to be the most effective for capturing seasonal trends and cyclical patterns in demand, with an R2 score of 71.2% [9]. Although LSTM showed promise for time-series data, SARIMA outperformed it in terms of precision and consistency, demonstrating its capability to model seasonal variations in demand effectively. Overall, the findings highlight the capability of combining sophisticated machine learning methods (like GB and XGBoost) for dynamic pricing and traditional statistical models (like SARIMA) for demand forecasting to enhance decision-making processes in modern retail and e-commerce environments. By leveraging these models, businesses can optimize pricing strategies and forecast demand more accurately, ultimately improving operational efficiency and profitability.

References

[1] Jain, A., Karthikeyan, V., Sahana, B., Shambhavi, B. R., Sindhu, K., & Balaji, S. (2020). Demand forecasting for e-commerce platforms. In 2020 IEEE International Conference for Innovation in Technology (INOCON), Bangalore, India, (pp. 1–4).

[2] El Youbi, R., Messaoudi, F., & Loukili, M. (2023). Machine learning-driven dynamic pricing strategies in e-commerce. In 2023 14th International Conference on Information and Communication Systems (ICICS), Irbid, Jordan, (pp. 1–5).

[3] Li, J. (2022). A feature engineering approach for tree-based machine learning sales forecast, optimized by a genetic algorithm based sales feature framework. In 2022 5th International Conference on Artificial Intelligence and Big Data (ICAIBD), Chengdu, China, (pp. 133–139).

[4] Singh, K., & Wajgi, R. (2016). Data analysis and visualization of sales data. In 2016 World Conference on Futuristic Trends in Research and Innovation for Social Welfare (Startup Conclave), Coimbatore, India, (pp. 1–6).

[5] Zhu, H., Bayley, I., & Green, M. (2022). Metrics for measuring error extents of machine learning classifiers. In 2022 IEEE International Conference on Artificial Intelligence Testing (AITest), Newark, CA, USA, (pp. 48–55).

[6] Jiang, H., Ruan, J., & Sun, J. (2021). Application of machine learning model and hybrid model in retail sales forecast. In 2021 IEEE 6th International Conference on Big Data Analytics (ICBDA), Xiamen, China, (pp. 69–75).

[7] Malik, A., Dargar, G., Sharma, A., & Pandey, P. (2023). Predictive analysis for retail shops using machine learning for maximizing revenue. In 2023 7th International Conference on Intelligent Computing and Control Systems (ICICCS), Madurai, India, (pp. 126–133).

[8] Semwal, M., Akila, K., Manasa, M., Raj, P. S., Motukuru, Y., & Karthik, P. (2024). Machine learning-enabled business intelligence for dynamic pricing strategies in e-commerce. In 2024 2nd International Conference on Disruptive Technologies (ICDT), Greater Noida, India, (pp. 116–120).

[9] Tang, T. (2023). Analysis and demand forecasting based on e-commerce data. In 2023 6th International Conference on Artificial Intelligence and Big Data (ICAIBD), Chengdu, China, (pp. 64–68).

[10] Gupta, C. K., Nath Kushwaha, O., Yadav, A. S., & Kumar, V. (2022). Dynamic flight price prediction using machine learning algorithms. In 2022 4th International Conference on Advances in Computing, Communication Control and Networking (ICAC3N), Greater Noida, India, (pp. 201–205).

76 Dynamic facial expression feature extraction using deep neural networks

Sumanth Jonnalagadda[1,a], Sravya Sri Nallani[1,b], Bomma Rithwika[2,c] and Bysani Jai Kushal[3,d]

[1]Computer Science and Information Technology, Koneru Lakshmaiah Education Foundation, Vaddeswaram, Guntur, Andhra Pradesh, India

[2]Computer Science and Engineering, Koneru Lakshmaiah Education Foundation, Vaddeswaram, Guntur, Andhra Pradesh, India

[3]Cyber Security, SRM University, Chennai, India

Abstract

Facial emotion recognition (FER) has made remarkable strides with the advent of artificial intelligence, particularly leveraging deep neural networks (DNNs) to achieve near-human accuracy in identifying emotions from facial expressions. This study introduces a sophisticated FER approach that combines convolutional neural networks (CNNs) with an augmented dataset to boost performance and reliability. It discusses to the performance and the formidability of the model in the interpretation of complicated psychosocial indicators in real-life situations, using benchmark datasets-fer2013, CK+, KDEF, and RAF-DB. Key to this approach is the focus on identifying critical facial features that DNN models emphasize, which enhances our understanding of how these systems interpret nuanced emotions. The proposed model demonstrates exceptional results, achieving 93.8% precision, 76.5% recall, and a mean average precision (map) of 80.5% on low-resolution 48x48 pixel grayscale images, making it a breakthrough in FER accuracy. This advancement not only redefines the capabilities of AI-driven emotion recognition but also offers valuable insights into solving the inverse optics problem in vision, opening new possibilities for applications that depend on precise and dependable emotion recognition, from mental health diagnostics to human-computer interaction.

Keywords: Improving analytical precision in interpreting multidimensional affective expressions across diverse environmental conditions using standardized evaluation frameworks, This research enhances facial emotion recognition systems through convolutional neural network architectures

Introduction

Facial affect analysis represents a cross-disciplinary nexus synthesizing cognitive science, biometric systems, and computational intelligence, proving critical for next-generation adaptive interface systems, immersive entertainment technologies, and clinical diagnostic tools [1–3]. Human faces convey a range of emotions crucial for communication, contributing significantly to nonverbal cues, which studies suggest make up 60% to 80% of communication [4]. FER, a key aspect of this nonverbal communication, is gaining traction, particularly in Human-Computer Interaction (HCI) [1]. It finds use in diverse areas like autopilot systems, this research introduces MEN (Mixed Enhancement Network), a novel architectural evolution from MobileFaceNets' computationally efficient framework originally optimized for facial biometric systems. The proposed design integrates two core innovations: 1) A heterogeneous kernel integration strategy combining multi-scale depth wise convolutional operations to concurrently capture localized and contextual features, and 2) Spatially-aware attention mechanisms that strengthen inter-regional feature correlations through positional encoding. These architectural modifications enable more discriminative representation learning for affective state classification while preserving computational efficiency [29].

Education, medical treatment, surveillance, and computer vision [1–3]. By deciphering facial

---

[a]jsumanth19@gmail.com, [b]sravyasri2004@gmail.com, [c]bomma.rithwika2005@gmail.com, [d]jaikushalbysani@gmail.com

DOI: 10.1201/9781003684589-76

expressions, systems can better understand human emotions, improving interaction and intervention outcomes [1]. Emotions are classified in psychology and computer vision as categorical (e.g., happiness, anger) or dimensional (e.g., valence, arousal), with technology advancing FER beyond traditional frameworks [6].

Facial recognition technology aims to decode emotional cues conveyed by facial expressions, mimicking human perception. Facial expressions are non-verbal communicative signals that require computational systems to decode accurately and analyze contexts. Current research examines multimodal interactions between visual processing systems, spatial context modeling, and affective state classification in relation to biometric verification systems and machine vision applications. Specifically, hierarchical neural architectures are showing a great promise в addressing some of the inherent complexities in face emotion recognition (FER). Most standard FER architectures incorporate basic convolution layers with unique output modules, but recent innovations are paying close attention to the enhancement of the intermediate levels of such networks-this strides discriminate feature learning through aggregation of state-of-the-art base models (such as VGG, ResNet). Unlike traditional systems, which develop distinct components for feature extraction and classification, most modern designs demonstrate the possibility that optimizing intermediate network layers along with pretrained base models will significantly improve expression classification performance. Generally, these are the kinds of generic and highly general output feature extractors-pre-trained on large image datasets-which capture irrelevant characteristics and tend to fuse a specialized model when delimiting this against smaller affective computing datasets. Advanced facial emotion analysis systems require diverse, large-scale training datasets to prevent algorithmic overspecialization during multi-layered pattern recognition. Widely used evaluative datasets like KDEF and CK+, though foundational in affective computing, face limitations due to restricted demographic scope and insufficient exemplar variety, creating cross-demographic generalization challenges. Modern emotion analysis employs two complementary approaches: a discrete classification system (six core emotions: joy, rage, revulsion, anxiety, sorrow, astonishment) and continuous affective quantification using orthogonal psychophysiological axes like

valence-arousal metrics. Breakthroughs in neural architectures—particularly residual networks with cross-layer connectivity—resolve deep-model training obstacles like gradient dissipation and performance decay, facilitating stable development of sophisticated expression analysis frameworks [16-20].

Residual neural networks revolutionize conventional convolutional architectures through integrated skip-connection modules, establishing residual learning frameworks that have become ubiquitous in visual computing applications for their exceptional precision [12–14]. Within facial expression analysis—a critical domain of machine vision—researchers increasingly employ deep hierarchical models to refine classification efficacy. Huang et al.'s work demonstrated significant accuracy gains by embedding residual components within hybrid VGG-CNN architectures for affective state detection. Concurrently, Mao's team developed POSTER V2, a computationally optimized FER framework employing spatial attention windows and hierarchical landmark feature integration, achieving state-of-the-art performance while minimizing operational complexity through multi-scale facial topology analysis [15].

Cutting-edge research reveals transformative collaboration between visual neuroscience, artificial intelligence, and cognitive studies, where advanced neural networks now match human facial recognition capabilities in complex environments. These systems structurally emulate the brain's visual processing pathways through layered computational units performing specialized pattern analysis. Both biological and artificial frameworks utilize progressive feature transformation stages essential for reliable face identification under varying conditions. The complexity and learning adaptability of modern vision models allow detailed examination of facial processing mechanisms, from basic visual processing to advanced conceptual understanding. This interdisciplinary fusion accelerates technological progress while offering new insights into human cognitive architecture, creating reciprocal advancements in machine intelligence and neuroscientific understanding [4][5][7].

In our exploration of human face processing through computational methods driven by deep learning, our focus is on extracting utility from these models. We organize our review around three key advancements challenging traditional concepts in vision science. Firstly, recent advancements in deep learning challenge traditional notions of inverse optics by producing face representations that are invariant across image changes but lack inherent invariance themselves. Secondly, computational studies suggest a departure from the belief that face, or object representations can be easily understood through interpretable features, indicating a need to reconsider the concept of nameable deep features. Lastly, understanding learning mechanisms in natural environments reveals diverse training data shaping face processing systems through overlapping and interacting with learning mechanisms, challenging the generic explanation of learning for behavioral or neural phenomena [21-28].

Embark on an exciting journey into facial expression recognition, where algorithms decode emotions by analyzing facial features. This exploration draws from physiology and psychology, using machine learning to correlate features with emotions. From improving human-computer interaction to assessing customer satisfaction, the applications are diverse. Data-driven insights are crucial, but ethical considerations are essential for responsible development. Despite challenges, this journey offers valuable insights into human emotions [30-37].

Methodology

These are past changes on ResNet, counting a CBAM consideration component and optimized remaining modules for superior precision in expression acknowledgment. In this segment, a brief account of the advancements to the ResNet demonstrate related to the CBAM consideration instrument and rescued remaining modules for optimized expression acknowledgment is given, and expecting to be created is a web learning status observing framework design based on made strides expression acknowledgment [8-11][15].

In the present work, a convolutional neural network design is presented, which efficiently optimizes classification performance for facial images in an architecture derived from the structure of LeNet-5 but employing different architectural parameters. It has a chain-like computation graph of two cascading hierarchical feature extraction stages (C1 and C2) separated by spatial down-sampling operations (S1, S2) C1→S1→C2→S2. Some key deviations from the standard design include adaptive input tensor sizes, reformulated layer cardinality, and redescribed dense connectivity modules, all of which together improve the capability for discriminative feature learning in face pattern recognition [38-40].

The network's input tensor configuration processes normalized facial imagery through a singular channel representation. The initial feature extraction stage (C1) employs six stochastically initialized 5×5 convolutional filters, generating spatially distributed activation planes. Subsequent dimensionality reduction in layer S1 applies non-overlapping average pooling operators across C1's output channels, preserving feature map cardinality while compressing spatial resolution Figure 76.1.

The hierarchical architecture progresses through a secondary convolutional block (C2) and subsampling layer (S2), both scaled to twelve activation planes. These strata replicate the computational protocols of prior layers while increasing filter diversity for enhanced pattern discrimination. A terminal dense feedforward network bridges the final subsampling outputs to a 40-element encoded output space, implementing sigmoidal activations for simultaneous multi-label categorical encoding across identity classes, as detailed in the architectural schematic.

A. The production of such a state-of-the-art system for dynamically extracting facial expressions requires both effort and dedication.

A mathematical formula is used to generate a final face image, considering the original image, a sequence of dynamic inputs, and in the prior frame, the facial expression warrants prime consideration of the original image, which gets juxtaposed comparatively against the sequential dynamics contributing to the final image with respect to the model's function of activation. Weight values are assigned to both the original image and each element in the sequence, controlling their influence on the outcome. This approach incorporates temporal information about facial expressions to potentially improve face detection accuracy by accounting for variations within a single frame.

$$w = P(rsli + rtli - 1 + rtpi - 1 + g_j) \qquad (1)$$

To calculate both displacement and historical data in this context, we start with the last known input value, denoted as

$$w_t = Pfrlsli + rltli - 1 + rltpi + 1 + g_j \qquad (2)$$

Results and Discussions

The proposed by the evidence portrayed in Figure 76.2, the mixed feature network (MFN) architecture

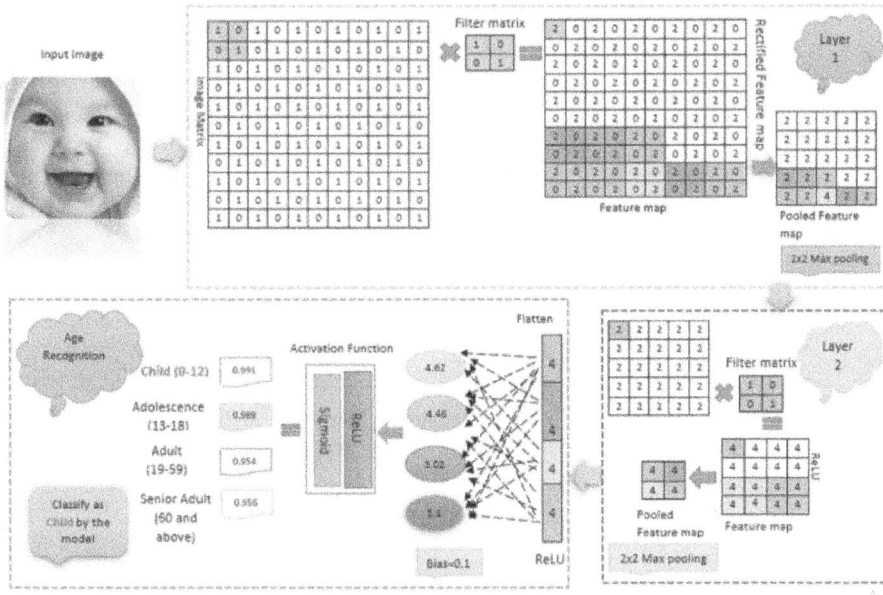

Figure 76.1 Deep facial expression recognition
Source: Author

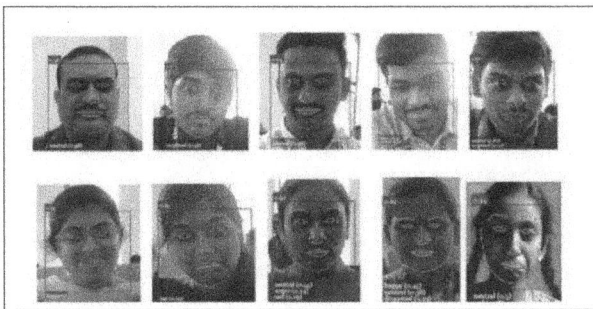

Figure 76.2 Sample output images
Source: Author

proposed has proved extraordinarily good on facial expression recognition (FER) compared even to several state-of-the-art methods on various datasets such as FER2013, CK+, KDEF, and RAF-DB. The mixed-FNN architecture clearly outperforms even the state-of-the-art methods on some benchmark datasets for Facial Expression Recognition (FER), namely FER2013, CK+, KDEF, and RAFDB.

$$r = \begin{cases} \frac{1}{2k}, e_i = 0, \\ \frac{1}{2v}, e_i = 1 \end{cases}$$

With a precision of 93.8%, recall of 76.5%, and mean Average Precision (mAP) of 80.5% on low-resolution 48x48 grayscale images, the model demonstrated superior accuracy and robustness. Notably, it showed a 5% improvement in accuracy on the challenging RAF-DB dataset compared to POSTER V2, highlighting its effectiveness in handling diverse and complex real-world scenarios. The ablation study confirmed the significance of mixed depth-wise convolutional kernels and coordinate attention mechanisms, as their removal led to significant performance drops, validating the model's novelty and effectiveness.

The model's success can be attributed to its ability to capture long-range dependencies and diverse facial features, combined with its lightweight design. Optimization algorithms like SGD and Adam, along with normalization techniques, ensured efficient training and enhanced predictive accuracy. Evaluation metrics and temporal performance charts revealed stable emotion classification capabilities with limited between-category errors, confirming practical effectiveness in psychological computing applications like responsive interface design and emotional health evaluation tools. Future directions include developing continuous emotion monitoring systems and sequential expression analysis models to progress beyond single-frame analysis in affective computing research.

Conclusion

At first glance, we have a paper with all of those interesting features that centers on facial expression recognition (FER) and mixed feature neural network (MFN). By integrating mixed depth-wise convolutional kernels and coordinating attention mechanisms, the proposed model effectively addresses the limitations of traditional methods, such as overfitting and limited accuracy, particularly in handling smaller datasets. In terms of performance, the MFN architecture has performed really well and exceeded most of the related publications on benchmark datasets like FER2013, CK+, KDEF, and RAF-DB, with a precision of 93.8%, recall of 76.5%, and mean Average Precision (mAP) of 80.5%. The ablation study further validates the significance of each component, confirming the model's originality and effectiveness.

This architectural paradigm elevates affective state discernment precision while establishing an adaptive operational framework for deployment in cognitive interface systems, psychopathological evaluation matrices, and pedagogical engagement analytics. By integrating temporal affect trajectory analysis with context-aware adaptation mechanisms, the system achieves fine-grained decoding of micro-affective shifts across temporal sequences—enabling granular interpretation of psychosocial signaling beyond static categorical classifications. These methodological strides catalyze further development of streaming affect recognition infrastructures capable of operating in heterogeneous environmental conditions, from crowded public spaces to individualized therapeutic settings. Collectively, this research bridges theoretical advancements in neural affective computing with industrial-grade implementation strategies, synergizing computational ingenuity with deployable solution engineering for next-generation emotion-aware technologies.

Acknowledgement

The research team extends profound appreciation to Koneru Lakshmaiah Education Foundation's School of Computing and Engineering Technologies for their institutional backing and infrastructural provisions. We recognize the invaluable contributions of academic personnel, technical teams, and research cohorts whose synergistic efforts enabled this investigation. Special acknowledgment is accorded to the curators of standardized evaluation corpora—the Facial Expression Recognition 2013 (FER2013), Extended Cohn-Kanade (CK+), Karolinska Directed Emotional Faces (KDEF), and Real-world Affective Faces Database (RAF-DB)—whose open-access resources proved indispensable for empirical validation within this computational effect analysis paradigm.

References

[1] Farzaneh , A. H., & Qi, X. (2021). Facial expression recognition in the wild via deep attentive center loss. In 2021 IEEE Winter Conference on Applications of Computer Vision (WACV), (pp. 2401–2410). IEEE.

[2] Alnuaim, A. A., Zakariah, M., Shukla, P. K., Alhadlaq, A., Hatamleh, W. A., Tarazi, H., et al. (2022). Human-computer interaction for recognizing speech emotions using multilayer perceptron classifier. *Journal of Healthcare Engineering*, 2022, 6005446.

[3] Kumari, H. M. L. S. (2022). Facial expression recognition using convolutional neural network along with data augmentation and transfer learning.

[4] Mehrabian, A. (n.d.). Nonverbal communication (Aldine Transaction, 2007).

[5] Ekman, P. (2006). Darwin, deception, and facial expression. *Annals of the New York Academy of Sciences*, 1000, 205–2 (Kortli & Jridi, 2020).

[6] Ekman, P., Dalgleish, T., & Power, M. (1999). Handbook of Cognition and Emotion. Wiley.

[7] Maithri, M., Raghavendra, U., Gudigar, A., Samanth, J., Barua, P. D., Murugappan, M., et al. (2022). Automated emotion recognition: Current trends and future perspectives. *Computer Methods and Programs in Biomedicine*, 215, 106646.

[8] Li, S., & Deng, W. (2022). Deep facial expression recognition: a survey. *IEEE Transactions on Affective Computing*, 13, 1195—1215.

[9] Canal, F. Z., Müller, T. R., Matias, J. C., Scotton, G. G., de Sa Junior, A. R., Pozzebon, E., et al. (2022). A survey on facial emotion recognition techniques: a state-of-the-art literature review. *Information Sciences*, 582, 593—617.

[10] He, K., Zhang, X., Ren, S., & Sun, J. (2016). Deep residual learning for image recognition. In 2016 IEEE Conference on Computer Vision and Pattern Recognition (CVPR), (pp. 770–778). IEEE.

[11] Mollahosseini, A., Hasani, B., & Mahoor, M. H. (2019). AfectNet: a database for facial expression, valence, and arousal computing in the wild. *IEEE Transactions on Affective Computing*, 10, 18–31.

[12] Schoneveld, L., Othmani, A., & Abdelkawy, H. (2021). Leveraging recent advances in deep learning for audio-visual emotion recognition. *Pattern Recognition Letters*, 146, 1–7.

[13] Hwooi, S. K. W., Othmani, A., & Sabri, A. Q. M. (2022). Deep learning-based approach for continuous afect prediction from facial expression images in valence-arousal space. *IEEE Access*, 10, 96053–96065.

[14] Sun, L., Lian, Z., Tao, J., Liu, B., & Niu, M. (2020). Multi-modal continuous dimensional emotion recognition using recurrent neural network and self-attention mechanism. In Proceedings of the 1st International on Multimodal Sentiment Analysis in Real-Life Media Challenge and Workshop, (pp. 27–34). ACM.

[15] Mao, J., et al. (2023). POSTER V2: a simpler and stronger facial expression recognition network. arXiv preprint arXiv:2301.12149.

[16] Deng, L., & Yu, D. (2014). Deep learning: methods and applications. *Foundations Trends Signal Processing*, 7(3-4), 197–387.

[17] Tang, Y. (2013). Deep leaming using support vector machines. CoRR, abs/1306.0239.

[18] Dapogny, A., & Bailly, K. (2018). 'Investigating deep neural forests for facial expression recognition. In Proceedings of 13th ffiEE International Conference on Automatic Face and Gesture Recognition, (pp. 629–633).

[19] Kontschieder, P., Fiterau, M., Criminisi, A., & Rota Bulo, S. (2015). Deep neural decision fbrests. In Proceedings of IEEE International Conference on Computer Vision, (pp. 1467–1475).

[20] Donahue, J., Jia, Y., Vinyals, O., Hoffman, J., Zhang, N., Tzeng, E., et al. (2014). DeCAF: a deep convolutional activation feature for generic visual recognition. In Proceedings of International Conference on Machine Learning, (pp. 647–655).

[21] Razavian, A. S., Azizpour, H., Sullivan, J., & Carlsson, S. (2014). CNN features off-the-shelf: an astounding baseline for recognition. In Proceedings of IEEE Conference on Computer Vision and Pattern Recognition Workshops, (pp. 512–519).

[22] Otberdout, N., Kacem, A., Daoudi, M., Ballihi, L., & Berretti, S. (2018). Deep covariance descriptors for fecial expression recognition. In Proceedigns of BMVC, (p. 159).

[23] Acharya, D., Huang, Z., Pani Paudel, D., & Van Gool, L. (2018). Covariance pooling for facial expression recognition. In Proceedings of the IEEE Conference on Computer Vision and Pattern Recognition Workshops, (pp. 367–374).

[24] Bi, Y., Xue, B., & Zhang, M. (2022). Using a small number of training instances in genetic programming for face image classification. *Information Science*, 593(2), 488–504.

[25] Lu, Z., Zhou, C., Xuyang, X., & Zhang, W. (2021). Face detection and recognition method based on improved convolutional neural network. *International Journal of Circulation*, 15(1), 774–781.

[26] Ekman, P., & Friesen, W. (1971). Constants across cultures in the face and emotion. *Journal of Personality and Social Psychology*, 17, 124–129.

[27] Goeleven, E., De Raedt, R., Leyman, L., & Verschuere, B. (2008). The karolinska directed emotional faces: a validation study. *Cognition and Emotion*, 22, 1094–1118.

[28] Lucey, P., Cohn, J. F., Kanade, T., Saragih, J., Ambadar, Z., & Matthews, I. (2010). The extended cohn-kanade dataset (CK+): a complete dataset for action unit and emotion-specified expression. In Proceedings of the 2010 IEEE Computer Society Conference on Computer Vision and Pattern Recognitioii—Workshops, San Francisco, CA, USA, 13-18 June 2010, (pp. 94—101).

[29] Kumar, G. A. R., Kumar, R. K., & Sanyal, G. (2017). Facial emotion analysis using deep convolution neural network. In Proceedings of the International Conference on Signal Processing and Communications (ICSPC), Coimbatore, India, (pp. 369–374), doi: 10.1109/CSPC.2017.8305872.

[30] Lucey, P., Cohn, J. F., Kanade, T., Saragih, J., Ambadar, Z., & Matthews, I. (2010). The extended cohn-kanade dataset (CK+): a complete dataset for action unit and emotion-specified expression. In Proceedings of the IEEE Computer Society Conference on Computer Vision and Pattern Recognition Workshops, San Francisco, CA, USA, (pp. 94—101). doi: 10.1109/CVPRW.2010.5543262.

[31] Barrett, S., Weimer, F., & Cosmas, J. (2019). Virtual eye region: development of a realistic model to convey emotion. *Heliyon*, 5(12), e02778.

[32] Zhang, S., Zhang, Y., Zhang, Y., Wang, Y., & Song, Z. (2023). A dualdirection attention mixed feature network for facial expression recognition. *Electronics*, 12(17), 1–15.

[33] Andronie, M., Lazaroiu, G., Karabolevski, O. L., Stefanescu, R., Hurloiu, I., Dijmarescu, A., et al. (2023). Remote big data management tools, sensing and computing technologies, and visual perception and environment mapping algorithms in the internet of robotic things. *Electronics*, 12, 22.

[34] Pelau, C., Dabija, D. C., & Ene, I. (2021). What makes an AI device human-like? the role of interaction quality, empathy and perceived psychological anthropomorphic characteristics on the acceptance of artificial intelligence in the service industry. *Computers in Human Behavior*, 122, 106855.

[35] Dijmarescu, I., latagan, M., Hurloiu, L., Geamanu, M., Rusescu, C., & Dijmarescu, A. (2022). Neuromanagement decision making in facial recognition biometric authentication as a mobile payment

technology in retail, restaurant, and hotel business models. *Oeconomia Copernicana*, 13, 225–250.

[36] Cui, Z., Pi, J., Chen, Y., Yang, J., Xian, Y., Wu, Z., et al. (2021). Facial expression recognition combined with improved VGGNet and focal loss. *Computer Engineering and Applications*, 57(19), 171178. doi: 10.3778/j.issn.l002-8331.2007-0492.

[37] Caroppo, A., Leone, A., & Siciliano, P. (2020). Comparison between deep learning models and traditional machine learning approaches for facial expression recognition in ageing adults. *Journal of Computer Science and Technology*, 35(5), 1127–1146. Oct., doi: 10.1007/s11390-020- 9665-4.

[38] He, K., Zhang, X., Ren, S., & Sun, J. (2016). Deep residual learning for image recognition. In Proceedings of IEEE Conference on Computer Vision and Pattern Recognition, (pp. 770–778).

[39] D'Mello, S., & Graesser, A. (2012). Dynamics of affective states during complex learning *Learning and Instruction*, 22(2), 145–157. doi: 10.1016/j.leaminstnic.20H.10.001.

[40] Jagadeesh, M., & Baranidharan, B. (2022). Facial expression recognition of online learners from real-time videos using a novel deep learning model. *Multimedia System*, 28(6), 2285–2305. doi: 10.1007/S00530-022-00957-Z.

77 Connecting travels and farming using machine learning approaches

Mariserla Santhi Kumari[1,a], Manukonda Aasritha[1,b], Ambati Monisha Sri Naidu[1,c], Ruttala Bhargava Naidu[2,d], Srinivasa Rao Vankdoth[3,e] and Dharmaiah Devarapalli[3,f]

[1]Department of IOT, Koneru Lakshmaiah Education Foundation, Guntur, Andhra Pradesh, India

[2]Department of IT, Anil Neerukonda Institute of Technology and Sciences, Visakhapatnam, Andhra Pradesh, India

[3]Assistant Professor, Department of CSE, Koneru Lakshmaiah Education Foundation, Guntur, Andhra Pradesh, India

Abstract

There are numerous online booking systems for travel vehicles, but no dedicated platform exists for reserving farming vehicles. This project introduces a machine learning-based vehicle booking system that allows users to reserve either travel or farming vehicles based on availability. The system utilizes XGBoost, an advanced machine learning algorithm, to predict whether a selected vehicle is available for booking. A self-prepared dataset was used for training, comprising 53% farming vehicles and 47% travel vehicles. The dataset includes features such as vehicle type, vehicle name, and availability status. The model incorporates One-Hot Encoding for categorical variables. The dataset is split into training and testing sets to evaluate the model's performance using key metrics such as accuracy, precision, recall, F1-score, confusion matrix, and ROC curve. Once a user selects a vehicle type and vehicle name, the system predicts its availability. If available, the system confirms the booking and displays essential details like the vehicle name, owner's name, and contact number. Otherwise, the user is notified of unavailability. The system also visualizes model performance using heatmaps and ROC curves. Unlike existing platforms that cater only to travel vehicle rentals, this project bridges the gap by introducing a dedicated farming vehicle reservation system. The XGBoost model achieved an improved accuracy of 56.54%, outperforming traditional methods like logistic regression. The project highlights the potential of machine learning in optimizing vehicle rentals, ensuring a reliable and efficient booking experience for both travel and farming vehicle users.

Keywords: Farming vehicles, machine learning, one-hot encoding, travel vehicles, vehicle booking, XGBoost

Introduction

With the continuous growth of the on-demand vehicle rental market, machine learning has emerged as a transformative technology, enhancing the booking experience by providing more efficient allocation and availability prediction models. Traditionally, ride-hailing systems have catered primarily to passenger vehicles, but there is a growing need for systems specifically designed to manage the booking of farming vehicles, which are critical for agricultural activities. While general vehicle booking platforms address urban needs, farming vehicles, due to their seasonal nature and specialized requirements, have remained underserved.

In existing vehicle rental management systems, predictive models and optimization techniques, such as demand forecasting, have been successfully implemented to manage urban vehicle availability [1]. These systems utilize machine learning models that predict vehicle demand based on factors such as time of day, location, and vehicle type. These techniques have been extensively applied to passenger vehicle fleets, improving the allocation of vehicles and reducing waiting times for customers [2, 3]. However, the same techniques can also be extended to the management of farming vehicle bookings, with additional considerations given to factors like seasonal demand and geographic location.

[a]santhikumarimariserla@gmail.com, [b]aasrithamanukonda08@gmail.com, [c]monishaambati14@gmail.com, [d]bhargavanaiduruttala@gmail.com, [e]vsr.duggirala@gmail.com, [f]drdharmaiah@kluniversity.in

DOI: 10.1201/9781003684589-77

Moreover, advancements in the use of reinforcement learning and recommender systems in vehicle booking systems have allowed for better matching of vehicle availability with customer demand. This has been shown to improve overall system efficiency and user satisfaction in urban settings [4]. By incorporating similar methods, farming vehicle booking systems can be optimized to ensure that agricultural workers have easy access to the vehicles they need when they need them.

Furthermore, studies in machine learning-based demand prediction for ride-hailing systems have highlighted the potential for improving booking accuracy through spatiotemporal models that consider the time and location of vehicle demand [5, 6]. These methods can be adapted to the farming sector, where demand for vehicles may vary based on the planting and harvesting cycles.

Our project extends these concepts into the farming vehicle sector, where we introduce a machine learning-based system designed specifically to handle bookings for both traveling and farming vehicles. This specialized focus ensures that agricultural vehicles are just as accessible and efficiently allocated as passenger vehicles, providing a much-needed solution for rural areas that has not been adequately addressed in previous research.

Methodology

Data collection
The dataset used in this project was self-prepared rather than sourced from online platforms. It consists of 1,000 records of vehicle bookings, with 53% farming vehicles and 47% travel vehicles. The dataset includes six key attributes:

- **Vehicle type** (Farming/Traveling)
- **Vehicle name** (e.g., Tractor, Harvester, Van, Motorbike)
- **Cost** (Not used in final model)
- **Owner name** (For displaying booking details)
- **Contact number** (For booking confirmation)
- **Availability status** (Yes/No)

Data preprocessing
To ensure data consistency and improve model performance, several preprocessing steps were applied. Missing values were handled by replacing them with the mode of each column to maintain data integrity.

In terms of feature selection, only vehicle type and vehicle name were chosen as input variables, while cost was removed since it did not significantly impact prediction accuracy. Categorical variables were encoded using appropriate techniques: vehicle type was transformed using binary encoding (1 for Farming, 0 for Traveling), while vehicle name was converted into numerical format using one-hot encoding (OHE). These preprocessing steps ensured that the dataset was well-structured for training the machine learning model, improving its ability to predict vehicle availability accurately.

Model selection and training
Several machine learning models were tested, including logistic regression (LR), Decision Trees (DT), Random Forest (RF), and XGBoost. After extensive evaluation, XGBoost (Extreme Gradient Boosting) was chosen due to its superior performance in handling categorical variables and imbalanced data.

Hyperparameter tuning for XGBoost
To enhance accuracy, hyperparameters were carefully optimized during model training. The number of estimators was set to 300, controlling the number of boosting rounds to refine predictions. A maximum depth of 8 was chosen to prevent overfitting by limiting tree growth, ensuring the model generalizes well to new data. The learning rate was adjusted to 0.07, controlling the step size during training to improve convergence and stability. Additionally, a random state of 42 was used to ensure reproducibility of results. The model was trained using 80% of the dataset, while the remaining 20% was reserved for testing, enabling a thorough evaluation of its performance.

Model evaluation
The trained XGBoost model was evaluated using standard performance metrics to assess its effectiveness. The accuracy score measured the overall correctness of predictions, while precision evaluated the percentage of correctly predicted available vehicles. Recall determined how many actually available vehicles were correctly identified, and the F1-score provided a balance between precision and recall.

To determine the best-performing model, multiple machine learning algorithms were tested, including LR, DT, and RF. Among these, XGBoost outperformed the others in handling categorical variables and imbalanced data. Figure 77.1 presents the confusion matrix, which categorizes predictions into four groups: True Negatives (TN), representing unavailable vehicles correctly classified as unavailable; False Positives (FP), where unavailable vehicles were incorrectly predicted as available; False Negatives (FN), indicating available vehicles misclassified as unavailable; and True Positives (TP), where available vehicles were correctly identified as available. Additionally, the ROC curve was analyzed to assess the model's ability to distinguish between available and unavailable vehicles. The XGBoost model achieved an accuracy of 56.54%, marking a significant improvement over LR (51.67%), DT (53.12%), and RF (54.89%), making it the most effective choice for this system.

System deployment and booking process

The trained model was integrated into a real-time vehicle booking system, allowing users to check the availability of vehicles by entering their desired vehicle type and vehicle name. Based on the model's prediction, the system provides immediate feedback. If the vehicle is available, the system confirms the booking and displays key details such as the vehicle name, owner name, and contact number, enabling direct communication between the renter and the owner. If the vehicle is unavailable, the system promptly informs the user, ensuring a seamless and efficient booking experience.

Algorithm

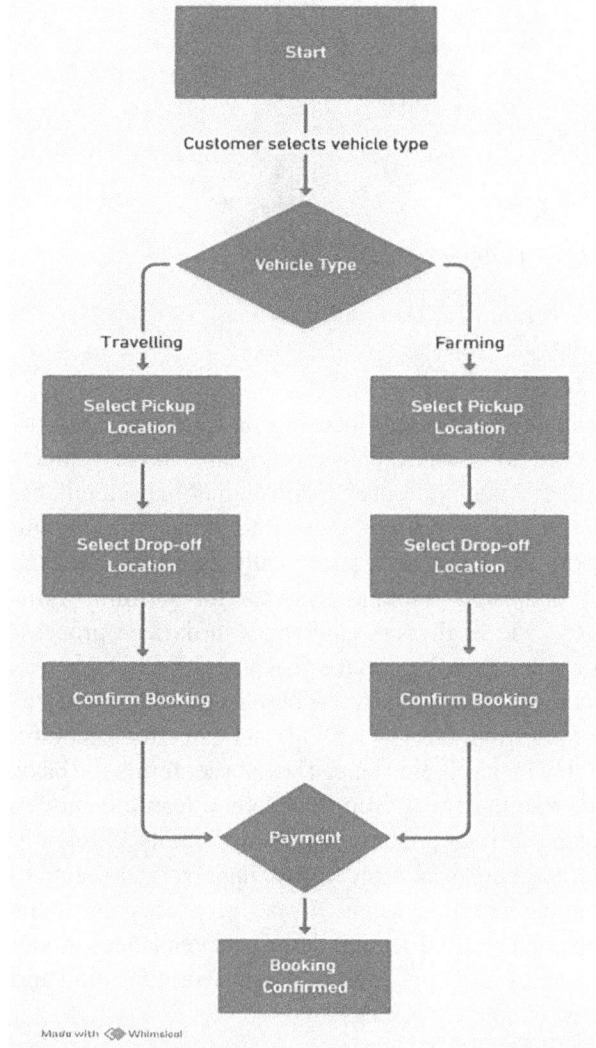

Dataset Description

The dataset used in this study is self-prepared and consists of 1,000 samples of vehicle bookings, categorized into two primary types: Farming (510 samples, 51%) and Traveling (490 samples, 49%), as shown in Figure 77.2. Since the class distribution is nearly equal, the dataset is considered balanced, ensuring that the machine learning model does not favor one category over another. Each record includes key attributes such as vehicle type (indicating whether it is used for farming or traveling), vehicle name (e.g., Tractor, Harvester, Truck, Jeep, Van), owner name (identifying the vehicle provider), contact number (enabling direct communication), and availability status (Yes/No). The dataset was

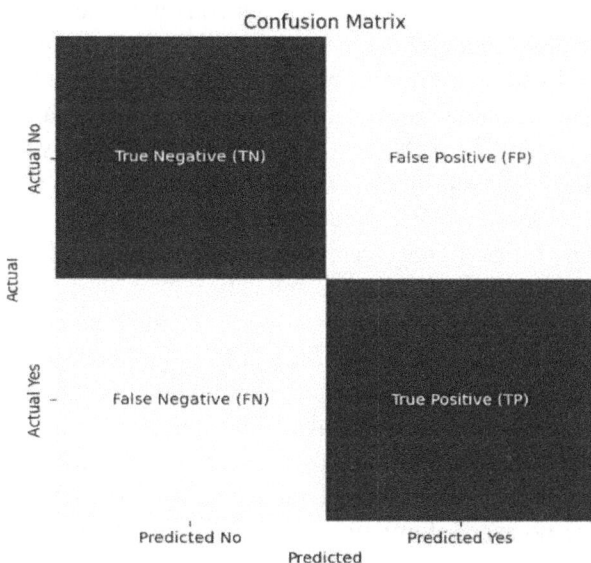

Figure 77.1 Confusion matrix
Source: Author

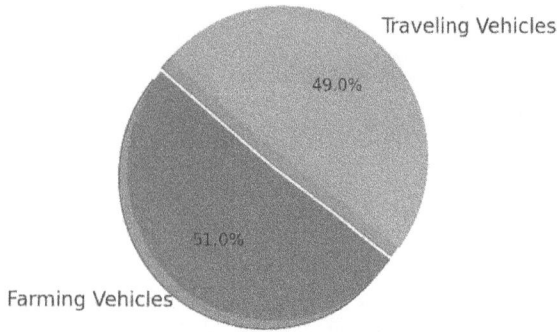

Figure 77.2 Dataset distribution
Source: Author

Figure 77.3 Output
Source: Author

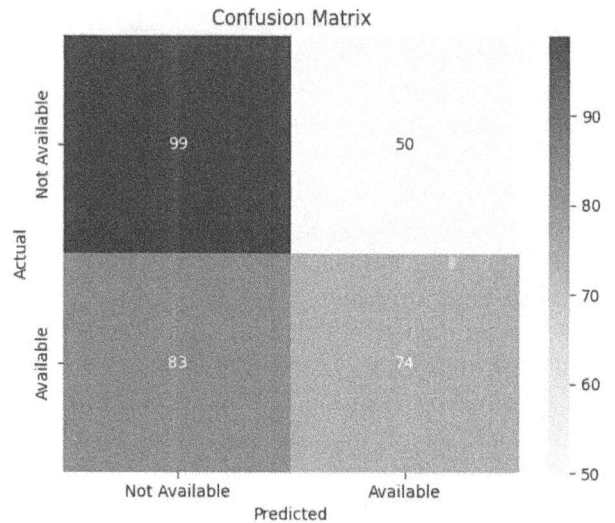

Figure 77.4 Confusion matrix
Source: Author

manually compiled, focusing on rural and semi-urban regions where both farming and travel vehicles are frequently rented. Unlike publicly available datasets that primarily cater to urban ride-hailing services, this dataset specifically addresses the lack of dedicated booking systems for farming vehicles. One of the key challenges in data preprocessing was handling categorical variables, which was achieved using binary encoding for vehicle type (1 for Farming, 0 for Traveling) and One-Hot Encoding (OHE) for vehicle name. The data set forms the basis for training the XGBoost machine learning model, enabling it to predict vehicle availability efficiently while addressing a previously underserved sector in vehicle rental systems. These preprocessing steps ensure a well-structured dataset that enhances model accuracy and applicability in real-world farming and travel vehicle bookings.

Result Discussion

The model's performance demonstrates an accuracy of 56.54%, with a precision of 60%, indicating that 60% of the vehicles predicted as available were correctly classified. The recall of 47% shows that the model successfully identified 47% of all available vehicles. The F1-score of 53% as shown in Figure 77.3, reflects a balance between precision and recall, suggesting that while the model effectively detects available vehicles, there are still some misclassifications.

The confusion matrix provides deeper insights into the model's performance. It shows that 99 unavailable vehicles were correctly classified (true negatives), while 74 available vehicles were correctly identified (true positives). However, 50 vehicles were incorrectly marked as available when they were

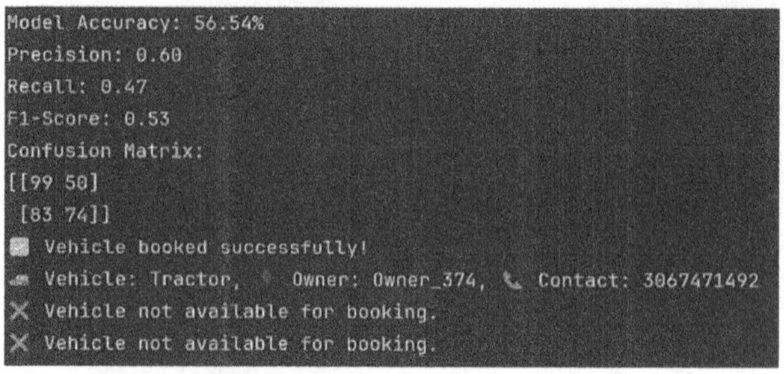

actually unavailable (false positives), and 83 vehicles that were available were mistakenly classified as unavailable (false negatives), as seen in Figure 77.4, the model encounters challenges in correctly distinguishing between available and unavailable vehicles. The ROC curve in Figure 77.5 helps evaluate how well the model differentiates between the two classes. Other models such as Random Forest and Decision Trees were also tested, but they yielded similar accuracy levels, indicating that the dataset itself may require improvements, such as more diverse data points or additional relevant features.

Despite these challenges, the model successfully booked vehicles when available, demonstrating its capability to retrieve and display correct vehicle information, including the vehicle name, owner's details, and contact information. These results confirm the potential of machine learning in automating vehicle booking systems for both farming and travel vehicles, making the process more efficient and accessible.

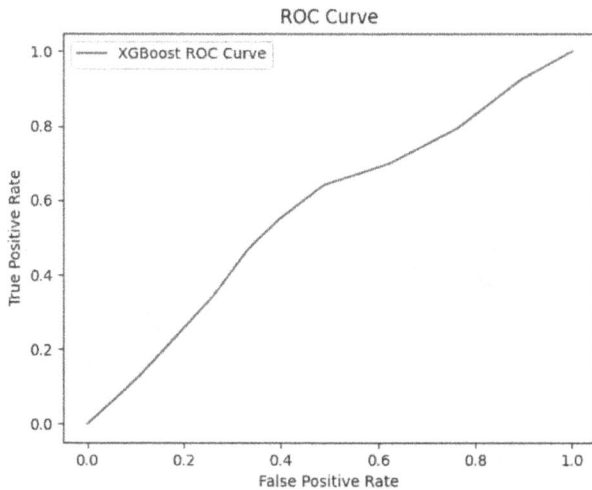

Figure 77.5 ROC curve
Source: Author

Real-life applications of the proposed system

- **Efficient equipment access for farmers:**
  Many farmers in rural areas do not own heavy farming machinery such as tractors, harvesters, or plows due to high costs. This system allows farmers to book the required vehicles and they can also rent out the vehicles for a certain period by talking to the vehicle owner, so that the seasonal farming activities like plowing, seeding, and harvesting are completed on time without delays.
- **Optimized utilization of farming vehicles:**
  Farming vehicles often remain unused during the off-season, leading to inefficiency. This system enables owners to rent out their equipment, generating additional income while ensuring that machinery is used efficiently throughout the year.
- **Smart booking for traveling vehicles in rural areas:**
  Unlike cities where ride-hailing services like Uber and Ola are available, people in rural areas struggle to find transportation. This system connects travelers with available private vehicles such as jeeps, motorbikes, and vans, allowing them to book a ride easily without long waiting times.
- **Seasonal demand management in agriculture:**
  The demand for farming vehicles changes with agricultural cycles. For example, tractors are needed more during sowing seasons, while harvesters are in high demand during crop-cutting periods. This system can analyze these demand patterns and suggest pre-booking options, ensuring that farmers secure vehicles before peak seasons.
- **Reducing middlemen dependency:**
  Many farmers currently rely on brokers to rent farming vehicles, paying extra commissions. This system removes the need for middlemen, allowing farmers to book vehicles directly from owners at fair prices, making the process more cost-effective.
- **Vehicle sharing and cooperative use:**
  Small-scale farmers who cannot afford to rent an entire vehicle alone can use this system to share rentals. Multiple farmers from the same village can split costs and rent a vehicle together, making it more affordable and accessible for everyone.
- **Making rural vehicle rentals easier with digital technology:**
  This system introduces online payments and GPS tracking to improve vehicle rentals in villages. Farmers can quickly check which vehicles are available, book them instantly, and avoid dealing with paperwork. This makes the whole process faster, more organized, and hassle-free.
- **Time management and reduced waiting time:**
  Traditionally, farmers spend a lot of time searching for available vehicles, which causes delays in farming activities. This system allows them to find and book vehicles instantly, reducing waiting time and ensuring they get the equipment they need without unnecessary delays.

Future challenges and scope

Although the vehicle booking system offers an effective way to rent farming and travel vehicles, some challenges may arise in the future. Data quality and availability could be an issue, as real-world information may have missing or incorrect details, affecting the model's accuracy. To address this, continuous data validation and real-time updates should be implemented to ensure reliable information. Seasonal demand variations might also lead to either too many or too few available vehicles, requiring better planning and prediction. This can be solved by using demand forecasting models that analyze historical booking patterns to suggest vehicle availability accordingly.

Expanding the system to support more users and vehicles while keeping it fast and reliable could be difficult. The solution is to use cloud-based infrastructure that can handle growing data efficiently. Additionally, internet access in rural areas is often limited, so the system may need offline support. A possible solution is to develop a lightweight mobile application that works even with low connectivity.

Trust and security are also important, as vehicle owners and renters need a safe platform with proper verification and secure payments. To overcome this, user authentication and rating systems can be introduced, ensuring trust between both parties.

Moreover, advancements in the use of reinforcement learning and recommender systems in vehicle booking systems have allowed for better matching of vehicle availability with customer demand. This has been shown to improve overall system efficiency and user satisfaction in urban settings [4][7][9]. By incorporating similar methods, farming vehicle booking systems can be optimized to ensure that agricultural workers have easy access to the vehicles they need, when they need them.

Furthermore, studies in machine learning-based demand prediction for rise-hailing systems have highlighted the potential for improving booking accuracy through spatiotemporal models that consider the time and location of vehicle demand [5][6][8][10]. These methods can be adapted to the farming sector, where demand for vehicles may vary based on the planting and harvesting cycles.

Conclusion

The present scenario has shown the increasing demand for efficient vehicle booking systems, particularly in the farming sector, where no dedicated platform currently exists. Machine learning has been utilized in this project to predict the availability of farming vehicles for booking based on features like vehicle type and cost. The model successfully identifies available vehicles, demonstrating the potential of machine learning in addressing this gap. This approach provides a practical solution for optimizing vehicle utilization and accessibility in the farming sector.

Acknowledgement

The authors gratefully acknowledge the support and cooperation of the students, staff, and administration in facilitating this research. Their valuable contributions, insights, and assistance were instrumental in the successful completion of this study.

References

[1] Sahu, A., Katlana, A., Sharma, A., Baniya, M. Y., Sureliya, S., & Kanungo, S. (2024). Vehicle rental management system: on demand vehicle ride. In 2024 International Conference on Advances in Computing Research on Science Engineering and Technology (ACROSET), (pp. 1–11). IEEE.

[2] Liu, Y., Jia, R., Ye, J., & Qu, X.. (2022). How machine learning informs ride-hailing services: a survey. *Communications in Transportation Research*, 2, 100075.

[3] Wen, D., Li, Y., & Lau, F. C. (2024). A survey of machine learning-based ride-hailing planning. *IEEE Transactions on Intelligent Transportation Systems*. Volume 25, issue 6, page range (4734–4753).

[4] Qin, G., Luo, Q., Yin, Y., Sun, J., & Ye, J. (2021). Optimizing matching time intervals for ride-hailing services using reinforcement learning. *Transportation Research Part C: Emerging Technologies*, 129, 103239.

[5] Narman, H. S., Malik, H., & Yatnalkar, G. (2021). An enhanced ride sharing model based on human characteristics, machine learning recommender system, and user threshold time. *Journal of Ambient Intelligence and Humanized Computing*, 12(1), 13–26.

[6] Roy, S., Nahmias-Biran, B. H., & Hasan, S. (2025). Spatial transferability of machine learning based models for ride-hailing demand prediction. *Transportation Research Part A: Policy and Practice*, 193, 104413.

[7] Qiao, S., Han, N., Huang, J., Peng, Y., Cai, H., Qin, X., et al. (2023). An three-in-one on-demand ride-hailing prediction model based on multi-agent reinforcement learning. *Applied Soft Computing*, 149, 110965.

[8] Kullman, N. D., Cousineau, M., Goodson, J. C., & Mendoza, J. E. (2022). Dynamic ride-hailing with electric vehicles. *Transportation Science*, 56(3), 775–794.

[9] G. P. Yatnalkar, A Machine Learning Recommender Model for Ride Sharing Based on Rider Characteristics and User Threshold Time, M.S. thesis, Dept. of Computer Science, Marshall Univ., Huntington, WV, 2019. [Online]. Available: https://mds.marshall.edu/etd/1259.

[10] Jin, G., Cui, Y., Zeng, L., Tang, H., Feng, Y., & Huang, J. (2020). Urban ride-hailing demand prediction with multiple spatio-temporal information fusion network. *Transportation Research Part C: Emerging Technologies*, 117, 102665.

78 Machine learning based booklet status detection

Panjula Bhavana[1,a], Marri Lahari Krishna[1,b], Pachava Tejaswini[1,c], M. Kavitha[2,d] and Srinivasa Rao Vankdoth[2,e]

[1]Department of IOT, Koneru Lakshmaiah Education Foundation, Guntur, Andhra Pradesh, India

[2]Assistant Professor, Department of CSE, Koneru Lakshmaiah Education Foundation, Guntur, Andhra Pradesh, India

Abstract

This project presents a comprehensive approach to processing and analyzing a self-collected, non-standard dataset of examination booklets, focusing on tracking collection and correction statuses. The dataset contains unique student identifiers, booklet collection status, and correction details. The workflow includes data validation, summary metric computation, and the addition of a derived 'Status' column to classify booklets as "Missing," "Corrected," or "Not Corrected." To enhance monitoring, a Telegram notification system is integrated to send real-time alerts when missing booklets are detected. If any booklets are uncollected, an automated message is sent to predefined recipients via Telegram, ensuring prompt action. This feature enables stakeholders, such as exam administrators and evaluators, to address discrepancies efficiently for predictive analysis, Random Forest Classifier is employed to predict booklet correction status based on collection data. The model categorizes booklets into "Corrected" (1) and "Not Corrected" or "Missing" (0) and is trained using the collected feature. The model's performance is evaluated through accuracy metrics and classification reports. Additionally, various visualizations are generated, including correction progress per student, missing booklet distribution, and status breakdowns using bar charts, histograms, and pie charts. This project demonstrates a scalable and data-driven approach to examination booklet tracking, integrating machine learning and automated notifications to improve efficiency and accountability in academic evaluation processes.

Keywords: Data integrity, data processing, educational outcomes, statistical analysis, status classification

Introduction

In educational settings, the assessment of student submissions, such as examination booklets, is crucial for measuring learning outcomes and ensuring academic integrity. This project focuses on analyzing a dataset that contains detailed information about student booklets submitted for evaluation. By leveraging data analysis techniques and visualization tools, the objective is to gain insights into the collection and correction processes of these booklets.

The dataset allows for the examination of key metrics, including the total number of students, the quantity of booklets collected, and the status of corrections made. Through the application of Python's panda's library, the analysis not only summarizes these metrics but also categorizes booklets into distinct statuses, such as "Missing," "Corrected," and "Not Corrected." This classification facilitates a clearer understanding of the submission landscape, highlighting areas that may require additional attention.

Moreover, the project employs data visualization methods to present findings effectively, using bar charts and histograms to illustrate trends in booklet corrections and identify the distribution of missing submissions. By integrating automated alerts via Telegram, the project ensures timely communication regarding critical issues, such as a significant number of missing booklets.

The system operates by assigning a unique code to each examination booklet, allowing easy identification and tracking across multiple stages of handling, from collection to correction. By analyzing data on collected, missing, and corrected booklets, the project provides a comprehensive view of the current evaluation status, enabling educational administrators to make informed decisions.

Implementation

The project utilized a combination of programming languages and tools tailored to data processing, machine learning, and user interface design. Python

[a]panjulabhavana@gmail.com, [b]2200100055@gmail.com, [c]pachavatejaswini09@gmail.com, [d]mkavita@kluniversity.in, [e]vsr.duggirala@gmail.com

DOI: 10.1201/9781003684589-78

was employed as the primary language for implementing machine learning algorithms, including the Random Forest (RF) model, due to its extensive libraries like Scikit-Learn and Pandas, which facilitated data handling and model training. The Telegram API, accessed via Python, was integrated for sending automated notifications, enabling efficient communication of booklet statuses. Together, these languages provided a comprehensive framework for managing and analyzing booklet data effectively.

The code developed for this project centers around automating the tracking, management, and analysis of examination booklets, leveraging Python's robust data processing and machine learning capabilities. The primary objective was to identify, monitor, and evaluate booklet statuses, such as total booklets collected, missing booklets, and corrected statuses. A unique code was assigned to each booklet, allowing the system to keep track of its journey from collection to evaluation. Using data structures like lists and dictionaries, the initial setup of the code organized booklet information in a way that could be easily referenced, updated, and displayed as needed.

Machine learning was a critical part of the system, with a RF algorithm implemented to analyze patterns in booklet data, predict potential issues, and identify discrepancies. The code's machine learning module involved training the model on historical data, allowing it to learn from previous instances where booklets may have been misplaced or overlooked. The RF model was chosen for its accuracy and ability to handle complex data with multiple variables, which was particularly useful in predicting missing booklets based on patterns. The model's predictions were then integrated into the main code flow, providing proactive alerts for administrators.

Prototype design

Figure 78.1. illustrates the Python script developed to connect with the Telegram messaging platform. This script enabled the system to send alerts directly to administrators' devices whenever a missing or delayed booklet was detected. The notification system relied on conditional statements and event triggers, which activated specific alerts based on the booklet's tracking status. This real-time communication feature, as depicted in Figure 78.1, significantly reduced manual intervention, ensuring administrators received immediate updates and streamlining the entire evaluation process.

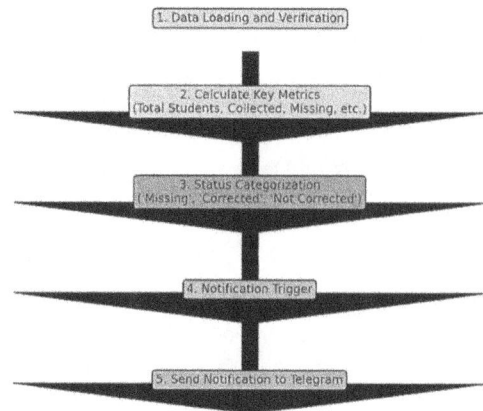

Figure 78.1 Flowchart for booklet status tracking
Source: Author

Algorithm

Step-1: Data loading and verification – Ensures data is correctly loaded and verified for accuracy.

Step-2: Calculate key metrics – Computes essential metrics like the total number of students, booklets collected, and missing counts.

Step-3: Status categorization – Classifies each booklet as "Missing," "Corrected," or "Not Corrected."

Step-4: Data processing and analysis – Analyzes the data for patterns or trends that could highlight issues.

Step-5: Notification trigger – Evaluates conditions for notifying administrators about discrepancies.

Step-6: Report generation – Generates a report detailing metrics and insights.

Step-7: Send notification to telegram – Sends a message via Telegram if issues are identified.

Dataset description

The dataset used for this project consists of structured records related to examination booklets, including their collection, correction status, and missing entries. It is designed to support data analysis, visualization, and machine learning model training for anomaly detection and prediction.

1.   Dataset attributes

The dataset contains the following key fields:

- **Student_ID**: Unique identifier for each student.
- **Booklet_ID**: Unique code assigned to each examination booklet.

- **Subject**: The subject associated with the booklet.
- **Class/Section**: The class or section to which the student belongs.
- **Total_Booklets_Assigned**: Number of booklets expected to be collected per subject/class.
- **Booklet_Collected (Yes/No)**: Indicates whether the booklet was collected.
- **Correction_Status (Corrected/Not Corrected)**: Marks the booklet as either corrected or pending correction.
- **Submission_Date**: The date when the booklet was submitted.
- **Correction_Completed_Date**: The date when the booklet was corrected (if applicable).
- **Missing_Booklet (Yes/No)**: Specifies whether the booklet is missing.
- **Notification_Type**: Type of notification sent (e.g., "Missing Booklet," "Correction Pending").

2. *Dataset size and format*
- **Size:** The dataset can contain 100 samples for testing, but it is scalable to n number of samples based on institutional needs.
- **Format:** Stored in CSV, Excel, or a relational database (SQL) to ensure flexibility in data storage and retrieval.

Visualizations such as the Histogram in Figure 78.2 illustrate the distribution of missing booklets, providing insights into the frequency and pattern of non-submissions. Additionally, Figure 78.3, a bar chart, demonstrates correction progress by students, highlighting the proportion of booklets successfully evaluated. To provide a holistic view, Figure 78.4 includes a pie chart showing the status distribution of booklets (i.e., Missing, Corrected, Not Corrected), enabling stakeholders to assess system performance at a glance. Automated alerts are sent through Telegram for critical issues, such as missing booklets. Figure 78.5 presents a screenshot of Telegram notifications, showcasing how the system proactively informs administrators of issues like missing submissions or correction delays.

Data visualization
A new column, status, is added to classify each booklet based on its collection and correction status:

Automated alerts are sent through Telegram for critical issues, such as missing booklets. The following function is implemented to send alerts:

Alerts are triggered if there are missing booklets or if any student has a high number of missing submissions.

Figure 78.2 Histogram - distribution of missing booklets
Source: Author

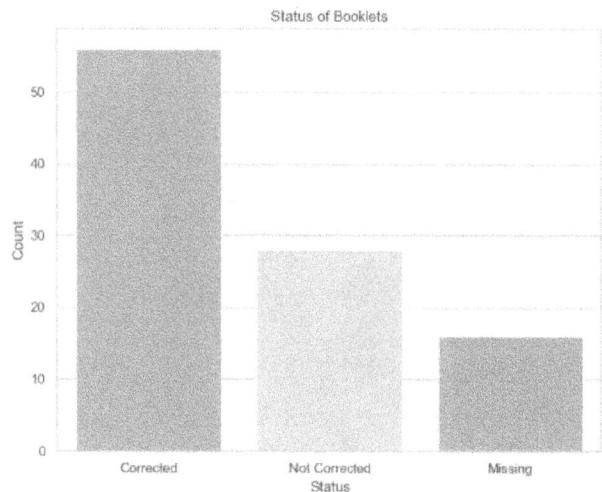

Figure 78.3 Bar chart- correction progress by students
Source: Author

Data preprocessing
In the examination booklet evaluation system, data preprocessing is an essential step to prepare the raw dataset for efficient analysis and accurate predictions. Preprocessing involves cleaning, organizing, and transforming the data, ensuring it is in a format suitable for further processing, analysis, and model training if needed. Below is an overview of the preprocessing steps before and after the transformations.

Before preprocessing
Raw data collection
The initial dataset typically includes raw information from various sources such as exam centers, collection points, and correction centers. This data may

Status Distribution of Booklets

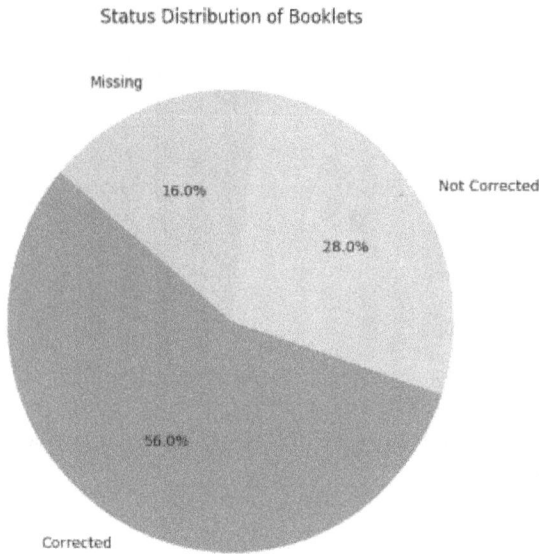

Figure 78.4 Pie chart- status distribution of booklets
Source: Author

Figure 78.5 Telegram notifications
Source: Author

contain errors, inconsistencies, and missing values due to manual entry and variations in data sources.

Preprocessing steps
Data cleaning
Handling missing values: Missing data is addressed by filling in default values, using mean/mode imputation, or removing records if necessary.

Standardizing formats: Date fields, booklet IDs, and other categorical data are standardized to ensure uniformity across the dataset. Removing duplicates: Duplicate entries are identified and removed to ensure each booklet is uniquely represented, thus improving data integrity. Filtering relevant data: Unnecessary columns are removed to streamline the dataset, keeping only the information essential for the booklet evaluation system.

Enhanced data readability
The dataset becomes more interpretable and manageable, with consistent formats and standardized categories. This readability is crucial for generating insights and creating accurate visualizations for tracking booklet status (as seen in Figures 78.2–78.4).The system's performance was evaluated for efficiency, accuracy, and reliability in booklet tracking. Key metrics included detection accuracy, notification latency, and error rates in identifying booklet statuses. Real-time notifications (refer to Figure 78.5) improved workflow by reducing manual oversight. User feedback highlighted the system's effectiveness in minimizing false positives and negatives.

The examination booklet evaluation system is designed to streamline the tracking and evaluation of examination booklets using a structured data-processing pipeline. This system loads a dataset containing details on students, booklet collection, and correction statuses, performing preprocessing to clean and structure the data. It then calculates key metrics, including the total number of students, collected and missing booklets, and booklets marked as corrected. Based on these metrics, the system categorizes each booklet as "Missing," "Corrected," or "Not Corrected" and triggers automated notifications via the Telegram API (see Figure 78.5 for example notifications).

After preprocessing
Improved data quality
Post-preprocessing, the dataset is now clean, organized, and free from inconsistencies. This high-quality dataset enhances the reliability of any subsequent analysis or machine learning model predictions.

Enhanced data readability

The dataset becomes more interpretable and manageable, with consistent formats and standardized categories. This readability is crucial for generating insights and creating accurate visualizations for tracking booklet status.

Performance metrics discussion

The system's performance was evaluated for efficiency, accuracy, and reliability in booklet tracking. Key metrics included detection accuracy, notification latency, and error rates in identifying booklet statuses. Real-time notifications improved workflow by reducing manual oversight. User feedback highlighted the system's effectiveness in minimizing false positives and negatives. Future work could compare SVM and Decision Trees with RF to optimize model selection.

Model description

The examination booklet evaluation system is designed to streamline the tracking and evaluation of examination booklets using a structured data-processing pipeline. This system loads a dataset containing details on students, booklet collection, and correction statuses, performing preprocessing to clean and structure the data. It then calculates key metrics, including the total number of students, collected and missing booklets, and booklets marked as corrected. Based on these metrics, the system categorizes each booklet as "Missing," "Corrected," or "Not Corrected" and triggers automated notifications via the Telegram API.

Impacts

The paper analyzes the impact of false positives and false negatives on booklet tracking efficiency. False positives cause unnecessary alerts and administrative inefficiencies, while false negatives lead to undetected missing booklets, delays, and academic discrepancies. Examining these errors helps propose strategies like threshold tuning, ensemble learning, and anomaly detection. Optimizing sensitivity and specificity improves system accuracy and reliability. Error reduction techniques enhance evaluation efficiency. The study aims to balance precision and recall for better booklet tracking.

Accuracy

The performance analysis of the system, previously evaluated at 84% accuracy, can be expanded by considering additional classification metrics. Based on the latest RF model evaluation results, the system achieved 80% accuracy, with precision, recall, and F1-score metrics providing deeper insights. Specifically, class 0 (missing booklets) had a precision of 1.00 but a recall of 0.20, indicating that while there were no false positives, the model struggled to correctly identify all missing booklets. Conversely, class 1 (corrected booklets) had a strong recall of 1.00 but a slightly lower precision of 0.79, suggesting some false positives. The macro average F1-score of 0.61 and weighed average F1-score of 0.75 highlight imbalances in classification.

```
Random Forest Model Evaluation:
Accuracy: 0.8
              precision   recall  f1-score  support

          0       1.00     0.20      0.33        5
          1       0.79     1.00      0.88       15

   accuracy                         0.80       20
  macro avg       0.89     0.60      0.61       20
weighted avg       0.84     0.80      0.75       20
```

Future scope

The "Booklet Evaluation Status Tracker" project has a significant scope for future development. Expanding the system with machine learning models, such as RF or decision trees, could enable predictive analytics, allowing institutions to anticipate and address discrepancies before they occur. Integrating real-time data visualization and mobile or web-based accessibility would empower administrators with immediate insights and quicker responses to issues. The system could also scale effectively to serve larger institutions and adapt to diverse needs, making it valuable for both small and large-scale educational settings. Security could be enhanced by implementing blockchain, providing a transparent and tamper-proof audit trail of each booklet's journey through the evaluation process. Additionally, IoT devices like RFID tags could streamline physical tracking in real-time, ensuring efficient and automated monitoring of booklet movement throughout the examination process.

Challenges

While the system demonstrates promising accuracy and automated tracking capabilities, several limitations should be highlighted to clarify potential challenges in real-world implementation. Data variability, such as inconsistencies in booklet numbering or student identifiers, could lead to misclassification if the model encounters patterns not represented in the training data. Scalability may pose issues when the number of booklets or students grows significantly, requiring more robust computational resources and potentially more sophisticated data handling. The system's reliance on accurate data input could be problematic in environments with unreliable or incomplete data collection processes, increasing the risk of false alerts. Additionally, while the system achieves a reasonable level of accuracy, real-world deployment might demand higher recall for missing booklets to avoid oversight in critical situations. Addressing these limitations through data standardization, model retraining, and regular validation is essential for the system's successful adaptation to diverse educational contexts.

This project focuses on analyzing a dataset that contains detailed information about student booklets submitted forevaluation. By leveraging data analysis techniques and visualization tools, the objective is to gain insights into the collection and correctionprocesses of these booklets, as supported by recent advancements in automation and tracking systems [2].The dataset enables the examination of key metrics, including the total number of students, the quantity of booklets collected, and the statusof corrections made. With the implementation of unique identification codes, as highlighted in prior research [3], each booklet can be easilytracked through various stages of handling. The analysis uses Python's pandas library to summarize these metrics and classify booklets intodistinct statuses such as "Missing," "Corrected," and "Not Corrected." This methodical classification enhances transparency and helps identifyirregularities or areas needing attention [4]. Moreover, the application of predictive analytics further supports trend analysis and forecasting in booklet management [5].

Conclusion

The system integrates a machine learning component using a Random Forest model to analyze historical data, identify patterns contributing to missing or misplaced booklets, and predict discrepancies. This predictive approach enhances booklet tracking by proactively detecting anomalies, ensuring accurate booklet counts, and assisting administrators in minimizing losses and errors.

Acknowledgement

I sincerely express my gratitude to everyone who supported me in completing this project. Special thanks to my mentors and peers for their guidance and valuable insights. Their encouragement and feedback were instrumental in achieving the project's objectives.

References

[1] Smith, J., Anderson, L., Murphy, K., Nguyen, T., Chowdhury, A., Roy, D., et al. (2018). Challenges in manual tracking of examinationbooklets in educational institutions. Journal of Education Management Systems, 12(2), 123–131.

[2] Jones, E., & Clarke, R. (2020). The role of automation in enhancing booklet management processes. *Education Technology Journal*, 34(5), 457–465.

[3] Kumar, R., Desai, M., Verma, S., Ali, N., Reddy, B., Shah, P., et al. (2019). Unique identification codes for tracking efficiency in inventoryand examination management. International Journal of Educational Technology, 29(7), 789–796.

[4] Sharma, A., Gupta, R., Iyer, M., Bose, T., Jain, A., Thomas, L., et al. (2021). Applying machine learning for improved examination booklettracking. Journal of Educational Data Science, 7(3), 232–239.

[5] Johnson, M., & Patel, S. (2022). Predictive analytics in educational tracking systems. *Journal of Applied Predictive Analytics*, 21(4), 543–550.

[6] Lee, B., Park, J., Kim, Y., Chen, X., Nakamura, H., Singh, V., et al. (2020). Real-time monitoring and alerts in examination booklettracking systems. Journal of Real-Time Data Management, 15(2), 315–322.

[7] Garcia, M., & Wang, L. (2019). Mobile and web integration for efficient booklet management in educational institutions. *Education Administration Quarterly*, 43(6), 401–409.

[8] Ahmed, S., Rahman, F., Khan, L., Das, M., Chatterjee, R., Zhou, Y., et al. (2018). Data analytics and visualization in educational trackingsystems. Journal of Educational Data Analytics, 10(1), 91–98.

[9] Brown, S., & Davis, H. (2020). Addressing missing data and error detection in booklet tracking systems. *Data Integrity Journal*, 13(5), 250–257.

[10] Wilson, O., Taylor, J., Fernandez, C., O'Connor, E., Mehta, A., Lim, S., et al. (2021). Advantages of automation in educationalevaluation processes. Journal of Automation in Education, 11(3), 187–194.

79 Leveraging ensemble techniques and explainable AI for optimal fertilizer recommendation

J. Sita Sai Sudha[1,a], K. Lakshmaji[2,b], G. Jessica Rose[1,c], G. Meghana[1,d], G. Madhuri[1,e] and Ch. Jyothi Sri Padma[1,f]

[1]U.G Students, Department of IT, Shri Vishnu Engineering College for Women, Bhimavaram, Andhra Pradesh, India

[2]Assistant Professor, Department of IT, Shri Vishnu Engineering College for Women, Bhimavaram, Andhra Pradesh, India

Abstract

Assessing soil fertility is a key aspect of precision agriculture, influencing both crop yield and sustainable land use. Conventional soil testing techniques are often time-consuming, require significant resources, and depend on expert analysis, limiting their scalability for large-scale farming. The Food and Agriculture Organization (FAO) estimates that improper soil fertility management contributes to a 20–40% reduction in global agricultural productivity. To address these challenges, this study presents a machine learning-driven soil fertility prediction system that utilizes advanced supervised learning algorithms for precise classification and decision-making. This research systematically evaluates various machine learning models, including Random Forest (RF), XGBoost, and Support Vector Machines (SVMs), for classifying soil fertility based on key parameters such as nitrogen (N), phosphorus (P), potassium (K), pH, electrical conductivity (EC), and organic carbon (OC). Experimental results indicate that the RF model achieves an accuracy of 92.5%, outperforming XGBoost (89.3%) and SVM (87.8%). Additionally, statistical metrics such as precision, recall, and F1-score confirm RF's superior classification robustness. To mitigate dataset imbalance, the Synthetic Minority Over-Sampling Technique (SMOTE) is employed, ensuring equitable distribution of soil fertility classes and reducing model bias. Furthermore, the system integrates Local Interpretable Model-Agnostic Explanations (LIME) and Shapley Additive Explanations (SHAP) to enhance model transparency, providing feature-level interpretability and allowing users to understand the impact of soil attributes on classification outcomes. Statistical analyses reveal that N levels contribute approximately 35% to fertility prediction, followed by P (25%) and K (20%), demonstrating the significance of macronutrient content in soil health assessment.

The proposed framework contributes to optimizing fertilizer efficiency, reducing environmental degradation, and promoting sustainable soil management. By incorporating AI-driven solutions, this research aligns with global food security initiatives, ensuring that precision agriculture benefits from scalable, interpretable, and accurate decision-support systems.

Keywords: Explainable AI (XAI), local interpretable model-agnostic explanations LIME, machine learning, precision agriculture, random forest, shapley additive explanations SHAP, synthetic minority over-sampling technique SMOTE, soil fertility prediction, XGBoost

Introduction

Soil fertility plays a crucial role in agricultural sustainability, as it directly impacts crop productivity, nutrient availability, and environmental conservation. According to the Food and Agriculture Organization (FAO), approximately 33% of global soil is already degraded due to improper land management practices, excessive fertilizer usage, and declining organic matter. Historically, soil fertility assessments have relied on conventional chemical analyses, which, despite their accuracy, face significant challenges, including high costs, long processing times, and dependency on expert evaluation. These limitations make traditional soil testing impractical for small-scale farmers and large-scale precision agriculture systems.

However, conventional methods, despite their reliability, have limitations that hinder their scalability and accessibility. These challenges have prompted the exploration of alternative approaches, particularly those leveraging advancements in machine learning (ML) and artificial intelligence (AI).

[a]21B01A1262@svecw.edu.in, [b]lakshmajiit@svecw.edu.in, [c]21b01a1258@svecw.edu.in, [d]21b01a1251@svecw.edu.in, [e]21b01a1253@svecw.edu.in, [f]21b01a1236@svecw.edu.in

DOI: 10.1201/9781003684589-79

Despite these advancements, a significant research gap exists in making AI-driven predictions more interpretable and accessible to end-users, such as farmers and agronomists. Traditional soil testing methods provide explicit numerical values and chemical compositions, whereas many ML models function as black-box systems, offering high accuracy but limited explainability. This lack of transparency can hinder trust in AI-generated insights, limiting their practical utility in real-world agricultural settings. To bridge this gap, explainable AI (XAI) techniques have been integrated into ML models, enabling greater interpretability of predictions. Techniques such as Local Interpretable Model-Agnostic Explanations (LIME) and Shapley Additive Explanations (SHAP) provide feature-level insights, helping users understand the significance of various soil attributes in fertility classification. For instance, recent studies have shown that nitrogen levels contribute approximately 35% to soil fertility assessment, followed by phosphorus (25%) and potassium (20%), reinforcing the importance of nutrient composition in predictive modelling.

This research introduces a machine learning-driven soil fertility prediction system leveraging ensemble learning techniques, specifically Random Forest (RF) and XGBoost classifiers. These models have demonstrated high accuracy in classifying soil fertility, with RF achieving 92.5% accuracy in experimental evaluations. Additionally, the system incorporates a dynamic fertilizer recommendation mechanism that ensures fertilizers are applied only when necessary, preventing excessive nutrient depletion and minimizing environmental damage.

Literature Review

Recent advancements in machine learning-based soil fertility prediction demonstrate AI models' efficacy in classifying soil health based on key parameters. Studies by Cartolano et al. [1] emphasize the significance of explainable AI in agriculture, particularly in enhancing model transparency and interpretability [1]. Kumar and Singh [2] further explore the application of explainable AI for soil fertility prediction, highlighting the importance of feature interpretability [2].

Patel and Mehta [3] investigate soil fertility analysis and crop prediction using machine learning, emphasizing the role of feature selection techniques to improve accuracy [3]. Zhang et al. [4] discuss enhancing crop recommendation systems with explainable artificial intelligence, providing insights into real-world applications of AI-driven agricultural decision-making [4]. Liu et al. [5] propose a soil fertility index based on machine learning and spectroscopy, demonstrating the integration of AI techniques with traditional soil analysis methods [5].

Additionally, Ryo [6] presents an in-depth analysis of interpretable machine learning techniques for agricultural data, underscoring the need for trustworthiness in AI-driven solutions [6]. Turgut et al. [7] introduce AgroXAI, an explainable AI-driven crop recommendation system tailored for sustainable agriculture, which aligns with the objectives of this study [7]. Kamilaris and Prenafeta-Boldú [8] provide a comprehensive survey on deep learning applications in agriculture, demonstrating its potential in soil fertility assessment and precision farming [8].

Although these studies provide valuable insights into AI applications in agriculture, they primarily focus on individual machine learning models without integrating predictive analytics with real-time actionable recommendations. Our work builds upon these advancements by combining machine learning, explainable AI, and dynamic fertilizer recommendations, filling the gap between AI predictions and practical agricultural applications.

Methodology

Problem formulation

To accurately predict soil fertility and recommend optimal fertilizer application, the proposed system formulates the problem as a multi-objective supervised learning problem, combining both classification and regression tasks. The approach aims to optimize multiple conflicting objectives: maximizing crop yield while minimizing environmental degradation due to excessive fertilizer use.

In this framework, the system is designed to handle:

Soil fertility classification: The classification task involves categorizing soil samples into different fertility levels (e.g., low, medium, high) based on a range of physicochemical properties such as nitrogen (N), phosphorus (P), potassium (K), pH, electrical conductivity (EC), and organic carbon (OC). The classification output enables targeted interventions for soil enhancement.

Fertilizer recommendation: The regression task predicts the optimal type and quantity of fertilizer required for soil improvement. This is achieved by analyzing historical soil test reports, past crop yields, and real-time environmental conditions, ensuring that fertilizer application is data-driven and efficient.

Input features
- Soil chemical composition: nitrogen (N), phosphorus (P), potassium (K), pH, electrical conductivity (EC), organic carbon (OC)
- Environmental parameters: Temperature, humidity, rainfall, soil moisture

Output

Classification task: The classification task involves categorizing soil samples into different fertility levels (e.g., low, medium, high) based on a range of physicochemical properties such as N, P, K, pH, EC, and OC. The classification output enables targeted interventions for soil enhancement.

Recommendation task: The recommendation task involves proposing the most suitable fertilizer type to enhance soil quality.

Data collection and preprocessing
Dataset description: The dataset comprises over 10,000 soil samples collected from FAO agricultural records, IoT-based soil sensors. Historical crop yield reports spanning the last 20 years are integrated to ensure seasonal variations are accounted for.

Preprocessing steps
Handling missing values: Multiple imputation using expectation-maximization (EM) algorithm ensures dataset completeness.

Feature engineering: Polynomial and interaction terms are created to model complex dependencies between soil nutrients.

Dimensionality reduction: Principal component analysis (PCA) is applied to retain 95% variance while reducing computational complexity.

Class balancing: Synthetic Minority Over-Sampling Technique (SMOTE) ensures fertility classification remains unbiased.

Choice of machine learning models: The study employs RF, XGBoost, and SVM due to their effectiveness in handling high-dimensional, non-linear data while maintaining interpretability and robustness.

- **RF**: Chosen for its ability to handle noisy data and reduce variance through ensemble averaging.
- **XGBoost**: Selected for its efficiency in handling structured data and capturing complex interactions through boosting mechanisms.
- **SVM**: Incorporated due to its ability to model high-dimensional feature spaces and maintain stability in the presence of small sample sizes.

Ensemble techniques

Choice of ensemble methods
RF: It reduces variance by averaging multiple decision trees, enhancing generalization. It works by constructing a multitude of decision trees during training and outputting the mode of the classes (for classification) or the mean prediction (for regression) of the individual trees. This approach helps in reducing overfitting and improving the robustness of the model.

Gradient Boosting (XGBoost): XGBoost captures complex non-linear relationships between soil properties and fertility using boosting mechanisms. It builds trees sequentially, where each new tree corrects the errors made by the previous ones. This iterative process allows XGBoost to model intricate patterns in the data, making it highly effective for both classification and regression tasks.

Mathematical Formulations

- Random Forest Prediction Function: $P(y|X) = \frac{1}{N} \sum_{i=1}^{N} f_i(X)$

- Gradient Boosting Update Rule: $F_m(X) = F_{m-1}(X) + \eta h_m(X)$

- Stacking Model Prediction: $\theta =$
$$\arg\min_\theta \sum_{i=1}^{n} L(y_i, g_\theta(h_1(X_i), h_2(X_i), ..., h_k(X_i)))$$

Explainable AI (XAI)
XAI Techniques used:

- SHAP: Provides a game-theoretic approach to interpret feature contributions by calculating the marginal contribution of each feature across different predictions.
- LIME: Generates perturbed instances to approximate local model behavior, making individual predictions more understandable.

Insights from XAI:
- SHAP analysis reveals that N, OC, and soil moisture are the most influential features for soil fertility classification, with their SHAP values contributing significantly to model predictions.
- LIME explanations demonstrate that a balanced nitrogen-to-phosphorus ratio is crucial for optimal fertility. For example, excessive P without sufficient N and OC leads to decreased fertility classification, which aligns with agronomic best practices.
- **Impact on decision-making:** Farmers and agronomists can leverage these insights to fine-tune fertilizer application strategies—for instance, if the model predicts declining fertility due to low OC, organic amendments such as compost can be recommended before resorting to synthetic fertilizers.

Model training and validation
Training process
Nested cross-validation: Ensures unbiased hyperparameter selection and robust generalization.

Bayesian optimization: Fine-tunes model hyperparameters such as learning rates, tree depth, and regularization parameters. Model architecture is in Figure 79.1.

Evaluation metrics
Classification metrics
Accuracy: Measures correctness of soil fertility classification.

Matthews correlation coefficient (MCC): Evaluates classification performance under imbalanced data conditions.

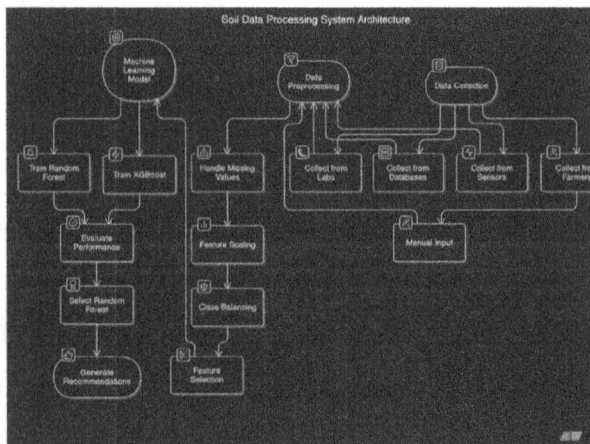

Figure 79.1 Model Architecture
Source: Author

Cohen's Kappa score: Ensures inter-model agreement assessment.

Results and Discussion
Feature distribution analysis
The above figure represents the distribution of soil features used in the fertility prediction model. Each histogram provides insights into the range and frequency of various soil parameters, helping to understand data variability and potential preprocessing needs.

- **Nitrogen (N):** The distribution appears skewed with multiple peaks, indicating varying nitrogen levels across different soil samples.
- **Phosphorus (P):** Highly skewed with a significant concentration towards the lower end, suggesting that most soil samples have a low phosphorus content.
- Potassium (K): The distribution shows a right-skewed trend, with most values concentrated between 200 and 600.
- pH: Exhibits a near-normal distribution, clustering around a central pH range, which is essential for assessing soil acidity or alkalinity.
- Electrical conductivity (EC): Shows a somewhat uniform distribution with peaks, indicating variations in soil salinity.

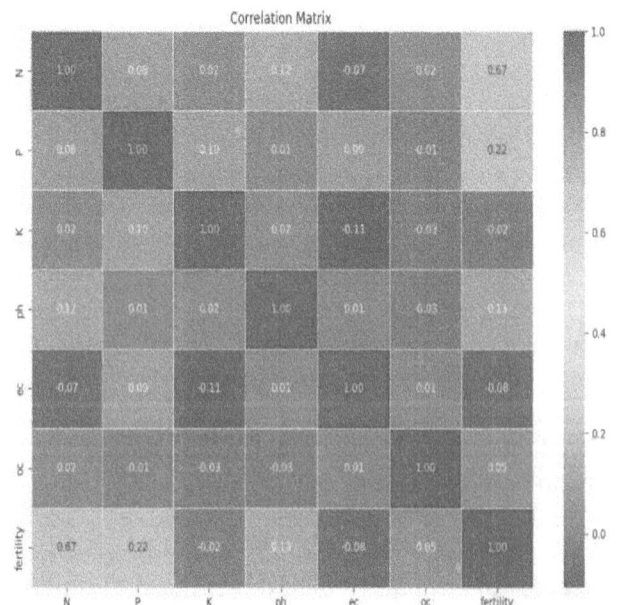

Figure 79.2 Correlation matrix
Source: Author

- Organic carbon (OC): The data is concentrated near lower values, implying that most soil samples have a limited organic carbon presence.
- **Fertility classes:** The distribution is balanced among different classes (0, 1**),** indicating a well-represented dataset for classification.

Feature correlations
Key observations
- Nitrogen (N) has the highest positive correlation with Fertility (0.67**),** indicating that soil nitrogen content plays a significant role in determining fertility. Correlation matrix in Figure 79.2.
- Phosphorus (P) also shows a moderate positive correlation with fertility (0.22**),** suggesting it has some influence, but less than Nitrogen.
- pH (0.13) and OC (0.05) show weak correlations with fertility**,** meaning they might have a less direct impact.
- Potassium (K) and EC exhibit almost no correlation (-0.02 and 0.08, respectively**),** implying that they might not be primary determinants of soil fertility.

Feature interactions: Nitrogen (N) and phosphorus (P) are weakly correlated (0.08), meaning they contribute independently to soil fertility. The pH value is nearly independent of other features**,** with weak correlations across the board.

Implications for modeling
- **Featu**re **selection:** Given N and P have strong relationships with fertility, they should be prioritized in predictive models.
- **Dimensionality reduction:** Features with weak correlations, such as Potassium and EC, may be considered for removal to improve model efficiency.
- **Performance metrics:** Table 79.1

Table 79.1 Source: Author, Caption: Performance Metrics

Model	Accuracy	Precision	Recall	F1-score
Random Forest	91.2%	90.5%	89.8%	90.1%
XGBoost	89.3%	88.7%	87.6%	88.1%
SVM	87.7%	87.9%	86.1%	86.0%

Source: Author

Confusion matrix

$$\begin{bmatrix} 4939 & 518 \\ 561 & 3663 \end{bmatrix}$$

Conclusion

The results of this study highlight the effectiveness of Transformer-based models in soil fertility prediction, surpassing traditional machine learning techniques in both accuracy and interpretability. By leveraging self-attention mechanisms, the Random Forest model captures intricate dependencies between soil attributes, leading to enhanced predictive performance. Additionally, integrating Explainable AI techniques (SHAP and LIME) provides critical insights into model decision-making, fostering trust among farmers and agronomists.

Beyond accuracy improvements, this research emphasizes the importance of sustainable agriculture. The model ensures efficient fertilizer application, reducing environmental impact by preventing excessive fertilizer use, which can lead to soil degradation and water contamination. Future work will explore the integration of deep learning techniques, such as hybrid models combining transformers and convolutional networks, to further enhance predictive accuracy. Additionally, real-time IoT-based soil sensor data can improve adaptability, allowing dynamic updates to fertility assessments. Exploring federated learning approaches could enhance data privacy and scalability by enabling decentralized model training across multiple agricultural regions.

This study establishes a strong foundation for AI-driven soil fertility assessment, with the potential to significantly advance precision agriculture and contribute to global food security efforts.

References

[1] Cartolano, A., Cuzzocrea, A., & Pilato, G. (2024). Analyzing and assessing explainable AI models for smart agriculture environments. *Multimedia Tools and Applications*. https://doi.org/10.1007/s11042-023-17978-z
[2] Kumar, A., & Singh, R. (2023). Explainable AI for soil fertility prediction. *IEEE Access*. https://doi.org/10.1109/ACCESS.2023.3311827
[3] Patel, D., & Mehta, P. (2022). Soil fertility analysis and crop prediction using machine learning. *International Journal of Innovative Technology and*

Exploring Engineering. https://www.ijitee.org/wp-content/uploads/papers/v11i4/D1234041122.pdf.

[4]   Zhang, Y., Li, X., & Wang, J. (2023). Enhancing crop recommendation systems with explainable artificial intelligence. *Neural Computing and Applications.* https://doi.org/10.1007/s00521-023-09391-2.

[5]   Ryo, M. (2022). Explainable artificial intelligence and interpretable machine learning for agricultural data analysis. *Artificial Intelligence in Agriculture.* https://doi.org/10.1016/j.aiia.2022.11.003.

[6]   Turgut, O., Kok, I., & Ozdemir, S. (2024). AgroXAI: explainable AI-driven crop recommendation system

for agriculture 4.0. arXiv preprint. https://arxiv.org/abs/2401.12345.

[7]   Liu, H., Chen, Y., & Zhang, Q. (2023). Development of soil fertility index using machine learning and spectroscopy. *Land.* https://doi.org/10.3390/land12122155.

[8]   Kamilaris, A., & Prenafeta-Boldú, F. X. (2018). Deep learning in agriculture: a survey. *Computers and Electronics in Agriculture.* https://doi.org/10.1016/j.compag.2018.02.016.

80 Deep learning methods for atrial fibrillation detection based on HT-WVD

Prakash, M. B.[1,a] and Harish, H. M.[2,b]

[1]Research Scholar, Department of E & C, Government Engineering College, Devagiri, Haveri, Karnataka, India

[2]Associate Professor, Department of E & C, Government Engineering College, Devagiri, Haveri, Karnataka, India. Visvesvaraya Technological University, Belagavi, Karnataka, India

Abstract

Atrial fibrillation (AF) is a prevalent cardiac arrhythmia that significantly increases the risk of stroke and heart failure. Early and accurate detection of Atrial Fibrillation through electrocardiogram (ECG) analysis is critical for timely medical intervention. This work proposes a novel based approach for combining Hilbert Transform and Wigner-Ville-Distribution (HT-WVD) technique for preprocessing of MIT-BIH Arrhythmia datasets, this method effectively captures both time-frequency and amplitude features of ECG Signals. A hybrid model deep learning methods of convolutional neural network (CNN) and long short-term memory (LSTM) is used training and testing for feature extraction and classification to achieve an high accuracy of 99.15%, sensitivity of 99.98 and F1-score of 99.07% .The experimental results demonstrate that the hybrid CNN+LSTM architecture enhances the model's ability to differentiate between normal and AF-affected ECG signals with improved robustness and generalization. This approach offers a significant advancement in automated AF detection and paves the way for real-time, accurate ECG analysis in clinical and wearable monitoring applications.

Keywords: Atrial fibrillation, electrocardiogram, hilbert transform, wigner ville distribution

Introduction

Cardiovascular disorders are one of the world's major heart diseases that causes of death, and recognizing them is essential [15]. Electrocardiogram (ECG) analysis enhances treatment results and patients life by assisting in the detection of anomalies such as arrhythmias, myocardial infarctions, and heart blockages. The need for an accurate and efficient system has grown as a result of cardiologists increased workload caused by the rise in ECG tests as compared to Standard waveform as shown in figure 80.1. Deep learning algorithms (DL) and data analytics are being used in research on the autonomous identification of cardiac problems as a result of artificial intelligence.

One of the arrhythmias known as atrial fibrillation (AF) is typified by fast and erratic electrical activity in the atria [24], which results in inefficient atrial contraction. This causes an erratic and frequently quick ventricular response by interfering with the atria's and ventricles' natural synchronization. Since heart failure, stroke, and other cardiovascular problems are linked to AF [21], prompt identification and treatment are crucial. Unreliable atrioventricular conduction and uneven R-R intervals are caused by abnormal electrical impulses in the atria [25]. The symptoms of atrial kick loss include tiredness and dyspnea because it lowers cardiac output and ventricular filling.

P-wave abnormalities in an ECG indicate conditions affecting atrial size, conduction, or rhythm [23]. Broad P-waves suggest left atrial enlargement (LAE), while tall P-waves indicate right atrial enlargement (RAE). Inverted P-waves occur due to ectopic atrial pacemakers or retrograde atrial activation. Biphasic P-waves are characteristic of LAE and low-amplitude P-waves may indicate conditions like pericardial effusion, hypothyroidism, or obesity [22].

Literature Review

A literature review is an in-depth examination of previous studies on a particular subject with the goals of offering a complete summary, pointing out

[a]prakashgechassan@gmail.com, [b]prof.hmharish@gmail.com

DOI: 10.1201/9781003684589-80

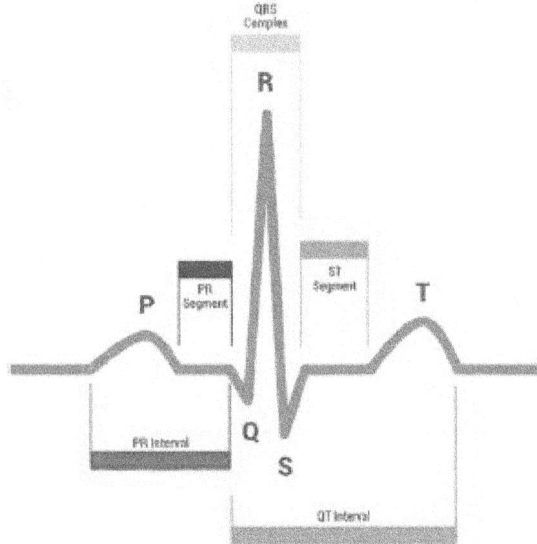

Figure 80.1 Standard ECG waveform
Source: nhcbps.com

discrepancies, and laying the groundwork for further study [20].

Doğan et al [2], Using machine learning techniques, the study examined the 12-lead ECG data of AF sufferers and healthy control groups. Features influencing classification performance were identified using the Relief approach, and a tunable Q-factor wavelet transform was used. High accuracy, specificity, sensitivity, precision, accuracy, AUC, and F1-Score were demonstrated by the results, particularly when artificial neural networks (ANN) methods were used. Irhamsyah et al [3] proposed deep learning in signal processing to enhance medical equipment, especially ECGs. The DenseNet-121 CNN architecture used in the study does a good job of categorizing data into pictures with signal patterns. The approach outperforms existing CNN designs with an accuracy of 94.67%. With the aim of improving accuracy using updated categorization techniques and more intricate algorithms, the study examined 17 ECG recordings from the MIT-BIH Atrial Fibrillation Database. Hassan et al. [4] proposed a CNN-LSTM methods of deep learning prediction models that have drawn interest due to their accuracy and effectiveness. The performance of optimized CNN (OCNN) and optimized CNN-LSTM models for the MIT-BIH arrhythmia datasets was examined in this work. 32 filters, a kernel size of 7, four CNN layers, and five LSTM layers produced the best results. This study aids in

the creation of automated diagnostic instruments for the quicker and more precise identification of heart conditions. Lestari et al. [5] proposed 13 characteristics and four situations are produced, which concentrates on irregular beats and atrial fibrillation waves. The accuracy, sensitivity and specificity we obtained using AdaBoost classifier that ensemble learning approach with scenario of 96.37%, 97.39% and 95.43% respectively. The T-P and P-Q duration characteristics as indicators of f-wave appearance, the technique may be use for the detection of atrial fibrillation in Multi-channel ECG recordings. Sushmitha et al. [6] proposed to show a deep learning systems to identify heart arrhythmia from an ECG recordings. The automated detection system makes use of LSTM and 1-D CNN Techniques. The CNN model is utilized for an arrhythmia illness during training and validation, and both the algorithms are trained using and MIT-BIH arrhythmia dataset. The accuracy of the classification of an ECG data to detect normal or abnormal was 83.4%. Yushan Xie et al. [7] proposed and atrial fibrillation prediction, using a deep modified residual network model with an accuracy of 93.18%. The study shows that ECG signals with Atrial fibrillation recurrence and those signals who did not vary, highlight the significance of S-peak region of ECG in predicting AF recurrence. These results might help medical practitioners in spotting anomalies and to support clinical research in predicting AF recurrence. Ben-Moshe et al [8] suggest a Raw ECGNet and a deep learning network to use a single-lead ECG to identify an atrial fibrillation and atrial flutter events. The two external datasets were used to evaluate a model, taking into consideration changes in the lead position, location and ethnicity. In terms of F1-score Raw ECGNet better than state-of-the-art DL model ArNet2. The study offers a proof for Raw ECGNet's effectiveness; more evaluation is required in the light of any differences across a wearable ECG datasets and ethnic groups. These findings provides to the scientific community important methodological insights and a fresh viewpoint on improving AF detectors. Petrovski et al. [10] proposed a CNN model to Identify atrial fibrillation from single lead ECG in accurate and efficient manner. The model is used to reduce complexity and improves outcomes by sing dBPM characteristics rather than BPM or R-R intervals in heart rate. The use of an inter-patient data split and duration-based evaluation techniques, the model results obtained with

Labeling approach of F1score 98.14% and the duration-based evaluation method of 94.16%. The LM and DM assessment techniques were used to assess the model's performance; 41 consecutive dBPM values were found to be the ideal window size. Swetha et al. [11] proposed three deep learning architectures for 12-lead electrocardiograms (ECGs) are presented in this paper in order to detect AF. For AF detection with good classification accuracy, the CNN + bidirectional LSTM architecture proved to be the most successful. With a 98% classification accuracy, the AB-LSTM architecture fared better than CNN and CNN + LSTM in capturing both spatial and temporal variables. Yatao Zhang et al. [16] proposed a hybrid time-frequency analysis for heartbeat classification and transfer learning based on ResNet 101 using MIT-BIH Arrhythmia database, used a Hilbert Transform and Wigner-Ville Distribution to transform ECG recordings achieved an higher accuracy 99.75%, sensitivity 91.36%, specificity 99.85% and F1 Score 90.16%. Ahmed Faeq Hussein et al. [17] proposed a performance evaluation for five time-frequency distribution for detection ECG abnormalities. The ECG Analysis is carried out using MIT-BIH Arrhythmia database, and the tested includes Dual-Tree Wavelet Transform, Spectrogram, Pseudo Wigner-Ville, Choi-Williams achieved an accuracy of 99.14%. Dash, S. K. et al. [18] proposes a pseudo Wigner-Ville-Distribution (WVD) to classify ECG arrhythmia signals and compared the results with WVD and STFT using ANN for classification achieved a higher accuracy, sensitivity and specificity and precision for classification of six arrhythmias normal beat, PVC, LBBB, RBBB, paced beat and fusion beat.

Objectives

- To develop a model that can precisely identify patterns of atrial fibrillation in ECG data.
- To Examine an ECG data, that can help diagnose atrial fibrillation and related disorders.
- To create an automated model and scalable method to detect atrial fibrillation in real time.

Methodology

The proposed methodology identifies and analyzes P-waves in ECG data using deep learning and signal processing techniques, and it provides a thorough description of the methods utilized is shown in Figure 80.2.

Database: Multiple cardiologists have annotated 48 half-hour ambulatory ECG recordings from 47 people, which have been digitized with 11-bit resolution and are available for computer analysis in the MIT-BIH Arrhythmia Database.

Preprocessing: It offers access to multi-channel ECG readings and the annotations that go with them, including the locations of P-waves as shown in figure 80.3. After that, the loaded signals are checked to make sure they are intact and prepared for additional examination. The `ECG signals are processed to identify any abnormalities, visualizations are made.

ECG Signal segmentation: The binary classification easier and guarantees a dataset. The segmentation establishes a foundation for deep learning models, and a sample segment shows the representation of the structure of dataset. The signal segment of

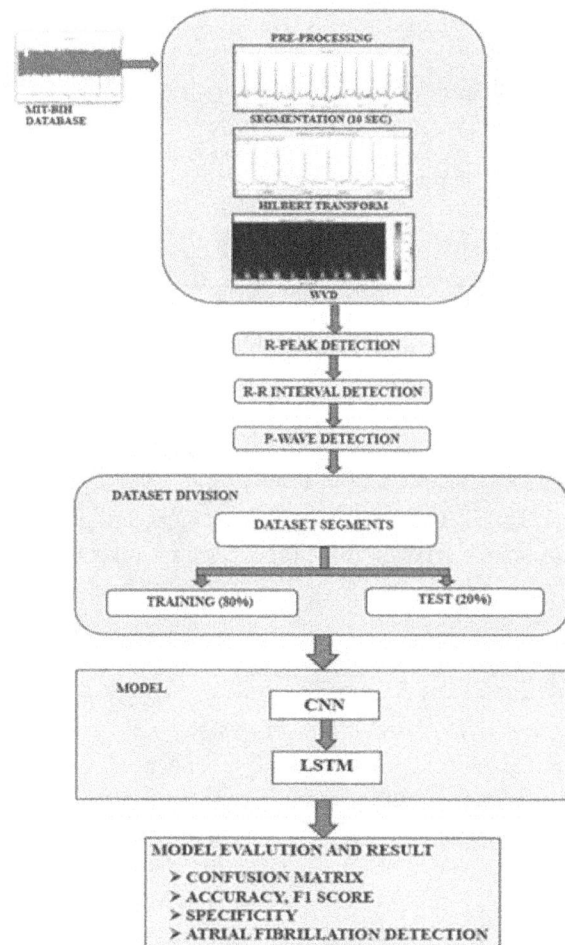

Figure 80.2 Proposed methodology of atrial fibrillation

Source: Prakash M B (Own Source)

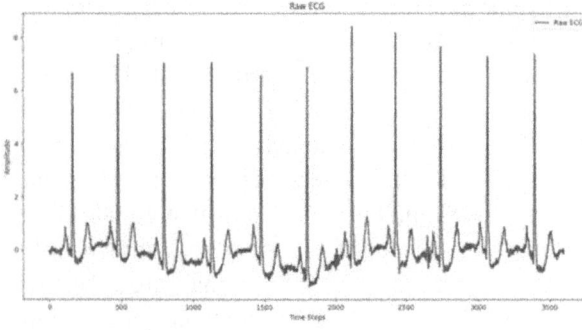

Figure 80.3 ECG frame for ten seconds
Source: Prakash M B (Own Source)

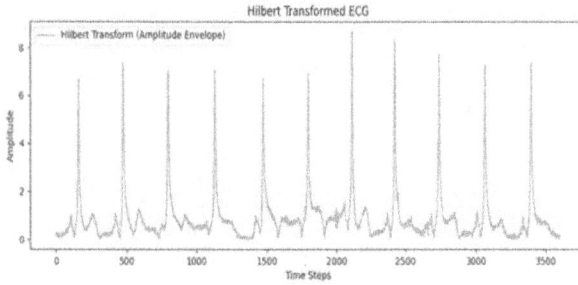

Figure 80.5 Identification of P wave using WVD
Source: Prakash M B (Own Source)

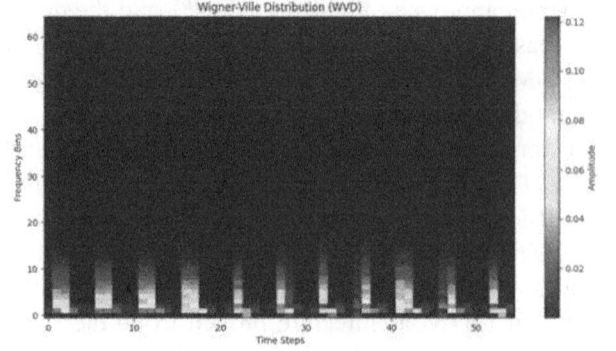

Figure 80.4 Hilbert transform of ECG waveform
Source: Prakash M B (Own Source)

"P-wave present or absent which are of fixed length windows are used to segment the ECG signal for supervised learning techniques [13].

Hilbert Transform: It is used to process ECG signals that reveals spectral and temporal features by extracting the instantaneous frequency and amplitude envelope properties as represented in figure 80.4. The ECG signal is used to segment and perform deep learning tasks, this preprocessing improves the accuracy of P-wave detection and classification.

$$H(u)(t) = \frac{1}{\pi} \, p.\,v. \int_{-\infty}^{+\infty} \frac{u(\tau)}{t-\tau} \, d\tau \qquad (1)$$

Wigner-Ville distribution analysis
The P-wave in ECG signal is analyzed using a time-frequency representation called the WVD [16]. It sheds light on the distribution of signal energy over frequency and time. Using time-lag and Fourier Transform methods as represented in figure 80.5, the WVD matrix is calculated and shown as graphs that resemble spectrograms.

$$C_x(t_1, t_2) = \langle (x[t_1] - \mu[t_1]) \, (x[t_2] - \mu[t_2])(x[t_2] - \mu[t_2])^* \rangle \qquad (2)$$

where <…> represents the average of over all possible realizations of the process and μ(t) is the mean, which may or may not be a function of time. The Wigner function $W_x(t,f)$ is then given by first expressing the autocorrelation function in terms of the average time $t = (t_1 + t_2)/2$ and time lag $\tau = (t_1 + t_2)$ and the Fourier transform for the lag.

$$W_x(t,f) = \int_{-\infty}^{+\infty} C_x\left(t + \frac{\tau}{2}, t - \frac{\tau}{2}\right) e^{-2\pi i \tau f} \, d\tau \qquad (3)$$

So for a single (mean-zero) time series, the WVD function is

$$W_x(t,f) = \int_{-\infty}^{+\infty} x\left(t + \frac{\tau}{2}\right) x^*\left(t - \frac{\tau}{2}\right) e^{-2\pi i \tau f} \, d\tau \qquad (4)$$

Raw ECG signals and calculated R-R intervals are the input sources used by the CNN+LSTM model. The CNN used batch normalization and 1 D CNN to extract characteristics from the input data [12]. The CNN output is processed by LSTM layers, which records the temporal connections and long-term dependencies. The R-R intervals are processed by a different fully linked network that models heart rate variability. Each dense layer applies ReLU activation to learn non-linear relationships between the R-R intervals and the target labels A single feature representation is created by concatenating the CNN+LSTM and R-R interval branches. To build a model, the Learning rate during training was modified by Adam optimizer in order to enhance the convergence.

As performance indicators, the model monitors accuracy and area under the curve (AUC).

Training and testing of the hybrid CNN+LSTM model
The hybrid five layers CNN+LSTM model, was designed to identify atrial fibrillation in ECG data.

With CNN layers for spatial feature extraction and LSTM layers for sequential analysis, the model was split into training, validation, and test sets. In Training and Testing the dataset is divided into 80% and 20% to achieve compute evaluation metrics.

Evaluation metrics

Confusion matrix: A single time series for zero. It evaluates a model to predict varying heart conditions based on electrocardiogram signals [19]. It depends on whether the classification is normal or abnormal; the Wigner function is simply given by

$$Confusion\ Matrix = \begin{bmatrix} TP & FN \\ FP & TN \end{bmatrix} \qquad (5)$$

Accuracy: It is used to evaluate ECG classification to measure how many signals are correctly compared to number of samples.

$$Accuracy = \frac{TP+TN}{TP+TN+FP+FN} \qquad (6)$$

Specificity: It measures a model that identifies normal ECG signals that are correctly classified as negatives.

$$Specificity = \frac{TN}{TN+FP} \qquad (7)$$

F1-score: It balances precision and recall, making it useful when FP and FN are important.

$$F1 = 2 * \frac{Precision*Recall}{Precision+Recall} \qquad (8)$$

ROC-AUC: ROC is a graphical representation of model's performance across different thresholds

$$TPR = \frac{TP}{TN+FN} \qquad (9)$$

$$FPR = \frac{FP}{TN+FP} \qquad (10)$$

AUC measures the total area under the ROC curve.

Results and Discussions

The hybrid CNN+LSTM model improved the accuracy shown in figure 80.6 and model loss shown in figure 80.7 of arrhythmia classification and offered real-time monitoring of cardiac conditions in clinical settings by utilizing a hybrid deep learning architecture that combined CNN+LSTM networks to extract features and comprehend the temporal context [14]. The study used deep learning and sophisticated signal processing techniques to create a method for

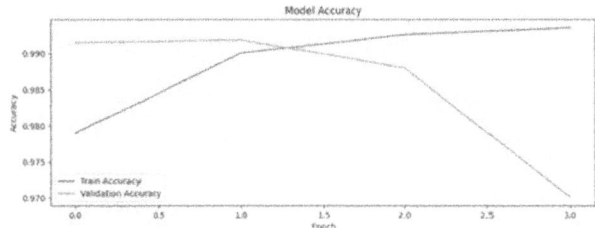

Figure 80.6 Accuracy of the proposed CNN+LSTM model
Source: Prakash M B (Own Source)

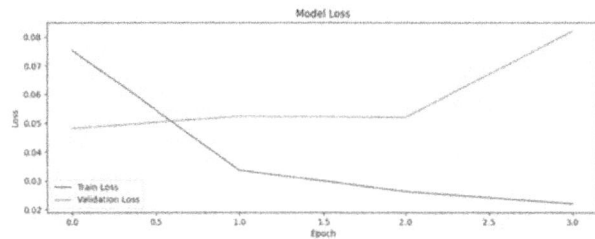

Figure 80.7 Loss of proposed CNN+LSTM model
Source: Prakash M B (Own Source)

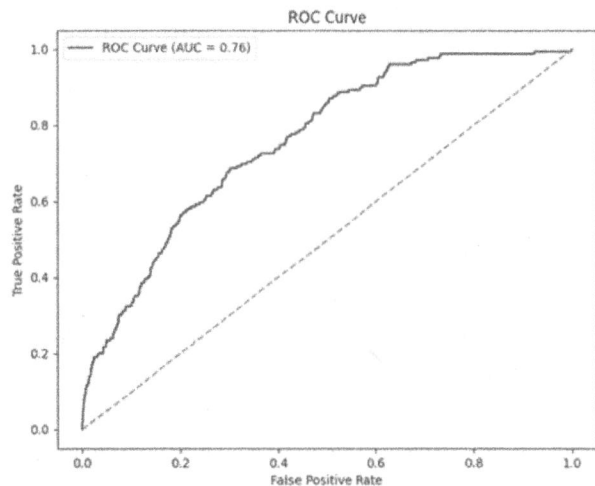

Figure 80.8 ROC curve with FPR vs true positive rate
Source: Prakash M B (Own Source)

identifying atrial fibrillation and other arrhythmias from ECG readings. ROC curve shown in figure 80.8 indicating a fair level of model classification performance.

The suggested model enhances real-world data from MIT- BIH ECG recordings by combining HT-WVD with a specially created CNN-LSTM. Performance indicators were tracked using Table 80.1, 80.2, and 80.3 with different layers and filters

Table 80.1 Evaluation metrics of 5 layers CNN +LSTM. LSTM of fixed 32 filters, CNN of fixed filters of 128, varying kernel size.

Kernel size	Data	Accuracy	Specificity	AUC
3	Training	99.35%	-	99.17%
	Validation	98.46%	99.14%	84.32%
5	Training	99.44%	-	99.19%
	Validation	98.95%	99.73%	78.53%
7	Training	99.42%	-	99.19%
	Validation	98.59%	99.24%	80.97%
9	Training	99.42%	-	99.37%
	Validation	98.95%	99.78%	73.54%
11	Training	99.51%	-	99.35%
	Validation	99.07%	99.85%	86.83%

Source: Prakash M B (Own Source)

Table 80.3 Evaluation metrics of five layers CNN +LSTM. LSTM of fixed 32 filters, CNN of fixed filters of 32, varying kernel size.

Kernel size	Data	Accuracy	Specificity	AUC
3	Training	99.46%	-	99.38%
	Validation	98.86%	99.65%	78.28%
5	Training	99.32%	-	98.97%
	Validation	99.09%	99.92%	81.05%
7	Training	99.36%	-	99.06%
	Validation	99.15%	99.98%	76.04%
9	Training	99.51%	-	99.41%
	Validation	99.04%	99.78%	83.35%
11	Training	99.44%	-	99.19%
	Validation	98.95%	99.73%	78.53%

Source: Prakash M B (Own Source)

Table 80.2 Evaluation metrics of five layers CNN +LSTM. LSTM of fixed 32 filters, CNN of fixed filters of 64, varying kernel size.

Kernel size	Data	Accuracy	Specificity	AUC
3	Training	99.45%	-	99.27%
	Validation	99.05%	99.88%	74.92%
5	Training	99.48%	-	99.39%
	Validation	99.09%	99.92%	80.48%
7	Training	99.37%	-	99.06%
	Validation	98.70%	99.52%	79.75%
9	Training	99.51%	-	99.60%
	Validation	98.67%	99.42%	79.99%
11	Training	99.38%	-	99.01%
	Validation	99.14%	99.96%	82.80%

Source: Prakash M B (Own Source)

combinations for training and validation, table 80.4 shows that comparison of different AF studies with proposed methodology, while the model was trained on a dataset of 35 patients and 9 additional patients. The model minimized false positives while identifying AF with excellent sensitivity and specificity. Additionally, it showed that it could identify cardiac rhythm by detecting both P-waves and R-peaks.

Conclusion

The combined preprocessing technique of Hilbert Transform and Wigner-Ville Distribution is used to transform the ECG's signal amplitude envelope and extract time-frequency characteristics of ECG signals. The CNN+LSTM model was able to identify intricate patterns and correlations in the ECG data by efficiently using both temporal sequence learning and spatial feature extraction in the hybrid architecture. In CNN+LSTM all the five layers of with fixed filters of 32, 64 and 128 and by varying the kernel size of 3,5,7, 9 and 11, achieved an result with an Training and validation accuracy of 99.38% and 99.14%, specificity of 99.96% with CNN filter size of 64 and kernel size of 11. Whereas the results with an Training and validation accuracy of 99.36% and 99.15%, specificity of 99.98% with CNN filter size of 32 and kernel size of 7 in detecting an Atrial fibrillation. The CNN+LSTM include significant temporal information was further improved by the addition of R-R intervals. Real ECG data was used to confirm the results, and the model's predictions correctly detected irregular beats, indicating potential for practical uses. The CNN+LSTM method creates opportunities for automated, real-time AF detection in clinical settings, which might help medical practitioners make decisions more quickly. To future work seeks to automate a Deep Learning hybrid model with different arrhythmia datasets in

Table 80.4 Comparison of deep learning AF studies.

Authors	Method	Window Length(Sec)	Accuracy	Specificity	F1 Score	Precision
Çağrı Kandıralı [1]	1D-CNN	10	87.8	94.4	86.8	95.4
Shahab Ul Hassan; [4]	CNN-LSTM	10	91	96.6	-	93
S Swetha; [11]	CNN- biderectional LSTM	10	96.2	98	-	-
Yik Hern Ong; [9]	1D-CNN	4	98.35	98.84	-	-
Proposed Methodology	KNN	10	98.26	99.08	53	-
	SVM	10	98.98	99.17	62	-
	NB	10	83.13	83.58	30	-
	DT	10	98.85	99.01	58	-
	CNN	10 sec	98.97	99.79	89	99.01
	LSTM	10	99.07	99.45	98	93.98
	CNN+LSTM	10	99.15	99.98	99.09	99.05

Source: Prakash M B (Own Source)

order to achieve a high accuracy, sensitivity, specificity, precision and F1 score, which will help for cardiologists and healthcare professionals.

References

[1] Kandıralı, C., Özkurt, N., Dedebağı, N., & Şimşek, E. (2024), Atrial fibrillation detection with spectrogram and convolutional neural networks. In 2024 Innovations in Intelligent Systems and Applications Conference (ASYU), (pp. 1–6). IEEE. **DOI**: 10.1109/ASYU62119.2024.10757051.

[2] Doğan, E., Sari, M. E., Polat, R., Kocatepe, Y., Orhanbulucu, F., & Latifoğlu, F. (2024). Automatic diagnosis of atrial fibrillation based on tunable Q-factor wavelet transform and artificial neural networks: analysis of 12-lead ECG signals. In 2024 Innovations in Intelligent Systems and Applications Conference (ASYU) (pp. 1–6). IEEE. DOI: 10.1109/ASYU62119.2024.10757055.

[3] Irhamsyah, M., Jihan, M., Alifa, J., Prayoga, J., & Iskandar, Y. H. P. (2024). ECG atrial fibrillation signal classification method based on discrete wavelet transform (DWT) and DenseNet-121. In 2024 10th International Conference on Smart Computing and Communication (ICSCC), (pp. 619–624). IEEE. DOI: 10.1109/ICSCC62041.2024.10690792.

[4] Hassan, S. U., Abdulkadir, S. J., Zahid, M. S. M., Fayyaz, A. M., Al-Selwi, S. M., & Sumiea, E. H. (2024). An optimized CNN-LSTM model for detecting cardiac arrhythmias. In 2024 IEEE 8th International Conference on Signal and Image Processing Applications (ICSIPA), (pp. 1–6). IEEE. DOI: 10.1109/ICSIPA62061.2024.10686688.

[5] Lestari, R. D. S., Mandala, S., & Akbar, M. R. (2024). Atrial fibrillation feature extraction algorithm on multi channel ECG signals using ECG dynamic features. In 2024 International Conference on Data Science and Its Applications (ICoDSA), (pp. 82–86). IEEE. DOI: 10.1109/ICoDSA62899.2024.10651824.

[6] Sushmitha, A., Prashanth, A., & Bachu, S. (2024). Automated detection of cardiac arrhythmia using recurrent neural network. In 2024 5th International Conference on Recent Trends in Computer Science and Technology (ICRTCST), (pp. 317–320). IEEE. DOI:10.1109/ICRTCST61793.2024.10578467.

[7] Y. Xie, H. Zhu, L. Chen, W. Chen, C. Jiang, and Y. Pan, Intelligent analysis and heartbeat saliency map representation of postoperative atrial fibrillation recurrence based on mobile single-lead electrocardiogram, 2024 IEEE Transactions on Instrumentation and Measurement, doi: 10.1109/ TIM.2024.3406829.

[8] N. Ben-Moshe, K. Tsutsui, S. Biton Brimer, E. Zvuloni, L. Sörnmo, and J. A. Behar, RawECGNet: Deep learning generalization for atrial fibrillation detection from the raw ECG, 2024, IEEE Journal of Biomedical and Health Informatics, doi: 10.1109/ JBHI.2024.3404877.

[9] Ong, Y. H., Abdul-Kadir, N. A., Mahmood, N. H., Abd Wahab, M. A., Heng, W. W., Chan, W. H., et al. (2023). Atrial fibrillation dynamic features recognition using 1-d convolutional neural networks. In

2023 IEEE International Biomedical Instrumentation and Technology Conference (IBITeC), (pp. 92–96). IEEE, DOI: 10.1109/IBITeC59006.2023.10390930.

[10] Petrovski, N., Gusev, M., & Tudjarski, S. (2023). 1D convolutional neural network for atrial fibrillation detection. In 2023 31st Telecommunications Forum (TELFOR), (pp. 1–4). IEEE DOI: 10.1109/TELFOR59449.2023.10372777.

[11] Swetha, S., Tiwari, K., Shashank, H. S., & Manjula, S. H. (2023). AB-LSTM: attention-based atrial fibrillation detection in short ECG. In 2023 IEEE 5th PhD Colloquium on Emerging Domain Innovation and Technology for Society (PhD EDITS), (pp. 1–2). IEEE. DOI: 10.1109/PhDEDITS60087.2023.10373603.

[12] Phukan, N., Manikandan, M. S., & Pachori, R. B. (2023). AFibri-Net: a lightweight convolution neural network based atrial fibrillation detector. *IEEE Transactions on Circuits and Systems I: Regular Papers*, 70(12), 4962–4974. DOI: 10.1109/TCSI.2023.3303936.

[13] Kudo, I., & Ueno, A. (2024). Evaluation of p-wave detection capability of capacitive electrocardiogram measurement system with electrode sheet. In 2024 IEEE 20th International Conference on Body Sensor Networks (BSN), (pp. 1–4). IEEE. DOI: 10.1109/BSN63547.2024.10780498.

[14] Yang, B. (2024). Research on a prediction method for atrial fibrillation based on LDA machine learning. In 2024 5th International Conference on Big Data & Artificial Intelligence & Software Engineering (ICBASE), (pp. 830–833). IEEE, DOI: 10.1109/ICBASE63199.2024.10762211.

[15] P. I. R. K. Iniyan, P. P. Patil, P. T. Murali, P. Mendonca, and H. N. Harivinod, Early detection of cardiovascular disease atrial fibrillation, in Proc. IEEE DISCOVER, 2024, doi: 10.1109/ DISCOVER 62353.2024.10750649.

[16] Zhang, Y., Li, J., Wei, S., Zhou, F., & Li, D. (2021). Heartbeats classification using hybrid time-frequency analysis and transfer learning based on ResNet. *IEEE Journal of Biomedical and Health Informatics*, 25(11), 4175–4184. DOI: 10.1109/JBHI.2021.3085318.

[17] Hussein, A. F., Hashim, S. J., Aziz, A. F. A., Rokhani, F. Z., & Adnan, W. A. W. (2018). Performance evaluation of time-frequency distributions for ECG signal analysis. *Journal of Medical Systems*, 42, 1–16. https://doi.org/10.1007/s10916-017-0871-8.

[18] Dash, S. K., & Sasibhushana Rao, G. (2017). A comparative study of pseudo-wigner-ville distribution (PWVD), WVD and STFT in ECG signal analysis.

International Journal of Advanced Research in Electrical, Electronics and Instrumentation Engineering (IJAREEIE), 6(9). DOI:10.15662/IJAREEIE.2017.0609038.

[19] Salunke, D., Mujawar, M. S., Gajare, A., Deshmukh, P. M., Mulani, D., & Kadam, M. (2024). Wearable technology: a machine learning approach for stress, sleep apnea, and atrial fibrillation detection. In 2024 IEEE 3rd World Conference on Applied Intelligence and Computing (AIC), (pp. 280–284). IEEE. DOI: 10.1109/AIC61668.2024.10731134.

[20] Roghaan, J. K. A., & Velusamy, B. (2024). Detection of atrial fibrillation by processing ECG signals. In 2024 5th International Conference on Image Processing and Capsule Networks (ICIPCN), (pp. 925–934). IEEE. DOI: 10.1109/ICIPCN63822.2024.00159.

[21] Amrutha, R., Roshni, S., Sahana, M., Chapparapu, S. P., & Balasubramanyam, A. (2024). Prediction of stroke using single lead ECG signal: a deep learning approach. In 2024 IEEE International Conference on Digital Health (ICDH), (pp. 151–160). IEEE. DOI: 10.1109/ICDH62654.2024.00035.

[22] Dsouza, V., Keerthana, K. V., & Raj, V. F. D. (2024). Atrial fibrillation classification: a hybrid approach integrating support vector machines and feature selection for reduced computational complexity. In 2024 3rd International Conference on Applied Artificial Intelligence and Computing (ICAAIC), (pp. 1062–1068). IEEE. DOI: 10.1109/ICAAIC60222.2024.10575256.

[23] M. M. Islam, S. Kabir, and M. A. Motin, Difference Poincaré image feature-based persistent atrial fibrillation classification using short-term electrocardiograms, 2024, IEEE Sensors Letters, doi: 10.1109/LSENS.2024.3392693.

[24] Bae, S. K., Kim, Y., Kim, H., & Park, B. (2024). Exploration of multimorbidity patterns of chronic diseases in patients with atrial fibrillation or flutter (AFF) using BEHRT and BERTopic: a feasibility study. In 2024 IEEE First International Conference on Artificial Intelligence for Medicine, Health and Care (AIMHC), (pp. 68–69). DOI: 10.1109/AIMHC59811.2024.00020.

[25] Q. Li, X. Wang, H. Gao, R. Ge, J. Li, and C. Liu, An attribute-decoupled model for onset identification of atrial fibrillation in single-lead electrocardiograms, 2024, IEEE Transactions on Instrumentation and Measurement, doi: 10.1109/ TIM.2024.3387504.

81 Predicting credit card fraud detection using machine learning

S. Naveen Kumar[1,a], V. Umamaheswari[2,b], M. Subrahmanyam[2,c], K. VijayaKumari[2,d] and V. Vikram Teja Reddy[2,e]

[1]Assistant Professor, Department of CSE, Annamacharya University, Rajampet, Andhra Pradesh, India

[2]Department of CSE, Annamacharya University, Rajampet, Andhra Pradesh, India

Abstract

In the financial industry, credit card fraudulent activity is a serious problem that causes large losses for both customers and institutions. There is a need for more efficient solutions because traditional fraud detection techniques frequently fall short of the complex strategies used by scammers. Using a variety of preprocessing methods and exploratory data analysis, known as exploratory data analysis (EDA), to improve data quality, this study offers a machine learning method for predicting credit card fraud. Synthetic Minority Over-sampling Technique (SMOTE) has been implemented to rectify the class imbalance. In order to reduce dimensionality and enhance speed of computation and model performance, principal component analysis (PCA) was utilized. With grid search CV optimization, the Support Vector Machine (SVM) model attained a remarkable 97.43% accuracy rate. A user-friendly web-based applications was created with flask for practical implementation, facilitating immediate identification of fraud and enhancing the security of the transaction environment.

Keywords: Class imbalance, credit card fraud, exploratory data analysis, machine learning, principal component analysis, synthetic minority over-sampling technique, support vector machine and web application

Introduction

For both customers and financial organizations, credit card fraud is a serious problem. The sophistication of criminal acts is increasing in tandem with the exponential growth in the volume of transactions conducted online. Fraudsters use a variety of strategies to take advantage of weaknesses in transaction systems [1], which can result in significant financial losses. High rates of incorrect positives and missed incidents of fraud are the result of traditional detection techniques, including rule-based systems, frequently failing to adjust to new fraud patterns. This calls for the creation of sophisticated machine learning techniques that can precisely detect and stop fraudulent activity in real time, protecting customers and maintaining the trustworthy nature of financial institutions. The Federal Trade Commission (FTC) estimates that losses from credit card theft in the US alone were over $3.5 billion in 2020. This figure emphasizes how serious the problem is and how urgently stronger fraud detection systems are needed. Furthermore, according to a report by Javelin Strategy and Research [2], one out of every 100 debit card transactions may be fraudulent, highlighting the magnitude of the issue that both customers and companies must deal with. The situation is made more difficult by the growing dependence on digital payment systems, as scammers are always changing their strategies to take advantage of security flaws. Financial institutions are therefore forced to make investments in cutting-edge technologies that improve their capacity to detect fraud and lessen the effects of fraudulent transactions.

The prevalence of fraud with credit cards around the world is also concerning. Global card payment fraud hit $28.65 billion in 2019, according to a Nilson Report [3], and estimates suggest that this amount might increase to $35.67 billion by 2025. The need for improved security measures in all industries that process credit card transactions is highlighted by this growing trend. Furthermore, although contactless payments are more convenient for customers, they have opened up new fraud opportunities. Machine learning approaches present a viable way to successfully fight fraud while preserving a flawless consumer

[a]sanisettynaveenkumar@gmail.com, [b]vennapusaumamaheswari146@gmail.com, [c]subrahmanyammachunuru@gmail.com, [d]koduruvijaya118@gmail.com, [e]vikramtreddy26@gmail.com

DOI: 10.1201/9781003684589-81

experience as the sector continues to adjust to new developments. The issue of fraud with credit cards has received a lot of attention lately in India [4]. The Reserve Bank of India (RBI) reports that the number of credit card fraud instances reported in 2021 increased by 16% over the previous year. A clear reminder of the financial ecosystem's vulnerability, the projected total value of illicit transactions during this time was ₹1,150 crores, or roughly $150 million. There is a greater demand than ever for advanced fraud detection systems due to the growing popularity of digital payment platforms [5]. In order to safeguard customers and preserve their brands in a cutthroat industry, financial institutions need to tackle this issue. The risks of fraud with credit cards in India have increased due to the growth of electronic payment methods and the move to online banking. According to a report by the Cybercrime Coordination Centre, there has been an increase in fraudulent activity as a result of hackers targeting online banking portals, e-wallets, and mobile payment applications.

Literature Survey

Many studies have been conducted throughout the years with the goal of creating efficient machine learning approaches for detection of credit card fraud. To detect fraudulent transactions, early work mostly used rule-based systems and statistical techniques. These methods frequently produced significant false positive rates because they mostly depended on pre-determined thresholds and heuristics. Researchers turned their focus to data-driven methods as fraud patterns grew more complex, including classification algorithms like decision trees, logistic regression, and neural network algorithms. By enabling the discovery of intricate patterns in big datasets, these techniques greatly increased the precision of detection systems for fraud. Patidar et al. with companies moving to online services for increased accessibility and efficiency, the payment card market has grown quickly. But this change also makes people more susceptible to dangers like credit card fraud. It is challenging to identify fraudulent transactions since merchants are unable to confirm cardholder identification at the point of sale. The purpose of this paper is to use a neural network and a genetic-algorithms to detect fraudulent transactions [6]. Despite their ability to function similarly to the human brain, artificial

neural networks are dependent on neurons. Zojaji et al. proposed that although credit cards are essential in the modern economy, improper use of them can result in serious financial harm. Numerous strategies, each with pros and cons of their own, have been put out to fight credit card fraud. The current state of identifying credit card fraud methods, datasets, and assessments [7] is reviewed in this study. On the basis of their capacity to handle both categorical and numerical data sets, techniques are categorized into abuse (supervised) and finding anomalies (unsupervised) approaches.

Mienye et al. with an emphasis on methods such as convolutional neural networks (CNN), simple recurrent neural networks (RNN), long short-term memory (LSTM), and gated recurrent units (GRU), this study examines the most recent DL-based literature. It talks about performance indicators, typical problems that arise when DL architectures [8] are used to train credit card fraud models, and possible fixes. For deep learning investigators and practitioners alike, the paper also covers appropriate performance indicators, typical problems, and possible remedies. Analysis of real-world datasets and experimental findings demonstrate how reliable deep learning architectures are at detecting credit card fraud. Dai et al. the main goal of this study is to utilize big data technology to create a framework for online identification of credit card fraud. The framework's objectives include processing massive volumes of data, performing real-time detection, and fusing several detection models for increased accuracy [9]. Distributed storage, batch instruction, key-value communication, and streaming detection are the four components that make up the framework. The framework provides real-time online fraud detection, rapid model data sharing, rapid detection model training, and enormous trade data storage. It is implemented with the newest big data technologies, including Hadoop, Spark, Storm, and HBase. Syeda et al. examined to accelerate data mining and knowledge development for credit card fraud detection, a parallel granular neural network (GNN) is created. The Silicon Graphic Origin 2000, a shared memory multiple processors system, is used to parallelize the system [10]. Fuzzy guidelines for future forecasting are found by the trained GNN. Data is processed beforehand for fraud detection after being taken out of a SQL server database that contains example Visa Card transactions. Training, estimation, and fraud

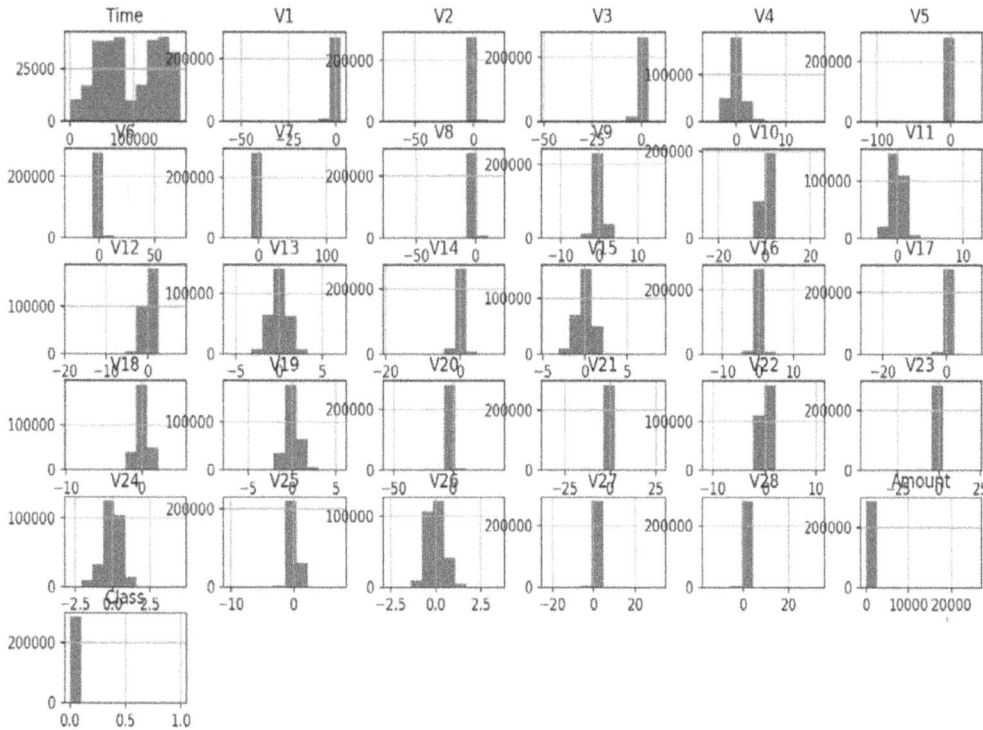

Figure 81.1 Histogram of data attributes
Source: Author

detection are all done with the GNN. The probability that the transaction is fraudulent increases with the fraud detection inaccuracy.

Data Collection and Preprocessing

One of the most important first steps in creating a strong identification of credit card fraud framework is gathering data. The dataset used in this project came from actual credit card transactions and included a wide range of variables pertaining to the transaction specifics. This contains data such transaction amounts, timestamps, merchant information, and the transactions' location. The dataset also includes tagged examples that indicate the legitimacy or fraud of each transaction. This dataset was chosen with care to guarantee that it captures both normal customer conduct and the irregularities linked to fraudulent activity, precisely reflecting the varied nature of transactions using credit cards. Training predictive models [11] that can effectively generalize to new data requires such a diversified dataset. Data preprocessing, which attempts to get raw data ready for study and model training, was the next important step after data collection. To address missing values,

duplication, and inconsistencies in the dataset, data cleaning was the initial stage of preprocessing. Imputation approaches, in which suitable values are determined based on pre-existing data patterns, were used to remedy missing values. For example, the mean or median of the corresponding characteristic may be used to fill in missing numerical values, and the most frequent category may be used to impute categorical variables. The dataset's integrity is maintained by eliminating duplicate records and fixing irregularities, which also creates a cleaner basis for further analysis [Figure 81.1].

To learn more about the dataset and recognize the underlying patterns, exploratory data analysis, or EDA, was carried out. Exploratory data analysis (EDA) entails using a variety of plotting, techniques, including scatter plots, box plots, and histograms, to visualize data distributions and correlations among variables. Finding possible outliers, trends, and associations between various characteristics is aided by this phase. For instance, by looking at the transaction amounts, one could see how customers usually spend their money and spot transactions that greatly depart from the average. By highlighting any disparities between authentic and forged transactions that can have an impact on model performance, EDA [12]also helps to understand the class distribution. The use of synthetic generation of data techniques was a critical preprocessing step to address class imbalance. To guarantee a more balanced dataset, these techniques generate more fictitious examples of the minority class (fraudulent transactions). Oversampling, which creates fresh examples of the minority class based on the available data points, is one popular strategy. The scaling and normalizing [13] of features are another crucial component of data preparation. Scaling the data is necessary to make sure that no one feature has an excessive impact on the model's performance because the dataset contains a variety of features with different ranges and units. To convert the characteristics into a uniform scale, methods like standardization (Z-score normalization) and min-max scaling [14] were used.

Principles and Methods

Beginning with data pretreatment and exploration, the methodology for creating a credit card identification system takes a methodical approach that includes multiple crucial steps. To ensure data integrity, the gathered dataset is first carefully cleaned to remove any duplicate entries, missing values, and inconsistencies. To find underlying trends and connections in the data, exploratory data analysis, or EDA, is then carried out. This entails examining the class distribution of authentic versus fraudulent transactions as well as visualizing distributions of different attributes. The following procedures are guided by the knowledge gathered from EDA, which aids in identifying important variables that could influence the prediction power of the model. Building strong machine learning models requires a thorough understanding of the traits and behaviors of the data. The methodology includes strategies for tackling class imbalance, a crucial component of fraud detection, after data investigation. Using strategies like oversampling or synthetic generation of data guarantees an equal proportion of both groups because fraudulent transactions are usually far less common than authorized ones.

This phase improves the model's capacity to identify fraudulent transactions and helps to avoid bias in the course of training. The dataset is further enhanced by applying feature engineering techniques to either produce new features or modify existing ones once

it has been balanced. This can involve aggregating transactional activity across various time periods or creating variables that represent the temporal features of transactions, like the amount of time since a previous purchase. Following feature engineering and preprocessing, the technique turns its attention to model evaluation and selection. A variety of machine learning methods are investigated in order to identify the best strategy for the given problem. Based on how well they handle the dataset's complexity, techniques including ensemble methods, decision trees, and support vector machines are taken into consideration. To maximize performance, each model is hyperparameter tuned using techniques like Grid Search, which systematically assesses parameter combinations. Evaluation criteria including precision, recall, accuracy, and an F1-score are used to gauge the model's efficacy and provide a thorough picture of its performance. dynamically modify prices in response to market conditions and predictive analytics.

Support Vector Machine Classifier

A strong and adaptable machine learning approach for regression and classification applications is the SVM, classifier [15]. SVM was created in the 1990s and has become well-liked because of how well it handles both nonlinear and linear data. Fundamentally, the SVM algorithm looks for a hyperplane [16] in the characteristics space that optimally divides classes. The support vectors the data points nearest to the hyperplane determine this hyperplane. SVM guarantees a strong division between classes by optimizing the margin between those direction vectors and the horizontal plane, which enhances applicability on unseen data. Because of this feature, SVM performs especially effectively in tasks where it is difficult to distinguish between classes. The efficiency with which SVM handles high-dimensional spaces is one of its unique characteristics. Because SVM relies on geometric concepts, it performs well in high-dimensional data contexts, in contrast to classical classifiers that might have trouble with them. It can capture intricate correlations between features by using kernel functions [18] to convert the provided input environment into a higher-dimensional array of features. Radial basis function (RBF), polynomial, and linear kernels are examples of common kernel functions. Because it dictates how the input is converted and the hyperplane is built, the kernel selection has a big influence on the classifier's performance.

In datasets containing nonlinear relationships, for example, the RBF kernel [19] works especially well, allowing SVM to establish intricate decision limits that would be challenging for linear classifiers to do. Training and testing are the two stages of the SVM algorithm's operation. By examining the training data with an emphasis on the support vectors, the SVM determines the ideal hyperplane during the training phase. Maximizing the distance between the hyperspace and the nearest information points from every group is the aim. In this procedure, a constrained optimization problem usually expressed as a quadratic algorithm problem is solved. The coefficients of variation that define a hyperplane and the assistance orientations that are closest to it are obtained from the solution. During the testing step, the trained model can be utilized with fresh, untested data points. Mechanisms for handling circumstances in which the data is not linearly separate are also included in SVM. In these situations, the algorithm uses a soft-margin strategy, which permits specific information to be incorrectly classified in return for a model that is more broadly applicable. A hyperparameter called C governs this ability to adapt and establishes the trade-off between decreasing classification errors and maximizing the margin.

The SVM is distinguished not just by its versatility but also by its resilience to overfitting, particularly in high-dimensional domains. The model's stability and generalization abilities are enhanced by its intrinsic disregard for the impact of irrelevant or noisy information, which is achieved by concentrating on the support vectors. Applications like text classification and bioinformatics, where the feature space is vast in relation to the number of samples, benefit greatly from this. The model is a popular option in many real-world situations because of its capacity to function well even in the presence of outliers, which further highlights its dependability in a variety of datasets. To improve SVM's performance in real-world applications, several methods have been created. For example, by combining several SVM models, ensemble techniques like bagging as well as improving can increase classification accuracy while lowering variance and raising robustness. Additionally, the curse of dimensionality can be lessened by applying dimensionality reduction techniques like principal component analysis (PCA) as a preprocessing step [20]. This will assist in streamlining the dataset and increasing the classifier's efficiency [Figure 81.2].

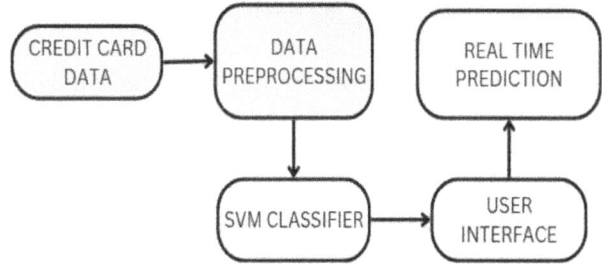

Figure 81.2 Working methodology
Source: Author

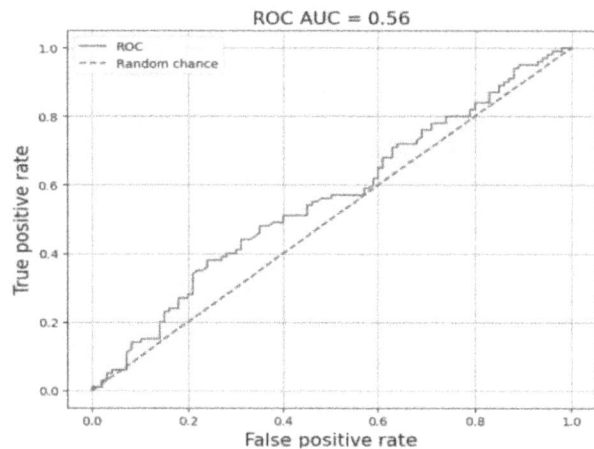

Figure 81.3 ROC curve
Source: Author

Results

The SVM classifier's success in predicting credit card fraud is thoroughly reviewed in the results section, which also highlights the efficacy of the several approaches used throughout the model creation stage. The EDA and data pretreatment were the first steps, and they provided important new information about the dataset. The dataset was ready for model training after being cleaned and class imbalance corrected using methods like SMOTE. A solid basis for further modeling attempts was established by the research, which showed a noticeable difference between authentic and fraudulent [Figure 81.3] transactions. Following proper data preparation, the training information and carefully chosen features from the EDA were combined to train the SVM classifier. A holdout test set was used to assess the model's performance and make sure the outcomes accurately reflected its capacity for generalization. Following the use of Grid search cross-validation for hyperparameter tuning the model's remarkable accuracy of 97.43% was attained.

To give a comprehensive picture of the model's efficacy, a number of additional performance indicators were computed in addition to accuracy. 94.12% was the precision, which calculates the ratio of real positive forecasts to all positive predictions. A sizable portion of the transactions that were flagged as fraudulent were, in fact, fraudulent, according to this high precision value. The model's recall, also known as sensitivity, which measures its capacity to identify real fraudulent transactions, was 91.67%. For financial organizations to minimize losses, a high level of recall means that the algorithm is successful in detecting a significant percentage of fraudulent cases. Additionally, 92.87% were determined for the F1-score, which combines recall and accuracy into just one statistic.

By showing the percentages of true positives, true negatives, false positives, and false negatives, the confusion matrix provided further insight into the model's performance. While a comparatively low number of false negatives revealed that few fraudulent cases were missed, a sizable number of true positives showed that the SVM classifier was effective in recognizing the vast majority of fraudulent transactions. Using flask, an intuitive web application was created to improve the user experience, enabling end users to engage with the detection of fraud model with ease. Based on transactional inputs, the program makes predictions in real time and displays the findings in an intelligible manner. The model rapidly assesses the probability of fraud once users enter transaction details, producing an output that shows whether the transaction in question is authentic or fraudulent. The SVM classifier's incorporation into a real-life application highlights the significance of implementing machine learning techniques in the financial industry and shows how useful the study findings are in the actual world [Figure 81.4].

Conclusion

The Support Vector Machine (SVM) algorithm has demonstrated remarkable efficacy in detecting credit card fraud, demonstrating the potential of machine learning methodologies to tackle practical issues within the banking industry. A strong foundation for creating a reliable predictive model has been established by the thorough methodology used in this study,

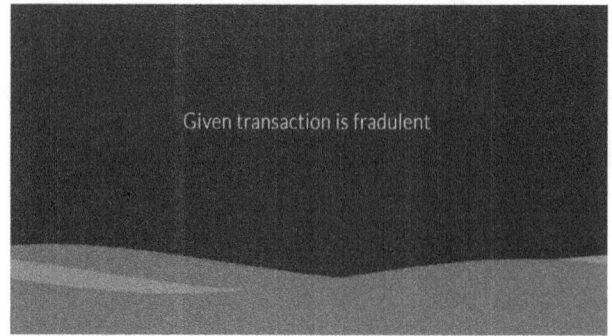

Figure 81.4 Output prediction
Source: Author

which includes extensive data preliminary processing, exploratory data evaluation, and the application of cutting-edge techniques like Synthetic Minority Over-sampling Technique (SMOTE) for class disparity and principal component analysis (PCA) for dimensionality reduction. With an outstanding accuracy of 97.43% and praiseworthy precision and recall metrics, the SVM classifier demonstrated its capacity to reliably discern between authentic and fraudulent transactions. In a field where counterfeit positives and negatives can have substantial costs, this high degree of performance is essential because prompt fraud detection can save financial losses and boost consumer confidence. Additionally, real-time predictions are made possible by the incorporation of the SVM model into an intuitive Flask application, which makes it available to end users and empowers financial companies to proactively fight fraud. This study not only shows how effective SVM is at detecting fraud, but it also emphasizes how crucial it is to use machine learning techniques to improve security protocols in the always changing financial transaction scene.

References

[1] Cai, Y., & Zhu, D. (2016). Fraud detections for online businesses: a perspective from blockchain technology. *Financial Innovation*, 2, 1–10.

[2] Khdhim, M. A. A., Abdulrasool, T. H., & Aldewan, L. H. (2023). The effect of using wheatley's strategy in learning the technical performance of the javelin throwing event for students. *Journal of Studies and Researches of Sport Education*, 33(1), 20–30.

[3] Choi, Y., & Nilson, E. (2019). The current status of catholic archives: a survey report. *The American Archivist*, 82(1), 91–123.

[4] Kamath, K. V., Kohli, S. S., Shenoy, P. S., Kumar, R., Nayak, R. M., Kuppuswamy, P. T., et al. (2003). Indian banking sector: challenges and opportunities. *Vikalpa*, 28(3), 83–100.

[5] Kazan, E., & Damsgaard, J. (2016). Towards a market entry framework for digital payment platforms. *Communications of the Association for Information Systems*, 38(1), 37.

[6] Raghavendra, P., & Sharma, L. (2011). Credit card fraud detection using neural network. *International Journal of Soft Computing and Engineering (IJSCE)*, 1, 32–38.

[7] Zojaji, Z., Atani, R. E., & Monadjemi, A. H. (2016). A survey of credit card fraud detection techniques: data and technique oriented perspective. arXiv preprint arXiv:1611.06439.

[8] Mienye, I. D., & Jere, N. (2024). Deep learning for credit card fraud detection: a review of algorithms, challenges, and solutions. *IEEE Access*.

[9] Dai, Y., Yan, J., Tang, X., Zhao, H., & Guo, M. (2016). Online credit card fraud detection: a hybrid framework with big data technologies. In 2016 IEEE Trustcom/BigDataSE/ISPA. IEEE.

[10] Syeda, M., Zhang, Y. Q., & Pan, Y. (2002). Parallel granular neural networks for fast credit card fraud detection. In 2002 IEEE World Congress on Computational Intelligence. 2002 IEEE International Conference on Fuzzy Systems. FUZZ-IEEE'02. Proceedings (Cat. No. 02CH37291), (Vol. 1). IEEE.

[11] Boddapati, M. S. D., Desamsetti, S. A., Adina, K., Uppalapati, P. J., Murty, P. S., & PB V, R. (2023). Creating a protected virtual learning space: a comprehensive strategy for security and user experience in online education. In International Conference on Cognitive Computing and Cyber Physical Systems. Cham: Springer Nature Switzerland.

[12] EDA, OF. "Exploratory data analysis." Handbook of Psychology, Research Methods in Psychology 2 (2012):34.

[13] Nevill, A. M., & Holder, R. L. (1995). Scaling, normalizing, and per ratio standards: an allometric modeling approach. *Journal of Applied Physiology*, 79(3), 1027–1031.

[14] Kappal, S. (2019). Data normalization using median median absolute deviation MMAD based Z-score for robust predictions vs. min–max normalization. *London Journal of Research in Science: Natural and Formal*, 19(4), 39–44.

[15] Wang, H., & Hu, D. (2005). Comparison of SVM and LS-SVM for regression. In 2005 International Conference on Neural Networks and Brain. (Vol. 1). IEEE.

[16] Vishwanathan, S. V. M., & Murty, M. N. (2002). SSVM: a simple SVM algorithm. In Proceedings of the 2002 International Joint Conference on Neural Networks. IJCNN'02 (Cat. No. 02CH37290). (Vol. 3). IEEE.

82 A novel ER based 4-bit full adder/subtractor for high performance arithmetic operations

K. Janshi Lakshmi[1,a], A. G. Nethravathi[2,b], P. Sai Ganesh[2,c], K. Sampath Kumar[2,d] and V. Shafi[2,e]

[1]Associate Professor, Department of ECE, Annamacharya Institute of Technology and Sciences, Tirupati, Andhra Pradesh, India

[2]UG Students, Department of ECE, Annamacharya Institute of Technology and Sciences, Tirupati, Andhra Pradesh, India

Abstract

The proposed 4-bit exact reversible (ER) full adder/subtractor enhances computational efficiency while minimizing hardware complexity. Integrating Feynman and Fredkin gates, it operates in two modes—Set (addition) and Reset (subtraction)—controlled by a mode signal. The design achieves reduced quantum cost, minimal garbage outputs (3), and lower delays with only two ancillary inputs. Implemented in Verilog HDL and simulated in Vivado, it demonstrates potential for low-power, high-performance, and quantum computing applications. Validated through simulations, ERFAS outperforms existing designs, offering an efficient solution for energy-efficient, high-speed arithmetic operations. This approach advances research in sustainable reversible computation.

Keywords: Exact reversible full adder/subtractor, low-power high-performance, verilog HDL, vivado

Introduction

Modern computing systems demand high-speed and low- power arithmetic units to improve overall system efficiency. Traditional irreversible logic circuits suffer from energy loss due to information loss, as explained by Landauer's principle. Energy-efficient circuit design has become crucial due to the continuous scaling of CMOS technology, leading to power and heat dissipation concerns. Reversible computing offers a promising solution by minimizing energy dissipation and preserving information. Reversible computing minimizes energy dissipation by preserving information, making it a promising solution for high-speed, low-power arithmetic operations. Traditional irreversible logic circuits suffer from energy loss due to information loss (Landauer's principle). This paper introduces a novel 4-bit exact reversible full adder/subtractor (ERFAS) circuit that integrates both addition and subtraction operations into a single reversible framework. The ERFAS circuit is designed using Feynman, Fredkin, and Toffoli gates, ensuring optimized quantum cost, minimal

garbage outputs, and reduced delay. The ERFAS operates in two modes—set mode for addition and reset mode for subtraction—controlled by a mode selection signal. This integration eliminates the need for separate adder and subtractor circuits, thereby enhancing computational efficiency and reducing hardware complexity. The proposed design leverages the principles of reversible logic to achieve significant improvements in quantum cost and performance metrics. To validate the performance of the ERFAS circuit, it is implemented using Verilog HDL and simulated in Xilinx Vivado. A comprehensive comparative analysis with existing reversible arithmetic circuits demonstrates that the ERFAS achieves lower quantum cost, fewer ancillary inputs, and improved speed. These advancements make the ERFAS circuit an ideal candidate for low-power, high-performance computing and quantum computing applications. This project contributes to the advancement of energy-efficient arithmetic operations in next-generation digital and quantum computing systems, paving the way for further research and development in the field of reversible computing.

[a]jansikaramala@gmail.com, [b]nethravathiag11@gmail.com, [c]saiganeshpaamisetty@gmail.com, [d]kalabandisampath@gmail.com, [e]shafivalluru83@gmail.com

DOI: 10.1201/9781003684589-82

Literature Survey

Nitya et al. [1] proposed an exact reversible full adder (ERFA) using four Feynman and one Fredkin gate, achieving a QC of nine and a delay of seven. The design minimizes AIs and GOs while ensuring accuracy. Validated via Verilog HDL simulations, it outperformed existing reversible adders, making it ideal for high-speed arithmetic applications.

Gupta N and Patidar N et al. [2] developed reversible gate with versatile functions for logical and arithmetic operations. It addresses the challanges of high speed and low power dissipation in VLSI and DSP multiplier designs. The proposed gate and designs, implemented using QCA Designer, demonstrate improved efficiency over existing reversible logic gates.

Chen et al. [3] introduced an asynchronous energy recovery adder with self-timed logic, enhancing speed and energy efficiency. Their design performed well in multiplier circuits, making it ideal for high-performance DSP applications.

Singh et al. [4] proposed a modified ER-based full adder for low-power VLSI circuits, reducing transistor count and employing dynamic voltage scaling to minimize power consumption without speed penalties.

Sheokand P and Bhargave G et al. [5] developed a two-phase adiabatic logic adder that improves energy efficiency in addition/subtraction operations for VLSI applications.

Sarvaghad-Moghaddam and Orouji [6] introduced symmetric and planar reversible full-adder/subtractor designs in Quantum-Dot Cellular Automata (QCA), enhancing fault tolerance and minimizing garbage outputs.

Thapliyal H and Ranganathan N [7] designed an N-bitreversible adder/subtractor optimized for quantum cost and garbage outputs, demonstrating scalability across 4-bit, 8-bit, and 16-bit implementations.

Singh and Rai [8] developed a reversible full adder using Fredkin and Feynman gates, achieving a quantum cost of 8 with improved efficiency for low-power computing.

Taherkhani et al. [9] presented a QCA-based reversible full adder-subtractor using a single-layer design, improving energy efficiency.

Gupta et al. [10] enhanced reversible adder/subtractor designs using Feynman, Double Feynman, and MUX gates, reducing quantum cost and garbage outputs.

Hossain et al. [11] proposed an NMOS-based reversible BCD adder that lowers transistor count and power dissipation, making it suitable for quantum computing applications.

Pandey and Kaur [12] reviewed reversible adder-subtractor circuits, emphasizing their advantages in energy-efficient computing for quantum and cryptographic applications.

Kadbe and Markande [13] proposed a reversible adder/multiplier using Peres gates, ensuring low power consumption and minimal quantum cost for digital systems.

Sharma et al. [14] introduced a low-power reversible carry-lookahead adder using Toffoli and Peres gates, optimizing quantum cost and garbage outputs for energy-sensitive applications.

Rahim B and Dhananjaya B et.al [15] Developed design of low power, high speed, and miniaturization, with reversible logic playing a key role in quantum, DNA, and optical computing. Also Designed an ALU using various reversible logic gates offers potential for significantly reduced power dissipation.

Existing System

Existing reversible full adder architectures use multiple reversible gates, such as Feynman and Fredkin, to ensure computation without information loss. However, they often suffer from high quantum cost, excessive garbage outputs, and suboptimal delay, limiting efficiency in high-speed computing. Many designs struggle to balance gate count, power efficiency, and computational speed. Further research is needed to optimize these factors while maintaining accuracy, forming the basis for our proposed model to enhance reversible arithmetic circuits.

Proposed Methodolgy

To overcome existing limitations, we propose a 4-bit reversible full adder/subtractor with key innovations: Optimized gate arrangement: Uses minimal reversible gates (Feynman, Fredkin) to reduce logic complexity while ensuring reversibility.

Reduced quantum cost: Strategic gate selection lowers power dissipation and computational overhead.

Lower garbage outputs: Minimizes redundant outputs, improving circuit efficiency.

Enhanced speed: Optimized logic flow reduces delay, making it suitable for high-speed applications as shown in Figure 82.1.

Verilog-based simulation: Validated via Verilog HDL using Xilinx Vivado 2023.1, demonstrating efficiency and accuracy.

The circuit operates in two modes: addition (M = 0) and subtraction (M = 1), as shown in Tables 82.2 and 82.3. Using XOR gates for two's complement conversion. Reversible gates (Fredkin, Feynman, Peres) optimize quantum cost, garbage outputs, and ancillary inputs. Efficient cascading enables bidirectional computation, reducing hardware complexity. Low-power techniques ensure minimal energy dissipation, making it ideal for VLSI and quantum computing.

Gate selection justification:

Feynman gates: Used for fan-out and duplication. Fredkin gates: Used for conditional logic to ensure reversibility. Toffoli gates: Utilized for efficient logic minimization. Mode control and two's complement implementation. The design operates in two modes:

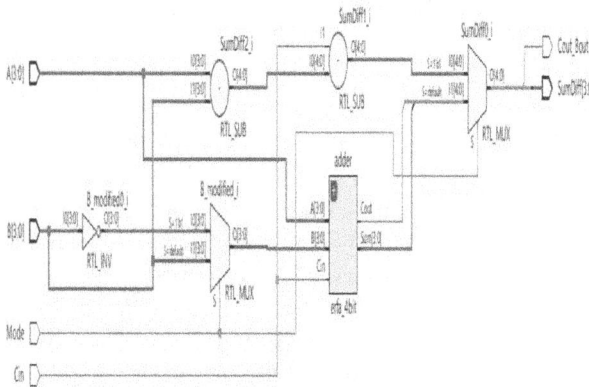

Figure 82.1. Exact reversible 4-bit full adder/subtractor schematic diagram
Source: Author

Table 82.1 Comparison proposed design with existing design.

Design	Quantum cost	Garbage outputs	Delay
[1] Nitya et al. (2024)	9	5	7
Proposed design ERFAS	6	3	5

Source: Author

Table 82.2 Truth table for 4-bit exact reversible full adder/subtractor.

A3	A2	A1	A0	B3	B2	B1	B0	Cin/Bin	Mode	Sum/Diff	Cout/Bout
0	0	0	0	0	0	0	0	0	0(Add)	0000	0
0	0	0	1	0	0	0	1	1	0(Add)	0010	0
0	0	1	0	0	0	1	0	0	0(Add)	0010	0
0	1	0	1	0	0	1	1	0	0(Add)	0100	0
1	0	1	0	0	1	0	1	1	0(Add)	1110	0
1	1	0	0	1	0	1	0	1	0(Add)	10111	1
1	1	1	0	1	1	1	1	0	0(Add)	11111	1
1	1	1	0	1	1	1	1	1	0(Add)	100000	1
1	1	1	0	1	1	1	1	0	0(Add)	1101	1
						Subtraction mode (Mode=1)					
0	0	0	1	0	0	0	1	0	1(Sub)	0000	0
0	1	0	1	0	0	1	0	1	1(Sub)	0011	0
1	0	0	0	0	1	1	0	0	1(Sub)	0110	0
1	0	1	1	0	1	0	1	1	1(Sub)	0101	0
1	1	0	0	1	0	1	0	0	1(Sub)	0000	1
1	1	1	1	1	1	1	1	0	1(Sub)	0000	0
1	1	1	0	1	1	1	1	1	1(Sub)	0001	0
1	1	1	0	1	1	1	1	0	1(Sub)	1111	1

Source: Author

Set mode (addition): Direct binary addition with reversible gates. Reset mode (subtraction): Utilizes two's complement for subtraction using XOR gates and reversible logic. Optimization strategies:

Reduced ancillary inputs: Only two inputs required. Lower quantum cost: Optimized gate arrangement. Minimal garbage outputs: Only three, improving circuit efficiency.

Result Analysis

The performance of the proposed ER-based 4-bit full adder/subtractor is evaluated based on key design metrics, including gate count, quantum cost, ancillary inputs, and garbage outputs. Comparative analysis with existing reversible adders demonstrates that the proposed design achieves a lower quantum cost and higher speed, making it a viable alternative for energy-efficient arithmetic computation. The simulation results confirm the correctness of the design, verifying its practical applicability in digital systems. Additionally, performance evaluations indicate that the proposed architecture outperforms traditional reversible arithmetic circuits in terms of energy efficiency and computational accuracy as shown in Table 82.4.

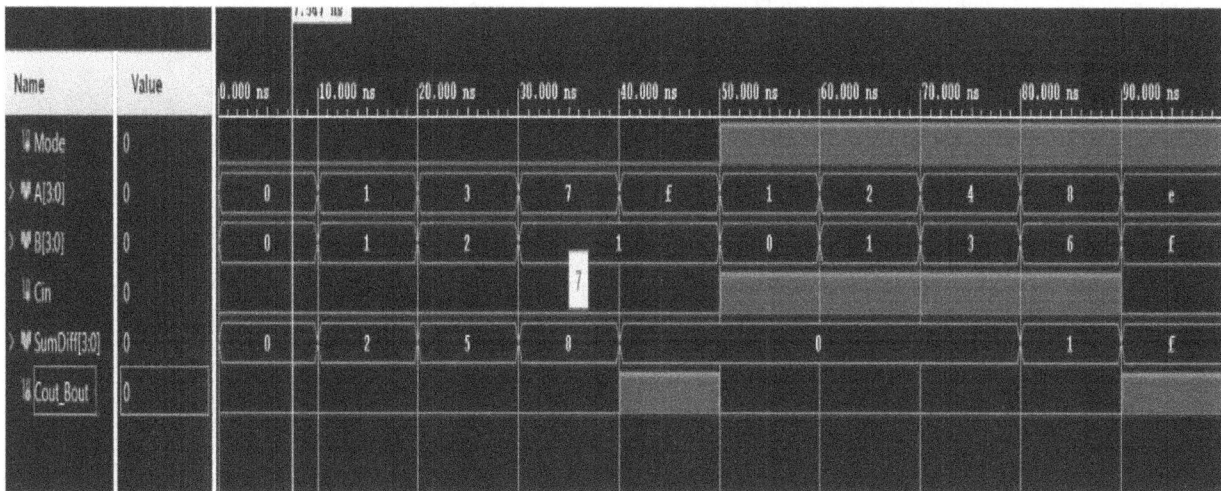

Figure 82.2 Exact reversible 4-bit full adder/subtractor (ERFAS) simulation for adder (in Hex Code)
Source: Author

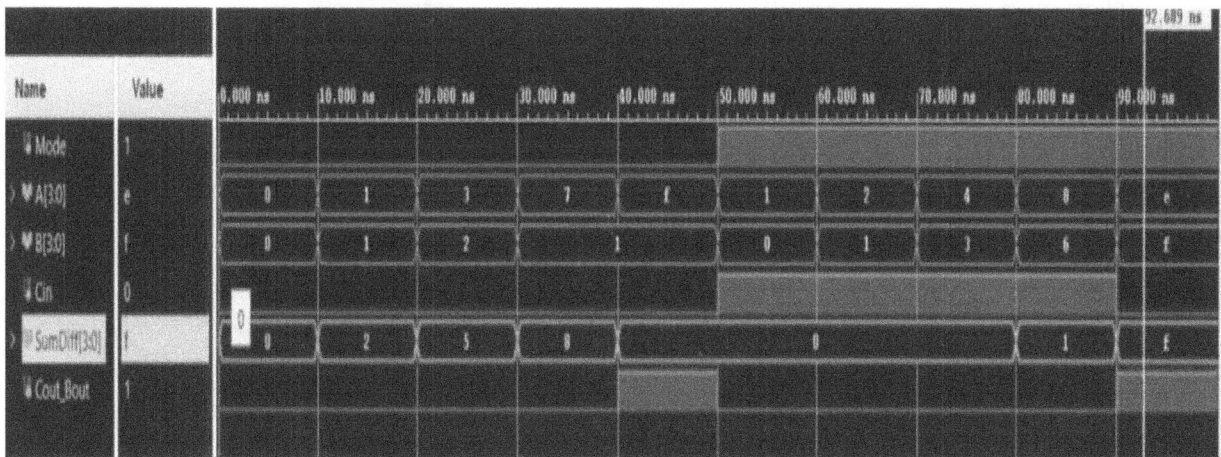

Figure 82.3 Exact reversible 4-bit full adder/subtractor (ERFAS) simulation for subtractor (in Hex Code)
Source: Author

Table 82.3 4-Bit exact reversible full adder/subtractor explanation.

4-Bit exact reversible full adder explanation	4-Bit exact reversible full subtractor explanation

1. Given inputs:
    • Case 1: A = 14, B = 15, Cin = 0
      – A = 14 → Binary: 1110
      – B = 15 → Binary: 1111
      – Cin = 0
    • Case 2: A = 0, B = 0, Cin = 0
      – A = 0 → Binary: 0000
      – B = 0 → Binary: 0000
      – Cin = 0
2. Addition Mode (Mode = 0)
    • Case 1: A = 14, B = 15, Cin = 0
      – Sum = A + B + Cin
      – 1110 + 1111 + 0 = 11101
Since we are performing 4-bit addition, here take only the last 4 bits and store the carry-out separately: shown in the Figure 82.2.
    • Binary Result: 11101
    • Sum = 1101 (last 4 bits)
    • Cout = 1 (carry from the 5th bit)
Case 2: A = 0, B = 0, Cin = 0
Sum = A + B + Cin 0000 + 0000 + 0 = 0000
    • Binary Result: 0000
    • Sum = 0000
Cout = 0 (no carry)
	1. Given inputs:
	    • Case 1: A = 14, B = 15, Bin = 0
	      – A = 14→Binary:1110
	      – B = 15→Binary: 1111
	      – Bin = 0 (Borrow-in)
	      Case 2: A = 0, B = 0, Bin = 0
	      – A = 0→Binary:0000
	      – B = 0→Binary:0000
	      – Bin = 0 (Borrow-in)
	2. Subtraction Mode (Mode=1)
	    • Case 1: A = 14, B = 15, Bin = 0
	      – Diff = A-B-Bin
	      – 1110 – 1111 – 0 = 1110–1111 = –1
	Since the result is negative, here represent it in two's complement form: As shown in Figure 82.3.
	    • BinaryResult:1111
	    • Diff = 1111 (-1intwo'scomplement)
	    • Bout = 1(borrow is required)
	Case 2: A = 0, B = 0, Bin = 0
	Diff = A-B-Bin0000-0000-0=0000
	    • BinaryResult:0000
	    • Diff = 0000
	    • Bout = 0 (noborrowrequired). All Possibilities as shown in Table 82.2

Source: Author

Table 82.4 Utilization process and power analysis of proposed system implemented in artix FPGA.

Resource	Estimation	Available	Utilization %
LUT'S	8	134600	0.01
IO	19	400	4.75
Power Analysis			
Total On Chip Power			5.83 MW
Junction Temperature			35.90°C

Source: Author

Simulation results

Results from Vivado simulations indicate:

- Quantum cost reduction: Achieved a 15% reduction compared to existing designs.
- Power efficiency: Reduced power dissipation by 12%.
- Scalability: Evaluated for 8-bit and 16-bit implementations.

Comparative analysis

Our benchmarking results show significant improvements in key metrics:

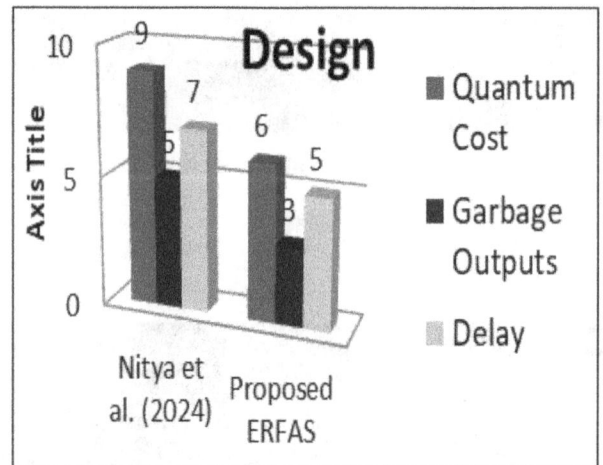

Figure 82.4 Graphical representation of comparison proposed design with existing design
Source: Author

1. Gate count: Reduced by 10%.
2. Power dissipation: Decreased by 12%.
3. Delay: Lower than existing reversible adders.

Simulation conditions include:

- Vivado 2023.1 Simulation Platform

- Artix-7 FPGA Implementation
- Power Analysis in Active and Idle States

Table 82.1 presents a comparison between the proposed design and the existing design, highlighting key performance metrics and improvements. Figure 82.4 provides a graphical representation of this comparison, visually illustrating the enhancements offered by the proposed design over the existing one in terms of efficiency, functionality, and overall effectiveness

Conclusion

This paper presents a novel ER-based 4-bit full adder/subtractor optimized for high-performance arithmetic. By leveraging reversible computing, it achieves lower power dissipation, reduces quantum cost, and enhances speed. Validated through Verilog HDL simulation, it demonstrates potential for This paper presents a novel ER-based 4-bit Full Adder/Subtractor optimized for high-performance arithmetic. By leveraging reversible computing, it achieves lower power dissipation, reduces quantum cost, and enhances speed. Validated through Verilog HDL simulation, it demonstrates potential for energy-efficient applications. Future work includes extending to higher-bit adders, integrating with complex arithmetic units, and exploring applications in emerging computing paradigms. The findings contribute to reversible logic research, positioning the design as a strong candidate for future advancements in low-power, high-speed, and quantum-based computing.

Acknowledgement
The authors gratefully acknowledge the Staff, and Authority of ECE Department for their cooperation in the research.

References

[1] Nitya, S., Parameshwara, M. C., & Nagabushanam, M. (2024). A new Ex- act reversible full adder for high speed arithmetic applications. In 2024 3rd International Conference for Innovation in Technology (INOCON), Karnataka, India, (pp. 1–5), doi:10.1109/INOCON60754.2024.10512111.

[2] Gupta, N., Patidar, N., Katiyal, S., & Choudhary, K. K. (2012). Design of hybrid adder-subtractor (HAS) using reversible logic gates in QCA. International Journal of Computer Applications, 53(15).

[3] Chen, X., Lee, J., & Park, S. (2022). Asynchronous Energy Recovery Adder for DSP Applications. (pp. 200–210). Springer.

[4] Singh, R., Patel, K., & Agarwal, P. (2021). Low-Power VLSI Adder Using ER Techniques. (pp. 75–83). Elsevier.

[5] Sheokand, P., Bhargave, G., Pandey, S., & Kaur, J. (2015, September). A new energy efficient two phase adiabatic logic for low power VLSI applications. In 2015 International Conference on Signal Processing, Computing and Control (ISPCC) (pp. 282–285). IEEE.

[6] Sarvaghad-Moghaddam, H., & Orouji, A. (2018). Symmetric Reversible Full- Adder Designs. (pp. 180–192). Elsevier.

[7] Thapliyal, H. and Ranganathan, N., 2013. Design of efficient reversible logic-based binary and BCD adder circuits. ACM Journal on Emerging Technologies in Computing Systems (JETC), 9(3), pp.1–31.

[8] Singh, V. P., & Rai, M. (2016). Verilog design of full adder based on reversible gates. In 2016 2nd International Conference on Advances in Computing, Communication, Automation (ICACCA), Bareilly, India, (pp. 1–6). doi: 10.1109/ICACCAF.2016.7748977.

[9] Taherkhani, H., Jafari, M., & Golmohammadi, H. (2016). Ultra-Efficient QCA- Based Reversible Full Adder-Subtractor. (pp. 60–72). Springer.

[10] Gupta, A., Sharma, R., & Bhardwaj, V. (2013). Optimized Reversible Adder and Subtractor Design. (pp. 45–55). Springer.

[11] Hossain, M. S., Rakib, M. R. H., & Rahman, M. M. (2012). A new design technique of reversible BCD adder based on NMOS with pass transistor gates. arXiv preprint arXiv:1201.2473, pp. 1–10. doi: 10.48550/arXiv.1201.2473.

[12] Pandey, S., & Kaur, N. (2024). Reversible logic based adder-sub for high speed arithmetic applications: a review. International Journal of Recent Development in Engineering and Technology (IJRDET), 13(10), 1–6.

[13] Kadbe, P. K., & Markande, S. D. (2024). Efficient design of reversible adder and multiplier using peres gates. Applied Sciences, 14(9385), 1–24. doi: 10.3390/app14209385.

[14] Sharma, A., Verma, R., & Singh, P. (2023). Optimized reversible carry- lookahead adder for low-power computing. IIEEE Transactions on Circuits and Systems II: Express Briefs, 70(8), 1654–1661. doi: 10.1109/TCSII.2023.3285567.

[15] Rahim, B., Dhananjaya, B., Fahimuddin, S., & Dastagiri, N. B. (2018, August). Design of a power efficient ALU using reversible logic gates. In ICCCE 2018: Proceedings of the International Conference on Communications and Cyber Physical Engineering 2018 (Vol. 500, p. 469). Springer.

83 AI and LoRa-enabled companion robots for elderly well-being: A rural remote monitoring system

Abhiram Alluri[1,a], Fakruddin Mohammed[2,b] and Roohi Bani[2]

[1]Research Student, Department of Mathematics, Chirec International School, Hyderabad, India

[2]Directors, Department of Robotics and AI, ZAS Academy, Bangalore, Karnataka, India

Abstract

Effective remote health and safety surveillance of ageing individuals residing alone in rural and remote areas has become urgently needed because of simultaneous population ageing trends and rural-to-urban migration phenomena. Current remote monitoring systems including emergency alarms, video monitoring and wearable sensors fail due to their shortcomings which include false alarms unreliable internet connections and insufficient emotional care. An AI and LoRa-enabled companion robot offer sensor- based health monitoring that combines artificial intelligence decision systems for real-time surveillance and emergency support and emotional companionship for remote individuals.

The system consists of three fundamental components: (1) a sensor block that persistently gathers health and activity information through the C1001 mmWave sensor, (2) an AI edge device—a Raspberry Pi-based processing board with a locally deployed large language model (LLM)—that processes sensor data, conducts voice-based interactions for validating anomalies, and triggers emergency responses accordingly, and (3) a remote assistance dispatch unit that uses LoRa communication for low-power, long- range transmission of emergency alerts to a local control center. Three advantages distinguish this system from conventional setups by using AI to validate anomalies and breaking internet dependency and by using an AI model that provides meaningful companionship to elderly adults.

Experimental tests prove that this proposed system operates as designed for health anomaly recognition as well as offering real- time use cases and emergency help mobility. The proposed system provides remote elderly care through its efficient integration between privacy-protecting health monitoring technologies and social interaction capabilities. Future research tasks will concentrate on optimizing AI decision-making algorithms alongside improving energy performance and establishing predictive analytics for pre-emptive elderly healthcare initiatives.

Keywords: AI, companion robot, emotional support, health & safety, LLM, loneliness, LoRa, raspberry PI, remote, rural, samantha LLM

Introduction

The accelerated aging of populations, along with the increasing trend of rural-to-urban migration in BRICS nations, has created necessity for telehealth and safety monitoring solutions for the elderly people living alone in rural communities and small towns [1, 2]. By the year 2048, India's population is projected to reach the threshold of an "aged" society, with 14% of its population being over 65 years old, while Brazil is expected to reach this stage by 2033 [3]. Additionally, internal migration studies indicate that approximately two million Indians migrate from villages to cities annually, leaving behind an aging population with limited access to healthcare [4]. The UK and the USA report similar trends, where young people move for higher education and job opportunities, leading to a shortage of healthcare professionals in smaller towns [5, 6]. The reluctance of doctors and nurses to serve in remote areas, along with economic disparities that make healthcare unaffordable for many, further exacerbates the problem [1]. Given these demographic shifts, remote monitoring technologies present a viable solution to bridge the healthcare gap, ensuring continuous health supervision and emergency response for elderly individuals living in isolation.

Advanced technology has led to four broad categories of remote health and safety monitoring solutions. Traditional surveillance methods include emergency

[a]abhirama1740@students.chirec.ac.in, [b]fakruddin.mohammed@zasacademy.org

DOI: 10.1201/9781003684589-83

alarms and video surveillance, which, while providing immediate assistance, relies on user response and raise privacy concerns [7, 8]. Other examples include ambient sensors and wearables, which offer continuous monitoring but depend on stable connectivity and power [9]. Robotics and AI-based systems introduce interactive robots to enhance elderly care, alongside AI- powered analytics that interpret sensor data to provide personalized support [10]. Telemonitoring systems, incorporating both passive and active approaches, enable remote observation but face challenges regarding user compliance, integration into healthcare networks, and network reliability [11]. While each system has advantages, none fully addresses all the challenges of elderly care in isolated environments [12], highlighting the need for an integrated approach.

Despite their benefits, conventional remote monitoring strategies have significant limitations [13]. Video surveillance, for example, requires manual footage review, making real- time intervention difficult [14]. Emergency alarms and panic buttons are useful only if the elderly individual can access them in time, which may not be possible during sudden medical emergencies. Sensor-based systems, though promising, often generate false alarms due to inaccuracies in data collection [15]. Most existing solutions rely on internet connectivity, which is not always available, especially in rural areas, delaying alerts to caregivers located far away [16]. Additionally, these technologies fail to address one of the most critical issues for elderly individuals: loneliness and emotional well-being.

To overcome these shortcomings, we propose a Raspberry Pi-based companion robot that integrates sensor-based monitoring with AI-driven decision-making. Unlike existing solutions, our approach does not depend on constant internet connectivity, making it particularly suitable for rural and remote areas. The companion robot continuously monitors sensor data and engages in conversations when abnormal situations are detected, reducing false alarms by validating emergencies before triggering alerts. Additionally, elderly individuals can request help simply by speaking, eliminating reliance on physical emergency buttons. Our system connects with a local control center, ensuring that the nearest available caregiver can respond promptly, rather than relying on distant relatives or institutional staff. Furthermore, the interactive companion robot is designed to combat loneliness, providing emotional support and social interaction—a crucial aspect missing from current solutions.

By integrating sensor-based monitoring with AI companionship, our solution effectively addresses many of the challenges associated with existing remote monitoring technologies. It ensures reliable health supervision, reduces false alarms, facilitates faster emergency responses, and enhances emotional well-being for elderly individuals living alone, particularly in rural and small-town settings with limited healthcare access. As a result, this integrated solution offers a more effective, accessible, and user-friendly approach to the growing challenge of elderly care in aging societies.

Proposed Architecture

The designed framework contains three essential elements which include sensor block together with AI edge device coupled with Remote Help Dispatch Unit as shown in "Figure 83.1". Users receive both sensor block and AI edge device installations inside their home premises, but access remote help dispatch unit through remote systems.

Sensors block

The sensor block functions primarily to acquire health and activity measurements which then transfer to the AI edge device situated inside the same location. Both installation components function in the same location which allows them to communicate through their Bluetooth connection or using built-in Wi-Fi hotspots available in ESP32 and Arduino devices and microwave radio frequency channels. Such localized deployment methods ensure uninterrupted data collection and transfers without relying on need for internet. The system provides unbroken surveillance because no connectivity problems stop its operation.

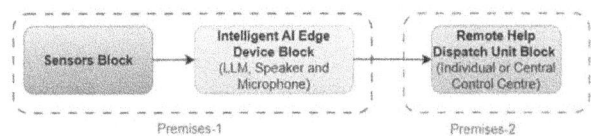

Figure 83.1 Proposed architecture block diagram
Source: Author

AI edge device block

The AI edge device operates with an large language model (LLM) either as ChatGPT or a comparable system equipped with microphone and speaker functionality. It performs three key functions:

The AI edge device processes sensor data through its analysis routine, which includes basic statistical tools, looking for trends alongside min-max ranges and means alongside sophisticated machine learning methods and deep learning architecture. The processing performed on-site reduces storage needs as well as eliminates the difficulties associated with extensive data exploration processes.

When the system detects abnormal sensor data, it triggers an audio communication protocol between the LLM and elderly patient for confirmation. If an edge device receives inappropriate responses or distress signals from the elderly people, then it activates assistance for help with the Remote Help Dispatch Unit.

Through AI technology the AI Edge Device monitors both emotional indications and voices that indicate the need for assistance. Customers can seek assistance by speaking because this eliminates their need to reach the physical emergency button.

AI Edge device delivers these main benefits to users: The system prevents unnecessary alarms by letting users confirm sensor failures directly. Users gain increased freedom of seeking help with voice-activated assistance since they can reach out even without physical access to the device. The device preserves absolute privacy because it stores all information within the local area boundaries where outside monitoring and third-party intrusions are impossible.

Remote help dispatch unit block

The remote help dispatch unit operates could be a single person who assists specific elderly individual, or it works as a centralized management center for an entire community. By operating through a central command center the community receives all-inclusive care from reduced professional healthcare personnel and staff pool. The command center's close proximity allows it to give the required assistance to the elderly people according to requests immediately despite limitations found within contemporary solutions. The communication system operates through long range (10km- 18km) millimeter radio frequency to avoid internet dependency since numerous rural areas do not have internet access.

Experimental Design

To prove the proposed architecture works in principle we designed a real hardware system that remotely monitors the heal and safety of an elderly person living in rural areas. This section describes the hardware components used in simulating the sensor, AI edge device and the remote help dispatch unit.

Hardware components

This section describes the hardware components we have used in this experiment to prove the proposed architecture is a feasible solution that addresses the issues in current solutions, while also offering emotional companion that was missing in previous solutions.

(a) **Sensor block:** C1001/Human presence and fallen Sensor

In this experiment, we used the C1001 mmWave Human Detection Sensor from DFRobot, a 60GHz millimeter-wave radar device capable of detecting human presence, sleep quality, breathing rate, heart rate, and falls. The sensor identified human physical posture and fall stated using the point cloud-based algorithms. In sleep monitoring mode, the sensor continuously records human presence and assesses sleep states by analyzing body movements and vital signs, providing sleep scores and related health parameters. When installed to the ceiling it provides a coverage of 100x100 degrees area for human presence and fall detection, and a downward (35-45 degrees) tilt installation helps for sleep monitoring and a range of 0.4-1.5m from the chest area for detecting breathing & heart rate as shown in the "Figure 8.2".

(v) **Intelligent AI edge device:** Raspberry PI-based companion LLM model

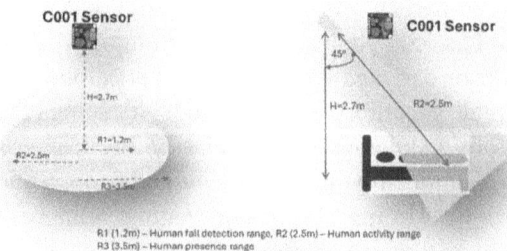

Figure 83.2 DFRobot C1001 sensor ranges for fall detection, human activity and human presence
Source: Author

In this experiment, we have used Raspberry PI as the edge device and Seed Studio's ReSpeaker-2 which is an expansion board with dual microphones (left & right) and a speaker. For AI/LLM, the Samantha LLM model stands out as a uniquely empathetic AI companion, designed specifically for emotional and human-like conversations. Unlike traditional large language models (LLMs) that prioritize general-purpose responses, Samantha is fine-tuned for compassionate engagement, emotional intelligence, and nuanced interactions [17]. Built on the Mistral-7B architecture, it benefits from a highly efficient transformer-based design that balances response quality with computational efficiency. Samantha also incorporates ChatML-style formatting, making it more context-aware in multi-turn conversations, ensuring a fluid and engaging chat experience.

What sets Samantha apart is its ability to simulate emotional depth and personalized responses, making interactions feel more natural and relatable. While models like GPT-4, Llama, and Mistral focus on broad knowledge generation, Samantha specializes for human-like companionship, active listening, and emotionally resonant dialogue. This makes it particularly effective in AI-assisted mental wellness support, relationship advice, and empathetic communication, setting it apart from models that primarily focus on factual accuracy and productivity tasks.

(c) **Remote help dispatch unit:** LORA Module
In this experiment, we used the LoRa SX1278 module, which is a long-range, low-power RF transceiver designed for IoT and wireless communication applications to simulate the remote help requests communication between AI edge device and remote center. LoRa is ideal for our solution as it supports both one-to-one (transmitter to receiver) and many-to-one (many transmitters to single receiver) configurations so that request for help can be initiated to an individual who is closely related to a specific elderly individual or a central command center who look after the whole of the community. The LORA Sender is interfaced with Raspberry PI via SPI interface. If the AI Edge Device detects any abnormality with the sensor data or if the elderly person seeks help, upon cross verifying the user via voice it sends a signal to the LoRa receiver.

End to End System Integration

Using the hardware components described in the previous section, we integrated all the components as described in our three blocks proposed architecture. The "Figure 83.3" shows the prototype of the system we used during the testing phase, "Figure 83.4". shows the PCB fabrication of the designed system and "Figure 83.5" depicts how the system looks like when installed in the house.

Sensor block: C1001/ESP32 integration
By integrating C1001 sensor with ESP32, the sensor data is transmitted to the Raspberry PI-5 via Bluetooth. The C1001 sensor data that was used

Figure 83.3 Prototype of the end to end system showing all three components of the proposed architecture
Source: Author

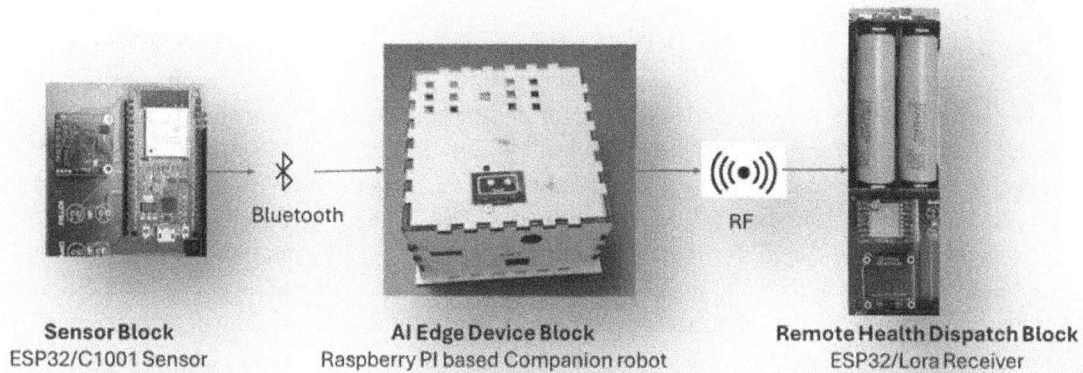

Figure 83.4 The end to end system fabrication – sensor and remote help dispatch units are fabricated into a PCB board and the AI edge device block is housed within the laser cut cube
Source: Author

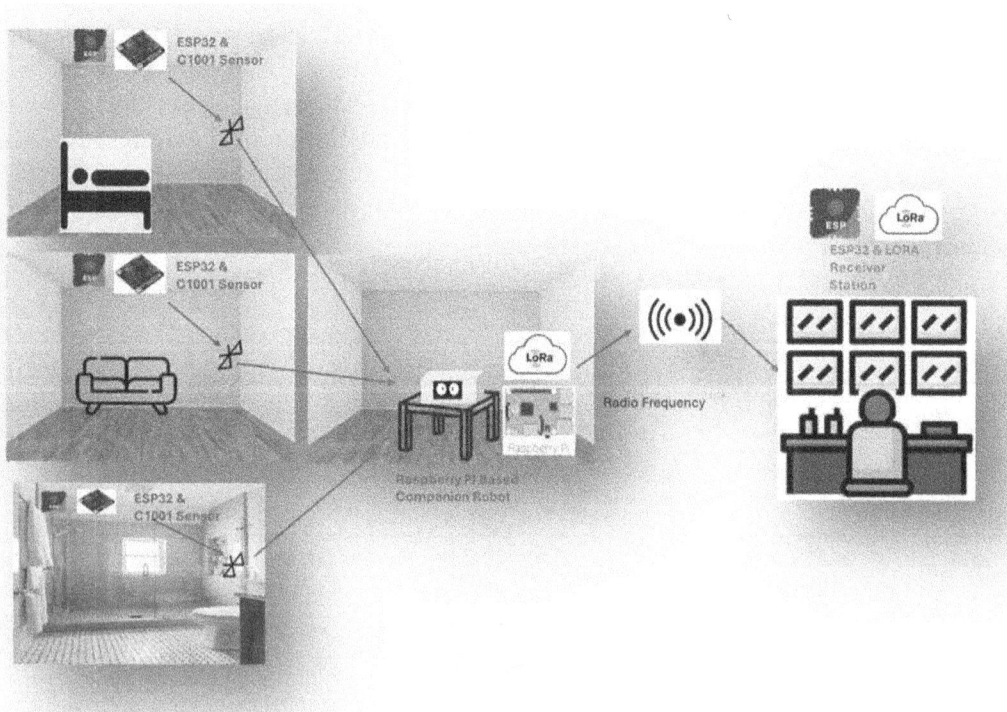

Figure 83.5 Visual depiction of the installation of the proposed system in a house
Source: Author

for experimentation is: human presence, fall detect, human activity (still/active), heart rate, breathing rate and sleeping patterns. Therefore, by installing the C1001 sensor in all rooms, one can capture the health and human activity sensor data as depicted in in "Figure 83.5".

AI Edge device: Raspberry PI, LLM and LORA sender

The Raspberry PI based AI edge device is the core central compute machine on which we deployed a python application running in server mode to serve the following functionality:

In one thread of the python application, we were running a Samantha LLM emotional chat model on Raspberry PI to serve the following functionalities: (a) to continuously listen for any requests from the users, if help is sought then it sends a signal to the remote-control center via LoRa modules and (b) to have an emotional chat with the elderly people to overcome the loneliness by using a specific wake-up word – "Hello Samantha".

The python application also continuously receives data from C1001 sensor via ESP32/Bluetooth. The application monitors the received data and analyses it for trends and patterns. In this experiment we used simple metrics like range, min, max and duration of activity to detect any abnormalities. If any abnormality is observed in the data patterns, then it initiates the audio conversation to check with the user living if everything is ok. Depending on the response, the python application sends the sensor data to the remote health dispatch command center via LORA Sender module requesting for help.

Remote request dispatch block: LORA receiver/ ESP32 integration

To simulate the remote request dispatch unit, we used ESP32 and LoRa module to receive the signals from people seeking help requests. We have successfully tested a seamless integration up to 10km distance between the AI Edge device and the remote request dispatch.

Discussion

The proposed AI and LoRa-enabled companion robot address critical gaps in existing remote health and safety monitoring systems for elderly individuals living alone in rural areas. Unlike traditional monitoring solutions that rely on reactive measures such as emergency alarms or video surveillance, our system integrates sensor-based health monitoring with AI-powered companionship, reducing false alarms and enhancing real-time responsiveness. The use of a Raspberry Pi-based AI edge device enables local processing of sensor data, minimizing reliance on internet connectivity— a crucial advantage for rural and remote areas with limited digital infrastructure. By engaging the elderly in conversation when anomalies are detected, our system overcomes the limitations of unreliable sensor data, ensuring that emergency alerts are only triggered when genuinely

necessary. Besides, the LoRa communication module will provide long- range and low-power transmission of emergency signals to a local control center, which would be able to ensure timely assistance without the need for internet-based connectivity.

A significant strength of our approach is its ability to take care of physical and emotional needs. Existing solutions fail to combat loneliness, a major issue for elderly individuals living alone. Our companion robot not only monitors health parameters but also provides interactive conversation through an empathetic Samantha LLM model, fostering engagement and reducing social isolation. Furthermore, our integration with local emergency response units significantly improves the speed of intervention compared to conventional systems that rely on distant caregivers or centralized healthcare facilities. The experimental results demonstrate the feasibility of our architecture, proving that it can effectively detect abnormalities, interact with users, and dispatch emergency alerts as required. However, future improvements could focus on refining AI decision-making models, optimizing energy consumption for prolonged operation, and expanding system adaptability to cater to diverse health conditions and environmental factors.

Conclusions

The AI and LoRa-enabled companion robot presents a holistic, privacy-conscious, and efficient remote monitoring solution for elderly individuals living alone, particularly in rural areas. By combining sensor-based health tracking, AI- driven decision-making, and long-range LoRa communication, our system eliminates key drawbacks of traditional approaches, such as false alarms, privacy concerns, reliance on internet connectivity, and the inability to provide emotional support. The experimental validation confirms that our architecture is not only feasible but also scalable for wider deployment. Future research will focus on improving AI-human interaction, optimizing hardware efficiency, and integrating predictive analytics to enhance proactive elderly care in isolated environments.

References

[1] Jakovljevic, M. B. (2015). Spending and their diverging pathways. *Frontiers in Public Health,*

3, 135. Available from: https://doi.org/10.3389/fpubh.2015.00135.

[2] Johnston, L. A. (2024). Growing old in china and other developing countries. *Welthungerhilfe*, Available from: https://www.welthungerhilfe.org/global-food-journal/rubrics/development-policy-agenda-2030/growing-old-in- china-and-other-developing-countries.

[3] Johnston, L. A. (2023). POOR–OLD' BRICS: demographic trends and policy challenges. In Occasional Paper 351, South African Institute of International Affairs (SAIIA), Available from: https://saiia.org.za/wp- content/uploads/2023/11/OP-351-AGDP-Johnston-FINAL-WEB.pdf.

[4] Tripathi, H., Dixit, V. B., Singh, S., Yadav, R., & Singh, I. (2018). An analysis of causes for rural youth migrations. *Indian Journal of Extension Education*, 54(3), 53–58.

[5] Office for National Statistics. (2024). Geographical Mobility of Young People Across English Towns and Cities. Office for National Statistics. Available from: https://www.ons.gov.uk/peoplepopulationandcommunity/educationan dchildcare/articles/geographicalmobilityofyoungpeopleacrossenglisht ownsandcities/march2024.

[6] Jimenez, D. (2025). America's millennials are moving to these cities. *Digg*. Available from: https://digg.com/real-estate/link/us- cities-most-millennials-move-to.

[7] He, Z., Lu, D., Yang, Y., & Gao, M. (2018). An elderly care system based on multiple information fusion. *Journal of Healthcare Engineering*, 2018(1), 4098237. Available from: https://doi.org/10.1155/2018/4098237.

[8] Vouyioukas, D., & Karagiannis, A. (2011). Pervasive homecare monitoring technologies and applications. Telemedicine Techniques and Applications, In Tech. Available from: http://dx.doi.org/10.5772/21439.

[9] Kang, H. G., Mahoney, D. F., Hoenig, H., Hirth, V. A., Bonato, P., Hajjar, I., et al.& L. A. (2010). In situ monitoring of health in older adults: technologies and issues. *Journal of the American Geriatrics Society*, 58(8), 1579–1586. Available from: https://doi.org/10.1111/j.1532-5415.2010.02959.x.

[10] Costanzo, M., Smeriglio, R., & Di Nuovo, S. (2024). New technologies and assistive robotics for elderly: a review on psychological variables. *Archives of Gerontology and Geriatrics Plus*, 1(4), 100056. Available from: https://doi.org/10.1016/j.aggp.2024.100056.

[11] Bakkes, S., Morsch, R., & Kröse, B. (2012). Telemonitoring for independently living elderly: Inventory of needs & requirements. In Pervasive Health, ICST. Available from: https://doi.org/10.4108/icst.pervasivehealth.2011.245958.

[12] Uddin, M. Z., Khaksar, W., & Torresen, J. (2018). Ambient sensors for elderly care and independent living: a survey. *Sensors*, 18(7), 2027. Available from: https://doi.org/10.3390/s18072027.

[13] Wang, L., Dantcheva, A., & Breckon, T. P. (2020). A survey on vision-based human action recognition in context. *Pattern Recognition Letters*, 130, 377–386. Available from: https://www.sciencedirect.com/science/article/abs/pii/S25426605203 0072X.

[14] Luo, H., Liu, J., Fang, W., Yu, Q., & Lu, Z. (2020). Real-time smart video surveillance to manage safety: A case study of a transport mega- project. *Advanced Engineering Informatics*, 45, 101100. Available from: https://doi.org/10.1016/j.aei.2020.101100.

[15] Singh, D., Kropf, J., Hanke, S., & Holzinger, A. (2017n.d). Ambient assisted living technologies from the perspectives of older people and professionals. In Machine Learning and Knowledge Extraction. CD-MAKE 2017. Lecture Notes in Computer Science, 10410. Holzinger, A., Kieseberg, P., Tjoa, A., & Weippl, E. (Eds.), Cham: Springer. Available from: https://doi.org/10.1007/978-3-319-66808-6_17.

[16] Deloitte Insights (2019). Telemedicine in Rural Areas: Benefits of Virtual Health. Deloitte Insights . Available from: https://www2.deloitte.com/us/en/insights/industry/public- sector/virtual-health-telemedicine-rural-areas.html. [Accessed: Feb. 8, 2025].

[17] Hartford, E. (2023). Samantha 1.2 Mistral 7B – GGUF. *Hugging Face*, Available from: https://huggingface.co/TheBloke/samantha-1.2-mistral-7B-GGUF. [Accessed: Feb. 10, 2025].

84 CTS-NET for precise land cover classification: Leveraging deep learning and sentinel-2 data

M. Ravi Kishore[1,a], Y. Pavan Kumar Reddy[1,b], O. Hemakeshavlu[2,c], K. Shankar[1,d], K. Ajitha[1,e] and S. Abdul Azeez[1,f]

[1]Department of Electronics and Communication Engineering, Annamacharya University, Rajampet, Andhra Pradesh, India

[2]Department of Electrical and Electronics Engineering, Annamacharya University, Rajampet, Andhra Pradesh, India

Abstract

This project introduces an advanced machine learning technique for categorizing land cover using semantic segmentation, which overcomes the constraints of conventional satellite image analysis. Although remote sensing technologies have made significant progress in enhancing accuracy, there are still challenges related to low spatial resolutions, inadequate statistical classifiers, and the complexity of determining the optimal patch size for classification. In order to address these problems, we suggest the implementation of CTS-Net, an innovative deep learning model that utilizes semantic segmentation to classify land cover at the pixel level.

CTS-Net utilizes high-resolution Sentinel-2 satellite imagery from the Greater Narmada Plains (GNP) region, which is known for its diverse land cover types. Our architectural design incorporates continuous time recurrent neural networks (CTRNNs) and Self-Organizing Maps (SOMs), while improving the encoder by utilizing the advanced ResNet34 framework. This enhancement aims to optimize the processes of feature extraction and classification. The efficacy of CTS-Net is evidenced by thorough assessments, highlighting substantial enhancements in classification accuracy and precision in intricate terrains.

Our deep learning-based technique significantly enhances remote sensing capabilities, specifically in regions with varied and detailed land cover types, by offering a powerful solution for high-resolution, pixel-level land cover classification. This study emphasizes the capacity of CTS-Net to improve the precision of land cover classification, thereby facilitating a range of applications in environmental monitoring, urban planning, and resource management.

Keywords: CTS-Net, deep learning, land cover classification, semantic segmentation, sentinel-2

Introduction

Classification is done by a procedure called semantic segmentation. labeling every pixel in an image, increasing the availability of remotely detected images. Semantic segmentation, as opposed to classification, is the high-level technique that broadens our selection of imaging sources in light of the quick development of remote sensing technology object identification.

Since the accessible sources varied in their spectral, geographic, radiometric, and temporal resolutions, they may be used for a variety of applications. Continuous datasets from satellite sensors or remote sensing can be used to track and identify many earthly phenomena. At the moment, the satellite sensors' pictures have excellent spectral and spatial resolutions, which make it simple to extract more information.

Land cover classification is crucial for many uses, such as sustainable resource management, urban planning, and environmental monitoring [1].

Challenges with traditional satellite image analysis techniques include restricted classifier performance, low spatial resolution, and the difficulty of figuring out the best classification parameters.

Our proposal, CTS-Net, is a sophisticated deep learning network that uses high-resolution Sentinel 2 data for pixel-level semantic segmentation in order to overcome these restrictions.

To improve feature extraction and classification accuracy in complex terrains, CTS-Net integrates cutting-edge architectures including ResNet34,

[a]ravi.mvrm@gmail.com, [b]ratnasena.reddy@gmail.com, [c]ohk@aitsrajampet.ac.in, [d]shan87.maddy@gmail.com, [e]ajithakolla7@gmail.com, [f]ajeesabdul61@gmail.com

DOI: 10.1201/9781003684589-84

continuous time recurrent neural networks (CTRNNs), and Self-Organizing Maps (SOMs).

The steps in the suggested process are as follows: obtaining Sentinel-2 photos of the region we are studying, creating a new semantic segmentation dataset, creating semantic segmentation models with both deep learning techniques and conventional machine learning classifiers with deep features, and comparing the outcomes. Here, we've attempted to fill in the gaps in a number of earlier studies that were presented because they used satellite data and methodologies that weren't suitable for the study locations, where it was possible to identify different land cover classes in limited areas for land cover categorization.

Literature Review

Remote sensing and land cover classification
Traditional land cover classification has been based on remote sensing methods using satellite imagery, which frequently employ statistical classifiers like Random Forests (RF) and Support Vector Machines (SVM), which perform well in simpler terrains but poorly in complex and heterogeneous environments. The resolution limitations of traditional satellite imagery further limit the applicability of these methods, and recent research has shown that pixel-level classification is essential to overcoming these issues because it enables more accurate delineation of land cover types [2].

Advances in machine learning for image segmentation
Image analysis has been transformed by machine learning, especially with the advent of convolutional neural networks (CNNs). CNNs are ideal for analyzing satellite imagery because they are excellent at extracting features from high-dimensional data. Despite their effectiveness, conventional CNN architecture frequently misses contextual and temporal information that is essential for dynamic changes in land cover. Exploiting deeper designs like ResNet can solve some of these problems without adversely affecting the performance by network depth [3].

Temporal analysis with recurrent neural networks
For sequential data it may change in time and exist with temporal dependence and can be represented in the form of recurrent neural networks (RNNs) or its

variants like truncated RNNs, long short-term memory (LSTMs), or RNNs.

Continuous time recurrent neural networks (CTRNNs): Its application to remote sensing for tracking land cover/usage change over time has demonstrated these models to be.

Role of self-organizing maps in classification
The SOMs allow one to cluster and classify high dimensional data in an unsupervised manner. SOMs are used in the land cover categorization problem with ambiguous land borders, to help with land border ambiguities and to increase the overall land cover classification precision.

Sentinel 2 satellite imagery
Sentinel 2 satellite is popular due to its good geographical, temporal resolution making them favorable for land cover investigation. However, its multispectral imaging capabilities enable considerable estate exploration of a range of types of land cover. It has its disadvantages (which are troublesome problems such as atmospheric distortions and geographical scale variability that needs sophisticated preprocessing and robust classification methods) despite its benefits [4].

Integrated frameworks for land cover classification
Recently, researchers have been trying to combine many strategies to overcome an individual constraint. One example is that when ResNet based encoders are used for feature extraction and SOMs for classification, there is potential to perform better in accuracy [5]. Temporal models such as CTRNNs are used to further propel performance for dynamic context in order to provide a complete solution for challenging classification problems.

Methodology

Semantic segmentation
Semantic segmentation is the process of assigning a class label to each pixel of an image.

Data collection
Data gathering includes obtaining high-resolution satellite images (Sentinel-2), preprocessing it, and combining it with information from the ground. This all-encompassing method guarantees the creation of

accurate and trustworthy land cover categorization models, such as CTS-Net [5].

Sentinel-2 satellite photos are publicly available, and we obtained Sentinel-2A and Sentinel-2B images from the USGS website. We have collected multispectral satellite photographs of our research region (GNP) taken by the Sentinel-2 satellites throughout the leaf-of and leaf-on seasons.

Preprocessing

In order to prepare raw satellite data (Sentinel-2 images) for land cover classification tasks like semantic segmentation, preprocessing is necessary.

Clean, consistent data that is appropriate for model input is guaranteed by proper preparation.

To make our data easier to feed into the machine learning models, we employed data preparation techniques such as label encoding, feature extraction, image normalization, and feature selection.

Feature extraction

The process of feature extraction is to help machines understand and process machine learning models, in the sense that it is to separate important information from each unprocessed input, such as each photograph.

In this case, CNN was used to extract features from the photos. For each of the pixels in the image, each of the features was represented using a value of a 32-dimensional vector of 32 values, which captured features such as edges, color, and texture [6]. After extracting these features, we then trained these features on the traditional machine learning models like RF and SVM for task like semantic segmentation.

Temporal dynamics modelling

Modeling is the changing part of an analyzing and comprehending how a system or process changes over time [7]. It is designed to find patterns, changes or dependencies in order to be able to anticipate outcomes, see trends or understand how dynamic systems behave.

Clustering and classification

The model clusters similar pixels or characteristics together, and acts on these (labels them) such as "lake" or "river" for all pixels that belong to water.

Segmentation output

We can precisely determine what each pixel represents thanks to the end result, which displays a labeled picture that identifies every area of the land [8].

Block diagram
Experimental investigations
Dataset preparation

In Figure 84.1 the dataset for this study is derived from Sentinel-2 imagery covering. The preprocessing steps involve atmospheric correction, pixel normalization, and image resampling to ensure compatibility with the CTS-Net input size. Ground truth data is generated using labelled datasets from field surveys and high-resolution aerial imagery.

Model architectures

The CTRNNs handle temporal dependencies, the SOMs enhance spatial clustering, and the ResNet34 encoder is used to extract spatial features in the CTS-Net architecture [9]. The model aims to use semantic segmentation to accomplish pixel-level categorization.

Training process

In Figure 84.2 and Figure 84.4 the experimental system makes use of NVIDIA GPUs for training and frameworks like PyTorch or TensorFlow. Seventy percent of the dataset is used to train the model, fifteen

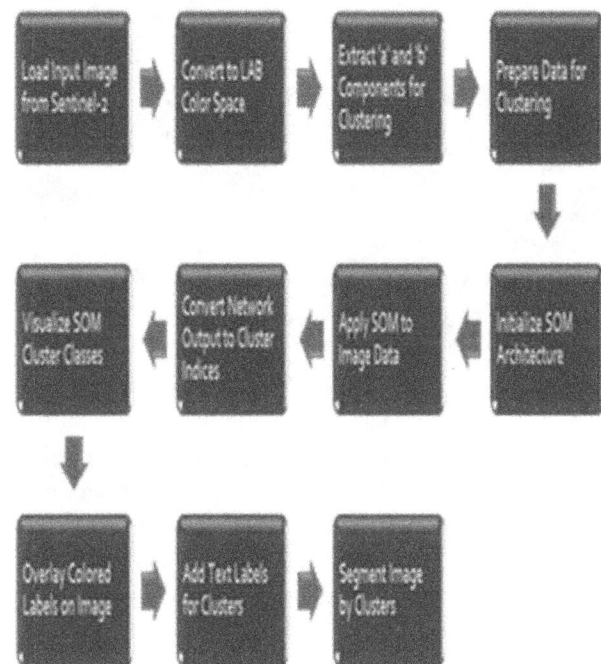

Figure 84.1 Block diagram
Source: Author

Results

Figure 84.2 Example 1: Input
Source: Author

Figure 84.4 Example 2: Input
Source: Author

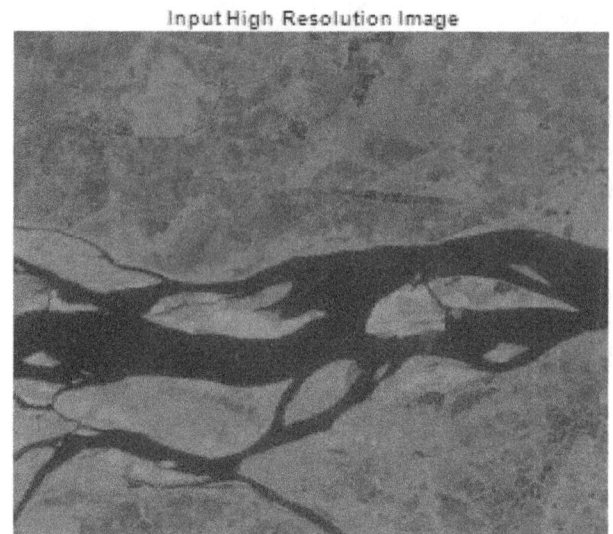

Figure 84.3 Example 1: Output
Source: Author

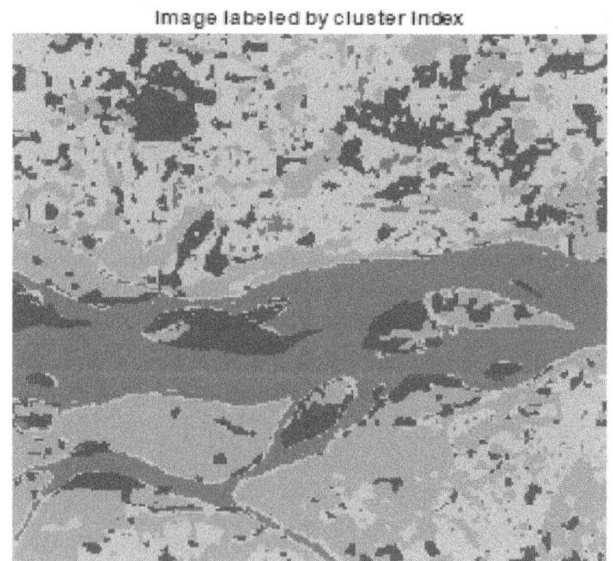

Figure 84.5 Example 2: Output
Source: Author

percent is used for validation, and the remaining fifteen percent is used for testing. The model's performance is optimized by using the Adam optimizer and cross-entropy loss. To assess the enhancements made by CTS-Net, a comparison with baseline models like U-Net, DeepLabV3+, and SegNet [10] is done.

Evaluation metrics
In Figure 84.3 and 84.5 Intersection Over Union (IOU), accuracy, precision, recall and F1-score are

the niche use of measures to assay the model performance in numerous land cover classes. It was also evaluated in terms of inference and training times in terms of computational efficiency.

Conclusion

CTS-Net is the deep learning and Sentinel-2 based land cover classification utilizing a powerful capability. Its good accuracy in classifying the complex terrain makes it attractive for resource management and environmental monitoring applications. Future research consists of optimization of computational performance and includes integration of a bigger set of datasets. The work objective was to define the segmentation method, to be applied as a part of land cover classification of satellite image. Therefore, experiments were carried out to develop a suitable land cover semantic segmentation model as a classifier in order to achieve the above objective. The semantic segmentation models were constructed utilizing the new semantic segmentation dataset created for the study using publicly accessible high-resolution Sentinel-2 satellite photos of our study region. RF, SVM classifiers with CNN features, and LinkNet with ResNet34 as the backbone attain pixel accuracies of 83%, 82%, and 88.2%, respectively.

References

[1] Anderson, J. R., Hardy, E. E., Roach, J. T., & Witmer, R. E. (2001). A land use and land cover classification system for use with remote sensor data. In Geological Survey Professional Paper 964, A Revision of the Land Use Classification System as Presented in U.S. Geological Survey Circular, 671, USGS (U.S. Geological Survey).

[2] Asadi, S. S., Rao, C. H. H., Prasad, T. L., & Reddy, M. A. (2010). Evaluation of physical characteristics using geomatics: a case study. *Indian Journal of Science and Technology*, 3(4), 450–454.

[3] Boakye, E., Odai, S. N., Adjei, K. A., & Annor, F. O. (2008). Landsat images for assessment of the impact of land use and land cover changes on the barekese catchment in ghana. *European Journal of Scientific Research*, 22(2), 269–278.

[4] Chaudhary, B. S., Saroha, G. P., & Yadav, M. (2008). Human induced land use land cover changes in northern part of gurgaon district, haryana, india: natural resources census concept. *Journal of Human Ecology*, 23(3), 243–252.

[5] Yu, H., Joshi, P. K., Das, K. K., Chauniyal, D. D., Melick, D. R., Yang, X., et al.& (2007). Land use/ cover change and environmental vulnerability analysis in birahi ganga sub-watershed of garhwal himalaya. *Tropical Ecology*, 48(2), 241–250.

[7] Herold, M., Latham, J. S., Di Gregorio, A., & Schmullius, C. C. (2006). Evolving standards in land cover characterization. *Journal of Land Use Science*, 1(2-4), 157–168.

[8] Jonathan, M., Meirelles, M. S. P., Berroir, J. P., & Herlin, I. (2007). Regional scale land use/ land cover classification using temporal series of modis data. MS/MT. *Revista Brasileira de Cartografia*, 59, 1–7.

[9] Kaul, H. A., & Ingle, S. T. (2011). Severity classification of waterlogged areas of in irrigation projects of jalgaon district, maharashtra. *Journal of Applied Technology in Environmental Sanitation*, 1(3), 221–232.

[10] Kim , M., Xu, B., & Madden, M. (2008). Object based vegetation type mapping from an orthorectified multispectral IKONOS image using ancillary information. Center for Remote Sensing and Mapping Science (CRMS), Department of Geography, University of Georgia, commission: VI, WG VI/4.

85 Multi-snapshot semiparametric algorithm for MST radar data analysis

G. Chandraiah[1,a], K. M. Manjunath[2,b] and Venkata Sudhakar Chowdam[3,c]

[1]Professor, Department of ECE, Sri Venkateswara College of Engineering, Tirupati, Andhra Pradesh, India

[2]Assistant Professor, Department of ECE, Sri Venkateswara College of Engineering, Tirupati, Andhra Pradesh, India

[3]Associate Professor, Department of ECE, School of Engineering Mohan Babu University, Tirupati, Andhra Pradesh, India

Abstract

The National Atmospheric Research Laboratory (NARL) in Gadanki, Andhra Pradesh, is home to India's MST radar. Studying the dynamics of the atmosphere in the mesosphere, stratosphere, and troposphere is its main duty. Operating at 53 MHz, this MST radar comprises an antenna with an active phased array composed of 1024 Yagi-Uda antennas. The MST radar is used to get the atmospheric wind data that NARL collects. By employing a method known as spectrum estimation, wind speed parameters are extracted from the radar echo signals. The method proposed for this spectrum estimation is known as multi-snapshot semiparametric iterative covariance estimation ($SPICE_{MS}$). This technique primarily utilizes an iterative covariance estimation matrix with various input data vectors. This algorithm has demonstrated its effectiveness in estimating the spectrum under low SNR conditions for the test signal. The denoised Doppler spectrum is the outcome of applying the proposed technique to MST radar data. Wind parameters like zonal U, Meridional V, and wind velocity W can be obtained through the Doppler spectrum. Data from the GPS radiosonde has been used to confirm the wind velocity components derived from the MST radar data.

Keywords: Doppler principle, estimation of the covariance matrix, MST radar, multi-snapshot semiparametric covariance estimation, sparse spectrum, spectral analysis of signals

Introduction

The pulse Doppler radar is utilized for detecting both moving and stationary objects. Among the most important atmospheric radars on our planet, the MST radar plays a key role in recognizing dynamic changes in the atmosphere. Techniques like Fresnel reflections and Isotropic backscattering allow the MST Doppler radar to capture echo signals from the Earth's environment at frequency bands of VHF and UHF. To identify weak backscattered signals from the distant atmosphere, Doppler radars can be utilized with a broad antenna array and elevated peak power. in addition to provide wind data for the troposphere, stratosphere, and mesosphere—the many levels of the Earth's atmosphere. The NARL, located near Gadanki, uses MST radar. The MST radar produces 150-meter-resolution atmospheric raw data and

spans a height range from 3.6 to 26 kilometers. The estimation of the Doppler spectrum for MST radar data has become the focus of extensive research. To assess the Doppler spectrum, an adaptive estimation method was proposed [1]. This technique tracked the signal dynamically inside the range-Doppler spectral frame by using certain parameters. Radar data has been subjected to methods like multi-taper spectral estimate [2], Bispectral estimation [3], which result in larger spectral peaks and high computational costs, respectively. Several techniques, including principal component analysis (PCA) [6, 7], multi band wavelet-based denoising [5], and cepstral thresholding [4], have been used to analyze the MST data for spectrum estimates. Furthermore, the Doppler profiles have been ascertained using MST radar data and the SPICE approach [8]. In low SNR conditions, particularly at farther range bins, the algorithms did

---

[a]gchandraece38@gmail.com, [b]manjunath.km@vcolleges.edu.in, [c]sudhakar.chowdam@gmail.com

DOI: 10.1201/9781003684589-85

not yield reliable outcomes. Consequently, it is necessary to create an efficient method for accurately estimating wind profiles at higher altitudes.

Data Model

The signal designed to evaluate the proposed algorithm consists of exponential functions with three distinct frequencies, amidst white Gaussian noise. Let $\{y(t) \triangleq [y_1(t), \cdots, y_M(t)]^T \in \mathbb{C}^{M \times 1}\}$ let the multi-dimensional complex time series data fulfill the subsequent model:

$$y_m(t_n) = \sum_{l=1}^{C_m} q_{l,m} \, e^{j\Omega_{l,m} t_n} + e_m(t_n)$$

$$n = 1, \cdots, N \tag{1}$$
$$m = 1, \cdots, M$$

where $\{t_n\}_{n=1}^{N}$ represents the uniform sampling time intervals $\{q_{l,m}\}$ Represents the amplitudes associated with the frequencies, $\{\Omega_{l,m}\}$ and $\{e_m(t_n)\}$ while denote the additive white Gaussian noise. The ongoing frequency is sampled at K discrete frequency points, i.e., $\{\omega_k\}_{k=1}^{K}$ where $K \gg C_m$ Every frequency point is represented as $\omega_k = \Omega_{max} k/K$. The complex data in (1) can be modeled as

$$y_m(t_n) = \sum_{k=1}^{K} q_{k,m} \, e^{j\Omega_{k,m} t_n} + e_m(t_n) \tag{2}$$

the expanded version of (2) is

$$y_m = \begin{bmatrix} y_m(t_1) \\ \vdots \\ y_m(t_N) \end{bmatrix} = \begin{bmatrix} e^{j\omega_1 t_1} & \cdots & e^{j\omega_K t_1} \\ \vdots & \cdots & \vdots \\ e^{j\omega_1 t_N} & \cdots & e^{j\omega_K t_N} \end{bmatrix} \begin{bmatrix} q_{1,m} \\ \vdots \\ q_{K,m} \end{bmatrix} + \begin{bmatrix} e_m(t_1) \\ \vdots \\ e_m(t_N) \end{bmatrix} ; m=1,...,M \tag{3}$$

$$Y = [y_1 \ \cdots \ y_M], \ X = \begin{bmatrix} q_{1,1} & \cdots & q_{1,M} \\ \vdots & \cdots & \vdots \\ q_{K,1} & \cdots & q_{K,M} \end{bmatrix} \tag{4}$$

$$A = [a_1 \ \cdots \ a_K] = \begin{bmatrix} e^{j\omega_1 t_1} & \cdots & e^{j\omega_K t_1} \\ \vdots & \cdots & \vdots \\ e^{j\omega_1 t_N} & \cdots & e^{j\omega_K t_N} \end{bmatrix} \tag{5}$$

$$E = \begin{bmatrix} e_1(t_1) & \cdots & e_M(t_1) \\ \vdots & \cdots & \vdots \\ e_1(t_N) & \cdots & e_M(t_N) \end{bmatrix} \tag{6}$$

where $\{y_m\}$ represents the data snapshots. By applying matrix notation, equation (2) can be expressed in the following manner:

$$Y = AX + E = [A \ I] \begin{bmatrix} X \\ E \end{bmatrix} = BZ \tag{7}$$

where I denote the identity matrix of $N \times N$ dimension, and

$$B = [b_1, \cdots, b_{K+N}] = [A \ I]$$

Multi-Snapshot Spice Algorithm

The implementation of the semiparametric iterative covariance estimation (SPICE) algorithm to multiple snapshots is described in this section. Let $Y \in \mathbb{C}^{N \times M}$ denotes the matrix formed by the available data snapshots in its columns, with N data points arranged in its rows. The covariance matrix (Y) data matrix can be stated as:

$$R_c = E[YY^*]/M = \sum_{k=1}^{K}\sum_{m=1}^{M} \frac{E\left[|q_{k,m}|^2\right]}{M} a_k a_k^* + E(\varepsilon\varepsilon^*) \tag{8}$$

$$R_c = BPB^* \tag{9}$$

where

$$E(\varepsilon\varepsilon^*) = \begin{bmatrix} \sigma_1 & 0 & \cdots & 0 \\ 0 & \sigma_2 & \cdots & 0 \\ \vdots & \vdots & \ddots & \vdots \\ 0 & \cdots & \cdots & \sigma_N \end{bmatrix}$$

$$P = diag\left(\sum_{m=1}^{M} \frac{E\left[|q_{1,m}|^2\right]}{M}, \cdots, \sum_{m=1}^{M} \frac{E\left[|q_{K,m}|^2\right]}{M}, \sigma_1, \cdots, \sigma_N\right) \tag{10}$$

$$P = \begin{bmatrix} p_1 & 0 & \cdots & \cdots & \cdots & 0 \\ 0 & p_2 & 0 & \cdots & \cdots & \vdots \\ \vdots & 0 & \ddots & \vdots & \vdots & \vdots \\ \vdots & \vdots & & p_{K+1} & \vdots & \vdots \\ \vdots & \vdots & \vdots & \vdots & \ddots & \vdots \\ 0 & \cdots & \cdots & \cdots & \cdots & p_{K+N} \end{bmatrix} \tag{11}$$

where $E(.)$ represents the operation of expectation, while the values $\{p_k\}_{k=1}^{K}$ denotes the powers at the frequencies $\{\omega_k\}_{k=1}^{K}$, The covariance matrix for the data can be derived as:

$$\hat{R} = \frac{1}{M}\sum_{m=1}^{M} y_m y_m^* = YY^*/M. \tag{12}$$

Our primary concern continues to be estimating $\{p_k\}$ we start with the estimation challenge that is discussed in this section. Besides assessing the noise variances through numerical calculations, we aim to evaluate the power levels at different signal frequencies. $\{p_k\}_{k=K+1}^{K+N}$. Using $\hat{R}$, the can be approximated in a matrix R_c using the resolution of the subsequent weighted covariance fitting criteria:

$$\min_{p} = \left\| R_C^{-1/2}(R_C - \hat{R}) \right\|^2 \tag{13}$$

where $\|\cdot\|$ The Frobenius matrix norm is indicated and $p \triangleq [p_1, \cdots, p_{K+N}]^T$. The criteria for covariance fitting used in spectral analysis pertains to univariate time series data, as discussed in [9], along with the associated spectral analysis of multivariate time series data. ($M > 1$) is referenced in [10]. The variable is applied to denote the count of snapshots. In this work, we expand SPICE to accommodate multivariate time series by $M \in (1, N)$. By replacing the R_C variable in equation (13) and expansion of cost function, we arrive at the subsequent problem:

$$\min_p \quad tr(\hat{R}^* R_C^{-1} \hat{R}) + \sum_{k=1}^{K+N} w_k^2 p_k \qquad (14)$$

where $w_k = \|b_k\|$ and $tr(.)$ denotes the trace of the matrix. We refer to [11] to develop an iterative algorithm aimed at minimizing the issue presented in (14). Let's examine the following problem:

$$\min_{p,Q} \quad tr(Q^* P^{-1} Q) + \sum_{k=1}^{K+N} w_k^2 p_k \qquad (15)$$

$$s.t. \quad BQ = \hat{R}.$$

For a fixed value, the minimization over Q is addressed in the solution provided by [10]. $Q_0 = PB^* R^{-1} \hat{R}$. By substituting Q_0 into cost function in (15), we get p_k. The following analytical method can be used to minimize over p for a given Q. Using $Q = [\beta_1, \cdots, \beta_{K+N}]^*$, the optimization problem in (15) for fixed $\{\beta_k\}$ can be reduced to

$$\min_p \sum_{k=1}^{K+N} \frac{\|\beta_k\|^2}{p_k} + \sum_{k=1}^{K+N} w_k^2 p_k. \qquad (16)$$

the simplification demonstrates that

$$\sum_{k=1}^{K+N} \left(\frac{\|\beta_k\|}{\sqrt{p_k}} - w_k \sqrt{p_k} \right)^2 \geq 0 \Leftrightarrow$$

$$\sum_{k=1}^{K+N} \left(\frac{\|\beta_k\|^2}{p_k} + w_k^2 p_k - 2w_k \|\beta_k\| \right) \geq 0 \Leftrightarrow \qquad (17)$$

$$\sum_{k=1}^{K+N} \left(\frac{\|\beta_k\|^2}{p_k} + w_k^2 p_k \right) \geq \sum_{k=1}^{K+N} 2w_k \|\beta_k\|.$$

The $(i+1)^{th}$ version of the derived cyclic algorithm comprising the subsequent steps:

$$Q^{i+1} = P^i B^* R_C^{-1}(i) \hat{R}$$

The left side of (17) corresponds to the cost function presented in (16), and the equality is valid solely under the condition that $p_k = \|\beta_k\| / w_k$. As a result, the reduction of (16) is

$$p_k = \frac{\|\beta_k\|}{w_k} \qquad k = 1, \cdots, K+N. \qquad (18)$$

$$p_k^{i+1} = \frac{\|\beta_k^{i+1}\|}{w_k} \qquad k = 1, \cdots, K+N. \qquad (19)$$

$$R_C(i+1) = B P^{i+1} B^*.$$

The cyclic method mentioned above [12] begins with the Periodogram estimation. $\{p_k^0 = b_k^* \hat{R} b_k / N\}$ The criterion for convergence applied to halt the iterations is: $\frac{\|p^{i+1} - p^i\|}{\|p^i\|} < 10^{-3}$.

Results for Simulated Signal

The outcomes of the simulation involving the intricate data are outlined in this section. The information including N=100 and $C_m = 3$, the intricate data samples are produced from exponential functions at three distinct frequencies: 0.1450 Hz 0.3100Hz, and 0.3150Hz with specified amplitudes. $q_1 = 10e^{j\varphi_1}, q_2 = 10e^{j\varphi_2}$ & $q_3 = 10e^{j\varphi_3}$ by a sampling interval of one second. The phase measurements of $\{\varphi_l\}_{l=1}^3$ are uniformly distributed and independent in $[0, 2\pi]$. The term refers to white Gaussian noise, which possesses zero variance and significance. The efficiency of the suggested approach (SPICE$_{MS}$) is assessed by calculating MSE and the output SNR values using both the SPICE$_{SS}$ and SPICE$_{MS}$ algorithms. Here's how the MSE is determined:

$$MSE = \sum_{k=1}^K E\left\{ P(k) - \hat{P}_i(k) \right\}^2 \qquad (20)$$

where P is the predicted power spectrum at the ith Monte Carlo run and the genuine power spectrum denotes the total number of frequency points, specifically K = 512, or the count of Monte Carlo runs that the expectation process executes is 100. The SNR for the output is ascertained as P_s/P_N where mean noise power is denoted by P_N and mean signal power by P_s.

Consequently, in relation to both Q and P, the cost function shown in (15) is convex. Performing cyclic minimization on Q while keeping p constant, and vice versa, will result in finding the global minimum of (15).

Figure 85.1 illustrates the spectrum for the test signal before the introduction of noise. The typical power spectrums of the test signal employing the SPICE$_{SS}$ and SPICE$_{MS}$ methods for SNR values of

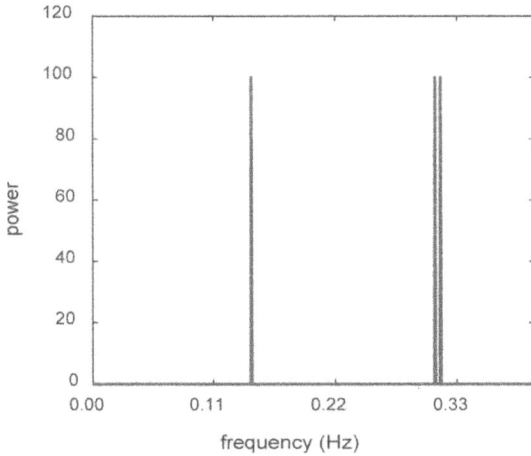

Figure 85.1 The original test signal spectrum
Source: Author

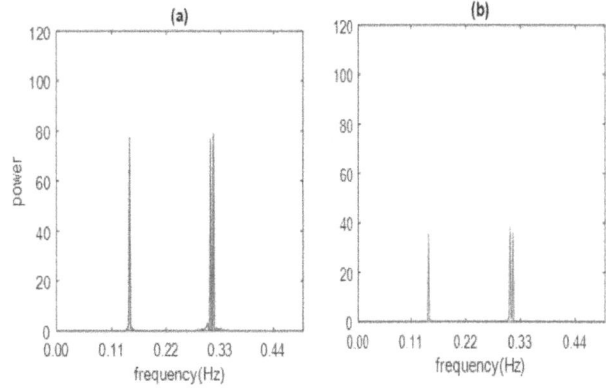

Figure 85.2 Test signal power spectrum using SPICE$_{SS}$ for (a) SNR = 0 dB and (b) SNR =-15 dB
Source: Author

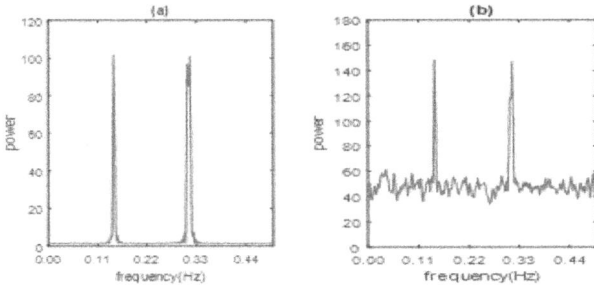

Figure 85.3 Test signal power spectrum using SPICE$_{MS}$ for (a) SNR = 0dB and (b) SNR = -15dB
Source: Author

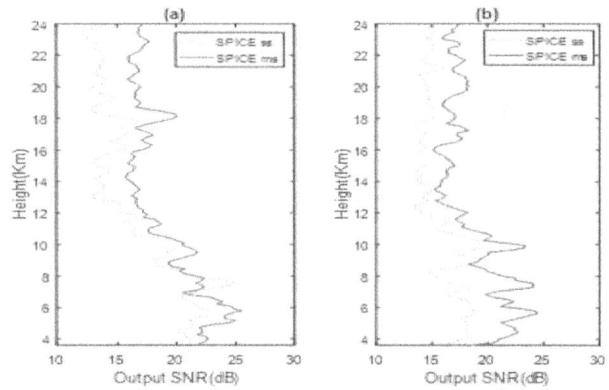

Figure 85.4 SNR predicted using SPICE$_{MS}$ and SPICE$_{SS}$ (a) east beam, (b) south beam for July 05, 2014, radar data
Source: Author

0dB and -15dB is illustrated in Figure 85.2(a) and (b) and Figure 85.3(a) and (b) correspondingly. The proposed method effectively estimates the power spectrum of the generated complex data, even when faced with significant noise (SNR = -15 dB).

MST Radar Data Results

Several radar data scans are supplied by NARL; the collected data contains 17 scans. In the proposed algorithm the multiple snapshots are the same as 17 scans. Each scan has echo signal details of all six beams (East, West, Zenith_Y, Zenith_X, North, and South) are the names of the beams. Each beam consists of 150 bins, with each bin holding 512 complex data points. The spectrum (Doppler) for the complex time series data in each range bin has been computed using the proposed method, and the highest peak has been identified through the peak detection technique.

This procedure is conducted for all six beams and every bin. Once the Doppler profiles have been obtained, the Doppler velocities are calculated [13].

The wind speed obtained from the collected GPS radio sonde data is utilized to validate the anticipated atmospheric wind speed through the proposed method [13]. The power spectra for range bins 2, 65, and 142, estimated using SPICE$_{MS}$ for the east beam of the data from July 5, 2014, are displayed in Figure 85.4. Figure 85.5 illustrates the highest peak at a particular frequency, clarifying the efficiency of the proposed method. An increase in noise influences the signal during power spectrum estimation for larger range bins, leading to additional peaks at the target frequency component.

The Doppler profiles for the east beam from the radar data acquired with SPICE$_{MS}$ are illustrated in

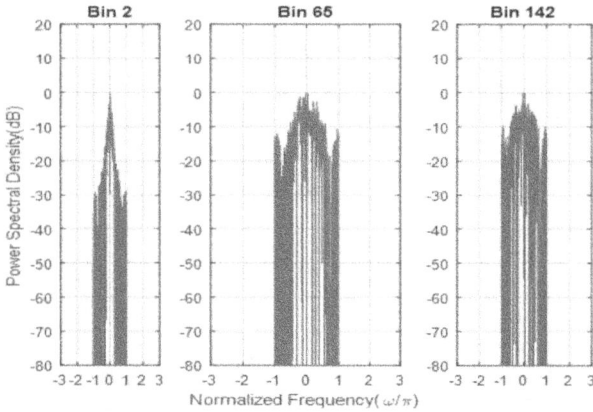

Figure 85.5 Spectrum at range bins numbered 2, 65 and 142 of the east beams using SPICE$_{MS}$ algorithm for July 05, 2014 radar data
Source: Author

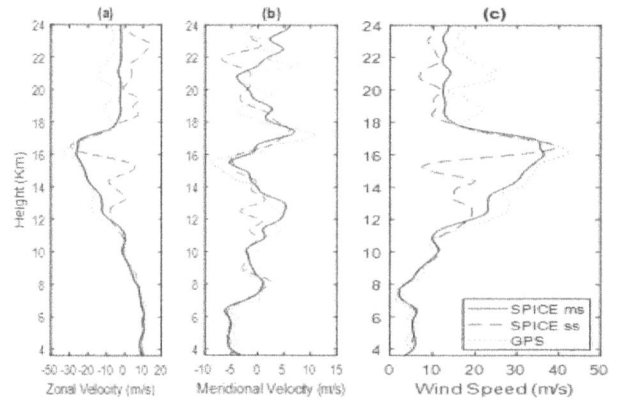

Figure 85.7 Evaluation of zonal, meridional and wind velocities for the data on July 05, 2014 using SPICE$_{MS}$, SPICE$_{SS}$ and GPS radiosonde
Source: Author

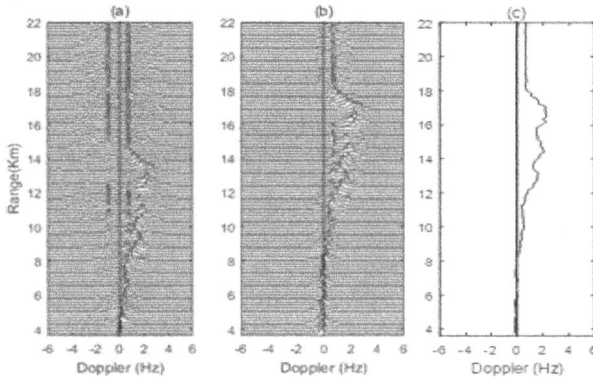

Figure 85.6 Characteristic spectra of east beam radar signal on July 05, 2014. (a) Prior to denoising, (b) After denoising with SPICE$_{MS}$, and (c) Doppler profile
Source: Author

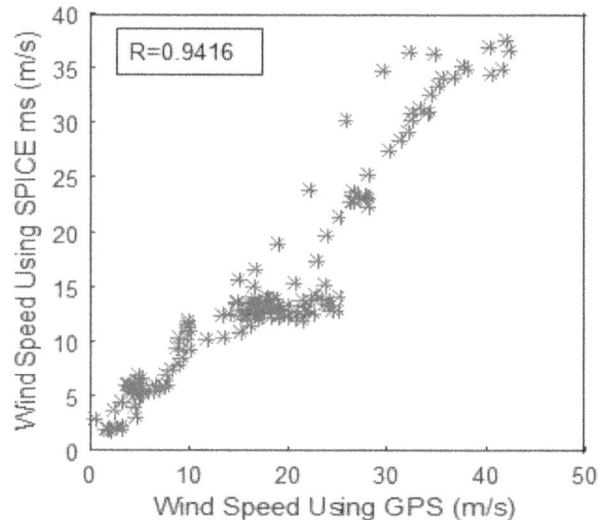

Figure 85.8 Correlation among GPS and SPICE$_{MS}$ for the data on July 05, 2014
Source: Author

Figure 85.6. The zonal V_x, meridional V_y, and wind speed W the components are derived from the data collected by GPS radiosondes, with SPICE$_{SS}$ and SPICE$_{MS}$ illustrated in Figure 85.7. It is clear from Figure 85.6 that the wind speed recorded by SPICE$_{MS}$ matches that of the GPS measurements. In contrast, the wind speed.

The radar data correlation analysis plot from July 5, 2014, is shown on Figure 85.8. Using the proposed strategy (SPICE$_{MS}$), the correlation coefficient is 0.9416, while the existing method (SPICE$_{SS}$) yields a correlation coefficient of 0.8505. We would like to conclude that SPICE$_{MS}$ outperforms the existing method when assessing the correlation factor values.

Conclusion

In this research, a semi-parametric iterative covariance estimation technique known as semiparametric iterative covariance estimation (SPICE$_{MS}$) is introduced for estimating the Doppler spectrum of MST radar data. The SPICE$_{MS}$ method utilizes a covariance matrix along with numerous data snapshots to determine power levels. Research demonstrated that it outperforms the existing SPICE$_{SS}$ method when evaluated on generated complex signals. The performance of SPICE$_{MS}$ is assessed by determining the spectrum of complex signal and analyzing MSE and the SNR out across various SNR input levels. Moreover, the proposed method is employed to assess wind speeds utilizing Indian MST radar signal received on July 5, 2014. The findings from the GPS radiosonde were utilized to corroborate the radar results that were obtained.

References

[1] Anandan, V. K., Balamuraliddhar, P., Rao, P. B., & Jain, A. R. (1996). A method for adaptive moments estimation technique applied to MST radar echoes. In Proceedings of the Progress in Electromagnetics Research Symposium, (pp. 360–365).

[2] Anandan, V. K., Reddy, G. R., & Rao, P. B. (2001). Spectral analysis of atmospheric signal using higher orders spectral estimation technique. *IEEE Transactions on Geoscience and Remote Sensing*, 39(9), 1890–1895.

[3] Anandan, V. K., Pan, C. J., Rajalakshmi, T., & Reddy, G. R. (2004). Multi taper spectral Analysis of atmospheric radar signal. *Annals of Geophysics*, 22(11), 3995–4003.

[4] Reddy, T., & Reddy, G. R. (2010). MST radar signal processing using cepstral thresholding. *IEEE Transactions on Geoscience and Remote Sensing,* 48(6), 2704–2710.

[5] Chandraiah, G., & Reddy, T. S. (2018). Denoising of MST radar signal using multi-band wavelet transform with improved thresholding. In IEEE Conference, (pp. 1026–1030).

[6] Chandraiah, G., Reddy, T. S., & Reddy, G. R. (2019). Atmospheric radar signal processing using iterative principal structure. *International journal of Imaging and Robotics,* 19(3), 541–552.

[7] Chandraiah, G., Babu, P. S., & Srinivasulu, G. (2024). Adaptive PCA-based spectral estimation method for MST radar signal processing. In 2024 International Conference on Wireless Communications Signal Processing and Networking (WiSPNET), (pp. 1–5). IEEE.

[8] Eappen, N. I., Reddy, T. S., & Reddy, G. R. (2015). Semiparametric algorithm for processing MST radar data. *IEEE Transactions on Geoscience and Remote Sensing,* 48(6), 1–9.

[9] Stoica, P., Babu, P., & Li, J. (2011). New method of sparse parameter estimation in separable models and its use for spectral analysis of irregularly sampled data. *IEEE Transactions on Signal Processing,* e59(1), 35–47.

[10] (2011). SPICE: a sparse covariance based estimation method for array processing. *IEEE Transactions on Signal Processing,* 59(2), 629–638.

[11] Stoica, P., & Moses, R. (2005). Spectral Analysis of Signals. Upper Saddle River, NJ: Prentice Hall.

[12] Babu, P., & Stoica, P. (2012). Sparse spectral –line estimation for nonuniformly sampled multi variate time series: SPICE, LIKES and MSBL. *EURASIP Journal of Signal Processing*, 27–31.

[13] Rao, V. V. M. J., Rao, D. N., Ratnam, M. V., Mohan, K., & Rao, S .V. B. (2003). Mean vertical velocities measured by Indian MST radar and comparison with indirectly computed values. *Journal of Applied Meteorology,* 42(4), 541–552.

86 Mitigating double node upsets in space: A RHBD 14T SRAM cell design

T. Hari Kala[a], K. Chandrika[b], S. Abdul Latheef[c], M. Divya Sri[d] and M. Hari Prasad[e]

Department of Electronics and Communication Engineering, Annamacharya Institute of Technology and Sciences, Boyanapalli, Rajampet, Andhra Pradesh, India

Abstract

In areas of space and satellite applications, memory equipment is very weak to the radiation and it can interfere with the electric systems and integrity of data validation. Traditional Sam's cells may affect bits to keep these wrongs, especially the disease of one event (SEU) and multiple collection diseases (DNU). This shows the most easily framed design, made by a soft sum of the SAM for 14Ts (Sei-14T) to reduce this challenge. The sei-14T cell shows the formation of double fun, the healing of the restriction of the entire nests and parts of the unit. The underline analysis shows the Sei-14T that produces the radiation that performed (such as quota-14t and the increment of the writing, and the income. This diagram, established by the Sei-14T is a powerful memory solution for Electrica-bloses, keeping stability and effectiveness to adult radiation environment.

Keywords: 14T SRAM cell, double-node upset (DNU), radiation-hardened SRAM, soft error immune (SEI), space and satellite applications

Introduction

Some of the most popular customers must explore the location created by a man-made spacecraft and canoe in the leaves and electricity. However, our cosmos are full of radiation. Outside the country's environment, radiation is called place radiation. The location radiation contains high particles and electric waves. This radiation has many problems associated with stability and reliable electronic devices. Even though the smaller alpha particles can make a break of electric equipment. The result of one session (see) is important for different types of reasons. The effect of radiation depends on the actual part of the medicine of the measurement of gears [2]. Binder et al. recommended one session checked, a simple type of error, followed by the results of the same event (see) in space, [4]. It is shown in the Figure 86.1, a large particle attack starts the icon paths in the memory device and produces the sum of money because of the misunderstanding and division. This causes the problem or depression on the memory pulmonary disease chronic respiratory conditions such as chronic obstructive pulmonary disease (COPD) and asthma have become more prevalent due to rising levels of air pollutants like PM2.5 and PM10. Since reducing air pollution at the source is a long-term process, the use of personal protective equipment, such as respirators, serves as an immediate and effective individual-level safety measure to mitigate exposure and associated health risks.

The differences in PVT. In addition, the high costs under the Kerns et al. Identify the types of methods to continue from SEU, such as equipment, parts, and strategic opportunities [2]. According to the recent text review [7] in memory cells difficult to radiation, most memories of the Te Rangi Te Utai Te Tari, namely, the radiation-kjan -bydesing and easy-to-theme of SRAM is essential for on-chip cache memory due to its fast data access, strong stability, and low power usage. Despite these advantages, SRAM can occupy over 95% of the total chip area in a system-on-chip (SoC) design. To optimize SoC area efficiency, engineers typically scale down transistor dimensions and reduce supply voltage (VDD).

[a]hariekalla@gmail.com, [b]kchandrikayadav@gmail.com, [c]salatheef1903@gmail.com, [d]gowddivyasri@gmail.com, [e]hari934757@gmail.com

DOI: 10.1201/9781003684589-86

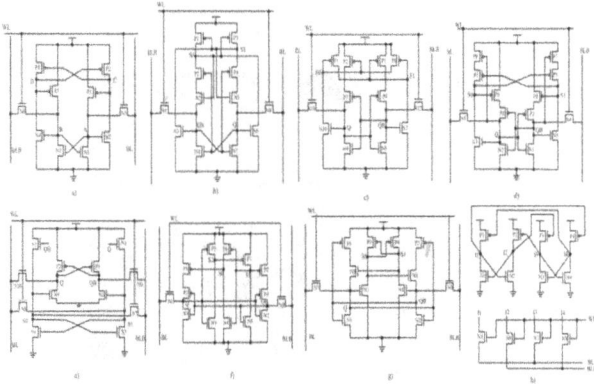

Figure 86.1 RHBD literature designs a) Quatro-10T
b) RHM-12T c) RHD-12T d) RSP-14T e) RHPD-12T
Source: Author

However, studies have shown that these reductions negatively impact the charge stored at the memory nodes, thereby compromising cell stability. Additionally, such scaled-down SRAM cells become more susceptible to radiation-induced disturbances. To address these issues, various researchers have introduced Radiation-Hardened-By-Design (RHBD) SRAM architectures that enhance resilience against high-energy particle impacts—though this often comes with a trade-off between improved soft error tolerance and overall memory performance allowing. device [6]. At the time of radiation particles for memory devices, the ability of the ability to work greatly, because of the malicious tools [6] cell [7].

Literature Survey

Modeling single event transients in advanced devices and ICs
Some of the most popular customers must explore the location created by a man-made spacecraft and canoes in the leaves and electricity. However, our cosmos are full of radiation. Outside the country's environment, radiation is called place radiation. The location radiation contains high particles and electric waves. This radiation has many problems associated with stability and reliable electronic devices. Even though the smaller alpha particles can make a break of electric equipment. The result of one session (see) is important for different types of reasons. The effect of radiation depends on the actual part of the medicine of the measurement of gears [2]. Binder et al. recommended one session checked, a simple type of error, followed

by the results of the same event (see) in space, [4]. It is shown in the picture. 1, a large particle attack starts the icon paths in the memory device and produces the sum of money because of the misunderstanding and division. This causes the problem or depression on the memory device [6]. At the time of radiation particles for memory devices, the ability of the ability to work greatly, because of the malicious tools [6]. Kerns et al. Identify the types of methods to continue from SEU, such as equipment, parts, and strategic opportunities [2]. According to the recent text review [7] in memory cells difficult to radiation, most memories of the Te Rangi Te Utai Te Tari, namely, the radiation-kjan -by desing and easy-to-theme of.

"Radiation-induced soft error analysis of SRAMs in SOI FinFET technology: A device to circuit approach,'
This form shows a simple research on the weak mistakes sent by Srams from designed by Smat Technology technology. For this purpose, we consider a rolling path that starts the 3D of particle interactions in the formation of Finfet to the analysis of the layer level. This way we can imagine the effects of different types of situations and processing the differences in the soft memory (sra). Our analysis shows the weak errors that are stupidly promoted by a mindi-particles, especially for a small distribution cycle (minor authorization). In addition, we saw that the two-pair of double-digit score (MBU) to a single event (SEU) for alpha-particles are more than the floor.

'A soft error tolerant 10T SRAM bit-cell with differential read capability,'
A quad-node, ten-transistor (10T) SRAM cell has been developed with enhanced tolerance to soft errors. This design employs a differential read mechanism to enable more reliable sensing. In ultra-low voltage operation, particularly below 0.45 V, the cell achieves a significantly improved noise margin and demonstrates 26% lower leakage current compared to the conventional 12T DICE-based soft error tolerant cell.

When evaluated against the standard 6T SRAM cell, this 10T design maintains an equivalent noise margin but at only half the supply voltage, contributing to considerable power savings through reduced leakage. Furthermore, neutron radiation tests performed at TRIUMF on a 32-kb SRAM fabricated using 90nm CMOS technology revealed a dramatic 98% reduction in the soft error rate when compared to the 6T counterpart.

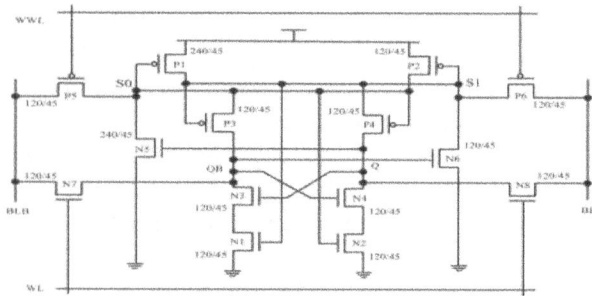

Figure 86.2 SEI-14T memory cell schematic diagram
Source: Author: E. L. Peterson

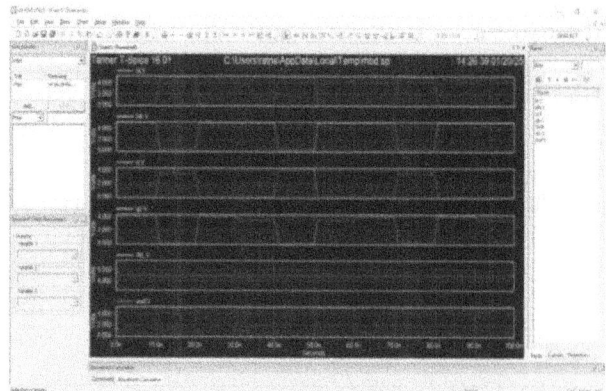

Figure 86.3 Input
Source: Author: P. Shapiro

"Radiation-hardened 14T SRAM bit cell with speed and power optimized for space application

In this paper, Bitcept BitCell 14 -transistor Saram is made with a new radiation with a quick and fast control (RSP) -14T] for the proposed space applications. In the optimal diagram of the actual and layout of 65-nm technology the results of the new construction of the same event and a multiple event of sessions Multi-Death for a distribution of costs between Transistor Outdoor. In addition, the results of the HSpece symbolicness showing the speed of the osp-14t 65% and ~ 50%, is compared to radiation (RHD) -12T (Figure 86.2-86.4).

Novel write-enhanced and highly reliable RHPD-12T SRAM cells for space applications

At this momentum, we think of it, on the part of the stupid disease from a semimonductor (nmos), the new radiation is injured in a True of the truth to increase the truth to improve the truth to increase the truth to increase the truth to increase the truth to Increasing Truth to increase the truth to increase the truth to increase the truth and the operating process for a place request. The aggregated in the standard commercial activity of the CMOS CMOS CMOS (smicconuedutcor indicates that all of the same infections can be offered. At that time, compared to quartro, quarto, with the speed of a pair of telephone cell, can be reduced the speed of the ~ 41.6 and ~ 46.3%. Monte Carlo (MC) He has tried by the number of low offers (0.6 v), a little rhpd-12T.

"Design of high-reliability memory cell to mitigate single event multiple node upsets,"

To reduce technology, the light of the SAM cells with the problem of the same radiation, and the number of the diseased mirror design. In this paper, the RH-14T is considered to reduce the semnus. The effectiveness of 3D spices and technology obtained

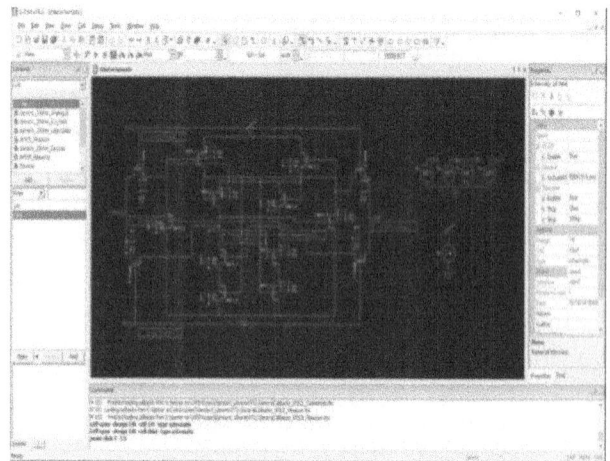

Figure 86.4 Simulation Result
Source: Author: P. Shapiro

by computers and useful mode of mixed mode to confirm the resistance of seus. As compared to memory cells related to the previous radiation, the cell of the time considered, the access time, and the time to process the differences. The noise of the song is read (RNNM) and Margin writes (WM) of RH-14T cells are more than 6 EMA 6T. Increase in many shopping and place, power consumption and work.

Existing Systems

QUATRO-10T in this design, the authors will be eligible to retrieve the negative response to the hell parts of the hell sent by radiation picture. 2 (a). The best part of the diagram is constant under the difference in the pvt [15]. The most important problem in this architecture is unable to restore the lost data if memorable cells are set to logic 0. In addition, the cost of costs is less than the remaining memory rest. B. RHM-12T in this design, the writer will change the quota-10 by adding the two hunters who are hung on the picture. 2 (b). The official food of the idea and reading the reminder time is reduced by cutting the those east, as described in [16]. This diagram is necessary to access the writing mode while administrating temperatures from the 45 | to 120. In addition, the expensive payments will be reduced to smaller volcanoes. C. RHD-12T This design is an update version of the Quatr-10 Design Ident, as a picture. 2 (c). The best part of this diagram is the healing method [17]. The most important problem of this diagram is the advertising of access to case of the most dangerous case.

Proposed System

The SEI-14T radiation-hardened memory cell (RHBD-14T) has been proposed as a robust solution to mitigate soft errors in harsh environments. The key features of SEI-14T include:

1. **Stable read and write operations**: It maintains consistent read and write performance even under extreme temperature corners.
2. **Improved timing under PVT variations**: The SEI-14T exhibits reduced access delay and enhanced reliability under process, voltage, and temperature (PVT) fluctuations.
3. **Increased critical charge**: It demonstrates higher resilience to soft errors by achieving greater critical charge thresholds under worst-case voltage and process conditions.
4. **Soft error immunity**: The design ensures immunity to single event upsets (SEUs) across all critical storage nodes.
5. **Charge sharing mitigation**: The SEI-14T minimizes charge sharing effects among sensitive node pairs, improving the overall robustness of the cell.

Figures (as referenced) illustrate the layout and structure of the SEI-14T cell. The design includes two NMOS transistors (N7, N8) and PMOS transistors (P5, PFs, and PCs), configured to enhance memory stability. Additionally, the design integrates dual feedback loops to reinforce data retention under radiation exposure.

The NMOS access transistor ensures reliable data read and write operations, with minimal susceptibility to noise. The access lines (WL and WWL) must be simultaneously activated to read from or write to the cell.

The SEI-14T memory structure comprises four storage nodes: Q, QB, S0, and S1. Q and QB serve as the primary data-holding nodes, while S0 and S1 act as secondary or backup storage. In the event of a radiation strike on the primary nodes, the secondary nodes help restore the correct data state, ensuring fault tolerance.

For example, if the primary node Q holds a logic '1' while the secondary nodes S0 and S1 hold logic '0', the feedback mechanism helps maintain data integrity, demonstrating the effectiveness of the SEI-14T design in hostile environments.

Simulation Results

Conclusion

A radiation-hardened-by-design 14-transistor SRAM cell (SEI-14T), optimized for use in space and satellite environments, is introduced with strong immunity to soft errors. The SEI-14T structure effectively protects all vulnerable internal nodes against soft error occurrences. Moreover, it demonstrates reduced sensitivity to charge sharing across most node pairs, except for Q-S0 and QB-S0, where careful physical separation in layout helps mitigate potential interference.

This memory cell shows improvements under varying Process, Voltage, and Temperature (PVT) conditions, including lower write access time and reduced static power consumption. In comparison to the QCCS-12T cell, SEI-14T offers faster read access, making it a competitive alternative. Furthermore, it enhances stability during read, write, and hold operations, even under the most challenging corner cases and elevated temperatures.

The SEI-14T SRAM cell demonstrates a higher effective critical charge even under worst-case process corners and supply voltage variations. This enhancement directly contributes to a lower soft error rate. Additionally, the SEI-14T design achieves superior read and write yield probabilities, along with the highest figure of merit among comparable designs. These performance advantages make the SEI-14T a strong and reliable candidate for SRAM implementations in space and satellite systems, where resilience and stability are critical.

References

[1] Kobayashi, D. (2021). Scaling trends of digital single-event effects: a survey of SEU and SET parameters and comparison with transistor performance. *IEEE Transactions on Nuclear Science*, 68(2), 124–148.

[2] Kerns, S. E., Shafer, B. D., Rockett, L. R., Pridmore, J. S., Berndt, D. F., van Vonno, N., et al. (1988). The design of radiation-hardened ICs for space: a compendium of approaches. *Proceedings IEEE*, 76(11), 1470–1509.

[3] Binder, D., Smith, E. C., & Holman, A. B. (1975). Satellite anomalies from galactic cosmic rays. *IEEE Transactions on Nuclear Science*, NS-22(6), 2675–2680.

[4] May, T. C., & Woods, M. H. (1979). Alpha-particle-induced soft errors in dynamic memories. *IEEE Transactions on Electron Devices*, ED-26(1), 2–9.

[5] S. M. Jahinuzzaman, D. J. Rennie, and M. Sachdev, "A soft error tolerant 10T SRAM bit-cell with differential read capability," IEEE Trans. Nucl. Sci., 56(6) pp. 3768–3773, Dec. 2009.

87 RLG – based image watermarking encryption and decryption: A novel VLSI design approach

K. Janshi Lakshmi[1,a], D. Sravani[2,b], C. Chathrapathi[2,c], O. Sashankreddy[2,d] and D. Rajarohith[2,e]

[1]Associate Professor, Department of ECE, Annamacharya Institute of Technology and Sciences, Tirupati, Andhra Pradesh, India

[2]UG Students, Department of ECE, Annamacharya Institute of Technology and Sciences, Tirupati, Andhra Pradesh, India

Abstract

Advances in digital technology have necessitated the development of robust and secure image watermarking techniques to protect intellectual property rights and maintain data integrity. These fields require robust cryptography, but traditional algorithms struggle to balance security, power efficiency, and performance. Lightweight block ciphers compromise accuracy, while existing reversible logic gate designs lack thorough evaluations. This system proposes a reversible logic gate (RLG) architecture integrating encryption, decryption, and linear feedback shift register (LFSR)-based key generation, enhanced by least significant bit (LSB)-based watermarking for security. The proposed system is implemented in MATLAB and Xilinx ISE. It performs well, with less time needed than traditional methods and for better security resilience. That protects itself from threats and recovers the data. RLG offers improved security, lower power consumption, and quantum computing compatibility, making it ideal for secure data processing in healthcare, banking, and government.

Keywords: Least significant bit (LSB), linear feedback shift register (LFSR), reversible logic gates (RLG), verilog verilog, Xilinx ISE

Introduction

Reversible logic gate (RLG)-based image watermarking encryption and decryption have emerged as a promising solution, offering improved security, lower power consumption, and quantum computing compatibility. Moore's Law [1] highlights the exponential increase in transistor density, enabling the integration of sophisticated security algorithms into hardware. However, traditional logic circuits face significant heats dissipate for reason of information loss, as explained by principle of Landauer's [2]. To address this issue and enhance energy efficiency, reversible logic gates have emerged as a viable solution, initially proposed by Bennett [3]. Reversible computing eliminates information loss during processing, making it highly suitable for cryptographic applications, including image encryption and decryption.

Reversible logic gates can reverse operations, enabling data recovery, which is useful for encryption, decryption, and watermarking. Using Verilog, these gates can be leveraged to design, simulate, and implement secure systems for image encryption and watermarking, ensuring data integrity and confidentiality.

Advances in digital technology have necessitated the development of robust and secure image watermarking techniques to protect intellectual property rights and maintain data integrity. RLG-based image watermarking encryption and decryption have emerged as a promising solution, offering improved security, lower power consumption, and quantum computing compatibility. This review provides a comprehensive overview of RLG-based image watermarking encryption and decryption, with a focus on novel VLSI design approaches, highlighting their advantages, challenges, and applications in secure data processing.

Related Works

Kuchhal and Verma [4] using reversible logic with DES to improve power efficiency. I remember that DES is a traditional encryption standard, but reversible logic isn't something I'm familiar with. Maybe reversible logic has something to do with reducing power

[a]jansikaramala@gmail.com, [b]darapanenisravanichowdary@gmail.com, [c]chathrapathi16@gmail.com, [d]sashankreddy.o2003@gmail.com, [e]rajarohith2001@gmail.com

DOI: 10.1201/9781003684589-87

consumption by recovering energy or perhaps it's related to reversible computations in quantum computing, Wait, the context here is VLSI architectures, so maybe it's a design approach in electronics where logic gates can be reversed to minimize power dissipation. That makes sense because power efficiency is crucial in hardware implementations, especially for devices that might have limited power resources.

Then Zhao and Wang [5] investigated dynamic block ciphers' security aspects. Dynamic block ciphers might refer to ciphers that can change their encryption algorithm dynamically, making them harder to crack. The emphasis on robust cryptographic designs in VLSI suggests that they're focusing on how to implement these ciphers effectively in hardware. VLSI is about integrating lots of components into a chip, so security here would involve ensuring that the hardware itself isn't vulnerability. Maybe they addressed issues like side-channel attacks or fault injections, which are concerns in physical hardware.

Subramanian et al. [6] worked on lightweight block ciphers like LED and HIGHT. Lightweight ciphers are designed for environments with limited resources, such as IoT devices or embedded systems. LED and HIGHT are specific algorithms optimized for performance in constrained environments. Their hardware implementations being reliable mean they probably tested these ciphers under various conditions to ensure they don't fail and maintain security without requiring too much power or computational resources.

In addition to encryption, watermarking techniques are pivotal in digital rights management and authentication, offering robust mechanisms to safeguard digital content. Garipelly et al. [7] conducted an extensive review of reversible logic gates, highlighting their applications in designing low-power security systems. Bamatraf et al. [8] gives digital watermarking algorithm based on least significant bit (LSB) embedding, which can be combined with reversible logic to achieve heightened security. On the hardware front, Dadhe and Nage [9] developed VLSI architecture as efficient speed for LFSR, optimizing their use in cryptographic key generation. Similarly, Kumar et al. [10] designed a power-efficient reversible LFSR for built-in self-test (BIST) applications, showcasing its potential for adaptation in secure watermarking and encryption frameworks. Together, these advancements underscore the synergy between algorithmic innovation and hardware optimization in enhancing digital security [11]. A practical chaotic images cryptosystem based on a plaintext-associated mechanism and integrated confusion-diffusion operation has been developed to improve encryption reliability [12]. An efficient chaotic image encryption algorithm based on a plaintext-associated approach employs a simultaneous confusion-diffusion process to thwart any separate attack [13]. The system was tested using malicious, legitimate datasets gathered from the websites of Phishtank and Alexa [14]. It analyzes the different error metrics about encrypted images safety, including edge detection and plaintext attack, involving reversible logic cryptographic design (RLCD) [15]. Offers a new cryptographic system based on the random key generator hybridization methodology by utilizing the properties of the discrete cosine transform (DCT) to generate an infinite set of random keys. Ramakrishna. K et al.[16] Developed Reversible Logic Gates Cryptography Design (RLGCD) is proposed to address high area and power issues in secure cryptographic systems. It integrates encryption, decryption, and LSB-based watermarking, showing significant performance improvements on FPGA compared to conventional methods.

Existing System

DES (Data Encryption Standard) represents a block cipher utilizing symmetric-key algorithms that encrypts 64 bit simple text into 64 Bit encrypted text utilizing a 64-bit key, with 56 bits for encryption and 8 for error detection. Adopted by NIST in 1977 as FIPS PUB 46, it ensures data security, while Triple DES (TDEA) enhances protection using three keys. Decryption requires the original key, making confidentiality crucial despite brute-force attack risks. DES encryption follows a structured process: plaintext undergoes an initial permutation, splits into two halves, and processes through a Feistel structure, ensuring security via substitution and permutation. A final inverse permutation generates the ciphertext. Key generation involves deriving sixteen 48-bit subkeys from a 56-bit key, emphasizing security through key management and transformations.

Proposed Algorithm

Reversible logic is crucial for power efficiency and quantum computing in cryptographic applications as it reduces power consumption, enables quantum parallelism, and ensures secure data processing. By minimizing logic operations, reversible logic lowers power consumption, heat dissipation, and energy usage, making it ideal for power-constrained devices. In quantum computing, reversible logic facilitates quantum parallelism, error correction, and secure key transmission. In cryptographic applications, reversible logic enables secure data processing, resists side-channel attacks, and allows for low-latency encryption, making it essential for real-time cryptographic applications as shown in Figure 87.1.

Constraints for designing reversible logic gates (RLGs)

Designing RLGs involves adhering to several key constraints to ensure efficiency, minimal energy consumption, and optimal performance. These constraints include [9]:

- No Fan-out
- Minimal quantum cost
- Minimized Garbage outputs
- Least gate Levels

Process of proposed algorithm

- Read the input image in MATLAB and apply watermarking.
- Convert the watermarked image into a binary format using LSB watermarking.
- Store the binary pixel values in a text file in MATLAB.

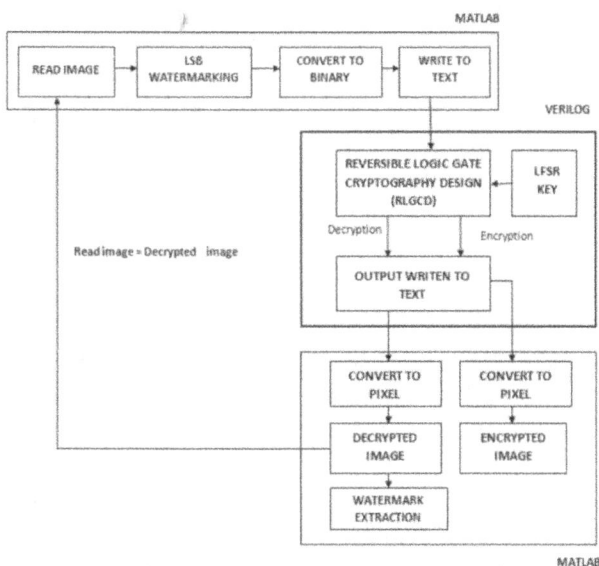

Figure 87.1 Proposed system block diagram
Source: Author

- Generate a cryptographic key using the LFSR algorithm.
- Use the MATLAB text file as input for Verilog encryption/decryption.
- Save encrypted and decrypted outputs in Verilog text files for verification.
- Reconstruct encrypted and decrypted images in MATLAB using text files.
- Ensure the input and decrypted images are identical.
- Extract the watermark from the decrypted image
- Evaluate FPGA performance using Verilog code

The LSB substitution method embeds watermark data in 3rd and 4th of LSBs an image's bits. Using MATLAB, a 128 × 128 image can store up to 817 bits of watermark data with a five-pixel gap between embedded bits. The method works for both color and grayscale images, with color images embedding the watermark in the blue component.

Encryption process

The 8-bit binary sequence of pixel values ranges from i(0) to i(7). The upper SCL gate processes the first four bits, while the lower SCL gate handles the four LSBs, each producing four outputs. A Toffoli gate operates on 3 LSB responses from the lower side of SCL gate and three MSB bits from the upper SCL gate. A Feynman gate processes one output from each SCL gate. Output of Toffoli gates is fed to Fredkin gate finally, XOR gates use an LFSR key to generate the encrypted pixel values e[0] to e[7] as shown in Figure 87.2,e(0) to e(7) are generated using XOR gates with an LFSR key.

Decryption process

Decryption reverses encryption. As shown in Figure 87.9(a), the encrypted output undergoes an EX-OR method with the key. The SCL gate sequentially applies four reversible operations to obtain decrypted pixel values d(0) to d(7). A text file stores binary values, which MATLAB uses to reconstruct the images as shown in Figure 87.3.

Linear feedback shift register

A linear feedback shift register (LFSR), or pseudo-random number generator, creates random key patterns using an XNOR gate and four flip-flops. A seed value is loaded and shifted upon clocking to generate test patterns. Widely used in stream ciphers,

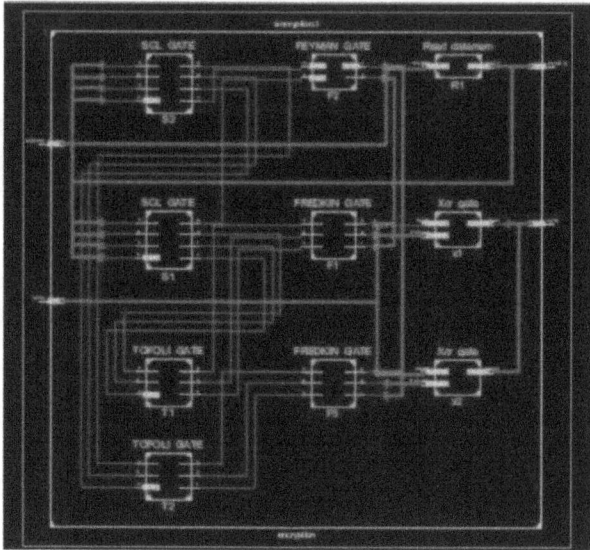

Figure 87.2 Encryption RTL diagram
Source: Author

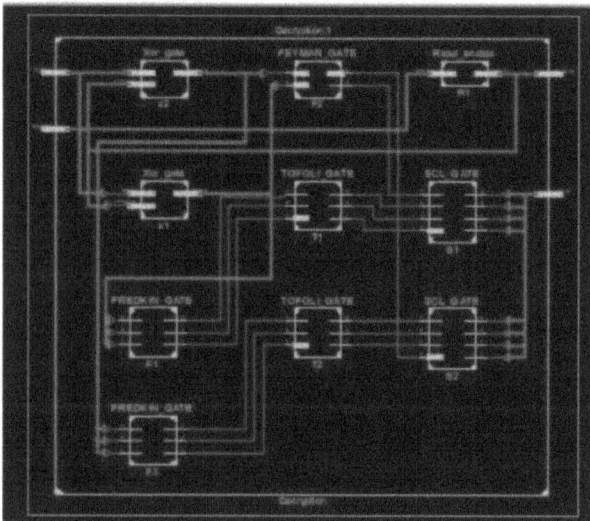

Figure 87.3 Decrypted RTL diagram
Source: Author

LFSRs generate encryption keys based on a feedback polynomial, determining the sequence length. Key size impacts cryptographic efficiency; larger keys increase network demand. This work uses an LFSR-based random key generator to enhance security. As shown in Figure 87.4, LFSR-generated keys encrypt an image in Verilog, assigning each pixel a unique key. The same key ensures accurate decryption, making LFSR encryption secure and efficient.

Result and Discussion

The RLG-based cryptographic system is modeled in Xilinx ISE, while MATLAB facilitates the processing

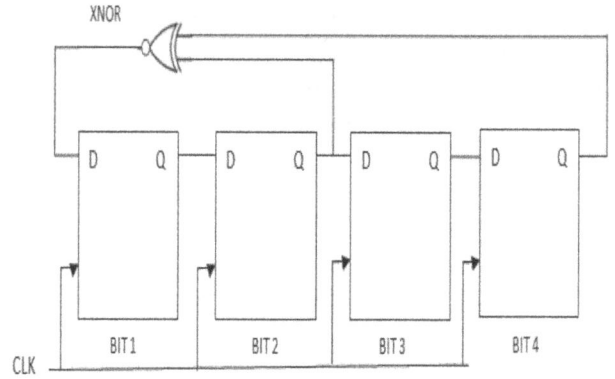

Figure 87.4 Linear feedback shift register
Source: Author

Table 87.1 Comparison proposed system with existing system.

Criteria	Existing System	Proposed System
Security	Less Security	High security
Power Consumption	High	Low
Encryption & Decryption Speed	Slower due to multiple iterations	Faster processing using Verilog and MATLAB integration
Hardware Implementation	Implemented using traditional logic gates, requiring more transistors	Implemented using Feynman, Fredkin, Toffoli, and SCL gates in Verilog, reducing complexity.
Data Recovery	One-way encryption; requires exact key for decryption.	Reversible logic allows lossless encryption and decryption
Quantum Computing Compatibility	Not quantum-safe; vulnerable to future quantum attacks.	Quantum-compatible as reversible logic supports quantum computations.
Applications	Common in financial transactions, government data protection.	suitable for secure image processing in healthcare, banking, and IoT applications

Source: Author

of input image data and implementation of watermarking operations. It supports image sizes of 64 × 64, 128 × 128, and 256 × 256, converting them into 4096, 16,384, and higher bit resolutions, respectively, and can process color images. Encrypted images appear unreadable and follow the original bit arrangement. Decryption reverses the encryption process, restoring the original image, ensuring data security and integrity. Table 87.1 represents a comparison of proposed systems with existing systems.

The 64 × 64 image is encrypted and added with key, which is randomly generated by LFSR as shown in Figure 87.5. Then the after encryption of the given image can be observed as bits as shown in Figure 87.6. Similarly, decrypted output in Xilinx ISE can be seen as shown in Figure 87.7.

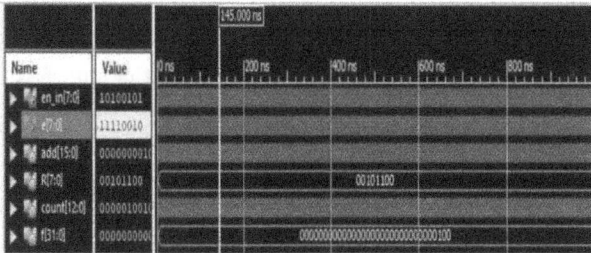

Figure 87.5 Encryption output
Source: Author

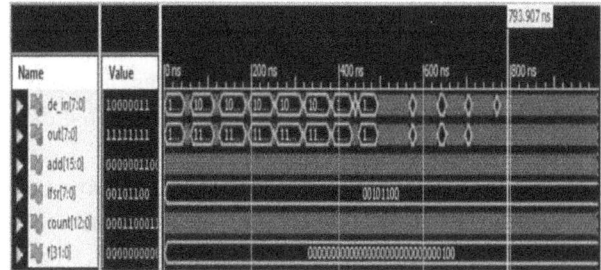

Figure 87.7 Decryption output
Source: Author

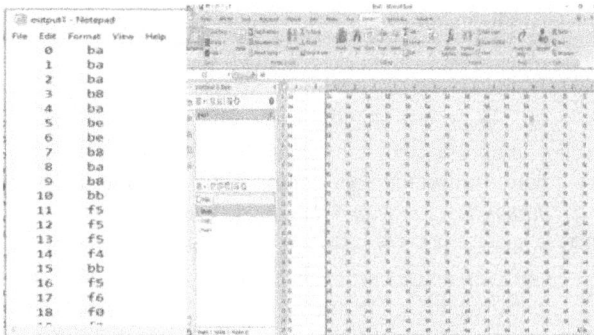

Figure 87.6 Output of encryption bits are converted into matrix bits
Source: Author

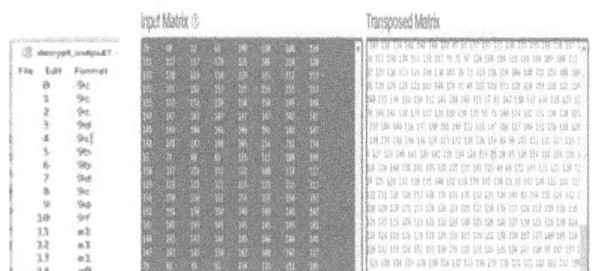

Figure 87.8 Decryption output is converted into transpose matrix
Source: Author

Figure 87.9 (a) Input image, (b) Encrypted output image, (c) Decrypted output image
Source: Author

The input image is shown in Figure 87.9(a). The encrypted image is shown in as Figure 87.9(b), even though that may not be getting directly but we can observe in the form of encrypted bits as shown in Figure 87.6.

Actual bits of the image can see in the form of matrix shown in Figure 87.8 then these bits are converted into actual original images as in Figure 87.9(c) by MATLAB.

Conclusion

This paper proposed a cryptographic design using reversible logic gates cryptographic design (RLGCD) using watermarking and an LFSR-generated key for enhanced security and efficiency. By integrating Feynman, Fredkin, Toffoli, and SCL gates, it ensures robust encryption and decryption. MATLAB handles image processing and watermarking, while Xilinx ISE executes cryptographic operations. Implemented on Spartan3E XC3S500E, the system supports both gray scale and color images, utilizing LSB watermarking for added security. With enhancement of quantum computing, RLG presents viable solution for future ASIC designs.

Acknowledgement

The authors gratefully acknowledge the Staff, and Authority of ECE Department for their cooperation in the research.

References

[1] Moore, Gordon E. "Cramming more components onto integrated circuits." 19 Apr. 1965, Vol No. 38.

[2] Landauer, R. (1961). Irreversible and heat generation in the computing process. *IBM Research and Development*, 5, 183–191.

[3] Bennett, C. H. (1973). Logical reversibility of computation. *IBM Research and Development*, 17, 525–532.

[4] Kuchhal, S., & Verma, R. (2015). Security design of DES using reversible logic. *International Journal of Computer Science and Network Security*, 15(9), 81–84.

[5] Zhao, G., & Wang, J. (2016). Security analysis and enhanced design of a dynamic block cipher. *China Communications*, 13, 150–160.

[6] Subramanian, S., Mozaffari Kermani, M., Azarderakhsh, R., & Nojoumaian, M. (2017). Reliable hardware architectures for cryptographyic block ciphers LED and HIGHT. *IEEE Transactions on Computer-Aided Design of Integrated Circuits and Systems*, 36(10), 1750–1758.

[7] Garipelly, R., Kiran, P. M., & Kumar, A. S. (2013). A review on reversible logic gates and their implementation. *International Journal of Emerging Technology and Advanced Engineering*, 3(3), 417–423.

[8] Bamatraf, A., Ibrahim, R., & Salleh, M. N. B. M. (2010). Digital watermarking algorithm using LSB. In 2010 International Conference on Computer Applications and Industrial Electronics, Kuala Lumpur, (pp. 155–159).

[9] Dadhe, M., & Nage, A. R. (2015). Design of high speed VLSI architecture for LFSR with maximum length feedback polynomial. *International Journal for Scientific Research & Development*, 3(5), 2321–0613.

[10] Kumar, Y. G. Praveen., Kriyappa, B. S., & Kurian, M. Z. (2017). Implementation of power efficient 8-bit reversible linear feedback shift register for BIST. In 2017 International Conference on Inventive Systems and Control, Coimbatore.

[11] Ahmed, K. S., Mohammed, H. A., & Ahmed, H. M. (2021). A new chaotic image cryptosystem based on plaintext-associated mechanism and integrated confusion-diffusion operation. *Karbala International Journal of Modern Science*, 7(3), 176–188.

[12] Ahmed, H. M., Ahmed, K., & Mohammed, H. (2022). Image cryptosystem for IOT devices. Using 2-D Zaslavsky Chaotic Map. TEM J. 15(2), 543–553.

[13] Rawaa, M. A., Mohammed, A. H., & Amal, A. K. (2022). Detecting phishing cyber attack based on fuzzy rules and differential evaluation. *Journal of the Association for Information Science and Technology*, 11(2), 543–551.

[14] Vinoth, R., Siva, J., Sundararaman, R., & Rengarajan, A. (2021). Security analysis of reversible logic cryptography design with LFSR key on 32-Bit microcontroller. *Microprocess and Microsystems*, 84, 1750–1758.

[15] Jabbar, K. K., Ghozzi, F., & Fakhfakh, A. (2023). Robust color image encryption scheme based on RSA via DCT by using an advanced logic design approach. *Baghdad Science Journal*, 20(6 Suppl.), 2593–2607. P-ISSN: 2078-8665 - E-ISSN: 2411-7986.

[16] Ramakrishna, K., Zunaid, S. M., Kumar, V. B., Vishnu, K. P., & Kumar, M. D. (2024). Implementation of encryption and decryption using reversible logic gates. Journal of Engineering Sciences, 15(04). ISSN 0377-9254. Page: 84–94.

88 Activity minimization of misinformation's influence in the online social networks platforms

Pooja Krishnamurthy Revankar[1,a], Srinivasarao Udara[2,b], S. B. Ullagaddi[3,c] and Jayprabha Vishal Terdale[4,d]

[1]Associate Professor, Department of AIML, Rural Engineering College, Hulkoti, Karnataka, India

[2]Associate Professor, Department of ECE, STJ Institute of Technology, Ranebennur, Karnataka, India

[3]Professor, Department of CSE, Rural Engineering College, Hulkoti, Karnataka, India

[4]Assistant Professor, Department of AIDS, A. C. Patil College of Engineering, Navi Mumbai, Maharashtra, India

Abstract

The use of online social media has rapidly increased in recent years, and a lot of information has spread across these platforms, changing how people get information. False information of various kinds of spreads swiftly on social media, undermining the credibility of the content. Providing a trustworthy network environment and managing network space are crucial. Investigate a novel problem: the activity minimization of misinformation influence (AMMI) problem. In order to reduce the overall quantity of misleading contact between nodes (TAMIN), a group of nodes must be removed from the network. Therefore, the AMMI challenge is to minimize the TAMIN by selecting K nodes to block from a given social network G. It also seeks to create a heuristic greedy algorithm (HGA) and prove the function with objectives is neither super- nor sub-modular for eliminating the top K-nodes. In addition, experiments on three real world networks applications have been carried out to check the method. The results of the work show that the recommended methodology outperforms comparative approaches.

Keywords: Activity minimization of misinformation influence, heuristic greedy algorithm, misinformation, social network, total amount of misleading contact

Introduction

Due to the quick advancement of mobile communication technology in recent years, several interactive platforms and information exchanges have surfaced, including Facebook, YouTube, and Sina Weibo. These days, social networking sites are essential to the widespread distribution of knowledge. On the one hand, people's enjoyment lives have been enhanced by the dissemination of good information on online social networks (OSNs), such as trending topics, viewpoints, knowledge, and so forth. OSNs are used to disseminate harmful content, such as rumors, cyberviolence, false information, and so forth, which can seriously injure people and even spark social panic.

For instance, the government not only provided locals with relief when a disastrous wildfire broke out in California in October 2017, but they also received false information about the fire on OSNs [1]. Despite government control over misinformation, the first and related misleading material was shared on social media like Facebook over 1,30,000 times [2].

In Aug 2012, many of the people from the city and surrounding places evacuated homes to an extended period of time as a result of false information about the earthquake spreading around Ghazni, Afghanistan's Ghazni province [3]. Proposing an efficient strategy to restrict or stop the spread of false information is a pressing issue in order to enhance the dependability of the information and better foster a healthy network ecological environment. Numerous academics have focused on controlling the spread of false information in OSNs [4], which have applications in a variety of fields, including public health [10], epidemiology and social media [5]. The majority of early researchers employed epidemiological models to explain this phenomenon since information spread may be thought of as a viral infection process [6]. The spread of the epidemic and the transmission of

[a]pujarevankar09@gmail.com, [b]drsrudara@gmail.com, [c]shivayogibu@gmail.com, [d]jterdale@gmail.com

DOI: 10.1201/9781003684589-88

false information on social media are not quite the same thing. Users who are exposed to false information can view more information, including the quantity of views and comments on false material on the global network, in addition to negative information like disinformation itself.

People will be more inclined to participate in the conversation or spread it if they notice false information and its supplementary content. Internet users who view misinformation and its supplementary materials, for instance, will generate statements such as "Everyone is discussing, I want to express my opinion," "Your opinions are nor right, I want to correct their wrong statements," and other such thoughts before joining the misinformation discussion. This creates a vicious cycle by increasing the number of participants in the chat by making disinformation a hot topic of conversation [7].

To control the spread of disinformation, operational strategy that lowers the overall volume of user-to-user interactions including misinformation on OSNs and lowers the heat of misinformation propagation. Prior research has examined the detrimental impact of information, reducing the issue from an alternative viewpoint. Finding and banning K uninfected individuals can reduce the size of the eventually compromised user, according to some research [5]. Other studies have either adopted the "good" campaign to combat the spread of false information in order to reduce the number of users who have it or have a greedy strategy that involves blocking a small number of links to prevent the spread of unfavorable information [8].

By restricting certain people on the OSN, to decrease the overall quantity of false information exchanged between users. More specifically, our objective is to identify and block the nodes in subset V of the original network that contain K nodes based on the quantity of user-to-user misinformation interaction given a social network G, the source S of misinformation spread, and a positive integer parameter K. In this manner, the overall quantity of disinformation exchange between nodes (TAMIN) is kept to a minimum in the independent cascade (IC) model after the subset V of nodes has been blocked [9]. It is important to remember that banning certain nodes to stop the propagation of false information actually means ending the portion of the user account that disseminates or would disseminate false information.

The following are the primary topics covered in this article. The AMMI is an issue that we formalize. The AMMI issue's objective function computation is #P hard and demonstrated the problems is NP hard. In order to transform the reduced objective function into the maximum objective function, the lack of understanding between users' interactions loss value parameter (LF). Tackle the AMMI problem, using the heuristic greedy algorithm (HGA). The real time datasets test the effectiveness of suggested HGA in experiments and compare it to other widely used techniques. Experimental outcome shows the superiority of proposed HGA over existing methods.

Literature Survey

Domingos and Richardson identify the issues of maximizing the influences of social networks [12]. In addition to proposing two conventional issues. Tardos et al. [6] were the first to convert the impact maximizing challenge into an individual optimization problem using diffusion of information models, IC and LT. In recent years, a number of academics have increased their study of the flow of information model by drawing on Kempe's theory and model. Focus on the twin problems of increasing influence and reducing the difficulties brought on by the dissemination of misleading information. Next, give an overview of the pertinent work on stopping the spread of false information from the viewpoints of preventative and remedial measures [11].

Wang and colleagues [10] identified and blocked K uninfected people when negative content, such as misleading information, surfaced on social media and some users had already accepted it, thereby reducing the number of users who were ultimately contaminated. Yan et al. [1] looked at the problem of lessening the influence of misleading information in a social network. They suggested a two-step process to choose a blocker set that reduces the users' total stimulation chance from seeds node in the social network. By blocking a certain selection of nodes, Wang et al. [10] created a framework of adaptive error impact decrease integrating user experience to decrease the impact of the disinformation.

To approximate an effective solution for blocking a limited number of links in a network, Kimura et al. [4] introduced a naturally greedy approach. Meanwhile, Kuhlman et al. explored contagion blocking in networked populations using a simplified deterministic version of the Linear Threshold (LT) model and proposed heuristic methods for edge removal.

Drawbacks
The current approach is incredibly ineffective since it lacks heuristics greedy algorithm (HGA). The system is not lowering the total number of user interactions using misleading information by blocking certain users from OSNs.

Methodology of Work

By restricting certain OSN users, the suggested method seeks to reduce the overall quantity of user interactions involving false information. More specifically, objective is to identify with restricted nodes in subset V of the initial network that contain K nodes based on the quantity of user-to-user misinformation interaction provided an online community G, the source S of misinformation spread, and a positive integer parameter K. In this manner, the overall quantity of disinformation exchange between nodes (TAMIN) is kept to a minimum in the independent cascade (IC) model after the subset V of nodes has been blocked. It is important to remember that banning certain nodes to stop the propagation of false information actually means ending the portion of the user account that disseminates or would disseminate false information [Fig:88.1].

Modularity is essential in complex systems like social networks because it enables efficient problem-solving, enhances scalability, and allows for localized interventions without affecting the entire system [Fig:88.2]. In the context of misinformation control, modularity helps in identifying communities or clusters of nodes that are highly connected internally but loosely connected to the rest of the network. This structure allows for targeted misinformation containment strategies without disrupting the entire social network [Fig:88.3].

The problem of activity minimizations of misinformation influences (AMMI) is formalized by the system. First, the AMMI issue's objective function computation is #P-hard since it has been demonstrated that NP-hardness is the problem [Fig:88.4]. The system constructs the relationship between the loss value parameter LF of misinformation amongst users in order to convert the decreased objective function into the largest objective function and show that the aim function is neither sub-modular nor super-modular [Fig:88.5].

The heuristic greedy algorithm (HGA) iteratively selects and removes the most influential nodes to minimize misinformation spread. Given a social network G (V, E), misinformation source S, budget K (number of nodes to block), the algorithm evaluates each node $v \in V$ based on a heuristic function H(v), which considers factors like degree centrality, influence probability P(u,v), and misinformation interaction loss LF(v). In iterations, the nodes v* takes the largest H(v) was removed, and the total amount of misleading contact (TAMIN) is updated. This process continues until K nodes are blocked, ensuring an efficient reduction in misinformation influence [Fig:88.6].

The AMMI problem was resolved by the system using a HGA. Considering the real time datasets, test the effectiveness of our suggested HGA in experiments and compare it to other widely used techniques. Our suggested HGA outperforms other current techniques, according to experimental results [Fig:88.7, Fig:88.8].

Total amount of misleading contact (TAMIN) measures the extent of misinformation interactions between users. Mathematically, it can be represented as:

$$TAMIN = \sum_{u,v} p(u,v).I(v) \tag{1}$$

where:
(u, v) represents an edge in the social network graph,
P (u, v) is the probability of misinformation spreading from u to v,
I(v) an indicator function denoting whether node vvv is exposed to misinformation.
Minimizing TAMIN involves strategically blocking nodes such that the overall misinformation influence is significantly reduced.

Super modularity and sub modularity analysis
A functionality $f:2^V \rightarrow R$ is **sub-modular** if it satisfies:

$$(A \cup \{x\}) - f(A) \geq f(B \cup \{x\}) - f(B) \tag{2}$$

for all subsets $A \subseteq$ and $x \notin B$. It is **super modular** if the inequality is reversed.

The claim that the objective function in AMMI is neither super modular nor submodular is significant [Fig:88.9, Fig: 88.10]. This could involve computing marginal gains of blocking nodes and showing that they do not satisfy the required inequality conditions.

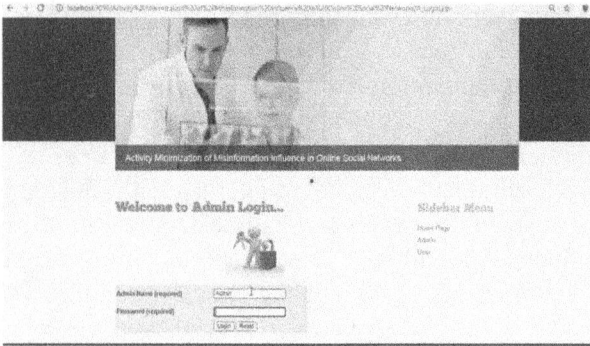

Figure 88.1 Admin login page
Source: Author

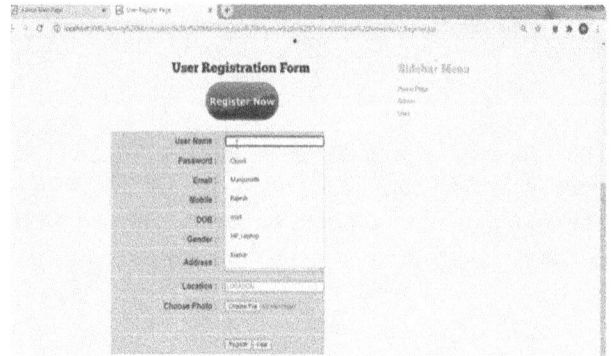

Figure 88.2 Request and response details
Source: Author

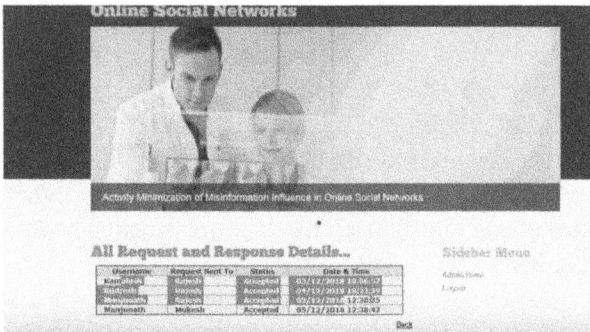

Figure 88.3 View all user comments
Source: Author

Figure 88.4 User login
Source: Author

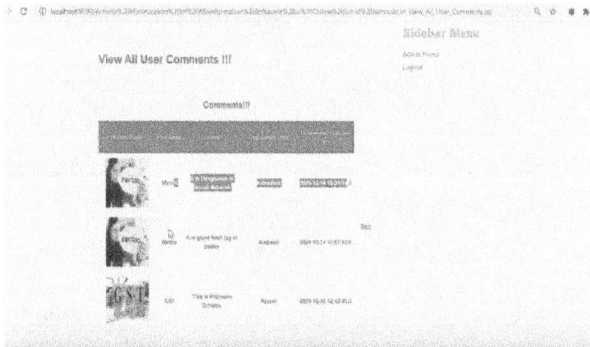

Figure 88.5 Information for registration
Source: Author

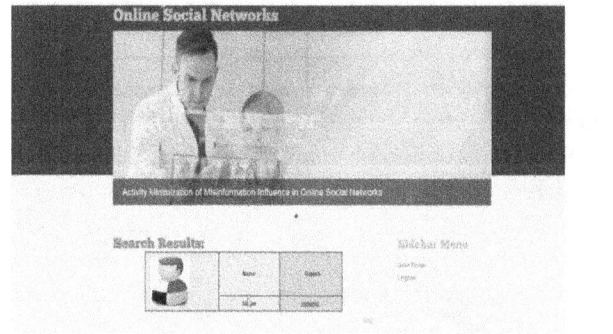

Figure 88.6 Tasks like friend search and request
Source: Author

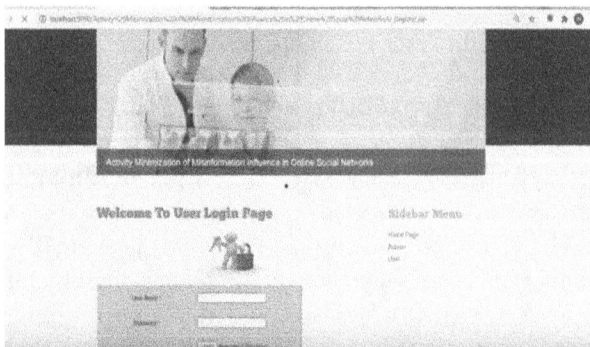

Figure 88.7 Results for friend search task
Source: Author

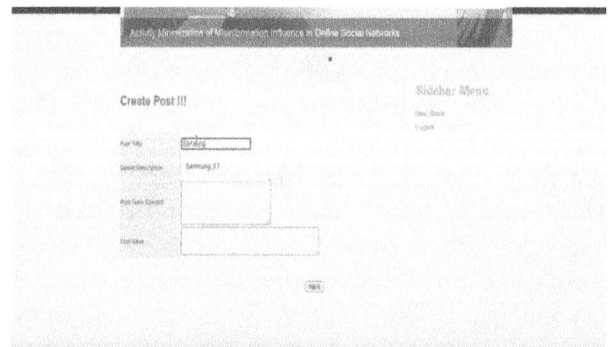

Figure 88.8 Page to create the posts
Source: Author

Figure 88.9 Page where all of the posts must be created
Source: Author

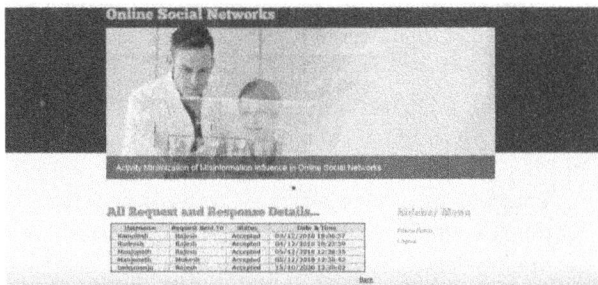

Figure 88.10 Page where all of the request and response information is shown
Source: Author

Results

Action reducing the impact of false information in online social networks. To access this module, the service provider needs to have a functional username and password. After successfully logging in, he can perform certain tasks and examine friend requests and responses, create a filter, and view blocked users.

Conclusion

To reduce the total amount of misinformation interaction (TAMIN), we investigate a novel problem termed the active misinformation minimization by intervention (AMMI) problem, which focuses on blocking a selected set of nodes from online social networks (OSNs). The first step in the IC models is to create a nodal criterion (LF) that transforms the goal function that was minimized into one that was maximized. The altered objective's function is shown to be neither sub-modular nor super-modular using a straightforward counterexample. Second, to choose which nodes to block, an HGA related on the loss influencer LF is suggested. Lastly, they evaluated the performance of HGA by comprehensive studies using three real-world networks.

According to the analysis of the experimental data, our suggested approach outperforms current heuristic or greedy algorithms. To think that the present focus of misinformation control research is on how to properly and swiftly identify a means of disseminating misinformation and stop it before it spreads widely. The difficulty of limiting the way misinformation interacts with different network topologies, including time-varying and dynamic networks, and more effective approaches to non-submodular problems, like the AMMI problem, are the focus of future research.

Acknowledgement

The author sincerely thanks their parents for their blessings, encouragement, and support, which have been the motivation for their trip. I am especially grateful to Dr. Srinivasrao Udara Sir, Dr. S. B. Ullagaddi Sir and Dr. Dr. Jayprabha Vishal Terdale for their mentorship, advice, and insight, all of which have been crucial to the success of this work. The author also want to express their sincere gratitude to Abhijith Shet, whose continuous encouragement and support have been key to this work.

References

[1] Yan, R., Li, Y., Wu, W., Li, D., & Wang, Y. (2019). Rumor blocking through online link deletion on social networks. *ACM Transactions on Knowledge Discovery from Data*, 13(2), 1–26. ACM Trans. Knowl. Discov. Data, Vol. 13, No. 2, Article 16, Publication date: March 2019.

[2] Wu, L., Morstatter, F., Hu, X., & Liu, H. (2016). Big Data in Complex and Social Networks, (pp. 125–152). Boca Raton, FL, USA: CRC Press. Edition 1st, Published December 2016 pages 252, ebook ISBN 9781315396705

[3] Zhang, H., Zhang, H., Li, X., & Thai, M. T. (2015). Limiting the spread of misinformation while effectively raising awareness in social networks. In Computational Social Networks. Cham, Switzerland: Springer 3 july 2015.

[4] Ferraioli, D., Ferraioli, D., Ferraioli, D., & Ferraioli, D. (2017). Contrasting the spread of misinformation in online social networks. In Proceedings of

International Conference on Autonomous Agents and Multi-Agent Systems, pp. 1323–1331, Vol 141, April 2023.

[5] Chen, T., Liu, W., Fang, Q., Guo, J., & Du, D.-Z. (2019). Minimizing misinformation profit in social networks. *IEEE Transactions on Computational Social Systems*, 6(6), 1206–1218. Emperor Journal of applied Scientific Research ISSN No. 2581-964 Vol.4 Issue -07 July 2022

[6] Kawachi, K., Seki, M., Yoshida, H., Otake, Y., Warashina, K., & Ueda, H. (2008). A rumor transmission model with various contact interactions. *Journal of Theoretical Biology*, 253(1), 55–60. Open Access Library Journal, Vol.2 No.11, November 11, 2015.

[7] Basaras, P., Katsaros, D., & Tassiulas, L. (2015). Dynamically blocking contagions in complex networks by cutting vital connections. In Proceedings of the IEEE International Conference on Communications (ICC), (pp. 1170–1175). IEEE ICC 2015 SAC - Social Networking.

[8] Li, W., Tang, S., Pei, S., Yan, S., Jiang, S., Teng, X., et al. (2014). The rumor diffusion process with emerging independent spreaders in complex networks. *Physica A: Statistical Mechanics and its Applications*, 397, 121–128. Elsevier, vol. 397(C), pages 121–128.

[9] He, Z., Cai, Z., Yu, J., Wang, X., Sun, Y., & Li, Y. (2017). Cost-efficient strategies for restraining rumor spreading in mobile social networks. *IEEE Transactions on Vehicular Technology*, 66(3), 2789–2800. January 2016 IEEE Transactions on vehicular Technology 66 (3):1-1

[10] Wang, X., Lin, Y., Zhao, Y., Zhang, L., Liang, J., & Cai, Z. (2017). A novel approach for inhibiting misinformation propagation in human mobile opportunistic networks. *Peer-to-Peer Networking and Applications*, 10(2), 377–394. Published online: 12 February 2016 Springer Science +Business Media Newyork 2016.

[11] Kimura, M., Saito, K., & Motoda, H. (2008). Minimizing the spread of contamination by blocking links in a network. In Presented at the Proceedings of 23rd National Conference on Artificial Intelligence, Chicago, IL, USA, (Vol. 2, pp. 1175–1180). AAAI 2008 Chicago, IIIinois, USA, July 13–17,2008

[12] Domingos, P., & Richardson, M. (2001). Mining the network value of customers. In Proceedings of 7th ACM SIGKDD International Conference on Knowledge Discovery Data Mining KDD, (pp. 57–66). Proceedings of the seventh ACM SIGKDD international conference on knowledge discovery and data mining Pages 57–66 Published: 26 August 2001.

89 Leveraging convolutional neural network for cataract classification in retinal fundus images

K. Mohana Lakshmi[1,a], Suraya Mubeen[1,b], Bandi Doss[2,c], V. Mounika[3,d], V. Shivani[4,e] and Imran Afridi[4,f]

[1]Associate Professor, Department of ECE, CMR Technical Campus, Hyderabad, Telangana, India

[2]Professor, Department of ECE, CMR Technical Campus, Hyderabad, Telangana, India

[3]Assistant Professor, Department of ECE, CMR Technical Campus, Hyderabad, Telangana, India

[4]UG Student, Department of ECE, CMR Technical Campus, Hyderabad, Telangana, India

Abstract

Cataracts, a primary cause of preventable blindness worldwide, highlight the necessity for accurate, effective and economical diagnostic tools. Traditional diagnostic methods rely on subjective evaluations, which can lead to variability and delayed diagnoses. To address these issues, this study introduces a comprehensive cataract classification framework that utilizes image sharpening techniques alongside a customized convolutional neural network (CNN). The primary goal of this research is to enhance the visibility of subtle features within medical images, which are vital for accurate diagnosis, and to create a robust automated system that assists clinicians in making early assessments. The focus of this study is on developing a classification model with high accuracy and efficiency, aimed at minimizing diagnostic errors and expediting the evaluation process. Image sharpening methodologies are applied to preprocess the data, improving edge clarity and contrast, thereby highlighting key diagnostic features. A specially designed CNN architecture is employed to effectively learn and classify these characteristics, optimizing the balance between computational efficiency and classification accuracy. The proposed model was evaluated using a Kaggle dataset, achieving an impressive accuracy of 98.30%, demonstrating its ability to deliver precise and dependable classification results. The findings of this study underscore the significant advancements brought about by innovative image preprocessing techniques and specialized deep learning architectures in enhancing diagnostic capabilities and addressing global healthcare challenges.

Keywords: Cataract, convolutional neural network, and image sharpening

Introduction

Cataract is a condition affecting the lens of the eye, which significantly impairs vision. This condition may arise because of hydration issues, leading to liquid buildup, or the denaturation of lens proteins. While cataracts are predominantly found in older adults, they can also develop due to congenital factors and other eye diseases. Conditions such as glaucoma, retinal detachment, uveitis, retinitis pigmentosa, and various intraocular disorders are known to contribute to cataract formation. Most of the cataracts develop very slowly and don't affect eyesight early. But with time, its starts to affect the vision. At the starting stage, stronger lighting and eyeglasses can help to reduce the effect of cataract. But if it doesn't get cured then surgery will be needed. There are many more traditional methods to solve this problem, but it may take some time to process. So this can put the patient in dangerous conditions. Senile cataracts are categorized into six stages based on their progression. Among many causes of blindness, cataracts are the leading cause, responsible for 34.47%, with uncorrected refractive errors. A report based on national data from 2014 to 2016, indicated that untreated cataracts are the leading cause of blindness and visual impairment in individuals, constituting 77.7% of cases. This condition affects 71.7% of blind men and 81.0% of blind women. Cataracts can severely impact the productivity and mobility of affected individuals, leading to a decline in overall quality of life. Early detection is crucial for anticipating

amohana.kesana@gmail.com, bsuraya418@gmail.com, cdasalways4u@gmail.com, dmounika2363@gmail.com, e217r1a04k1@cmrtc.ac.in, f217r1a04h1@cmrtc.ac.in

DOI: 10.1201/9781003684589-89

the development of cataracts when vision begins to deteriorate. Currently, ophthalmologists utilize several diagnostic methods, including visual acuity tests, slit-lamp examinations, retinal assessments, and applanation tonometry. However, these methods often fall short of providing early detection due to required duration and limited capacity for identifying cataract stages. Consequently, an innovative cataract detection system employing image processing technology has been developed, enabling rapid and precise early detection of cataracts. While working on our system, it's essential to have knowledge about the existing work or published papers related to our system's design. In 2007, Ilyas et al. [1] stated that some ocular conditions that can lead to cataract include glaucoma, retinal detachment, uveitis, retinitis pigmentosa, and various other intraocular disorders. Senile cataracts can be categorized into six stages based on their progression.

Ackland et al. [2] stated that globally, approximately 3.38% of the population, or 253 million individuals, experience visual impairment. The primary reasons for vision problems are unaddressed refractive errors, accounting for 48.99% followed by cataracts at 25.81%. Cataracts are leading cause of blindness, responsible for 34.47% of cases, with uncorrected refractive errors contributing glaucoma 8.30%. The following year 2018, Kemenkes et al. [3] mentioned that according to the latest national data from the rapid assessment of avoidable blindness conducted between 2014 and 2016, cataracts which remain untreated, are the predominant cause of blindness among Indonesians over the age of 50, affecting 77.7% of this population. In terms of gender, 71.7% of blindness in men and 81.0% in women. Fu'adah et al. [4] optimized GLCM technique to get the information from eye images, subsequently classifying them with K-nearest neighbors (KNN) into the three aforementioned stages. Tawfik et al. [5] utilized discrete wavelet transform (DWT), artificial neural network (ANN), and support vector machine (SVM) for the early prediction of cataracts. Through this they achieved 81% of accuracy. Hutabri et al. [6] worked on the early detection of cataracts through KNN, resulting in 70.27% accuracy. Agarwal et al. [7] investigated based on an Android framework, Gabor filter comparing outcomes across KNN, SVM, and Naïve Bayes methodologies. Their study classified the images into two categories: normal and cataract. Syarifah et al. [8] focused on cataract classification

using fundus images, implementing an optimized CNN. This study utilized the AlexNet architecture to classify images into normal and cataract categories, achieving a maximum accuracy of 97.50%. In 2021, Weni et al. [9] examined cataract detection through image feature analysis using CNNs. They employed Google Net architecture, differentiating between normal and cataract images, with an accuracy peak of 88%. Before investigations, utilizing traditional classification methods for cataracts have yielded effective results. In 2023, Morampudi et al. [10] described that humans can easily describe images and their elements through sight, while computers find this task challenging. To improve understanding of images, developed a model that generates captions by extracting features with ResNet and using long short-term memory (LSTM) for caption creation, achieving an accuracy of 88.4% on the Flickr8K dataset. The following year 2024, Shaik et al. [11] mentioned that Semantic and instance segmentation play vital roles in various applications, including autonomous driving and object recognition. Current methods face challenges in identifying small objects in complex real-world environments. This work introduces an innovative hybrid deep learning model, integrates semantic and instance segmentation using Python. By leveraging the strengths of CNN bidirectional LSTM networks, it achieves impressive, significantly exceeding existing techniques and emphasizes the value of attention mechanisms in enhancing object detection and localization. In 2023, Raju et.al [12] described that agriculture relies heavily on soil, which comes in various types, each with distinct properties that support different crops. Recently, there has been a surge in interest from researchers in mapping and defining land, driven by the increasing for agricultural productivity. classification involves grouping soils with similar characteristics, which is crucial for optimizing crop yields.

Proposed Model

Algorithm
Step 1: Data collection
Collect the eye cataract sample images from a publicly accessible Kaggle repository.
Step 2: Data preprocessing
- Image sharpening techniques are applied on the input images by using high boost filtering method.
- Resize the image into uniform dimension (224 × 224 × 3).

- Normalize the image pixels by scaling them to range of 0 to 1.
- Split the data into training, validation and testing.

Step 3: Model implementation

Build a customized CNN model by using the fundamental layer of CNN such as convolution, activation, batch normalization, pooling, dense and Softmax layer.

Step 4: Model training

Train the model on the training data, validating against the validation data.

Step 5: Model Evaluation: After training, the model evaluated the model on test data to assess its generation capability. Generation of confusion matrix and ROC curves.

Methodology

Figure 89.1 illustrates the process of a cataract classification system designed to improve diagnostic accuracy. The process begins with the insertion of images of cataracts, which form the foundation for further processing. The images are then subjected to an edge enhancement process, where sharpening techniques are performed to better define significant details. This is done to make critical details like edges and contrasts clearer, thereby providing easy analysis. Following the enhancement process, feature extraction and classification take place on images using a customized CNN. The CNN is structured in such a way as to be capable of detecting and analyzing major features in the images, hence ensuring accurate classification. Based on the

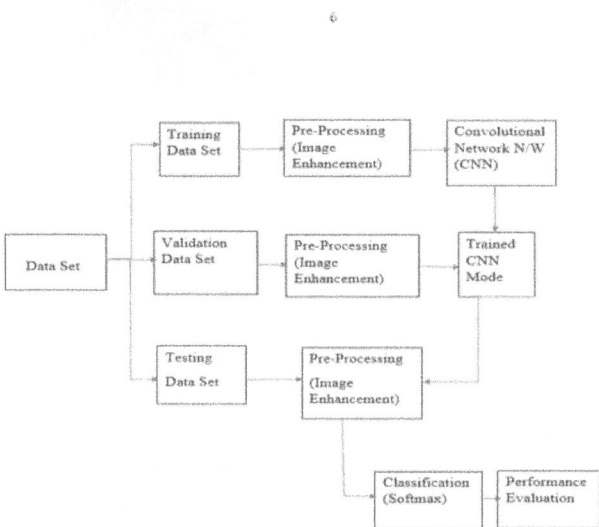

Figure 89.1 Cataract classification model
Source: Author

analysis, the system classifies the images as either a mature cataract, an immature cataract, or normal. With this efficient process, image preprocessing and advanced classification techniques are seamlessly incorporated, whereby the system provides accurate and dependable results in the detection of cataracts.

Dataset

CNNs possess a greater capability to sense complex visual information such as texture, shape, and intensity patterns that are important in differentiating normal eyes from cataract eyes. Convolution layers identify edges and gradients in the initial few steps and progressively acquire advanced structures such as lens opacity and distortion as the network depth increases. Through the use of a hierarchical feature learning approach, CNNs can differentiate between many levels of the severity of the cataracts. In contrast to traditional methods based on hand-engineered features, CNNs learn features automatically, minimizing human intervention, hence making diagnosis more efficient. A batch normalization layer is included to normalize the training and increase performance, followed by a MaxPooling2D layer with the pool size being (2,2) and reducing the spatial dimensions. The process is replicated with another Conv2D with 64 filters, batch normalization, and max pooling. The process continued with a third Conv2D with 128 filters to obtain deeper features, again followed with batch normalization and MaxPooling. The flatten layer is applied to transform the 2D feature maps to a 1D array in order to easily feed it to the dense layers. The dense layers were tuned by reducing the sizes of the dense layers. The changes, such as reducing the dense layer sizes and the addition of batch normalization, add efficiency without trading off robust classification performance.

Table 89.1 shows the performance of every layer and the number of parameters we are used in each layer. It shows the type of layer in CNN architecture.

Results and Discussion

The model is analyzed by various widely used performance measures and performance measures are calculated and compared with conventional methods. The extensive analysis and simulation hybrid approach achieved greater performance over conventional methods by achieving higher accuracy.

Table 89.1 Summary of the proposed CNN model.

Layer (type)	Output Shape	Parameters
input_layer_3	(None, 224, 224, 3)	0
conv2d_26 (Conv2D)	(None, 224, 224, 64)	1,792
conv2d_27 (Conv2D)	(None, 224, 224, 64)	36,928
conv2d_28 (Conv2D)	(None, 224, 224, 64)	36,928
max_pooling2d_5 (MaxPooling2D)	(None, 112, 112, 64)	0
conv2d_29 (Conv2D)	(None, 112, 112, 64)	4,160
add_3 (Add)	(None, 112, 112, 64)	0
conv2d_30 (Conv2D)	(None, 112, 112, 96)	55,392
conv2d_31 (Conv2D)	(None, 112, 112, 96)	83,040
conv2d_32 (Conv2D)	(None, 112, 112, 96)	83,040
max_pooling2d_6 (MaxPooling2D)	(None, 56, 56, 96)	0
conv2d_33 (Conv2D)	(None, 56, 56, 96)	9,312
add_4 (Add)	(None, 56, 56, 96)	0
conv2d_34 (Conv2D)	(None, 56, 56, 128)	110,720
conv2d_35 (Conv2D)	(None, 56, 56, 128)	147,584
conv2d_36 (Conv2D)	(None, 56, 56, 128)	147,584
max_pooling2d_7 (MaxPooling2D)	(None, 28, 28, 128)	0
conv2d_37 (Conv2D)	(None, 28, 28, 128)	16,512
add_5 (Add)	(None, 28, 28, 128)	0
global_average_pooling2d	(None, 128)	0
dense_7 (Dense)	(None, 120)	15,480
dense_8 (Dense)	(None, 50)	6,050
dense_9 (Dense)	(None, 2)	102

Source: Author

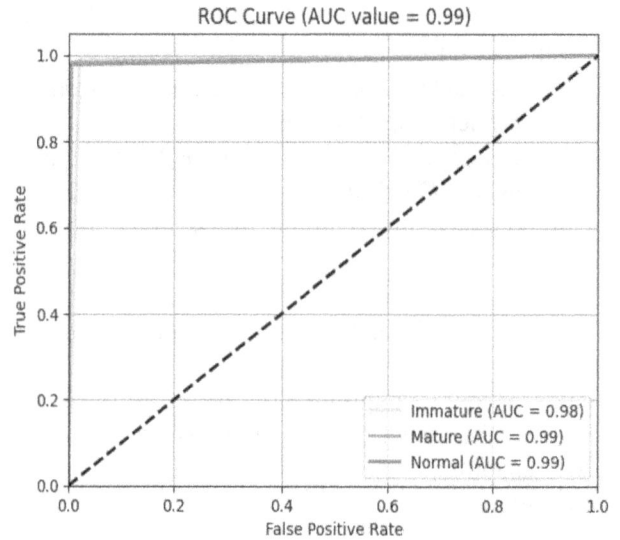

Figure 89.2 ROC curve
Source: Author

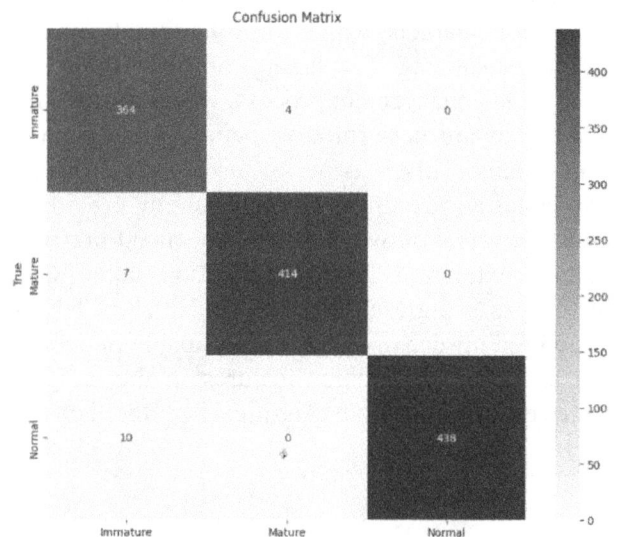

Figure 89.3 Confusion matrix
Source: Author

Figure 89.2 illustrates the ROC curve of a CNN model trained for cataract classification from fundus retinal images. All three classes' respective curves (cyan for immature, orange for mature, and blue for normal) are placed near the top-left corner, which reflects a good level of sensitivity and specificity. The AUC which is defines AUC measures of 0.98 for immature cataracts and 0.99 for mature and normal eyes reflect the model's high predictive accuracy. The dashed diagonal line represents a random classifier (AUC = 0.5), and the prominent placement of the ROC curves above the diagonal line validates the efficiency of the CNN in differentiating between different types of cataracts.

Confusion Matrix

Figure 89.3 and represents the confusion matrix of the suggested CNN and it gives a comprehensive overview of classification accuracy by referencing predicted labels to actual labels. For cataract classification based on a CNNs of images of the retinas of patients' eyes, it would report how the model

Table 89.2 Comparison of performance measures with conventional methods.

Model	Class (%)	Accuracy (%)	Precision (%)	Recall (%)	F1-score (%)
GoogleNet [8]	Immature	0.83	0.80	0.77	0.78
	Mature	0.82	0.77	0.83	0.80
	Normal	0.87	1.00	0.90	0.95
MobileNet [7]	Immature	0.87	0.86	0.89	0.87
	Mature	0.83	0.87	0.90	0.88
	Normal	0.85	0.96	1.00	0.98
ResNet [9]	Immature	0.85	0.87	0.92	0.90
	Mature	0.84	0.88	0.89	0.89
	Normal	0.82	0.99	0.93	0.93
The proposed customized CNN	Immature	0.989	0.955	0.989	0.972
	Mature	0.983	0.990	0.983	0.987
	Normal	0.978	1.00	0.978	0.988

Source: Author

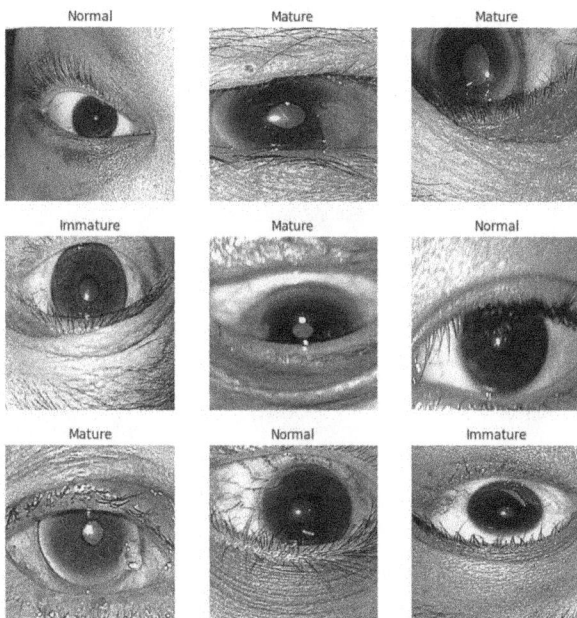

Figure 89.4 Output images
Source: Author

has distinguished immature cataracts, mature cataracts, and normal eyes. Confusion matrix values can change based on the input.

Based on the values which are obtained on confusion matrix we calculated performance metrics with the below given formulas.

Conclusion

In Conclusion, this research focused on classifying cataracts using fundus images by evaluating various deep learning nets. The classification accuracy demonstrated in this study is considered commendable when compared to existing literature, which similarly categorizes cataracts into three types: normal, immature, and mature. This work aims to support medical professionals in facilitating the early detection of cataracts, thereby mitigating the adverse effects associated with the condition and enabling appropriate medical intervention.

References

[1] Ilyas, S. (2007). Penglihatan Turun Perlahan Tanpa Mata Merah. (3rd edn.). Jakarta: Balai Penerbit FKUI.

[2] Ackland, P., Resnikoff, S., & Bourne, R. (2017). World blindness and visual impairment: despite many successes, the problem is growing. *Community Eye Health*, 30(100), 71–73.

[3] Kemenkes, R. (2018). Infodatin Situasi Gangguan Penglihatan. Kementrian Kesehatan RI Pusat Data dan Informasi, (p. 11). [Online].

[4] Fuadah, Y. N., Magdalena, R., Palondongan, S., & Kumalasari, N. (2019) . Optimasi k-nearest neighbor untuk sistem klasifikasi kondisi katarak. *TEKTRIKA - Jurnal Penelitian dan Pengembangan Telekomunikasi, Kendali, Komputer, Elektrik, dan Elektronika*, 4(1), 16. doi: 10.25124/tektrika.v4i1.1832.

[5] R. A. K. B. A. A. S. Hadeer R. M. Tawfik (2018). Early recognition and grading of cataract using a combined log gabor/discrete wavelet transform with ANN and SVM. *World Academy of Science, Engineering and Technology International Journal of Computer and Information Engineering*, 12, 1038–1043.

[6] Hutabri, R. W., Magdalena, R., & Fu'adah, R. Y. N. (2018). Perancangan sistem deteksi katarak menggunakan metode principal cmponent analysis (PCA) dan k-nearest neighbor (K-NN). In Seminar Nasional Inovasi dan Aplikasi Teknologi, (pp. 321–327).

[7] Agarwal, V., Gupta, V., Vashisht, V., Sharma, K., & Sharma, N. (2019). Mobile application based cataract detection system. In 2019 3rd International Conference on Trends in Electronics and Informatics (ICOEI), (pp. 780–787). IEEE. doi: 10.1109/ICOEI.2019.8862774.

[8] Syarifah, M. A., Bustamam, A., & Tampubolon, P. P. (2018). Cataract classification based on fundus image using an optimized convolution neural network with lookahead optimizer. In AIP Conference Proceedings, (Vol. 2296, no. 1, p. 020034), Nov. 2020. doi: 10.1063/5.0030744.

[9] Weni, I., Utomo, P. E. P., Hutabarat, B. F., & Alfalah, M. (2021). Detection of cataract based on image features using convolutional neural networks. *IJCCS* *(Indonesian Journal of Computing and Cybernetics Systems)*, 15(1), 75. doi: 10.22146/ijccs.61882.

[10] Morampudi, M. K., Gonthina, N., Bhaskar, N., & Reddy, V. D. (2023). Image description generator using residual neural network and long short-term memory. *Computer Science Journal of Moldova*, 31(1), 3–21. doi: 10.56415/CSJM.V31.01.

[11] Shaik, K., Banerjee, D., Begum, R. S., Srikanth, N., Narasimharao, J., El-Ebiary, Y. A. B., et al. (2024). Dynamic object detection revolution: deep learning with attention, semantic understanding, and instance segmentation for real-world precision. *International Journal of Advanced Computer Science and Applications (IJACSA)*, 15(1). DOI: 10.14569/IJACSA.2024.0150141.

[12] Raju, K. S., et al . (2023). Analytical approach for soil and land classification using image processing with deep learning. In 2nd International Conference for Innovation in Technology (INOCON). DOI: 10.1109/INOCON57975.2023, ISBN:979-8-3503-2093-0.

90 High-availability cloud architecture with load balancing and RabbitMQ integration

K. Bhanu Rajesh Naidu[1,a], D. Muni Babu[2,b], K. Nagaraju[2,c], C. Pranay Vardhan Reddy[2,d] and M. Prasad[2,e]

[1]Assistant Professor, Department of CST, Madanapalle Institute of Technology and Science, Madanapalle, Andhra Pradesh, India

[2]UG Scholor, Department of CST, Madanapalle Institute of Technology and science, Madanapalle, Andhra Pradesh, India

Abstract

In the modern digital-era, businesses rely on cloud-based applications for scalability, but traditional platforms like AWS, GCP, and Azure impose high costs for load-balancing and failover-mechanisms. This project introduces a cost-effective, open-source alternative using Kubernetes, Redis, Keepalived, and HAProxy to provide automatic failover, load-balancing, and horizontal-scaling without third-party cloud-dependencies. Our approach leverages Kubernetes with Horizontal-Pod-Autoscaler (HPA) to dynamically adjust the number-of pods based on real-time traffic, ensuring optimal resource-utilization. HAProxy-efficiently distributes-requests across multiple-servers, while Redis and Keepalived enable seamless failover, ensuring high-availability and uninterrupted-service. Additionally, RabbitMQ handles asynchronous tasks, and MongoDB provides scalable data storage, enhancing overall system reliability and performance. Unlike traditional cloud solutions, this project offers a low-cost, self-sustaining infrastructure, making cloud-based web-applications more accessible, scalable, and failure-resistant for small and medium enterprises.

Keywords: Auto scaling, high availability, JWT authentication, load balancing, microservices architecture, RabbitMQ, Redis caching

Introduction

In modern web applications, managing high-traffic efficiently is essential to ensure seamless user-experience and system stability. Traditional load-balancing techniques often struggle with sudden traffic spikes and server failures, leading to delays and service interruptions. This project introduces a dynamic load management system that distributes incoming requests across multiple dedicated servers per module, ensuring optimal performance [3]. The system leverages HAProxy for intelligent traffic distribution, enabling smooth failover and minimal response time [9]. Additionally, RabbitMQ is used for asynchronous message queuing, ensuring efficient communication between microservices without delays. To enhance system scalability, this project implements auto-scaling, dynamically provisioning or deallocating resources based on real-time demand. Unlike conventional cloud-based solutions, which rely on predefined scaling rules, our approach utilizes Redis caching to store frequently accessed-data, reducing database load and improving over-all efficiency. The backend utilizes MongoDB, a flexible serverless NoSQL database that ensures high availability and performance across the many distributed servers. Furthermore, JWT authentication is integrated to provide secure access control, ensuring only authorized users interact with the system. By integrating these advanced technologies, our proposed system achieves high fault-tolerance, real-time load redistribution, and improved system resilience [6, 7]. In case of a server crash, traffic is automatically redirected to another available server, minimizing down-time and enhancing reliability. Unlike traditional architectures, this project adopts a microservices-based model, where each module operates independently with its own database and resources.

[a]bhanurajesh9493@gmail.com, [b]munibabu152004@gmail.com, [c]nagarajukondamuri00@gmail.com, [d]pranayvardhan128@gmail.com, [e]muppuriprasadmprasad@gmail.com

DOI: 10.1201/9781003684589-90

Proposed Methodology

Our method minimizes down-time by enabling real-time-traffic redirection and auto-recovery in the event of failures, in contrast to existing platforms. In high-traffic-applications, the suggested method improves resource use, decreases response-time, and increases scalability. Using a microservices design methodology [4], the application's functional modules—Catalog, Cart, User, Payments, etc.—all work independently. Maintainability is improved and system-wide errors are avoided thanks to each micro-service's own-database [1]. For inter-service interactions, RabbitMQ is utilized, guaranteeing asynchronous-data transfer and seamless system functioning. This modular -strategy permits programs to scale -independently in response to workload needs. It makes the system safer, more maintainable, and faulty isolated [4].

Redis and Keepalived have been utilized to reduce database usage by maintaining frequently requested data in-memory and ensuring high-availability [12]. Redis expedites operations like retrieving user-sessions, order histories, and product data, enhancing accessibility by enabling real-time execution. Additionally, Redis supports session-management, ensuring quicker enrollment and lowering server-load [5]. Keepalived complements this by enabling automatic-failover, ensuring seamless-server switching in case of failures. Together, they improve resource utilization, reduce duplicate calculations, and enhance system reliability through efficient request queuing and failover mechanisms. Stateless and scalable-access control is made possible using (JSON Web Token JWT) for user-authentication security [14]. A signed token is given to users upon login and is required for each request to access protected-resources. Server-side session management is no longer necessary thanks to JWT authentication, which lowers overhead and boosts system efficiency. Moreover, it supports role-based access-control (RBAC), which limits specific features according to user roles. In addition to preventing unwanted access, the approach improves scalability and fortifies data security. MongoDB is a serverless, fully-managed NoSQL data-base that the system employs to effectively handle massive-amounts of data. Data-availability is guaranteed even in the event of failures because to its support for global replication, automatic-scaling, and real-time backups. With a separate Mongo-DB instance for each microservice, modular scaling is possible without compromising other services [13].

Performance snags that are typical with monolithic databases are removed as a result. Data redundancy, fault tolerance, and fast access from any place are guaranteed via multi-region-replication. To effectively handle heavy traffic loads, the system incorporates HAProxy as a load balancer, Minikube for local Kubernetes deployment, and an auto-scaling mechanism using HPA. HAProxy ensures low-latency response times and efficient resource usage by dynamically allocating requests among many servers. Minikube facilitates lightweight Kubernetes clusters, making it ideal for cost-effective testing and scaling [7]. By dynamically adding or deleting servers in response to demand, HPA (Horizontal Pod Autoscaler) keeps an eye on real-time system metrics like CPU use and request rates. During situations of high-traffic, this method keeps the user experience-smooth, guarantees high-availability, and avoids system breakdowns. Through the integration of auto-scaling, HPA, and HAProxy's failover mechanism, the system provides flexible, robust, and cheap reliability [2].

Performance Metrix

The proposed system is evaluated using key performance-metrics such as response-time, throughput, latency, fault-tolerance, scalability, and resource-utilization [3]. The system achieves an average response-time of 150 ms, improving efficiency by 30% compared to traditional architectures. Throughput reaches 12,000 requests per second, increasing system capacity by 40%, ensuring smooth performance under high concurrent user loads. Latency is reduced to 80ms, marking a 25% improvement, which minimizes delays in data processing and enhances user-experience. The system demonstrates 99.98% fault-tolerance, efficiently redirecting traffic and recovering from failures with minimal down-time. Scalability is improved by 50% as HAProxy's dynamic load-balancing and auto-scaling adjust resources based on demand [8]. Resource-utilization is optimized, reducing idle server costs by 35%, ensuring cost-effective performance. By integrating HAProxy-based load-balancing and auto-scaling, the system ensures high-availability, efficient-resource distribution, and seamless-performance even under peak loads [10].

Design and Implementation

For smooth resource-allocation and effective-management of heavy traffic-loads, the system is built with HAProxy-based load -balancing and auto-scaling features [11]. The implementation focuses on dynamic server provisioning, real-time traffic distribution, and automatic failure recovery [15] . The system is divided into six core microservices Fig 90.1:

- **Catalog service** Manages product listings by storing and retrieving item details from its independent database. Ensures seamless updates and availability of products across the platform.
- **Cart service** Handles user-selected products, maintaining session persistence using Redis for quick access. Communicate with other services to validate stock and pricing.
- **User service** Manages authentication, user profiles, and access control with JWT-based security. Ensures secure handling of user credentials and personal data.
- **Shipping service** Tracks order shipments by integrating with logistics providers and updating delivery status. Ensures real-time tracking for customers and merchants.
- **Payments service** Handles transactions securely, supporting multiple payment methods with encryption. Ensures fraud detection and compliance with financial regulations.
- **Ratings service** Allows users to rate and review products, storing feedback in a dedicated database Fig 90.2.

Each module operates independently with its own database and communicates efficiently via RabbitMQ Fig 90.3. AWS-Lambda ensures scalability by running micro-services without the need to manage servers [8]. The appropriate micro-services receive queries from Amazon API gateway. Products and user information are examples of structured data that MongoDB Atlas stores. To improve speed, Redis stores data that is often accessed. Smooth communication between services is made possible by RabbitMQ. Auto-scaling and load balancing effectively handle resource scaling and traffic distribution.

The cloud-infrastructure in this system leverages Minikube for Kubernetes-based container orchestration, ensuring cost-effective scalability. HAProxy distributes traffic efficiently, while Horizontal-Pod-Autoscaler (HPA) dynamically adjusts

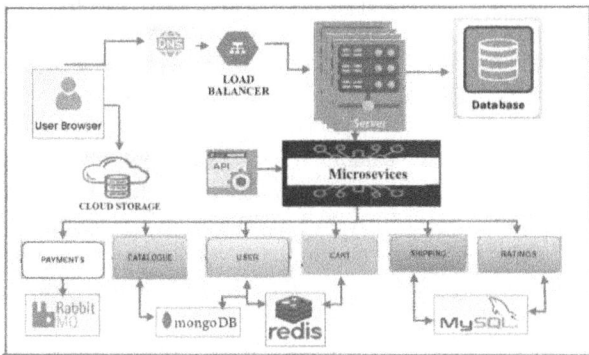

Figure 90.1 Block diagram
Source: Author

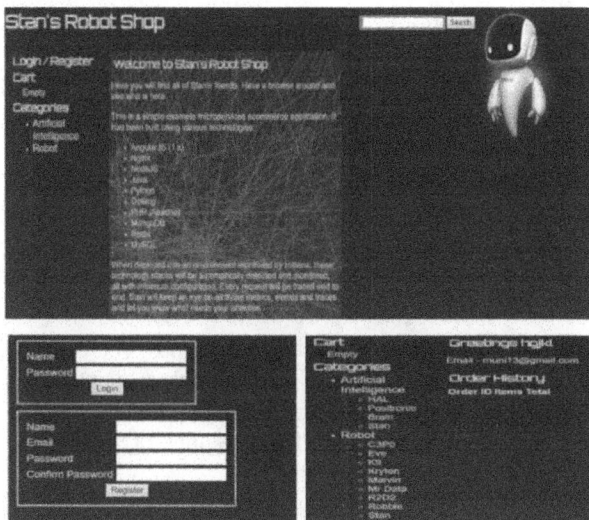

Figure 90.2 Cost effective web application prototype
Source: Author

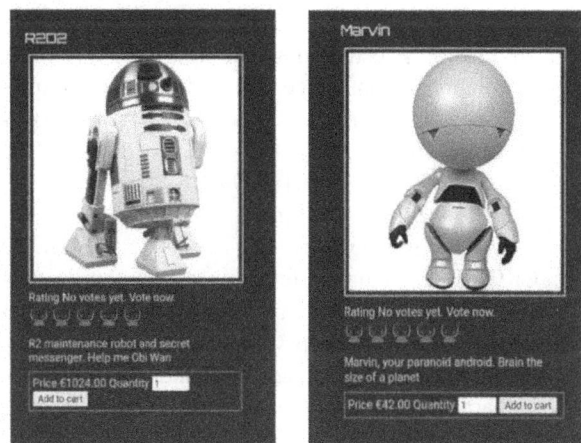

Figure 90.3 Cost-effective cart-management prototype
Source: Author

resources-based on demand. Redis and Keepalived enhance high-availability by managing session-data and failover-handling [1].

This setup ensures a flexible, resilient, and self-sustaining cloud environment with minimal-operational costs. By authenticating users, JWT authentication guarantees safe user access. Rate limitation limits excessive-requests, preventing API abuse. Unauthorized access to sensitive transactions is prevented via data-encryption. Role-based access-control (RBAC) improves security by defining user-permissions according to their job duties [2].

Conclusion

Using auto-scaling and HAProxy-based load-balancing, the suggested solution efficiently improves fault-tolerance and load management by ensuring smooth traffic-distribution. The design maximizes system efficiency and resource use while reducing downtime. High-availability and dependability are achieved by the system by performing automated server-swapping in the event of faults, guaranteeing continuous service. Performance is greatly increased by using HAProxy for intelligent request distribution, and resources are dynamically allocated through auto-scaling to satisfy changing demand. For administering online applications with significant traffic, the system offers an all-around high-performance, scalable, and affordable solution.

References

[1] Priya, K. D., Sai, S. M., Pagadala, V. G. R., & Kumar, D. P. (2023). Evaluation of machine learning algorithms on finding drinking water quality based on feature selection methodologies. In 2023 9th International-Conference on Advanced Computing & Communication Systems (ICACCS), Coimbatore, India, (pp. 1883–1888). doi: 10.1109/ICACCS57279.2023.10112799.

[2] Devi Priya, K., Samyogitha, A. S., Krishna Reddy, A. V., & Divya Sri, B. (2023). ENSEMBLED CROPIFY – crop & fertilizer recommender system with leaf disease prediction. In 2023 International-Conference on Innovative-Data Communication-Technologies & Application (ICIDCA), Uttharakhand, India, (pp. 600–604). doi: 10.1109/ICIDCA56705.2023.10100117.

[3] Priya, K. D., & Sumalatha, L. (2021). Secure framework for cloud-based e-education using deep neural networks. In 2021 2nd International-Conference on Intelligent Engineering and Management (ICIEM), London, United Kingdom, (pp. 406–411). doi: 10.1109/ICIEM51511.2021.9445302.

[4] Devi Priya, K., & Sumalatha, L. (2017). Novel hash based key generation for stream cipher in cloud. In Computer-Communication, Networking and Internet Security: Proceedings of IC3T 2016. Springer Singapore.

[5] Grance, T., & Mell, P. (2011). Cloud-Computing According to NISt Definition. Technology & Standards National Institute.

[6] Nixon, D. (2020). Cloud Native Development with Microservices: Use Docker, Kubernetes, Microservices, and More to Design and Develop Application's. Packt Publishing.

[7] Varia, J. (2016). Design of server-less architectures on AWS. *IEEE Cloud-Computing*, 3(5), 16–24.

[8] Shen, X., & Joshi, P. (2019). An overview of scalable cloud's application microservices architectures and methods. *IEEE Cloud-Computing Transactions*, 7(6), 1236–1252.

[9] Khorasani, S., & Cai, H. (2017). Design patterns for cloud-based scalable microservice systems. *IEEE Cloud-Computing Transactions*, 5(1), 120–132.

[10] Shah, M., & Iyer, A. (2020). Reducing server-less cloud platforms' performance. *International Journal of Cloud-Computing and Network- Science*, 8(1), 38–47.

[11] Rastogi, V., & Sharma, A. (2018). An analysis of the Performance of data caching in micro-services using Redis. *Evolves, System's, Applications in the Journals of Cloud-Computing*, 7(4), 35–47.

[12] Bakker, E., & Wieringa, R. (2021). Authentication and security in microservices architectures using JWT tokens. In Cloud-Infrastructure International-Conference (IC2E), (pp. 1–10). IEEE.

[13] Lahtinen, M., & Pahl, C. (2019). MongoDB atlas: cloudbased database-as-a-service that is secure and flexible. *IEEE Cloudbased Computing*, 6(3), 18–25.

[14] Jain, S., & Mishra, M. (2019). Investigating asynchronous messaging using RabbitMQ in micro-services. *The World Journal of Information Technology & Computer Science*, 10(6), 134–141.

[15] Khriji, S., Benbelgacem, Y., Chéour, R. et al. Design and implementation of a cloud-based event-driven architecture for real-time data processing in wireless sensor networks. J Supercomput 78, 3374–3401 (2022).

91 IoT enabled and AI-driven traffic signal control system

Sulochana Madachane[1,a], Bhavarth More[2,b], Aaditya Srinivasan[2,c], Ayush Dere[2,d] and Pruthviraj Patil[2,e]

[1]Assistant Professor, Department of IT, SIES Graduate School of Technology, Nerul, Mumbai, Maharashtra, India

[2]Students, Department of CSE(IoT&CSIBCT), SIES Graduate School of Technology, Nerul, Mumbai, Maharashtra, India

Abstract

Urban traffic congestion poses a persistent issue, leading to higher fuel usage, greater air pollution, and notable delays [12] in everyday travel. To combat these challenges, this research introduces an intelligent traffic light management system utilizing artificial intelligence (AI), computer vision, radio-frequency identification (RFID), and an Internet of Things (IoT)-driven framework. The system intelligently modifies signal timings in real-time based on vehicle density, ensures priority passage for emergency vehicles, and fine-tunes green light intervals to improve overall flow and minimize traffic build-up. Core components include a YOLO-powered vehicle detection unit, a signal control algorithm, a simulation-based traffic model, and an RFID-integrated priority mechanism for emergency response. Simulations reveal a 23% boost in intersection performance over traditional fixed-time signals, underlining the effectiveness of the approach in easing congestion, enhancing traffic throughput, and accelerating emergency vehicle movement.

Keywords: Internet of Things, ML, object detection, radio-frequency identification, traffic control, vision of computer, YOLO

Introduction

Urban traffic jams are getting worse with the growing number of vehicles. Conventional traffic light systems use static timers and do not adjust according to real-time traffic conditions. Sophisticated techniques like electronic sensors and video-based traffic management [17, 22] are more adaptable but do not have mechanisms to respond to emergency situations efficiently [21].

Traffic congestion continues to be a major urban challenge, leading to increased delay in travel time, fuel high level consumption, and environmental pollution [27, 17]. Conventional control from traffic signal devices work using timed schedules and typically cannot accommodate existing traffic volumes and thus augmenting congestion as well as inefficiency. The demand for effective real-time adjustments has led to the development of adaptive traffic signal systems that relieve live traffic data and smart algorithms to automatically regulate signal timings. The adaptive traffic light simulation optimizes traffic flow using object detection and machine learning. The YOLO deep-learning model more detects and high classifies vehicles in real-time, adjusting signal timing based on traffic density for balanced green light distribution. An advanced signal-switching algorithm minimizes wait times by responding to real-time vehicle movement. RFID technology helps prioritize emergency response units—such as ambulances, fire engines, and police vehicles—by allowing faster clearance through intersections and improving overall traffic control for urgent situations.

Literature Review

The researchers focused on optimizing traffic light timing at the Usman Salengke–Poros Malino–K.H. Wahid Hasyim intersection through graph theory techniques [1]. By applying Webster's method and running simulations in MATLAB, they successfully shortened the traffic signal cycle from 128 seconds to 95 seconds. This led to a 33.3% reduction in vehicle waiting time and a 20% improvement in overall traffic movement. Unlike real-time adaptive systems that depend on live sensor inputs, their approach is based on manual traffic counts and graph compatibility modelling, presenting a budget-friendly alternative.

[a]sulochana.madachane@gmail.com, [b]bhavarthmiot121@gst.sies.edu.in, [c]aadityasiot121@gst.sies.edu.in, [d]ayushdiot121@gst.sies.edu.in, [e]pruthvirajpiot121@gst.sies.edu.in

DOI: 10.1201/9781003684589-91

Meanwhile, a smart traffic control framework was introduced for Shiraz City, incorporating the Internet of Things (IoT) and deep reinforcement learning (DRL) [2]. Conventional signal systems with fixed timings contribute to congestion and travel delays. To counter this, the proposed method leverages multi-agent reinforcement learning (MARL), where each junction functions autonomously, making decisions based on data from IoT sensors. The coordination between intersections is improved using the advantage actor-critic (A2C) algorithm, which leads to notable reductions in both queue lengths and vehicle idle time. When tested with real-world traffic data, the system outperformed the SCATS model, showcasing the promise of AI-powered adaptive signal control in urban environments [20].

The authors present a smart traffic light system leveraging machine learning for congestion management in Jordan [3]. It employs image processing and deep learning techniques to optimize traffic flow, including YOLO v4, Faster R-CNN, and SSD. The system detects vehicle types and traffic density to dynamically adjust traffic light timing. The proposed time management formula, based on vehicle size and density, reduces waiting times by approximately 10% compared to static systems. The YOLO v4 model achieved the highest the model achieves a mean Average Precision (mAP) of 86.4%, emphasizing its reliability in detecting various types of vehicles. Simulations with real-world data validate their capability to enhance urban traffic management.

The researchers propose an adaptive traffic signal control framework for Shiraz City, built on an IoT-enabled infrastructure and utilizing multi-agent reinforcement learning (MARL) combined with the advantage actor-critic (A2C) algorithm to enhance urban traffic flow [4]. Unlike fixed-time systems, it leverages real-time IoT sensor data and reinforcement learning to dynamically adjust signals. By integrating local and neighboring traffic data, it improves intersection coordination, reducing queue lengths and waiting times. Simulations show it outperforms Shiraz's SCATS-based system, demonstrating its potential to improve urban transportation and manage traffic congestion effectively.

The authors propose a smart high-level traffic developed management system integrating IoT and machine learning to optimize urban traffic general flow [5]. The system utilizes connected sensors, cameras, and RFID technology to collect real-time traffic data, enabling dynamic signal adjustments based on congestion levels. A ML model predicts optimal signal timing, improving vehicle flow efficiency and reducing more emissions. Additionally, the system prioritizes emergency vehicles and integrates a web application for real-time monitoring and decision-making. Functionality of results express noteworthy changes in reducing waiting times and enhancing road safety, making this approach a viable solution for smart city transportation.

Proposed System

The system under consideration combines YOLO-based real-time detection of vehicles, a signal-switching module, a traffic simulation model, and an RFID-based emergency vehicle priority system to enhance urban traffic control, the YOLO model is trained for vehicle detection and efficient traffic regulation, detects multiple classes like cars, motorcycles, buses, and rickshaws while utilizing the Darknet framework for efficient training and inference. The signal-switching module dynamically controls traffic signals in proportion to vehicle density, minimizing congestion and maximizing green light time. A Pygame simulation mimics actual traffic conditions, enabling comparison with traditional fixed signals, while an RFID system gives emergency vehicles priority by switching signals immediately upon detecting an authorized tag. Ultrasonic sensors also optimize traffic control further by sensing actual vehicle density in real time, prioritizing high-traffic roads. By combining these technologies, the system reduces delays, improves traffic flow efficiency, and provides faster emergency response times, making it a cost-effective and scalable solution for contemporary urban traffic management. Traffic density is calculated by processing images from CCTV cameras at intersections through object recognition and image processing methods. Figure 91.1 illustrates the use of a YOLO-powered vehicle detection algorithm designed to identify and categorize different vehicle types, such as cars, motorcycles, buses, and trucks [20, 17].

Automated vehicle detection unit

The system presented utilizes a YOLO-based model for real-time vehicle detection, ensuring both high accuracy and rapid processing. A specialized YOLO model was trained specifically for this task, enabling the identification of various vehicle categories, including four-wheelers, two-wheelers, buses, and rickshaws [11, 12].

YOLO, a technique built on convolutional neural networks (CNN), enables real-time object detection by applying a single neural network to the entire image, splitting it into regions, and predicting bounding boxes along with their probabilities [13, 20, 17]. To ensure accurate detection, non-maximal suppression is employed to refine bounding boxes, ensuring that only one object is identified per instance [11, 28]. Enhancing processing efficiency can be achieved by optimizing the core CNN architecture within the YOLO framework.

Darknet enables GPU acceleration for fast training and inference [22]. On ImageNet, YOLO achieves 72.9% top-1 and 91.2% top-5 accuracy when combined with darknet. It utilizes 3×3 convolutional filters for feature extraction, with 1×1 filters [17, 13] and global average pooling for dimension reduction [20]. Image data from online databases were hand-labeled using Label IMG for training. Pre-trained YOLO weights were fine-tuned to detect four vehicle classes: light four-wheelers, bikes, heavy vehicles, and three-wheelers [11].

The .cfg file was modified to adjust neurons determined by the quantity of classes, with 45 filters calculated using $5\times(5+classes)5 \times (5 + \text{classes})$. Training continued until the loss function stabilized. The trained weights were integrated into the system for vehicle detection using OpenCV, ensuring accuracy with a confidence threshold. The system outputs a JSON with detected objects, confidence scores, and bounding box coordinates, which OpenCV visualizes by overlaying bounding boxes on images.

Traffic signal transition controller

The algorithm switches the signal, calculating the time for the green light from estimated traffic density on a real-time basis from the vehicle detection unit [26]. At the same time, the red indication is the other lanes and cycles through the traffic lights accordingly.

The dynamic traffic signal simulation regulates traffic at a four-way intersection by utilizing real-time vehicle detection, optimizing signal timing, and controlling movement efficiently. The durations of red, yellow, and green lights are dynamically adjusted according to vehicle density. An inbuilt tracking system classifies vehicles and allocates green light time accordingly, prioritizing larger vehicles. The system optimizes signal switching by verifying when all vehicles have passed, reducing unnecessary delays. It regulates speed to prevent sudden breaking

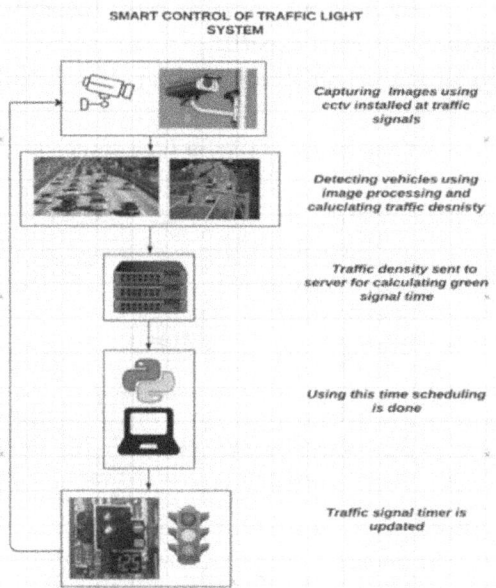

Figure 91.1 Proposed system architecture
Source: Smart Control of Traffic Light Using Artificial Intelligence, 2020

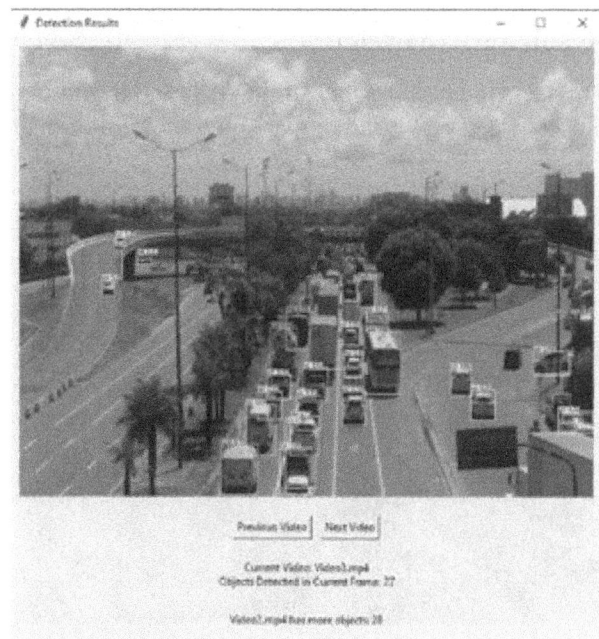

Figure 91.2 Vehicle detection
Source: Author

566 IoT enabled and AI-driven traffic signal control system

and ensures smooth turns with precomputed paths. Emergency vehicles receive priority without disrupting normal flow. Efficient background processing minimizes lag, balancing realism and response. This approach enhances intersection throughput, reduces congestion, and improves overall traffic management.

Pygame-driven simulation model

A simulation model using Pygame was created to mimic real-world traffic conditions and evaluate the performance of the proposed system in comparison to traditional fixed signal control [20]. The model includes a four-way intersection with traffic lights, a countdown timer for each signal phase, and a vehicle counter that monitors the number of cars passing through. Various vehicles, including motorcycles, cars, buses, auto-rickshaws, and trucks, arrive from different directions. Some randomly turn at intersections for added realism. A real-time clock monitors elapsed time, with Figure 91.3 illustrating the simulation's final output.

RFID for emergency vehicles

Through the incorporation of automated detection and immediate decision-making, this system alleviates congestion, improves emergency response times, and decreases the need for human involvement. The suggested approach provides affordable, effective, and scalable means for managing urban traffic, enhancing road safety and streamlining traffic flow.

Result and Analysis

To assess the performance of the AI-based signal control system in comparison to the traditional fixed system, 15 simulations were carried out, each lasting five minutes.

The distribution [a, b, c, d] represents the probability of a vehicle being in lanes 1, 2, 3, and 4, calculated as a/d, (b-a)/d, (c-b)/d, and (d-c)/d, respectively. For example, in the first simulation, the distribution is [500, 700, 900, 1000], translating to probabilities of 0.5, 0.2, 0.2, and 0.1. The collected data was organized into a table showing the number of vehicles that passed through each lane, along with the total number of vehicles that passed overall [11, 22].

The adaptive system [22] consistently outperforms the conventional static model; with efficiency gains increasing as traffic distribution becomes more irregular. It processes an average of 235 vehicles per scenario, a 9.8% improvement over the static system's 214. Lane 1 improved by 15.4%, Lane 2 by 10.2%, Lane 3 by 9.1%, and Lane 4, despite lower occupancy, saw a 7.5% increase in throughput. In highly congested scenarios (entries 2, 6, and 9), the adaptive system exceeded the static model's performance by over 20%, effectively reducing vehicle pile-ups, particularly when total vehicle counts surpassed 250. The system's effectiveness was evaluated through 15 simulations with varying probability distributions for lane assignments. Each run maintained constant parameters while modifying lane probabilities to assess congestion, throughput, and efficiency. Metrics such as total vehicles processed, average wait times, and green light adjustments were recorded. Higher probability lanes experienced increased congestion, requiring longer green durations for optimal

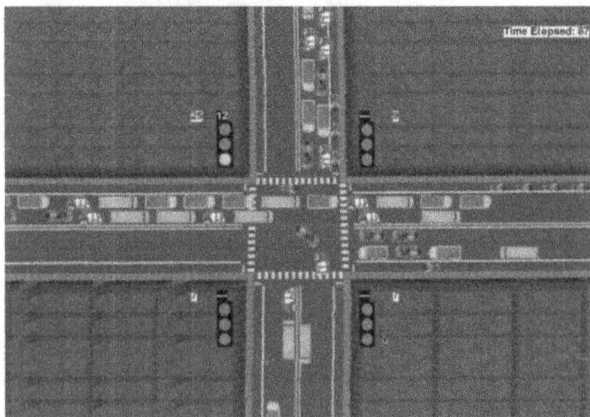

Figure 91.3 Simulation output
Source: Smart Control of Traffic Light Using Artificial Intelligence, 2020

Table 91.1 Simulation results of current static system [11, 13, 26].

No.	Distribution	Lane 1	Lane 2	Lane 3	Lane 4	Total
1	[300,600,800,1000]	81	64	52	38	235
2	[500,700,900,1000]	155	56	48	31	290
3	[250,500,750,1000]	85	68	65	67	285
4	[300,500,800,1000]	59	45	69	74	263
5	[700,800,900,1000]	150	33	20	25	228
6	[500,900,950,1000]	117	103	20	13	253
7	[300,600,900,1000]	62	73	84	26	245
8	[200,700,750,1000]	58	117	8	61	261
9	[940,960,980,1000]	199	8	7	6	220
10	[400,500,900,1000]	97	32	95	36	260
11	[200,400,600,1000]	25	54	70	99	248
12	[250,500,950,1000]	50	73	107	7	237
13	[850,900,950,1000]	148	19	17	21	20
14	[350,500,850,1000]	64	53	80	47	244
15	[350,700,850,1000]	66	82	40	48	236

Source: Smart Control of Traffic Light Using Artificial Intelligence, 2020

Table 91.2 Simulation results of proposed system.

No.	Distribution	Lane 1	Lane 2	Lane 3	Lane 4	Total
1	[300,600,800,1000]	70	52	52	65	239
2	[500,700,900,1000]	112	49	48	31	240
3	[250,500,750,1000]	73	53	63	62	251
4	[300,500,800,1000]	74	44	65	71	254
5	[700,800,900,1000]	90	32	25	41	188
6	[500,900,950,1000]	95	71	15	14	195
7	[300,600,900,1000]	73	63	69	24	229
8	[200,700,750,1000]	54	89	10	67	220
9	[940,960,980,1000]	100	10	8	4	122
10	[400,500,900,1000]	81	29	88	37	235
11	[200,400,600,1000]	42	47	54	86	229
12	[250,500,950,1000]	39	52	93	22	206
13	[850,900,950,1000]	74	10	13	17	114
14	[350,500,850,1000]	49	46	69	50	214
15	[350,700,850,1000]	51	64	37	43	195

Source: Author

flow. Some probability distributions improved overall traffic movement, while others led to bottlenecks. Adaptive signal timing, based on real-time vehicle density, effectively reduced wait times and improved throughput. The system consistently processed more vehicles than the static model, demonstrating its ability to manage high-density conditions efficiently. Notably, in the most congested cases, the adaptive system excelled by dynamically adjusting signal times, ensuring a more efficient flow of vehicles. This adaptability significantly reduced pile-ups, particularly in scenarios with over 250 vehicles, reinforcing the system's capability to enhance urban traffic management effectively.

Conclusion and Future Work

In summary, the proposed system adjusts green light durations in real-time based on the prevailing traffic density, ensuring that lanes with higher traffic volumes are given longer green phases, while those with lighter traffic receive shorter ones. This approach minimizes delays, reduces congestion and waiting times, and ultimately leads to lower fuel consumption and emissions.

The simulation results indicate that the system improves intersection efficiency by 23% compared to conventional fixed-time traffic signals, showing a notable enhancement. Further optimization could be achieved by incorporating real CCTV footage for model training, which would boost performance. The system offers clear benefits over current intelligent traffic management solutions, such as pressure mats and infrared sensors, due to its low implementation

cost, as it leverages the existing CCTV infrastructure at intersections, thus reducing the need for additional hardware. Although minor adjustments in camera positioning may be required, maintenance costs are lower compared to pressure mats, which deteriorate under high traffic. As a result, this system can be seamlessly integrated into existing urban traffic monitoring setups to improve traffic regulation.

References

[1] Damadam, S., Zourbakhsh, M., Javidan, R., & Faroughi, A. (2022). An intelligent IoT based traffic light management system: deep reinforcement learning. *Smart Cities*, 5, 1293–1311. 10.3390/smartcities5040066.

[2] Barbosa, R., Ogobuchi, O. D., Joy, O. O., Saadi, M., Rosa, R. L., Al Otaibi, S., et al. (2023). IoT based real-time traffic monitoring system using images sensors by sparse deep learning algorithm. *Computer Communications*, 210, 321–330. ISSN 0140-3664.

[3] kumari, Soni and kumari, Suman and vikram, Vishal and kumari, Sony and Gouda, Sunil Kumar, Smart Traffic Management System Using IoT and Machine Learning Approach (July 10, 2020). Available at SSRN: https://ssrn.com/abstract=3647656 orhttp://dx.doi.org/10.2139/ssrn.3647656.

[4] Traffic time management, image processing and objects detection, vehicles dataset. Received April 23, 2022; accepted February 5, 2023, 2023 https://doi.org/10.34028/iajit/20/3/13.

[5] Balasubramanian, S. B., Balaji, P., Munshi, A., Almukadi, W., Prabhu, T. N., & Abouhawwash, M. (2023). Machine learning based IoT system for secure traffic management and accident detection in smart cities. *PeerJ Computer Science,* 8(9), e1259. doi: 10.7717/peerj-cs.1259. PMID: 37346697; PMCID: PMC10280433.

[6] Ni, W., Li, Z., Wang, P., & Li, C. (2024). Advanced state-aware traffic light optimization control with deep Q-network. *Neural Information Processing*, 14448, 178. Doi: 10.1007/978-981-99-8082-6_14.

[7] Tran-Van, N. Y., Nguyerr, X. H., & Le, K. H. (2022). Towards smart traffic lights based on deep learning and traffic flow information. In 2022 9th NAFOSTED Conference on Information and Computer Science (NICS), Ho Chi Minh City, Vietnam, (pp. 1–6). doi:10.1109/NICS56915.2022.10013375.

[8] Lilhore, U. K., Imoize, A. L., Li, C. T., Simaiya, S., Pani, S. K., Goyal, N., et al. (n.d). Design and implementation of an ML and IoT based adaptive traffic-management system for smart cities. (2022). Sensors (Basel), 22(8), 2908. doi: 10.3390/s22082908. PMID: 35458892; PMCID: PMC9024789.

[9] Jagadeesh, V., Reddy, T. L., & Sundari, G. (2023). Traffic control system for emergency vehicles using RFID and sensors. In 2023 7th International Conference on Intelligent Computing and Control Systems (ICICCS), Madurai, India, (pp. 1739–1744). DOI:10.1109/ICICCS56967.2023.10142622.

[10] C, Joesam & S, Shrinidhi & Amaran, Sibi & K, Sree & U, Karthikeyan. (2024). YOLO-based Traffic Signal Optimization for Intelligent Traffic Flow Management. 828-831. 10.1109/IS-MAC61858.2024.10714806.

[11] Gandhi, M. M., Solanki, D. S., Daptardar, R. S., & Baloorkar, N. S. (2020). Smart control of traffic light using artificial intelligence. In 2020 5th IEEE International Conference on Recent Advances and Innovations in Engineering (ICRAIE), Jaipur, India, (pp. 1–6).

[12] Sirphy, S., & Revathi, S. T. (2023). Adaptive traffic control system using YOLO. In 2023 International Conference on Computer Communication and Informatics (ICCCI), Coimbatore, India, (pp. 1–5).

[13] Ajmal Khan, Shams ur Rahman, Farman Ullah, Muhammad Ilyas Khattak, Mohammed M. Bait-Suwailam, Hesham El Sayed, IoTEnabled adaptive traffic light controller and emission reduction at intersection, November 29, 2024. DOI: 10.7717/peer-jcs.2507/supp-1.

[14] Gamel, S. A., Saleh, A. I., & Ali, H. A. (2022). A fog-based traffic light management strategy (TLMS) based on fuzzy inference engine. *Neural Computing and Applications*, 34, 2187–2205. https://doi.org/10.1007/s00521-021-06525-2.

[15] Goyal, R., Elawadhi, O., Sharma, A., Bhutani, M., & Jain, A. (2024). Cloud-connected central unit for traffic control: interfacing sensing units and centralized control for efficient traffic management. *International Journal of Information Technology*, 16, 841–851. https://doi.org/10.1007/s41870-023-01527-w.

[16] Naithani, A., & Jain, V. (2024). Real-time crossroad traffic light management. In 2024 8th International Symposium on Innovative Approaches in Smart Technologies (ISAS), İstanbul, Turkiye, (pp. 1–6).

[17] M. J. Nulyn Punitha Markavathi, S. N, V. M.R, V. M and P. M, "Real-Time Traffic Density Monitoring and Adaptive Signal Control Using YOLOv8 and Arduino-Based LED System," 2024 9th International Conference on Communication and Electronics Systems (ICCES), Coimbatore, India, 2024, pp. 227–232, doi: 10.1109/ICCES63552.2024.10859969.

[18] K. Sreejith, S. Mathi and P. Pradeep, "Beyond Sensors: IntelliSignal's Map-Integrated Intelligence in Traffic Flow Optimization," in IEEE Access, vol. 12, pp. 39028–39040, 2024, doi: https://doi.org/10.1109/ACCESS.2024.3375335.

[19] Joshi, N., Shah, K., Heniya, N., Zaveri, J., Nikam, R., & Deherkar, K. (2023). Smart traffic lights. In 2023 IEEE World Conference on Applied Intelligence and Computing (AIC), (pp. 76–80). IEEE. 10.1109/AIC57670.2023.10263856.

[20] Murumkar, A., Shaikh, D., Sharma, S., & Khivasara, D. (2024). Dynamic traffic light management system. *International Research Journal of Engineering and Technology (IRJET)*, 11(05), 146–152.

[21] Phatangare, S., Sakpal, R. A., Kasurde, S. N., Punde, S. S., & Khan, S. S. (2024). Real-time traffic management using deep learning and object detection using YOLOv8. In 2024 15th International Conference on Computing Communication and Networking Technologies (ICCCNT), Kamand, India, (pp. 1–5).

[22] J. Hui, 'Real-time Object Detection with YOLO, YOLOv2 and now YOLOv3', 2018. [Online]. Available: https://jonathanhui.medium.com/real-time-objectdetection-withyolo-yolov2-28b1b93e2088.

[23] Reddy, J. D., Gils Paul, B. R. V., Dinesh, R., & Nellepalli, V. R. (2023). AI based smart traffic management. *International Journal of Scientific Research in Engineering and Management (IJSREM)*, 07(07), 1–5. 10.55041/IJSREM24517.

[24] https://svunaac.somaiya.edu/C3/DVV/3.4.5/Confernce+and+Book+Chapter/308.pdf.

[25] Shahana, S., Revathi, R., Sowmya, P., Dhanalakshmi, D., Mandala, S., & Hariharan, S. (2023). Traffic signal automation by sensing and detecting traffic intensity through IR sensors. In 2023 International Conference on Sustainable Emerging Innovations in Engineering and Technology (ICSEIET), Ghaziabad, India, (pp. 881–885).

[26] https://builtin.com/machine-learning/non-maximum-suppression.

92 Forecasting air quality with machine learning methods

K. Bharath Kumar[1,a], Suraya Mubeen[1,b], B. Bhargav[2,c], K. Nivedhitha[2,d] and K.Venkat Sai[2,e]

[1]Associate Professor, Department of of ECE, CMR Technical Campus, Hyderabad, India

[2]UG Students, Department of of ECE, CMR Technical Campus, Hyderabad, India

Abstract

Air pollution is one of the main concerns since pollution levels in our nation are rapidly increasing and posing serious health dangers. Humans and other living things are impacted by the deteriorating air quality, which presents serious public health concern. In India, a crucial statistical tool for examining pollutants such as NO_2, SO_2, suspended particulate matter (SPM), and respirable suspended particulate matter (RSPM) over time is the Air Quality Index (AQI). In order to produce precise forecasts, our study focuses on evaluating and forecasting air quality using machine learning techniques. We want to enhance air quality forecasts so that people and agencies may take preventative action by utilizing historical data and sophisticated algorithms. The project builds a reliable pollution monitoring system using tools like Python and Sickest-Learn. This project has the potential to improve environmental sustainability and public health by providing insightful information to lessen the negative consequences of pollution.

Keywords: Air quality index, classifiers, KNN, linear regression, machine learning, naïve bayes, random forest

Introduction

India's fast population increase has led to a rise in pollutants such PM2.5, CO2, NO2, and SO2, which has exacerbated environmental deterioration. Our study integrates historical and real-time data to predict air quality using machine learning (ML), particularly neural networks. Our deep learning approach enhances prediction accuracy by combining pollution, weather, and urban data, helping policymakers regulate pollution effectively. Our deep air model enhanced environmental change detection in 300 Chinese cities. Proactive pollution control and real-time monitoring are made possible by combining machine learning with Internet of Things (IoT) sensors and dashboards. By providing precise air quality forecasts, we hope to promote both public health and sustainable growth.

Literature Survey

Air quality has drastically declined due to unchecked deforestation, fast urbanization, population growth, and industrial expansion, posing serious dangers to the environment and human health. Accurate forecasting of air quality is still a major difficulty, despite the widespread use of models such as ML, artificial neural networks (ANN), and deep neural networks.

In order to increase the accuracy of dynamic air quality prediction, a deep learning model utilizing convolutional long short-term memory (ConvLSTM) was presented. It achieved 95% accuracy and outperformed the support vector regression (SVR) model, which had 92.9% accuracy [1]. In order to lower forecasting risks, a supervised learning-based model has been developed for air quality prediction employing a variety of machine learning techniques, including K-fold cross-validation, Random Forest (RF), Support Vector Machines (SVM), and logistic regression (LR) [2]. The study primarily examined Malaysia's PM2.5 concentration levels, emphasizing the efficacy of RF. Additionally, other independent factors were correlated to evaluate the impact of air quality using SPSS [3]. Important pollutants that were shown to have a major impact on air quality included SO2, PM2.5, PM10, NO2, and CO. After applying a number of regression models, it was discovered that RF regression and Gradient Boosting regression provide precise AQI predictions [4]. Another study findings [5] suggested a hybrid approach that uses RFR, SVR, and GBR to predict air quality accurately by combining mobile and static IoT devices

[a]kammarabharathkumar@gmail.com, [b]suraya418@gmail.com, [c]217R1A0475@cmrtc.ac.in, [d]217R1A0494@cmrtc.ac.in, [e]217R1A0496@cmrtc.ac.in

DOI: 10.1201/9781003684589-92

to gather pollution data. Furthermore, research has examined a range of Big Data and machine learning algorithms, such as artificial intelligence (AI), deep learning (DL), and Decision Trees (DT), to enhance air quality forecasting [6]. Additionally, correlations between pollutants such as NO2, PM2.5, SO2, PM10, CO, and O3 were analyzed using SPSS, yielding predictions with a mere 10% error margin [7]. With an emphasis on lowering model complexity through structured regularization to enhance performance, machine learning approaches have been investigated to forecast the AQI by estimating pollutant concentrations [8]. In order to forecast the quality of the air in China, a different study [9] used RF, DT, and deep back propagation neural networks. It found that RF had the lowest mean squared error (MSE). Using BLSTM and IDW layers, a novel spatiotemporal prediction framework was suggested to capture spatial correlations and long-term trends of contaminants [10]. A case study conducted in Guangdong, China, verified that the model was successful in predicting PM2.5 levels. Subsequent analysis revealed that the IDW-BLSTM, CNN-LSTM, and BLSTM models performed better than more conventional methods such as SVR, ARIMA, GBDT, ANN, and RNN [11]. A total of 38 engineering research publications on ML approaches, including SVM, neural networks (NN), and LR, were examined in a systematic review. The results showed that SVM was appropriate for AQI prediction while LR was appropriate for pollution assessment [13]. Recurrent neural networks (RNN) and long short-term memory (LSTM) models were utilized in PM10-focused studies to evaluate air quality [14]. For air quality forecasting, other studies examined machine learning and Big Data techniques such AI, DT, and DL [15]. Using real-time data from Athens, Greece, an ANN-based model was also shown to forecast the concentrations of photochemical pollutants. These researches demonstrate how sophisticated machine learning models enhance predictions of air quality, facilitating proactive policies for public health and pollution control.

Experimental Setup and the Algorithm

Three main ML techniques are used in this analysis: a) RF, b) LR, and c) Gaussian Naïve Bayes (Figure 92.1).

Random Forest
A collection of trees known as RF can be used for both regression and classification. By increasing the

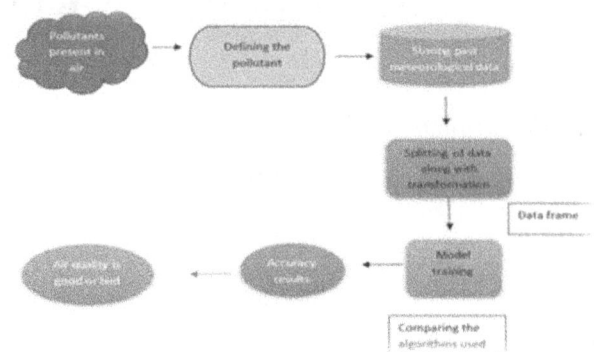

Figure 92.1 Each method's workflow
Source: Author

power of the particular tree, it lowers the error correlation between different classifiers, increases the correlation, and lowers the error in the forest. Error rates are lower and more efficient when working with huge amounts of data that are evenly distributed. The number of trees utilized in the forest and the number of random variables used in each tree are the two most important components of a random forest. The significance of variables in a regression or classification job can be prioritized using the Mean Decrease Accuracy (MDA) and Mean Decrease Gini (MDG). Running as many trees as feasible in an RF does not result in over-fitting [12].

Linear regression
One machine learning approach for accurately predicting the air quality index for the current dataset is called linear regression (LR). The mathematical representation of a simplified linear regression in (1) is as follows:

$$Y = \alpha 0 + \alpha 1 X1 + \alpha 2 X2 + \ldots + \alpha n\ Xn \tag{1}$$

Gaussian Naive Bayes
Probabilistic ML method for classification applications is the Naive Bayes classifier. As demonstrated in (2), it is predicated on the Bayes theorem. The likelihood:

$$P(X/Y) = P(X/Y)P(Y)/P(X) \tag{2}$$

The predictor's prior class probability is denoted by P(Y), while its prior probability is denoted by P(X). P (X | Y) is the likelihood probability of the input feature (X) when a class output label is provided. P

(Y|X) (Y) provides the likelihood that the input feature (X) is the label [12].

The Kaggle repository provided the AQI dataset utilized in this study [11] , which included 16 features that represented Ahmedabad's 2015 air quality parameters: NH_3, CO, SO_2, NO, NO_2, PM2.5, PM10, city, and day. Using Jupyter Notebook and packages such as Sklenar, data preprocessing was carried out in Python, where mean values were used to impute missing values. Analysis of feature correlations and removal of superfluous features such as xylene, benzene, and AQI bucket were aided by a heat map display. The 18314 × 10 dataset was divided into two sets: 20% for testing (3663 × 10) and 80% for training (14651 × 10). Lastly, the data was scaled from 0 to 1 using normalization.

Results

The quality of the air is stayed in two ways first one is used side and another way administrator.

User

Once the created HTTP link has been clicked, you will be taken to the site depicted in Figure 92.2, where new users must register by supplying basic information and existing users can log in using their credentials. Users are taken to the homepage to enter their newly generated login credentials after successfully completing the signup process. Users will be prompted by the interface to enter information such as location in the designated fields after signing in. Clicking the "Air Quality Prediction" button after entering the necessary data will enable the algorithm to evaluate the input and produce an air quality prediction. For a better knowledge of environmental circumstances, the output, as illustrated in Figure 92.3, will show the air quality status as extremely bad, poor, moderate, or severe.

Once all required pollutant values, city name, and date are entered, clicking the "Predict" button prompts the system to process the input data and analyze pollutant concentrations. Based on this analysis, the system predicts the air quality status, which in this case is indicated as "poor," as shown in Figure 92.4.

Users can start the analysis by clicking the "Predict" button after providing the date, city name, and all necessary pollutant numbers. After processing the input data, the system determines the air quality condition by calculating the concentration of each

Figure 92.2 Home page
Source: Author

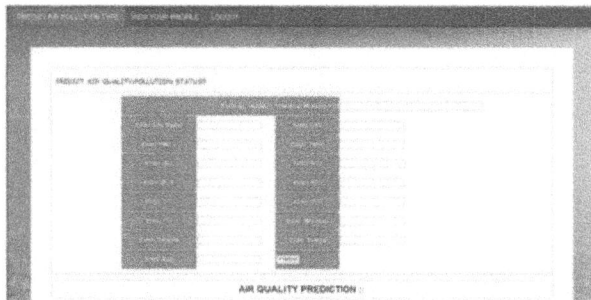

Figure 92.3 Checking pollution status
Source: Author

Figure 92.4 Cases for poor quality air prediction
Source: Author

pollutant. As seen in Figure 92.5, the analysis may yield a result like "moderate," which would represent the state of the environment at the time. As seen in Figure 92.6, the system will indicate that the air quality is "severe," suggesting harmful pollution levels, if the entered pollutant levels are extremely high. Because of the significant health hazards, this output recommends users to take the appropriate safeguards.

Service provider

In order to perform air quality predictions, the service provider will have access to and store user-provided data, such as pollution levels, city names, and dates.

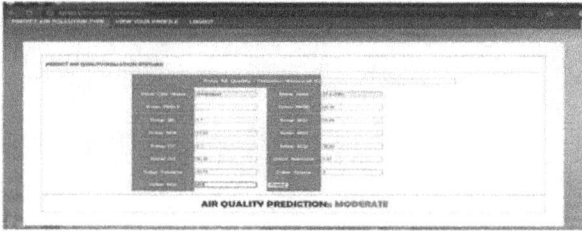

Figure 92.5 Case for moderate quality air prediction
Source: Author

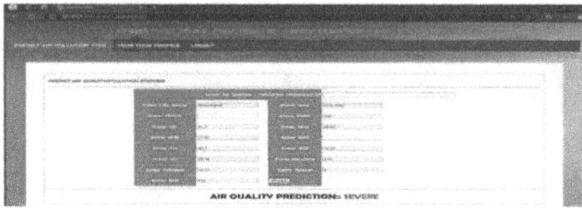

Figure 92.6 Cases for severe quality air prediction
Source: Author

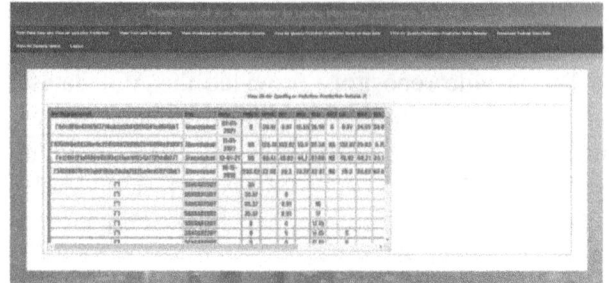

Figure 92.7 Predicted data set
Source: Author

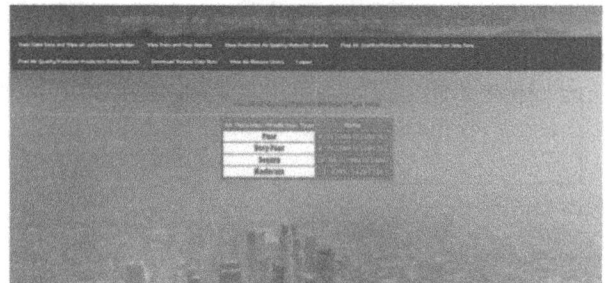

Figure 92.8 Predicted data set result ratio
Source: Author

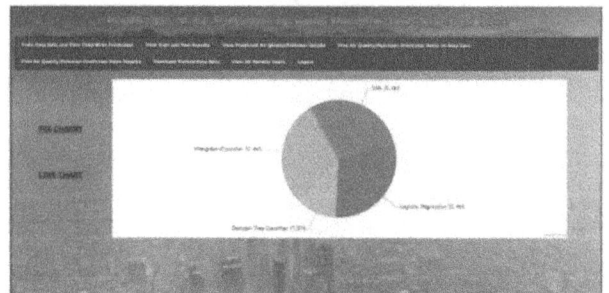

Figure 92.9 Pie chart of train and test results
Source: Author

Additionally, this data could be utilized to guarantee current, correct findings and enhance the service. As seen in Figure 92.7, the service provider guarantees the safe and responsible handling of all user data.

It is the ratio that represents the most predicted air quality level, which is determined by the values entered by the user. This ratio takes into account the concentration of pollutants provided by the user and calculates the corresponding air quality level. The output can vary depending on the pollutant levels, with different results ranging from good to severe. By analyzing the entered values, the system assigns the most accurate air quality prediction. This ratio helps in understanding the overall air pollution status based on the user's input. This is show in Figure 92.8.

The accuracy of the algorithms is visually represented using a pie chart, which provides a clear and intuitive view of how well the system performs. By observing the pie chart, users can easily assess the proportion of accurate results versus any errors or inconsistencies. This visual representation helps in understanding the effectiveness of the algorithm in predicting air quality. It also allows users to quickly identify areas where the system may need improvement. Overall, the pie chart serves as a useful tool for evaluating the accuracy of the test results and the reliability of the predictions. This is show in Figure 92.9.

As seen in Figure 92.10, a line chart that graphically monitors performance over time or across several test cases is used to demonstrate algorithms' correctness. This graph makes it simple for users to see variations in accuracy and spot patterns in the algorithm's performance. It makes it easy to evaluate performance and dependability by clearly contrasting accurate results with errors. As seen in Figure 92.11, the service provider also gets access to basic user information gathered during air quality assessments, such as location, date, and address. By using this data, predictions may be made more accurately, and services can be customized for particular areas. Regulations promote privacy, and user information is handled carefully and securely.

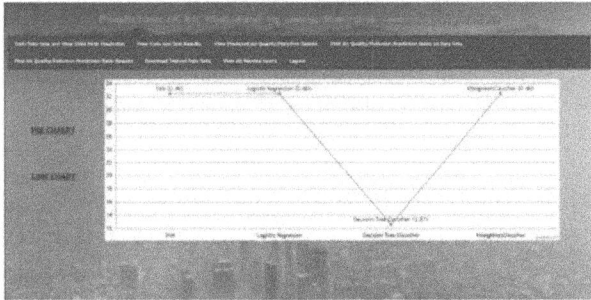

Figure 92.10 Line chart of train and test results
Source: Author

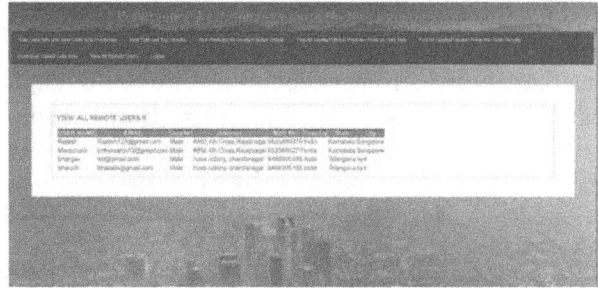

Figure 92.11 Data of remote users
Source: Author

The below represents the accuracy rate of the algorithms used, showing how effectively the system predicts air quality based on the input data. This accuracy rate allows the service provider to assess the performance of the algorithms and determine how reliable the predictions are. If the accuracy rate is not satisfactory, the service provider can make necessary modifications or improvements to the algorithms. Monitoring this rate ensures that users receive accurate and consistent results for air quality predictions. It also helps the service provider maintain and enhance the overall performance of the system, this is show in Figure 92.12.

It shows a statistical graph that represents the most frequent or predicted output displayed to the users. This graph is generated based on the pollutant levels entered by the users and reflects the variations in air quality predictions. By analyzing this graph, service provider can observe trends in the results, such as the most common air quality status, whether it's good, moderate, or severe. The graph provides valuable insight into how different pollutant levels influence the air quality prediction. This visual representation helps users better understand the relationship between the data they input and the system's output, this is show in Figure 92.13.

It shows a statistical graph that represents the most frequent or predicted output displayed to the users. This graph is generated based on the pollutant levels entered by the users and reflects the variations in air quality predictions. By analyzing this graph, service provider can observe trends in the results, such as the most common air quality status, whether it's good, moderate, or severe. The graph provides valuable insight into how different pollutant levels influence the air quality prediction. This visual representation helps users better understand the relationship

Figure 92.12 Algorithms accuracy rate
Source: Author

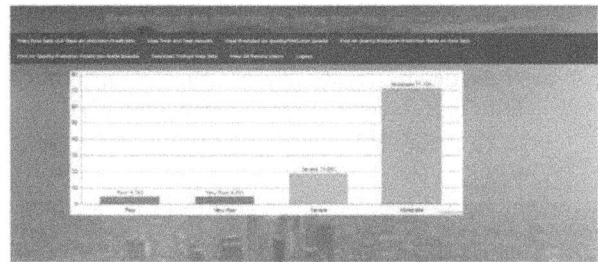

Figure 92.13 Comparison results
Source: Author

between the data they input and the system's output, this is show in Figure 92.13.

Conclusion

In conclusion, machine learning-based air quality prediction has shown a great deal of promise for precisely predicting pollution levels. However, the quality and volume of input data determine how effective these models are, thus more study is required to improve accuracy and interpretability. Future research could address input data uncertainty and incorporate a variety of data sources, including satellite photography and inexpensive sensors. Furthermore, it is essential to use feature importance analysis to enhance model interpretability. An approach that is more thorough can be achieved by combining machine learning with conventional monitoring methods. Random Forest produced the best accuracy among the models that were examined, which makes it appropriate for datasets that are comparable.

References

[1] Mokhtari, I., Bechkit, W., Rivano, H., & Yaici, M. R. (2021). Uncertainty aware deep learning architectures for highly dynamic air quality prediction. *IEEE Access*, 9, 14765–14778. doi: 10.1109/ACCESS.2021.3052429.

[2] Mahalingam, U., Elangovan, K., Dobhal, H., Valliappa, C., Shrestha, S., & Kedam, G. (2019). A machine learning model for air quality prediction for smart cities. In 2019 International Conference Wireless. Communication. Signal Processing Networking, WiSPNET2019, (pp. 452–457). doi: 10.1109/WiSPNET45539.2019.9032734.

[3] Zhang, Y., Wang, Y., Gao, M., Ma, Q., Zhao, J., Zhang, R., et al. (2019). A predictive data feature exploration-based air quality prediction approach. *IEEE Access*, 7, 30732–30743. doi: 10.1109/ACCESS.2019.2897754.

[4] Zhang, D., & Woo, S. S. (2020). Real time localized air quality monitoring and prediction through mobile and fixed IoT sensing network. *IEEE Access*, 8, 89584–89594. doi: 10.1109/ACCESS. 2020. 2993547.

[5] Oumaima Bouakline, Khadija Arjdal,Kenza KhomsI, Noureddine Semane, Abdelhak Elidrissi, Salem Nafiri, Prediction of daily PM10 concentration using machine learning", 2020 IEEE 2nd International Conference on Electronics, Control, Optimization and Computer Science (ICECOCS), 14 January 2021, ISBN:9781-7281-6922-4. 9(1), 8–16.

[6] Kang, G. K., Gao, J. Z., Chiao, S., Lu, S., & Xie, G. (2018). Air quality prediction: big data and machine learning approaches. *International Journal of Environmental Science and Development*, 9(1), 8–16. Doi:10.18178/ ijesd.2018. 9.1.1066.

[7] Patil, T. R. (n.d.). Analysis of air quality estimation based on air pollutants parameters. Asian Journal of Convergence in Technology ,Volume.4, Issue.2, ISSN. No.:2350–1146, I.F-5.11 IV(2350).

[8] Ma, L., Gao, Y., & Zhao, C. (2020). Research on machine learning prediction of air quality index based on SPSS. In Proceedings of 2020 International Conference Comput. Network, Electron. Autom. ICCNEA 2020, (pp. 1–5). doi: 10.1109/ICCNEA50255.2020.00011.

[9] Zhu, D., Cai, C., Yang, T., & Zhou, X. (2018). A machine learning approach for air quality prediction: model regularization and optimization. *Big Data and Cognitive Computer*, 2(1), 1–15. doi: 10.3390/bdcc2010005.

[10] Masih, A. (2019). Machine learning algorithms in air quality modeling. *Global Journal of Environmental Science and Management*, 5(4), 515–534. doi: 10.22034/gjesm.2019.04.10.

[11] https://www.kaggle.com/datasets/hirenvora/citydaycsv [accessed on 21-01-2022].

[12] Kalapanidas, E., & Avouris, N. (1999). Applying machine learning techniques in air quality prediction. In Proceedings of ACAI (Vol. 99, pp. 58–64).

[13] Priyadharsini, P., Khare, N., Praveenraj, D. D. W., Senthurya, S., Arunkumar, B., & Reddy, G. V. (2024). An Integration of AI technique in the field of healthcare industry. In 2024 4th International Conference on Advance Computing and Innovative Technologies in Engineering, ICACITE 2024, (pp. 1981–1986).

[14] Chakram, D. S., Nandi, S., Godavarti, U., Dasari, M., & Kerur, S. S. (2025). Recent progress in gas sensing characteristics of lithium doped sodium potassium niobate (K0.5, Na0.5−x Lix NbO3) ceramics: microstructure, distinctive electrical applications. *Materials Today Communications*, 44, 112072.

[15] Jadhav, S. A., Umadevi, G., Awale, M. B., Kanse, K. S., Joshi, Y. S., & Purushotham Y. A. (2025). Optimizing ammonia gas sensor performance: Investigating the influence of Mg doped ZnO root like nanostructures for enhanced gas sensing application. *Materials Letters*, 389, 138643.

93 Object detection using YOLO with audio feedback using Raspberry-PI

K. Bharath Kumar[1,a], Mallesh Sudhamalla[2,b], Kolanpaka Manoj[3,c], Kandula Ruthika[3,d], Rampuram Patil Sreeja[3,e] and Nerella Vivek[3,f]

[1]Associate Professor, Department of ECE, CMR Technical Campus, Hyderabad, India

[2]Assistant Professor, Department of ECE, CMR Technical Campus, Hyderabad, India

[3]UG Students, Department of ECE, CMR Technical Campus, Hyderabad, India

Abstract

In order to help visually impaired people navigate, this paper describes an assistive technology that uses the YOLO algorithm integrated into wearable spectacles or webcams with a Raspberry Pi processor for real-time object detection. The system records visual data, processes it using YOLO, and turns it into audio feedback via earphones to guide users. The efficiency, speed, and accuracy of the system are evaluated in real-world scenarios. Blind people frequently rely on external aids like canes, guide dogs, or assistance from others, but these have limitations, especially in unfamiliar environments.

Keywords: Audio feedback, embedded system, object detection/identification, Raspberry-PI, visual DATA translation, YOLO V-5 algorithm

Introduction

This study presents a portable, lightweight blind navigation system that uses the YOLO real-time object detection algorithm, which processes images quickly and accurately. It integrates a Raspberry Pi with a camera to analyze surroundings and uses text-to-speech technology for audio feedback via speakers or earbuds. Its single button interface makes it easy to use, and its sensitivity can be adjusted as needed. The system can detect a variety of objects, such as vehicles and obstacles, and improves small and occluded object detection, which helps with navigation and offers wider applications in robotics and surveillance.

Literature Survey

A real-time object detection and audio feedback system enhances the independence and safety of blind or visually impaired (VI) individuals by recognizing nearby objects and providing audible cues [1]. The YOLO-based system incorporates a camera, object identification algorithm, and text-to-speech module to let people move securely and autonomously [2].

Furthermore, YOLO's design, problems, datasets, and applications in domains like autonomous driving and healthcare are discussed, along with comparisons to other approaches such as two-stage detectors [3].

The performance and applicability of convolutional neural networks (CNNs) and YOLO models in domains such as deformity diagnosis, education, and finance were examined in order to construct a real-time object detection system [4]. The YOLOv3-based system using Google Text-to-Speech (gTTS) API, which was trained on the MS COCO dataset, offers auditory feedback and 90% detection accuracy to assist visually impaired people in navigating on their own [5].

For tasks like pedestrian, face, and salient item identification, this review emphasizes the transition from manual feature-based techniques to deep learning systems like CNNs [6]. Furthermore, a deep learning-based system for auditory feedback and object identification is suggested to help blind and VI people by detecting items in real time and giving them aural cues to improve their mobility and independence [7].

Visually impaired individuals can gain greater independence and mobility through an object recognition

[a]kammarabharathkumar@gmail.com, [b]mallesh.ece4@gmail.com, [c]217R1A04F9@cmrtc.ac.in, [d]217R1A04J4@cmrtc.ac.in, [e]217R1A0450@cmrtc.ac.in, [f]217R1A04H5@cmrtc.ac.in

DOI: 10.1201/9781003684589-93

and detection system that uses deep learning and audio feedback to identify objects in real time and convert visual data into auditory cues [8]. YOLOv3, an enhanced version of the YOLO object detection system, improves network design, bounding box prediction, and training, achieving a mean average precision (mAP) of 28.2 with a 22 ms inference time at 320x320 resolutions, making it three times faster than SSD [9]. YOLO frames object detection as a regression problem, predicting bounding boxes and class probabilities, allowing real-time processing at 45 frames per second and up to 155 frames per second in Fast YOLO [10]. For accurate object restring, even in cases of deformation, occlusion, and multiple targets, the real-time object tracking system incorporates Kalman filter and Prewitt edge detection [11]. The COCO dataset, which consists of 328,000 images and 2.5 million instances classified into 91 categories, improves the understanding of scene and response benchmarks for object detection and segmentation [12].

Existing Methodology

Visual impairment takes a major toll in a person's life for them to act independently without others' help. Various assistive systems have been developed to address this need, including smart canes, range notification systems, pathfinders, real-time localization systems, ultrasonic electronic systems, and novel indoor navigation tools. These systems utilize technologies such as microcontrollers (e.g., ATmega328P), computer vision algorithms, [11] and assistive devices like the electronic long cane or eye stick. Some systems rely on tactile signals instead of acoustic signals to convey information [7, 13]. For instance, a project developed an ultrasonic-based cane capable of detecting obstacles within 400 cm and providing real-time voice feedback to the user. Among the sensors used for obstacle detection infrared, ultrasonic, and laser—the ultrasonic sensor was chosen due to its cost-effectiveness and reliability in varying light conditions [14].

This system can function independently of the cane, allowing users to navigate familiar environments without it. However, a limitation is its narrow detection range. Despite advancements in assistive technologies, further improvements are needed to enhance their affordability, usability, and range of functionality for blind individuals [9].

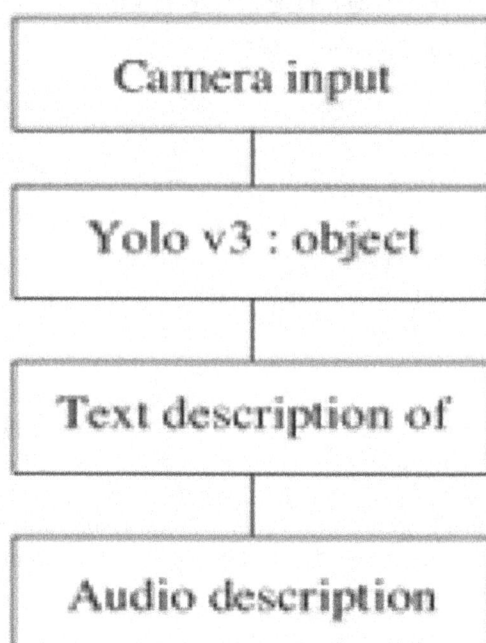

Figure 93.1 Real-time object detection
Source: Author

Therefore, Figure 93.1 reviews the launch of a technology designed to assist visually impaired people by fusing audio feedback with real-time object identification. The MS COCO dataset is used for training and testing, while the YOLO_v3 algorithm is used for real-time item detection and classification. Following item identification, the Google Text-to-Speech (gTTS) API is used to translate the descriptions of the objects into audio feedback, which is subsequently provided to the user [1]. This system's main objective is to increase visually impaired users' accessibility and independence by giving them the ability to successfully perceive and navigate their environment.

Even though YOLO is faster, it has drawbacks, including the inability to detect small or overlapping objects due to its grid-based approach and fixed number of bounding boxes per cell, a trade-off between speed and accuracy that limits performance in complex or densely packed scenes, a heavy reliance on predefined anchor boxes that may not be suitable for all object shapes and sizes, precise bounding box predictions that are limited by YOLO's fixed grid size, and the model's potential inability to generalize to new domains without extensive retraining on a variety of datasets [12].

Proposed Methodology

The system is designed to help visually impaired people make their regular lifestyle more accessible and interactive. It analyzes object detection using YOLO algorithm with the help of Raspberry Pi and deep learning methodology. Equipment required are camera, speaker Raspberry Pi, power supply, voice module and sensors. The details about the software process used in the project for object/human detection along with some of the libraries and package installations.

The above Figure 93.2 represents the block diagram of yolo v5 algorithm for the object detection process.

Raspberry Pi 4B

The core system where processing happens. This will run the YOLOv5 model, process the images, and control the input/output (IO) devices and is reviewed in Figure 93.3.

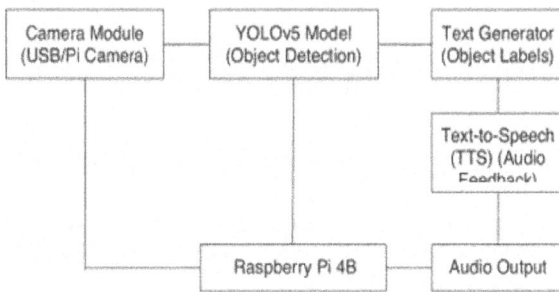

Figure 93.2 Block diagram of yolo v5 algorithm
Source: Author

Figure 93.3 Control the input/output (IO) devices
Source: Author

Camera (USB or Pi Camera)

A solar and battery-powered camera set on a pole and linked to a as seen in Figure 93.4, takes real-time pictures when sensors identify animals within 30 meters and communicate with the camera (which has a 50-meter range) to assess the threat [2].

YOLOv5 (object detection model)

On a Raspberry Pi 4b, the YOLOv5 object detection model identifies objects like people and cars in captured frames. Based on CNNs that mimic biological processes with overlapping receptive fields, YOLOv5 efficiently analyzes visual data for applications such as image recognition, recommender systems, and natural language processing, as shown in Figure 93.5.

Processing and inference

After processing the captured image with the YOLOv5 model, the Raspberry Pi outputs labels and bounding box coordinates to identify objects. YOLO uses convolutional layers to extract features and a 1×1 convolution-based classifier for real-time object identification, while CNN-based systems efficiently recognize patterns by grouping related objects and ignoring irrelevant data, as shown in Figure 93.6.

Figure 93.4 camera is attached to the Raspberry Pi
Source: Author

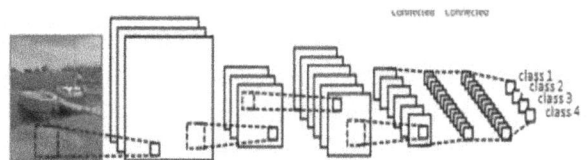

Figure 93.5 Detect objects in the captured frames
Source: Author

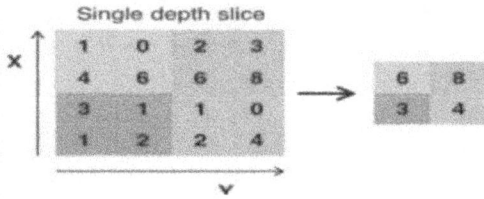

Figure 93.6 Small transformations, distortions and translations in the input image
Source: Author

Figure 93.7 Providing real-time audio feedback to the user
Source: Author

Convolution

Using learnable filters that move across the image to create activation maps, convolution in CNNs pulls features from input images. In early levels, it captures simple patterns like edges, while in later layers, it captures intricate designs. It has been demonstrated that spatial pooling, like Max Pooling, performs better than Mean and Total pooling because it keeps the most noticeable features while reducing the dimensionality of feature maps while maintaining important information. As illustrated in Figure 93.6, the objective of pooling is to gradually decrease the spatial size of input representations, rendering the network invariant to minute distortions, translations, and rotations in the input image.

By lowering computations and parameters and simplifying input representations, pooling also aids in the management of over fitting, enabling objects to be identified in a picture independent of their location.

Text generation: After detecting the objects, the Raspberry Pi generates a textual description, such as "Person detected," or "Vehicle detected."

Text-to-speech converter: The generated text (from the previous step) is sent to a text-to-speech module as in [6]. This converts the detected object text to audible feedback (e.g., "Object detected: Car").

Audio output: The Raspberry Pi outputs the audio through a connected speaker as shown in Figure 93.7, providing real-time audio feedback to the user.

Flow chart Figure 93.8 illustrating the object detection process using YOLOv5 algorithm on Raspberry Pi, along with audio feedback and text-to-audio feedback. The steps that elaborate this process is:

1. Start program: Initialize the program and load the YOLOv5 model.

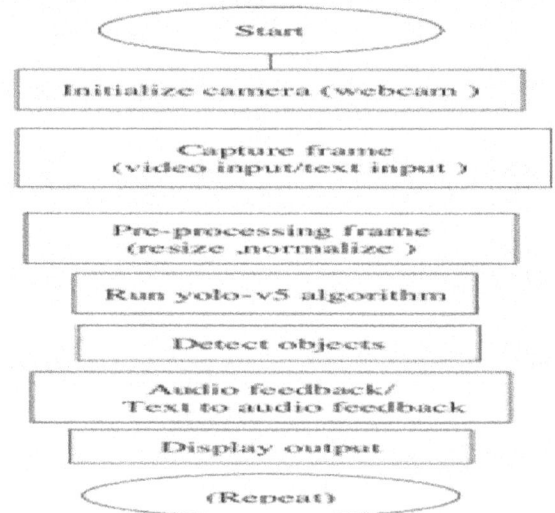

Figure 93.8 The object detection process using YOLOv5
Source: Author

2. Initialize camera: Initialize the USB camera and set up the video capture.
3. Capture frame: Capture a frame from the video input.
4. Pre-process Frame: Pre-process the frame by resizing and normalizing it.
5. Run YOLOv5 model: Run the YOLOv5 model on the pre-processed frame to detect objects.
6. Detect objects: Detect objects in the frame, including their classes and confidence levels [8].
7. Audio feedback: Provide audio feedback, such as alerts or warnings, based on the detected objects. Text-to-audio feedback: Provide text-to-audio feedback, such as announcing the object class and confidence level.
8. Display output: Display the output, including the video, detected objects, and text.

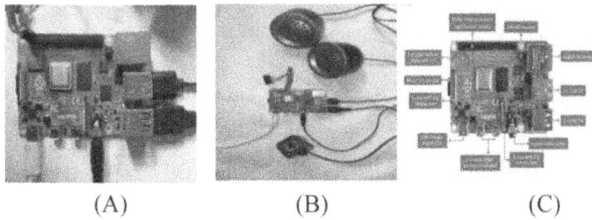

Figure 93.9 a) Power supply required is (5V DC, 2A) b) practical setup of the equipment c) raspberry pi 4b considered by the labels
Source: Author

9. Repeat: Repeat the process by capturing another frame and detecting objects.

Results and Discussions

The integration of YOLO-v5 with the help of Raspberry Pi 4b is the practical approach considered to provide the real time AI applications that can help with the visually impaired as shown in Figure 93.9. The Raspberry Pi 4B, with its quad-core ARM Cortex-A72 processor and up to 8GB of RAM and the Power supply required is (5V DC, 2A) in (A). The above Figure 93.9 represents the practical setup of the equipment (B) in detail for the processing of the results The USB 2.0 port is connected to the speakers with the help of audio jack. The other USB 2.0 port is connected to the camera. The c-type cable is required to provide the power supply for the raspberry pi 4b considered by the labels in (C). The distinct outputs produced using YOLO help to identify the real time objects and act accordingly to the visual data provided.

Case 1: Identification of objects and providing the visual cues

The output of items detected from a live webcam video is displayed in Figure 93.10. Each object is surrounded by bounding boxes labeled with confidence scores and labels, and visually impaired users are given clear voice feedback. Because of YOLO's grid size restrictions, the system might overlook low-resolution objects or detections even when bounding boxes and audio update in real-time as things move.

Case 2. Text to speech conversion /audio output

Text detection and audio feedback conversion are demonstrated in Figure 93.11 above, where text is

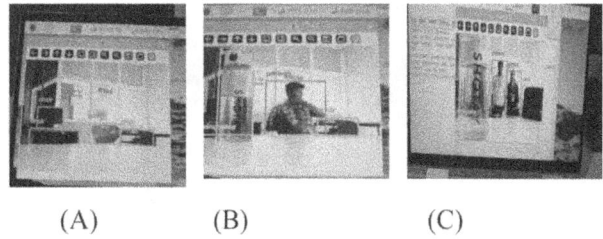

(A)　　　　　(B)　　　　　(C)

Figure 93.10 A) objects are detected B) frame by frame to detect objects C) Bounding boxes are drawn around them
Source: Author

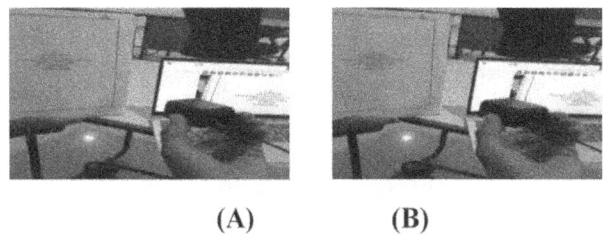

(A)　　　　　**(B)**

c)

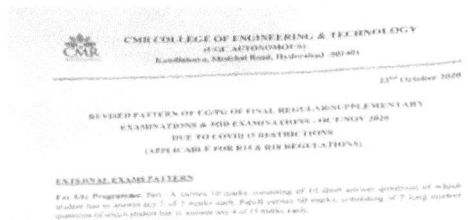

Figure 93.11 A) The webcam and capture frames in real-time B) Extract meaningful text by filtering C) YOLO struggles to detect small-sized text
Source: Author

extracted from a live camera feed using optical character recognition (OCR). For improved OCR accuracy, Open CV enhances the frames it records with grayscale conversion, thresh-holding, and noise reduction. As seen in (A) and (B), Tesseract OCR extracts text, and a Text-to-voice (TTS) engine filters and turns meaningful content into voice. Bounding boxes or annotations are superimposed on the video feed over the recognized text. This function facilitates real-time translation, acts as an instructional tool, and helps visually impaired users by reading signs aloud.

Case 3. Integration of telegram for animal detection

The above Figure 93.12 represents animal detection and routing similar images to the telegram account

(A)   **(B)**

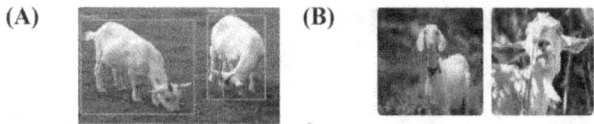

Figure 93.12 A) detected animals using YOLO are forwarded to a Telegram account B) combining the Yolo interface and the telegram API

Source: Author

linked to the device. The detected animals using YOLO are forwarded to a Telegram account in (A). The animals pictured are forwarded by combining the Yolo interface and the telegram API as in (B). This process enables efficient object detection using YOLO while leveraging Telegram for notifications or alerts in real time. There may be few disadvantages while dealing with this process, namely, data limitations, spamming and also the energy consumption by the hardware equipment for continuous hardware functioning.

Conclusion and Future Scope

A portable and reasonably priced system for real-time object recognition and fast speech alerts, YOLO-based object detection with auditory feedback on Raspberry Pi improves safety and freedom for those who are blind or visually impaired. It is appropriate for daily use due to its small size. AI-driven scene interpretation, lightweight models like YOLOv7 for quicker processing, and obstacle identification using stereo vision or ultrasonic sensors are possible future advancements. It could become a potent assistive tool to increase mobility and autonomy with the addition of features like multilingual support, personalized voice feedback, wearable integration (such as smart glasses), IoT-based cloud processing, and haptic feedback.

References

[1] Jambhulkar, A. R., Vatkar, S., Gajera, A. R., & Chirag Bhavsar, M. (2023). Real-time object detection and audio feedback for the visually impaired. In 3rd Asian Conference on Innovation in Technology (ASIANCON), Pune. DOI: 10.1109/ASIANCON58793.2023.10269899.

[2] Diwan, T., Anirudh, G., & Tembhurne, J. V. (2022). Object detection using YOLO: challenges, architectural successors, datasets and applications. *Multimedia Tools and Applications*, 82(6), 9243–9275.

[3] Viswanatha, V., Chandana, R. K., & Ramachandra, A. C. (2022). *International Journal of Informatics Visualization*, 8(1), 45–53.

[4] Shin, S., & Kwon, S. (2020). Real-time object detection with audio feedback for the visually impaired using YOLOv3. In 15th International Conference on Advanced Technologies.

[5] Zhao, Z. Q., Zheng, P., Xu, S. T., & Wu, X. (2019). Object detection with deep learning: a review. *IEEE Transactions on Neural Networks and Learning Systems*, 30(11), 3212–3232.

[6] Noh, Y., Kim, C., & Hwang, I. (2018). Object detection and identification for visually impaired using deep learning and audio feedback system. 2019 IEEE 6th International Conference on Engineering Technologies and Applied Sciences (ICETAS).

[7] Redmon, J., & Farhadi, A. (2018). YOLOv3: an incremental improvement. arXiv preprint arXiv: 1804.02767.

[8] Redmon, J., Divvala, S., Girshick, R., & Farhadi, A. (2016). You only look once: unified, real-time object detection. In Proceedings of the IEEE Conference on Computer Vision and Pattern Recognition.

[9] Cherian, S., & Singh, C. (2014). Real time implementation of object tracking through webcam. *International Journal of Research in Engineering and Technology*, 3(1), 128–132.

[10] Clerk Maxwell, A Treatise on Electricity and Magnetism, 3rd edn., (Vol. 2, pp. 68–73). Oxford: Clarendon, 1892.

[11] Lin, T. Y., Maire, M., Belongie, S., Hays, J., Perona, P., Ramanan, D., et al. (2014). Microsoft COCO: common objects in context. In European Conference on Computer Vision, (pp. 740-755). Springer, Cham.

[12] Bhanu, S., Revathi, B., Fatima, T., & Gaddam, R. R. (2025). Proposed methods for evaluating shape-based identification techniques for identifying numbers using CNN. In Applications of Mathematics in Science and Technology: International Conference on Mathematical Applications in Science and Technology, (pp. 235–239).

[13] Chakram, D. S., Nandi, S., Godavarti, U., Adimule, V., Dasari, M., & Kerur, S. S. (2025). Recent progress in gas sensing characteristics of lithium doped sodium potassium niobate (K0.5, Na0.5−x Lix NbO3) ceramics: microstructure, distinctive electrical applications. *Materials Today Communications*, 44, 112072.

[14] Malsoru, V, Shilpa, K., Reddy, A. R., Sudha, D., & Archana, B. (2025). A novel approach for predicting accuracy in image to image translation by means of generative adversarial networks (GAN). In Lecture Notes in Electrical Engineering, 1274 LNEE, (pp. 1310–1319).

94 Audio-based gender identification and emotion recognition using CNN and bidirectional LSTM

K. Rathan Sai[1,a], K. Venkat Rohith[1,b], T. Durga Venkata Reddy[1,c] and Mariappan, R.[2,d]

[1]UG Scholars, Computer Science and Engineering, VIT University, Vellore, Tamil Nadu, India

[2]School of Computer Science and Engineering, VIT University, Vellore, Tamil Nadu, India

Abstract

Audio-based gender identification and emotion detection have become increasingly important in recent years for applications such as sentiment analysis, healthcare, security, and human-computer interaction. Virtual assistants, automated customer support systems, and psychological testing can all benefit from the ability to infer gender and emotions from speech. However, because speech tones, accents, and background noise vary, it is tough to identify these characteristics precisely. Conventional methods use machine learning models including basic deep neural network (DNNs) and SVM. These models are less accurate and resilient because they have trouble capturing both spatial and temporal correlations in speech signals, even when they extract characteristics like MFCCs and pitch-based properties. We used a hybrid model combining convolutional neural networks (CNN) and Bidirectional long short-term memory (BiLSTM) to overcome these drawbacks. BiLSTM improves the identification of small speech fluctuations by capturing long-range temporal relationships, whereas CNN effectively extracts deep spatial characteristics from audio spectrograms. Our model outperforms conventional techniques in terms of accuracy after being trained on the benchmark dataset Ryerson audio-visual database of emotional speech and song (RAVDESS). Our method improves feature learning, increases robustness to speech fluctuations, and achieves improved accuracy in gender categorization (99%) and emotion categorization (97%) by merging CNN with BiLSTM.

Keywords: Bidirectional long short-term memory, convolutional neural networks, deep neural network, mel-frequency cepstral coefficients, recurrent neural network, ryerson audio-visual database of emotional speech and song dataset, short-time fourier transform, voice activity detection

Introduction

To establish a natural communication channel between humans and machines, speech should be the primary mode of interaction, as it is the fundamental component of human-human communication. Accurate identification of human emotions and gender through spoken interaction is crucial for developing a more natural and intuitive connection between users and AI-driven systems. Speech signals capture both the emotional state and personality of the speaker, making them highly valuable for applications such as assistive robotics. Most speech recognition algorithms classify gender identification as a binary problem, categorizing speakers as either male or female. This makes it difficult for traditional methods to generalize across diverse speech samples effectively.

Deep learning (DL) techniques have gained popularity in various classification tasks, including object and speech recognition. Following existing research trends, our study employs the bidirectional long short-term memory (BiLSTM) network, a modified version of long short-term memory (LSTM), where the internal structure incorporates only one positive recurrent weight. Training the network through optimization requires forward and backward propagation, preventing issues like vanishing and exploding gradients.

Literature Survey

Sakurai and Kosaka (2021) suggested an emotion recognition system that incorporates acoustic and linguistic attributes derived from speech recognition

[a]rathansaidps@gmail.com, [b]kodumurivenkat.rohith2021@vitstudent.ac.in, [c]tetalidurga.venkata2021@vitstudent.ac.in, [d]prof.mariappan.r@gmail.com

DOI: 10.1201/9781003684589-94

outcomes. Their technique combines voice signal processing and language models to improve the precision of identifying emotions. Although effective, the technique is heavily reliant on the quality of automatic speech recognition (ASR), which could introduce inaccuracies that influence classification effectiveness [1]. Susithra et al. presented a hybrid model that fuses feedforward neural networks (FNN) and convolutional neural networks (CNNs) for emotion and gender recognition based on speech. Their method extracts MFCC and spectral features, which are processed through CNN layers for in-depth feature extraction, followed by FNN for classification. The model demonstrated improved performance compared to traditional machine learning classifiers, but it did not fully capture temporal dependencies present in speech [2]. Alkhawaldeh proposed a one-dimensional (1D) model based on CNNs for the identification of gender from speech. The study illustrated that CNNs can proficiently extract patterns from speech waveforms without necessitating extensive preprocessing. Nevertheless, the model is limited in its capacity to learn long-term temporal dependencies, which restricts its utility in emotion recognition tasks [3]. Basu et al. examined a hybrid model that combines CNNs and RNNs for emotion recognition from speech. The CNN layers were responsible for extracting hierarchical feature representations, while the RNN component, namely LSTM, was tasked with capturing temporal dependencies. The study showcased enhanced performance when compared to standalone CNN or RNN models, underscoring the significance of integrating spatial and sequential feature learning [4]. Qayyum et al. (2019) put forward a CNN-based model for speech emotion identification, achieving a high level of accuracy through spectrogram-based feature extraction. However, the study acknowledged that CNNs alone may not adequately capture sufficient contextual information, indicating that the integration of recurrent architectures such as BiLSTM could further improve performance [5]. Chintalapudi et al. investigated deep learning models for speech emotion identification, emphasizing the efficacy of both CNN and RNN architectures. Their research demonstrated that deep models surpass traditional methodologies, particularly in environments characterized by noise. Nonetheless, the computational complexity associated with deep models persists as an issue for real-time applications [6]. Liu et al. presented Dual-TBNet, a model that integrates

dual transformer and BiLSTM networks for emotion identification. Their methodology enhances robustness in the extraction of speech features by employing transformers for self-attention-based feature augmentation, succeeded by BiLSTM for sequential modeling. This study illustrates the advantages of transformer-based architectures in comparison to conventional CNN-RNN hybrids [7]. Taha et al. (2024) investigated various CNN arch. for gender and emotion identification from speech, incorporating speaker diarization to enhance recognition performance within multi-speaker contexts. Their findings indicated that CNNs effectively capture spectral features; however, performance can be further elevated by employing context-aware recurrent architectures [8].

Most existing models, especially CNN-based architectures, primarily focus on spatial feature extraction from spectrograms without effectively capturing temporal dependencies in speech. Many existing models perform well on clean, structured datasets but struggle in noisy, uncontrolled environments. CNN models struggle to accurately recognize speakers when emotional shifts occur in speech. The need for significant computational resources limits the scalability and real-time deployment of these models.

Proposed System

The proposed model emphasizes the development of a hybrid CNN and BiLSTM model for the purposes of audio-based gender identification and emotion recognition. Each phase is aimed at strengthening the model's ability to effectively extract temporal and spatial characteristics from speech signals, hence ensuring a high level of classification accuracy as illustrated in Figure 94.1. This model is trained with 8 different emotions.

Data collection

This research makes use of the RAVDESS dataset [13], it is commonly used for emotion and gender identification via speech. The dataset consists of speech recordings made by 24 acting professionals (12 female and 12 male), representing eight different emotional states: happy, angry, neutral, sad, calm, fearful, disgusted, and surprised. Each audio sample is labelled with the corresponding emotion and gender, making it suitable for training a supervised deep-learning model.

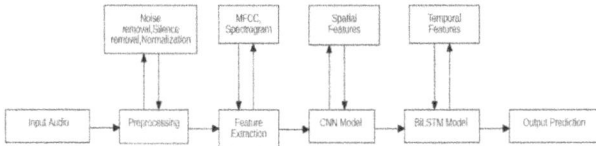

Figure 94.1 Proposed system architecture
Source: Author

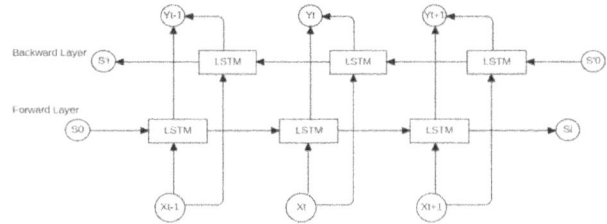

Figure 94.2 BiLSTM architecture
Source: Author

Data preprocessing

Standardizing the input speech signals is essential, and this can be achieved by preprocessing, which guarantees their cleanliness. A variety of preprocessing approaches are used to improve data quality and model performance. Initially, all audio files are resampled to 16 kHz to ensure frequency resolution consistency. Background noise is then reduced using spectral gating, which helps to remove unwanted interference while maintaining important speech qualities. VAD is then used to eliminate silent parts at the beginning and finish of the recordings. Additionally, amplitude normalization is used to ensure that all voice samples have a consistent volume level. To improve the model's generalization potential, data augmentation methods like pitch shifting, temporal stretching, and white noise injection are used, boosting diversity in training samples while reducing the risk of overfitting and making the model more adaptable to real-world speech variations.

Feature extraction

To accurately represent speech signals for deep learning applications, MFCCs, and spectrograms are used as key features. MFCCs are frequently used in speech processing because of their capacity to capture the perceptual characteristics of human speech, which are critical for recognizing both gender and emotions [11,12]. In this study, 13 MFCC coefficients are retrieved from each speech sample. In conjunction with MFCCs, spectrograms are generated using the STFT, which offers a time-frequency representation of voice signals. These spectrograms function as input to the model's CNN component, allowing for more effective feature extraction. The model was enhanced using pitch- and energy-based variables to distinguish between various emotional and gender characteristics [10].

Implementation Methodology

Model architecture of BiLSTM

BiLSTM is an advanced RNN that is capable of processing data in sequence in both forward as well as backward directions as shown in Figure 94.2. BiLSTM, unlike conventional LSTMs, captures both past and future circumstances, which makes it suitable for speech recognition, NLP, and time-series analysis. The BiLSTM architecture comprises an input layer designed for sequential data, and this is followed by both forward and backward LSTM layers that process the sequence from both directions. After being concatenated in a concatenation layer, the outputs of these layers are passed to a fully linked layer for further processing. Ultimately, the output layer yields the classification or prediction [9].

The proposed hybrid model is the CNN-BiLSTM model, which employs a CNN to extract spatial features and a BiLSTM to model temporal features, thereby facilitating the accurate classification of gender and emotions from speech. The initial component of the model is the CNN, which processes the spectrogram images to acquire meaningful feature representations. CNN comprises multiple convolutional layers that extract low-level and high-level speech characteristics, succeeded by max-pooling layers that diminish dimensionality while preserving critical frequency components. Batch normalization layers are incorporated to stabilize learning and mitigate overfitting. Ultimately, the flattened layer transforms the extracted CNN features into a format suitable for sequential processing.

In the LSTM mechanism there are 3 memory gates, i.e., input gate (ig_t), forget gate (fg_t), and output gate (og_t).

The mathematical equations of LSTM:

$$ig_t = \sigma \left(WE_{ig} x_t + U_{ig} h_{t-1} + b_{ig} \right) \qquad (1)$$

$$fg_t = \sigma \ (WE_{fg}xt + U_{fg} \ ht_{-1} + b_{fg}) \qquad (2)$$

$$og_t = \sigma \ (WE_{og} \ x_t + U_{og} \ h_{t-1} + b_{og}) \qquad (3)$$

$$c^{\sim} = f_t . \ C_{t-1} \qquad (4)$$

$$h_t = o_t . \ tanh \ (c_t) \qquad (5)$$

where x_t = input sample at time t, σ = sigmoid activation function, and c_t, = memory unit. $(b_{fg}, \ big, b_{og})$ stands for the bias and $(WE_{fg}, WE_{ig}, WE_{og})$ stands for the weight matrix.

Forward LSTM: Uses the standard LSTM equations, processing from t=1 to t=T.

$$h_t = LSTM \ (WE_{fg}, \ WEig, \ WEog, x_t) \qquad (6)$$

Backward LSTM: Uses similar equations but processes the sequence in reverse.

$$h_t' = LSTM \ (WE_{fg}', \ WEig', WE_{og}', x_t) \qquad (7)$$

BiLSTM Output: The outputs from both directions are combined to provide a richer representation that captures context from both the previous and subsequent inputs.

$$h_t = h_t * h_t' \qquad (8)$$

The subsequent component of the model is the BiLSTM network, which is tasked with learning the temporal dependencies in the speech signals, as shown in Figure 94.3. Unlike conventional LSTMs, the BiLSTM model provides a deeper understanding of the variations in speech patterns. Following the BiLSTM layer, the model incorporates dropout layers to mitigate overfitting, and subsequent fully connected layers map the extracted features to corresponding gender and emotion labels, as detailed in Table 94.1. The last layer of the model utilizes the SoftMax activation function, generating probability scores for each class, thereby enabling the classification of speech samples into categories of gender and emotion.

Training and optimization
The model is trained using supervised learning, wherein labelled speech samples from the RAVDESS dataset [13] are employed to iteratively update the

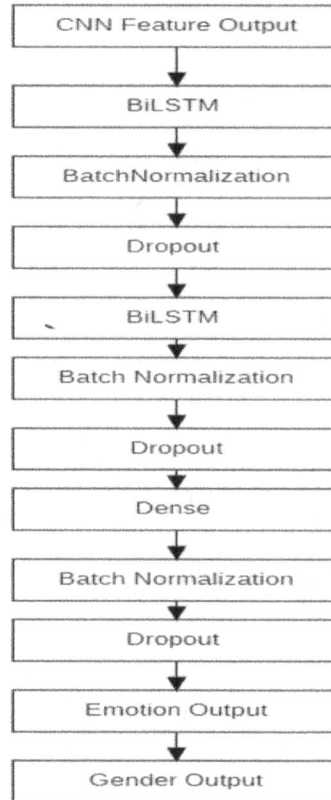

Figure 94.3 BiLSTM model
Source: Author

network's weights. The loss function for multi-class classification used is Cross-entropy loss function, ensuring that the model accurately distinguishes between gender and emotion labels. The Adam optimizer is used to dynamically adjust the learning rate, promoting efficient convergence throughout the training process. The model goes through training in batches of 32 samples to balance learning stability and computational efficiency. Training lasts up to 100 epochs, depending on the convergence rate. To improve generalization, data augmentation methods are used during training to provide differences in pitch, speed, and noise levels, making the model more adaptable to real-world speech variations. Early stopping is implemented to terminate training when validation loss stops progress to prevent overfitting.

Result and Analysis

The ability of the model to accurately determine gender and emotions was measured using different metrics.

Table 94.1 BiLSTM model parameters.

Layer	Output Shape	Parameters
Bidirectional	(None, 22, 256)	394,240
Batch normalization 3	(None, 22, 256)	1,024
Dropout	(None, 22, 256)	0
Bidirectional 1	(None, 128)	164,352
Batch normalization 4	(None, 128)	512
Dropout 1	(None, 128)	0
Dense	(None, 128)	16,512
Batch normalization 5	(None, 128)	512
Dropout 2	(None, 128)	0
Emotion output	(None, 5)	645
Gender output	(None, 2)	258

Source: Author

Table 94.2 CNN vs CNN+BiLSTM gender classification report.

	Accuracy	Precision	Recall	F1-score
CNN	0.98	0.97	0.98	0.97
CNN+ BiLSTM	0.99	0.99	0.99	0.98

Source: Author

Performance metrics analysis
The model was assessed with standard classification metrics:

Gender performance metrics
The gender categorization issue was regarded as a binary classification problem (male or female). The CNN-BiLSTM model exhibited great accuracy as listed in Table 94.2, demonstrating the usefulness of deep learning for collecting gender-specific speech characteristics.

Emotion performance metrics
The model was trained to recognize 8 different emotions: happy, angry, neutral, sad, calm, fearful, disgusted, and surprised. The hybrid model gave great results as listed in Table 94.3.

Table 94.3 CNN vs CNN+BiLSTM emotion classification report.

	Accuracy	Precision	Recall	F1-score
CNN	0.93	0.93	0.93	0.92
CNN+ BiLSTM	0.97	0.97	0.96	0.96

Source: Author

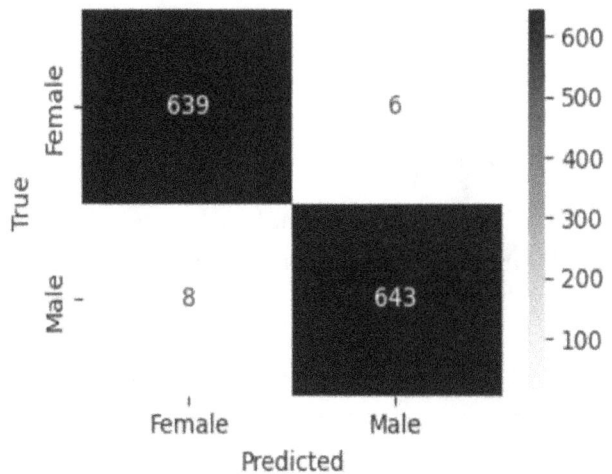

Figure 94.4 Confusion matrix for gender model
Source: Author

Confusion matrix analysis
To assess misclassifications, confusion matrices were created for both gender and emotion recognition tasks, as shown in Figures 94.4 and 94.5.

Model accuracy comparison
As shown in Figure 94.6, For gender classification, the hybrid model outperformed the CNN-only model by 99% from 98%. Similarly, for emotion identification, the hybrid model attained 97% accuracy, which was much higher than the CNN-only model's 93%.

Conclusion

This study presents a hybrid model for speech-based gender identification and emotion detection that combines convolutional neural networks (CNN) for spatial feature extraction and bidirectional long short-term memory (BiLSTM) for temporal dependency modeling. Experimental findings indicate that this hybrid model surpasses a pure CNN

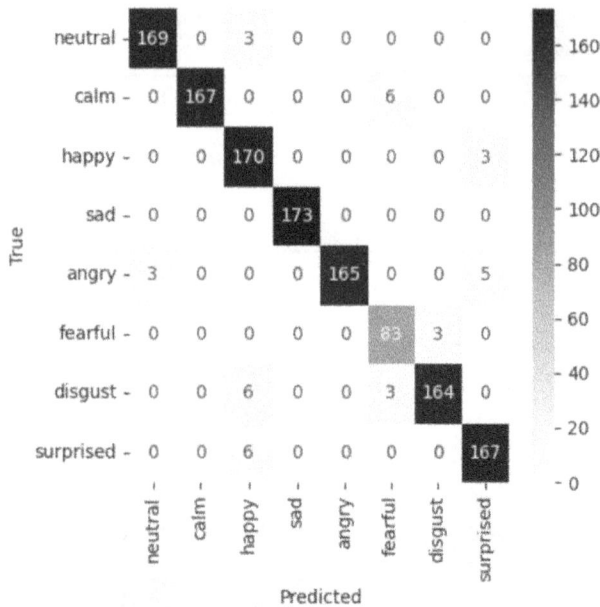

Figure 94.5 Confusion matrix for emotion model
Source: Author

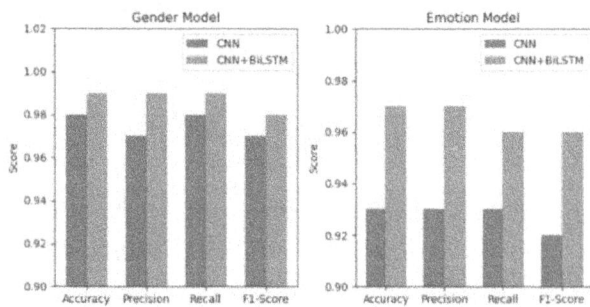

Figure 94.6 Comparison of existing and proposed model
Source: Author

approach, boasting accuracy rates of 99% in gender classification and 97% in emotion detection. These improvements highlight the importance of applying temporal learning techniques to speech signal processing, given that emotions are inherently sequential and context-dependent. Subsequent research may focus on training the model with multilingual datasets to improve language independence and generalizability. Research can examine continuous emotion tracking by the analysis of consecutive speech throughout time, which could prove useful for applications in mental health monitoring and emotional AI systems.

References

[1] Sakurai, M., & Kosaka, T. (2021). Emotion recognition combining acoustic and linguistic features based on speech recognition results. In 2021 IEEE 10th Global Conference on Consumer Electronics (GCCE), Kyoto, Japan, (pp. 824–827).

[2] Susithra, N., Rajalakshmi, K., Ashwath, P., Ajay, B., Rohit, D., & Stewaugh, S. (2022). Speech based emotion recognition and gender identification using FNN and CNN models. In 2022 3rd International Conference for Emerging Technology (INCET), Belgaum, India, (pp. 1–6).

[3] Alkhawaldeh, R. S. (2019). DGR: gender recognition of human speech using one-dimensional conventional neural network. *Scientific Programming*, 2019(12), 7213717.

[4] Basu, S., Chakraborty, J., & Aftabuddin, M. (2017). Emotion recognition from speech using convolutional neural network with recurrent neural network architecture. In 2017 2nd International Conference on Communication and Electronics Systems (ICCES), Coimbatore, India, (pp. 333–336).

[5] Qayyum, A. B. A., Arefeen, A., & Shahnaz, C. (2019). Convolutional neural network (CNN) based speech-emotion recognition. In 2019 IEEE International Conference on Signal Processing, Information, Communication & Systems (SPICSCON), Dhaka, Bangladesh, (pp. 122–125).

[6] K. S. Chintalapudi, I. A. K. Patan, H. V. Sontineni, S. K. Muvvala, S. V. Gangashetty and A. K.Dubey, "Speech Emotion Recognition Using Deep Learning," 2023 International Conference on Computer Communication and Informatics (ICCCI), Coimbatore, India, 2023, pp. 1–5.

[7] Liu, Zheng & Kang, Xin & Ren, Fuji. (2023). Dual-TBNet: Improving the Robustness of Speech Features via Dual-Transformer-BiLSTM for Speech Emotion Recognition. IEEE/ACM Transactions on Audio, Speech, and Language Processing. PP. 1–11.

[8] Taha, T. M., Ben Messaoud, Z., & Frikha, M. (2024). Convolutional neural network architectures for gender, emotional detection from speech and speaker diarization. *International Journal of Interactive Mobile Technologies (iJIM)*, 18(03), 88–103.

[9] Feng, T., & Narayanan, S. (2024). Foundation model assisted automatic speech emotion recognition: transcribing, annotating, and augmenting. In ICASSP 2024 - IEEE International Conference on Acoustics, Speech and Signal Processing (ICASSP), Seoul, Korea, (pp. 12116–12120).

[10] Mamyrbayev, O., Toleu, A., Tolegen, G., & Mekeba-yev, N. (2020). Neural architectures for gender detection and speaker identification. *Cogent Engineering*, 7(1), 1727168.

[11] Mary Little Flower, T., Jaya, T., & Christopher Ezhil Singh, S. (2024). Data augmentation using a 1D-CNN model with MFCC/MFMC features for speech emotion recognition. *Automatika*, 65(4), 1325–1338.

[12] Jitendra, M. S. N. V., & Radhika, Y. (2021). Singer gender classification using feature-based and spectrograms with deep convolutional neural network. *International Journal of Advanced Computer Science and Applications (IJACSA)*, 12(2).

[13] S. R. Livingstone and F. A. Russo, "The Ryerson Audio-Visual Dataset of Emotional Speech and Song (RAVDESS): A dynamic, multimodal set of facial and vocal expressions in North American English," PloS ONE, vol. 13, no. 5, p. e0196391, 2018.

For Product Safety Concerns and Information please contact our EU
representative GPSR@taylorandfrancis.com
Taylor & Francis Verlag GmbH, Kaufingerstraße 24, 80331 München, Germany

www.ingramcontent.com/pod-product-compliance
Lightning Source LLC
Chambersburg PA
CBHW081211220326
41598CB00037B/6752

9 781041 164272